P9-BYV-876

THE HYDROGEN ECONOMY

Opportunities and Challenges

In the light of ever-increasing global energy use, the increasing cost of energy services, concerns over energy supply security, climate change and local air pollution, this book centres on the question of how growing energy demand for transport can be met in the long term. Given the sustained interest in and controversial discussions on the prospects of hydrogen, this book highlights the opportunities and challenges of introducing hydrogen as an alternative fuel in the transport sector from an economic, technical and environmental point of view. Through its multi-disciplinary approach, the book provides a broad range of researchers, decision makers and policy makers with a solid and wide-ranging knowledge base concerning the hydrogen economy. The geographical scope of the book is global.

The Hydrogen Economy: Opportunities and Challenges is the first book to cover hydrogen in a holistic manner from a technical, environmental and socioeconomic perspective. Particular highlights include:

- Assessment of the benefits and downsides of hydrogen compared with other alternative fuels;
- Strategies and scenarios for a hydrogen infrastructure build-up;
- Interactions between hydrogen production and the electricity sector;
- Long-term global hydrogen supply scenarios and their impact on resource availability;
- The potential of hydrogen for decarbonising the transport sector;
- Macroeconomic impacts of introducing hydrogen as alternative fuel.

MICHAEL BALL studied Industrial Engineering at the University of Karlsruhe, where he also received his Ph.D. in 2006 in the field of energy-system modelling, developing a model for hydrogen infrastructure analysis, which served as a tool for producing the European Hydrogen Energy Roadmap. After a stay as a researcher at the Fraunhofer Institute for Systems and Innovation Research (ISI) in Karlsruhe, he joined Shell in the Netherlands in 2006 as project CO_2 advisor.

MARTIN WIETSCHEL is the co-ordinator of the 'Energy Economics' business unit at the Fraunhofer Institute for Systems and Innovation Research, Karlsruhe. He is a Professor at the University of Karlsruhe and a Lecturer at the ETH in Zurich. His work focuses on socioeconomic and technical research, mainly in the field of energy and the environment.

Reviews of this book

'The world is facing a severe energy and environmental challenge, a challenge that is particularly acute for Europe – how to secure competitive and clean energy for its citizens against a backdrop of climate change, escalating global energy demand and future supply uncertainties. Hydrogen and fuel cell technologies have the potential to play a significant role in the development of a low-carbon, high efficiency energy system in Europe. This multidisciplinary book significantly broadens the perspective on the prospects of hydrogen as a universal energy vector and fuel, and provides a very important addition to the policy debate over future sources of transportation energy and the role hydrogen can play herein for the decades to come.'

Herbert Kohler, Chair of the European Hydrogen and Fuel Cell Technology Platform

'Sustainability of energy is one of the most important subjects in today's world. Our civilisation still relies almost entirely on fossil fuels to cover its energy needs. Their use has caused harmful consequences for the environment, from air pollution to global warming and climate change. What's more, fossil fuels are being depleted fast, with oil ranking first. All this should lead us to a transition away from today's petrol-based paradigm towards cleaner and ultimately renewable fuels. In this context, hydrogen is an ideal energy carrier: clean, efficient and safe, and as a synthetic fuel that can be produced from any primary energy source, it has the potential to address most energy needs of a sustainable transport system. In this book, the authors have carefully outlined the possible energy dilemma that could occur in the near future, and the particular challenges of the transport sector. The book is an important contribution to the discussion about the role of hydrogen in the future energy system, and should be of great interest to a broad readership, from policy makers to the general public.'

Mustafa Hatipoglu, Managing Director of the International Centre for Hydrogen Energy Technologies of the United Nations Industrial Development Organization (UNIDO-ICHET)

'The price of petroleum is rising continuously, as oil resources are being depleted fast. This is followed by price increases in natural gas and coal. In the meantime, the effects of global warming – such as stronger typhoons, floods and droughts – are becoming more prominent and destructive. The total cost of environmental damage last year alone is estimated to be six trillion dollars worldwide. The hydrogen economy is the permanent solution to these intertwined problems. *The Hydrogen Economy: Opportunities and Challenges* is a timely book outlining the opportunities presented by the hydrogen economy, as well as the challenges posed. I strongly recommend this excellent book to energy engineers, environmentalists and decision makers, as well as those interested in the future of humankind and the welfare of planet Earth.'

T. Nejat Veziroglu, President of the International Association for Hydrogen Energy (IAHE)

'Europe has the unique opportunity to lead the world and to create a low carbon energy economy, by boosting the development and deployment of cleaner and more efficient energy technologies. Hydrogen and fuel-cell-based energy systems hold great promise for achieving this vision. This book helps to understand the options around future mobility and stands out by its holistic approach in critically addressing the prospects of hydrogen in the transport sector from a technical, environmental and socioeconomic perspective. This book should be read by anyone involved in shaping the mobility mix of the future.'

Gijs van Breda Vriesman, Chair of the Governing Board of the European Joint Technology Initiative on Fuel Cells and Hydrogen

THE HYDROGEN ECONOMY

Opportunities and Challenges

Edited by

MICHAEL BALL
Shell
Den Haag, The Netherlands

MARTIN WIETSCHEL
Fraunhofer Institute for Systems and Innovation Research
Karlsruhe, Germany

CAMBRIDGE
UNIVERSITY PRESS

CAMBRIDGE UNIVERSITY PRESS
Cambridge, New York, Melbourne, Madrid, Cape Town, Singapore, São Paulo, Delhi

Cambridge University Press
The Edinburgh Building, Cambridge CB2 8RU, UK

Published in the United States of America by Cambridge University Press, New York

www.cambridge.org
Information on this title: www.cambridge.org/9780521882163

First published 2009

Printed in the United Kingdom at the University Press, Cambridge

A catalogue record for this publication is available from the British Library

Library of Congress Cataloging-in-Publication Data

The hydrogen economy : opportunities and challenges / edited by Michael Ball, Martin Wietschel.
p. cm.
Includes bibliographical references.
ISBN 978-0-521-88216-3 (hardcopy)
1. Hydrogen as fuel. 2. Hydrogen–Research–Economic aspects. 3. Hydrogen industry.
4. Alternative fuel vehicles. I. Ball, Michael. II. Wietschel, Martin, 1962– III. Title.

TP359.H8H858 2009
665.8′1–dc22

2009010742

ISBN 978-0-521-88216-3 hardback

On n'hérite pas de la terre de nos parents, on ne fait que l'emprunter à nos enfants.

Antoine de Saint-Exupéry

There are risks and costs to a programme of action, but they are far less than the long-range risks and costs of comfortable inaction.

John F. Kennedy

Contents

Main contributors

Dr Clemens Cremer works as an analyst with EnBW, Karlsruhe. Until 2008, he was affiliated with the Fraunhofer Institute for Systems and Innovation Research (ISI) as senior scientist and co-ordinator of the business unit 'Energy Efficiency'. He is a lecturer at the Swiss Federal Institute of Technology, Zurich (ETH Zurich).

Dr Maximilian Fichtner is leader of the 'Energy Storage Group' at the Institute of Nanotechnology Research Center, Karlsruhe (FZK). He is also co-ordinator of the 'HyTecGroup' at FZK, which is one of the largest activities on hydrogen research worldwide. His scientific interests are the structural and kinetic aspects of nanocomposite systems for hydrogen storage and for high-performance batteries.

Dr Eberhard Jochem is Professor Emeritus for Economics and Energy Economics at the Centre of Energy Policy and Economics (CEPE) at ETH Zurich, which he founded in 1999. He has been senior executive at the Fraunhofer Institute for Systems and Innovation Research (ISI), Karlsruhe, Germany, since 2000. He is an internationally acknowledged expert in technical, socioeconomic and policy research, mainly in the field of energy efficiency and climate change.

Dr Frank Marscheider-Weidemann is a technical chemist and manages the business unit 'Fuel cells' at the Fraunhofer Institute for Systems and Innovation Research (ISI), Karlsruhe. His work includes technical and socioeconomic analyses. One example is the assessment of fuel cell technology impacts on tradesmen, the automobile industry and other industrial sectors.

Dr Joan Ogden is Professor of Environmental Science and Policy at the University of California, Davis, and Director of the Sustainable Transportation Energy Pathways Program at the campus's Institute of Transportation Studies. Her primary research interest is the technical and economic assessment of new energy technologies, especially in the areas of alternative fuels, fuel cells, renewable energy and low carbon energy systems. She received a B.S. in Mathematics from the University of Illinois, and a Ph.D. in Physics from the University of Maryland.

Dr Gustav Resch is leader of the business unit 'Renewable Energy Policy' at the Energy Economics Group, Vienna University of Technology. Within several European and international research activities, he has been acting as senior expert in the area of energy policy and energy modelling with a focus on renewable energy technologies.

Werner Weindorf received his Engineering Degree in Physics at the Munich University of Applied Sciences. He has been with Ludwig-Bölkow-Systemtechnik since January 1999 as technology and policy consultant. His major activities are life-cycle analysis (LCA) and technoeconomic analysis of alternative and conventional fuels, hydrogen infrastructure and renewable energies.

Dr Christopher Yang is a researcher at the Institute of Transportation Studies at the University of California, Davis. He is a co-leader of Infrastructure Systems Analysis within the Sustainable Transportation Energy Pathways (STEPS) Program and his work focuses mainly on the analysis of hydrogen infrastructure, the grid impacts of electric vehicle charging and the reduction of greenhouse-gas emissions from transportation systems.

Preface

The world is facing a severe energy and environmental challenge – how to provide competitive and clean energy for its citizens in light of an escalating global energy demand, concerns over energy supply security, climate change and local air pollution. More specifically, with soaring crude oil prices and with conventional oil becoming harder to find and produce, and its production eventually declining, there is a growing imperative to develop alternative fuels. At the same time, governments are stepping up their efforts to address the challenges of sustainable mobility and to foster the expansion of low-carbon fuels. Against this backdrop, this book centres around the question on how the growing energy demand for transport services can be met in the long term, while adhering to the aforementioned external framework conditions.

While the road-transport sector is expected to witness a much broader portfolio of fuels in the future, the context for considering alternative fuels is dynamic and uncertain. However, there is a growing consensus that electric mobility (i.e., whereby the vehicle drive is provided by an electric motor) is going to play a significant role in transforming the transport sector and could experience a substantial uptake in the future. Under such a scenario, hydrogen-powered vehicles could capture a noticeable market share. Hydrogen is particularly promising as it has the potential to address simultaneously all the major energy policy objectives in the transport sector, i.e., greenhouse-gas emissions reduction, energy security and reduction of local air pollution.

We have been involved in various hydrogen-related R&D projects, most notably aiming at developing strategies and roadmaps for the introduction of hydrogen in the transport sector. Given the sustained interest and controversial discussion on the prospects of hydrogen, this book intends to highlight not only the opportunities, but also the challenges of introducing hydrogen as an alternative fuel in the transport sector. The possible transition to a largely hydrogen-based transport system is placed in the context of the development of the global energy scene in the coming decades and analysed in a holistic manner from a technical, environmental and economic perspective.

Avoiding excessive technical jargon and technological details, the book aims to be of interest to a fairly broad readership (academia, policy makers and industry, as well as the interested reader) and to provide decision makers – through its multidisciplinary approach – with a comprehensive and up-to-date reference and knowledge base about hydrogen. We hope that this book will broaden the perspective on the prospects of hydrogen as a universal energy vector and fuel, and that it will contribute positively to the policy debate over future sources of transportation energy and the role hydrogen can play herein for the decades to come. Areas covered include, among others:

- The benefits and downsides of hydrogen compared with other alternative fuels;
- Strategies and scenarios for a hydrogen infrastructure build-up;
- Interactions between hydrogen production and the electricity sector;
- Long-term global hydrogen-supply scenarios and their impact on resource availability;
- The potential of hydrogen for decarbonising the transport sector; and
- Macroeconomic impacts of introducing hydrogen.

While hydrogen and fuel-cell technologies are progressing, there is also continuing technical progress in a variety of other alternative fuels and efficient vehicle technologies, such as hybrid, plug-in hybrid, and pure electric vehicles, and liquid biofuels. In this respect, hydrogen should be seen as one option available in a broad move towards a lower-carbon energy system.

This book does not intend to pretend that hydrogen will solve all of our energy and environmental problems; nor does it intend to make forecasts about how the energy system in general and the transport sector in particular will evolve in the coming decades. Rather, this book is about presenting the choices at hand. In this sense, it strives to reflect critically on the various alternatives and strategies available to respond to the global energy challenge, in particular how to secure sustainable energy for transportation, as one of the pillars of our globalised world. Hydrogen and fuel-cell technologies are certainly very well positioned to become a major part of the solution.

For a long time, hydrogen has been the fuel of the future. The coming decade will be critical to prove the commercial viability of hydrogen and fuel-cell technologies. It will be interesting to look back in 20 or 30 years time to see how the *Future of Hydrogen* will have unfolded.

Michael Ball and Martin Wietschel
Stepanakert and Karlsruhe

Acknowledgements

This book being an edited volume, we would like to thank again all the lead authors and various co-authors for their valuable contributions, discussions and critical feedback.

We thank our editor Matt Lloyd at Cambridge University Press for his positive feedback and support from the first time he read the book synopsis. We also thank Diya Gupta, Anna-Marie Lovett and Alison Lees, who took us through the final stages of editing and production.

Last but not least, we would like to thank our families for their support and the time they gave us, which have allowed the realisation of this book. A special thanks goes to Ainhoa for her enduring patience and support during the writing of this book as well as for the critical reading and editing of parts of the manuscript.

Abbreviations

ABM	Agent-based models
ACEA	European Automobile Manufacturers' Association
AFC	Alkaline fuel cell
AFS	Alternative fuel standard
APEC	Asia-Pacific Economic Cooperation
API	American Petroleum Institute
APU	Auxiliary power unit
AR4	IPCC Fourth Assessment Report
ASPO	Association for the Study of Peak Oil
ATR	Autothermal reforming
Balmorel	Baltic Model of Regional Energy Market Liberalisation
BAU	Business as usual
BCO	Bio crude oil
BEV	Battery-electric vehicle
BGR	Bundesanstalt für Geowissenschaften und Rohstoffe (Federal Institute for Geosciences and Natural Resources)
boe	Barrel of oil equivalent
BTL	Biomass-to-liquids
BTU	British thermal unit
BWR	Boiling-water reactor
CAES	Compressed-air energy storage
CAFE	Corporate average fuel economy
CARB	California Air Resources Board
CBM	Coal-bed methane
CCGT	Combined cycle gas turbine
CCS	Carbon/carbon dioxide capture and storage
CDM	Clean development mechanism
CFC	Chlorofluorocarbon
CGE	Computable general equilibrium
CGH_2	Compressed gaseous hydrogen
CHP	Combined heat and power

cif	Cost, insurance, freight
CIS	Commonwealth of Independent States
CMG	Compressed methane gas
CMM	Coal-mine methane
CNG	Compressed natural gas
CONCAWE	The Oil Companies' European Association for Environment, Health and Safety in Refining and Distribution
COP	Conference of the Parties
cP	centipoise
CRW	Combustibles, renewables and waste
CSM	Coal-seam methane
CTL	Coal-to-liquids
DCL	Direct coal liquefaction
DDGS	Distillers dried grains with solubles
DEFC	Direct ethanol fuel cell
DICI	Direct-injection compression ignition
DISI	Direct-injection spark ignition
DME	Dimethylether
DMFC	Direct methanol fuel cell
DOE	Department of Energy (USA)
DP	Dynamic programming
DPF	Diesel particulate filter
ECBM	Enhanced coal-bed methane
EGR	Enhanced gas recovery
EIA	Energy Information Administration (US DOE)
EOR	Enhanced oil recovery
EPR	European Pressurised Water Reactor; Evolutionary Power Reactor
EROEI	Energy returned on energy invested
ETBE	Ethyl tertiary butyl ether
ETS	Emission trading scheme
EU	European Union
EUCAR	European Council for Automotive Research and Development
EUR	Estimated ultimate recovery
FAME	Fatty acid methyl ester
FAO	Food and Agriculture Organization of the United Nations
FBR	Fast-breeder reactor
FC	Fuel cell
FCV	Fuel-cell vehicle
FFV	Flexible-fuel vehicle
FOB	Free on board
FPFC	Fuel-processor fuel cell

FSU	Former Soviet Union
FT	Fischer–Tropsch
GAMS	General Algebraic Modeling System
GDP	Gross domestic product
GH_2	Gaseous hydrogen
GHG	Greenhouse gas
GIS	Geographical information system
GTL	Gas-to-liquids
GWP	Global-warming potential
HDV	Heavy-duty vehicle
HEV	Hybrid-electric vehicle
HFC	Hydrofluorocarbons
HFP	Hydrogen and Fuel Cell Technology Platform
HHV	Higher heating value
HOV	Highly occupied vehicle
HVDC	High-voltage direct current
IAEA	International Atomic Energy Agency
ICE	Internal-combustion engine
ICL	Indirect coal liquefaction
IEA	International Energy Agency
IEF	International Energy Forum
IET	International emissions trading
IGCC	Integrated-coal gasification combined-cycle
IMF	International Monetary Fund
IO	Input–output
IOC	International Oil Company
IPCC	Intergovernmental Panel on Climate Change
IR	Inferred resources
ITER	International Thermonuclear Experimental Reactor
JI	Joint implementation
JODI	Joint Oil Data Initiative
JRC	Joint Research Centre
LCA	Life cycle analysis
LCFS	Low Carbon Fuel Standard
LDV	Light-duty vehicle
LEV	Low-emission vehicle
lge	Litre of gasoline equivalent
LH_2	Liquid hydrogen
LHV	Lower heating value
LNG	Liquefied natural gas

LP	Linear programming
LPG	Liquefied petroleum gas
LULUCF	Land use, land use change, and forestry
MCFC	Molten-carbonate fuel cell
mD	millidarcy
MEA	Membrane electrode assembly; Mono-ethanolamine
MFC	Microbial fuel cell
MMV	Measurement, monitoring and verification
MOREHyS	Model for Optimisation of Regional Hydrogen Supply
MOx	Mixed oxide
mpgge	Miles per gallon gasoline equivalent
MTBE	Methyl tertiary butyl ether
MTG	Methanol-to-gasoline
MTO	Methanol-to-olefins
NAFTA	North American Free Trade Agreement
n.a.	Not available
NEA	Nuclear Energy Agency
NEDC	New European Driving Cycle
NG	Natural gas
NGC	Natural gas from coal
NGL	Natural-gas liquids
NGPL	Natural gas plant liquids
NMVOC	Non-methane volatile organic carbons
NOC	National Oil Company
NUTS	Nomenclature of Territorial Units for Statistics
OGJ	Oil & Gas Journal
OLADE	Latin American Energy Organization
OECD	Organisation for Economic Cooperation and Development
OPEC	Organization of Petroleum Exporting Countries
OSPAR	Oslo Paris Commission for the Protection of the Marine Environment of the North East Atlantic
PAFC	Phosphoric-acid fuel cell
PEMFC	Proton-exchange-membrane fuel cell; Polymer-electrolyte membrane
PFC	Perfluorocarbons
PGM	Platinum-group metals
PHEV	Plug-in hybrid-electric vehicle
PISI	Port-injection spark ignition
PM	Particulate matter
POX	Partial oxidation
ppm	Parts per million
PPP	Purchasing power parity

PSA	Pressure-swing adsorption
PV	Photovoltaic
PWR	Pressurised water reactor
R&D	Research and development
RAR	Reasonably assured resources
RCS	Regulations, codes and standards
RD&D	Research, development and demonstration
RES	Renewable energy sources
RES-E	Renewable energy sources for electricity generation
RME	Rapeseed methyl ester
SCO	Synthetic crude oil
SCPC	Super-critical pulverised coal
SD	System dynamics
SEC	US Securities and Exchange Commission
SMR	Steam methane reformer
SNG	Synthetic natural gas
SOFC	Solid-oxide fuel cell
SPE	Society of Petroleum Engineers
SRES	IPCC Special Report on Emissions Scenarios
SULEV	Super-ultra-low-emission vehicle
SUV	Sport utility vehicle
TAR	IPCC Third Assessment Report
tce	Tons of coal equivalent
TDM	Transport demand management
toe	Tons of oil equivalent
TPES	Total primary energy supply
TSA	Temperature swing adsorption
TTW	Tank-to-wheel
UCG	Underground-coal gasification
UCTE	Union for the Coordination of Transmission of Electricity
ULEV	Ultra-low-emission vehicle
UN	United Nations
UNDP	United Nations Development Programme
UNECE	United Nations Economic Commission for Europe
UNEP	United Nations Environment Programme
UNFC	United Nations Framework Classification
UNFCCC	United Nations Framework Convention on Climate Change
UNSD	United Nations Statistics Division
URR	Ultimate recoverable resources
USEPA	United States Environmental Protection Agency
USGS	United States Geological Survey

VOC	Volatile organic compounds
vol.%	Per cent by volume
WEC	World Energy Council
WEO	World Energy Outlook
WNA	World Nuclear Association
WRI	World Resources Institute
WTT	Well-to-tank
WTW	Well-to-wheel
wt.%	Per cent by weight
XTL	X-to-liquids
ZEV	Zero-emission vehicle

1

Scope of the book

Michael Ball and Martin Wietschel

Already in 1874, Jules Verne, in his novel *The Mysterious Island*, lets the engineer Cyrus Harding reply when asked what mankind will burn instead of coal, once it has been depleted:

> water decomposed into its primitive elements, . . . and decomposed doubtless, by electricity . . . Yes, my friends, I believe that water will one day be employed as fuel, that hydrogen and oxygen which constitute it, used singly or together, will furnish an inexhaustible source of heat and light, of an intensity of which coal is not capable.

Today's energy and transport system, which is based mainly on fossil fuels, can in no way be evaluated as sustainable. In the light of the projected increase of global energy demand, concerns over energy supply security, climate change, local air pollution and increasing prices of energy services are having a growing impact on policy making throughout the world.

At present, oil, with a share of more than one third in the global primary energy mix, is still the largest primary fuel and covers more than 95% of the energy demand in the transport sector. With continued growth of the world's population and industrialisation of developing nations, such as China and India, accompanied by an increasing 'automobilisation', a surge in global demand for oil is expected for the future. A growing anxiety about the economic and geopolitical implications of possible shortcomings in the supply of oil as a pillar of our globalised world based on transportation is increasingly triggering the search for alternative fuels. However, this search is not only motivated by possible oil shortcomings, but also in response to the climate change issue, because worldwide CO_2 emissions from the transport sector have been growing for decades and most projections show a further increase for the future.

At the heart of the book stands the question of how the growing energy demand in the transport sector can be met in the long term, when conventional (easy) oil will be running out. Among the principal options are unconventional oil from oil sands or oil shale, synthetic Fischer–Tropsch fuels on the basis of gas or coal, biofuels,

The Hydrogen Economy: Opportunities and Challenges, ed. Michael Ball and Martin Wietschel. Published by Cambridge University Press.

electricity as 'fuel' for fully or partially battery-powered electric vehicles, and hydrogen. Unconventionals have a huge potential, but their extraction is very energy intensive and bears higher environmental impacts than conventional production. Synthetic liquid fuels from fossil energy sources can rely on the existing distribution infrastructure, but also come with a higher environmental footprint; moreover, from an energy-efficiency point of view, the syngas route is more favourable for the production of hydrogen than for Fischer–Tropsch fuels (neglecting infrastructure build-up and vehicle availability). 'Sustainable' biofuels increasingly face resource constraints and growing competition with electricity and heat generation, as well as with food production. Electric mobility on the basis of battery electric vehicles is, apparently, the most energy-efficient solution, but major technical and economic breakthroughs for vehicle batteries have to be realised first to bring this path into the market.

The need to modify the present trend, characterised by the unsustainable development of energy systems, requires that effective solutions are found and widely applied. Hydrogen is increasingly seen as offering a range of benefits with respect to being a clean energy carrier (if produced by 'clean' sources), which are receiving ever greater attention as policy priorities. Creating a large market for hydrogen as an energy vector offers effective solutions to both the aspects of emission control and the security of energy supply: hydrogen is emission-free at the point of final use, it is a secondary energy carrier that can be obtained from any primary energy source and it can be utilised in different applications (mobile, stationary and portable).

While the emergence of the so-called 'hydrogen economy', where hydrogen plays a major role as energy vector in the energy system, has been forecast by a range of experts from industry, policy and research and less-than-experts alike, for a number of decades, the discussion about hydrogen as future energy carrier or fuel has only in recent years been increasingly taken up by relevant stakeholders in the field, not least because of breakthroughs in fuel cell technology in the last decade. Despite the attention that hydrogen is receiving, in particular from policy makers and research communities, there is still a lack of publicly available literature about hydrogen for a broader expert group that covers all relevant topics. This book aims to close this gap and provide a synthesis of the latest, most important and interesting research findings and facts regarding the possible transition to a hydrogen-based energy and transport system.

The book intends to highlight both the opportunities and the challenges of introducing hydrogen as a potential energy vector from an economic, technical and environmental point of view. The focus is on the use of hydrogen as alternative fuel in the transport sector, which is generally considered the major driver for its introduction. Given the current controversy and popularity of the hydrogen issue, the book aims to provide – through its multidisciplinary approach – a broad range of decision makers (policy makers, academia, industry) as well as the interested reader with a solid and comprehensive knowledge base about hydrogen. The analysis focuses primarily on the time horizon until 2030. The geographic scope of the book is global, with the exception of a few chapters that are confined to a more European perspective.

The book at hand is the first book to cover hydrogen in a holistic manner from a technical, environmental and economic perspective. Particular highlights include:

- Assessment of the virtues and downsides of hydrogen compared with other alternative fuels in the transport sector;
- Strategies and scenarios for a hydrogen infrastructure build-up;
- Long-term global hydrogen supply scenarios and their impact on resource availability and contribution to CO_2-emissions reduction in the transport sector; and
- Macroeconomic impacts of introducing hydrogen as alternative fuel.

The book is organised as follows.

Chapter 2 addresses why hydrogen has recently been receiving increased attention. First, the challenges of today's energy system – security of supply and reduction of greenhouse-gas emissions – are discussed and existing and emerging energy policies to cope with them addressed. This sets the scene for the introduction of hydrogen, which needs to be seen in the context of the development of the global energy scene. The possible emergence of a hydrogen economy is then reflected from the perspective of historical transitions of energy sources. Next, the chapter outlines which are the major drivers for the possible transition to a hydrogen economy and what potential benefits could be expected from using hydrogen as an energy vector.

One major driver for hydrogen is concern about energy supply security due to shortcomings in the supply of fossil fuels, particularly oil. This aspect is dealt with in Chapter 3. First, the development of the past and present global energy supply are briefly analysed. The focus is on the development of oil and natural-gas production and consumption, as these fuels are expected to be most sensitive with respect to resource-economic constraints in the coming decades. Regarding the remaining reserves of oil and gas, both the 'pessimistic' and the 'optimistic' views are discussed. A special emphasis is placed on the potential availability of unconventional oil and gas deposits and the possible implications resulting from their extraction. Based on these assessments, the interdependency of fossil-fuel resources on the one hand and the development of global energy demand on the other is analysed, and scenarios are derived for the future availability of oil and gas at a global level. The chapter continues with a brief description of the supply situation for coal.

In the light of the projected growth of demand for energy services, particularly electricity, there is a renewed interest in the extension of nuclear power in some countries. With uranium being a finite resource as well, Chapter 4 focuses primarily on the question of a future expansion of nuclear power in the context of the availability of nuclear fuels. Moreover, the evolution of the next generation of nuclear reactors, such as breeder reactors or reactors suitable for hydrogen production, is addressed.

Renewables are often seen as *the* future feedstock for hydrogen, if hydrogen is to make a real contribution to energy security and CO_2-emission reduction. However, 'cheap' renewable potentials are also limited and will be increasingly in competition with heat and electricity generation. An overview of the renewable potentials worldwide and with a particular focus on the situation in the European Union is at the centre of Chapter 5.

There is an urgent need for deploying carbon-dioxide capture and storage (CCS). This is a vital part of a portfolio of technologies and strategies, besides renewable energies and energy efficiency, that are required to reduce and eventually reverse CO_2-emission growth worldwide. With fossil fuels to remain a major primary energy source in the world for several decades to come – not least for the production of hydrogen during the initial phase – it is the only technology that could, potentially, directly achieve very large and rapid reductions in fossil-fuel emissions, although significant challenges still lie ahead. The various technical, economic and legal aspects of CCS are dealt with in Chapter 6.

Given the continuing growth of transport energy demand and showing the limits of fossil fuels in Chapter 3, Chapter 7 focuses on the potentials of alternative transportation fuels, including electricity. The chapter starts with a general overview of the different fuel supply options available. In the following, the major fuel pathways and their technical characteristics are described in more detail. The chapter concludes with a comparison of alternative fuels and drive systems based on a well-to-wheel analysis. This analysis accounts for the entire pathway (from feedstock to the drive system), the energy efficiency, CO_2 emission and costs and allows the important advantages and disadvantages of alternative fuels to be compared. On this basis, the major competitors for hydrogen are identified.

Chapter 8 first provides a brief overview of the evolution of hydrogen vehicles and points out major hydrogen demonstration projects around the globe. The development of a roadmap for hydrogen is essential because a widely accepted and harmonised hydrogen roadmap will give investors more planning security, will stimulate private and public R&D and is necessary for the establishment of an industry policy. To reflect the views of today's stakeholders about the introduction of hydrogen and fuel cells in the next decades, the status of roadmap development in the EU, the USA and Japan is thus another main issue in Chapter 8. In addition, the critical aspect of social acceptance of hydrogen, as a prerequisite for its introduction as vehicle fuel, is discussed.

Chapter 9 addresses the fundamental chemical and physical properties of hydrogen and how they play out when using hydrogen as vehicle fuel.

In Chapter 10, the most important (commercial) hydrogen production processes available today are described and analysed from the perspective of technology and economics, including their parameterisation for the hydrogen infrastructure model discussed in Chapter 14. Future development goals necessary to reach a market breakthrough of these processes, as well as novel hydrogen production technologies that still require basic research, are also addressed. The chapter finishes by discussing the use of hydrogen as industrial gas and the availability of industrial surplus hydrogen for fuelling hydrogen vehicles during the transition phase.

Chapter 11 addresses the critical question of hydrogen storage on board the vehicles. For hydrogen vehicles to reach competitive driving ranges, storage is crucial. There are still significant technical challenges to be overcome, which are discussed in this chapter.

Chapter 12 discusses and analyses the different options for hydrogen distribution – pipelines and trailers (including liquefaction) – from a technical and economic point of view, in the same way as the hydrogen production technologies in Chapter 10. Further, different hydrogen refuelling station concepts are described and the necessity for the development of codes and standards addressed.

Some of the most important benefits of hydrogen can only be realised if hydrogen is used in fuel cells; for example, the high overall conversion efficiency compared with the internal combustion engine as well as the reduction of local pollution and noise. Therefore, the market success of fuel cells plays a key role in a hydrogen economy. In Chapter 13, the fuel cell as a technology is introduced and its strategic role outlined. The chapter describes the various types of fuel cell and their potential uses in mobile, stationary and portable applications. As preparing for the structural changes in industry is just as important as the technical optimisation of fuel cells, the remainder of the chapter is devoted to this aspect.

Constructing a hydrogen infrastructure with user centres, a mix of hydrogen production technologies, plant sizes and locations, as well as related transport choices, is crucial and constitutes a challenging task for its introduction as vehicle fuel. In Chapter 14 different hydrogen infrastructure scenarios are developed and analysed. For the hydrogen infrastructure analysis, a model-based approach is described to assess its schedule and geography. What this build-up could look like, what it might cost and what the resulting CO_2-emission reductions in the transport sector are, are shown in a detailed case study for Germany, followed by more general strategies and conclusions at a European level. Closing the loop to the resource analyses in Chapters 3 to 5, this chapter concludes with some global hydrogen supply scenarios and their impacts on primary resource requirements.

While Chapter 14 focuses on a hydrogen infrastructure analysis for Europe, Chapter 15 addresses the build-up of a hydrogen infrastructure in the USA.

If hydrogen production is to be fully integrated into the energy system, a more holistic view needs to be applied with respect to its interactions with the electricity sector. The various aspects of the interplay between hydrogen production and electricity generation are addressed in Chapter 16. For instance, with growing capacities of wind power or photovoltaic generators, hydrogen could become a promising storage medium for surplus electricity from these intermittent renewable energies. On the other hand, with fossil fuels remaining the prevalent energy supply in the foreseeable future, despite their drawbacks with regard to climate change, routes to large-scale cost-effective hydrogen production with integrated CO_2 management for use in either power generation or as transport fuel are investigated. A special focus is on the technological and strategic aspects of co-production of hydrogen and electricity in integrated gasification combined-cycle (IGCC) power plants. Finally, from the perspective of overall CO_2-emissions reduction in the energy system, the question is addressed: whether renewable energies are better deployed in the transport sector or the power sector.

In the long run, hydrogen corridors offer, among other things, the chance to manage the energy resource limitations for hydrogen production within the EU and to improve energy supply security. Therefore, Chapter 17 deals with the assessment of possible hydrogen corridors between the EU and neighbouring countries, using consistent hydrogen scenarios, cost and potential calculations. Barriers for hydrogen corridors are also identified.

Often, only technical aspects are considered when looking at the deployment of hydrogen technologies. However, the introduction of hydrogen could have relevant implications for GDP, welfare and job development in a nation or region. The competitiveness of a nation could be one major driver for hydrogen use as an energy carrier. These issues are discussed in Chapter 18. Among others, possible economic effects are shown on the basis of a quantitative model analysis and assessed for relevant EU member states.

The results of Chapter 18 form the basis for the question of whether hydrogen technologies might be able to contribute to sustainable development by promoting both economic growth and environmental protection. The environmental issue is handled in Chapter 19, which integrates two debates: one on sustainable transport and the other on the future of hydrogen-powered transport technologies. Transport systems perform vital social functions, but in their present state cannot be considered 'sustainable'. Particular areas that need to be addressed in this respect include emissions, safety, land use, noise and social inclusion. Vehicle technologies will play a key role in addressing several of these areas. This chapter examines the role of hydrogen, and fuel-cell vehicle technologies in particular, in contributing to a future sustainable transport system and also shows the limitation of such an approach.

However, the question is whether our two major energy challenges of the future, climate change and shortcomings of conventional energy resources, can be solved by technical developments alone. The high losses at each level of energy conversion and use indicate that energy has to be used much more efficiently than is currently done. The importance of energy efficiency is at the centre of Chapter 20.

Chapter 21 summarises the major findings and conclusions of the book and reflects critically on the perspectives of a transition to a hydrogen-based energy and transport system.

Note on economic figures presented in the book

The majority of the economic figures, such as energy prices or capital costs of hydrogen technologies, presented in this book are expressed in euros (€), as they have been taken largely from European studies and literature sources. Past cost data or future projections have generally been converted into money of today. On the other hand, in various literature sources the cost data are indicated in US dollars instead of euros. The correct approach would be to apply the industrial-sector-specific inflation rate for each process to convert past cost data into today's costs and subsequently to convert the cost figures from US dollars to euros using today's exchange rates. This would have to be done for each process where the cost figures are indicated in US dollars.

The industry-specific inflation rate is different from sector to sector; also the inflation rate in the USA is different from that in the EU. In addition, the exchange rate between US dollars and euros has historically fluctuated to a large extent. It is assumed that the different inflation rates in the USA and the EU partly compensate for the error, which would be made if the exchange rate were assumed to be one euro per US dollar. To avoid modifying the figures indicated in the original literature sources, in this book the exchange rate is thus assumed to be one euro per US dollar.

The energy price scenarios and economic figures for hydrogen technologies and infrastructure build-up presented in the book date back mostly to the years 2005 and 2006. Across all industrial sectors, in the recent past, an unexpected and lasting surge in energy prices (above all, oil) has been experienced worldwide. For instance, the oil, gas and coal prices projected for 2030 in the high-energy price scenario in this book have already partially been exceeded by today's market prices of these commodities. However, the cost competitiveness of renewable energies as well as of hydrogen and fuel cell vehicles is positively influenced by this development, and hydrogen, on the basis of renewable energy sources, is the winner of such a development.

Steel and metal prices have escalated as well; this is manifested, for instance, in the drastic increase of capital costs for plant equipment or pipelines. For technologies with a high share of energy costs in total production costs, such as steam methane reformers, the impact of higher feedstock prices can significantly influence their economic attractiveness. This recent increase in energy and material or commodity prices has not been factored into the hydrogen supply costs. Hence, the absolute costs presented in this book have to be taken with caution, as they represent a rather optimistic estimate. Nevertheless, it has to be taken into account that the prices of conventional energy and vehicle technologies are also affected by the price increase.

2

Why hydrogen?

Michael Ball

The world is facing a new era of energy anxiety with complicated choices regarding fuel sources, new technologies, and government regulations and actions. There is also a growing global consensus that greenhouse-gas emissions need to be managed, resulting in the challenge to search for the best way to rein in emissions while also providing energy to sustain economies. The projected increase in global energy demand, and the economic and geopolitical implications of possible shortcomings in the supply of oil have been major drivers stirring the debate about the future energy supply. Supply disruptions of oil would primarily hit the transport sector, since this is still almost entirely dependent on oil worldwide. This situation is increasingly triggering the search for alternative automotive fuels. In this respect, hydrogen has in recent years been gaining increased attention.

This chapter addresses why. First, the challenges of today's energy system – security of supply and reduction of greenhouse-gas emissions – will be discussed and existing and emerging energy policies to cope with them will be addressed. This sets the scene for the introduction of hydrogen, which needs to be seen in the context of the development of the global energy scene. The possible emergence of a hydrogen economy is then reflected from the perspective of historical transitions of energy sources. Next, the major drivers for the possible transition to a hydrogen economy will be outlined as well as the potential benefits that could be expected from using hydrogen as an energy vector.

2.1 The challenges of today's energy system

There are two major concerns about the future of the energy sector: *security of energy supply* and *climate change* (due to greenhouse-gas emissions, mainly CO_2).[1] Figure 2.1 demonstrates why. Global primary energy use per capita has increased

[1] Another issue is local air pollution, which is also linked to fossil-fuel combustion, not least in the transport sector. Especially in the world's growing megacities, road traffic has an increasingly negative impact on urban air quality (see also Chapter 19).

The Hydrogen Economy: Opportunities and Challenges, ed. Michael Ball and Martin Wietschel. Published by Cambridge University Press.

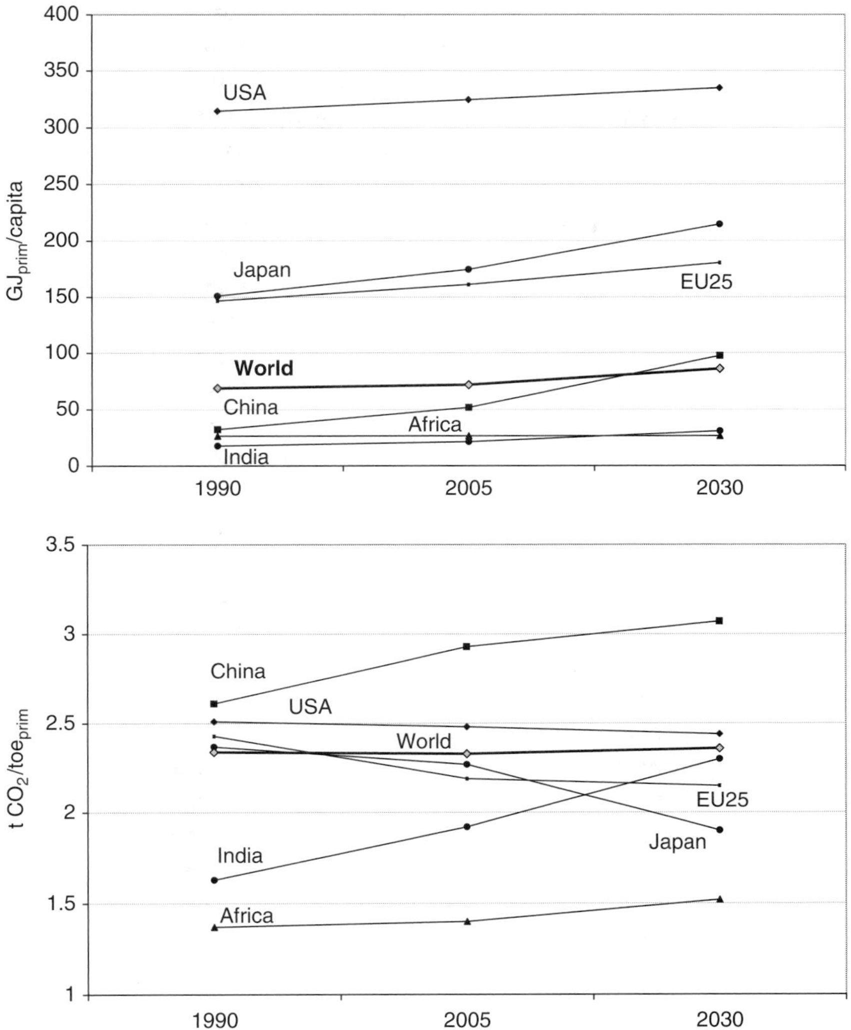

Figure 2.1. Development of global energy use per capita and carbon intensity (IEA, 2006; UNPD, 2006).

from 69 GJ/capita in 1990 to 72 GJ/capita in 2005 (a growth of 0.3% per year) and global carbon intensity has remained almost constant at 2.33 t CO_2/toe_{prim} (0.64 t C/toe_{prim}) (see also Fig. 2.10); both trends are projected to continue in the business-as-usual case until 2030 (IEA, 2006).

In 2005, the United States had, with 325 GJ/capita, the highest per-capita energy use, followed by Japan with 175 GJ/capita and the EU25 with 161 GJ/capita; the per-capita use of India and Africa amounted to 21 and 26 GJ/capita, respectively. It is not expected that energy use per capita will decrease. According to the

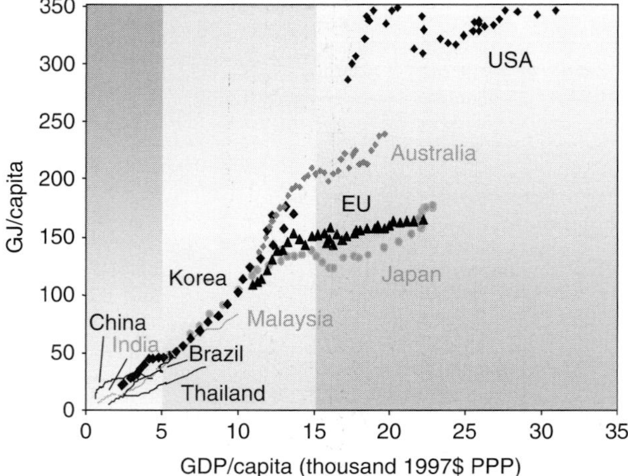

Figure 2.2. Climbing the energy ladder (Shell, 2001).

International Energy Agency (IEA), the average global energy use is projected to grow to 86 GJ/capita in 2030, according to their Reference Scenario (IEA, 2006).[2] The biggest growth rate, with more than 85%, is expected for China.

Energy is essential for economic development and rising living standards. Changes in the energy system mark transitions in the economic and social development of countries and societies, as they climb the energy ladder (Fig. 2.2). The first and most important step is substituting commercial for traditional fuels, such as biomass fuels and animal wastes, which are still the major energy source for people in many developing countries. After that there is a strong – but constantly changing – relationship between income and energy demand, with economic growth becoming increasingly decoupled from energy demand (Shell, 2001). When per capita GDP (on a purchasing power parity basis) reaches some:

- $3000 – demand explodes as industrialisation and personal mobility take off.
- $10 000 – demand slows as the main spurt of industrialisation is completed.
- $15 000 – demand grows more slowly than income as services dominate economic growth and basic household energy needs are met.
- $25 000 – economic growth requires little additional energy.

As basic energy needs are met, consumer priorities shift to other, often less energy-intensive, goods and services, pointing to eventual saturation of energy needs. However, in spite of a reduction in energy intensity (more energy saving, more efficiency in end uses), the rising demand for higher levels of comfort may still lead to a higher per capita consumption, as for instance projected for the US and EU25 in Fig. 2.1. Newly industrialising countries are able to climb the

[2] The IEA World Energy Outlook 2006 forms the basis for the analysis in this book (IEA, 2006). However, there are no fundamental changes in trends projected by the International Energy Agency (2008).

Table 2.1. *Development of energy use per capita and GDP*

	Energy use per capita (GJ/capita)		Energy use per GDP (GJ/1000 US$PPP(2000))	
	1990	2005	1990	2005
USA	314.5	324.5	11.3	9.6
China	32.2	51.8	20.3	9.2
India	17.6	21.2	10.6	8.4
EU25	146.7	161.2	7.2	6.7
Africa	26.3	26.4	14.9[a]	14.9[a]
Japan	151.1	174.5	6.5	6.6
World	**69**	**72**	**10.7**	**9.2**

Note:
[a] Sub-Saharan Africa.
Source: own calculations according to (IEA, 2007c; UNSD, 2007; WRI, 2007).

ladder more quickly, thanks to proven technologies with lower costs from earlier development elsewhere. But more efficient technologies mean they also require less energy at every stage. Overall, however, Fig. 2.2 underlines the fact, that while countries are further industrialising, their energy intensity will grow. On the other hand, despite the underlying economic growth, the energy use per unit GDP is falling, showing that energy is generally being used more efficiently (Table 2.1).

For the last three decades, GDP/capita and population growth were the main drivers of the increase in global greenhouse-gas or CO_2 emissions. To decompose the main driving forces of GHG emissions, the *Kaya identity* is used, which expresses the level of energy-related CO_2 emissions as the product of four indicators: carbon intensity (CO_2 emissions per unit of total primary energy supply (TPES)); energy intensity (TPES per unit of GDP); gross domestic product per capita (GDP/cap); and population (CO_2 emissions $=$ population $\times$ (GDP/population) $\times$ (energy/ GDP)$\times$ (CO_2/energy)). The average global growth rate of CO_2 emissions between 1970 and 2004 of 1.9% p.a. is the result of 1.6% p.a. population growth, 1.7% p.a. $GDP_{PPP(2000)}$/cap, -1.2% p.a. energy intensity and -0.2% p.a. carbon intensity (IPCC, 2007c). At the global scale, declining carbon and energy intensities could not offset income effects and population growth, and consequently carbon emissions have risen in absolute terms.[3] Under the Reference Scenario of the IEA (IEA, 2006), these trends are expected to remain valid until 2030, as population rises, developing countries

[3] In the future, with the scope and legitimacy of controlling population subject to ongoing debate, the two technology-oriented factors, energy and carbon intensities, have to bear the main burden for emission reduction. This is all the more daunting in the light of the expected increase in global population: from around six billion today to about eight or nine billion in the period between 2030 and 2050.

expand their economies and energy is not expected to be further decarbonised (see Chapter 3).

While the carbon intensity of industrialised nations is projected to fall slightly, the energy supply of the developing economies, e.g., China and India, will become much more carbon intensive (see Fig. 2.1). It appears that rising carbon intensities accompany the early stages of the industrialisation process, which is closely linked to accelerated electricity generation, based mainly on fossil fuels (primarily coal). In addition, the emerging but rapidly growing transport sector is fuelled by oil, which further contributes to increasing carbon intensities. Stepped-up fossil fuel use, GDP/capita growth and, to a lesser extent, population growth, result in the dramatic increase in carbon emissions in India and China. Worldwide, carbon intensity is projected to increase from 2.33 t CO_2/toe_{prim} in 2004 to 2.36 in 2030 in the IEA Reference Scenario (average growth rate of CO_2 emissions of 1.7% p.a.).

Differences in terms of per-capita income, per-capita emissions and energy intensity among countries remain significant (see Fig. 2.3). In 2004, UNFCCC Annex I countries held a 20% share in world population, produced 57% of world gross domestic product based on purchasing power parity (GDPppp), and accounted for 46% of global GHG emissions.

The need to modify the present trend of energy use, characterised by the unsustainable development of global energy systems, requires that effective solutions are found and widely applied. Such solutions are characterised by a high degree of complexity, as they can have far-reaching effects on the quality of life of citizens and depend on a large set of variables. They should handle the fact that the energy supply relies at present, and also in the medium term, on fossil fuels, often with reduced security for energy imports, and threats to the equilibrium of the ecosystem. There is, therefore, a need to enlarge and diversify the energy supply using clean and renewable sources and to increase the efficiencies and cost effectiveness of the energy-conversion technologies.

2.1.1 Security of energy supply

Energy insecurity can be defined as the loss of welfare that may occur as a result of a change in the price or availability of energy (Bohi and Toman, 1996). Global demand for oil has reached new heights, led mainly by industrialising countries in Asia. With crude oil prices reaching new record highs exceeding $140 per barrel in July 2008 (although falling again since then!) and with prices of internationally traded natural gas, coal and uranium following suit (see Figs 2.4 and 2.5), energy-supply-security concerns are back on the agenda of policy makers (in the Western world).[4] Part of the surge of oil prices in recent years can be explained by a shrinking margin between oil production capacity and demand.

[4] There is a growing number of studies that investigate the possible impacts of high oil prices on the global economy, see for instance (IMF, 2000; IEA, 2004; Greene and Ahmad, 2005; Hirsch *et al.*, 2005).

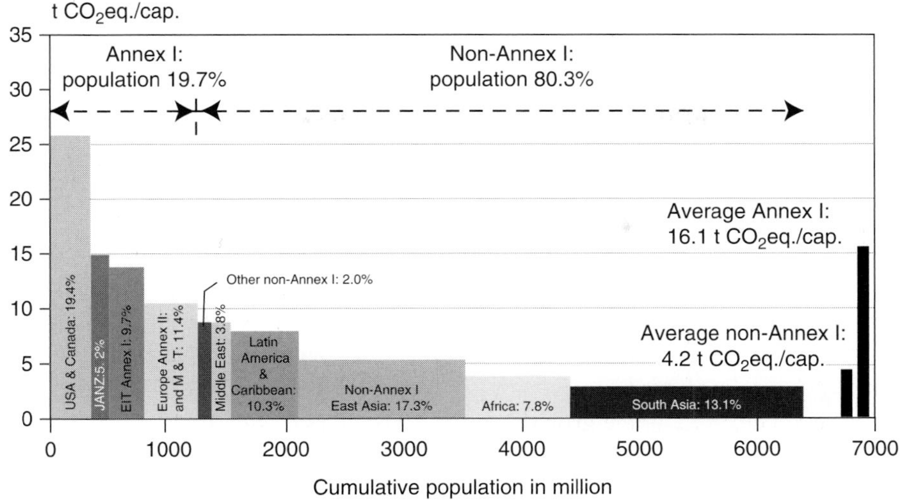

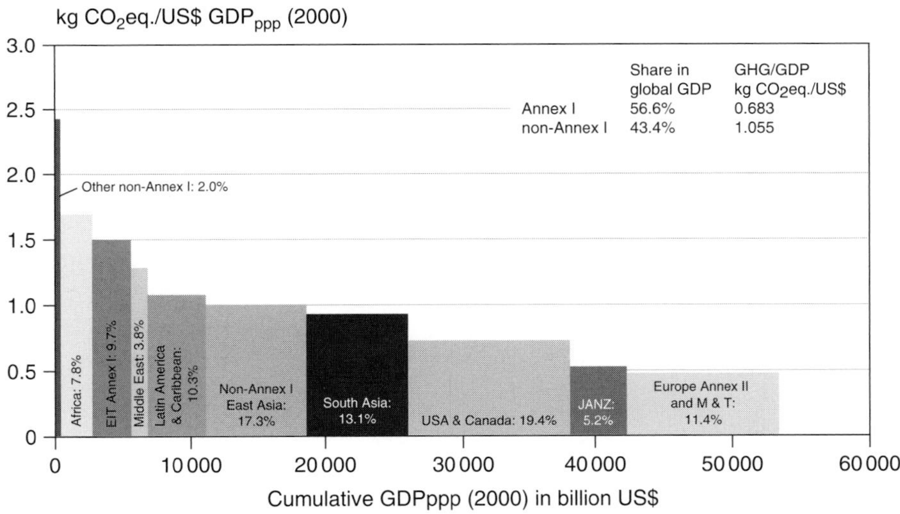

Figure 2.3. Distribution of CO_2 emissions per capita and GDP (IPCC, 2007c).

Given a continued growth of the world's population to more than eight billion in the coming decades as well as a progressing industrialisation and economic development of developing nations, particularly in Asia, it has to be expected that the global demand for energy is likely to continue to grow in the coming decades. According to the IEA, world primary energy demand is projected to expand by more than 50% until 2030, and could more than double by 2050: fossil fuels will continue to dominate global energy use, accounting for more than 80% of the increase. (While this 'traditional' energy-growth prognosis is a commonly accepted 'given' of other long-term global energy scenarios as well, they generally fall short of discussing its actual

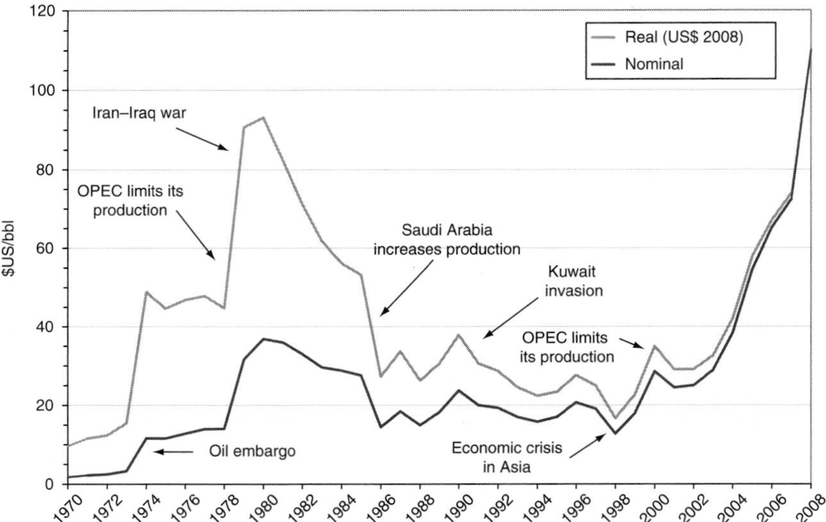

Figure 2.4. Historic development of crude oil prices since 1970 (BP, 2008).

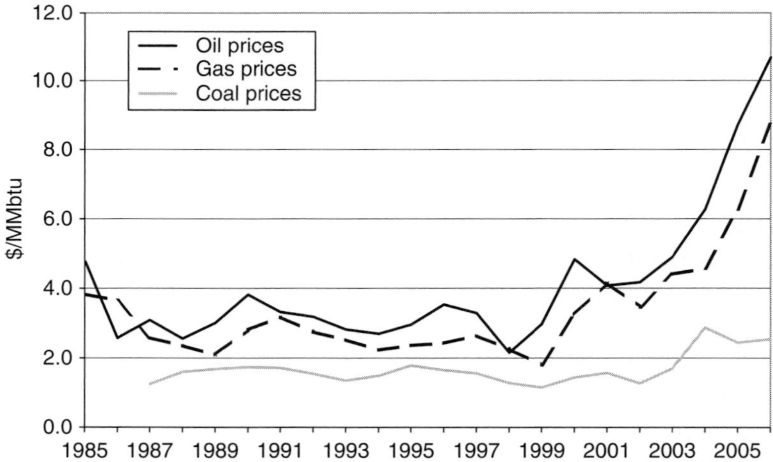

Figure 2.5. Development of crude oil, gas and coal prices (BP, 2008). Crude oil prices: OECD (cif); gas prices: EU (cif); coal prices: EU.

dimension and impact on the planet, such as accompanying resource depletion and ecological consequences.) Shortcomings in the supply of fossil fuels are likely to occur first with crude oil, as it is the most depleted fossil fuel today. In addition, with oil being expected to remain the single largest fuel in the primary energy mix – having a share of 35% today – according to the above projection almost a doubling of cumulative oil production would be required until 2030 to meet the rising demand (see also Chapter 3).

Moreover, there is a high geographic concentration of oil and gas occurrences as well as a growing import dependence from few (often politically unstable) countries. For instance, the oil import dependence in 2004 in the USA amounted to 64%, in the EU25 to 79%, and in Japan even to 100%; while China was still a net exporter of oil in 1990, its net imports amounted to 46% of its consumption in 2004 (IEA, 2006). The oil import dependence in all major importing regions is projected to grow further in the future, especially from OPEC countries as they hold around 75% of the remaining reserves. There is also a growing import dependence on natural gas. In 2005, the EU25 for instance, imported 53% of its gas, with Russia accounting for some 41% of the imports.[5] On top of the fact that reserves of oil and natural gas are restricted to only a few countries in the world, their global resources are also limited. The rising global energy demand is aggravating this situation and sooner or later we are likely to face a gap between supply and demand.

Consequently, there is renewed public interest in alternatives to fossil fuels, especially to oil, resulting in new technology initiatives to promote renewables, biofuels, nuclear power and hydrogen. Higher oil prices also tend to open up larger markets for more carbon-intensive liquid-fuel production processes, such as from oil shale or oil sands, or synthetic liquid fuels derived from coal or gas. However, energy-security concerns tend first of all to invigorate a higher reliance on indigenous energy supplies and resources. Regions where coal is the dominant domestic energy resource tend to use more coal, especially for electricity generation, which increases greenhouse-gas emissions. Coal has been enjoying a revival in recent years for a variety of reasons – among others, high gas prices – in Asian developing countries, the USA and some European countries.[6] In some countries, the changing relative prices of coal and natural gas have changed the dispatch order in power generation in favour of coal.

Energy security also means access to affordable energy services by those people, largely in developing countries, who currently lack such access. Energy security plays an important role in mitigating climate change. Striving for enhanced energy security can affect GHG emissions in opposite ways. On the one hand, GHG emissions may be reduced by stimulating rational energy use, efficiency improvements, innovation and the development of alternative energy technologies with inherent climate benefits. On the other hand, measures supporting energy security may lead to higher GHG emissions due to stepped-up use of indigenous coal or the development of lower-quality and unconventional oil resources (IPCC, 2007c).

[5] In total, the EU25 currently imports around 50% of its primary energy and if no additional measures are taken, the import dependence will rise to around 70% by 2030 (EC, 2006a). The EU also imports 35% of its coal as well as 100% of its uranium.

[6] In the USA, some 150 new coal-fired power stations are planned. In China, two 500 MW coal-fired power plants are currently starting up every week, and each year the country's coal-fired power generating capacity increases by the equivalent of about half the entire German grid (The Economist, 2007).

With regard to energy security policy, a distinction can be made between government actions to mitigate the short-term risks of physical unavailability occurring in case of a supply disruption and efforts to improve energy security in the long term. In the first case, actions include establishing strategic reserves, dialogue with producers, and determining contingency plans to curtail consumption in times of important supply disruptions. In the second case, policies tend to focus on tackling the root causes of energy insecurity. These can be broken down into four broad categories (IEA, 2007a):

- energy-system disruptions linked to extreme weather conditions or accidents;
- short-term balancing of demand and supply in the electricity markets;
- regulatory failures; and
- concentration of fossil-fuel resources.

The uneven distribution of fossil-fuel resources around the world is the most long-lasting cause of energy insecurity. Policies addressing concerns linked to resource concentration may have the most significant implications for climate-change mitigation and vice versa, as in both cases policies are likely to affect fuel and associated technological choices (IEA, 2007a). Besides energy conservation, the best guarantee of the security of energy supply is clearly to maintain a diversity of energy sources and supplies. In this respect, the advantage of hydrogen is that it can help in the long term to diversify the energy supply, particularly in the transport sector, as it can be produced from any primary energy source, unlike other alternative fuels (except electricity). As hydrogen can also be produced at various places in the world, the dependence on only a few countries is low.

The number of policy measures to enhance energy security is growing worldwide at an unprecedented pace. Actions resulting from the EU's energy policy, for instance, include (CEU, 2007):

- an increase in renewable energy to 20% of EU supply by 2020,
- an increase in use of biofuels in road transport to 10% by 2020 (5.75% by 2010), and
- a reduction in energy use of 20% by 2020 to improve energy efficiency.

Similar actions are to be observed in other parts of the world, increasingly with the objective of diversifying the fuel supply in the transport sector. Examples are in Brazil, which has the world's most developed biofuel industry, and where a 25% blend (mainly ethanol) is mandatory, or the Alternative Fuel Standard (AFS) at federal level in the USA, or various biofuel mandates being introduced at state level (see also (EC, 2006b)).

2.1.2 Climate change

Global primary energy use more than doubled from 225 EJ in 1970 to 470 EJ in 2004 (at an average annual growth of 2.2%), with more than 80% of the energy today

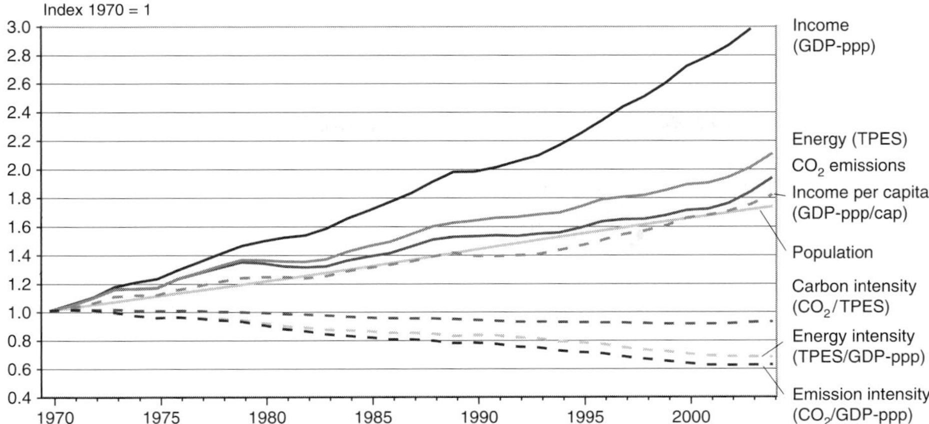

Figure 2.6. Relative global development of gross domestic product measured in PPP (GDPppp), total primary energy supply (TPES), CO_2 emissions (from fossil-fuel burning, gas flaring and cement manufacturing) and population (Pop). In addition, in dotted lines, the figure shows income per capita (GDPppp/Pop), energy intensity (TPES/GDPppp), carbon intensity of energy supply (CO_2/TPES), and emission intensity of the economic production process (CO_2/GDPppp) for the period 1970–2004 (IPCC, 2007c).

being supplied by fossil fuels. As a consequence, emissions of carbon dioxide (CO_2), the main greenhouse gas from human activities, are the subject of a worldwide debate about energy sustainability and the stability of the global climate. Evidence that human activities are causing the planet to warm up is now unequivocal, states the United Nations Intergovernmental Panel on Climate Change (IPCC) in its Fourth Assessment Report (AR4); this is supported by observations of increases in global average air and ocean temperatures, widespread melting of snow and ice, and rising global average sea level (IPCC, 2007a).[7] The IPCC concludes that it is at least 90% certain ('very likely') that anthropogenic emissions of greenhouse gases rather than natural variations are responsible for the observed increase in global average temperatures. It is, however, important to note that the *natural* greenhouse effect is very important for life on Earth, as the average temperature at the Earth's surface would otherwise be roughly 30 °C lower than the current global average of 15 °C. For a comprehensive analysis of the science of climate change, its impacts and mitigation strategies see (IPCC, 2007a; b; c).

2.1.2.1 GHG emission trends – review and future outlook

Total annual greenhouse-gas emissions are continuously rising. Figure 2.6 shows that the effect on global emissions of the decrease in global energy intensity during 1970 to

[7] Current concentrations of GHG have already caused the mean global temperature to increase by 0.76 °C in the period from 1850 to 2005; owing to the inertia of the climate system this will lead to at least a further half-degree warming over the next few decades. Eleven of the twelve years from 1995 to 2006 rank among the 12 warmest years in the instrumental record of global surface temperature (since 1850).

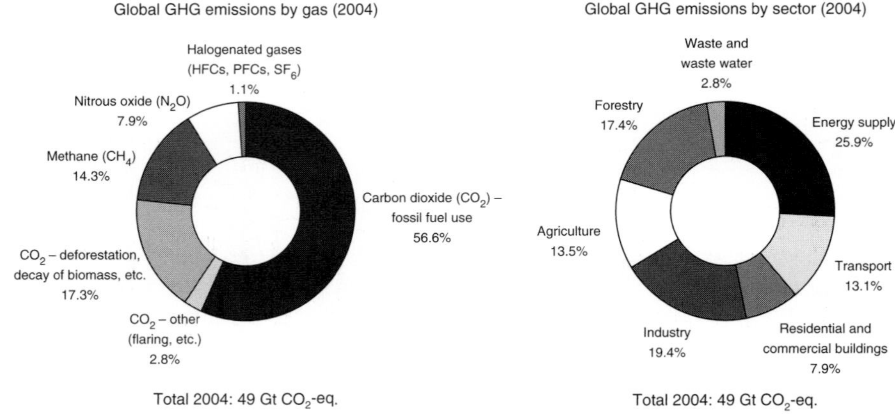

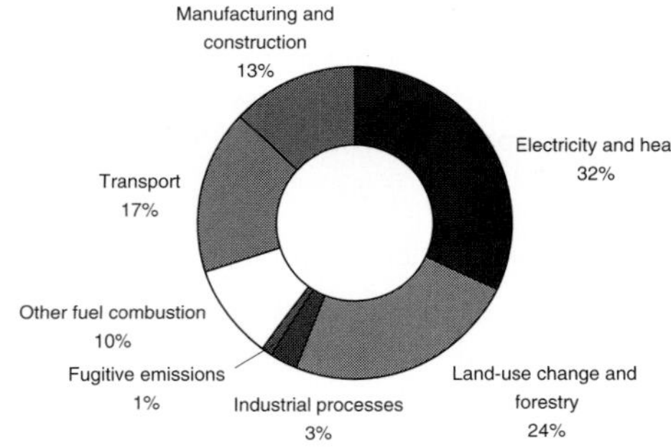

Figure 2.7. Breakdown of GHG emissions by gas and sector (IPCC, 2007c; WRI, 2006).

2004 has been smaller than the combined effect of global income growth and global population growth; both drivers of increasing energy-related CO_2 emissions. The long-term trend of a declining carbon intensity of energy supply reversed after 2000.

Over the last three decades, GHG emissions increased by an average of 1.6% per year, with CO_2 emissions from fossil-fuel use growing at 1.9% per year. Total GHG emissions in 2004 (Kyoto gases)[8] amounted to 49.0 Gt CO_2-equivalent (CO_2-eq.), a 70% increase since 1970 and a 24% increase since 1990. Total CO_2 emissions in 2004 amounted to 26.1 Gt (see Fig. 2.7). Emissions of CO_2 have grown by approximately 80% since 1970 (28% since 1990) and represented 77% of total anthropogenic emissions in 2004. Total methane (CH_4) emissions rose by about 40% from 1970; sectorally there was an 84% increase from combustion and the use of fossil fuels,

[8] Carbon dioxide (CO_2), methane (CH_4), nitrous oxide (N_2O), hydrofluorocarbons (HFCs), perfluorocarbons (PFCs) and sulphur hexafluoride (SF_6).

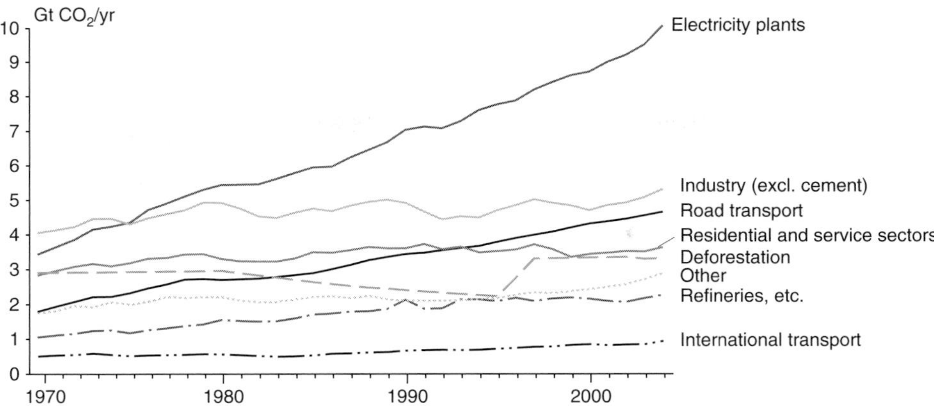

Figure 2.8. Sectoral breakdown of global CO_2 emissions, 1970–2004 (IPCC, 2007c).

while agricultural emissions remained roughly stable, owing to compensating falls and increases in rice and livestock production, respectively. Nitrous oxide (N_2O) emissions grew by 50% since 1970, mainly owing to the increased use of fertiliser and the aggregate growth of agriculture (IPCC, 2007c).

The largest growth in CO_2 emissions has come from the power-generation and road-transport sectors, with industry, households and the service sector remaining at approximately the same levels between 1970 and 2004 (Fig. 2.8). The transportation sector accounts for about 13% of global GHG emissions (17% of global CO_2 emissions), with most of these emissions (72%) resulting from road transport, followed by aviation (about 12%), and marine transport (8%) (WRI, 2005; 2006).

Since 1970, GHG emissions from the energy-supply sector have grown by over 145%, transport emissions by over 120%; industry sector's emissions grew by almost 65%, land use, land-use change, and forestry (LULUCF) by 40%, while the agriculture sector (27%) and the residential and commercial sectors (26%) saw the slowest growth (IPCC, 2007c). On a geographical basis, there are important differences between regions. North America, Asia and the Middle East have driven the rise in emissions since the 1970s. The former region of the Soviet Union has shown significant reductions in CO_2 emissions since 1990, reaching a level slightly lower than in 1972. Developed countries (UNFCCC Annex I countries) hold a 20% share in world population, but account for 46% of global GHG emissions. In contrast, the 80% of world population living in developing countries (non-Annex I countries) account for the remaining 54% of GHG emissions. Based on the metric of GHG emission per unit of economic output (GHG/GDP_{ppp}), Annex I countries display generally lower GHG intensities for the economic production process than non-Annex I countries (see also Fig. 2.3).

The single biggest impact on the climate resulting from human activities comes from emissions of CO_2, the most important anthropogenic greenhouse gas. Global atmospheric CO_2 concentrations have increased by just over one third of their pre-industrial level of about 280 ppm (around 1750, before the Industrial Revolution), reaching 379 ppm in 2005.[9] The atmospheric concentration of CO_2 in 2005 exceeded by far the natural range over the last 650 000 years (180 to 300 ppm) as determined from ice cores. The total CO_2 equivalent concentration of all long-lived greenhouse gases is about 455 ppm CO_2-eq., although the effect of aerosols, other air pollutants and land-use change reduces the net effect to around 375 ppm CO_2-eq. (IPCC, 2007a). (Global atmospheric concentrations of methane (CH_4) and nitrous oxides (N_2O) in 2005 were at 1774 ppb and 319 ppb, respectively.) The primary source of the increased atmospheric concentration of CO_2 since the pre-industrial period results from fossil-fuel use, with land-use change providing another significant but smaller contribution.

Despite ongoing improvements in energy intensities, global energy use and supply is projected to continue to grow, especially as developing countries pursue industrialisation. In the absence of further policy action, GHG-emission trends are expected to continue as well, if global energy demand and associated supply patterns continue to rely on fossil fuels. All world energy scenarios project that the global energy mix until 2030 will essentially remain unchanged, with fossil fuels remaining the world's dominant energy source, and with consequent implications for GHG emissions (see also Section 3.1). For instance, according to the SRES (non-mitigation) scenarios, GHG emissions are set to grow by between 25% and 90% between 2000 and 2030 under the business-as-usual scenario (IPCC, 2000). Developing countries are expected to account for a large share of this rise. The fastest growth in emissions comes from road transport and electricity generation. A new trend seen in recent years is a slight rise in the global share of energy generated from coal, largely in China.

The consequence is that the greenhouse effect, which has maintained an average temperature on the Earth's surface at about 15 °C, is intensifying, further warming the Earth and changing the climate system. While this may entail both beneficial and adverse effects on the environment and socioeconomic systems, the larger the changes and the rate of change in climate, the more the adverse effects are likely to predominate. The question of precisely how much the world will warm in the future is still an area of active research. A measure of the response of the climate system to changes in radiative forcing is the equilibrium climate sensitivity, which is defined as the global average surface warming following a doubling of *equivalent* atmospheric CO_2 concentrations from pre-industrial levels (roughly equivalent to 550 ppm CO_2-eq.). The equilibrium climate sensitivity is likely to be in the range of 2 °C to 4.5 °C, with a best estimate of about 3 °C, and is very unlikely to be less than 1.5 °C (IPCC, 2007a). For illustration, a warming of 4.5 °C on a global scale would

[9] The annual CO_2 growth rate was larger during the last ten years (1995–2005 average: 1.9 ppm/year) than it has been since continuous direct atmospheric measurements began (1960–2005 average: 1.4 ppm/year), although there is a year-to-year variability in growth rates.

be comparable to the change in average temperatures from the last ice age to today (Stern, 2006).

2.1.2.2 Impacts of climate change

The IPCC report concludes that climate change is already having major impacts. The number and severity of climate effects will further increase as climate change progresses, e.g., melting glaciers, coastal flooding due to rising sea levels, declining crop yields, water shortages, spread of diseases, impacts on ecosystems (such as extinction of species), etc. For more details on the potential effect of climate change, refer to (IPCC, 2007a; b).

The impact of these effects will vary. The availability of water will increase in regions where it is already good, but decline in many areas that are already dry. Glacial retreat will reduce access to water in areas that are dependent on meltwater from the mountain chains around the world. If global temperature increases by more than 1.5 to 2.5 °C, there is increasing likelihood of extinction of 20–30% of plant and animal species on the planet. Major changes may also take place in the functioning of ecosystems, having negative effects on biodiversity and ecosystem services. The most exposed population centres are densely populated coastal areas and communities and businesses that are sensitive to extremes or dependent on climate-sensitive resources.

Climate change could impact on growth and development very seriously. The four areas of the world thought to be the most vulnerable to climate change are: the Arctic, where temperatures are rising fast and ice is melting; sub-Saharan Africa, where dry areas are forecasted to get dryer; small islands, because of their inherent lack of capacity to adapt, and Asian mega-deltas, where millions of people will be at increased risk of flooding.

2.1.2.3 Definition of climate-change targets

Limits to climate change that are deemed as prevention of dangerous anthropogenic interference with the climate system, as defined in Article 2 of the United Nations Framework Convention on Climate Change (UNFCCC), (see Section 2.1.2.5), can be defined with respect to different criteria, such as concentration stabilisation at a certain level, global mean temperature or sea-level rise, or levels of ocean acidification.[10] Defining what is dangerous anthropogenic interference with the climate system, and hence what are limits to be set for policy purposes is a complex task and can only partially be informed by science, as it inherently involves normative judgments (IPCC, 2007c).

Because Article 2 of UNFCCC states as its objective the 'stabilisation of greenhouse-gas concentrations in the atmosphere at a level that would prevent dangerous anthropogenic interference with the climate system', a commonly used target has been the stabilisation of the atmospheric CO_2 concentration. Even though it is very

[10] The choice of different targets is not only relevant because it leads to different uncertainty ranges, but also because it leads to different strategies. Stabilisation of one type of target, such as temperature, does not imply stabilisation of other possible targets, such as sea-level rise, radiative forcing, concentrations or emissions.

difficult to assess the global climate changes related to the increase of greenhouse gas concentration, there is a general consensus that a 'sustainable' situation can be reached if GHGs remain below 550 ppm CO_2-eq. (Socolow and Pacala, 2006; Stern, 2006). This is generally considered by scientists to be the maximum acceptable value to avoid significant climate changes.

If more than one GHG is included, most studies use the corresponding target of stabilising radiative forcing, thereby weighting the concentrations of the different gases by their radiative properties. The advantage of radiative-forcing targets over temperature targets is that the consequences for emission trajectories do not depend on climate sensitivity, which adds an important uncertainty. The disadvantage is that a wide range of temperature impacts is possible for each radiative forcing level. Temperature targets, by contrast, provide a more direct first-order indicator of potential climate-change impacts, but are less practical to implement in the real world, because of the uncertainty about the required emission reductions (IPCC, 2007c).

However, several other climate change targets may be chosen, e.g., rate of temperature change, radiative forcing, or climate change impacts. In general, selecting a climate policy target early in the cause–effect chain of human activities to climate-change impacts, such as emissions stabilisation, increases the certainty of achieving required reduction measures, while increasing the uncertainty on climate change impacts (see Table 2.2). Selecting a climate target further down the cause–effect chain (e.g., temperature change, or even avoided climate impacts) provides for greater specification of a desired climate target, but decreases certainty on the required emission reductions to reach that target.

As mentioned above, targets relating to Article 2 of UNFCCC would determine the level of greenhouse gas concentrations in the atmosphere (or the corresponding climate change). Choosing a stabilisation level implies the balancing of the risks of climate change (risks of gradual change and of extreme events, risk of irreversible change of the climate, including risks for food security, ecosystems and sustainable development) against the risk of response measures that may threaten economic sustainability (IPCC, 2007c). However, for any stabilisation target, deep emission reductions are unavoidable, to achieve stabilisation. The lower the stabilisation level, the earlier these deep reductions have to be realised.

2.1.2.4 Mitigation – targets, strategies and economic impacts

Several studies have identified a temperature rise of 2 °C above the pre-industrial level as being the critical threshold, above which severe changes affecting humans and biodiversity are to be expected. A 2 °C maximum increase is also the EU climate target (EC, 2007a). However, only the current level of around 380 ppm CO_2 would already mean an increase of about 2 °C when equilibrium is reached (see Table 2.3). Respecting this constraint is already outside the range of scenarios considered by the IPCC, if the higher value of the 'likely' climate sensitivity range of 2–4.5 °C

Table 2.2. *Advantages and disadvantages of using different stabilisation targets*

Target	Advantages	Disadvantages
Mitigation costs	Lowest uncertainty in costs.	Very large uncertainty in global mean temperature increase and impacts. Either needs a different metric to allow for aggregating different gases (e.g., GWPs) or forfeits opportunity of substitution.
Emissions mitigation	Lower uncertainty in costs.	Does not allow for substitution among gases, thus losing the opportunity for multigas cost reductions. Indirect link to the objective of climate policy (e.g., impacts).
Concentrations of different greenhouse gases	Can be translated relatively easily into emission profiles (reducing uncertainty in costs).	Allows a wide range of CO_2-only stabilisation targets due to substitutability between CO_2 and non-CO_2 emissions.
Radiative forcing	Easy translation to emission targets, thus not including climate sensitivity in cost calculations. Allows for full flexibility in substitution among gases. Connects well to earlier work on CO_2 stabilisation. Can be expressed in terms of CO_2-eq. concentration target, if preferred, for communication with policy makers.	Indirect link to the objective of climate policy (e.g., impacts).
Global mean temperature	Metric is also used to organise impact literature; and has shown to be a reasonably proxy for impacts.	Large uncertainty in required emission reductions as result of the uncertainty in climate sensitivity and thus costs.
Impacts	Direct link to objective of climate polices.	Very large uncertainties in required emission reductions and costs.

Source: (IPCC, 2007c).

Table 2.3. *Characteristics of different stabilisation scenarios*

Anthropogenic addition to radiative forcing at stabilisation (W/m^2)	Stabilisation level for CO_2 only (ppm CO_2)	Multi-gas concentration level (ppm CO_2-eq.)	Global mean temperature °C increase above pre-industrial at equilibrium,[a] using 'best estimate climate sensitivity' of 3 °C	Peaking year for CO_2 emissions	Change in CO_2 emissions in 2050 (% of 2000 emissions)[b]
2.5–3.0	350–400	445–490	2.0–2.4	2000–2015	−85 to −50
3.0–3.5	400–440	490–535	2.4–2.8	2000–2020	−60 to −30
3.5–4.0	440–485	535–590	2.8–3.2	2010–2030	−30 to +5
4.0–5.0	485–570	590–710	3.2–4.0	2020–2060	+10 to +60
5.0–6.0	570–660	710–855	4.0–4.9	2050–2080	+25 to +85
6.0–7.5	660–790	855–1130	4.9–6.1	2060–2090	+90 to +140

Notes:

[a] The global mean temperature at equilibrium is different from expected global mean temperature at the time of stabilisation of GHG concentrations owing to the inertia of the climate system. In most scenarios, stabilisation of GHG concentrations occurs between 2100 and 2150.

[b] Ranges correspond to the 15th to 85th percentile of the post-TAR scenario distribution.

Source: (IPCC, 2007c).

(see Section 2.1.2.1) is taken into account. Projections of future warming depend on projections of global emissions. If annual GHG emissions were to remain at today's levels, GHG concentrations would be close to double pre-industrial values – about 550 ppm CO_2-eq. – by 2050, assuming an average increase of 2 ppm per year.[11] But the annual flow of emissions is accelerating, as fast-growing economies invest in high-carbon energy sources and as demand for energy and transport increases around the world. The level of 550 ppm CO_2-eq. could be reached as early as 2035 (Stern, 2006).

The timing of emission reductions depends on the stringency of the stabilisation target. To prevent concentrations rising above today's level, reductions of up to 85% will be necessary by 2050 (compared with 2000). In the most stringent stabilisation category, i.e., stabilisation *below 490 ppm CO_2-eq.*, emissions are required to decline before 2015 and need to be further reduced to less than 50% of today's emissions by 2050. For somewhat higher stabilisation levels (e.g., a stabilisation level *below 590 ppm CO_2-eq.*), global emissions have to peak between 2010 and 2030, followed by a return to 2000 levels on average around 2040. For high stabilisation levels, e.g., *below 710 ppm CO_2-eq.*, the emissions should peak around 2040.

The costs of stabilisation crucially depend on the choice of the baseline, related technological change and resulting baseline emissions; stabilisation target and level; and the portfolio of technologies considered. In 2030, macroeconomic costs for multi-gas mitigation (multi-gas emissions reduction scenarios are able to meet climate targets at substantially lower costs than CO_2-only strategies (IPCC, 2007c)), consistent with emissions trajectories towards stabilisation between 445 and 710 ppm CO_2-eq., are estimated at between a 3% decrease of global GDP and a small increase of 0.6%, compared with the baseline (IPCC, 2007c). The strictest goal, limiting concentrations of GHG to 445 ppm CO_2-eq. in the atmosphere, would reduce annual GDP growth rates by less than 0.12% per year until 2030.[12] For the same stabilisation targets, the corresponding global average macroeconomic costs in 2050 are estimated at between a 5.5% decrease of global GDP and a 1% gain.[13] For a stabilisation level between 535 and 590 ppm CO_2-eq. the costs range from a 0.2–2.5% GDP loss in 2030, and from slightly negative to 4% GDP loss in 2050. The ranges reflect uncertainties over the scale of mitigation required, future rates of technological innovation, reductions in the cost of low-carbon technologies, improvements in energy efficiency and degree of policy support.

The *Stern Review* estimates that in case of inaction, the overall costs and risks of climate change will be equivalent to losing at least 5% of global GDP each year, now

[11] For the approximate CO_2-eq. concentrations corresponding to the computed radiative forcing due to various anthropogenic GHG emission scenarios refer to the SRES Report (IPCC, 2007a). SRES scenarios do not include explicit mitigation policies.

[12] For comparison, world military expenditure in 2006 was estimated at US $1204 billion (at 2006 prices), which amounted to 2.6% of global GDP (SIPRI, 2007; *Die Zeit*, 2007).

[13] According to Stern (2006), the annual costs of emissions reduction, consistent with a 550 ppm CO_2-eq. stabilisation level, are likely to be in the range of –1.0 (net gain) to +3.5% of GDP by 2050, with an average estimate of approximately 1%, if strong action is taken now.

and forever; if a wider range of risks and impacts is taken into account, the estimates of damage could rise to 20% of GDP or more.[14] But it would be possible to 'decarbonise' both developed and developing economies on the scale required for stabilisation, while maintaining economic growth in both: an annual cost rising to 1% of GDP by 2050 would pose little threat to standards of living, given that economic output in the OECD countries is likely to rise in real terms by over 200% by then, and in developing regions as a whole by 400% or more (Stern, 2006).

Climate change will have an impact on economic activity, on the environment and on the life of people around the world by affecting access to water, food production and health (UNDP, 2008). The poorest countries and populations are most susceptible to adverse impacts, even though they have contributed least to the causes of climate change. This is because developing countries are heavily dependent on agriculture, which is the most climate sensitive of all economic sectors. For adaptation, it is, hence, essential that climate change be fully integrated into development policy.

Achieving any stabilisation target requires the deployment of a portfolio of technologies and actions.[15] Policy options for significant reductions in greenhouse-gas emissions imply substantial modifications in the conversion and utilisation of different energy sources and strong efforts being made in the following directions:

- energy efficiency improvement, with reduction of fossil fuel use (short term);[16]
- switch to low-carbon or carbon-free energy sources (natural gas, renewables, nuclear);
- policy support for innovation and the deployment of low-carbon technologies;
- capture and storage (CCS) of the CO_2 produced from fossil fuels (see Chapter 6);
- reduction of non-energy GHG emissions, such as avoiding deforestation;
- pricing of carbon, implemented through tax, trading or regulation; and
- inform, educate and persuade individuals about what they can do to respond to climate change.

The costs of taking action are not evenly distributed across sectors or around the world. Even if developed countries take on responsibility for absolute cuts in emissions of 60–80% by 2050, developing countries must take significant action too (EC, 2007a). Emission reductions have to take place in various sectors, such as transport,

[14] It should, however, be mentioned that economic forecasting over such long periods is difficult and imprecise, as the results are specific to the applied model and its assumptions. Hence, they should not be endowed with a precision and certainty that is not possible to achieve (Stern, 2006).

[15] Socolow and Pacala (2006), for instance, propose a so-called 'wedge concept'. The concept is based on defining a stabilisation triangle, which represents the difference between today's carbon emissions of around 7 Gt C and a doubling of carbon emissions over the next 50 years to 14 Gt C (assuming CO_2 emissions to continue growing at the pace of the last 30 years), resulting in a doubling of pre-industrial CO_2 concentrations by 2056 (a level considered capable of triggering severe climate changes). The stabilisation triangle can be divided into seven 'wedges', each a reduction of 25 billion tonnes of carbon emissions over 50 years. The wedge has proved to be a useful unit because its size and time frame match what specific technologies can achieve. Many combinations of technologies can fill the seven wedges, such as increasing the fuel economy of cars, raising energy efficiency of coal-fired power plants or stopping deforestation.

[16] Technological improvements can increase the efficiency of power plants and energy-using equipment such as appliances, cars, lighting equipment, as well as buildings. In addition, behavioural change towards more economical utilisation can also contribute in reducing overall energy use (EC, 2005b; IEA, 2007a) (see also Chapter 19).

buildings, industry, agriculture or forestry, with each sector having a different set of options and potentials for reduction. The power sector around the world would need to be at least 60% decarbonised by 2050 for atmospheric concentrations to stabilise at or below 550 ppm CO_2-eq., and deep emissions cuts will also be required in the transport sector (Stern, 2006). Particular options for the transport sector include improved vehicle efficiency measures, biofuels or modal shifts (e.g., from road to rail). Implementing emission reduction in the transport sector often comes with the co-benefits of reducing traffic congestion and improving air quality (see also Chapter 19). The loss of natural forests around the world contributes more to global emissions each year than the transport sector. Curbing deforestation is a highly cost-effective way of reducing emissions: recent stabilisation studies indicate that land-use mitigation options could provide 15 to 40% of total cumulative abatement over the century (IPCC, 2007c).

Mitigation opportunities with net negative costs (so called 'no-regret opportunities') have the potential to reduce emissions by around 6 Gt CO_2-eq./year in 2030, if implementation barriers can be overcome, mainly in the building sector (Enkvist *et al.*, 2007; IPCC, 2007c) (see also Chapter 20). The IPCC reckons that a carbon price of \$20–50/tCO_2-eq. by 2020–30 is needed to stabilise atmospheric GHG concentrations at 550 ppm CO_2; it would limit the increase in temperature to 2.8–3.2 °C. To achieve this, it would have to be applied globally.[17]

Stabilisation at 550 ppm CO_2-eq., for instance, requires that annual emissions be brought down to more than 30% below current levels before 2030. This is a major challenge, especially given today's heavy reliance on fossil fuels, which are also widely forecast to supply the major part of energy till the middle of the century, even with very strong expansion of the use of renewable energy and other low-carbon energy sources. Among fossil fuels, coal will continue to be important in the energy mix around the world, particularly in fast-growing economies in Asia. Extensive carbon capture and storage will be necessary to allow the continued use of fossil fuels without contributing to a major rise of GHG emissions. However, the *Stern Review* concludes that sustained long-term action can achieve reduction at costs that are low in comparison to the risks of inaction; the costs of stabilising the climate are significant but manageable; delay would be dangerous and much more costly.

While mitigation measures need to be implemented to reduce GHG emissions, adaptation measures are crucial as well. Adaptation practices refer to actual adjustments, or changes in decision environments, which might ultimately enhance resilience or reduce vulnerability to observed or expected changes in climate; these can include crop diversification, irrigation, water management, disaster risk management, or insurance (IPCC, 2007b). (The EU also recently produced a Green Paper on adaptation to climate change (see EC, 2007b).)

[17] For a detailed analysis of sector-specific global mitigation options and their economics, as well as of the sectoral economic mitigation potential as a function of carbon price (see Enkvist *et al.*, 2007 and IPCC, 2007c).

2.1.2.5 International policy framework

Scientific evidence pointing towards anthropogenic climate change only started accumulating over the course of the 1970s and 1980s. In the late 1980s, increased public awareness of international environmental issues and concerns about the possibility of global warming due to anthropogenic emissions of greenhouse gases moved the climate change debate from the scientific to the political arena, leading to the first international policy response in 1992, with the adoption of the United Nations Framework Convention on Climate Change (UNFCCC) (for more information, see http://unfccc.int). The UNFCCC was signed by 155 countries in 1992 at the so-called 'Earth Summit' in Rio de Janeiro and came into force in 1994. (In August 2007, 192 countries out of the 194 UN member states have ratified the convention.) The ultimate objective of the UNFCCC

> is to achieve . . . *stabilisation of greenhouse-gas (GHG) concentrations* in the atmosphere at a level that would prevent dangerous anthropogenic interference with the climate system. Such a level should be achieved within a time-frame sufficient to allow ecosystems to adapt naturally to climate change, to ensure that food production is not threatened and to enable economic development to proceed in a sustainable manner.
>
> *(Article 2) (UN, 1992)*[18]

The implementation of the convention is shaped by the Conference of the Parties (COP), which convenes at regular intervals. The third Conference of the Parties (COP-3) was held in Kyoto, Japan in December 1997, where the parties adopted the Kyoto Protocol. To date, 175 Parties have ratified the Protocol (August 2007), but many, such as the United States and Australia, remain opposed to ratification. With the ratification of Russia, the Kyoto Protocol entered into force on 16 February 2005.

Under the Kyoto Protocol, the greenhouse-gas reduction commitments apply to six gases or groups of gases: carbon dioxide (CO_2), methane (CH_4), nitrous oxide (N_2O), and the fluorinated compounds hydrofluorocarbons (HFCs), perfluorocarbons (PFCs) and sulphur hexafluoride (SF_6). (Those substances that contribute to ozone depletion, namely chlorofluorocarbons (CFCs) and halons, are covered by the *Montreal Protocol*.) The contributions of the different gases (with their different atmospheric lifetimes and radiative properties) are weighted according to their *global warming potentials (GWP)*,[19] with CO_2 being the most significant contributor to climate change. The Kyoto Protocol has as its target a total cut in overall greenhouse-gas emissions of developed countries (so-called Annex I countries) of at least 5% below 1990 levels in the commitment period 2008 to 2012.

[18] It should be noted that enabling economic development to proceed in a sustainable manner has two sides. While projected anthropogenic climate change appears likely to affect sustainable development adversely, with adverse effects tending to increase with higher levels of climate change and GHG concentrations, costly mitigation measures could have adverse effects on economic development. This tension gives rise to the debate over the right scale and balance between climate policy (mitigation and adaptation) and economic development.

[19] The GWPs are an index for estimating the time-integrated relative global warming effect due to the atmospheric emission of a kilogram of a particular greenhouse gas compared to the emission of a kilogram of carbon dioxide. The time horizon used for the GWP index is typically 100 years.

Under the terms of the protocol, parties with legally binding obligations may meet them by applying three flexible mechanisms: *joint implementation (JI)*, *clean development mechanisms (CDM)*, and *international emissions trading (IET)*. These mechanisms were created to enable governments to meet part of their greenhouse-gas reduction commitments by developing emission-reduction projects in other countries. Joint implementation projects are undertaken in industrialised countries that have quantitative emissions reductions targets, and CDM projects are hosted by developing countries that have no quantitative targets. Both JI and CDM will transfer environmentally sound technologies to the host countries, which will assist them in achieving their sustainable development objectives. The concept behind all three mechanisms is that a proportion of the required reductions in GHG emissions should be achieved at the lowest possible costs (IEA, 2007b).

- *Joint implementation*: an Annex-I country (or an entity within an Annex-I country) can receive emission-reduction units (ERUs) generated by emission-reduction projects in another Annex-I country.
- *Clean development mechanisms*: non-Annex I parties can create certified emissions reductions (CERs) by developing projects that reduce net emissions of greenhouse gases. Annex-I parties (both governments and private entities) can help to finance these projects and purchase the resulting credits as a means of meeting their own reduction commitments.
- *International emissions trading*: Annex-I parties may trade their emission allowances with other Annex-I parties. The aim is to improve the overall flexibility and economic efficiency of emissions cuts.

International emissions trading has already been implemented through the European Emissions Trading Scheme (ETS) since January 2005, while Australia and the USA are planning to introduce cap-and-trade systems in 2011 and 2012, respectively. There are also a growing number of projects underway based on the application of JI and CDM. Clean development mechanism projects can take many forms and include those based on achieving improvements in energy efficiency (both end use and supply side), increased use of renewable energy sources, methane reduction (e.g., from gas capture from landfills or flaring reduction), fuel switching, enhanced industrial processes, and the application of sequestration techniques and CO_2 sinks (afforestation and reforestation). For more information, see http://cdm.unfccc.int/index.html.

The Kyoto Protocol represents the first international joint action towards greenhouse-gas emission controls. Because climate change is a global problem, the response to it must be international. Many countries and regions are taking action already: the EU, California, Australia and China are among those with the most ambitious policies to reduce greenhouse-gas emissions. The UNFCCC and the Kyoto Protocol provide a basis for international co-operation, along with a range of partnerships and other approaches. However, as the Kyoto Protocol ends in 2012, the international community faces the challenge of creating a workable post-Kyoto framework to reduce global greenhouse-gas emissions.

There is a growing number of policy measures to curb increasing GHG emissions. In particular, the European Union has set ambitious goals to reduce greenhouse-gas emissions. The EU has agreed to limit global warming to within 2 °C, which means progressively reducing global greenhouse-gas emissions by up to 50% by 2050 compared with 1990, implying reductions in developed countries of 60–80% (EC, 2007a). The EU Council of Ministers decided in March 2007 an EU-wide reduction in CO_2 emissions (from their 1990 levels) by 20% in 2020, and by 30% if international agreement can be reached (CEU, 2007).[20] The EU further recognises the need to accelerate its international lead in CCS and has an ambition to deploy 10–12 full-scale demonstration projects within Europe by 2015, testing various ways of integrating carbon capture and storage (CCS) in coal and gas-fired power generation; it also desires to require all new fossil-fuelled power plants built from 2020 onward to have CO_2 capture (i.e., to ensure that CCS is commercially viable for all new fossil fuel power plants by 2020), with existing plants progressively retrofitted (EC, 2007c).

2.1.3 Air pollution

This section briefly outlines the extent of air pollution as of today and its effects on the climate system, however neglecting a discussion about impacts, mitigation strategies and future emissions projections. The main sources of air pollution are transport, power generation, industry, agriculture and heating. All these sectors emit a variety of air pollutants – sulphur dioxide (SO_2), nitrogen oxides (NO_x), ammonia (NH_3), volatile organic substances (VOC) and particulate matter (PM) – many of which interact with others to form new pollutants, such as ozone.[21] These are eventually deposited and have a whole range of effects on human health, biodiversity, buildings, crops and forests. Air pollution is also closely linked to climate change, both in terms of common sources and mitigation strategies.

Sulphur emissions are relevant for the climate system as they contribute to the formation of aerosols, which affect precipitation patterns and reduce radiative forcing. Sulphur emissions also contribute to regional and local air pollution. Global SO_2 emissions have grown approximately in parallel with the increase in fossil-fuel use. Since about the late 1970s, however, emission growth has slowed considerably. Implementation of emissions controls, a shift to lower-sulphur fuels in most industrialised countries, and the economic transition process in Eastern Europe and the former Soviet Union have contributed to the lowering of global sulphur emissions.

[20] Under the Kyoto Protocol, in recognition of their different circumstances, countries agreed different reduction targets. For example, the European Union (EU15) agreed an 8% reduction, while Norway and Australia were actually allowed to increase their emissions by 1 and 8% respectively, relative to their 1990 levels. The Kyoto protocol target for the EU15 is to reduce GHG emissions by 8% below the 1990 level by 2008–2012.

[21] Ozone occurs naturally in the stratosphere and in the troposphere but ozone concentrations close to ground level are harmful to ecosystems and human health. Ground-level ozone is formed from the complex chemical reactions between volatile organic compounds (VOC) and NO_x in the presence of sunlight. Volatile organic compounds are emitted from many different sources, including petrol stations, tailpipe emissions from cars, and the use of solvents and solvent-containing products, such as paints and varnishes.

Conversely, with accelerated economic development, the growth of sulphur emissions in many parts of Asia has been high in recent decades, although growth rates have moderated recently. Global anthropogenic sulphur emissions are estimated for the year 2000 at between 55 Mt S and 62 Mt S, a decline from around 75 Mt S in 1990 (IPCC, 2007c).

The most important sources of NO_x emissions are fossil-fuel combustion from power generation and the transport sector, with emissions largely being related to the combustion practice. Together with other sources, such as natural and anthropogenic soil release, biomass burning, lightning and atmospheric processes, they amount to around 25 Mt N per year. In recent years, emissions from fossil-fuel use in North America and Europe are either constant or declining. In most parts of Asia and other developing parts of the world, emissions have been increasing, mainly from the growing transport sector.

Black- and organic-carbon emissions (BC and OC) are mainly formed by incomplete combustion. The main sources of BC and OC emissions include fossil-fuel combustion in industry, power generation, traffic and residential sectors as well as biomass and agriculture-waste burning. Natural sources, such as forest fires and savannah burning, are other major contributors. There has recently been some research suggesting that carbonaceous aerosols may contribute to global warming. However, the uncertainty concerning the effects of BC and OC on the change in radiative forcing and hence global warming is still high (IPCC, 2007c).

As noted above, some air pollutants, such as sulphur aerosols, have a significant effect on the climate system. Considerable uncertainties still surround the estimates of anthropogenic aerosol emissions. Data on non-sulphur aerosols are sparse and highly speculative. Sulphur emissions from fossil-fuel combustion led to the formation of aerosols, which affect regional climate and precipitation patterns and also reduce radiative forcing. There has been a slowing in the growth of sulphur emissions in recent decades. Other air pollutants, such as NO_x and black- and organic-carbon, are also important climatologically and adversely affect human health.

Quantitative analysis on a global scale for the implications of climate mitigation for air pollutants is relatively scarce. Air pollutants and greenhouse gases are often emitted by the same sources, and changes in the activity of these sources affect both types of emission. Many of the traditional air pollutants and greenhouse gases have common sources, offering a cost-effective potential for simultaneous improvements of traditional air-pollution problems and climate change. For instance, climate change measures that aim at reduced fossil-fuel combustion will have ancillary benefits for regional air pollutants. In contrast, some ammonia abatement measures can lead to increased nitrous oxide (N_2O) emissions, while structural measures in agriculture could reduce both regional air pollution and climate change. Methane (CH_4) is both an ozone (O_3) precursor and a greenhouse gas. Hence, CH_4 abatement will have synergistic effects and some cheap abatement measures may be highly cost effective.

The transportation sector is a major contributor to environmental problems, being the source of over 30% of global NO_x emissions and 18% of CO emissions (IEA, 2005). Air pollution, especially from road transport, is becoming an increasingly critical issue for urban air quality, particularly in the world's growing megacities (see also Chapter 19). In 2000, road transport in the EU25 contributed to 4.6% of all land-based SO_2 emissions, 61% of NO_x emissions, 39% of VOC emissions, 20% of $PM_{2.5}$ emissions and 75% of CO emissions (EC, 2005a). In contrast to the expected reductions in emissions from land-based sources, the maritime sector is becoming an even larger source of air pollution in the EU. It is projected that SO_2 emissions from the maritime sector will increase by around 45% and NO_x emissions by approximately 67%; with these growth rates, emissions of SO_2 and NO_x from the maritime sector should surpass total emissions from land-based sources by 2020.

The emissions from the transport sector have a particular importance because of their rapid rate of growth; for instance, goods transport by road in Europe has increased by 54% since 1980, passenger transport by road by 46% in the past ten years in the EU and passenger transport by air has increased by 67% in the past ten years. Whilst emission levels in the economically more developed countries have increasingly stabilised, they are continuing to rise in the less developed countries. The establishment of stricter standards for the emission of pollutants by motor vehicles has had positive results, but the progress achieved to date is threatened by the rising number of vehicles on the road and vehicle use: the number of vehicles worldwide is expected to increase from around 800 million today to more than 2 billion in 2050. The use of hydrogen as vehicle fuel would contribute significantly to a reduction of urban air pollution. While hydrogen fuel-cell vehicles only emit water, burning hydrogen in combustion engines produces NO_x emissions.

2.2 Historical transitions of energy sources

The discussion of the possible transition to a hydrogen-based energy economy can be better assessed if a closer look is taken at the historical development of the energy supply. Since about the middle of the nineteenth century, there has been a gradual change in the form of the energy supply from solid via liquid to gaseous energy carriers. Figure 2.9 shows the change in the composition of the global primary energy mix over time from industrialisation around 1850 up to the present day. It is interesting that the substitution of the various primary energy sources seems to have been independent of their availability, at least at global level; up to now, the transition has taken place long before a certain resource was exhausted (Marchetti, 1975).[22] The figure further exemplifies that transitions in the energy system are very slow and that it takes several decades to bring new energy technologies (or new energy sources) to the point of materiality, commonly defined as contributing 2% to the relevant energy mix.

[22] Cesare Marchetti was one of the first to work intensively on the development of the primary energy mix and the determining factors and causes for the transition between primary energy sources.

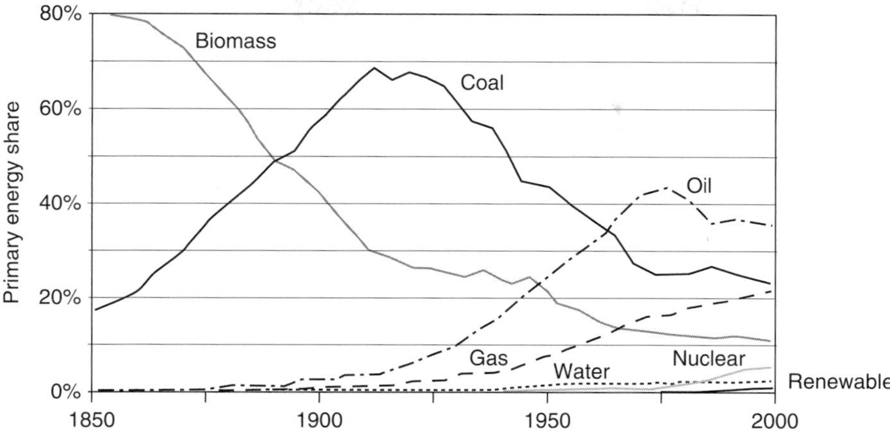

Figure 2.9. Development of the global primary energy mix since 1850 (own diagram based on (Shell, 2001)). Hydropower was evaluated according to the partial substitution method (see Chapter 3).

Wood (biomass) was the principal energy source for mankind for the longest period, until about the middle of the nineteenth century. The constantly growing demand for timber for iron making and shipbuilding and the associated deforestation led to an increase in the price for timber in England in the seventeenth century: by 1630, wood was already two and a half times more expensive than at the end of the fifteenth century (Rifkin, 2002). At around 1700, coal began to replace wood in England, where population density and energy consumption increased rapidly as well. This went hand in hand with the industrial revolution, a phase of profound upheaval in production, working and living conditions which began around 1760 in Great Britain and which characterised the nineteenth century in Europe and beyond. Its beginning is often linked to the invention of the steam engine by James Watt in 1769. In the wake of industrialisation, coal finally appeared on the scene in large parts of Europe by the middle of the nineteenth century and helped countries like Great Britain and Germany to economic prominence (Dunn, 2001). The spread of coal accompanied the massive expansion of railways and shipping lines, the growth of steel production and the electrification of the factories, and was the dominating fuel for the remainder of the nineteenth century and well into the twentieth century.

At the beginning of the twentieth century, oil gradually appeared as a fossil energy source alongside coal in the USA and other industrialised countries. The industrial extraction of crude oil took off in the middle of the nineteenth century almost simultaneously in the USA, Azerbaijan, Poland, Romania but also in Germany (Wietze) and the Alsace region (Pechelbronn, France). However, oil's triumphant advance began without doubt in the USA with the start of commercial production in 1859 in Pennsylvania. Oil gradually replaced coal in trains and ships as well as domestic heating systems and was there at the onset of automobilisation. The spread of oil went together with the inventions of the combustion engine at the end of the nineteenth century and the automobile by Karl Benz and Gottlieb Daimler in 1885. However, it was only possible for cars to achieve a breakthrough because of Henry Ford's mass production. The year 1911 saw the opening of the first filling station in the USA in Detroit (Rifkin, 2002). The founding of the *Standard Oil Company* by John D. Rockefeller in 1868 should also be mentioned in this context; his empire controlled more than 90% of the oil industry in the USA in the 1880s. The transport system's transition from railways to cars made the advantages of oil as a liquid energy carrier over coal generally accepted, such as, e.g., its higher energy density and the possibility of pipeline transport. Not least because of its rich oil reserves, the USA became the largest industrial nation in the twentieth century and was the world's largest oil producer right up to the beginning of the 1990s. After the first oil crisis in 1973 and 1974, however, the share of oil in total world primary energy use began to drop.

Oil also found its way into the chemical industry. In 1957, in Germany for instance, almost 80% of primary chemicals were still being manufactured based on coal products (Herz, 1983). Coal tar and benzene were some of the most important raw materials of organic chemistry. However, since both were mainly formed as by-products of coke produced for the steel industry, their supply did not meet the growing demand. This resulted in a structural change in organic chemistry from coal to petrochemicals. Today, about 90% of chemicals are produced using oil products. Global automobilisation has eventually made oil an indispensable element of private and commercial life. At the same time, oil has also been the trigger for conflicts and wars and the two World Wars made the strategic significance of oil especially clear. Despite this, almost half the global population still use wood as a fuel today (Rifkin, 2002). In the wake of the diversification of the energy sources and not least because of its environmentally-friendly features, a gaseous energy source, natural gas, has been substituting the liquid energy source of oil more and more in recent years. Currently, natural gas is in the process of ousting oil from domestic heating and other primary energies from electricity generation.

In the current energy policy discussion, hydrogen is assigned a special role by many with regard to the future energy supply, especially in mobile applications. Hydrogen is the most frequent element in space and represents about 75% of the total mass of the Universe as well as more than 90% of all atoms. Hydrogen was discovered by the Englishman Henry Cavendish in 1766; the Frenchman Antoine Lavoisier gave it its name in 1787. In 1839, William Grove published his work on fuel cells, which converted hydrogen into electricity and heat in an electrochemical process. From about 1800 onwards, hydrogen was only used in a single significant application, for energy purposes in the form of 'town gas' and 'water gas', which both comprised 50% hydrogen (Weber, 1991). Water gas, which contained about 40% carbon monoxide besides hydrogen, was produced by passing steam over hot coal. It was mainly used for soldering and welding or to drive commercial or industrial gas engines. Town gas or coal gas, in contrast, served as street lighting and was used to supply energy to residential buildings for cooking, heating and lighting. Town gas was produced by the dry distillation of coal and contained about 30% methane, 10% carbon monoxide and other gases, besides 50% hydrogen. These gases with a high proportion of hydrogen were important energy carriers for a long period which only ended in the 1950s and 60s when town gas was increasingly supplanted by natural gas; in Germany, for instance, the last substitution took place at the beginning of the 1990s in Saarbrücken (Germany) (Geitmann, 2002).

Hydrogen actually has a longer history as an energy carrier than as a chemical raw material. Hydrogen has only been used and produced for industrial purposes since about 1920. This can be traced back to the first commercial scale ammonia synthesis using the Haber–Bosch method which *BASF* began operating in 1913. Ammonia replaced saltpetre as the basic material for manufacturing explosives and artificial fertilisers. Since the 1950s, hydrogen has been used in space travel to power fuel cells. Today, hydrogen is mainly used for ammonia synthesis to manufacture artificial fertilisers, in refineries to process crude oil and for various chemical interim products or in the food industry to produce solid or semisolid edible fats.

In the 1960s, several scientists had the idea of using solar energy to separate water into hydrogen and oxygen and then to recombine these in fuel cells; this idea is currently being discussed again under the term 'solar hydrogen energy economy'. At the beginning of the 1970s, first studies were also being made of the use of nuclear energy for the commercial production of hydrogen as an energy supply, to reduce the reliance on fossil energy sources (Marchetti and de Beni, 1970). In 1970, the term 'hydrogen economy' was first coined by *General Motors* in connection with the future fuel supply in the transport sector. Scientific interest in hydrogen as an energy source was given a particular push by the first oil crisis in 1973–74. In 1974, the first international hydrogen conference took place in Miami; this has been repeated every two years since 1976. The *Hydrogen Implementing Agreement* of the IEA also dates from the same period. The Chernobyl reactor catastrophe in the spring of 1986 pushed the use of solar energy into the limelight and, in its wake, also the idea of a solar hydrogen economy and the first research programmes were launched, especially in Germany.[23] Significant progress in fuel cell technology in the late 1990s as well as the growing concern about the security of supply of the fossil energy sources, oil and natural gas, have refocused the interest in hydrogen in the energy policy debate in recent years, mainly as an alternative to oil in the transport sector. At present, there are lots of national and international research activities and co-operation as well as numerous pilot projects with regard to fuel cells and hydrogen, especially in the EU, the USA and Japan (for details see Chapter 8).

[23] In 1986, the Federal Republic of Germany and Saudi Arabia signed an agreement to co-operate in research, development and application in the field of solar hydrogen production (HYSOLAR).

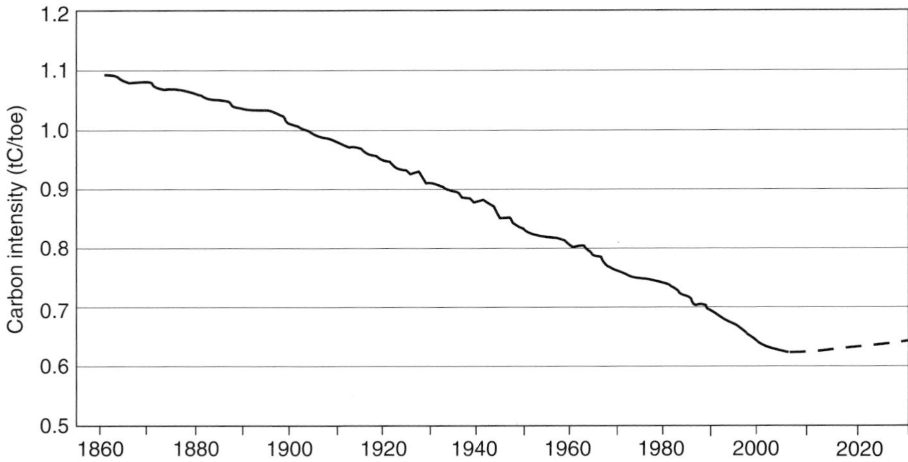

Figure 2.10. Decarbonisation of the energy supply (own diagram based on (Nakicenovic, 1996) and (IEA, 2006)).

With the transition from solid to liquid to gaseous energy sources, the energy raw materials have become less carbonaceous, richer in hydrogen and thus lighter. Whereas wood has a ratio of hydrogen atoms to carbon atoms of between 1:3 and 1:10, for coal this is 1:2 or 1:1, depending on the type (Dunn, 2001). For oil, the ratio is already 2:1, and 4:1 for natural gas (methane). Pure hydrogen is, of course, carbon-free. This process of replacing carbon atoms with hydrogen atoms is referred to as *decarbonisation*. This means that each new energy source emits less CO_2 or carbon than its predecessor. If the composition of the primary energy mix is known, its carbon intensity can be calculated using fuel-specific emission factors (see IPCC, 2006). Figure 2.10 shows the trend in carbon intensity since 1860. Since this date, the carbon emissions due to global primary energy use have dropped by more than 30% as a result of the transition from wood to coal to oil to natural gas to rational energy use and to carbon-free energies; this is equivalent to a reduction rate of 0.3% per year (Nakicenovic, 1996). In spite of this, the absolute worldwide primary energy use and the absolute carbon emissions have still increased, owing to the growth in population and the global economy.

According to IPCC (2007c), the year 2000 marked a tipping point as the long-term trend of a declining carbon intensity of energy supply reversed after about 150 years. The decarbonisation of energy sources might further be reversed in this century because of an increased use of coal to satisfy the future demand for energy, especially in China and India, as depicted in Fig. 2.10, which displays the projected carbon intensity according to the IEA Reference Scenario until 2030 (without CCS). How long this trend will continue will depend on the pace of expansion of renewable energies and low-carbon fuels, such as biofuels or renewable hydrogen, relative to the projected revival of coal use.

2.3 Prospects for a hydrogen economy

In consequence of the continuing growth of the world's population and the increasing industrialisation of developing nations, such as China and India, it has to be expected that the demand for energy, especially for fossil fuels, will continue to rise in the coming decades. According to the world energy scenarios of the *International Energy Agency* and other organisations, global primary energy demand is projected to increase by more than 50% until 2030. Such a surge in energy use in this period can, by and large, only be supplied by fossil energies, despite a growing share of renewable energies. In this context, the complex of problems relating to the limited availability of fossil fuels is becoming of utmost importance, particularly with respect to (conventional) oil. Currently, oil covers more than one third of global primary energy demand and is expected to remain the largest primary fuel in the coming decades.

The finiteness of fossil fuels, above all of oil and natural gas, raises the question of whether the current world energy projections, which rely on the assumption of a continued economic growth and, hence, a continued growth of consumption, take the effectively available remaining supply adequately into consideration. If the above world energy scenarios are to materialise, cumulative oil production would have to be doubled until 2030. However, conventional oil is – beyond doubt – the most depleted fossil fuel today. A decline of global oil production as a result of the depletion of oil occurrences, and, hence, possible shortcomings in its supply would primarily hit the transport sector, since this is still 95% dependent on oil worldwide. The growing concern about energy security in Western societies, an increasing import dependence from the Middle East and the economic and geopolitical implications of possible shortcomings in the supply of oil (and gas) are stirring the current debate about the future supply of energy and are key drivers for future energy policies, in particular in the transportation sector. While energy security has been a pillar of energy policy for a long time, recently concerns about climate change are gaining momentum in shaping energy policies. Hence, policy makers will have to develop cost-effective policies ensuring the security of the energy system, while at the same time reducing greenhouse-gas emissions. This situation is increasingly triggering the search for alternative automotive fuels and drive trains.

While in the short and medium term, lightweight construction, improved conventional internal-combustion engines (ICEs), hybridisation and dieselisation can improve the fuel economy of vehicles, longer-term strategies must focus on developing alternative fuels. Unless there will be a breakthrough in battery technology for either pure electric vehicles or plug-in hybrids, with electricity itself (largely) becoming the 'fuel' for automotive power – which is the most energy-efficient solution on a well-to-wheel basis and would make the discussion about hydrogen largely obsolete – four broad alternative fuel categories are available: unconventional oil from oil sands or oil shale, synthetic oil from natural gas or coal, biofuels and *hydrogen* (for a detailed discussion of the pros and cons of the different alternative fuels, see Chapter 7). Hydrogen can be used as fuel in vehicles with internal-combustion engines (ICEs) or with fuel cells. In an ICE, hydrogen is burned very similarly to petrol or gas-fired engines to produce mechanical energy. Fuel cells are devices that use a chemical reaction to generate electricity (heat and water) by combining hydrogen with oxygen (air); the electricity is then converted electromechanically in an electric motor into torque in the wheels, which drive the vehicle.

Among the above choices, hydrogen seems especially promising: as a secondary energy carrier that can be produced from any primary energy source, it can contribute to a diversification of automotive fuel sources; in addition, the production of hydrogen from synthetic gas shows a higher thermal process efficiency than, for example, synthetic fuels produced via the Fischer–Tropsch route. While hydrogen can also be used directly in modified ICEs, considering the whole well-to-wheel chain, hydrogen offers significant advantages in combination with fuel-cell vehicles, owing to their high conversion efficiencies – as compared with conventional gasoline or diesel vehicles – particularly at partial load, such as in urban traffic. Furthermore, as hydrogen is nearly emission-free at final use, hydrogen vehicles can contribute to a reduction of transport-related emissions of both CO_2 and air pollutants, the latter making it attractive for improving urban air quality. In the case of the hydrogen being produced from fossil fuels in centralised plants, CCS eventually offers an advantage by allowing the capture of CO_2 from a single point source, thus providing a 'clean' fuel, instead of otherwise having to reduce the CO_2 emissions from the tailpipe of each vehicle. Creating a large market for hydrogen thus offers the prospect of providing effective solutions to both the aspects of emission control and the security of energy supply. As an energy vector, hydrogen can also be utilised in different applications (transportation, electricity production, etc.).

However, it must be stressed that hydrogen is not an energy source but a secondary energy carrier, in the same way as electricity. As for electricity, the advantage of using hydrogen as a fuel, as far as security of supply or greenhouse-gas emissions are concerned, depends on how the hydrogen is produced. If produced from coal, it adds to security of supply but gives rise to higher CO_2 emissions (unless the CO_2 is captured and stored). If produced by non-fossil fuel (nuclear or renewable), it adds to security of supply and reduced CO_2 emissions, but only in so far as the non-fossil fuel source is additional to what would otherwise be used in electricity production. This means that any assessment of the virtues of switching to hydrogen as a transportation fuel involves a number of assumptions on long-term future energy policy developments. Unlike electricity, hydrogen allows storage over time and thus can offer buffering capacity for a decentralised non-fossil fuel-based energy system, e.g., for intermittent wind power.

Hydrogen production using renewable (locally available) energy sources is seen as an aspirational nearly zero-emission means of energy provision, offering at the same time the possibility of reducing dependence on fossil fuels and the depletion of finite resources, and of enhancing security of supply – though it is widely thought to be a distant and only partial solution. Under this scenario, hydrogen production using fossil fuels is an interim option for bridging the distance between today's carbon economy and a future hydrogen economy based on renewables. In particular, hydrogen produced from fossil fuels with CO_2 capture and storage is considered to be the cleanest way to continue using these fuels that will continue to play an important role in our societies in the future. Another zero-emission option for hydrogen production is the use of nuclear energy. Production from fossil fuels could be considered a 'technological bridge' to new production processes from renewables and 'new' nuclear fuels, which are expected for the second half of this century. In any case, the development of technologies for the transportation and final use of hydrogen produced from fossil fuels in the next decades will form the basis of the introduction of CO_2-free production technologies in the long term.

Thereby, neither the use of hydrogen as energy vector nor the vision of a hydrogen economy are new. Until the 1960s, hydrogen was used in many countries in the form of town gas for street lighting as well as for home energy supply (cooking, heating, lighting), and also the idea of a hydrogen-based energy system was already formulated in the aftermath of the oil crises of the 1970s. Moreover, hydrogen is an important chemical feedstock, for instance for the hydrogenation of crude oil or the synthesis of ammonia. Mainly breakthroughs in fuel-cell technology in the late 1990s have revived the interest in hydrogen.

Since then there has been a continuously growing number of national and international research activities and demonstration projects aimed at the use and promotion of hydrogen and fuel cells, to prove the feasibility of hydrogen solutions and gain experience. Besides its use in the transport sector, hydrogen may also be used in stationary applications for decentralised heat and electricity generation as well as for portable applications of fuel cells, such as in camcorders, mobile phones or laptops. As stationary fuel cells – unlike the polymer-electrolyte-membrane (PEM) fuel cells typically used in vehicles – can also be fuelled directly by natural gas and thus do not necessitate the use of pure hydrogen and because the hydrogen volumes needed for portable applications are negligible, the transport sector can be considered the major driver for the introduction of hydrogen. Given that policy makers around the globe are increasingly focusing on reducing CO_2 emissions from power generation, particularly coal-fired power stations, integrated-gasification combined-cycle (IGCC) plants with carbon capture and storage (CCS) might turn out to be an attractive option, as they offer not only the possibility of generating 'clean' electricity, but also 'clean' hydrogen. Thus, the power sector might become another potential user (and possibly supplier) of hydrogen, depending on the economic dispatch of the power plants (see also Chapter 16).

Hydrogen is increasingly seen as offering a set of benefits that are not generally offered by fossil-fuel combustion and that are rising as policy priorities (see also Lovins (2003)). Concerns over *climate change* (greenhouse-gas emissions) are having a growing impact on worldwide policy making, as are concerns over local air quality, *security of supply* and energy dependency. For these reasons, the hydrogen economy is one of the long-term priorities for the energy system of the European Commission. Apart from Europe, the hydrogen vision is also being investigated, in the USA and in Japan, especially and several national hydrogen energy activities have been or are currently being developed. As a consequence, hydrogen and fuel-cell research has been receiving increased funding recently, both at a national and an international level.[24]

The potential benefits of a hydrogen economy are recognised to differing degrees by national governments and supranational institutions, though the pathways and timeframes to achieve such a transition remain highly contended. In particular, there are various factors that are very critical for the transition towards a hydrogen economy, in particular the build-up of a hydrogen infrastructure. Under the premise that cost-efficient hydrogen vehicles are available – which certainly requires a significant cost reduction of fuel-cell-based drive trains (among other technical challenges, such as hydrogen storage on board to achieve acceptable driving ranges) – a crucial prerequisite for the introduction of hydrogen as alternative fuel is the implementation of a supply infrastructure, that comprises its production (including feedstock preparation), its distribution and the installation of refuelling stations. The implementation of an operational infrastructure will require considerable investments over several decades and especially involves a high investment risk regarding the future increase of hydrogen demand. In addition, the supply of hydrogen needs to be integrated in the context of the energy system as a whole, as its production may affect the conventional energy system – especially the electricity sector – in various ways: examples are the competing use of renewable energies regarding electricity, heat and hydrogen production, the dispatch of electrolysers or the possible co-production of electricity and hydrogen in IGCC plants. In addition, hydrogen offers the possibility of use as a storage medium for electricity from intermittent renewable energies, e.g., wind energy, thus facilitating load levelling.

Whether it is viable that hydrogen can solve most of the energy issues in the long term needs to be evaluated through well defined deployment scenarios that can provide quantitative information on the opportunities and risks related to large market introduction. In particular the large investments required for hydrogen take-off and the uncertainties about the development of key technologies, like fuel cells or hydrogen storage systems, must be known and accepted as affordable by all the stakeholders involved in such a critical transition. Therefore, the main objective is to identify reasonable future hydrogen scenarios, extract the relevant information from them and, at the end, if the hydrogen technologies can be judged really effective, propose specific action plans able to create a safe and economically viable transition

[24] Total public spending for hydrogen and fuel-cell RD&D in the OECD in 2003–2005 amounted to some US$ 1 billion per year, of which 30% was by Japan, 32% by the EU25 and 24% by the USA (IEA, 2005; Roads2HyCom, 2007). This represented some 12% of all public energy RD&D spending. Moreover, major car manufacturers are spending around US$100 million each year on fuel-cell vehicle development.

to the new society. The above issues plus the fact that there is no clearly outstanding hydrogen pathway in terms of economics, primary energy use and CO_2 emissions show that it is vital for all the stakeholders involved to start defining a strategic orientation as soon as possible. Moreover, fuel cells and hydrogen technologies exhibit a great innovation potential and offer promising economic prospects for export-oriented economies. There are definitely many unresolved problems and challenges, but also clear advantages and opportunities related to the use of hydrogen. This book will try to shed light on some of the major aspects related to the possible transition to a hydrogen economy.

2.4 Summary

The world is facing a new era of energy anxiety with complicated choices regarding fuel sources, new technologies, and government regulations and actions. Security of supply as well as climate change as a result of anthropogenic greenhouse-gas emissions have become a major concern of policy making in the energy sector.

Several studies have identified a temperature rise of 2 °C above the pre-industrial level as being the critical threshold, above which severe changes affecting human beings and biodiversity are to be expected. A 2 °C above pre-industrial limit on global warming is already almost out of reach, as it would imply global emissions to peak within the next decade and be reduced to less than 50% of today's level by 2050. Stabilisation at 550 ppm CO_2-eq. or below is also extremely challenging, given that global emissions would need to peak within the next 20 years and be followed by significant decline thereafter. Achieving these drastic CO_2 reduction targets, as assumed in some mitigation scenarios, requires an increase in the rates of improvement of energy intensity and carbon intensity by two to three times their historical levels. This will require a portfolio of technologies and mitigation actions, such as improving energy efficiency, carbon capture and storage (CCS) and the use of renewable energies or nuclear power. It also requires long-term changes in energy production and consumption patterns (power plants, buildings, transport, etc.).

The challenge – an absolute reduction of global GHG emissions – is daunting. This is also reflected in the various world energy *reference* scenarios, that project a further increase of GHG emissions until 2030, owing to fossil fuels remaining the major energy source in the global energy mix. In particular, the impacts of population growth, economic development, patterns of technological investment and of consumption continue to eclipse the improvement in energy intensities and decarbonisation. An absolute emissions reduction presupposes a reduction of energy and carbon intensities at a faster rate than income and population growth together, and revolutionary paths in decoupling energy use and economic development in newly industrialising countries.

However, the major conclusion of the *Stern Review*, is that the benefits of strong and early action far outweigh the economic costs of not acting. Mitigation, i.e., taking strong action to reduce emissions, should be viewed as an investment, a cost incurred now and in the coming few decades to avoid the risks of very severe consequences in the future. The less mitigation is done today, the greater the difficulty of continuing to adapt in future.

With the discovery of oil and the invention of the automobile, the transportation system started to shift away from railways to cars, and opened the way to oil, which became the world's leading energy source by around the middle of the twentieth century. Today, oil is, with a share of more than one third in the global primary energy mix, still the largest primary fuel and covers more than 95% of the energy demand in the transport sector, making the latter particularly vulnerable to potentially decreasing supplies. A projected increase in global energy demand, the economic and geopolitical implications of possible shortcomings in the supply of oil, and consequent concerns about energy supply security have put the discussion about hydrogen as a future energy carrier back on the agenda.

Of all the alternative fuels being discussed for meeting the future energy demand of mobility, hydrogen seems particularly promising. As hydrogen is a secondary energy carrier that can be produced from any primary energy source, it can contribute to a diversification of automotive fuel sources; hydrogen also offers the long-term possibility of being solely produced from renewable energies, thus allowing renewables other than biomass to enter the transport sector. In addition, hydrogen can contribute to a reduction of transport-related emissions of both CO_2 and air pollutants, the latter making it especially attractive for improving urban air quality.

References

Bohi, D.R. and Toman M.A. (1996). *The Economics of Energy Security*. Norwell, Massachusetts: Kluwer Academic Publishers.

BP (British Petroleum) (2008). *Statistical Review of World Energy 2008*. www.bp.com.

Council of the European Union (CEU) (2007). *Presidency Conclusions – Brussels, 8–9 March 2007*. European Commission. http://europa.eu/european_council/conclusions/index_en.htm.

Die Zeit (2007). Einsatz aller Kräfte. *Die Zeit*, **18**, 28.

Dunn, S. (2001). *Hydrogen Futures: Toward a Sustainable Energy System*. Worldwatch Paper No 157. Washington: Worldwatch Institute.

EC (European Commission) (2005a). *Annex to the Communication on Thematic Strategy on Air Pollution and The Directive on 'Ambient Air Quality and Cleaner Air for Europe'* (COM(2005), 446 final, COM(2005), 447 final). Commission of the European Communities, SEC (2005) 1133.

EC (European Commission) (2005b). *Doing More with Less*. Green Paper on energy efficiency. COM(2005), 265 final.

EC (European Commission) (2006a). *A European Strategy for Sustainable, Competitive and Secure Energy*. Green Paper. Commission of the European Communities, COM(2006), 105 final.

EC (European Commission) (2006b). *An EU Strategy for Biofuels*. Communication from the Commission, Commission of the European Communities, COM(2006), 34 final.

EC (European Commission) (2007a). *Limiting Global Climate Change to 2° Celsius. The way ahead for 2020 and beyond*. Communication from the Commission to the Council, the European Parliament, the European Economic and Social Committee and the Committee of the Regions, Commission of the European Communities, COM(2007), 2 final.

EC (European Commission) (2007b). *Adapting to Climate Change in Europe – Options for EU Action*. Green Paper. Communication from the Commission to the Council, the European Parliament, the European Economic and Social Committee and the Committee of the Regions, Commission of the European Communities, COM(2007), 354 final.

EC (European Commission) (2007c). *Sustainable Power Generation from Fossil Fuels: Aiming for Near-Zero Emissions from Coal after 2020*. Communication from the Commission to the Council and the European Parliament, COM(2007), 843 final.

Enkvist, P. A., Nauclér, T. and Rosander, J. (2007). A cost curve for greenhouse gas reduction. *The McKinsey Quarterly*, **1**, 35–45.

Geitmann, S. (2002). *Wasserstoff & Brennstoffzellen*. Berlin: Hydrogeit Verlag.

Greene, D. L. and Ahmad, S. (2005). *Costs of US Oil Dependence: 2005 Update*. Oak Ridge National Laboratory, Report No. ORNL/TM-2005/45. www.osti.gov/bridge.

Herz, H. (1983). *Analyse optimaler Prozessstrukturen zur Herstellung von Kraftstoffen und Grundchemikalien in der Bundesrepublik Deutschland im Jahr 2000*. Dissertation, VDI-Verlag Fortschrittsberichte Reihe 16, No 22.

Hirsch, R. L., Bezdek, R. H. and Wendling, R. M. (2005). *Peaking of World Oil Production: Impacts, Mitigation and Risk Management*. US Department of Energy (DOE), National Energy Technology Laboratory (NETL).

IEA (International Energy Agency) (2004). *Analysis of the Impact of High Oil Prices on the Global Economy*. Paris: OECD/IEA.

IEA (International Energy Agency) (2005). *Prospects for Hydrogen and Fuel Cells*. IEA Energy Technology Analysis Series. Paris: OECD/IEA.

IEA (International Energy Agency) (2006). *World Energy Outlook 2006*. Paris: OECD/IEA.

IEA (International Energy Agency) (2007a). *Energy Security and Climate Policy. Assessing interactions*. Paris: OECD/IEA.

IEA (International Energy Agency) (2007b). *IEA Greenhouse Gas R&D Programme*. www.ieagreen.org.uk.

IEA (International Energy Agency) (2007c). *Key World Energy Statistics*. Paris: OECD/IEA.

IEA (International Energy Agency) (2008). *World Energy Outlook 2008*. Paris: OECD/IEA.

IMF (International Monetary Fund) (2000). *The Impact of Higher Oil Prices on the Global Economy*. IMF. www.imf.org.

IPCC (2000). *IPCC Special Report on Emissions Scenarios. A Special Report of Working Group III of the Intergovernmental Panel on Climate Change*. Cambridge: Cambridge University Press.

IPCC (2006). *2006 IPCC Guidelines for National Greenhouse Gas Inventories, Prepared by the National Greenhouse Gas Inventories Programme*, eds. Eggleston H. S., Buendia L., Miwa K., Ngara T. and Tanabe K. Hayama, Japan: Institute for Global Environmental Strategies (IGES).

IPCC (2007a). *Climate Change 2007: The Physical Science Basis. Contribution of Working Group I to the Fourth Assessment Report of the Intergovernmental Panel on Climate Change*, eds. Solomon, S., Qin D., Manning, M., *et al.* Cambridge: Cambridge University Press.

IPCC (2007b). *Climate Change 2007: Impacts, Adaptation and Vulnerability. Contribution of Working Group II to the Fourth Assessment Report of the Intergovernmental Panel on Climate Change*, eds. Parry, M. L., Canziani, O. F., Palutikof, J. P., van der Linden, P. J. and Hanson, C. E. Cambridge: Cambridge University Press.

IPCC (2007c). *Climate Change 2007: Mitigation. Contribution of Working Group III to the Fourth Assessment Report of the Intergovernmental Panel on Climate Change*, eds. Metz, B., Davidson, O., Bosch, P. R., Dave, R. and Meyer, L. A. Cambridge: Cambridge University Press.

Lovins, A. B. (2003). *Twenty Hydrogen Myths*. Boulder, Colorado: Rocky Mountain Institute. www.rmi.org.

Marchetti, C. (1975). Primary energy substitution models: on the interaction between energy and society. *Chemical Economy and Engineering Review*, **7** (8), 9–14.

Marchetti, C. and de Beni, G. (1970). Hydrogen, key to the energy market. *Scientific and Technical Review of the European Communities, Eurospectra*, **IX** (2), 14–18. www.cesaremarchetti.org.

Nakicenovic, N. (1996). Freeing energy from carbon. *Daedalus, The Journal of the American Academy of Arts and Sciences*, **125** (3), 95–112.

Rifkin, J. (2002). *The Hydrogen Economy*. New York: Penguin Putnam, Inc.

Roads2HyCom (2007). *R&D Expenditure for Hydrogen and Fuel Cells as Indicator for Political Will*, eds. Lako, P. and Ros, M. E., Roads2HyCom. www.roads2hy.com.

Shell (2001). *Energy Needs, Choices and Possibilities: Scenarios to 2050*. www.shell.com.

SIPRI (2007). *SIPRI Yearbook 2007: Armaments, Disarmament and International Security*. 38th edn. Oxford: Oxford University Press, on behalf of Stockholm International Peace Research Institute.

Socolow, R. H. and Pacala, S. W. (2006). A plan to keep carbon in check. *Scientific American*, **295** (3), 50–57.

Stern, N. (2006). *The Economics of Climate Change. The Stern Review*. New York: Cambridge University Press.

The Economist (2007). Cleaning up: a special report on business and climate change. *The Economist*, **383** (8531).

UN (United Nations) (1992). *United Nations Framework Convention on Climate Change*. New York: UN.

UNDP (United Nations Development Program) (2008). *Human Development Report 2007/2008. Fighting Climate Change: Human Solidarity in a Divided World*. New York: UNDP.

UNPD (United Nations Population Division) (2006). *World Population Prospects. The 2006 Revision. Population Database. Online database*. Department of Economic and Social Affairs, Population Division, http://esa.un.org/unpp.

UNSD (United Nations Statistics Division) (2007). *Statistical Databases (Common Databases)*. Online database. http://unstats.un.org/unsd/databases.htm.

Weber, R. (1991). *Wasserstoff*. Frankfurt: Informationszentrale der Elektrizitätswirtschaft (IZE) e.V.

WRI (World Resources Institute) (2005). *Navigating the Numbers: Greenhouse Gas Data and International Climate Policy*. WRI Report prepared by Baumert, K. A., Herzog, T. and Pershing, J. Washington D.C.: WRI.

WRI (World Resources Institute) (2006). *Hot Climate, Cool Commerce: A Service Sector Guide to Greenhouse Gas Management*. WRI Report prepared by Putt del Pino, S., Levinson, R. and Larsen, J. Washington D.C.: WRI.

WRI (World Resources Institute) (2007). *Climate Analysis Indicators Tool (CAIT)*. Online Database version 4.0. Washington D.C.: WRI. http://cait.wri.org.

3

Non-renewable energy resources: fossil fuels – supply and future availability

Michael Ball

Today, the world's energy supply still depends to around 90% on non-renewable energy sources, which are largely dominated by fossil fuels. As the global energy mix is widely expected to continue relying predominantly on fossil fuels in the coming decades, the question arises to what extent and how long fossil fuels will be able to sustain the supply. The projected increase in global energy demand, particularly in the developing nations of Asia (such as China and India), as well as the economic and geopolitical implications of future shortcomings in the supply of oil and gas, are already creating serious concerns about the security of energy supply. Especially, the transport sector, which is still almost entirely dependent on oil worldwide and would be most vulnerable to supply shortages, is increasingly triggering the search for alternative fuels. The following chapter thus focuses primarily on the future availability of fossil fuels in the context of the development of global energy demand and sets the scene for the possible introduction of hydrogen.

3.1 Projections on the future development of global energy demand

In the following, the past and future development of global energy demand and its composition will be briefly analysed. The energy balance methodology for primary energy demand of oil, gas, coal and biomass is normally based on the calorific content of the energy commodities. Depending on the statistical methodology applied, however, figures about world primary energy demand can vary greatly, not only with respect to the absolute demand, but also with respect to the shares of the different fuels, particularly renewable energies, whose share in 2002, for instance, ranged from 6.3% to 13.5% according to different statistics. This is mainly because of the quantification of *electricity* generated from sources that do not have a calorific value, such as hydropower, wind or solar energy.[1] Differences also result from the

[1] To calculate the primary energy equivalent of electricity generated from these renewable sources two methods are used. In the *partial substitution method*, the primary energy equivalent represents the amount of energy that would be necessary to generate an identical amount of electricity in conventional thermal power plants, assuming an average

The Hydrogen Economy: Opportunities and Challenges, ed. Michael Ball and Martin Wietschel. Published by Cambridge University Press.

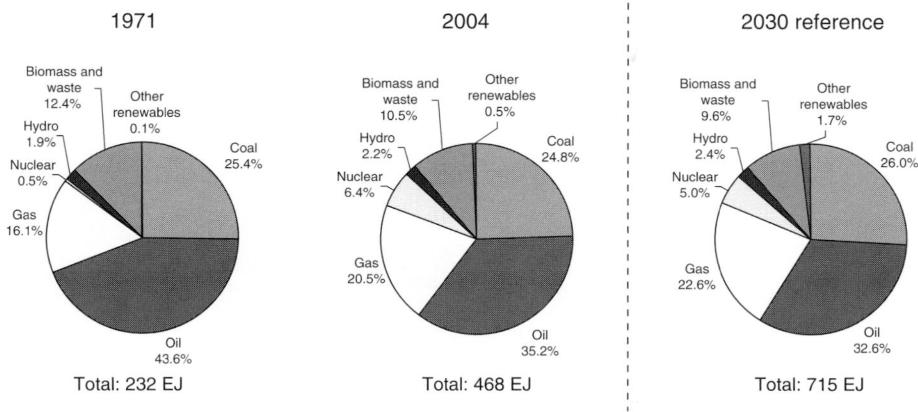

Figure 3.1. World primary energy supply 1971, 2004 and projection 2030 (IEA, 2006), including traditional biomass for developing countries.

evaluation of biomass and waste, especially *traditional biomass* (also referred to as non-commercial biomass) which includes fuels that are not traded commercially: fuel wood, charcoal, dung and farm residues. As the consumption of traditional biomass is very hard to assess and available figures show large discrepancies, it is often not considered in energy statistics, even though it can play an important role in developing countries.

Figure 3.1 shows a comparison of the composition of world primary energy supply in 1971 and in 2004 according to the *International Energy Agency (IEA)*. It can be seen that the total world primary energy consumption has more than doubled since 1971, from 232 EJ to 468 EJ. Today, oil is still the dominant energy source: although its share decreased from 44% in 1971 to 35% in 2004, its absolute consumption increased from 102 EJ to 165 EJ. While the share of natural gas has increased from 16% to 21%, the share of coal has almost remained constant. In total, fossil fuels account for around 80% of today's global energy supply. While nuclear energy was still insignificant in 1971, it contributes more than 6% to total supply today; this surge was mainly a result of the enforced construction of nuclear power plants following the first oil crisis in 1973 and 1974. The share of biomass and waste has slightly dropped, but still amounts to almost 11%; two thirds of this come from traditional biomass, which represents about 20% of total primary energy supply in developing countries, even though the shares differ from country to country (IEA, 2004). Other renewables, such as solar, wind or geothermal energy, still play a negligible role globally today with only 0.5% of total supply.

efficiency of ca. 38%. The *physical energy content method* uses the physical energy content of the primary energy source as its primary energy equivalent, which for electricity from hydropower, wind or solar energy is 100%; the share of renewables is accordingly smaller. For electricity from nuclear power both methods assume an average efficiency factor of 33%. Nowadays most international organisations (IEA, UN, Eurostat) have adopted the physical-energy-content method.

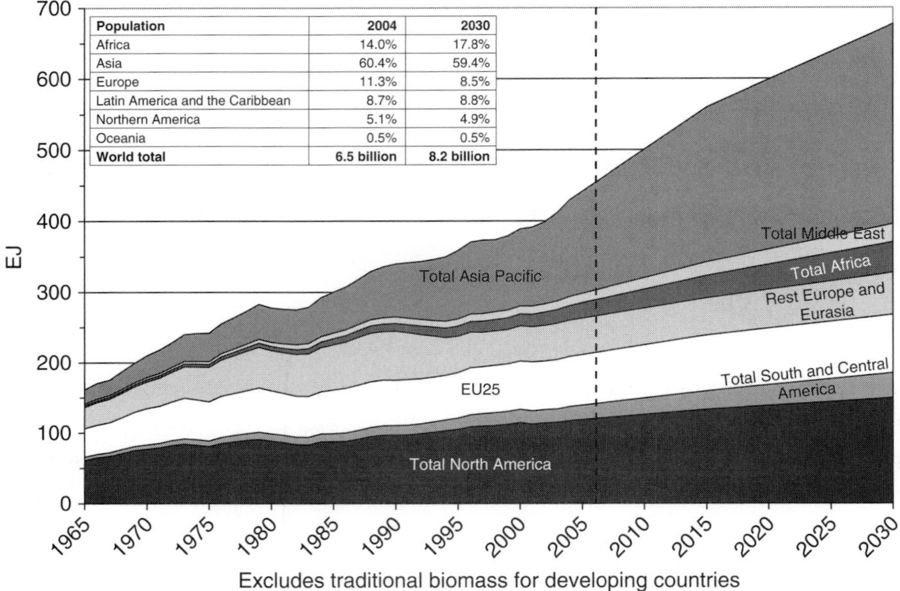

Population	2004	2030
Africa	14.0%	17.8%
Asia	60.4%	59.4%
Europe	11.3%	8.5%
Latin America and the Caribbean	8.7%	8.8%
Northern America	5.1%	4.9%
Oceania	0.5%	0.5%
World total	**6.5 billion**	**8.2 billion**

Figure 3.2. Development of primary energy demand for different world regions since 1965 and demand projections until 2030 (BP, 2006; IEA, 2006; UNPD, 2006), excluding non-commercial biomass for developing countries.

In 2004, the transport sector accounted for 47% of global primary oil consumption (58% of final energy consumption of oil and 18% of total primary energy use), compared with 33% in 1971. The share of oil in global transport energy demand has remained constant over the considered time period, at 95%. As for natural gas, the power generation sector has the highest share in the world gas market, amounting to 38% in 2004; for coal, this share was, with 68%, even higher (IEA, 2006).

Figure 3.2 shows the development of global primary energy demand since 1965, broken down into different world regions, as well as the projection until 2030 according to the *IEA Reference Scenario* (IEA, 2006). Between 1965 and 2005, global demand has been steadily increasing by about 2.5% per year on average, showing only a slowdown during the first and second oil crises (1973–74 and 1978–79) and the Asian economic crisis (1997–98).

Until 2002, North America used to have the highest share in total world primary energy demand. In 2004, this share amounted to 27%, of which the USA had nearly 23%; the share of North America in world population, however, is only about 5%. North America and the EU25 (17%) together made up around 44% of total demand in 2004, representing 16% of total population. The country with the highest share in primary energy demand in the EU25 was Germany with 3.1%, followed by France with 2.5%. As can be seen, around the 1990s, the breakdown of the Eastern bloc led to a strong decline in energy demand in that area. The increase in global energy use in recent years has mainly been caused by the growing demand in the Asia Pacific region, where more than half of the world population lives, and which accounted for 32% of total energy use in 2004. China today is responsible for more than 40% of total demand in the area, followed, with a significant difference, by Japan, India and South Korea, altogether representing almost 80% of regional demand; the average annual growth rate in the region has been 4.3% since 1990, largely driven by China. In contrast, the shares of Africa and South and Central America in world primary energy use amounted to 3.0% and 4.6% respectively only, while representing around 23% of world population. The Middle East region, with its major reserves of oil and natural gas, only had a share of 4.8%.

The most important recently published world energy scenarios looking at the future development of global primary energy demand with a time horizon 2030 are the *World Energy Outlook* (WEO) of the IEA (IEA, 2006; 2008a) and the *International Energy Outlook* of the US Department of Energy (EIA, 2008); the *World Energy Technology Outlook (WETO* H_2*)* of the European Commission (WETO, 2003; 2006), the *Energy Technology Perspectives* of the IEA (IEA, 2008b) and the *Shell Energy Scenarios* (Shell, 2008) cover the time horizon until 2050. A complete comparability of these scenarios is not possible, as country groupings and other boundaries are not uniform. All in all, however, the scenarios depict a largely similar picture of the projected development of world energy demand as well as on the composition of the global energy mix. In the following, the *IEA WEO 2006 Reference Scenario* until 2030 will be representatively investigated in more detail. (The IEA WEO 2008 (IEA, 2008a) was published after the analysis in this book had been completed. Nevertheless, as the WEO 2006 and 2008 Reference Scenarios are not substantially different, (WEO 2008 forecasts a slightly lower growth in energy demand until 2030) the validity of the conclusions drawn in this chapter is not affected.)

As Fig. 3.2 displays, global energy demand is expected to continue to grow until 2030, with 1.6% p.a.; however, to a lesser extent than in the period 1965 to 2005, with an average annual growth of 2.5%. This means that world energy use will increase by 53% until 2030. The rationale behind this forecast is the further growth of world population (from 6.5 billion today to more than 8 billion in 2030) and an assumed continuing growth of world GDP by an average of 3.4% p.a. (between 1971 and 2002, world GDP grew by 3.3% p.a.), particularly in transition countries such as China, India and Brazil (expressed in US$2005 purchasing power parity (PPP) terms). Over 70% of the increase in demand over the projection period is expected to come from developing countries, with China alone accounting for some 30%, thus shifting the centre of gravity of global energy use.

According to the *IEA Reference Scenario*, almost half of the increase in global primary energy use goes to generating electricity and one-fifth to meeting transport needs – almost entirely in the form of oil-based fuels. Regarding the relative shares of the different fuels in the energy mix, only minor shifts are expected (Fig. 3.1). Fossil fuels will remain the dominant source of energy, accounting for some 83% of the overall increase in demand until 2030; the total share of fossil fuels is even projected to increase to 81%. The demand for oil is assumed to grow by 1.3% per year (from 81 million b/d to 116 million b/d in 2030), for natural gas by 2.0% and for coal by 1.8% per year. (The *IEA World Energy Outlook 2004* assumed an increase of 1.6% for oil, 2.3% for gas and 1.4% for coal.) The share of renewables other than biomass is expected to remain marginal.

Global oil demand will increasingly focus on the transportation sector, which is responsible for two thirds of the demand growth, and in 2030 52% of primary oil use will be for transport (compared to 47% in 2004). More than 70% of the increase in oil demand comes from developing countries, which see an average annual demand growth of 2.5%. The *IEA Reference Scenario* further assumes that oil will remain the largest single fuel in the global energy mix until 2030 and continue to provide more than 90% of the energy demand for transportation. While the demand for natural gas grows the fastest in Africa, the Middle East and Asia, notably China, North America and Europe are going to remain the largest markets.

The power sector accounts for more than half of the increase in primary gas demand, increasing its share in global electricity generation from 21% in 2004 to 24% in 2030. Coal sees the biggest increase in demand in absolute terms, remaining the second-largest primary fuel. Power generation accounts for 81% of the increase in coal use, boosting its share of total coal demand from 68% in 2004 to 73% in 2030; accordingly, coal is also going to keep its high share of more than 45% in global electricity generation. Most of the growth in demand comes from Asia, particularly China and India, which alone account already for almost 80% of the entire increase in coal use until 2030 (see Fig. 3.3).

Figure 3.4 shows the past and expected future development of global energy-related CO_2 emissions for selected years, from 1971 until 2030 for the above-described *IEA Reference Scenario*. According to this scenario, global CO_2 emissions will increase by 1.7% per year over the projection period, from 26.1 Gt in 2004 to 40.4 Gt in 2030 (see also Chapter 2).

On a country basis, the United States had, with 22%, the highest share in global CO_2 emissions in 2004, followed by China, with 18%, and Russia, with 6%; the EU25 share was about 15%. Developing countries account for over three-quarters of the rise in global CO_2 emissions between 2004 and 2030, with China alone being responsible for nearly 40% of the increase. Developing countries will become the biggest emitter, as their share in total emissions rises from 39% at present to 52% by 2030. Today, the power sector accounts for 41% of total CO_2 emissions, followed by the transport sector with 20% and industry with 18%. Power generation is projected to contribute almost half the increase in global emissions; transport around 20%. While the power sector is expected to account for 44% of total emissions by 2030, transport remains the second largest sector, with its share of total emissions stable at around 20% throughout the projection period. Coal recently overtook oil as leading contributor to global energy-related CO_2 emissions (41% in 2004) and is likely to consolidate that position through to 2030.

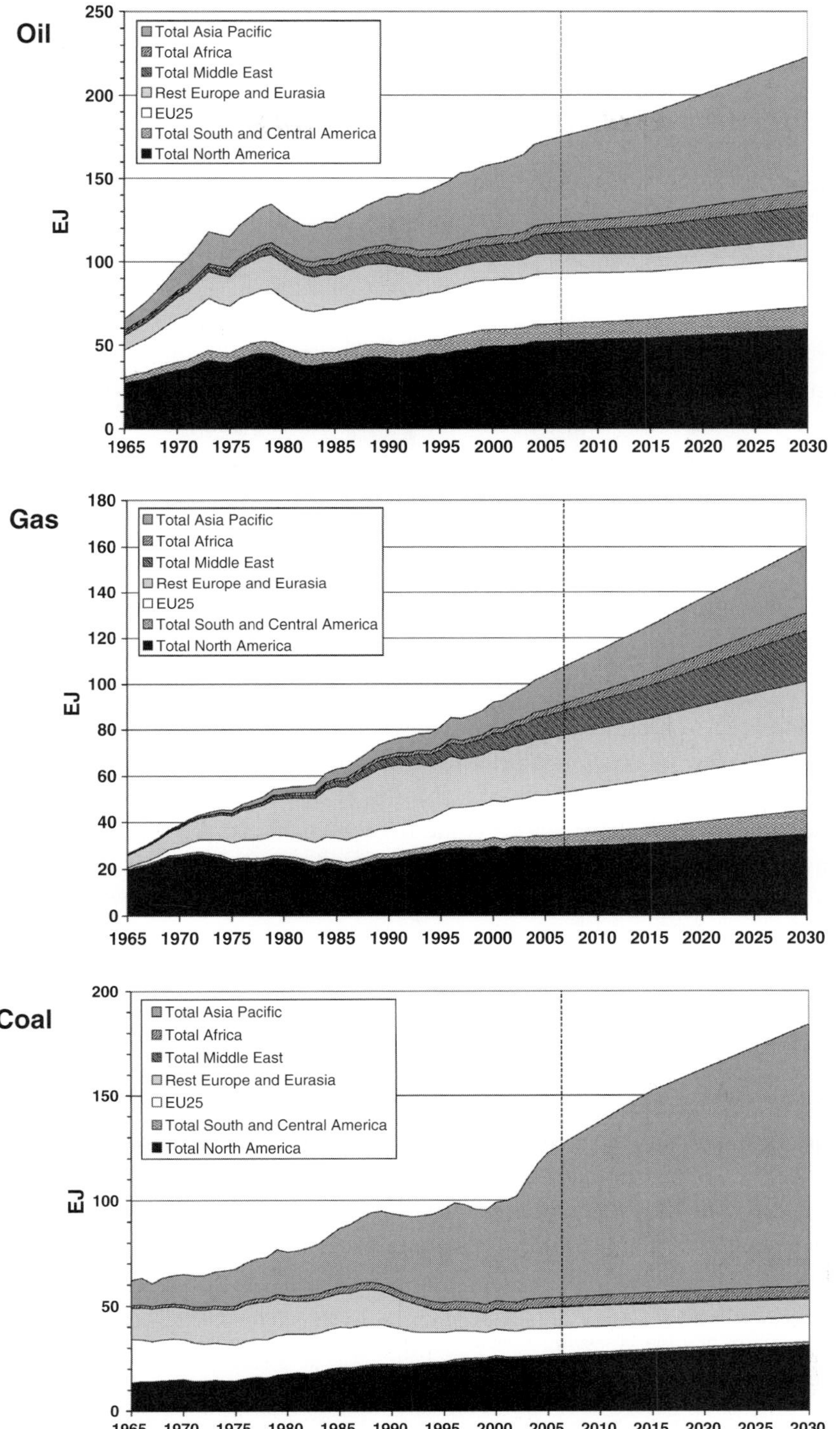

Figure 3.3. Development of oil, gas and coal consumption for different world regions since 1965 and demand projections until 2030 (BP, 2006; IEA, 2006).

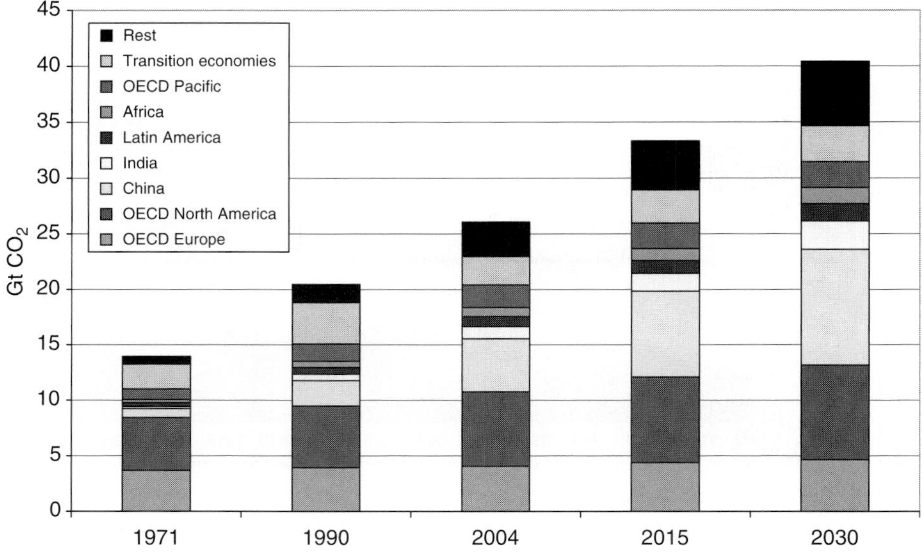

Figure 3.4. Development of energy-related CO_2 emissions (IEA, 2006).

3.2 General classification of reserves and resources

The previous section outlined the future development of global energy demand until 2030, as expected by various world energy scenarios, of which the *IEA Reference Scenario* has been exemplarily discussed in more detail. With the major part of the increase in energy use projected to come from fossil fuels, the aspect of their long-term availability needs to be addressed. Given the natural limits of fossil energy resources, the question arises: to what extent and how long a worldwide steadily growing energy demand can be met, particularly by oil and gas, which are the most depleted fuels today. Hence, the consequences of the above demand scenario for the future availability of oil and gas must be investigated.[2]

To assess the future availability and lifetime of fossil fuels, their occurrences are categorised as *reserves* and *resources*. However, a wide variety of terms is used to describe energy reserves and resources, and different authors and institutions have different meanings for the same terms; meanings also vary for different energy sources (WEA, 2000). Among the ways resources can be categorised are the degree of certainty that they exist and the likelihood that they can be extracted profitably. For explaining the differences and the boundaries between reserves and resources in a schematic way,

[2] The assessment and quantification of the remaining reserves and resources of fossil fuels is a very complex and broad field, characterised by a lack of internationally harmonised definitions and standards, great data uncertainties and discrepancies and, consequently, the potential danger of data abuse for political purposes. Within the scope of this publication, only an overview of the range of the currently available estimates of fossil resources is provided and the focus is rather on the general discussion of potential sources of uncertainty, than on a detailed assessment of the different methodological and statistical approaches and discrepancies at country or even field level.

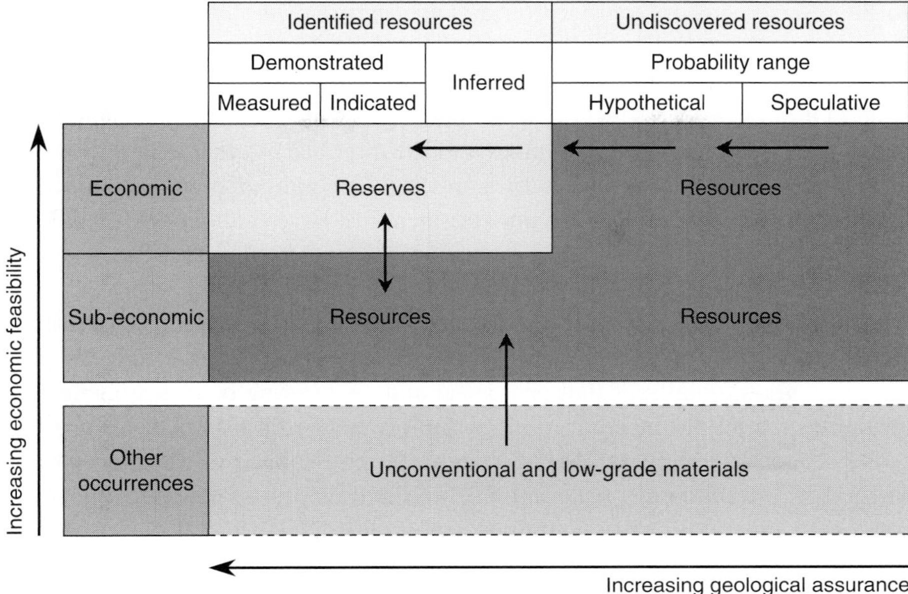

Figure 3.5. McKelvey diagram of reserves and resources (WEA, 2000).

the *McKelvey* diagram can be used, which presents resource categories for *finite raw materials* in a matrix with increasing degrees of geological assurance and economic feasibility (see Fig. 3.5). For the classification of renewable energy sources, see Chapter 5.

In the above classification system, resources are defined as concentrations of naturally occurring solid, liquid or gaseous material in or on the Earth's crust in such form that economic extraction is potentially feasible. The geological dimension is divided into identified and undiscovered resources. *Identified resources* are deposits that have *known* location, grade, quality and quantity or that can be estimated from geological evidence. Identified resources are further subdivided into demonstrated (measured plus indicated) and inferred resources, to reflect varying degrees of geological assurance and the ease of extraction of reserves. *Reserves* are identified resources that are economically and technically recoverable at the time of assessment. *Undiscovered resources* are quantities *expected* or *postulated to exist* under analogous geological conditions and which could be recovered economically today or in the future. Other occurrences are materials that are too low-grade or for other reasons not considered technically or economically extractable. For the most part, unconventional resources are included in 'other occurrences'.

The boundary between reserves, resources and other occurrences is current. For several reasons, reserve and resource quantities and related supply–cost curves are subject to continuous revision. Production inevitably depletes reserves and eventually exhausts deposits, while successful exploration and prospecting add new reserves and resources. Price increases and cost reductions expand reserves by moving resources into the reserve category and vice versa. The *dynamic nature of the reserve–resource relationship* is illustrated by the arrows in Fig. 3.5. Technology is the most important force in this process. Technological improvements are continuously pushing resources into the reserve category by advancing knowledge and lowering extraction costs. The outer boundary of resources and the interface to other occurrences is less clearly defined and often subject to a much wider margin of interpretation and judgement. Other occurrences are not considered to have economic potential at the time of classification. But over the very long term, technological progress may upgrade significant portions to resources.

Reserves of oil and gas are reported according to the *deterministic* or *probabilistic approach*. The *deterministic approach* refers to reserves (sometimes also called *proved reserves*) as those quantities that geological and engineering data demonstrate with reasonable certainty to be recoverable in future years from known reservoirs under existing economic and operating conditions, i.e., on the basis of assumptions about cost, geology, technology, marketability and future prices; estimates of reserves change over time as those assumptions are modified. There is, however, no internationally agreed benchmark or legal standard on how much proof is needed to demonstrate the existence of a discovery, and everyone seems to have his own definition of what is reasonably certain;[3] nor are there established rules about the assumptions to be used to determine whether discovered oil or gas can be produced economically (IEA, 2004). This has created inconsistency and confusion about the amount of oil and gas that can be extracted economically in the long run, and attempts have been made to harmonise definitions and methodologies and to improve the transparency in the reporting of reserves.

The Society of Petroleum Engineers (SPE) and the World Petroleum Congress (WPC) developed a *probabilistic* hydrocarbon-resource classification scheme, that takes into account the probability with which a reserve can be produced (SPE, 2007);[4] but such a probabilistic assessment is also subject to a potential level of misinterpretation.[5] Finally, as for *resources*, very few estimates exist, and those estimates that do exist are also subject to considerable uncertainty and the speculative character is even more pronounced than for reserves.[6] BGR (2003) refers to resources as those quantities that are geologically demonstrated, but at present

[3] There is also no single, commonly accepted technical definition of (proved) reserves, and in the above definition many words are ambiguous and without any quantification; a major drawback of the deterministic approach.

[4] In the SPE/WPC scheme, reserves are classified according to the probability with which they can be produced into 'proved', 'probable' and 'possible' reserves. Under these definitions, 'proved' reserves are those with a probability of at least 90% (P90) that the estimated volumes can be produced profitably; 'proved plus probable' reserves are required to have at least 50% probability (P50), while 'proved plus probable plus possible' reserves are based on a probability of at least 10% (P10).

[5] For a discussion of the advantages and disadvantages as well as the statistical consequences of the above reporting methods of reserves see (IEA, 2004; Kägi *et al.*, 2003; Schindler and Zittel, 2000).

[6] The most widely quoted *primary* sources of global *reserves* data of oil and gas, which compile data from national and company sources, are the journals *Oil & Gas Journal* and *World Oil*, the *US Geological Survey (USGS)* and the industry database of the *IHS Energy Group*. Specifically for oil, *OPEC* compiles and publishes data of its member states; for gas there is *Cedigaz*. Other publishers, such as *BP*, *BGR* or the *IEA*, base their estimates on the above-mentioned primary sources. Resource estimates at country level are reported by the *USGS* and *BGR*.

cannot be recovered for technological or economic reasons (but might be recoverable in the future), as well as quantities that are geologically possible but not demonstrated.

Another common criticism of oil reserves statistics relates to *backdating of reserves*, as new discoveries and revisions of previous reserve calculations are often not distinguished. When oil companies replace earlier estimates of the reserves left in many fields with higher numbers, it is a common practice that those revisions are backdated to the year in which a company or country corrected an earlier estimate and *not* to the year in which the field was discovered first. This practice leads to distortions of statistics regarding new discoveries, as the *reserve growth*, resulting from new discoveries and revisions of earlier estimates, is biased towards the present and often generally interpreted as discovery rate (Campbell and Laherrère, 1998).

Finally, there are both tendencies to deliberately over- and understate reserve figures, as outlined by BGR (2003). The following reasons can lead to an *overstating* of reserves:

- OPEC production quotas are directly tied to reserves (see also OPEC reserves jump in Section 3.3.2.2);
- intention of developing nations to attract investors;
- temptation of companies listed on stock exchanges to boost the share price.

Possible reasons for *understating* reserves can be:

- in many countries, reserves are considered as assets and taxed, so oil companies may tend to report lower figures;
- intention of developing nations to attract financial aid and obtain loans;
- unstable political and economic situations in a country that may jeopardise exploration and production activities.

Any statistics depend on their sources and what is included therein; oil statistics from different sources rarely match up. The correctness of the reported data depends to a large extent on the seriousness of the transmitter, and figures are often published without requiring proof of their credibility (for example, many OPEC countries have been reporting constant reserves figures for many years, see Section 3.3.4). International attempts have been made to improve the transparency in the reporting of reserves.[7] Financial reporting standards are the strictest and lead to the lowest estimates: the US Securities and Exchange Commission (SEC) lays down the most restrictive and most detailed reserve-reporting standards for companies quoted on

[7] As for oil statistics, in 2001 the *Joint Oil Data Initiative (JODI)* has been launched, a common effort made by APEC, Eurostat, IEA, IEFS, OLADE, OPEC and UNSD to harmonise oil data from different statistic sources (www.jodidata.org). The UN Economic Commission for Europe (UNECE) has further developed a *Framework Classification for Fossil Energy and Mineral Resources (UNFC)*. In 2006, the Society of Petroleum Engineers and the UNECE agreed to further efforts to develop one globally applicable harmonised standard for reporting fossil energy reserves and resources, that will ensure greater consistency and transparency in financial reporting and enhance energy resource management, energy studies and business processes (UNECE, 2007).

US stock exchanges based on the probabilistic approach, requiring the report of only proved reserves; they, nonetheless, also allow scope for some discretion. In conclusion, it can be said that, so far, there are no international requirements or standards in place regarding reserve classification and reporting, and the methodologies used for reserve estimates seem to vary according to their purpose; as a result, reserve data are often referred to as 'political data'. The consequences are inconsistency and controversy about the future supply of oil and gas.

The following terms are often used in the context of quantifying reserves and resources of fossil fuels: the *Estimated Ultimate Recovery (EUR)*, also called *Ultimate Recoverable Resources (URR)*, is the sum of past cumulative production, proved reserves at the time of estimation and the possibly *recoverable* fraction of undiscovered resources. The *remaining potential*, i.e., the sum of reserves and resources, is the total amount of an energy source that is still to be recovered. The *mid-depletion point* is the point of time when approximately 50% of the EUR (at field, country or world level) has been produced.

To measure the lifetime of hydrocarbon reserves, often the *static lifetime of reserves* (also called the *reserve:production ratio*) is calculated as the ratio between reserves and the annual production of the current year. A criticism of this approach is that it seems unrealistic to assume that the production of, e.g., crude oil will remain constant in future at today's level, and then suddenly fall from a high level down to zero; it can rather be expected that there will be an increase to a peak production and then a gradual decline in production.

3.3 Oil

3.3.1 Classification of conventional and unconventional oil

Crude oil is generally categorised as *conventional* and *unconventional*. However, there is no exact definition of these two terms and the meaning depends on whether the distinction is made according to physical or economic/technical criteria (Kägi *et al.*, 2003).

The *physical approach* categorises the different oils with regard to their density or specific gravity (see Fig. 3.6). According to this approach, crude oil is termed *conventional* if its specific gravity does not exceed 0.934 g/cm^3 (or is greater than 20 °API).[8] This definition also comprises *natural-gas liquids (NGL)* with a specific gravity of less than 0.8 g/cm^3, a liquefiable fraction separated from the methane during the production of natural gas or oil (i.e., non-associated or associated gas, respectively).[9] Crude oils with a specific gravity higher than 0.934 g/cm^3 are called

[8] The API gravity is an arbitrary metric and computed as °API = 141.5 / specific gravity − 131.5, with the specific gravity measured at a temperature of 60 °F (15.5 °C).

[9] Those hydrocarbons that exist in the reservoir as constituents of natural gas, but which are – according to their vapour pressures – recovered as liquids (propane and butane) at the surface, are generally referred to as *natural gas liquids* and comprise condensate as well as *liquefied petroleum gas (LPG)*.

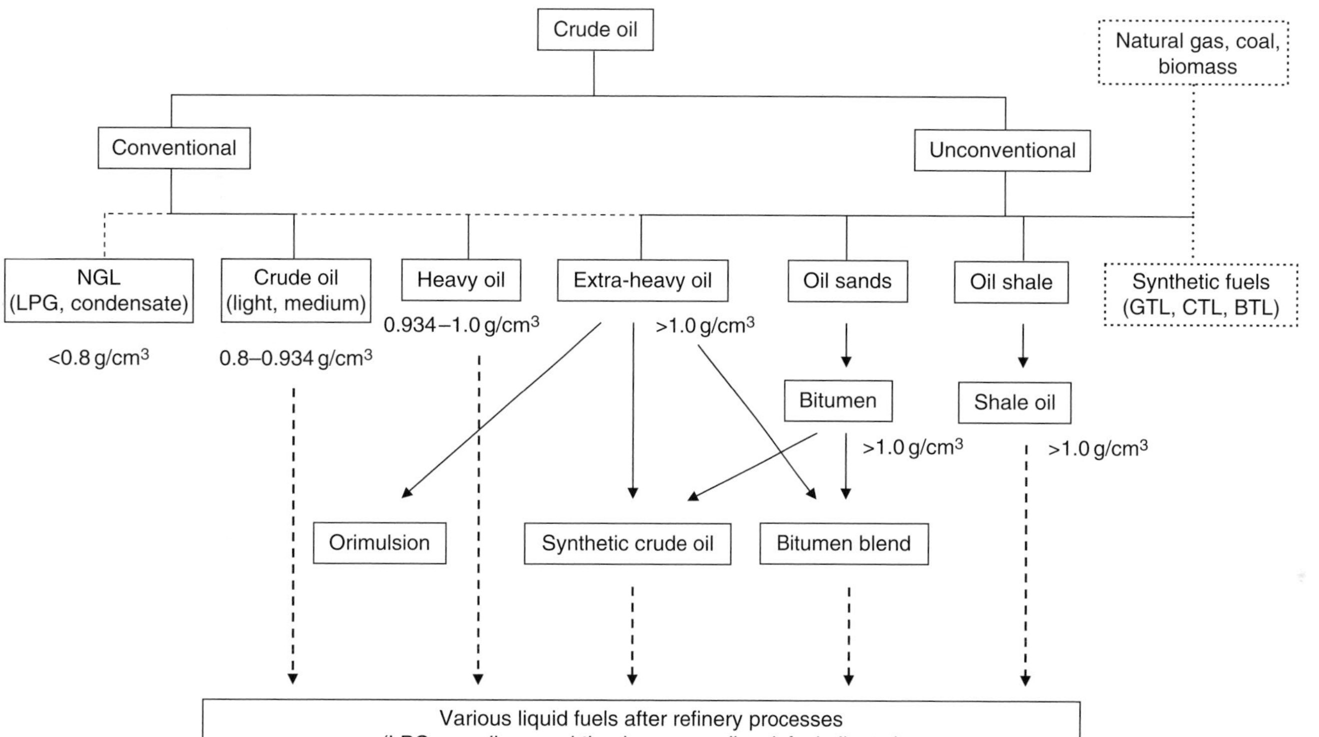

Figure 3.6. Classification of conventional and unconventional oil.

unconventional. The unconventional oil in this respect includes heavy and extra heavy oil, oil or tar sands (crude bitumen) and oil shale (shale oil). In addition, there are synthetic fuels derived from natural gas, coal or biomass, which are also often included under unconventional oil.

The physical approach is not undisputable, as it does not take into account the conditions of the occurrences or of the production of the crude oil (e.g., onshore or offshore, water depths, climatic conditions, etc.). Therefore, some authors give a narrower definition of conventional oil. Campbell (2006), for instance, considers crude oil that is found under deep-water conditions (water depths greater 500 m) or in arctic regions, as well as NGL as unconventional oil. As a consequence, the remaining potential of conventional oil is estimated to be lower.

The *economic/technical approach* focuses on the question of which occurrences can be extracted economically with today's technology. According to this approach, all occurrences that at present are economically extractable with existing available technology are classified as *conventional*, whereas occurrences that are not extractable at present owing to economic or technical hurdles are considered *unconventional*. The physical properties of the oil are not taken into account. It is important to note that, in this approach, the boundary between conventional and unconventional crude oil is current, as it is subject to the market price of oil, the production cost and the available technology.

The *physical* and *economic or technical approaches* made no big difference with respect to the oil that has been produced until now. According to the *economic or technical approach* – as advancing technology and rising prices will facilitate the economic production of new resources – the boundary will increasingly be shifted from unconventional oil towards conventional oil. This is, for instance, the case in Venezuela and Canada, where extra heavy oil and oil sands have already been economically produced for several years. According to the *physical approach*, however, this leads to a rise in the production volumes of unconventional oil. In this publication, the distinction between conventional and unconventional occurrences will be made according to the *physical approach*.

3.3.2 Conventional oil

3.3.2.1 Production, consumption and trade

As in the case of reserves, there are also different statistical methodologies for reporting production and consumption figures of oil, which can mainly be attributed to two factors: first, a lack of definitions regarding what should be reported under (conventional) oil, and second, a lack of consensus regarding how the amount of oil is being measured. The first point refers to the inclusion of NGL or unconventional oil (such as Canadian oil sands), the latter concerns the units of measurement. Depending on the statistic, oil can be measured as volume (in barrels) or as weight (in tons); there are, however, no uniform conversion factors, as the density of

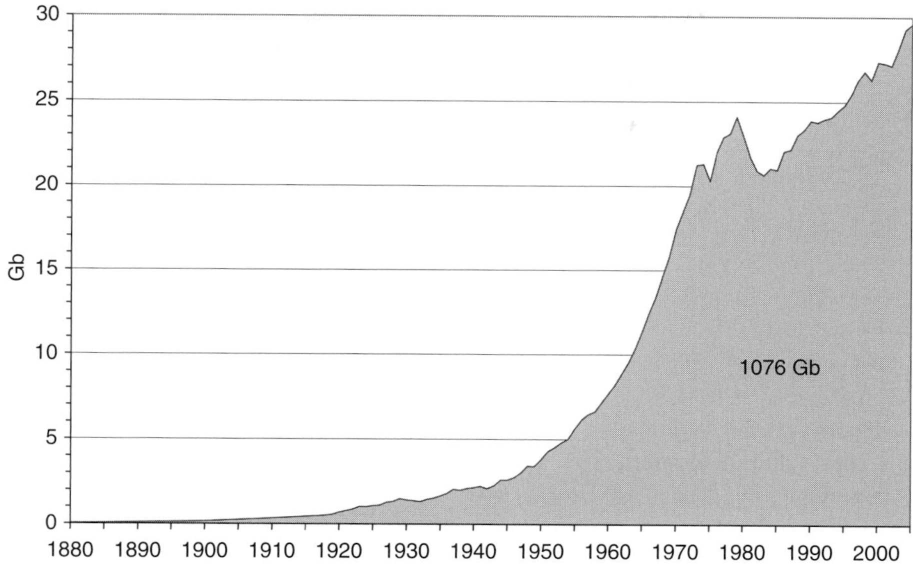

Figure 3.7. Development of annual oil production since 1880 (Bartlett, 2000; BP, 2006).

different crudes is not constant, and it usually varies with time and between countries; in addition, oil may also be measured as energy equivalent (Laherrère, 2001). As a result, different statistics are not directly comparable.

Since the beginning of industrial oil production, around 1880, almost 1100 Gb of crude oil have been extracted, of which more than two thirds were extracted in the last 20 years (Fig. 3.7). Annual production has been steadily increasing ever since, notably interrupted only by the first and second oil crises (1973–74 and 1979) and the Asian economic crisis in 1998.

Table 3.1 shows the ten major producers and consumers of oil in 2005. Total production in 2005 amounted to 81 million barrels per day (29.6 Gb per year). The most important producer countries were Saudi Arabia, Russia and the USA; the ten countries with the highest production are responsible for more than 60% of world production. Although the USA has been the biggest oil producer until the beginning of the 1990s, its production has been continuously declining since passing a peak in 1970; Russia's production, on the other hand, has significantly grown since the beginning of the 1990s. While OPEC's share[10] in total production amounted to around 50% in the mid 1970s, this fell to about 30% in the mid 1980s and today accounts for some 42%.

About one third of total crude oil production today comes from offshore fields (mainly the Persian Gulf, North Sea, Gulf of Mexico and West Africa). The oil production of the European Union (this includes the 25 member states as of

[10] The OPEC was founded in 1960 and, at the time of writing, comprises the following members: Algeria, Iran, Iraq, Kuwait, Libya, Nigeria, Qatar, Saudi Arabia, United Arab Emirates, Venezuela, Angola and Ecuador.

Table 3.1. *Worldwide oil production and consumption 2005*

Oil production		Oil consumption	
Country	(%)	Country	(%)
Saudi Arabia	13.6	USA	25.0
Russia	11.8	China	8.5
USA	8.4	Japan	6.5
Iran	5.0	Russia	3.3
Mexico	4.6	Germany	3.1
China	4.5	India	3.0
Canada	3.8	South Korea	2.8
Venezuela	3.7	Canada	2.7
Norway	3.7	Mexico	2.4
United Arab Emirates	3.4	France	2.4
Sum	**62.4**	**Sum**	**59.8**
OPEC share	41.7	EU25 share	17.9
World (Gb)	**29.6**	**World (Gb)**	**30.1**
World (EJ)	**170**	**World (EJ)**	**172**

Note:
Includes crude oil, shale oil and NGL, as well as oil sands and extra-heavy oil for Canada and Venezuela, respectively.
Source: (BP, 2006).

31st December, 2006), which relies largely on the offshore potential of the North Sea, contributed no more than 6.6% to world production in 2005, of which Norway had a share of 55% and the UK of 34%. The production in the UK, however, has been rapidly declining since passing the *mid-depletion point* in 1999; also Norwegian oil production has been declining since 2001, and this trend is likely to continue, unless new fields in the Barents Sea can successfully be developed. The average quality of crude oil is expected to become heavier and more sour (higher sulphur content) in the future, owing to a continuing decline in production from existing sweet fields, increased production of heavier oils in Russia and the Middle East, and an expected growth of unconventional oil production, such as from Canadian oil sands.

The geographical distribution of oil consumption depicts a very different picture from production. The USA alone is responsible for one quarter of total consumption, followed by China and Japan; the European Union accounts for 18%. In particular, consumption in China and India more than doubled in the last decade, with China showing a surge in demand of 17% between 2003 and 2004. The transportation sector today is responsible for 47% of total primary oil consumption,

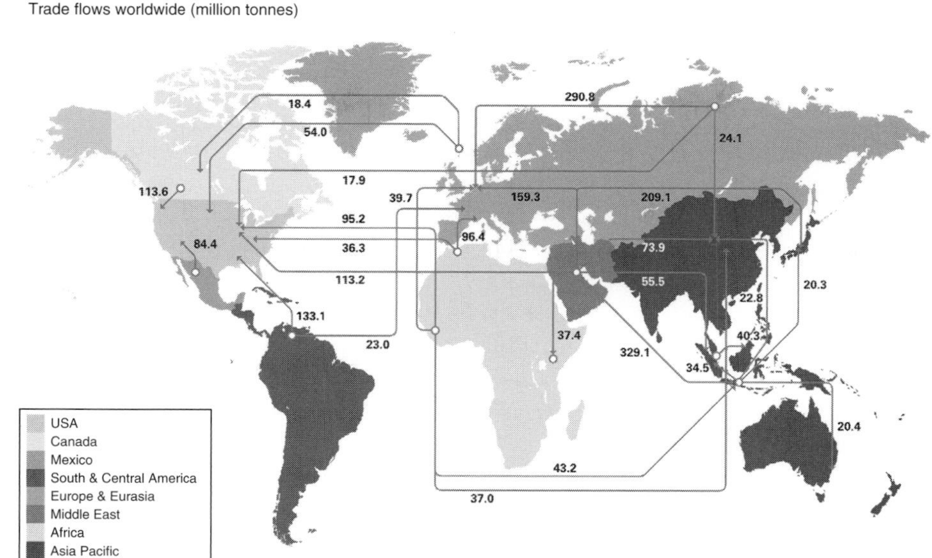

Figure 3.8. Major oil trade movements (Mt) (BP, 2007).

followed by the industry sector with 17%, the residential, services and agriculture sectors with 13% and the power sector with 7% (IEA, 2006).

Figure 3.8 shows the major oil trade movements (imports and exports) in 2006: 13.8 Gb of crude oil and 4.4 Gb of oil products were traded worldwide, around 60% of world crude oil production. In 2005, the USA and Europe each accounted for 27% of total imports (crude oil plus products), followed by Japan with 10% and China with 7%. The most important export regions were the Middle East with 40% of total exports, followed by Russia with 14% and West Africa (Nigeria) with 9%. While pipelines dominate the transport of oil within the continents, and countries, tanker transport dominates intercontinental transport (Middle East to Europe, Asia and America, Africa to Europe and America, and South America to North America) with a share of 75% to 80% (BGR, 2003).

Today, the USA has an oil-import dependence (net imports as percentage of consumption) of 64%, the European Union of 79%, China of 46%, India of 69% and Japan of 100%; as non-OPEC production declines, the import dependence of these countries will further rise in the coming decades until 2030.

3.3.2.2 Reserves and resources

As outlined in the previous section, oil statistics, particularly reserve statistics from various sources may differ significantly – both globally and at country level – as different methodologies for measuring the reserves and different statistical

Table 3.2. *Remaining potential conventional oil, end 2005*

Remaining potential		Reserves	Resources
Country	(Gb)	(Gb)	(Gb)
Saudi Arabia	328.8	264.9	63.9
Russia	173.8	74.6	99.2
Iran	161.1	132.5	28.7
Iraq	142.9	115.0	27.9
Kuwait	106.9	101.8	5.1
United Arab Emirates	105.1	97.8	7.4
Venezuela	101.8	79.7	22.1
USA	88.2	29.4	58.8
Kazakhstan	59.5	30.1	29.4
Nigeria	52.0	35.9	16.2
Sum	**1 320**	**962**	**359**
Others	471	226	244
World	**1 791**	**1 188**	**603**
World (EJ)	**10 262**	**6 807**	**3 455**
OPEC share (%)	61.4	75.6	33.3
EU25 share (%)	1.1	0.8	1.8

Source: (BGR, 2007).

boundaries are applied. Numerous estimates made by various institutions, such as the World Energy Council, IHS Energy, OPEC, the US Geological Survey, *Oil & Gas Journal*, Campbell and others, report current oil reserves in the range from 900 to 1300 Gb. Within the limits of this publication, however, it is not possible to investigate the differences between the various statistics in great detail. Instead, the estimates of the German *Federal Institute for Geosciences and Natural Resources (BGR)* are used in the following, as they are derived from an assessment of the most important primary sources and, thus, considered to reflect a representative average; in addition, *BGR* also publishes estimates of resources at country level. When calculating scenarios for the mid-depletion point of oil, the entire range of estimates will be taken into consideration.

Table 3.2 shows the remaining potential (the sum of reserves and resources) of *conventional oil* at the end of 2005, which amounts to around 1800 Gb, made up of 1200 Gb reserves and 600 Gb resources. In line with the definition in Section 3.3.1, these figures do *not* include unconventional oil, such as crude bitumen from oil sands production in Canada or extra heavy oil from Venezuela.[11] Almost three-quarters of

[11] The *Oil & Gas Journal*, for instance, reports total proved oil reserves at the end of 2006 of 1317 Gb, including Canadian oil sand reserves of 179 Gb (OGJ, 2006a). Taking these oil sand reserves into account, Canada ranks second to Saudi Arabia in terms of oil reserves.

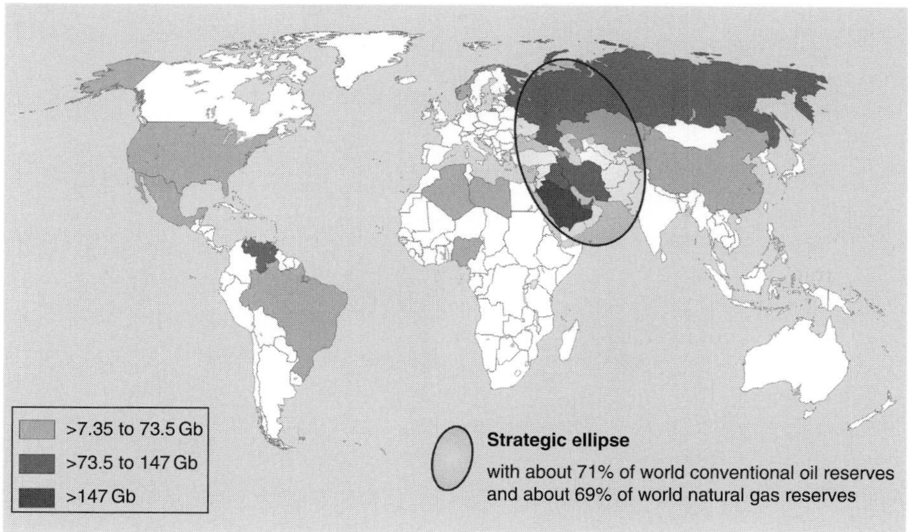

Figure 3.9. The strategic ellipse for oil (BGR, 2007).

the remaining oil is concentrated in ten countries only, headed by Saudi Arabia with 18% and Russia with 10%. Around 70% of conventional oil (and gas) reserves are located within the so-called '*strategic ellipse*' around the Caspian Sea, as illustrated in Fig. 3.9. Moreover, around 60% of the remaining potential and 75% of reserves are located in OPEC countries.

The remaining potential of Europe, mainly of Norway and Great Britain, has a share of 2.6% of the global potential only and, in 2005, already 56% of the EUR of Europe had been produced; while Great Britain, with 62%, has already passed its mid-depletion point and Norway is close to half of its EUR being depleted (BGR, 2007). According to Schindler and Weindorf (2003), 23 countries (excluding OPEC and Russia) have already passed their mid-depletion point. In Europe, offshore reserves exceed onshore reserves; worldwide, offshore reserves account for 24%. The fact that the estimated conventional oil resources are about only half the reserves, reflects the high degree of exploration and may indicate that no major new discoveries of conventional oil are expected. Figure 3.10 shows the distribution of EUR, i.e., cumulative production, reserves and resources, for different world regions.

Of the approximately 47 500 oil fields known today, only around 500 – the so-called 'giant' fields – produce about half of the world's oil and contain almost 75% of all reserves (Bahorich, 2006; Robelius, 2007; Schindler and Zittel, 2000); this is less than 1% of the total number of fields. The concentration of oil production in only a few fields can very well be exemplified by Saudi Arabia: the largest oil field in the world, the *Ghawar* field in Saudi Arabia, has been producing since 1951 and still

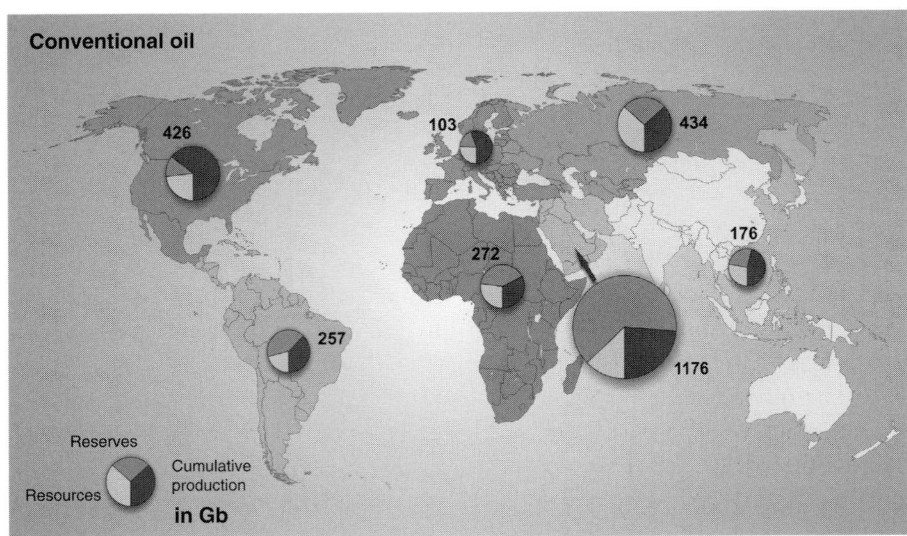

Figure 3.10. Geographical distribution of the EUR of conventional oil (BGR, 2007).

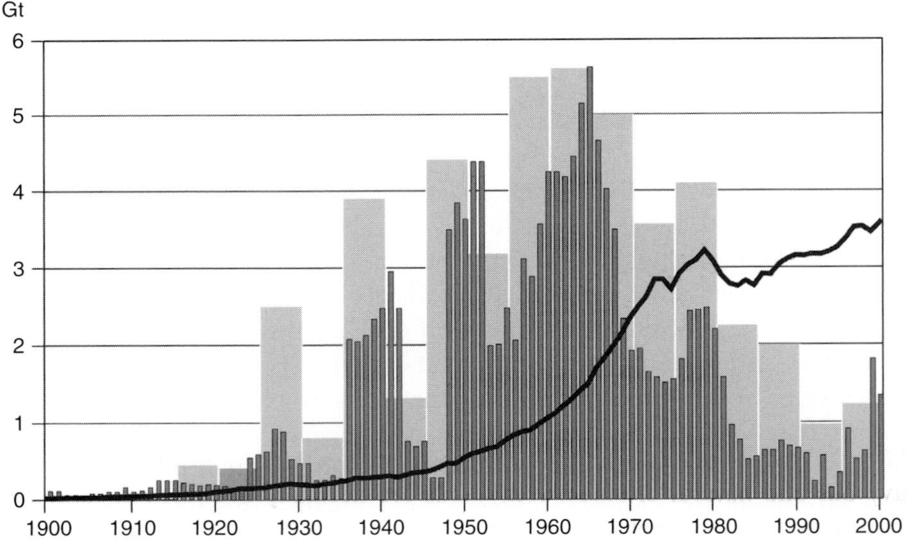

Figure 3.11. New discoveries and yearly production of crude oil from 1900 to 2000 (BGR, 2007).

contributes to about 50% of today's national oil production. The majority of new discoveries of oil fields took place between the 1950s and the 1970s, as shown in Fig. 3.11, which displays new discoveries (on a five-year average) and discoveries of giant fields, as well as the annual crude oil production since 1900. The discovery volumes of oil peaked in the 1960s (of gas in the 1970s); the actual number of discoveries peaked in the 1980s (Bahorich, 2006). Since the 1980s, annual production has been exceeding new discoveries, which today, are replacing about one-quarter of annual production only.

With the average replacement rate of oil reserves continuing to decline and a further fall in the average size of new fields discovered (as in the past four decades), the gap between production and new discoveries is likely to grow even further in the future. According to Campbell and Laherrère (1998), of all the oil being produced today, about 80% comes from fields that were discovered almost 40 years ago. The major potential for future discoveries is seen offshore (Sandrea and Sandrea, 2007a; b).

As much as 90% of the world's oil (and gas) reserves are in the hands of partially or wholly state-owned and operated companies, the so-called National Oil Companies (NOC), through which governments retain the profits from oil production. The three largest oil and gas firms are *Saudi Aramco*, followed by the *National Iranian Oil Company* and *Gazprom*; *ExxonMobil*, the international oil company (IOC) with the highest proved reserves, ranks fourteenth only (*The Economist*, 2006). Today, almost 60% of the world's oil production comes from non-OPEC countries, which hold about one-quarter of total reserves only; hence, the dependence on the production from OPEC countries can be expected to increase in the future. In addition, many of the world's remaining potential new sources of oil and natural gas are in countries with relatively high political and legal instability, such as Nigeria and Russia, or technically challenging regions, such as the Arctic and Asia–Pacific region.

3.3.3 Unconventional oil

According to the physical approach, the distinction between conventional and unconventional oil is made with respect to its specific gravity (see Section 3.3.1). Heavy crude oils often result from a bacterial oxidation of conventional oils inside the reservoir rock and have generally degraded physical and chemical properties, such as much higher viscosity, and higher heavy metal, sulphur and nitrogen contents. Different categories of heavy crude oil are defined according to specific gravity: the heavy oils (between 10 and 20 °API),[12] and the extra-heavy oils and bitumen (°API <10), where the '*in-situ*' level of viscosity makes the distinction between extra-heavy oils and natural bitumen. Heavy oils are normally not recoverable in their natural state through a well or by ordinary production methods; most require heat or dilution to flow into a well or through a pipeline. In the case of oil shale, the reservoir rock and source rock are the same because the oil has not migrated, and permeability is practically zero.

[12] Heavy crude oil will not be explicitly addressed, as there are often no clear boundaries with conventional oil and extra-heavy crude oil within a heavy oil reservoir. To avoid double counting, reserves of heavy oil will not be considered either, as they are already partially or completely included in conventional or extra-heavy oil reserves (BGR, 2003).

3.3.3.1 Oil sands

Properties and extraction processes *Oil-sand* or *tar-sand* deposits are naturally occurring mixtures of quartz sand, silt and clay, water and *natural bitumen* (also called *crude* or *natural bitumen*), along with minor amounts of other minerals. Each particle of oil sand is coated with a layer of water, which is surrounded by a thin film of bitumen.

Although there can be considerable variations, oil sands typically contain 10 to 12 wt.% bitumen, 75 to 80 wt.% inorganic material (of which 90% is quartz sand), and 3 to 5 wt.% water. With a density range of 8 to 14 °API and a viscosity greater than 10 000 centipoise (cP) (for comparison, light oil is defined to have a viscosity less than 100 cP), bitumen is a thick, black, tar-like substance that pours extremely slowly, and needs to be heated or diluted to be produced or to be transported in a pipeline. Bitumen can be recovered from oil sands either by *surface mining* or *in-situ extraction* methods (see Fig. 3.12).

In *surface mining*, the overburden is removed to expose the oil sands, and is then stockpiled for later use in reclamation; the maximum overburden thickness that can be removed economically is about 70 metres. The mined oil sands are then transported by trucks to crushers that break up lumps and remove rocks. In a process called *hydrotransport*, the oil sand is mixed with hot water to create a slurry mixture, and is piped to the extraction plant. During hydrotransport, the bitumen already begins to separate from the sand, water and minerals. In the *extraction plant*, the bitumen forms a thick froth at the top of the separation vessel and the sand settles out to the bottom. The bitumen froth is skimmed off, and later mixed with solvent and spun in a centrifuge to remove water and clay solids. The bitumen can then be processed in an *upgrader*. Surface-mining operations and subsequent water-based extraction of the oil sands produce large volumes of tailings. Tailings slurry from the extraction plants contains water, residual bitumen, sand, fine clay particles and solvent. The traditional method of dealing with tailings is to pump them into large holding ponds with reclamation to occur after the tailings slurry solidifies.

The *in-situ recovery* method is used where the oil sands occur below 50 to 70 metres of overburden. In (thermal) *in-situ* recovery, thermal energy is applied to heat the bitumen and allow it to flow to the well bore. Steam is often used as uplifting gas, i.e., to facilitate production by softening the bitumen, diluting and separating it from sand grains, and enlarging or creating channels and cracks through which the diluted bitumen can flow.[13] Existing *in-situ* technology uses natural gas-fired boilers to generate steam. Around 90% of the water used for steam to recover bitumen can be recycled, but for every barrel of bitumen produced, about 0.3 barrels of additional ground water must be used. There is less land disturbance from *in-situ* extraction than from surface mining; however, the recovery factor is, with currently 25% to

[13] There are numerous *in-situ* recovery methods that include steam injection, the most common ones being *cyclic steam stimulation ('huff & puff')* and *steam-assisted gravity drainage*. A comprehensive overview of oil sands production methods can be found in Gruson *et al.* (2005).

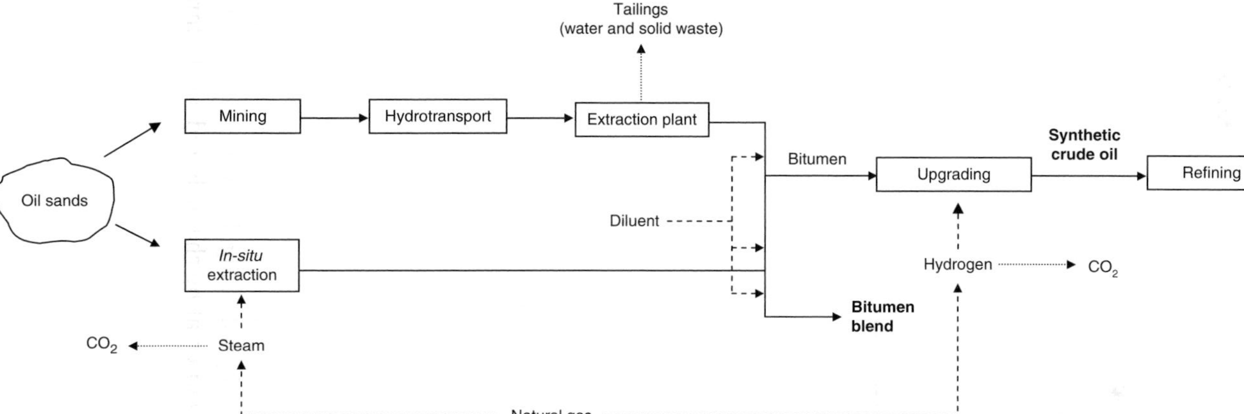

Figure 3.12. Schematic representation of oil-sands production.

Table 3.3. *Crude bitumen – reserves, resources and cumulative production, end 2006*

Country	Crude bitumen in-place (Gb)	Reserves (Gb)	Resources (Gb)	Cumulative production (end 2006) (Gb)
Canada	1 694	173.6	136.5	5.4
Mineable	*101*	*31.8*	*34.0*	*3.6*
In-situ	*1 593*	*141.8*	*102.5*	*1.8*
World	**2 515**	**220**	**220**	**5.4**
World (EJ)	**17 102**	**1 496**	**1 496**	**37**

Sources: (AEUB, 2007; BGR; 2003; 2007; NEB, 2004; 2006).

60%, significantly lower than for mining operations, with more than 90% (Babies, 2003; Gruson *et al.*, 2005). This yields about 0.56 to 0.7 b bitumen/ton oil sand for mining projects, and between 0.16 and 0.56 b bitumen/ton oil sand for *in-situ* projects.

To be processed by refineries, the bitumen generally needs to be upgraded. To transport the bitumen to the upgrader (or for direct use as bitumen), it must be blended with a diluent, normally gas condensate, to meet pipeline specifications for density and viscosity. The upgrading is first achieved through the addition of hydrogen to crack the large hydrocarbon molecules into smaller compounds (*hydrocracking*), or through the removal of carbon (*coking*); during a second stage, hydrogen is added to remove impurities such as sulphur (*hydro-treating*). Upgraders produce solid by-products that need to be handled: coke (for upgraders using coking) and sulphur (for all upgraders). During the upgrading process, the bitumen is converted from a viscous, tar-like oil into a high-quality synthetic crude oil (SCO), with an API degree of between 29 and 36 and a very low sulphur content; this synthetic crude oil can then be further processed in conventional refineries. The average liquid yield factor for upgrading is around 0.86. *In-situ* projects are generally on a smaller scale than mining projects and cannot accommodate the cost of a dedicated upgrader. In almost all such projects, the bitumen is blended with a lighter, less-viscous hydrocarbon (diluent) and sold as bitumen blend, with a density of around 21 °API (Gruson *et al.*, 2005).

Resource estimates and current production The vast majority of the world's oil-sand deposits is located in Canada. Of the total in-place volume of around 1700 Gb of bitumen, only slightly less than 20% is assumed to be recoverable. The EUR of Canadian crude bitumen, i.e., reserves, resources and cumulative production as of the end of 2006, amounted to around 316 Gb. Taking the remaining reserves of around 174 Gb into account, Canada ranks second following Saudi Arabia in global oil reserves (see Table 3.3). However, the practice of including oil sands in official reserve statistics, as for instance by the *Oil & Gas Journal*, is not without controversy, as will be discussed later. More than 80% of the initial established reserves are considered to be only recoverable by *in-situ* methods, since they are too deep below the surface to use open-pit mining (AEUB, 2007).

Canada's bitumen resources are situated almost entirely within the western province of Alberta (see Fig. 3.13). These deposits are distributed among three regions: *Athabasca*, *Cold Lake* and *Peace River*. Approximately 76% of crude bitumen is produced in the Athabasca region, 22% in the Cold Lake region and 2% in the Peace River region.

All commercial oil-sands projects are currently located in Canada. Conventional Canadian oil production peaked in 1973. Mainly as a result of declining production costs and state incentives, Canadian crude bitumen production has been continuously increasing since the end of the 1980s and exceeded conventional crude oil production for the first time in 2001; from 1990 to 2006 bitumen production has more than trebled (see Table 3.4). Crude bitumen is produced by mining, extraction and *in-situ* recovery (thermal and non-thermal).[14] In 2006, Canada produced a total of 458 Mb bitumen (1.25 Mb/d), 59% from the mineable area and 41% from the *in-situ* area. Bitumen produced from mining was upgraded, yielding 200 Mb of synthetic crude oil; whereas *in-situ* production was mainly marketed as non-upgraded crude bitumen (bitumen blend). This split between end uses for mining-based and *in-situ*-based bitumen production is historical. The Canadian government has developed an ambitious roadmap, which aims at almost fivefold production by 2030, raising bitumen production to 5 Mb/d (ACR, 2004). The industry is currently dominated by two companies, *Syncrude* and *Suncor*.

Overall production costs at the time of writing are in the range of US$(2005) 12–20 per barrel bitumen for *in-situ* projects (excluding upgrading), US$ 15–17 per barrel bitumen for oil-sands surface-mining or extraction projects, and US$ 30–34 per barrel SCO for integrated mining projects, i.e., mining or extraction and upgrading operations (NEB, 2006) (including capital costs, operating costs, taxes, royalties and a 10% real rate of return to the producer). Integrated mining and *in-situ* operations are estimated to be economic at US$ 30 to $ 35 per barrel. Capital expenditures for oil-sand projects have been growing tremendously over the last decade and far surpassed earlier projections. While in 1998, capital spending amounted to C$ 1.5 billion, C$ 10.4 billion have been spent in 2005; total oil-sand investment from 1996 to 2005 was around C$ 47 billion. Estimates of capital expenditures to construct all announced projects over the period 2006 to 2015 total C$(2005) 125 billion (US$ 105 billion).

[14] Around 25% of *in-situ* production currently comes from *primary (non-thermal) production*, so-called *cold heavy-oil production with sand (CHOPS)*, where the bitumen is co-produced with sand through the use of specialised pumps (the same technology is also used for conventional heavy oil production). A significant difference between primary bitumen and conventional heavy-oil production, however, is the amount of sand that is co-produced, which can be two to three times higher. Primary production has the advantage of being cheap, but recovery rates are, at 5% to 10%, very low. The share of primary production is projected to decline in the future.

Table 3.4. *Development of Canadian crude bitumen production*

	1990	2006	2015	2030
Total (Mb)	**123**	**458**	**1047**	**1825**
Mined crude bitumen	74	278	657	–
In-situ crude bitumen	49	180	390	–

Sources: (AEUB, 2007; ACR, 2004; BGR, 2003).

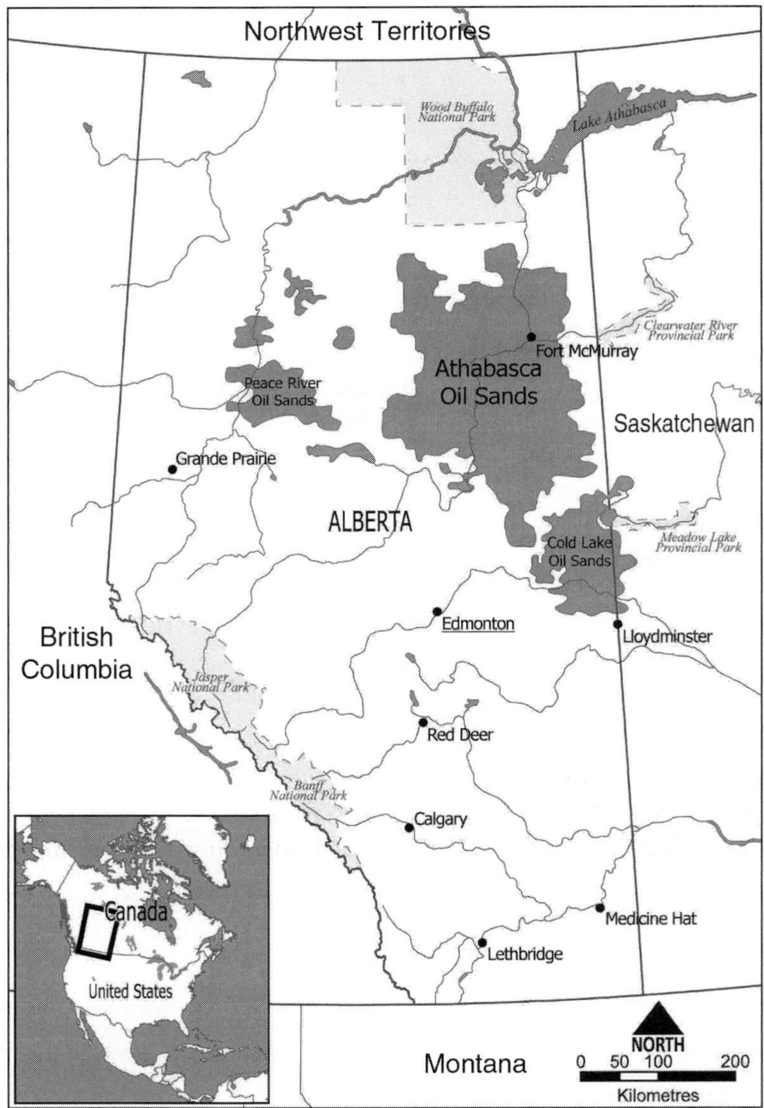

Figure 3.13. Geographical distribution of Canadian oil sands (Wikipedia Commons, 2007).

Despite the considerable growth of the Canadian oil-sand industry in recent years, there are still several difficulties that could impede the future development of the industry. Some key challenges are discussed next.[15]

Upgrading and refinery capacity While essentially all of the mined bitumen is upgraded in Alberta, the majority of *in-situ* production is shipped as bitumen blend with a light diluent to refineries in the United States that are suitably equipped to handle such feedstock. This historical split needs to be overcome in the future and further upgrading capacities will have to be installed in Canada, especially to reduce the need for diluents. In addition, the proposed extension of synthetic-crude-oil supply will require new refinery capacities, either in Canada or the United States.

Diluent supply To facilitate the transportation of bitumen from *in-situ* production to market by pipelines, it is generally necessary to add a diluent, such as gas condensate, to improve the mobility.[16] Typically, raw bitumen requires the addition of 30 to 40 vol.% of diluents. This suggests that a shortage of Canadian condensate may become a major bottleneck for *in-situ* projects, since condensate supply is directly tied to natural gas supply. Alternatives to reduce dependence on diluent might include heated pipelines, or blending with alternative viscosity reducers, such as conventional light crude oil, synthetic crude oil or refinery naphthas. The installation of additional local upgrading capacity in the field can further reduce or even eliminate diluent demand; as most bitumen blend is currently transported to markets outside Alberta (mainly the USA) for further processing, the diluents remain there, since their return to Alberta or Canada is economically not viable. Recycling of diluents might be a solution, but it requires a large investment to install an additional pipeline.

Natural gas dependency Oil-sand operators have historically depended on natural gas as their main source of energy, and thus, as oil-sand production has grown, so has the related demand for gas. A large part of the energy requirement for oil-sands mining, extraction and upgrading operations, as well as for *in-situ* operations, is met through natural gas. Natural gas-fired turbines generate electricity to operate equipment and facilities, provide heat that is used to generate steam (for *in-situ* recovery) and also provide process heat for bitumen extraction and upgrading. Some *in-situ* processes also use natural gas as uplifting gas. Natural gas is also the feedstock for the production of hydrogen for upgrading processes (hydro-cracking and hydro-treating).

[15] A comprehensive assessment of the environmental implications of Canadian oil sands production and how they can be mitigated is found in (ACR, 2004; NEB, 2004; 2006; WWF, 2008; Woynillowicz *et al.*, 2005).

[16] There are four options for transporting bitumen or extra-heavy oil by pipeline: heating, blending, mixing with water or mixing with a diluent (see also Gruson *et al.*, 2005; Saniere *et al.*, 2004). As the latter is most economic, it is this option that is most widely used today. To adapt to different refinery specifications, the bitumen is blended with either condensate, SCO or both (then called *DilBit*, *SynBit* or *DilSynBit*).

The use of natural gas as the principal energy source for the production of oil sands will decrease the amount of gas available for export to the United States, which currently receive around 90% of their gas imports from Canada. Even though there have been some efforts to reduce this dependence on natural gas, any increase in natural gas prices (e.g., due to an increase of demand in the North American gas market) or a sharp reduction in natural gas supply would have critical repercussions for the oil-sands industry. The major alternatives to natural gas being discussed and investigated are the use of coal or nuclear energy, especially for steam generation. In addition, both coal and nuclear energy can be used as alternative hydrogen feedstock. The gasification of oil-sand residues (e.g., coke from upgrading) is further being investigated as an option for hydrogen production. Another alternative is the development of non-thermal *in-situ* recovery methods, which use solvents instead of steam to facilitate the extraction of the bitumen.

Water consumption An *in-situ* facility requires freshwater to generate steam and for various utility functions throughout the plant; mining operations require water to separate the bitumen from sand and hydrotransport bitumen slurry. In addition, water is needed for upgrading the bitumen into lighter forms of oil for transport. The water requirements for mining projects range from 2 to 4.5 barrels of water per barrel of SCO produced (NEB, 2006). Despite some recycling, almost all of the water withdrawn for oil-sand operations ends up in tailings ponds. The principal environmental threats from tailings ponds are the migration of pollutants through the ground-water system and the risk of leaks to the surrounding soil and surface water. Because of their extremely low rate of consolidation, the management of fine tailings – fine clay and silt particles that settle out to form a layer after the coarse sand has separated from the tailings slurry – represents a major reclamation challenge. Even though in *in-situ* projects an average of 90% of the water is recycled, the process still requires large volumes of water. The development of non-thermal *in-situ* recovery methods would help reduce the need for water.

Environmental impacts Oil-sand development processes can be loosely divided into the categories of mining, extraction, upgrading and *in-situ* operations, of which each has an impact on the environment. During *surface mining*, the primary environmental issues relate to land disturbance during exploration, site preparation and active mining, productivity and stability of reclaimed lands after the mine is decommissioned, surface and ground-water use and quality, and air emissions from mining equipment and the open pit. Environmental issues related to *extraction* include air emissions from the extraction facilities and vehicles, storage and disposal of tailings, and waste water from processed sand and tailings. For *upgrading*, the main environmental issues relate to air emissions and waste materials, such as coke, wastewater and sulphur. The issues associated with *in-situ recovery* include land disturbance and

Table 3.5. *Specific natural gas demand for oil-sands production and upgrading*

	Fuel (Nm^3/b bitumen)	Hydrogen production (Nm^3/b bitumen)
Mining and extraction	7.0	–
In-situ process (thermal)	28.3	–
Upgrading	2.3	11.3

Source: (ACR, 2004).

habitat fragmentation during exploration, site preparation and operation (due to linear disturbances, such as seismic lines, roads, pipelines, and the presence of well pads, facilities and utility corridors), air emissions, and surface and ground-water use and quality.

Another critical issue facing the industry is that of CO_2 emissions, particularly as Canada has ratified the Kyoto Protocol and as the government aims to reduce the intensity of the sector's CO_2 emissions. Today, the oil-sand industry contributes to around 5% of Canada's CO_2 emissions; with oil-sand production boosting, total emissions are rising as well. Specific CO_2 emissions from oil sands production amount to 30–40 kg CO_2/b bitumen for mining, 60–80 kg CO_2/b bitumen for *in-situ* production and 70–90 kg CO_2/b bitumen for upgrading (ACR, 2004); this results in 100–130 kg CO_2/b for mining and 130–170 kg CO_2/b for *in-situ* operations. For comparison, the CO_2 intensity of conventional light crude oil production using primary recovery methods is between 15 and 20 kg CO_2/b, whereas the production of heavier crude oils or the application of tertiary recovery methods, such as EOR, which is generally more energy-intensive, can result in a CO_2 intensity of as much as 80–110 kg CO_2/b. As conventional crude oil is becoming heavier, the average CO_2 intensity of conventional crude oil production will also increase in the future.

Of the above-mentioned challenges of oil-sands production, the heavy dependence on natural gas is among the most critical. Table 3.5 shows the specific natural gas demand per barrel of bitumen for mining and extraction, (thermal) *in-situ* recovery and upgrading operations, as well as for the production of hydrogen. Depending on the recovery process, up to 25% of the energy content of the SCO is used in the form of natural gas.

Assuming the ratio of 60% mining and 40% *in-situ* production from today until 2030, the resulting natural gas demand to produce the envisaged 5 Mb/d can be calculated as 27 Gm^3, or 52 Gm^3 if natural gas consumption for upgrading and hydrogen production is included; this results in a cumulative gas consumption of 410 Gm^3 excluding upgrading and 780 Gm^3 including upgrading. Under this scenario, between 26 and 49% of Canada's remaining gas reserves of 1.6 Tm^3 (BP, 2007) would have been used up for oil-sands production by 2030. Given that more than 80% of bitumen reserves are only recoverable by *in-situ* production, a total of around 5000 Gm^3 of natural gas (including upgrading) would be necessary to produce those reserves; this would exceed Canadian gas reserves by about a factor of three. Including the Canadian bitumen reserves of 174 Gb in conventional oil reserves thus seems questionable as long as no alternative energy sources for steam generation and hydrogen production are developed.[17]

3.3.3.2 Extra-heavy oil

Properties and extraction processes Extra-heavy oil is usually defined as that portion of heavy oil having an API gravity of less than 10°. Extra-heavy oil is very similar to the natural bitumen of the oil sands, i.e., it has also only a limited capability of flowing in the reservoir and, therefore, requires measures to reduce its viscosity. Bitumen and extra-heavy oil only differ from each other in the degree by which they have been degraded from the original crude oil, with extra-heavy oil having a viscosity below 10 000 cP. The production methods applicable for extra-heavy oil are basically the same as for the *in-situ* recovery of oil sands. To meet pipeline specifications for the transportation of heavy crude oil, it is also generally necessary to add a diluent.

Resource estimates and current production According to the *USGS*, total resources of extra-heavy oil in place worldwide are estimated at around 1350 Gb, of which about 90% are located in the Orinoco Belt in Venezuela. It is estimated that between 240 and 270 Gb of the Venezuelan resources in place are ultimately recoverable. The synthetic crude produced from heavy oil is considered to be refined oil and is, therefore, not subject to OPEC quotas, unlike Venezuela's conventional oil production.

In the Orinoco Belt, all the heavy crude oil is currently extracted by cold production and transported by pipeline via dilution to an upgrader, where it is upgraded to a 26–32 °API crude, which can be exported and used as feedstock in common refineries. The diluent, which is added upstream, is recovered and sent back to the production plant in a dedicated pipeline, to be reused for the same purpose. Recycling the diluent reduces operating costs, but investment is higher, as a return pipeline has to be constructed. Cold production is the cheapest and the most environmentally friendly method, but has the lowest recovery rates, of only 5 to 10% of the oil in place. Thermal recovery methods, such as steam injection, could increase the product yield significantly. At the time of writing, the production of synthetic crude oil in the Orinoco Belt amounted to almost 0.6 Mb/day (Gruson *et al.*, 2005). Extraction and processing costs range from US$ 8–11 per barrel (EIA, 2006; Saniere *et al.*, 2004).

[17] However, it has to be noted that, from the perspective of providing mobility, by using that natural gas as feedstock – and taking into account fuel production and vehicle conversion efficiencies – the majority of passenger cars can be fuelled with the production of liquid fuels from oil sands, followed (with about a factor of three less) by its conversion into hydrogen and subsequent use in fuel-cell vehicles, and its direct use in CNG vehicles.

Owing to a lack of upgrading facilities, the state oil company *PDVSA* developed in the early 1980s a method for using the extra-heavy oil by creating an emulsion, made up of approximately 70% extra-heavy oil and 30% water. This branded product *Orimulsion®* is marketed as fuel for power generation and sold under long-term contracts to utilities in, among others, Japan, Italy, Denmark, Canada and China. The emulsion solves the production-transportation problem related to extra-heavy oil and eliminates refining by permitting the emulsion to be burned directly in boilers using conventional equipment with only minor modifications. The current production capacity is 5.2 Mt per year. The future of Orimulsion® is unclear, as with high oil prices it might be more profitable to upgrade the extra-heavy oil to refinery feed.

3.3.3.3 Oil shale

Properties and extraction processes The term oil shale is actually misleading, as it does not contain any free oil nor is it commonly shale. Oil shale is a sedimentary rock that is quite rich in organic matter, called *kerogen*. This kerogen, a solid bituminous material, is the precursor of oil and has, therefore, not gone through the 'oil window' of heat, the range of temperature at which oil forms naturally.[18] To convert the solid organic kerogen into a raw oil, the processes normally taking place in nature under geological times need to be artificially accelerated. To achieve this and to yield petroleum-like liquids, the shale must be heated to a high temperature, a process called *retorting*. There are two basic approaches for producing shale oil: *mining and surface retorting* and *in-situ retorting* (see Fig. 3.14). With conventional surface processes, the shale is brought to the heat source, namely the retort; with *in-situ* retorting, the heat source is placed within the oil shale itself.

The kerogen content of different oil shales, i.e., the average yield of oil of an oil shale can vary greatly. Most oil shales have oil yields between 50 and 150 l oil/t oil shale, rich shales may even yield more than 200 l/t. A particular characteristic of oil shale is its high areal density, which can exceed 1 million b/acre at its thickest (Bunger *et al.*, 2004). Oil shale can also be used directly as fuel for electricity generation, as, for instance, in Estonia. However, oil shale is characterised by a low calorific value between 7.5 and 9 MJ/kg (based on LHV), which is about one fifth of the energy content of crude oil (Porath, 1999).

Oil shale can be mined by surface or underground mining, with the first generally being the more efficient method, as more oil shale can be recovered (up to 80%). The mined oil shale is transported to the retort facility. Surface retorting involves crushing the oil shale and heating it in a vessel (the retort) to temperatures between 480 and 540 °C in the absence of air, to achieve a pyrolysis, which converts the kerogen into gaseous hydrocarbons, which can be separated by boiling point and condensed into a

[18] Geologists often refer to an 'oil window', which is the temperature range between 60 and 150 °C that oil forms in – below the minimum temperature, oil remains trapped in the form of kerogen; above the maximum temperature, the oil is converted to natural gas through the process of thermal cracking. This temperature corresponds to a burial depth of at least 2 km.

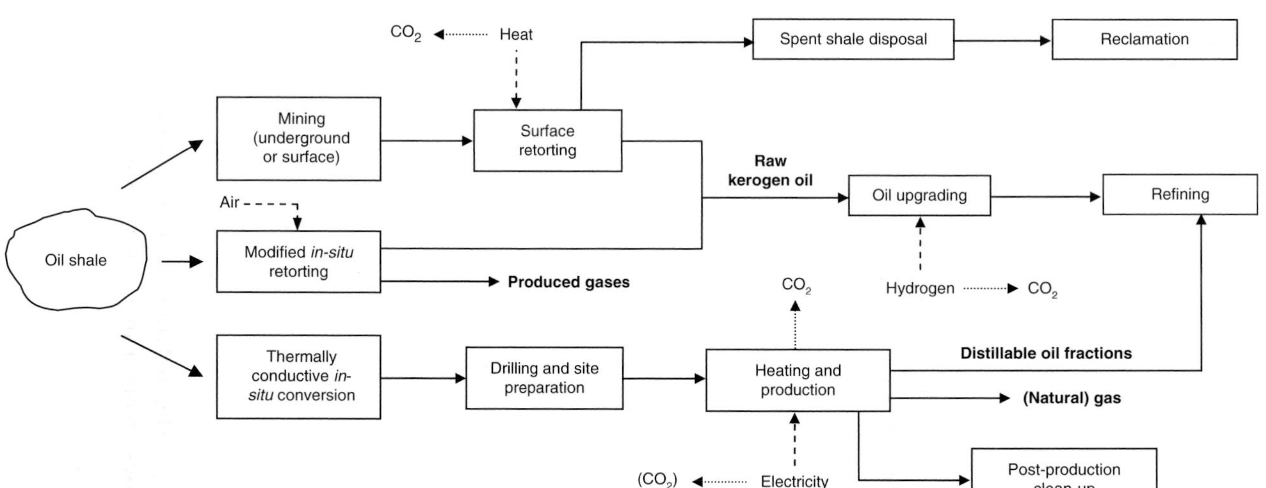

Figure 3.14. Schematic representation of oil-shale production.

variety of liquid fuels; this process, called 'retorting', typically converts 75 to 80% of the kerogen into oil. The hot raw kerogen oil leaving the retort is not stable and must be sent directly to an upgrading plant for catalytic processing with hydrogen (hydro-treating and hydro-cracking) to remove impurities and produce a stable product. This stable shale oil can then be used as a conventional refinery feedstock. After retorting, the spent shale is cooled and disposed of, awaiting eventual reclamation.[19]

In principle, mining and surface retorting is a technically viable approach for producing shale oil; however, except for the *Alberta Taciuk Processor* applied in the Australian *Stuart Shale Oil Project*, no significant development work in surface retorting has occurred for more than 20 years (Bartis *et al.*, 2005). Significant problems must be overcome concerning land use and ecological impact from mining operations, the large amounts of heat needed to process the oil shale, impact on local air quality and the great amounts of expanded shale rock left after extraction. A major constraint might further be the high water consumption of about three barrels per barrel of shale oil, which is needed for dust control, cooling and reclaiming spent shale, upgrading raw shale oil and power generation. While technical viability has been demonstrated by various developments of surface retorts during the 1970s and 1980s, the economics look less promising. According to Bartis *et al.* (2005), a first-of-kind conventional mining and retorting process would only be profitable with real crude prices in the range of US$ (2005) 75 to 95 per barrel.

For deeper, thicker deposits, not suitable for surface or underground mining, the kerogen oil can be produced by *in-situ* technology. *In-situ processes* introduce heat to the kerogen while it is still embedded in its natural geological formation to turn it into oil underground, which is then extracted from the ground and transported to an upgrading or refining facility. There are two general *in-situ* approaches: *(modified) in-situ retorting* and *in-situ conversion.*

Pure in-situ retorting processes drill access shafts to reach the shale layers, apply process heat to the shale by heaters or direct combustion, and move the resulting shale oil and gases to the surface through conventional oil and gas wells. Various approaches were investigated during the 1970s and 1980s. Much of this prior work was not successful, encountering serious problems in maintaining and controlling the underground combustion process and avoiding subsurface pollution. *Modified in-situ* processes attempt to improve performance by exposing more of the target deposit to the heat source and by improving the flow of gases and liquid fluids through the rock formation. These processes create a much larger shaft and involve direct blasting or partial mining beneath the target oil-shale deposit prior to heating, to create void space, needed to allow produced gases and pyrolised shale oil to flow towards

[19] When oil shale is retorted, the inorganic portion of the shale expands considerably. The spent shale remaining has no direct commercial value and, ideally, it is placed back in the mine. However, because of the *popcorn effect*, the volume of spent shale is greater than the volume of the mine from which it was taken.

production wells. A portion of the shale resource is fractured and rubblised, and the rubblised shale is then ignited to generate heat for pyrolysis. According to Bartis *et al.* (2005), no companies have recently shown interest in pursuing any type of *in-situ* retorting process.

Among the most promising *in-situ* processes is a new approach developed and patented by Shell. This thermally conductive conversion process, known as the *in-situ conversion process (ICP)*, involves placing electric heaters in vertically drilled wells and gradually heating the oil shale interval over a period of two to three years until the targeted volume of the deposit reaches a temperature of 340 to 370 °C. The kerogen is then converted to hydrocarbon gases and kerogen oil, which are produced through conventional recovery means. This very slow heating to a relatively low temperature is sufficient to cause the chemical and physical changes required to release oil from the shale. To keep ground water away from the process (which would affect the heating), and to keep hydrocarbons and by-products of the process away from the ground water, it is planned to freeze the ground water to establish an underground barrier – a 'freeze wall' – around the perimeter of the extraction zone. Owing to the slow heating and pyrolysis process, the product quality is improved and subsequent product treating is less complex, as compared with oil produced by surface retorting or conventional *in-situ* approaches. The produced oil is chemically stable and consists solely of lighter distillable oil fractions (i.e., no heavy residuum will be created), which should be a premium feedstock for refineries, without, in contrast to oil from surface retorting, the need for near-site upgrading with hydrogen. On an energy basis, about two thirds of the released product is liquid and one third is a gas similar in composition to natural gas. It is expected that 65% to 70% of the oil in place can be recovered (Bartis *et al.*, 2005).

A critical measure of the viability of oil shale recovery – as for any alternative fuel production process – is the ratio of energy used to produce the oil to the energy returned, expressed as *energy returned on energy invested (EROEI)*. According to Bartis *et al.* (2005), the *ICP* process requires about 250 to 300 kWh electricity for down-hole heating per barrel of extracted product. With the electricity supplied by a combined-cycle gas power plant, the EROEI is about 3.5:1, i.e., 3.5 units of energy are returned for every unit of fuel consumed; a conventional coal-fired power plant would result in a EROEI of around 2.5:1 (see also Udall, 2005). This compares with a figure of typically 5:1 for conventional oil extraction.

At the time of writing, oil-shale commercialisation (in the United States) was still in the research and development phase. Shell has successfully tested its *in-situ* process at a very small scale in Colorado's Piceance Basin, but larger operations are required to establish technical viability. The two major technological challenges are controlling ground water during production and preventing subsurface environmental problems. To date, Shell estimates that the technology should be profitable at crude oil prices around 25 \$/b. However, as it will take several years before the product stream reaches steady-state production (well after hoped-for first generation commercial start-up early in the next decade) and because the process is so capital intensive, the economic risk is very high. Although initial results are very promising, a commercial-scale operation will depend on overcoming certain technical hurdles and perceptions of future market conditions and investment risks.

Resource estimates and current production To estimate shale-oil resources, two measures are commonly used: *resources in-place* and *recoverable resources*. Resources in place are distinguished according to their grade – the gallons of oil that can be produced from a ton of shale. The rich ores that yield 95 to 200 l (25 to more than 50 gallons) per ton are the most attractive for early development. Deposits with grades below 10 gallons per ton are generally not counted as resources in-place because it is commonly assumed that such low yields do not justify the costs and energy expended in extraction and processing. However, estimating shale-oil resources is complicated by several factors. First, the amount of kerogen contained in oil-shale deposits varies considerably. Second, some countries report the total amount of oil in-place, and do not account for what fraction might be recoverable. Third, shale-oil recovery technologies are still developing, so the amount of recoverable kerogen can only be estimated. Fourth, no standard grade is used to define shale-oil resources and different resource estimates include different minimum grades, which complicates the process of summing up or comparing various estimates.

Oil shale exists in great quantities worldwide, with the world total amounting to about 2600 Gb of shale oil in-place, of which the vast majority – an estimated 2000 Gb – is located in the United States (based on a grade of >10 gallons per ton) (US DOE, 2004). By far the largest known oil-shale deposits in the world are found in the *Green River Formation* in the Rocky Mountains of the Western United States, which covers parts of Colorado, Utah and Wyoming (Fig. 3.15); assuming a grade greater than 15 gallons per ton, Bartis *et al.* (2005) estimates the resources of oil in-place at 1500 to 1800 Gb and the recoverable resources at 500 to 1100 Gb. A more conservative estimate by BGR (2003) puts the world's recoverable resources of shale oil at 1160 Gb, of which 380 to 500 Gb are allocated to the United States; total shale-oil reserves are estimated at 6 Gb. Other countries with important oil-shale deposits are Australia, Jordan, Morocco, China, Brazil and Estonia.

The utilisation of oil shale can be traced back to the seventeenth century. The modern use of oil shale to produce oil dates to Scotland in the 1850s. Towards the end of World War II the extraction of oil shale gained importance for the production of synthetic fuels, particularly in Germany. During and following the oil crises of the 1970s, oil companies such as Exxon, Tosco and Unocal, working on some of the richest oil-shale deposits in the western United States, spent several billion dollars in various unsuccessful attempts to extract shale oil commercially. In the 1980s, with oil prices bottoming out and increasing oil supplies from non-OPEC countries, the

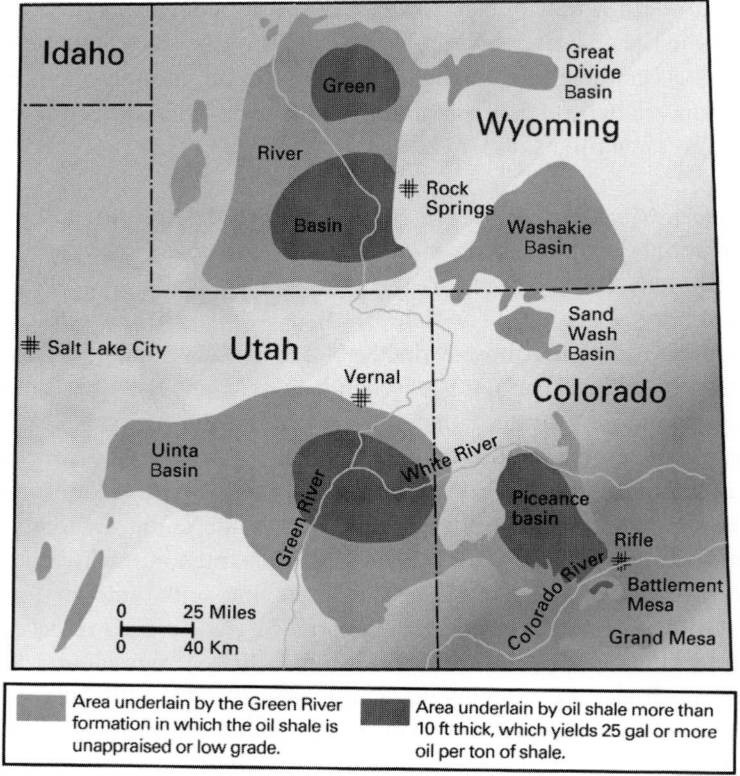

Figure 3.15. Geographical distribution of US oil-shale occurrences (Bunger *et al.*, 2004).

oil-shale boom collapsed and efforts to commercialise oil shale in the United States were abandoned; Unocal was the last to do so in 1991.[20]

World oil-shale production fell from its peak of 46 Mt in 1980 to about 16 Mt in 2000 (Brendow, 2003). At present, about 69% of world oil-shale production is used for the generation of electricity and heat, about 6% for cement production and other industrial uses, and 25% is processed into shale oil. Oil shale has been burned directly as a very low-grade, high-ash-content fuel in a few countries, such as Estonia, which is also the only country in Europe where oil shale is of any importance. With a yearly production of about 14 Mt, oil shale still generated more than 90% of the electricity in Estonia at the time of writing, and also most of Estonia's 7000 b/d oil production comes from oil shale. Other countries where surface retorting of oil shale has been used for many years to yield shale oil are Brazil and China, which produced 3100 b/d and 1500 b/d in 2002,

[20] A comprehensive description of the history of oil shale development in the United States is found in CRS (2006) and Laherrère (2005); for other countries, see WEC (2007). An overview of companies newly looking into oil shale and oil sands development in the USA is provided by US DOE (2007a).

respectively (Fischer, 2005; Laherrère, 2005). The *Stuart Shale Oil Project* in Queensland (Australia), which has been generating the most publicity, as the project consortium ran into financial and environmental compliance problems and because Greenpeace has permanently been opposing the project, was stopped in 2004 (QLD, 2007; Snyder, 2004; WEC, 2007). Further oil-shale projects are planned in China (WEC, 2007).

In 2004, the US Department of Energy, recognising a peak of world oil production, published a study carried out to review the potential of shale oil as a strategic resource to help meet the nation's needs for liquid fuels. The study announces a new generation of oil shale projects, although no commercial-scale retort has been demonstrated in the USA yet. The US DOE (2004) states: 'Looking to what is possible with a co-ordinated industry-government effort, it is possible that an oil-shale industry could be initiated by 2011 (*with initial production of 0.2 million b/d*), reaching an aggressive goal of 2 million b/d by 2020. Ultimate capacity could reach 10 million b/d, a comparable capacity to the long-term prospects for Alberta's oil sands.' Other sources are less optimistic regarding the future production of shale oil in the United States: the *Annual Energy Outlook 2006* published by the US DOE (EIA, 2006) estimates the annual shale-oil production in 2030 at between 0.05 and 0.4 million b/d (depending on different oil price scenarios), and Bartis *et al.* (2005) states: 'Under high growth assumptions, an oil-shale production level of 1 million barrels per day is probably more than 20 years in the future, and 3 million barrels per day is probably more than 30 years into the future.'

In contrast, increasing today's average fuel efficiency of personal vehicles in the USA of around 21 miles/gallon (11.2 l/100 km) by only two miles per gallon would save around 0.8 million b/d (assuming an average annual driving distance of 12 000 miles and a passenger car fleet of 243 million vehicles); for comparison, the average fuel efficiency of personal cars in the EU is about 7.8 l/100 km (30 miles/gallon; Woynillowicz *et al.*, 2005). A dieselisation of the entire US vehicle fleet could save as much as 3 million b/d. The current US oil production amounts to 6.8 million b/d.

Several technical, economic and environmental problems must be overcome for successful large-scale use of shale oil in the future. While mining and surface retorting is generally seen as a technically viable approach for producing shale oil, the current commercial readiness of surface retorting technology is more questionable. Potential environmental problems particularly associated with surface retorting comprise surface impacts from oil shale mining, air pollution, contamination of surface and ground water from mining operations and disposal of large quantities of spent shale; in addition, significant volumes of water are required for large-scale oil shale development. The *in-situ* recovery process developed by Shell is anticipated to be economically competitive in the mid-$20/b range and reduces the environmental impact resulting from previous shale-oil recovery techniques, as it does not involve open-pit or subsurface mining, creates no spent shale to be disposed of, and minimises unwanted by-products and water use. However, other subsurface impacts, including ground-water contamination, are possible and must be controlled. In addition, heating oil shale for retorting – whether above ground or *in-situ* – requires significant energy inputs, and, where the energy is supplied by fossil fuels, releases large amounts of CO_2.

3.3.4 Scenarios for determining the mid-depletion point of oil

Given a globally steadily growing demand for oil and, hence, the need for the production to keep pace, questions arise about the long-term availability of oil and to what extent and how long an ever-growing demand can be met. This question gains even more importance as future shortcomings in the supply of oil are likely to have significant economic and geopolitical implications. Having discussed the projected development of global oil demand and the supply situation for both conventional and unconventional oil in the previous sections, this section focuses on how a growing demand will affect the point in time of the mid-depletion point of oil – assuming a natural limitation of supply[21] – and what consequences can be drawn for the prospects of a hydrogen economy.

The current debate about peak oil (and gas) is characterised by the two controversial views of the 'optimists' and the 'pessimists'.[22] The 'pessimists' advocate the position that the world's resources are limited, expect a decline of world oil production within a few years from today, followed by shortages in supply and sharp price increases, and project that a large part of the demand cannot be met any more in the near future. They further argue that – owing to the time and money needed – the ramp-up of production of unconventional oil will not be quick enough to make up for the gap and point to the high environmental price that comes along with these substitutes. The 'optimists' deny the near-term existence of a global peak in oil production, argue that the volumes of exploitable oil are closely correlated to technological advances in exploration and production, economics of production and the market price of oil, and especially point out the large potentials of unconventional occurrences. They further stress that previous estimates of the EUR have all been revised upwards and refer to the long history of failed forecasts regarding the peaking of oil production (Maugeri, 2004). The risk of supply disruptions is considered geopolitical rather than geological in nature.

There is a long history of failed forecasts regarding the peaking of oil production and experience shows that reserves are usually underestimated. However, there are

[21] The most widely accepted theory as to how oil was formed is the *organic theory*, which maintains that oil (and also natural gas) are the product of compression and heating of ancient organic matter over geological time. There is an alternative theory about the formation of oil and gas deposits, the *abiotic theory*, which was developed in the 1950s by Russian and Ukranian scientists. According to this theory, hydrocarbons are continuously formed from *inorganic materials* under high pressures and temperatures found at upper-mantle to lower-crust depths of the Earth (10 to 30 km) from the reaction of carbonate rocks with iron oxide and water, and have no biological origin, such as dead plants or animals. This theory, which is very controversially discussed among geologists and supported by few people only, especially *Thomas Gold*, has recently become popular, as it is closely linked to the concept of peak oil, which sees oil as a finite resource. However, even assuming that oil is continuously forming on its own, this would not be at a rate that could compete with current depletion.

[22] The most prominent pessimists are *C. Campbell*, *J. Laherrère*, *R. Bentley* and *K. Deffeyes*; there is also a society called the Association for the Study of Peak Oil (ASPO), see www.peakoil.net. So-called optimists include *M. Adelman*, *C. Watkins*, *M. Lynch*, *J. Ryan* and *P. Odell*. (For further reading about the peak oil debate, see (CERA, 2006; Deffeyes, 2001; EWG, 2007; Jaccard, 2005; Kägi *et al.*, 2003; OGJ, 2004 or Robelius, 2007).)

compelling reasons why current projections might be more reliable than previous ones (Hirsch *et al.*, 2005):

1. Extensive drilling for oil and gas has provided a massive worldwide database and current geological knowledge is much more extensive than in years past.
2. Seismic and other exploration technologies have advanced dramatically in recent decades; nevertheless, the oil reserves discovered per exploratory well began dropping worldwide over a decade ago. Global production is exceeding new discoveries since the 1980s and the size of new discoveries is also decreasing (see Section 3.3.2.2).[23]
3. Many credible analysts have recently become much more pessimistic about the possibility of finding the huge new reserves needed to meet growing world demand, and even the most optimistic forecasts suggest that world conventional oil peaking will occur in less than 25 years.
4. The peaking of world oil production could create enormous economic disruption, as only glimpsed during the 1973 oil embargo and the 1979 Iranian oil cut-off.

Often the static lifetime is calculated as a measure for the remaining amounts of fossil fuels; however, methodologically, this approach is not considered appropriate and realistic for the reasons outlined in Section 3.2 and also does not contribute to the objective of determining the point in time of the mid-depletion point. From an economic point of view, it is not the temporal endpoint of the resource utilisation and thus the depletion of a resource that is of interest, but the time of maximum production, when half of the EUR has been produced and the demand – given an ultimately declining production rate – will probably start exceeding the supply.[24] The consequences could be supply disruptions and price explosions, accompanied by negative macroeconomic impacts if no alternatives to conventional oil can be developed. To derive a time window for the mid-depletion point, different estimates of the EUR of oil and different growth rates of oil demand are taken into account.

Different estimates of the ultimately recoverable resources lead to different time windows for the mid-depletion point of oil. Estimates of the EUR at country level can differ, for instance, because of different boundaries between conventional and unconventional occurrences, and depend on assumptions about recovery factors,

[23] However, most drilling in recent years has been concentrated in North America, a mature region with limited potential for new discoveries. Less than 2% of new wildcat wells were drilled in the Middle East alone, which holds the major part of the world's undiscovered oil (IEA, 2006).

[24] It is often assumed that the mid-depletion point and the peak, i.e., the time of maximum production, are close together, so that after the depletion of half the EUR a decline of production must be expected. However, this correlation is only true if a symmetrical single-peak profile of global oil production is assumed and the estimations of the EUR are correct and not influenced by political or economic impacts. The theoretical background for this theory is derived from a model developed in the 1950s by M. King Hubbert, a Shell geoscientist. According to Hubbert, oil production in any given field follows a bell-shaped trajectory, the Hubbert curve. Hubbert's basic assumption is that the production of oil grows exponentially at the beginning and starts declining at around that point of time when half of the ultimate recoverable resources have been depleted, after one or several production peaks. Some say Hubbert's curve was validated when he correctly predicted the peak of US Lower 48 oil production in 1970. However, there are some constraints to the Hubbert model and its application for projecting future oil production is not uncontroversial. For a discussion of the underlying assumptions and critics of the Hubbert model refer to CERA (2006), Kägi *et al.* (2003), Laherrère (2000), Maugeri (2004) and Robelius (2007).

which also reflect reserve growth resulting from improvements in drilling, exploration and production technologies.[25] Additions to world oil resources come from three sources: recovery growth from existing fields, undiscovered fields and unconventional sources. Upward revisions from existing fields, not new discoveries, have been the major contributor to world oil production in the last 25 years. Growth in recovery from existing fields is also likely to be the largest source of future additions to world oil supply, as is already the case in the United States, where new fields are providing less than a quarter of reserve additions.

To determine the mid-depletion point, the estimates of the *US Geological Survey (USGS)* (USGS, 2000) have been chosen, as they cover almost the entire bandwidth of recently published estimates of the EUR of conventional oil: the estimate of 3843 Gb for a 5% probability of discovery (P5) has been selected as an upper bound ('optimistic estimate') and the estimate of 2193 Gb with a 95% (P95) probability of discovery as a lower bound ('pessimistic estimate'); the mean estimate places world EUR at 3021 Gb. In the optimistic case, *reserve growth* amounts to 1107 Gb, in the mean case to 688 Gb, and in the pessimistic case to 268 Gb.[26] The absolute lower bound represents the latest estimate of 1900 Gb (ASPO, 2007; Campbell, 2006). This EUR is low because NGL, heavy, deepwater and polar oils and reserve growth are not included, as in the case of the *USGS* estimates. The main reason, however, is the low evaluation of oil reserves of some OPEC countries, primarily Saudi Arabia, as parts of them are considered political reserves.[27] The estimate of the Federal Institute for Geosciences and Natural Resources of 2845 Gb is in the middle of the USGS bounds (BGR, 2007). To assess the impact of *unconventional oil* on the peaking of production, *only the reserves* of 415 Gb have been considered, owing to the high uncertainties relating to resource estimates and the great variations with respect to the recoverable quantities of unconventional occurrences. The cumulative oil production as of today amounts to 1076 Gb.

Figure 3.16 shows the time window of the mid-depletion point for two different growth rates of annual oil demand – 0% and 1.3% according to the *IEA Reference*

[25] Oil production techniques are generally divided into three categories: primary, secondary and tertiary recovery. Traditional primary and secondary production methods usually recover between 15% and 40% of the oil in-place; tertiary recovery – also called *enhanced oil recovery (EOR)* – offers prospects for ultimately producing 30% to 60%. Enhanced oil recovery increases the recovery rate by changing the natural chemo-physical conditions in the reservoir and comprises three major categories: *thermal recovery* (e.g., steam injection), *gas injection* (e.g., natural gas, nitrogen or CO_2) and *chemical injection* (e.g., polymers). However, EOR techniques, especially steam injection, increase the energy and CO_2 intensity per barrel produced by about a factor of three. Different sources indicate the world average *recovery factor* of oil fields between 35% and 40%.

[26] This reserve growth is largely due to *enhanced oil recovery*. In 2004, EOR contributed with 1.8 million barrels per day to about 2% of world production, with more than a third coming from the US (Moritis, 2006). The injection of CO_2 as a means to increase oil production is discussed in more detail in the context of CCS in Chapter 6. The *Oil & Gas Journal* (2006b) estimates that more than 500 Gb can be produced with EOR methods. Enhanced oil recovery methods are also incentivised by high oil prices.

[27] Campbell further excludes the surge in reserves of 300 Gb caused by the so-called 'OPEC jump'. After Venezuela had newly included its extra-heavy oil reserves in 1987, which would have resulted in a doubling of its reserves and a corresponding increase of production quota at the expense of other OPEC countries, other OPEC members subsequently also increased their reserves to keep their production quotas. Moreover, at present, many OPEC countries have been reporting constant reserves figures for many years, despite ongoing production: out of the 97 countries covered by OGJ estimates at the end of 2003, the reserves of 38 countries were unchanged since 1998 and 13 more were unchanged since 1993 (IEA, 2004).

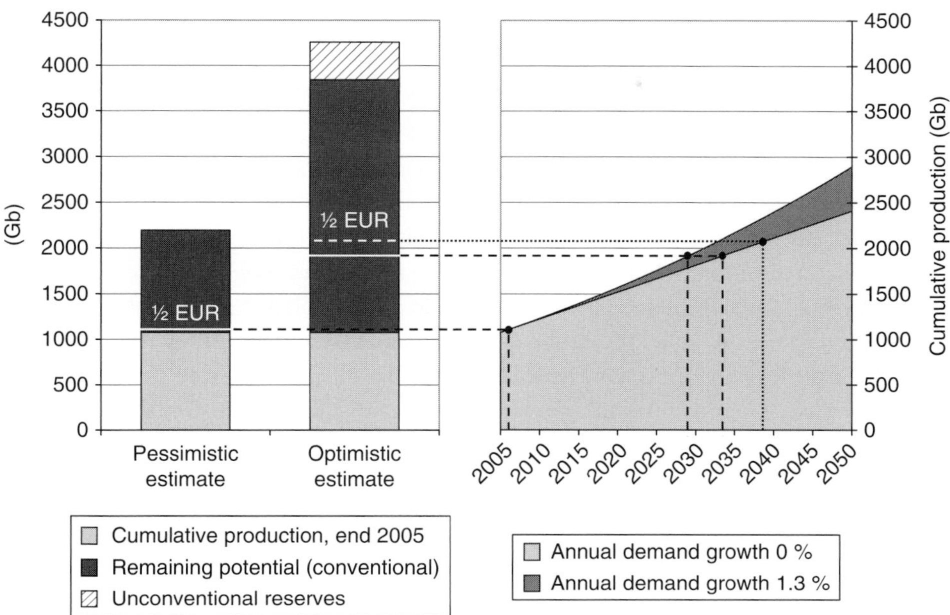

Figure 3.16. Scenarios for the mid-depletion point of oil.

Scenario (Section 3.1; between 1965 and 2005, global oil demand has grown by an average of 2.4% per year (BP, 2006)) – and the above estimates of ultimately recoverable resources. According to the 1.3% growth scenario, the cumulative oil production will have doubled by 2035. According to the pessimistic estimate, the mid-depletion point of conventional oil production is about to be passed today (2007); the optimistic estimate sees this time between 2028 and 2033. According to the *Campbell* scenario, the peak of conventional oil production *was* passed in 2007. The inclusion of unconventional oil reserves will shift the mid-depletion point to 2038. The conventional oil *BGR scenario* sees the peak of production around 2015. Unconventional oil reserves will prolong the mid-depletion point by five to ten years only. This means that an increase of the Canadian oil-sands production or US oil-shale production will have a negligible impact on the peak of conventional oil production until 2030: oil sands can shortly prolong the maximum of oil production; however, the global decline of production cannot be prevented (in the short or medium term).

The IEA (2006) estimates the necessary capital investments for the oil industry until 2030 at US$(2005) 4.3 trillion, or $164 billion per year; exploration and production account for almost three quarters of the total ($120 billion per year). For comparison, the IEA (2004) estimated the capital requirements until 2030 at only US$3 trillion. During the last five years an average of $100 billion was spent annually.

3.4 Natural gas

3.4.1 Classification of conventional and unconventional natural gas

Just as oil, natural gas is also categorised as conventional and unconventional. Unlike crude oil, however, natural gas deposits are normally classified according to the economic or technical approach, i.e., all occurrences that are currently extractable under economic conditions are considered conventional, whereas the rest are termed unconventional. Conventional natural gas includes '*non-associated gas*' from gas reservoirs in which there is little or no crude oil, as well as '*associated gas*', which is produced from oil wells; the latter can exist separately from oil in the formation (*free gas*, also known as cap gas, as it lies above the oil), or dissolved in the crude oil (*dissolved gas*). Unconventional gas is the same substance as conventional natural gas, and only the reservoir characteristics are different and make it usually more difficult to produce. Unconventional gas comprises natural gas from coal (also known as coal-bed methane), tight gas, gas in aquifers and gas hydrates (see Fig. 3.17). It is important to mention in this context so-called '*stranded gas*', a term which is applied to occurrences whose extraction would be technically feasible, but which are located in remote areas that at the moment cannot (yet) be economically developed (see Section 3.4.3.1).

The methane content of natural gas, which largely determines its calorific value, varies from reservoir to reservoir and can range from 75 to 99 vol.%. (Natural gas with a methane content between 80% and 87% is termed *L (Low)-Gas*; between 87% and 99%, *H (High)-Gas*.) Besides methane, the raw natural gas from both gas and oil wells often contains larger fractions of higher hydrocarbons (mostly pentanes and larger fractions), which are gaseous under reservoir conditions, but partially condense into liquid hydrocarbons when brought to the surface, and accordingly are

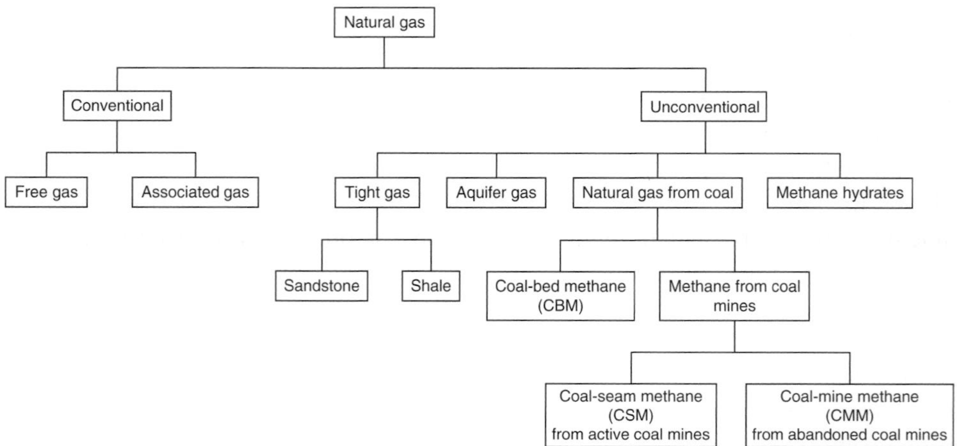

Figure 3.17. Classification of conventional and unconventional gas.

called *condensates*. In addition, propane and butane, in particular, are separated from the raw gas and marketed as separate fractions or in the form of *liquefied petroleum gas (LPG)*, as these *natural gas plant liquids (NGPL)* have a higher value as separate products. Statistically, the liquid hydrocarbons recovered from natural gas are treated inconsistently: while condensate and NGPL are usually reported under the generic term *natural gas liquids (NGL)*, some statistics report these fractions separately, or include in NGL only condensate, or add one or both fractions directly to crude oil production. Owing to the inconsistent definition and statistical treatment of NGL, significant differences in the reporting of oil production data can be observed, especially in countries like the USA, where NGL production plays an important role.

3.4.2 Conventional natural gas

3.4.2.1 Production, consumption and trade

As in the case of oil, natural gas statistics are also not directly comparable, owing to different statistical valuation methods. Natural gas is usually measured in volume units: cubic metres or cubic feet. As for production figures, it is often not clearly specified whether it is total (=gross) gas production or marketed production (total production less own consumption of production facilities, reinjected and flared gas); moreover, no clear distinction is made between dry gas (almost entirely methane) and wet gas (including NGPL). Particularly in the USA, NGPL production depends on the natural gas price: if prices are low, NGPLs are recovered from the natural gas, as it is more economic to sell them as separate products in the form of LPG. If natural gas prices are high, NGPLs are left in the natural gas: as a consequence NGPL production decreases and the reported natural gas production volumes apparently increase, even though the actual production has not changed; and vice versa.

Since 1900, almost 81 Tm^3 of natural gas have been extracted, of which about 80% was extracted in the last 20 years (Fig. 3.18).

Table 3.6 shows the production and consumption of natural gas of the ten most important countries in 2005. Total world (marketed) natural gas production amounted to 2760 Gm^3, of which Russia and the United States together had a share of around 40%; however, the United States has already produced 58% of its EUR. The production from offshore fields has steadily grown in recent years, and, at the time of writing, represented about 27% of global natural gas production, of which a third comes from the North Sea. The most important gas producers in Europe are Great Britain, Norway and the Netherlands, with 3.2%, 3.0% and 2.3% of world production, respectively. According to the Federal Institute for Geosciences and Natural Resources (BGR, 2007), the Netherlands – which was the biggest producer in Europe until the mid 1990s – and Great Britain have already passed their mid-depletion point. Consequently, European natural gas production is getting shifted more and more towards the offshore fields of Norway, where the largest

Table 3.6. *Worldwide natural gas production and consumption 2005*

Natural gas production		Natural gas consumption	
Country	(%)	Country	(%)
Russia	21.6	USA	23.0
USA	19.0	Russia	14.7
Canada	6.7	Great Britain	3.4
Great Britain	3.2	Canada	3.3
Algeria	3.2	Iran	3.2
Iran	3.1	Germany	3.1
Norway	3.1	Japan	2.9
Indonesia	2.8	Italy	2.9
Saudi Arabia	2.5	Ukraine	2.7
The Netherlands	2.3	Saudi Arabia	2.5
Sum	**67.5**	**Sum**	**61.9**
World (Gm^3)	**2763**	**World (Gm^3)**	**2750**
World (EJ)	**94**	**World (EJ)**	**94**

Source: (BP, 2006).

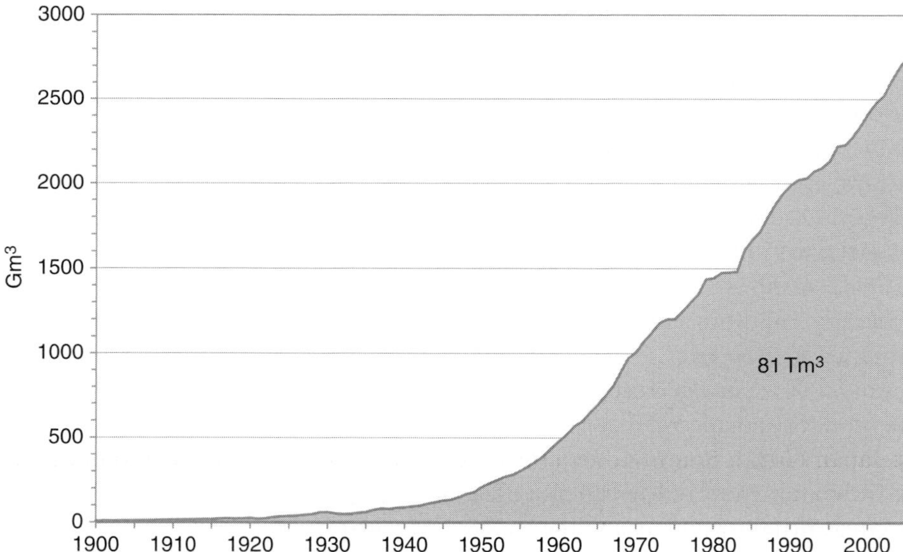

Figure 3.18. Development of annual natural gas production since 1900 (BP, 2006; Laherrère, 2004b).

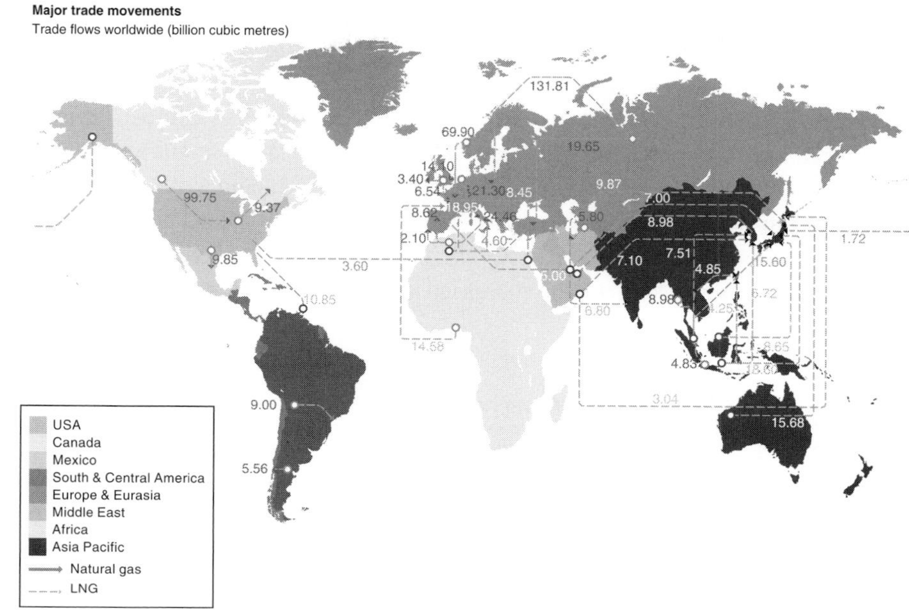

Figure 3.19. Major gas trade movements in Gm^3 (BP, 2007).

reserves are located. With the exception of the United States, *unconventional gas* has so far only a small share in production: in 2003, CBM (coal-bed methane) and tight gas amounted to 35% of total US gas production or 7.4% of world production (see Sections 3.4.3.3 and 3.4.3.4). The global gas consumption is – as in the case of oil – dominated largely by the United States, with a share of almost one quarter. Moreover, ten countries account for more than 60% of total consumption; the highest increases in absolute terms in recent years have been observed in Asia, particularly China.

Natural gas can be transported either in gaseous form in pipelines or in tankers as liquefied natural gas (LNG), which is produced by gas compression and cryogenic cooling. Figure 3.19 displays the major gas trade movements for pipeline and LNG transport. Cross-border trade amounted to almost 750 Gm^3 in 2006 (26% of production), of which a little more than one quarter was traded as LNG (BP, 2007). More than three quarters of global LNG trade (import) today relates to four countries: Japan (40%), South Korea (16%), Spain (12%) and the United States (8%). Liquified natural gas trade has more than doubled in the last decade and is predicted to grow further. According to the IEA (2006), LNG accounts for 70% of the projected increase in gas trade by 2030, thereby multiplying the current volume by a factor of five. North America is expected to see the biggest increase. Thus, LNG trade, which has until now been largely focused on the Asia-Pacific region, will become much more widespread, although OPEC countries will continue to dominate the supply.

Unlike oil, there is no global natural gas market, but four main regional markets in which producers and distributors have long-term contracts: the European market, with the main exporters Russia, North Africa, Norway and the Netherlands; the North American market (NAFTA); the Asian market, which is almost entirely a LNG market and characterised by large distances between the main consumers (mainly Japan, South Korea, and Taiwan) and the producing countries (mainly Indonesia, Malaysia and Brunei); and the South American market, which has been developed in the last few years. As LNG expands, these markets will become more integrated, which might lead to an increased competition for natural gas supply among these regions.[28]

3.4.2.2 Reserves and resources

In the following, the remaining potential of conventional natural gas will be addressed. As it is beyond the limits of this publication to investigate data discrepancies between different statistics in more detail, the estimates of the Federal Institute for Geosciences and Natural Resources (BGR) will be used, as they are derived from an assessment of the most important primary sources, and also include resource estimates at country level. Nevertheless, the consequences of different estimates of the EUR of natural gas on the time window of the mid-depletion point will be analysed in Section 3.4.4.

Table 3.7 shows the top ten countries with regard to conventional natural gas reserves and resources at the end of 2005. These countries concentrate 74% of the total remaining potential (reserves plus resources), of which one third is located in Russia. Around 70% of conventional gas reserves are located within the so-called '*strategic ellipse*' that extends from the Middle East to Western Siberia (see also Fig. 3.9). OPEC countries hold 34% of the remaining potential and 50% of reserves; the share of the EU25 amounts to 1.7% each. The European gas market, however, has access to one third of the global remaining potential for natural gas, owing to the accessibility to Russian fields; if the Middle East is considered as a potential supplier, this figure rises to about two thirds. The European gas market, therefore, is in a better position than other gas markets. A quarter (46 Tm^3) of the world's gas reserves is found offshore. In Europe, offshore reserves dominate onshore reserves with 4.2 vs. 2.5 Tm^3; globally, the Middle East has the highest offshore reserves, with the majority found in the *South Pars/North Field* of Iran and Qatar in the Persian Gulf.

As for oil, a correlation also exists for natural gas between new discoveries and the development of production, which usually follows the discoveries with some delay. Most gas fields were discovered at the end of the 1960s and during the 1970s; another major discovery was the *South Pars/North* field in Iran and Qatar at the beginning of

[28] Owing to the difficulty of finding new coastal sites for import or export, transport vessels are being developed that can regasify on ship and then deliver gas directly into coastal grid networks (so-called floating LNG plants). The introduction of such on-board gasification technology will allow LNG to be sold into markets where expensive receiving terminals do not exist, potentially broadening the market for LNG.

Table 3.7. *Remaining potential for conventional natural gas at the end of 2005*

Remaining potential		Reserves	Resources
Country	(Tm^3)	(Tm^3)	(Tm^3)
Russia	130.3	47.3	83.0
Iran	38.5	27.5	11.0
Qatar	28.3	25.8	2.5
USA	20.6	5.6	15.0
Saudi Arabia	17.8	6.8	11.0
China	12.4	2.4	10.0
Canada	9.6	1.6	8.0
Turkmenistan	8.8	2.8	6.0
Nigeria	8.7	5.2	3.5
United Arab Emirates	7.6	6.1	1.5
Total	**283**	**131**	**152**
Others	103	48	55
World	**386**	**179**	**207**
World (EJ)	**13 126**	**6 092**	**7 034**
OPEC share (%)	34.2	49.9	20.6
EU25 share (%)	1.7	1.7	1.6

Source: (BGR, 2007).

the 1990s, which is the world's biggest gas field. In analogy to oil, of the more than 26 000 known gas fields globally, only slightly more than 100, i.e., less than 1% are 'giants' and concentrate around three-quarters of all known gas reserves (BGR, 2003). Unlike oil, however, new natural gas discoveries still significantly exceed production volumes.

Figure 3.20 shows the distribution of the EUR, i.e., cumulative production, reserves and resources of conventional natural gas for different world regions.

3.4.3 Unconventional natural gas

3.4.3.1 'Stranded gas'

'Stranded gas' is an all-encompassing term, which refers to any gas whose production would be technically feasible, but that is uneconomic to deliver to the market and is thus wasted or unused. Stranded gas usually comprises gas from occurrences that are located in remote areas that (currently) cannot economically justify the construction of pipelines or are too far from commercial markets; factors that determine the profitability of a pipeline include resource volume,

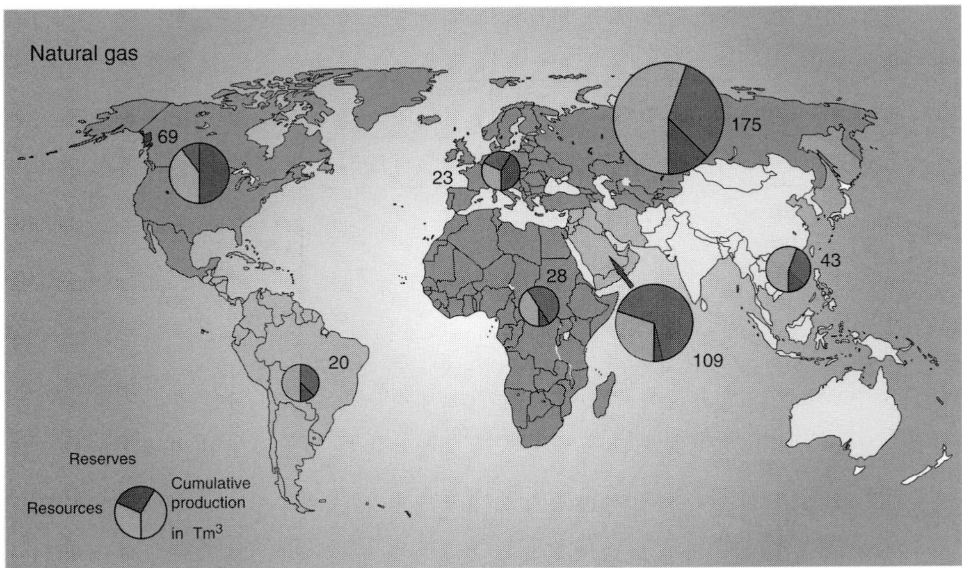

Figure 3.20. Geographical distribution of the EUR of conventional natural gas (BGR, 2007).

transport route, regulatory environment, market size and demand growth. Alternatives to pipeline transport, such as LNG projects, the development of LNG tankers that allow liquefaction and regasification on board, or gas-to-liquid (GTL) technology, if further commercialised, could change the status of these occurrences.[29] The choice of technology to develop stranded gas depends on many factors. Among those factors are the scale (capacity, size) of development and distance to markets.

Depending on how one defines 'reserves' and 'stranded', estimates of stranded gas vary from 25 to 250 Tm^3 (Fischer, 2001). Most publications estimate that between 50% and 60% of the world's proven conventional gas reserves can be considered stranded, i.e., between 90 and 110 Tm^3. According to the IEA (2005), the distribution of stranded gas by region is as follows: 53% Middle East, 20% CIS, 8% Africa, 7% Central and South America, 12% other areas.

Stranded gas may further include *gas flaring and venting* from oil production. Most developing countries that produce oil also flare and vent large volumes of associated gas, as long as it is not used by consumers nearby or reinjected. This practice of burning gas or releasing it into the atmosphere not only harms the environment by adding to greenhouse-gas emissions,[30] it also deprives developing

[29] To be economically viable, the LNG option usually requires minimum gas field sizes of 90–130 Gm^3, for a train size of 3–4 Mt/a over a lifetime of 20 to 25 years. For a production of 10 000 b/day of GTL, field sizes of about 15–20 Gm^3 are required over 20 years. Further options to bring stranded gas to market include gas-to-hydrate (which is still basic research) or gas-by-wire technologies.

[30] According to the GAO (2004), worldwide flaring and venting is estimated to contribute about 4% of the total methane and about 1% of the total CO_2 emissions caused by human activity.

countries of an energy source that is cleaner and often cheaper than others available and reduces potential tax revenues and trade balances. Data collected and reported on flaring and venting are limited as international reporting is voluntary, and no single organisation is responsible for collecting these data. Although the practice of flaring and venting has been diminished, the annual volume of natural gas being flared and vented worldwide is estimated at between 150 Gm^3 and 170 Gm^3, enough to satisfy the annual gas consumption of Central and South America or that of Germany and the Netherlands (Elvidge, 2007; Gerner *et al.*, 2004). The gas flared annually in Africa (around 37 Gm^3) could produce 200 TWh of electricity, about half the power consumption of the continent and more than twice that of sub-Saharan Africa (excluding South Africa) (Gerner *et al.*, 2004). According to Elvidge (2007), Russia flares about 52 Gm^3 of gas annually, followed by Nigeria with 21 Gm^3 and Iran with around 13 Gm^3.

3.4.3.2 Gas-to-liquids (GTL)

Gas-to-liquids (GTL) is the chemical conversion of natural gas into petroleum products. Gas-to-liquid plants use Fischer–Tropsch technology, which first converts natural gas into a synthesis gas, which is then fed into the Fischer–Tropsch reactor in the presence of a catalyst, producing a paraffin wax that is hydro-cracked to products (see also Chapter 7). Distillate is the primary product, ranging from 50% to 70% of the total yield.

At the time of writing, there were only three commercial-scale GTL plants in operation: the 25 000 b/d *PetroSA* facility in South Africa, Shell's 15 000 b/d Bintulu plant in Malaysia and, since 2006, the 34 000 b/d *Oryx* plant in Qatar by Qatar Petroleum, Chevron and Sasol. Several other GTL plants are currently under construction or planned, most of them in the Middle East, especially in Qatar using gas from its huge *North Field*. The IEA (2006) estimates another 280 000 b/d to be added by 2010; total capacity by 2030 could reach 2.2 million b/d, with global demand for gas from GTL producers surging from just 8 Gm^3 in 2004 (0.3% of world gas production) to about 200 Gm^3 in 2030. The EIA (2006) estimates global GTL capacity by 2030 at between 1.1 and 2.6 million b/d, depending on different energy price scenarios.

Gas-to-liquid technology is at the same time an economically viable option for the recovery of stranded gas *and* an option to produce clean fuels or chemical feedstocks. Besides the financial incentive to monetise otherwise worthless gas, GTL has received added impetus in recent years, especially with regard to diesel fuel; also, the trend in industrialised nations to reduce sulphur and particle contents in fuels is likely to accelerate. However, GTL competes with LNG for reserves of inexpensive, stranded natural gas; further declines in LNG supply costs could undermine the attraction of GTL. The future of GTL further hinges on the reduction of production costs, lowering the energy intensity of the process and the ratio of gas to oil prices.

Estimates of capital costs of GTL plants display a wide range: while the EIA (2006) indicates capital costs at US$ 25 000–45 000 per barrel of daily capacity, depending on production scale and site selection, the IEA (2006) reports capital costs of GTL plants currently completed or under construction with US$ 84 000 per barrel. By comparison, the costs of a conventional refinery are around $15 000 per barrel per day. Gas-to-liquid is assumed profitable when crude oil prices exceed $25 per barrel and natural gas prices are in the range of $0.5–1.0/GJ (EIA, 2006). The economics of GTL are extremely sensitive to the cost of natural gas.

3.4.3.3 Natural gas from coal

Properties and extraction processes As its name implies, natural gas from coal (NGC), commonly referred to as coal-bed methane, is natural gas that is formed and remains trapped in coal beds. Natural gas from coal is generated during the coalification process, wherein organic matter is transformed into coal. Natural gas from coal actually comprises three types of gas: coal-bed methane (CBM), coal-seam methane (CSM) and coal-mine methane (CMM), depending on whether the gas comes from *virgin coal deposits* (CBM) and is thus exclusively produced for the methane, or whether it is a by-product of underground coal mining, either from *active coal mines* (CSM) or *abandoned coal mines* (CMM).[31] These gases have different chemical compositions, especially with respect to their methane content: CBM has the highest methane content with 90–95 vol.%; the methane contents of CSM and CMM are with 25–60% and 60–80%, respectively, significantly lower (BGR, 2003). Regional resources of CBM are generally associated with the geographic distribution of bituminous coal and anthracite deposits, while lignite reservoirs are not relevant, owing to their low maturity. Deeper coals generally also contain more methane.

Because of NGC's co-occurrence with coal, the targeted coal seam locations and their geographical distribution are typically well known from coal assessments. Natural gas from coal is produced by reducing the natural pressure within the coal seam by creating fracture systems (so-called 'fracs') to allow the gas to release from the coal and then flow through a well to the surface.

Resource estimates and current production Natural gas from coal is present wherever coal is found and, as coal is found in great quantities throughout the world (see Section 3.5), natural gas from coal may represent a large energy source. As for all types of unconventional gas, the published reserve and resource figures show great variations and are often based on estimations from incomplete data. In addition, as

[31] In fact, coal-bed methane is an explosive hazard in underground mining operations and for safety reasons has traditionally been vented with mines' fresh air circulation. Since the 1970s, methane captured from underground mining has increasingly been used to supplement local gas supplies (WEA, 2000).

the production of NGC is regionally economic already, some countries include NGC in their conventional gas reserves. Even though NGC comprises coal-bed methane (CBM), coal-seam methane (CSM) and coal-mine methane (CMM), in the literature it is mostly referred to as CBM only, and it is generally not indicated to which extent this also includes CSM and CMM. For long-term methane supplies from coal beds, however, dedicated drilling in coal beds is more important than the methane from underground coal mines (WEA, 2000).

Major occurrences of NGC are found in Russia, China, Canada and the USA; worth mentioning in Europe are Bulgaria, Germany and Poland. While according to the Federal Institute for Geosciences and Natural Resources (BGR, 2003), the resources of natural gas from coal show a great variation between 92 and 195 Tm^3 (on average 143 Tm^3), the economically recoverable reserves are estimated at 1.1 Tm^3 only (around 0.6% of conventional gas reserves); the United States hold around 40% of these reserves (about one-third of the world's hard coal reserves are also located in the USA, see Section 3.5.2.2), followed by China with 17%. The low figure for NGC reserves compared with the average resources (around 0.75%) can partially be explained by the fact, that for CBM, only a small fraction of the gas in-place can be recovered, while for CSM and CMM, the majority of the gas released from coal mining escapes unused to the atmosphere.

World NGC production in 2001 was 42.3 Gm^3 (BGR, 2003). Owing to tax incentives, NGC has been used in the USA since 1975, and its exploitation has been steadily progressing. The United States are still the dominant producer of NGC (CBM): in 2001, production amounted to 40 Gm^3, which was 95% of world production; in 2006, US CBM production reached 51 Gm^3, 9% of total national production (Kuuskraa, 2007). Other countries with NGC production, however, with a negligible share as of today, are Australia, China, Canada, Germany, the UK and Poland. Natural gas from coal is usually used very near to the production site only; accordingly, there is no global NGC market and prices are regionally different. It can be expected that NGC production will grow in the future, owing to declining mining activities in many countries and governmental incentives, but the consumption is likely to remain regionally limited.

3.4.3.4 Tight gas

Properties and extraction processes Tight-formation gas is natural gas trapped in low-porosity (7 to 12%), low-permeability reservoirs with an average *in-situ* permeability of less than 0.1 millidarcy (mD), regardless of the type of the reservoir rock: tight gas usually comprises *gas from tight sands* (i.e., from sandstone or limestone reservoirs) and *shale gas*. Sometimes tight gas also comprises natural gas from coal and 'deep gas' from reservoirs below 4500 m. Shale gas is produced from reservoirs predominantly composed of shale rather than from more conventional sandstone or limestone reservoirs; a particularity of shale gas is that gas shales are often both the source rock for the generation of natural gas and the reservoir rock for the storage of the gas.

The production of tight gas is technically very demanding. The major differences from conventional production arise because of the poor permeability of tight reservoirs, where the natural gas cannot flow as quickly to the well or in sufficient volumes to be economic, and where production rates are usually quite low. The principal prerequisite for economically producing tight gas is, therefore, to improve reservoir permeability, e.g., by artificial stimulation techniques, such as hydraulic fracturing (i.e., the generation of artificial fracture systems).

Resource estimates and current production The size, location and quality of tight-gas reservoirs varies considerably. As with CBM, the USA is also the leading producer of tight gas: in 2006, 161 Gm^3 of gas from tight sands and 31 Gm^3 of shale gas have been produced; thus, total production from unconventional gas – including CBM – amounted to around 43% of total US natural gas production (a more than three-fold increase since 1990) (Kuuskraa, 2007). Tight gas is further being produced in small quantities in Canada and in Europe (Germany, UK, France and the Netherlands). Although the CIS countries have a considerable potential of tight gas resources, they are not expected to be produced in the near future, since there exist sufficient conventional gas reservoirs.

Although tight gas reservoirs exist in many regions, only the ones in the USA, Canada and Russia have been assessed. The Federal Institute for Geosciences and Natural Resources applies these estimates to extrapolate tight-gas resource potential (reserves plus resources) for other countries and regions, arriving at a speculative global potential of around 90 Tm^3; the range of these estimates is indicated with $\pm$ 50%. The highest potential is found in the CIS countries, followed by the Middle East and North America. The BGR (2007) indicates global tight gas reserves of 1 Tm^3 and resources of 90 Tm^3.

3.4.3.5 Aquifer gas

Properties and extraction processes Aquifer gas, also referred to as *geo-pressured gas* or *brine gas*, is natural gas found dissolved in aquifers, primarily in the form of methane. The solubility of natural gas, and thus the methane content of the water, can vary significantly, and depend on factors, such as the total pressure, temperature, salt content of the water and amount of other gases dissolved. The amount of gas dissolved in underground liquids increases substantially with depth. A general rule is that the deeper the aquifers and the higher the pressure, the higher the gas content. At depths down to 5 km, up to 5 m^3 of methane can be dissolved per m^3 of water in aquifers under normal hydrostatic pressure (load of water); under lithostatic pressure (load of water and rocks), this factor may increase to more than

10 m^3/m^3 water, while in zones of extreme pressure gas contents of up to 90 m^3/m^3 water are known.

The productivity of a reservoir depends on the characteristics of the ground water and the properties of the rocks of the aquifer. The most important characteristic of the ground water is its saturation with gas. The techniques to produce gas from aquifers already exist; however, they work less effectively than for conventional gas and are therefore less economic. High porosity and permeability of the rocks of the aquifer facilitate a high production of water, and can lead to significant land subsidence. This problem can by addressed by reinjecting the degasified ground water.

Resource estimates and current production Aquifer gas is expected to occur in nearly all sedimentary basins. The ranges of estimates of both *resources in-place* and *recoverable resources* of aquifer gas are enormous. While no detailed assessment of aquifer gas resources is available, the Federal Institute for Geosciences and Natural Resources (BGR, 2003) derives potential aquifer gas in-place from the ground water volume contained in high-permeability sandstones in the hydrosphere. This approach leads to an estimate of 2400–30 000 Tm^3 of aquifer gas in place, with a mean estimate of 16 200 Tm^3; in the absence of a more detailed assessment, a regional breakdown has been obtained by weighting the global mean estimate of gas occurrence in place with regional shares of total sedimentary area. The estimates of possibly recoverable resources show a tremendous scope, ranging from 24 to 1500 Tm^3; 800 Tm^3 are considered as resources (BGR, 2007).

While these estimates of aquifer gas occurrences are highly speculative, the potential quantities are vast. Even a future recovery factor of 5% of the mean aquifer gas in place implies a resource volume of more than four times the conventional gas reserves. However, considering the enormous amounts of aquifer gas is not relevant, as long as it is not known how much and at what cost this gas can be recovered. Aquifer gas is already produced in small quantities from shallow reservoirs in Japan, China and the USA. But in all cases aquifer gas recovery has been motivated by the production and economic use of trace elements (such as iodine) rather than by the gas itself. Production of aquifer gas in Italy was stopped in the 1960s because of land subsidence. With increasing primary energy prices and technological progress in the production techniques, aquifer gas might become economically attractive in the future. Until 2030, however, it is estimated that neither is the production of aquifer gas going to increase significantly nor will some occurrences be categorised as reserves.

3.4.3.6 Gas hydrates

Properties and extraction processes At high pressures and low temperatures, water and gas form an ice-like mixture, called gas hydrate, also known as clathrate or simply hydrate. Hydrates are a crystalline, solid substance composed largely of water that looks and behaves like dry ice, and in which the gas molecules are trapped within a framework of cages of water molecules. Most gas hydrates are *methane hydrates*, although many other gases also have molecular sizes suitable for forming hydrates. Hydrates are a gas concentrator: 1 m^3 of methane hydrate releases at atmospheric pressure about 164 m^3 methane, which makes gas hydrates a very interesting energy resource (BGR, 2003).[32]

Gas hydrates occur naturally where combinations of temperature and pressure favour the stability of gas hydrate over a gas–water mixture. Such conditions are present in oceanic sediments along continental margins and in polar continental settings (mainly of the northern hemisphere). Because of the low-temperature, high-pressure requirements for hydrate stability, hydrates are primarily found in two environments: as *sub-sea marine accumulations* downwards from the sea floor, where the water depth is greater than about 500 metres and pressures are sufficiently high and temperatures above those for ice stability, and as *terrestrial accumulations* associated with *permafrost* in polar regions and in shallow arctic seas where temperatures are sufficiently low (CfE, 2007). For marine gas hydrates, the stability zone extends from the sea floor down to depths of 300 to 1000 m beneath the sea floor, where the base of the layer is limited by increasing temperature. Permafrost hydrates range from some 200 m within the water–ice permafrost zone to a depth that is also determined locally by the rising temperature (down to 2000 m) (BGR, 2003); in association with permafrost, gas hydrates are stable both in onshore and offshore sediments. In addition, there may be *sea floor accumulations* of gas hydrates, as in the deep waters of the Gulf of Mexico, which result from seeps of deeper hydrocarbon deposits. Gas hydrates may also act as seals for underlying free natural-gas reservoirs.

Resource estimates and current production All gas hydrate occurrences are classified as resources. The quantity of gas in the form of hydrates remains very speculative and highly uncertain, and varies within the limits of 10^{13} to 10^{19} m^3 of methane; there are further discrepancies, depending on whether marine or continental resources are dominating. Various studies estimate the amount of gas hydrates below the ocean floor (marine gas hydrates) at a range of 100 to 10^7 Tm^3 and on land (continental gas hydrates) at 10 to 10^5 Tm^3 (BGR, 2003); this would be up to 50 000 times the world's known conventional gas reserves. Resource estimates for permafrost regions seem generally more reliable, as there is more knowledge due to many drillings for the exploration of conventional gas. But none of these geological resource estimates allows one to derive statements about how much can be practically and affordably recovered in the end. Although hydrates occur throughout the world, only those

[32] Hydrates are also of research interest as a source of greenhouse gases resulting from the decomposition of the trapped methane as well as for their role as a submarine geohazard, as the destabilisation of gas hydrates may initiate submarine landslides, which may cause tsunamis. The formation of gas hydrates further poses problems for gas pipelines at low temperatures, such as in deep and cold waters.

from northern Russia (the Messoyakha gas field), the Alaska North Slope, the Mackenzie-Delta and Beaufort-Sea Region (north-western Canada), offshore North Carolina, the Gulf of Mexico and the Nankai Trough offshore of Japan have been studied (CfE, 2007). To date there has been no well documented commercial production of gas hydrates. (For an overview of hydrate-related research projects, see www.netl.doe.gov/technologies/oil-gas/FutureSupply/MethaneHydrates/main content.htm.)

While the enormous resources of gas hydrates make them a very attractive choice for eventually replacing supplies of conventional natural gas, the knowledge required to produce gas hydrates economically is currently lacking. Attempts to produce gas hydrates focus on destabilising the hydrates in the reservoir by changing the composition from gas hydrates to gas plus water. There are three main processes envisaged for production: *change the pressure* conditions, *inject steam or hot water* to destabilise the gas hydrate, or *change the chemical conditions* for gas hydrate formation. A danger with offshore production is that the destabilisation of gas hydrates in the sediments can lead to submarine landslides. Occurrences of gas hydrates in permafrost areas have the advantage that in some regions (such as in the Mackenzie-Delta in Canada, Alaska or western Siberia) an infrastructure for the production of conventional gas already exists. According to the BGR (2003), the production of marine hydrates is expected to start much later (between 2030 and 2060), as the technical and scientific challenges are significantly more demanding.

The key to establishing gas hydrates as a significant energy resource is whether the methane gas will ever be economically and safely producible. The current state of knowledge is still too limited to allow reliable estimates on the start of an economic gas hydrate production. The BGR (2003) estimates gas hydrate resources at 500 Tm^3.

3.4.4 Scenarios for determining the mid-depletion point of natural gas

In analogy to oil, in the following it is analysed how the projected increase in natural gas consumption will affect the time of the mid-depletion point of gas production. The methodological approach is the same as described for oil in Section 3.3.4. The discussion about a possible peaking of gas production is (still) less controversial than for oil.

To represent the range of estimates of the EUR of natural gas, a very pessimistic estimate of 283 Tm^3 by Laherrère (2004a) and a very optimistic estimate of 558 Tm^3 by Chabrelie (2002) have been selected. For comparison, the USGS (2000) estimates the P95 EUR of natural gas at 301 Tm^3, the P5 EUR at 604 Tm^3 and the mean at 436 Tm^3. The estimate of the Federal Institute for Geosciences and Natural Resources (BGR, 2007) of 466 Tm^3 is in the middle. As for unconventional gas, only *reserves* have been taken into account, owing to the large uncertainties related to unconventional gas occurrences. As of today, 81 Tm^3 of natural gas have already been produced.

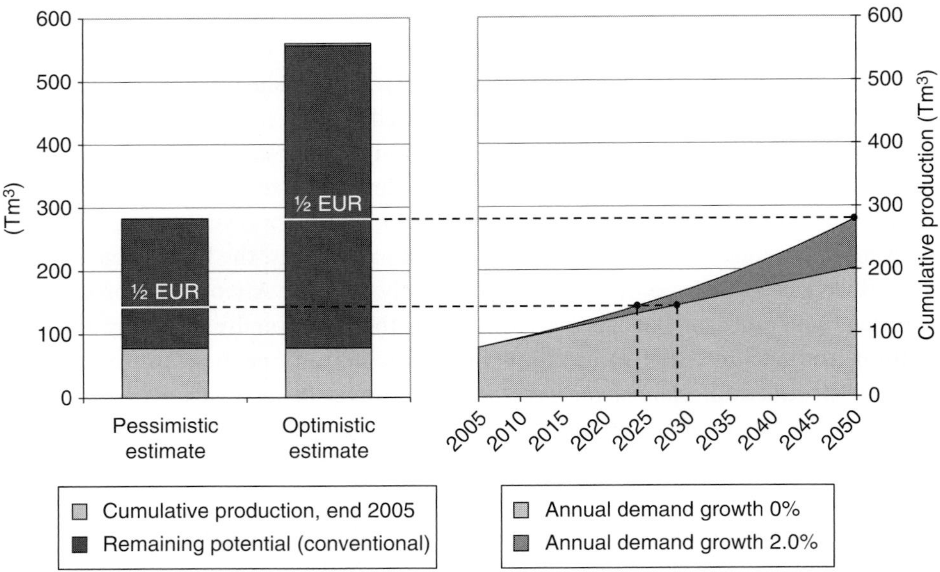

Figure 3.21. Scenarios for the mid-depletion point of natural gas.

Figure 3.21 shows the time window of the mid-depletion point for two different growth rates of annual gas demand – 0% and 2.0% according to the *IEA Reference Scenario* (Section 3.1; between 1965 and 2005, global natural gas demand has grown by an average of 2.9% per year (BP, 2006)) – and the above estimates of ultimately recoverable resources. According to the *IEA Reference Scenario*, the cumulative gas production will have doubled by around 2030. According to the pessimistic estimate, the peak of conventional gas production will be passed between 2024 and 2028; the optimistic estimate sees the production peak around 2050. The influence of unconventional reserves of 2.1 Tm^3 is negligible. According to the BGR's estimate, the mid-depletion point will be passed around 2042 for the high-growth scenario.

The projected growth scenarios for conventional gas seem justified from the point of view of ultimately recoverable resources, and unlike for oil, no major discrepancies between projected demand and supply are to be expected in the coming decades. An important aspect of the future availability of natural gas, however, is the creation of the necessary infrastructure for the production and subsequent transport of the gas to the customer. The cumulative investments for the gas-supply infrastructure until 2030 are estimated to amount to US$3.9 trillion (IEA, 2006).

3.5 Coal

With the beginning of industrialisation around 1850 (first in Great Britain), wood, the main energy source by then, began to lose out to coal, which was the dominating fuel for the remainder of the nineteenth century and well into the twentieth century.

For more than 200 years, coal has been an important feedstock for the supply of energy as well as for iron and steel manufacturing. At 25%, coal still ranks second in terms of global primary energy supply, and coal-fired power stations provide more than 45% of global electricity generation. With a further growing worldwide energy demand and the expectation of possible declines in the supply of (cheap) oil and gas, coal is increasingly getting back on the energy agenda. Given its relative abundance and geographical distribution, coal might well play an important role in the future again, not only in the electricity sector (for the future role of coal, particularly in the power sector, see MIT (2007)), but also as source for transportation fuels, such as hydrogen or synthetic liquid fuels (CTL), especially if clean coal technologies, especially with respect to carbon capture and storage, become economically viable.

3.5.1 Classification of coals

Coal originates from the transformation of terrestrial plants into carbon through burial. During this process, the organic matter was transformed sequentially through the different stages or ranks of coal – lignite, subbituminous, bituminous and anthracite. Depending on their rank, which is a function of time, temperature and pressure by which the organic matter was transformed, the various coal types differ in their carbon and water content: the higher the rank, the lower the water content and the higher the proportion of pure carbon.

Coals are usually subdivided into several broadly defined types according to their calorific value. However, almost every coal-producing country has its own coal classification scheme and it is common practice that coals with the same properties are categorised differently. In particular, the boundaries between hard coal (usually anthracite, bituminous and subbituminous coal) and soft brown coal (usually also subbituminous coal and lignite) overlap with respect to the allocation of high- and low-energy subbituminous coal. As in the case of oil, it is thus difficult to compare national and international coal statistics, and there is an international attempt to introduce a common categorisation scheme according to the UNECE standard, which distinguishes six subgroups (UNECE, 2007).

As for the quantification of coal reserves and resources, the classification of the Federal Institute for Geosciences and Natural Resources will be used, which distinguishes hard coal (anthracite, bituminous and subbituminous coal) and soft brown coal (lignite): lignite comprises coals with a calorific value between 6700 kJ/kg and 16 500 kJ/kg, hard coal comprises coals with a calorific value greater than 16 500 kJ/kg (the totals are generally in line with figures published by the World Energy Council or the US Department of Energy). The energy content of the different coals, and hence the conversion factors from ton to tce, are country-specific and typically range from 0.3 to 0.4 tce/t coal for lignite and 0.75 to 0.85 tce/t coal for hard coal. In the following, countries are ranked according to the energy content of their coals.

Table 3.8. *Worldwide hard coal production and consumption, 2005*

Hard coal production		Hard coal consumption	
Country	(%)	Country	(%)
China	36.3	China	35.9
USA	20.8	USA	20.7
India	6.4	India	6.5
Australia	6.2	South Africa	3.6
South Africa	4.9	Japan	3.1
Russia	3.3	Russia	2.3
Indonesia	2.3	Poland	1.6
Poland	2.0	Germany	1.5
Canada	1.6	Canada	1.5
Kazakhstan	1.3	South Korea	1.5
Sum	**85.1**	**Sum**	**78.3**
World (Gtce)	**4.18**	**World (Gtce)**	**4.19**
World (EJ)	**122.6**	**World (EJ)**	**122.7**

Source: (BGR, 2003; 2007).

3.5.2 Hard coal

3.5.2.1 Production, consumption and trade

Hard coal is produced both in surface and underground mining. Table 3.8 shows the world hard coal production in 2005: in total around 5 Gt (or 4.18 Gtce) were produced, with almost two thirds of total production (more than half of total energy content) coming from China and the United States; the share of the EU25 amounted to 3.5% of total production. The situation for hard coal consumption is similar, with China being by far the biggest consumer, followed by the United States; here the share of the EU25 was 7.4%.

Of the 25% share of coal in world primary energy supply, hard coal makes up around 90% (BGR, 2007). Owing to its high calorific value, the main use of hard coal is for electricity generation (60% of world hard coal production), 16% is used for steel making, and the rest in other industries and households. While in the EU25 around one third of the electricity is produced by (hard) coal, this share amounts to 50% in the United States, 70% in India, 80% in China, 85% in Australia, 90% in South Africa and 93% in Poland.

Because of its high calorific energy content, hard coal is internationally traded (unlike lignite). Total trade in 2005 amounted to 790 Mt (16% of production), of which around 90% was traded by sea transport; from the harbours, the coal is further distributed either by inland waterways or rail. The most important exporters of hard coal are Australia, South Africa and Indonesia.

Table 3.9. *Worldwide hard coal reserves and resources, 2005*

Hard coal reserves		Hard coal resources	
Country	(%)	Country	(%)
USA	30.0	Russia	34.2
China	11.0	China	18.2
India	10.4	USA	10.1
Russia	10.3	Australia	4.0
Australia	9.6	India	3.3
South Africa	6.6	South Africa	2.8
Kazakhstan	3.0	Pakistan	2.1
Ukraine	1.7	Colombia	1.5
Poland	1.1	Canada	1.3
Colombia	1.0	Poland	1.2
Sum	**84.8**	**Sum**	**78.7**
World (Gtce)	**626**	**World (Gtce)**	**3511**
World (EJ)	**18352**	**World (EJ)**	**102898**

Source: (BGR, 2003; 2007).

3.5.2.2 Reserves and resources

Table 3.9 shows the distribution of world hard coal reserves and resources in 2005. Total reserves amounted to 728 Gt (626 Gtce), of which the vast majority are located in the USA and China, followed by India and Russia. The top ten countries represent 85% of total reserves. Considering the production of 2005, the static lifetime of hard coal can be calculated at around 150 years; however, we should acknowledge the simplicity of this approach, as coal use is expected to increase significantly in the future. As for hard coal resources, whose quantification is more uncertain, Russia is leading, followed by China and the United States. Figure 3.22 shows the geographical distribution of cumulative production, reserves and resources of hard coal.

3.5.3 Lignite

Table 3.10 shows the world lignite production in 2005, which amounted to 936 Mt or 317 Mtce. In the last decade, world production has been quite constant. Since 1990, Germany has been the biggest producer of lignite (178 Mt), followed by the USA. The calorific values of lignite from Germany and the USA, however, are significantly different (0.31 tce/ton vs. 0.50 tce/ton). Lignite is produced in surface mining only. Because of its low energy content and high water content (between 35 and 75 wt.%) transportation costs are quite high and only allow for an economic utilisation of lignite close to the mining site (up 50 km, seldom up to 100 km). Ninety per cent of lignite worldwide is used in power plants for electricity and heat generation.

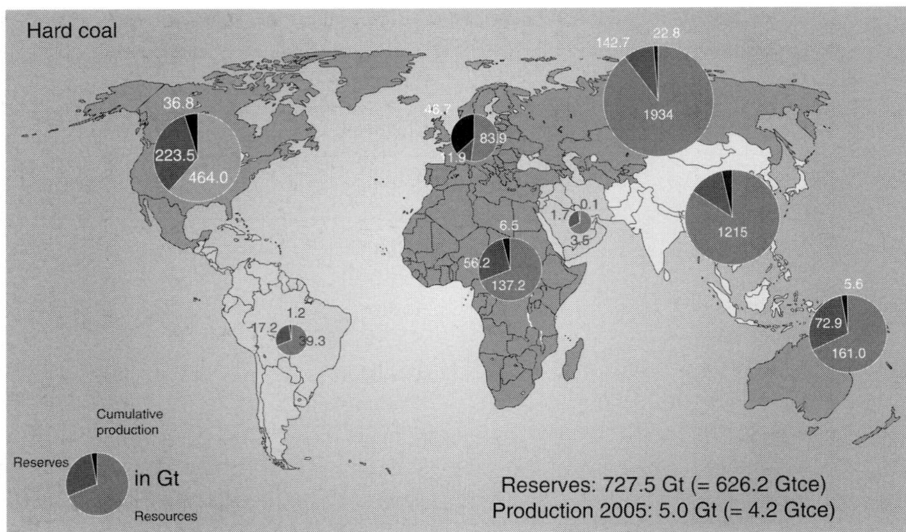

Figure 3.22. Geographical distribution of the EUR of hard coal (BGR, 2007).

Table 3.10. *Worldwide production, reserves and resources of lignite, 2005*

Production		Reserves		Resources	
Country	(%)	Country	(%)	Country	(%)
Germany	17.4	USA	23.7	USA	45.8
USA	11.8	Australia	18.7	Russia	20.2
Russia	10.6	India	16.2	China	8.1
Australia	7.0	China	10.6	Germany	5.5
Turkey	6.7	Serbia and Montenegro	7.2	Australia	3.2
China	6.1	Russia	6.7	Kazakhstan	2.8
Poland	5.5	Germany	2.9	Poland	2.7
Czech Republic	4.6	Brazil	2.2	Indonesia	2.4
Greece	4.1	Turkey	1.8	Serbia and Montenegro	1.7
Serbia and Montenegro	3.4	Indonesia	1.7	Brazil	0.9
Sum	**77.3**	**Sum**	**91.7**	**Sum**	**93.2**
World (Gtce)	**0.32**	**World (Gtce)**	**70.4**	**World (Gtce)**	**430**
World (EJ)	**9.3**	**World (EJ)**	**2063**	**World (EJ)**	**12611**

Source: (BGR, 2007).

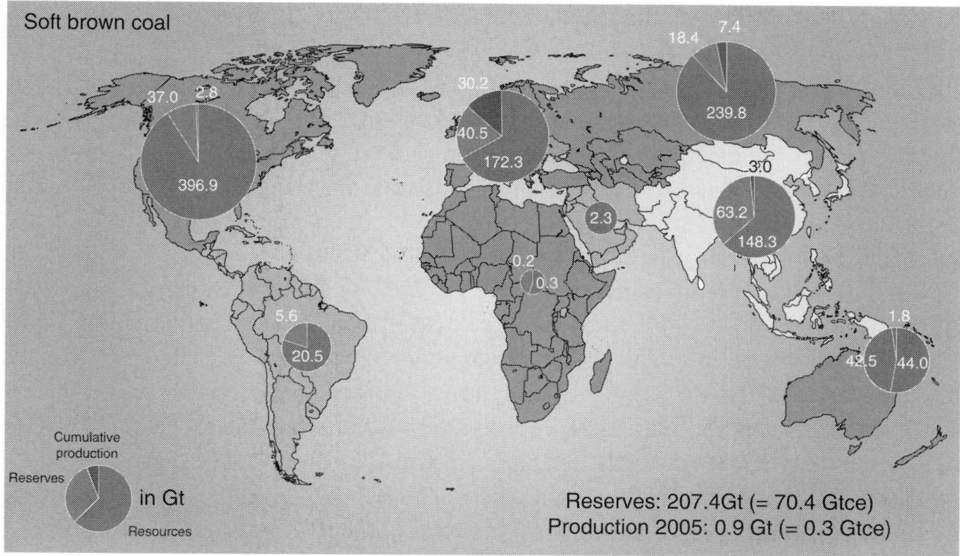

Figure 3.23. Geographical distribution of the EUR of lignite (BGR, 2007).

Table 3.10 further shows the global distribution of lignite reserves. Worldwide, the USA has the highest reserves in terms of energy content (16.7 Gtce), followed by Australia and India. The world total amounts to around 70 Gtce. It should be noted, again, that the calorific values of lignite show great variations among different countries. On the basis of the 2005 production, the static lifetime of lignite can be calculated at around 230 years. Table 3.10 also displays the world lignite resources. The ten most important countries concentrate around 93% of the total resources of 430 Gtce. The majority of resources are located in the United States, Russia and China. In general, the quantification of resources is far more uncertain than for reserves; this is especially the case for Russia, China and Kazakhstan. Figure 3.23 shows the geographical distribution of cumulative production, reserves and resources of lignite.

3.5.4 Coal-to-liquids (CTL)

With high oil prices, nations endowed with rich coal resources, such as the USA and China, might increasingly consider filling part of the resulting fuel gap with the generation of synthetic fuels from coal. Coal can be used to produce liquid fuels either by removal of carbon, a process known as *carbonisation* or *pyrolysis*, or by addition of hydrogen, a process called *liquefaction*. The disadvantage of all pyrolysis and carbonisation processes is that they have very low liquid yields (< 20%) and the liquids produced are of low quality; it has also not been successfully demonstrated to date that these processes can be economically viable (Gruson *et al.*, 2005). Coal-to-liquids (CTL) comprises two very different approaches: indirect coal liquefaction (ICL) and direct coal liquefaction or hydrogenation (DCL). For both approaches, a major challenge is to increase the hydrogen:carbon ratio. Another process under development is underground coal gasification (UCG), i.e., the *in-situ* gasification of coal in the seam. It is achieved by injecting oxidants, gasifying the coal and bringing the product gas to the surface through boreholes drilled from the surface, which can then be used for power generation or as chemical feedstock.

Indirect coal liquefaction is based on *coal gasification* and syngas production, with subsequent production of FT fuels (as in the case of GTL, see also Section 3.4.3.2 and Chapter 7); those technologies are either commercially proven or made up of proven modules. Today, Sasol in South Africa is the only commercial producer of coal-based synthetic fuels, with a total liquid fuel production from two plants of around 160 000 barrels per day (30% of total national oil consumption).[33] The output is 80% synthetic diesel and 20% synthetic naphtha. The EIA (2006) indicates capital costs for an ICL plant at US$50 000 to 70 000 per barrel of daily capacity and estimates the synthetic fuels to be economically competitive at an oil price of about $40 per barrel and a coal price between $1 to $2/GJ, depending on coal quality and location (e.g., the USA or China). Indirect coal liquefaction processes are very energy-intensive and CO_2 emissions are more than ten times higher per unit of output than from conventional oil refineries (without carbon capture and storage). New CTL projects are currently planned in China and the USA, with the USA aiming at up to 2.6 Mb/d until 2035 (US DOE, 2007b).

Direct liquefaction processes aim to add hydrogen to the organic structure of the coal, breaking it down only as far as is necessary to produce distillable liquids. Many different processes have been developed, with the common features being the dissolution of a high proportion of coal in a solvent at elevated temperature and pressure, followed by hydro-cracking of the dissolved coal. (For details of the current status of direct coal liquefaction, see Gruson *et al.* (2005) or Williams and Larson (2003).) Direct coal liquefaction is currently the most efficient route, with liquid yields in excess of 70% by weight. The world's first commercial direct coal liquefaction facility is under construction in China. The first train is expected to start operation in 2008, producing about 20 000 b/d of oil products (including gasoline and diesel); by 2020 it is planned to increase total capacity to 600 000 b/d. Gruson *et al.* (2005) estimate capital costs for a DCL plant at $60 000 per barrel of daily capacity.

While the IEA (2006) considers CTL to remain a niche activity until 2030, the EIA (2006) estimates global CTL capacity by 2030 at between 1.8 and 2.3 million b/d, depending on different energy price scenarios.

[33] The production of FT fuels from coal in South Africa started in the mid 1950s and was extended in the early 1980s when South Africa was embargoed for its apartheid regime.

3.6 Summary

Oil and gas still make the world work. Fossil fuels still account for about 80% of today's world primary energy supply, with the transport sector almost entirely depending on oil. Global demand for oil has reached new heights, led by China and other rapidly industrialising countries. A shrinking margin between oil production capacity and demand was largely responsible for the rapid rise in oil prices in recent years. Owing to a declining production coupled with a growing demand, the import dependency of the EU27 for instance is expected to grow from around 81% today to up to 97% in 2030. Given the extent to which the industrialised world has come to depend on oil as a pillar of its economy, possible shortages in the supply of oil as a consequence of declining production are likely to result in abrupt and disruptive changes.

Energy projections are highly sensitive to the underlying assumptions of GDP growth, the main driver of demand for energy services. Hence, global energy demand is projected to grow by more than 50% until 2030. The analysis of this chapter shows a mismatch of growth scenarios with fossil-fuel resources, particularly for oil: if we continue with business as usual, we are very likely to face shortcomings in the supply of oil in the coming decades.

There will always be considerable uncertainty concerning how much oil exists under the Earth's surface and how much can be recovered. There is a long history of failed forecasts regarding the peaking of oil production and experience shows that reserves are usually underestimated. However, there are compelling reasons why current projections might be more reliable than previous ones. For instance, global production has been exceeding new discoveries since the 1980s and the size of new discoveries has also been decreasing. The fact of peak oil production in the short to medium term has widely been accepted and the analysis in this chapter suggests that the world conventional oil production will peak around 2015.

The present level of oil prices and growing concerns about the ability of world oil supplies to meet increasing demands, especially from the developing economies of Asia, as well as increasing numbers of countries experiencing declines in conventional oil production are prompting significant investments in oil sands and a renewed interest in oil shale, as well as in synthetic fuels from gas and coal (GTL and CTL). The potential resources of unconventional fuels are vast, but they come at much higher costs and higher environmental penalty, as their production is much more energy intensive, and therefore much more CO_2 intensive, than conventional oil production. Growth prospects for any unconventional oil will depend on the prices of conventional hydrocarbons and on environmental constraints. When the price of producing unconventional oil is competitive with the price of oil from conventional sources – either by technological improvements or higher oil prices – and the environmental problems can be overcome, then unconventional fuels will find a place in the fossil-fuel market in the future. A further degree of uncertainty

Table 3.11. *Projections of the production of unconventional oil until 2030*

(Mb)	Extra-heavy oil	Natural bitumen[a]	Shale oil	CTL	GTL	Total
2006	157[b]	458	3–4	58	27	**703**
2030[c]	840–1130 (2.3–3.1 Mbd)	1060–1825 (2.9–5 Mbd)	18–730 (0.05–2 Mbd)	274–840 (0.75–2.3 Mbd)	400–950 (1.1–2.6 Mbd)	**2592–5475** (7.1–15 Mbd)
Break even price (US$/b)	n.a.	30–35	70–95	40–50 and coal prices from $1–2/GJ	>25 and gas prices from $0.5–1.0/GJ	

Notes:
[a] For simplicity reasons, a liquid yield factor of one is assumed for bitumen upgrading operations.
[b] 2003.
[c] According to EIA (2006).

generally associated with the production of unconventional oil (or any alternative fuel) is the potential response of OPEC nations to various market and technological developments.

Table 3.11 summarises the current projections of the production of unconventional oil, including synthetic fuels from coal and gas, until 2030. Today, unconventional fuels account for around 2% of world oil production of 81 Mb/day. Their future will depend on the oil price. If prices stay at relatively high levels, unconventional fuels could reach between 2.6 and 5.5 Gb (7 and 15 Mb/day) in 2030. According to the IEA (2006) *WEO Reference Scenario*, total oil production in 2030 will amount to 42.3 Gb (116 Mb/day). Hence, unconventional fuels would make up between 6% and 13% of total oil production in 2030, of which around one-third comes from oil sands. Unconventional fuels are not a silver bullet: they can briefly delay the maximum rate of oil production, however, the global decline of production cannot be prevented in the short to medium term, if demand for oil continues to surge.

Despite the considerable growth of the Canadian oil-sands industry in recent years, there are still several difficulties that could impede the future development of the industry; for instance, the heavy reliance on natural gas and water, which are necessary in both the extraction of bitumen from oil sands and the upgrading of bitumen to synthetic oil, as well as increasing CO_2 emissions.

For nearly a century, the oil shale in the western United States has been considered as a substitute source for conventional crude oil. If a technology can be developed to recover oil from oil shale economically, the quantities would be in the range of today's conventional oil reserves. But the economics of shale-oil production have persistently remained behind conventional oil. The prospects of oil-shale development are uncertain and many issues related to technology performance, and environmental and socioeconomic impacts remain unsolved. It is unlikely that shale oil recovery can be expanded to make a major contribution any time soon towards meeting the growing demand for oil.

As for gas, it can be said, that the projected growth scenarios for conventional gas seem justified from the point of view of ultimately recoverable resources and that, unlike oil, no major discrepancies between projected demand and supply are to be expected in the coming decades. An important aspect of the future availability of natural gas, however, is the creation of the necessary infrastructure for the production and subsequent transport of the gas to the customer. An increased demand competition between Europe, the United States and Asia is likely to be expected; this concerns, for instance, a possible supply of Asia, particularly China, with gas from Russia, or an extended competition for LNG between the EU and the USA.

The production of unconventional gas is mainly of importance in the United States, with tight gas being the largest of the unconventional gas resources, followed by coal-bed methane and shale gas. In 2006, production of unconventional gas in the USA represented about 43% of the total gas output. Although the production of unconventional gas in the USA could further grow if advanced technologies are developed and implemented, no significant production of unconventional gas at a global scale is expected until 2030. Estimates of unconventional gas occurrences indicate enormous energy sources, but uncertainties around those estimates are equally large, especially with respect to aquifer gas and gas hydrates, for which there is still fundamental research needed for both production techniques and reliable resource estimates.

As for coal, no resource constraints are expected in the coming decades.

Table 3.12 summarises the global reserves and resources of fossil fuels.

The analysis of resource potential vs. demand growth for oil shows that it is time to develop alternatives to oil as major fuel for the transport sector. Simply from a resource point of view – and neglecting adverse environmental impacts – no preference for hydrogen over unconventionals (oil sands, oil shale) can be concluded in the short to medium term, as the primary energy expended for their production – although being significantly higher than for the recovery of conventional oil – yields more 'mobility' than when used for hydrogen production. However, in the longer term, hydrogen can contribute to diversify fuel supply and help renewable energies (other than biomass) enter the transport sector.

Both the production of hydrogen from fossil fuels and unconventional fuels result in high CO_2 emissions. While the capture of the CO_2 from a central point source is equally possible for unconventional fuels and hydrogen production, in the case of hydrogen a clean fuel is provided, unlike in the case of liquid hydrocarbon fuels. There are no advantages for GTL and CTL over hydrogen, except for them being able to use the existing liquid-fuel infrastructure. If hydrogen vehicles and

Table 3.12. *Summary of global reserves and resources of fossil fuels*

	Reserves		Resources	
	Individual unit	EJ (LHV)	Individual unit	EJ (LHV)
Conventional oil and gas				
Oil	1 188 Gb	6 807	603 Gb	3 455
Natural gas	179 Tm^3	6 092	207 Tm^3	7 034
Unconventional oil				
Extra-heavy oil	189 Gb	1 083	189 Gb	1 083
Crude bitumen (Oil sands)	220 Gb	1 261	220 Gb	1 261
Shale oil (Oil shale)	6 Gb	34	1 160 Gb	6 647
Total	**415 Gb**	**2 378**	**1 569 Gb**	**8 991**
Unconventional gas				
Natural gas from coal	1 Tm^3	34	143 Tm^3	4 862
Tight gas	1 Tm^3	34	90 Tm^3	3 060
Aquifer gas	–	–	800 Tm^3	27 200
Gas hydrates	–	–	500 Tm^3	17 000
Total	**2 Tm^3**	**68**	**1 533 Tm^3**	**52 122**
Coal				
Hard coal	626 Gtce	18 352	3 511 Gtce	102 898
Lignite	70 Gtce	2 063	430 Gtce	12 611
Total	**696 Gtce**	**20 415**	**3 941 Gtce**	**115 509**

infrastructure are available, the syngas route would be better used for hydrogen, as it has a higher thermal process efficiency. Any investments needed for an initial hydrogen infrastructure must also be reflected in the context of the investments in the oil and gas sector, which are cumulatively projected to amount to $8.2 trillion until 2030 (IEA, 2006).

References

ACR (Alberta Chamber of Resources) (2004). *Oil Sands Technology Roadmap*. www.acr-alberta.com.

AEUB (Alberta Energy and Utilities Board) (2007). *Alberta's Energy Reserves 2006 and Supply/Demand Outlook 2007–2016*. Statistical Series EUB ST98–2007: Calgary (Alberta): AEUB.

ASPO (Association for the Study of Peak Oil and Gas) (2007). *Newsletter* No. 83 (November 2007). www.peakoil.net.

Babies, H. G. (2003). Ölsande in Kanada – eine Alternative zum konventionellen Erdöl? *Commodity Top News*, **20** (October 2003). Federal Institute for Geosciences and Natural Resources (BGR) www.bgr.bund.de.

Bahorich, M. (2006). End of oil? No, it's a new day dawning. *Oil & Gas Journal*, **104** (31).

Bartis, T. B., La Tourrette, T., Dixon, L., Peterson, D. J. and Cecchine, G. (2005). *Oil Shale Development in the United States. Prospects and Policy Issues*. RAND Corporation. www.rand.org.

Bartlett, A. (2000). An analysis of US and world oil production patterns using Hubbert-style curves. *Mathematical Geology*, **32** (1), 1–17.

BGR (Bundesanstalt für Geowissenschaften und Rohstoffe) (Federal Institute for Geosciences and Natural Resources) (2003). *Reserven, Ressourcen und Verfügbarkeit von Energierohstoffen 2002*. Rohstoffwirtschaftliche Länderstudien, Heft XXVIII. Hanover.

BGR (Bundesanstalt für Geowissenschaften und Rohstoffe) (Federal Institute for Geosciences and Natural Resources) (2007). *Reserves, Resources and Availability of Energy Resources 2005*. www.bgr.bund.de.

BP (British Petroleum) (2006). *Statistical Review of World Energy 2006*. www.bp.com.

BP (British Petroleum) (2007). *Statistical Review of World Energy 2007*. www.bp.com.

Brendow, K. (2003). Global oil shale issues and perspectives. *Oil Shale*, **20** (1), 81–92.

Bunger, J. W., Crawford, P. M. and Johnson, H. R. (2004). Is oil shale America's answer to peak-oil challenge? *Oil & Gas Journal*, **102** (30).

Campbell, C. J. (2006). *Regular Conventional Oil Production to 2100 and Resource Based Production Forecast*. (August 2006). www.oilcrisis.com/campbell.

Campbell, C. J. and Laherrère, J. H. (1998). The end of cheap oil. *Scientific American*, (March 1998).

CERA (Cambridge Energy Research Associates) (2006). *Why the 'Peak Oil' Theory Falls Down*. Decision Brief, November 2006. Cambridge, MA: CERA.

CfE (Canadian Centre for Energy Information) (2007). *Centre for Energy*. Calgary, Alberta. www.centreforenergy.com.

Chabrelie, M. F. (2002). *Prospects for Growth of the Gas Industry – Trends and Challenges*. The International Association for Natural Gas (Cedigaz). www.cedigaz.org.

CRS (Congressional Research Service) (2006). *Oil Shale: History, Incentives and Policy*. CRS Report for Congress (April 13). US Department of State. http://fpc.state.gov/fpc/65955.htm.

Deffeyes, K. S. (2001). *Hubbert's Peak: The Impending World Oil Shortage*. Princeton, NJ: Princeton University Press.

DOE (US Department of Energy) (2004). *Strategic Significance of America's Oil Shale Resource. Volume I: Assessment of Strategic Issues. Volume II: Oil Shale Resources Technology and Economics*. Washington, DC: Office of Petroleum Reserves, Office of Naval Petroleum and Oil Shale Reserves. www.fossil.energy.gov/programs/reserves/npr/publications.

DOE (US Department of Energy) (2007a). *Secure Fuels from Domestic Resources. The Continuing Evolution of America's Oil Shale and Tar Sands Industries*. Washington, DC: Office of Petroleum Reserves, Office of Naval Petroleum and Oil Shale Reserves. www.fossil.energy.gov/programs/reserves/npr/publications.

DOE (US Department of Energy) (2007b). *America's Strategic Unconventional Fuels. Volume I – Preparation Strategy, Plan and Recommendations*. Task Force on Strategic Unconventional Fuels. www.unconventionalfuels.org.

Elvidge, C. D. (2007). *A Twelve Year Record of National and Global Gas Flaring Volumes Estimated Using Satellite Data*. Final Report to the World Bank, Earth Observation Group. Boulder, Colorado: NOAA National Geophysical Data Centre. www.ngdc.noaa.gov/dmsp/interest/DMSP_flares_20070530_b.pdf.

EIA (Energy Information Administration) (2006). *Annual Energy Outlook 2006 with projections until 2030*. US Department of Energy. www.eia.doe.gov/oiaf/archive.html#aeo.

EIA (Energy Information Administration) (2008). *International Energy Outlook 2008*. US Department of Energy. www.eia.doe.gov/oiaf/ieo.

EWG (Energy Watch Group) (2007). *Crude Oil: The Supply Outlook*. EWG Series **3** (2007). www.energywatchgroup.org.

Fischer, P. A. (2001). Natural gas: how operators will bring 'worthless' gas to market. *World Oil*, **222** (11).

Fischer, P. A. (2005). Hopes for shale oil are revived. *World Oil*, **226** (8).

GAO (United States Government Accountability Office) (2004). *Natural Gas Flaring and Venting. Opportunities to Improve Data and Reduce Emissions*. Report GAO-04–809. Washington, DC. www.gao.gov/cgi-bin/getrpt?GAO-04-809.

Gerner, F., Svensson, B. and Djumena, S. (2004). *Gas Flaring and Venting: A Regulatory Framework and Incentives for Gas Utilization*. Public Policy Journal Note No. 279. Washington, DC: World Bank.

Gruson, J. F., Gachadouat, S., Maisonnier, G. and Saniere, A. (2005). *Prospective Analysis of the Potential Non-conventional World Oil Supply: Tar Sands, Oil Shales and Non-conventional Liquid Fuels from Coal and Gas*. Technical Report EUR 22168. European Commission, Joint Research Centre, Institute for Prospective Technological Studies (IPTS) and Institut Français du Pétrole (IFP).

Hirsch, R. L., Bezdek, R. H. and Wendling, R. M. (2005). *Peaking of World Oil Production: Impacts, Mitigation and Risk Management*. US Department of Energy (DOE). National Energy Technology Laboratory (NETL).

IEA (International Energy Association) (2004). *World Energy Outlook 2004*. Paris: OECD/IEA.

IEA (International Energy Association) (2005). *Prospects for Hydrogen and Fuel Cells*. IEA Energy Technology Analysis Series. Paris: OECD/IEA.

IEA (International Energy Association) (2006). *World Energy Outlook 2006*. Paris: OECD/IEA.

IEA (International Energy Association) (2008a). *World Energy Outlook 2008*. Paris: OECD/IEA.

IEA (International Energy Association) (2008b). *Energy Technology Perspectives 2008. Scenarios and Strategies to 2050*. Paris: OECD/IEA.

Jaccard, M. (2005). *Sustainable Fossil Fuels*. New York: Cambridge University Press.

Kägi, W., Siegrist, S., Schäfli, M. and Eichenberger, U. (2003). *Versorgung mit fossilen Treib- und Brennstoffen*. Bern: Bundesamt für Energie (BFE).

Kuuskraa, V. A. (2007). *A Decade of Progress in Unconventional Gas*. White Paper, Unconventional Gas Series, July (2007). Advanced Resources International, Inc. www.adv-res.com.

Laherrère, J. H. (2000). *The Hubbert Curve: Its Strengths and Weaknesses*. http://dieoff.com/page191.htm.

Laherrère, J. H. (2001). *Estimates of Oil Reserves*. Paper presented at the EMF/IEA/IEW meeting at IIASA, Laxenburg (Austria), June 19, 2001. www.oilcrisis.com/laherrere.

Laherrère, J. H. (2004a). *Future of Natural Gas Supply*. Paper presented at the ASPO Third International Workshop on Oil and Gas Depletion, Berlin, May 25, 2004. www.oilcrisis.com/laherrere.

Laherrère, J. H. (2004b). www.oilcrisis.com/laherrere/discovery/ProductionDiscovery.xls.

Laherrère, J. H. (2005). *Review on Oil Shale Data*. www.oilcrisis.com/laherrere.

Maugeri, L. (2004). Oil: Never cry wolf – why the petroleum age is far from over. *Science*, **304** (5674), 1114–1115.

MIT (Massachusetts Institute of Technology) (2007). *The Future of Coal*. An Interdisciplinary MIT Study. Cambridge: MIT. http://web.mit.edu/coal.

Moritis, G. (2006). CO_2 injection gains momentum. *Oil & Gas Journal*, **104** (15).

NEB (National Energy Board of Canada) (2004). *Canada's Oil Sands: Opportunities and Challenges to 2015*. www.neb-one.gc.ca.

NEB (National Energy Board of Canada) (2006). *Canada's Oil Sands: Opportunities and Challenges to 2015; an Update*. www.neb-one.gc.ca.

Oil & Gas Journal (OGJ) (2004). *Hubbert Revisited*. Different authors; series of six articles, *Oil & Gas Journal*, **102** (26) – **102** (31).

Oil & Gas Journal (OGJ) (2006a). Special report: oil production, reserves increase slightly in 2006. *Oil & Gas Journal*, **104** (47).

Oil & Gas Journal (OGJ) (2006b). Shell's interest in enhanced oil recovery grows. *Oil & Gas Journal*, **104** (43).

Porath, S. (1999). *Erzeugung von Chemierohstoffen aus Kukersit durch Pyrolyse*. Dissertation, University of Hamburg. www.sub.uni-hamburg.de/opus/volltexte/1999/23.

Queensland Government, Department of Infrastructure and Planning (QLD) (2007). *Projects: Stuart Oil Shale – Stage 2* www.dip.qld.gov.au/projects/energy/oil/stuart-oil-shale-stage-2.htm

Robelius, F. (2007). *Giant Oil Fields – The Highway to Oil*. Dissertation. Uppsala: University of Uppsala.

Sandrea, I. and Sandrea, R. (2007a). Global offshore oil-1: exploration trends show continued promise in world's offshore basins. *Oil & Gas Journal*, **105** (9).

Sandrea, I. and Sandrea, R. (2007b). Global offshore oil-2: growth expected in global offshore crude oil supply. *Oil & Gas Journal*, **105** (10).

Saniere, A., Hénaut, I. and Argillier, J. F. (2004). Pipeline transportation of heavy oils, a strategic, economic and technological challenge. Institut Français du Pétrole (IFP). *Oil & Gas Science and Technology*, **59** (5), 455–466.

Schindler, J. and Zittel, W. (2000). *Fossile Energiereserven (nur Erdöl und Erdgas) und mögliche Versorgungsengpässe aus Europäischer Perspektive*. Ottobrunn: Ludwig Bölkow Systemtechnik (LBST). www.lbst.de.

Schindler, J. and Weindorf, W. (2003). *'Well-to-Wheel' – ökologische und ökonomische Bewertung von Fahrzeugkraftstoffen und -antrieben*. Ludwig Bölkow Systemtechnik (LBST). www.HyWeb.de/Wissen/pdf/Nuernberg2003.pdf.

Shell (2008). *Shell Energy Scenarios to 2050*. www.shell.com.

Snyder, R. E. (2004). Oil shale back in the picture. *World Oil*, **225** (8).

SPE (Society of Petroleum Engineers) (2007). *Petroleum Resources Management System*. www.spe.org/spe-site/spe/industry/reserves/Petroleum_Resources_Management_System_2007.pdf.

The Economist (2006). *Special Report: National Oil Companies. The Economist*, (April 12, 2006).

Udall, R. (2005). *The Illusive Bonanza: Oil Shale in Colorado: Notes, References, Further Reading*. Association for the Study of Peak Oil & Gas (ASPO) – USA. www.aspo-usa.com/proceedings/News.cfm.

UNECE (United Nations Economic Commission for Europe) (2007). *United Nations Framework Classification for Fossil Energy and Mineral Resources*. www.unece.org/ie/se/reserves.html.

UNPD (United Nations Population Division) (2006). *World Population Prospects: The 2006 Revision; Population Database, Online database*. Department of Economic and Social Affairs, Population Division. http://esa.un.org/unpp.

USGS (United States Geological Survey) (2000). *World Petroleum Assessment 2000*. http://pubs.usgs.gov/dds/dds-060.

WEA (2000). Chapter 5 Energy resources. In *World Energy Assessment. Energy and the Challenge of Sustainability*, ed. Goldemberg, J., New York: United Nations Development Programme (UNDP), United Nations Department of Economic and Social Affairs (UNDESA), World Energy Council (WEC).

WEC (World Energy Council) (2007). *Survey of Energy Resources 2007*. www.worldenergy.org/publications.

WETO (2003). *World Energy, Technology and Climate Policy Outlook (WETO) 2030*. Report EUR 20366. Brussels: European Commission, DG Research.

WETO (2006). *World Energy Technology Outlook 2050. WETO* H_2. Report EUR 22038. Brussels: European Commission, DG Research.

Wikipedia Commons (2007). http://commons.wikinedia.org/wiki/Category:Oil_sands.

Williams, R. H. and Larson, E. D. (2003). A comparison of direct and indirect liquefaction technologies for making fluid fuels from coal. *Energy for Sustainable Development*, **VII** (4). Princeton: Princeton Environmental Institute, Princeton University.

Woynillowicz, D., Severson-Baker, C. and Raynolds, M. (2005). *Oil Sands Fever – The Environmental Implications of Canada's Oil Sands Rush*. The Pembina Institute: www.pembina.org/pubs.

WWF (World Wildlife Fund) (2008). *Unconventional Oil – Scraping the Bottom of the Barrel?* WWF, UK. http://assets.panda.org/downloads/unconventional_oil_final_lowres.pdf.

Further reading

BGR (Bundesanstalt für Geowissenschaften und Rohstoffe) (Federal Institute for Geosciences and Natural Resources) (2003). *Reserven, Ressourcen und Verfügbarkeit von Energierohstoffen 2002*. Rohstoffwirtschaftliche Länderstudien, Heft XXVIII. Hanover.

BGR (Bundesanstalt für Geowissenschaften und Rohstoffe) (Federal Institute for Geosciences and Natural Resources) (2007). *Reserves, Resources and Availability of Energy Resources 2005*. www.bgr.bund.de.

WEA (2004). *World Energy Assessment – Overview: 2004 Update*. New York: United Nations Development Programme (UNDP), United Nations Department of Economic and Social Affairs (UNDESA), World Energy Council (WEC).

WEC (World Energy Council) (2007). *Survey of Energy Resources 2007*. www.worldenergy.org/publications.

4

Non-renewable energy resources: nuclear fuels

Michael Ball and Felipe Andrés Toro

4.1 Nuclear fuels

4.1.1 Nuclear power today and outlook until 2030

Civilian use of nuclear power started with the opening of the first nuclear reactor in 1957 in the United Kingdom, generating approximately 50 MW_{el} in its first year. This picture has changed considerably since the beginning of the 1970s. In 2006, nuclear power contributed around 2700 TWh to 16% of global electricity generation (6% of primary energy use) (WNA, 2007). Worldwide, some 440 nuclear power plants are in operation in 30 countries, using the energy released by nuclear fission of the natural uranium radionuclide ^{235}U. (All commercial nuclear plants today use uranium as fuel (Olah *et al.*, 2006).) The total installed nuclear-generation capacity amounts to around 370 GW_{el}.

Three countries, namely the USA (104 plants), France (59 plants) and Japan, account for approximately 58% of the worldwide generation capacity, followed by Germany and the Russian Federation. These three countries also dominated the historical development of nuclear power expansion (see Fig. 4.1). The three countries with the highest nuclear energy share in their electricity mix today are France, with around 75%, followed by Lithuania, with 70%, and Slovakia, with 55%. While nuclear power contributes some 20% to power generation in the United States, the share in the EU25 is around 36%.

The historic growth of nuclear power can be divided into three broad periods: early growth (1957–1973), major expansion (1973–1990), and slow growth (1990 until today) (NEA/IAEA, 2006a). Accelerated by the first oil crisis, the period from 1973 until 1990 was the boom era for nuclear power, when over 300 plants were built (over 80% of the current nuclear capacity in the world), expanding capacity at an average of 16 GW_{el} per year. This rapid growth ended abruptly, mainly as a consequence of the Three Mile Island accident in the USA in 1979 and the Chernobyl disaster in the Ukraine in 1986. The annual increase in generating capacity between

The Hydrogen Economy: Opportunities and Challenges, ed. Michael Ball and Martin Wietschel. Published by Cambridge University Press.

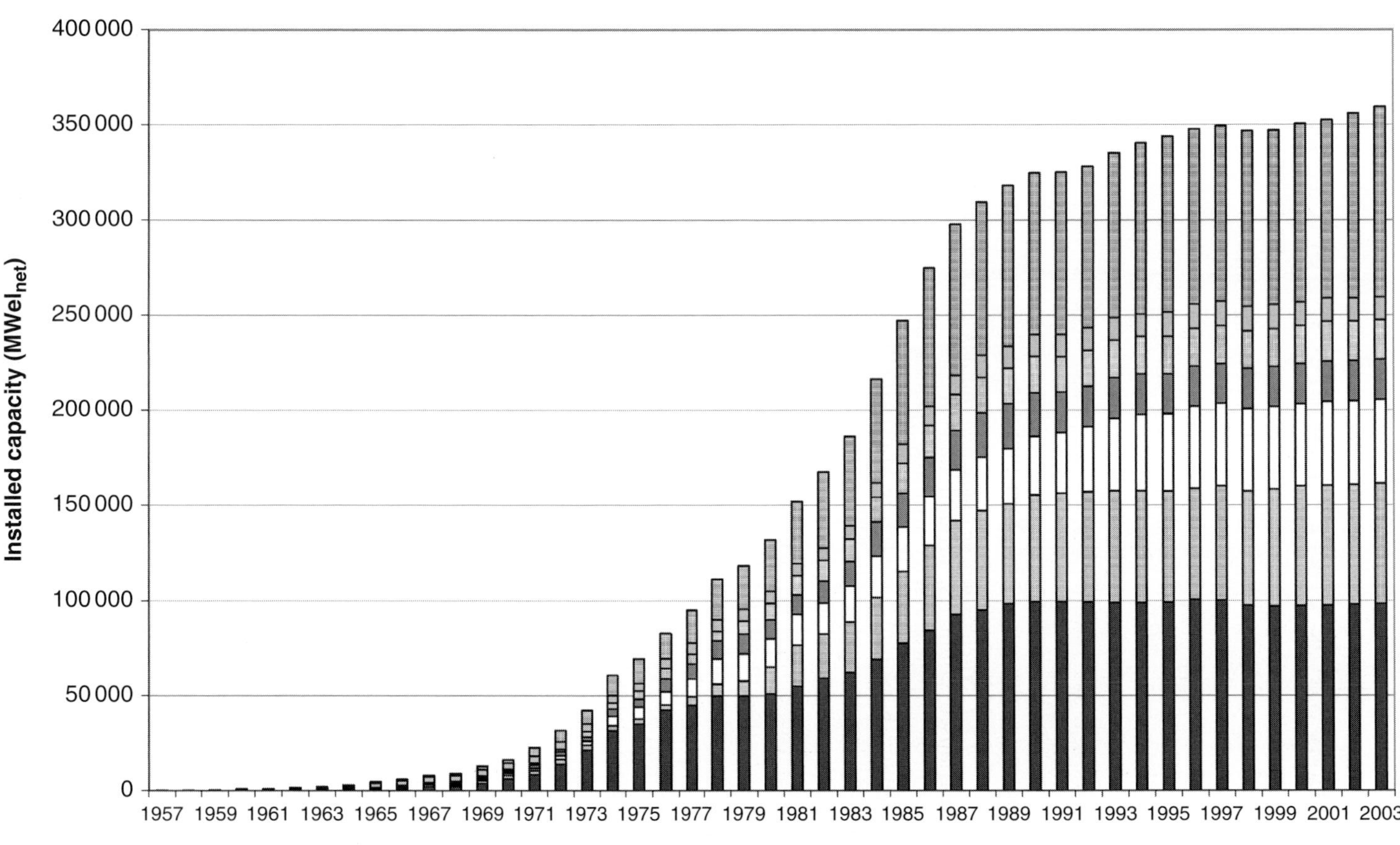

Figure 4.1. Development of nuclear capacities in selected countries (NEA/IAEA, 2006a).

1990 and 2003 averaged 2.3 GW_{el} only. Market liberalisation and cheap fossil fuels contributed to reducing nuclear power's attractiveness in the 1990s.

Owing to worldwide economic growth, coupled with a continuous strong increase in power consumption, the IEA projects in its Reference Scenario (see Chapter 3) an increase of world electricity demand of 2.6% per year, almost doubling the demand from 17 400 TWh in 2004 to 33 750 TWh by 2030 (IEA, 2006a). Transformation countries in Asia, especially China and India, are the main drivers for this growth. This is leading to a search for alternatives including, among others, the expansion at world scale of nuclear-energy programmes. As a result, various nuclear-power and uranium-exploration programmes and renewed production are taking place in some Western economies as well as in India, China and Central Asia. There are currently about 30 reactors under construction in more than 10 countries, notably China, India, South Korea, Japan and Russia.[1] While nuclear power is currently mainly used in industrialised countries, around half of the reactors under construction worldwide are in Asia (Sokolov and McDonald, 2006). Total lead time between the policy decision and commercial operation is between 7 and 15 years (IEA, 2006a).

Figure 4.2 illustrates the required installation of new nuclear-generating capacity according to different growth scenarios. The dark grey bars (1) show the present trend of the annual construction start of three new reactors on average, with 3 GW_{el}. Assuming an average construction time of five years and a decommissioning of old reactors after 40 years of operation,[2] the net capacity will decline by about 70 % until 2030 if present trends continue (EWG, 2006). The grey bars (2) indicate the annual construction start-ups necessary to maintain the present capacity of around 370 GW, which is represented by the grey line. The very light grey (3) bars indicate the annual construction start-ups necessary to meet the projection of the IEA Reference Scenario of 416 GW_{el} by 2030 (IEA, 2006a); the very light grey line provides the corresponding total capacity. In the Reference Scenario, nuclear capacity is projected to increase primarily in China, Japan, India and South Korea; however, the contribution of nuclear energy to the total electricity generation drops from 16% today to 10% in 2030. The light grey bars (4) indicate the annual construction start-ups necessary to meet the projection of the IEA Alternative Policy Scenario of 519 GW by 2030, with the light grey line showing the corresponding total capacity. (The MIT (2003) even describes a scenario where worldwide nuclear power generation could almost treble, to 1000 GW by 2050.)

Realising these growth scenarios seems very ambitious, particularly in the short term. At present, only three or four new reactors per year are completed. According

[1] The *World Nuclear Association* (WNA) provides frequent updates on reactors in operation, under construction, planned and proposed (for details see www.world-nuclear.org).

[2] Most nuclear power plants originally had a nominal design lifetime of up to 40 years, but engineering assessments of many plants over the last decade have established that many can operate longer. In the USA, nearly 50 reactors have been granted licence renewals that extend their operating lives from the original 40 to 60 years; in Japan, plant lifetimes up to 70 years are envisaged (WNA, 2007). At the end of 2005, eight nuclear power plants had been completely decommissioned and dismantled worldwide, with the sites released for unconditional use (IEA, 2006a).

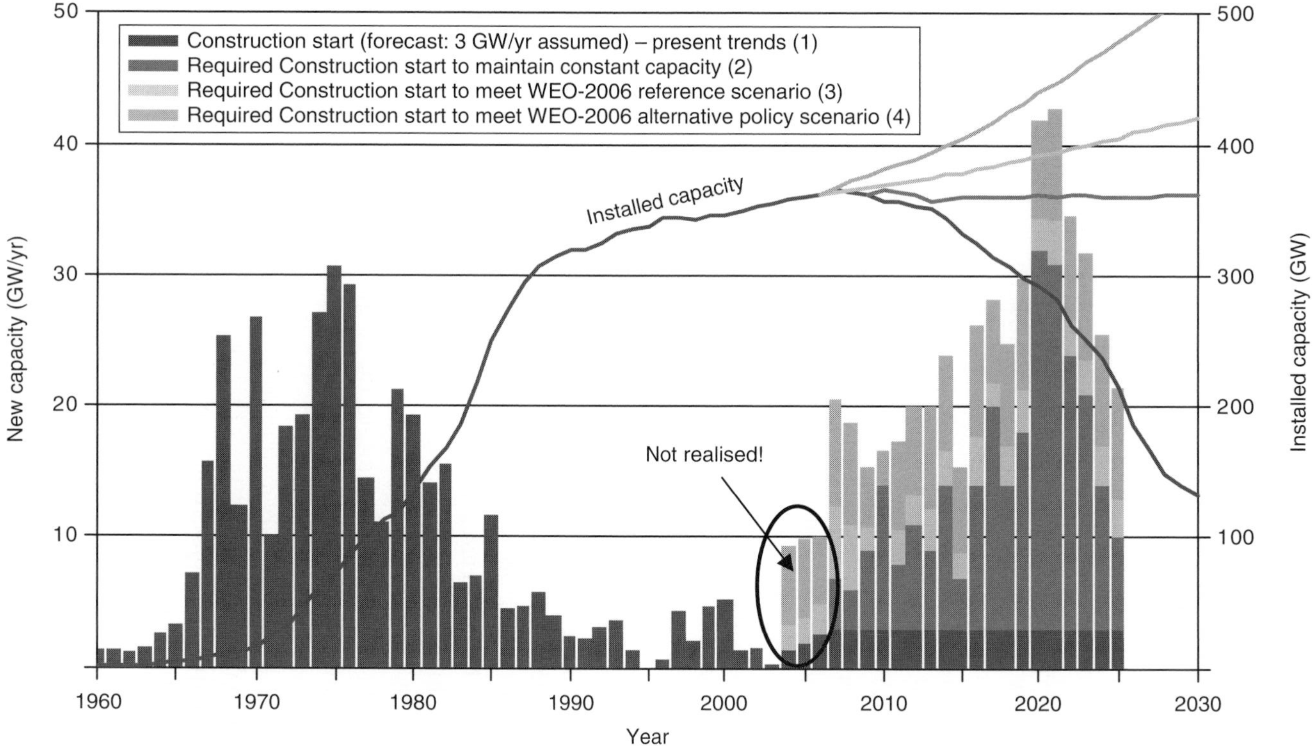

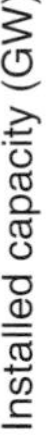

Figure 4.2. Installation of new nuclear capacity for different growth scenarios (EWG, 2006).

to the EWG (2006), this trend will continue at least until 2012, as over the last few years too few reactors started their construction to meet, for instance, the IEA Reference Scenario by then. For the IEA scenarios to materialise beyond 2012, between five and ten times more reactors than today must be constructed annually. Until about 2015, it is estimated that the long lead times of new reactors and the decommissioning of aging reactors are a barrier for fast extension. Merely maintaining the present reactor capacity will require the completion of 15–20 new reactors per year until 2030. In addition, current uranium production capacities would have to be increased by at least 30%.

4.1.2 New reactor technology

Nuclear reactors are classified by their neutron energy level (thermal or fast reactors), by their coolant (water, gas, liquid metal) and by their neutron moderator (light water, heavy water, graphite). Most existing plants are thermal reactors using pressurised (PWR) or boiling water (BWR) as a coolant and moderator; PWR and BWR together represent more than 80% of the commercial nuclear reactors today, of which PWR accounts for 60% alone (Olah *et al.*, 2006).

Reactor designs are broadly divided into four generations (Deutch and Moniz, 2006). The earliest prototype reactors, built in the 1950s and early 1960s, were often one of a kind. Most of the existing nuclear power plants today are Generation II light-water reactors, built in large numbers from the late 1960s to the early 1990s. The third generation, developed in the 1990s, includes evolutionary reactors with passive safety (reactor shutdown in accident conditions without active intervention), longer lifetime and modular design to reduce costs, licensing and construction time, and high fuel burn-up to optimise fuel use and minimise waste (IEA, 2007). Several such reactors have been built, mainly in East Asia. In the European Union, two third-generation EPRs (European Pressurised Water Reactor, also referred to as Evolutionary Power Reactor), of 1600 MW_{el} each, are planned in Finland and France. The Finnish reactor is expected to begin operation from 2011; construction of the French unit started at the end of 2007, with commissioning in 2012.

Future reactors – the fourth generation of nuclear reactors is expected to enter the market after 2030 (Abram and Ion, 2008). Generation IV reactors are being developed in an international co-operation framework – the Generation IV International Forum[3] – to improve safety and economic performance, to minimise nuclear waste and to enhance reliability and proliferation resistance. Six nuclear reactor technologies are being developed, believed to represent the future shape of nuclear energy. They are based on three general classes of reactor: gas-cooled, water-cooled

[3] More details can be found at http://gif.inel.gov. There is also the Global Nuclear Energy Partnership (GNEP) promoted by the US Department of Energy (see www.gnep.energy.gov) and the International Project on Innovative Nuclear Reactors and Fuel Cycles (INPRO) co-ordinated by the International Atomic Energy Agency (IAEA) (see www.iaea.org).

and fast-spectrum (Lake *et al.*, 2002). Most of the six designs employ a closed fuel cycle to maximise the resource base and minimise high-level wastes to be sent to a repository. Three of them are fast reactors. All of the new designs operate at higher temperatures up to 1000 °C, compared with less than 330 °C for today's light-water reactors. In particular, four of them are designed and can be used for thermochemical hydrogen production (see also Chapter 10, WNA, 2007). Between Generation III and IV reactors there are small-scale gas-cooled reactors, such as pebble-bed modular reactors, which introduce the interesting prospect of small-sized, modular nuclear-power plants, able to adapt to different needs.

With respect to the future availability of nuclear fuels, among the above concepts, fast breeder designs are of particular interest. In the following, their characteristics as compared with conventional reactor designs are described. For a better understanding, the nuclear fuel options are addressed first.

Naturally occurring uranium (U) consists primarily of a mixture of two isotopes: ^{235}U and ^{238}U. However, only ^{235}U, which makes up merely 0.7% of natural uranium, is fissile.[4] Although some reactors are able to use natural uranium directly, the vast majority of reactors require a higher concentration of ^{235}U, typically in the 3–5% range, which is achieved through enrichment of natural uranium. (In nuclear weapons, by contrast, the enrichment level generally needs to be greater than 90%.) Besides ^{235}U, other fissile materials usable for practical nuclear energy production are plutonium (^{239}Pu) and ^{233}U, which must be made from ^{238}U and thorium (^{232}Th), respectively. The latter are far more abundant than naturally occurring fissile ^{235}U, but not fissile themselves. The process of converting these 'fertile' materials (by means of neutron absorption) into 'fissile' materials is called 'breeding'.

Principally, open and closed nuclear fuel cycles can be distinguished. In an *open cycle*, also referred to as a once-through cycle, the uranium is burned once in a reactor and the spent fuel discharged from the reactor is treated as waste and – after interim storage – eventually disposed of in a geological repository.[5] This approach, which is applied by the majority of commercial reactors, uses only 1% of the uranium's energy content, as only ^{235}U contributes by fission to the production of energy (Lake *et al.*, 2002). The spent fuel consists of about 95% ^{238}U (which is left almost untouched), but still contains about 1% ^{235}U as well as about 1% ^{239}Pu (plutonium), both fissile material, which can further produce energy (the rest being mainly fission products, Olah *et al.*, 2006). To make better use of these resources and to reduce the amount of radioactive material to be stored, a *closed fuel cycle* can be used. Fuel can be recycled in thermal reactors, or in fast-breeder reactors.[6]

4 The energy released by one gram of ^{235}U that undergoes fission is equivalent to about 2.5 million times the energy released in burning one gram of coal (MIT, 2003).

5 At present, no country in the world has yet implemented a system for permanently disposing of the spent fuel (Deutch and Moniz, 2006). Since 1979, a salt dome in Gorleben (northern Germany) has been under investigation for final storage of nuclear waste. In 2000, a moratorium stopped the work for a period of three to ten years. To date, around € 1.3 billion have been invested in the Gorleben project.

6 A thorough analysis and evaluation of different fuel cycles with regard to economics, environmental impacts, nuclear waste management and proliferation risk is given by the MIT (2003).

Thermal-reactor recycle, which is the closed fuel cycle currently practised, requires a reprocessing plant. In a reprocessing facility, the useful ^{235}U is separated and sent back to the enrichment plant. The ^{239}Pu can also be separated and made into mixed oxide (MOx) fuel, in which uranium and plutonium oxides are combined. The use of MOx leads to a reduction of new fuel required of up to 30%, compared with a once-through fuel cycle (MIT, 2003). However, the plutonium produced from reprocessing represents a potential risk as it can be diverted for use in nuclear weapons (Deutch and Moniz, 2006; it takes about 10 kg of nearly pure ^{239}Pu to make a nuclear bomb). France, Japan, Russia, and the UK have reprocessing plants in operation. Mixed oxide as fuel is commonly used in reactors in Germany, France, Belgium, the UK, Russia and Japan; the total capacity of reactors using MOx amounts to about 27 GW_{el} (MIT, 2003). The economic viability of thermal reactors with reprocessing in a closed fuel cycle generally depends strongly on the amount of uranium resources available at economically attractive prices.

In recent years, *fast-breeder reactors* (FBR) have received renewed attention because their fast neutrons can convert ^{238}U into ^{239}Pu and produce fuel in excess of their own consumption. A fast-breeder reactor is a fast-neutron reactor[7] capable by design to breed fuel by producing more fissile isotopes than it consumes. In a thermal reactor, the fast (high-energy) neutrons generated in the fission reaction are slowed down by moderators to increase the probability of collision between these slow neutrons and the fissile ^{235}U, and thus to increase the amount of energy generated by fission. Fast neutrons, however, have the ability to convert ^{238}U, which does not directly undergo fission and represents around 95% of the nuclear waste, to ^{239}Pu. With plutonium as fuel, fast reactors produce more neutrons per fission than from uranium, sufficient not only to sustain the chain reaction but also to convert ^{238}U in a 'fertile blanket' around the core into fissile plutonium. In other words, the fast-breeder reactor 'burns' and can 'breed' plutonium.[8] To recover the plutonium, reprocessing of the blanket material is required. The produced plutonium is made into MOx fuel to be used as fuel in thermal reactors or fast breeders. Fast-breeder reactors make it possible to provide a growing energy resource that does not require a continuing supply of ^{235}U or ^{239}Pu after an initial input of fissile fuel at the start-up; after the initial fuel charge, the reactor can be refuelled by reprocessing. This could increase the energy extracted from natural uranium by a factor of 30 or more (IEA, 2007).

[7] In fast (neutron) reactors, the fission chain reaction is sustained by fast neutrons, unlike in thermal reactors. Thus, fast reactors require fuel that is relatively rich in fissile material: highly enriched uranium (> 20%) or plutonium. As fast neutrons are desired, there is also the need to eliminate neutron moderators; hence, certain liquid metals, such as sodium, are used for cooling instead of water. Fast reactors more deliberately use the ^{238}U as well as the fissile ^{235}U isotope used in most reactors. If designed to produce more plutonium than they consume, they are called fast-breeder reactors; if they are net consumers of plutonium, they are called 'burners'.

[8] As mentioned before, besides ^{238}U, ^{232}Th, another 'fertile' isotope existing in nature, can also be transformed into fissile atoms, when bombarded with fast neutrons. By doing so, ^{232}Th eventually transforms into ^{233}U, which has similar properties to ^{235}U and can also be used a nuclear fuel. This route is particularly promising, as thorium is considered to be about three times more abundant in the Earth's crust than uranium (see Section 4.1.4).

Fast-breeder reactors were originally conceived to extend the world's uranium resources. However, significant technical and material problems were encountered, and also geological exploration showed in the 1970s that scarcity was not going to be a concern for some time. Owing to both factors, by the 1980s it was clear that FBRs would not be commercially competitive with existing light-water reactors. Also, the separated plutonium (from reprocessing used light-water reactor fuel), which was originally envisaged for FBRs, is now being used as MOx fuel in conventional reactors. Fast-breeder reactors operate in Russia, Japan and France (IEA, 2006b). Prototypes have also been built in India, the USA and the UK. China intends to built a prototype; India and Russia are building FBRs that might be described as commercial.

By recycling the fuel from fast reactors, FBRs can deliver much more energy from uranium while reducing the amount of waste that must be disposed of for the long term. They can also be used as plutonium burners to dispose of excess plutonium from dismantled weapons. Today there has been progress on the technical front, but the economics of FBRs (which also include reprocessing) still depend on the value of the plutonium fuel that is bred, relative to the cost of fresh uranium. A concern about a fast-reactor power-generation economy is that it would also bring reprocessing and large amounts of fissile material with weapons potential into commercial use. If the use of nuclear energy is to grow significantly, breeder reactor designs are necessary and will be one of the keys to increasing the sustainability of future nuclear energy systems (Lake *et al.*, 2002). However, substantial research and development is needed to work through daunting technical and economic challenges to make this scheme work (Deutch and Moniz, 2006).

4.1.3 Uranium

4.1.3.1 Production, consumption and trade

The production of uranium started in 1938 and was mainly for military purposes. Twenty-five years later after a series of transformations, which led to the change in use for military purposes, uranium became a strategic energy resource. The historic production of uranium dates back to 1945 when approximately 500 tonnes of uranium where extracted. By 1965, the production amounted to 31 500 tonnes, according to the first statistics published at that time (NEA/IAEA, 2006a). During the 1980s, uranium production had a peak of over 70 000 tonnes coming from 22 countries. Production in 2005 amounted to almost 42 000 tonnes. From 1945 until 2005 about 2.3 million tonnes have been produced worldwide. Figure 4.3 shows the historic worldwide uranium production of various countries from 1950 until 2005. The black line shows the demand from nuclear reactors, which increased considerably since the 1980s. Countries at the bottom of the graph have already exhausted their reserves (EWG, 2006).

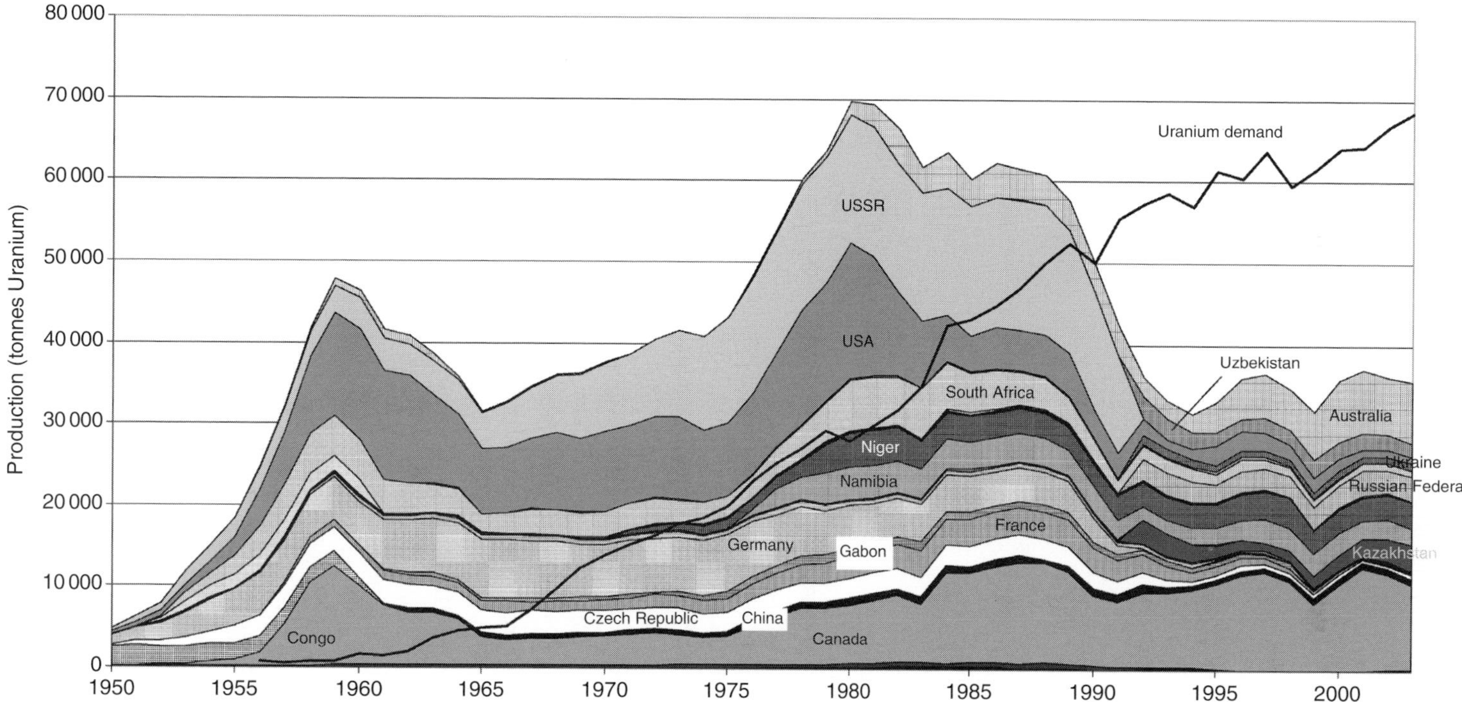

Figure 4.3. Development of worldwide uranium production and demand (NEA/IAEA, 2006a).

Table 4.1. *Worldwide uranium production in 2005*

Country	Uranium production (t)	Percentage of world production (%)
Canada	11 600	27.7
Australia	9 510	22.7
Kazakhstan	4 360	10.4
Russia	3 431	8.2
Namibia	3 147	7.5
Niger	3 093	7.4
Uzbekistan	2 300	5.5
USA	1 219	2.9
Ukraine	1 039	2.5
China	750	1.8
South Africa	674	1.6
Czech Republic	400	1.0
India	230	0.5
Others	199	0.5
Total	**41 952**	**100**

Source: (NEA/IAEA, 2006b).

At present, most of the uranium mine production is dominated by a small number of countries. In global terms, Canada produced roughly 28% of the total production, followed by Australia with around 23% and Kazakhstan with 10%; these three countries together account for more than 60% of global production. Very few production sites remain in countries like Germany, the USA and France, which are the biggest consumers. France terminated production in 2001, as economic resources were depleted; there is no uranium production in Japan. According to the WNA (2007), in 2006, 41% of the uranium were produced by open-pit mining, 26% by the *in-situ* leach process (ISL), 24% by underground mining, and 9% as by-products of the mining of gold, copper or other minerals (e.g., in South Africa). The time to bring a deposit into production after its discovery takes about 20 to 30 years. Table 4.1 summarises the worldwide uranium production in 2005.

Unlike other primary energy sources, the global consumption of uranium exceeds its production. Worldwide consumption in 2005 amounted to 66 500 t U (NEA/IAEA, 2006b).[9] As Fig. 4.3 shows, newly mined and processed uranium (*primary supply*) exceeded reactor-related uranium requirements until 1991. Since 1991, the gap between primary supply and uranium demand has been filled by *secondary supply*, i.e., material that has been held in inventory (both civilian and military in origin) or has been reprocessed. Besides reprocessed uranium and plutonium of spent

[9] The uranium consumption per power plant in Europe varies between 120 and 170 t U per year. One tonne of uranium yields between 40 000 and 50 000 MWh of electricity.

reactor fuels, secondary supply largely consists of weapons-grade uranium and plutonium declared surplus to military requirements in the USA and the Russian Federation and made available for use as civil fuel.[10] Secondary supply started to become an important resource to satisfy demand in the early 1980s. In 2003, total uranium requirements were met almost equally by primary and secondary supply; in 2005, primary supply made up 63% of total consumption.

The most important uranium consumers are depicted in Fig. 4.4. The USA account for almost 30% of the global consumption, followed by France with 16%, Japan with 11%, Germany with 6% and Russia with 5%.

The two main producers, namely Canada and Australia, which represent half of the uranium produced worldwide, export mostly to USA, Japan, France and South Korea. Canada, Russia, Niger and Australia remain the largest suppliers of nuclear materials to the EU (ESA, 2007). Uranium spot market prices have increased tremendously in the last couple of years, from around US\$20/kg U_3O_8 at the beginning of 2003 to almost \$300/kg at the beginning of 2007; at the time of writing (December 2007), prices were at around \$200/kg U_3O_8 (\$230/kg U).[11] It is difficult to identify an explanation for this behaviour. As mentioned before, there is a gap between supply and demand for uranium and therefore this situation contributes to create a very strong speculative attitude reflected in soaring prices. In addition, the expectations that demand will increase with new reactors being built in the coming years is also contributing to the development of prices in this way. But high uranium prices are likely to trigger new exploration and production.

4.1.3.2 Resources

The classification of uranium resources is not uniform across national and international organisations and definitions also change from time to time. These definitions are also different from other energy resources, such as oil, gas and coal. The (latest) reference system introduced by the *Nuclear Energy Agency (NEA)* and the *International Atomic Energy Agency (IAEA)*, which is frequently used across the world, is also used in the following analysis.[12] In addition to a differentiation based on the availability and geological certainty of the existence of the resource, cost classes are

[10] Uranium coming from the decommissioning of nuclear weapons under disarmament pacts as well as uranium and plutonium from the reprocessing of fuel rods will continue to play a role in the future. But the extent of deployment of these sources will depend on political decisions. According to the NEA estimates, the highly enriched uranium from the dismantling of nuclear weapons amounts to 249 500 tonnes; the Federal Institute for Geosciences and Natural Resources (BGR, 2003) indicates the total uranium available from military stockpiles at 358 000 to 408 000 tonnes. Another major potential secondary supply source is depleted uranium or enrichment tails (the fraction remaining after the uranium enrichment), which could provide some 450 000 tonnes of uranium, if it were enriched (NEA/IAEA, 2006a). However, this potential will only be realised if there is surplus enrichment capacity with relatively low operating costs; at present, only Russia is re-enriching tails.

[11] Most uranium, however, is bought on long-term contracts, and between 2000 and 2006 medium- and long-term uranium prices under existing contracts only increased by 20%–45% (WEC, 2007).

[12] There are other classification systems in use, such as in Canada, Germany (BGR), Australia, United States (DOE) and the Russian Federation; there is further the United Nations Framework Classification (UNFC) (EWG, 2006; NEA/IAEA, 2006a).

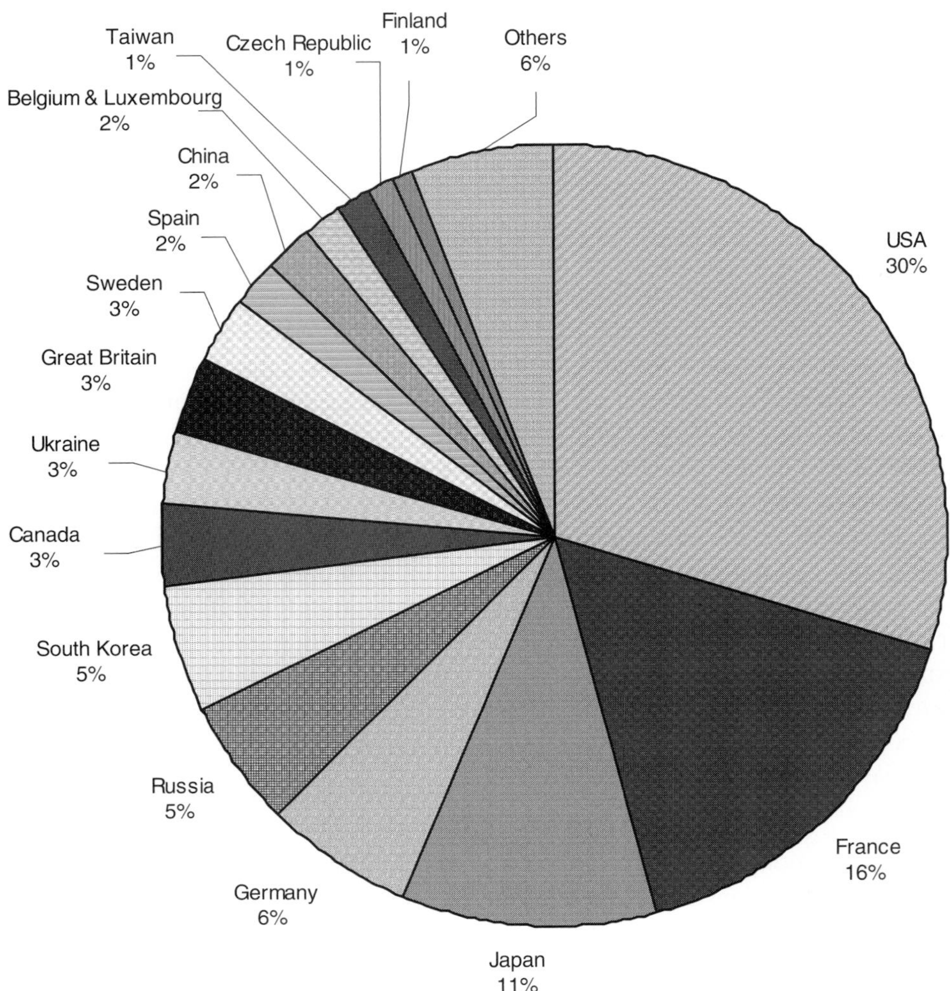

Figure 4.4. Worldwide uranium consumption in 2005 (BGR, 2007).

added as an indicator of the extraction costs, which are likely to be different for each kind of resource.

The NEA/IAEA system splits resources into 'known resources' and 'undiscovered resources'. 'Known resources' are further divided into 'Reasonably assured resources (RAR)' and 'inferred resources (IR)'. The categories are internally divided into various cost classes, according to suggested extraction costs as shown in Table 4.2. The definition of these classes also changed from time to time. The classes 'below \$40/kg U', 'below \$80/kg U' and 'below \$130/kg U' are the most widely used. Undiscovered resources are further subdivided into two categories, namely 'prognosticated' and 'speculative' resources. 'Prognosticated' resources are subdivided into

Table 4.2. *Global uranium resources*

Resource category	Extraction cost ranges ($/kg U)	Uranium resources (kt) Individual	Total	Data reliability
Reasonably assured resources (RAR)	<40	1 731	1 731	High
	40–80	727	2 458	↑
	80–130	711	**3 169**	
Inferred resources (IR)	<40	793	3 962	
	40–80	275	4 237	
	80–130	321	**4 558**	Low
Undiscovered resources				
Prognosticated	<80	1 475	6 033	
	80–130	780	6 813	
Speculative	<130	4 437	11 250	↓
	Unassigned	847	**12 097**	Very low

Sources: (NEA/IAEA 2006a; EWG, 2006).

two classes: 'below $80/kg U' and 'below $130/kg U'; 'speculative' resources are split into 'below $130/kg U' and 'unassigned'. Table 4.2 displays the total uranium resources according to this classification.

Reasonably assured resources below $80 per kg amount to 2.5 million tonnes of uranium, while RAR resources below $130 per kg amount to around 3.2 million tonnes. (This equals 1230 PJ and 1585 PJ, respectively, assuming that 1 tonne of uranium yields around 0.5 PJ (BGR, 2007).) IR below $80 per kg amount to roughly 1.1 million tonnes and below $130 per kg to approximately 1.4 million tonnes. Total RAR and IR sum up to almost 4.6 million tonnes (2280 EJ). Total undiscovered resources are estimated at 7.5 million tonnes. On top of these resources comes uranium from sources such as energy companies' stocks, nuclear arms uranium, etc.

Figure 4.5 illustrates the distribution of uranium resources among the ten major countries. Approximately 90% of the resources in all RAR and IR categories are to be found in those ten countries, with the leading ones being Australia, Kazakhstan, Canada and South Africa.

The quality and uranium content of the ore plays an important role when analysing uranium availability, as energy demand for uranium extraction increases steadily with lower ore concentrations. About 90% of world resources have ore grades below 1%, more than two thirds below 0.1% (1000 ppm). Today, only Canada is left with uranium deposits having an ore grade of more than 1%. Australia has, by far, the largest resources, but the ore grade is very low, with 90% of its resources containing less than 0.06%; Kazakhstan exhibits a similar situation with an ore concentration far below 0.1% (EWG, 2006). With

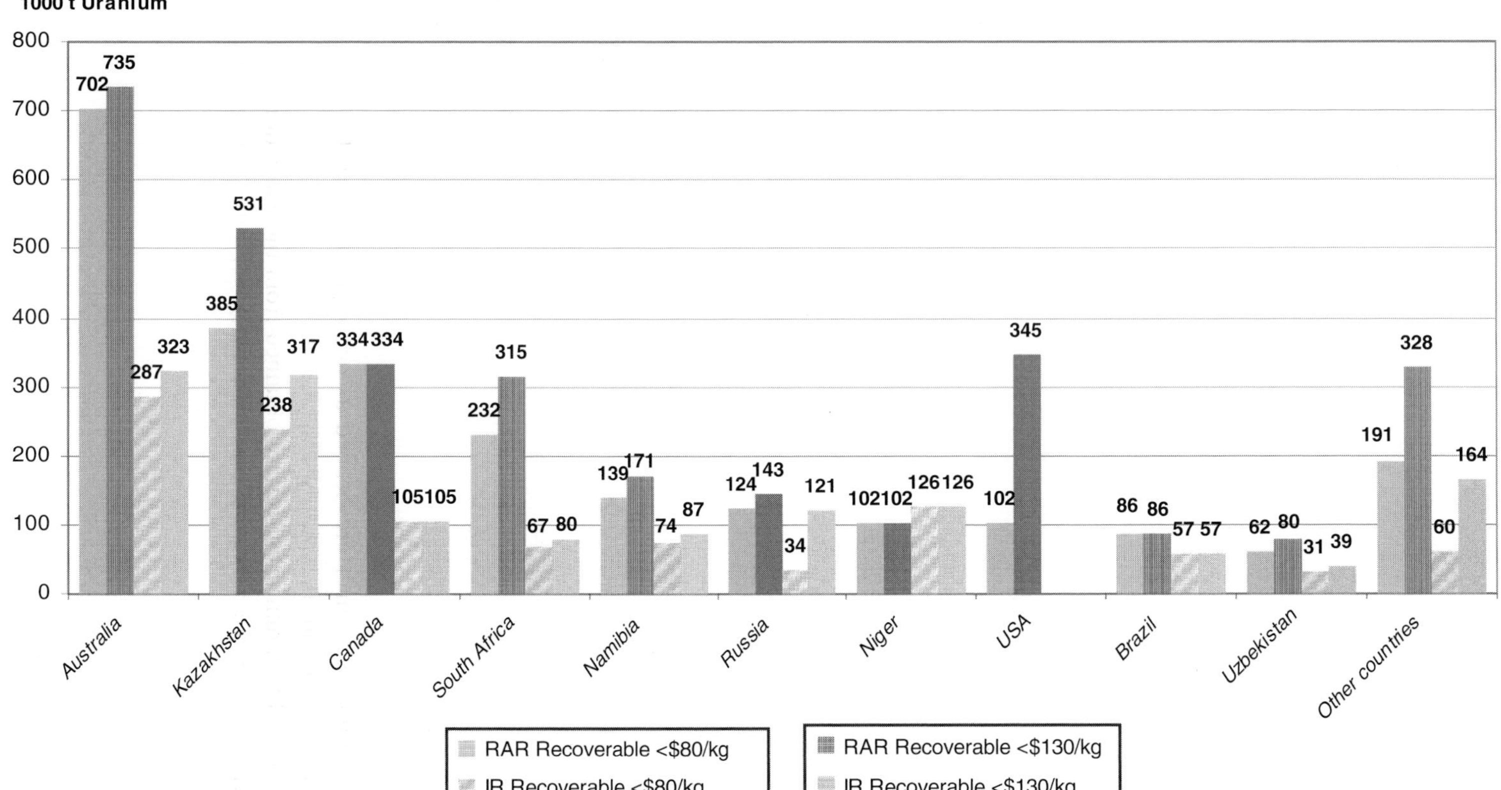

Figure 4.5. Top ten countries' uranium resources in 2003 (NEA/IAEA, 2006a).

concentrations below 0.01–0.02%, the energy needed for uranium extraction and processing is so high that the energy needed for supplying the fuel, operation of the reactor and waste disposal comes close to the energy that can be gained by burning the uranium in the reactor (EWG, 2006).[13]

With respect to the question of how long will uranium resources and reserves last, three demand scenarios are analysed. The first one assumes the current uranium demand of 66.5 kt U to be constant until 2030 (and hence no extension of nuclear power capacity), summing up to around 1700 kt over the entire period (starting in 2005); assuming a remaining 40 year lifetime of all current reactors, total consumption would add up to about 2600 kt U. A second scenario corresponds to the IEA Reference Scenario, which projects an increase in demand until 2030 to approximately 80 kt U per year, accumulating to 2000 kt until 2030 and to 4200 kt U over the lifetime of the reactors (IEA, 2006a). The third scenario is the IEA Alternative Policy Scenario, which for 2030 projects an annual demand of 100 kt U, 2200 kt in sum until 2030 and 5100 kt over the reactor lifetimes.

The first scenario would use up the entire RAR resources below US$80 /kg, leaving some scope for further use. The IEA Reference Scenario would use up all the RAR and IR resources, while the Alternative Policy Scenario would even need to make use of some of the prognosticated resources. A significant expansion of nuclear power beyond the lifetimes of the current reactors and the ones projected to be constructed until 2030 would, therefore, have to rely on tapping uranium resources that are still classified as undiscovered today. For the coming decades, the supply of uranium from nuclear weapons decommissioning and reprocessed nuclear waste will also continue to play an important role; however, this is limited to the political willingness to move forward on these decisions. To sustain nuclear power in the long term, closed fuel cycles, such as plutonium recycle, or fast-breeder reactors would have to be further developed and implemented, for instance, to make use of thorium as a nuclear fuel.

4.1.3.3 Unconventional uranium

Besides the conventional uranium resources, there are also the so-called 'unconventional uranium resources', which are defined as deposits with very low uranium content, from which uranium is typically only recoverable as a minor by-product. These unconventional uranium resources are obtained from the extraction of phosphates, non-ferrous ores and carbonatites, as well as black schist and lignite. It has to be noted that the distinction between conventional and unconventional resources is not entirely clear cut, but is, instead, somewhat transitional.

Historically, phosphate deposits with an average uranium content of 0.01% are the only type of deposit from which significant quantities have been recovered, mostly in

[13] It is very likely that most of the undiscovered prognosticated and speculative resources might refer to ore grades of below 0.02% (200 ppm) (EWG, 2006).

the USA and Morocco. Initial estimates indicate that with current high uranium prices unconventional resources obtained as a by-product from phosphate deposits could again become important for exploitation. According to the NEA, with high prices, the production of uranium from phosphate deposits could even exceed the production of uranium from gold production and, therefore, possibly become a conventional resource. The *Red Book 2003* indicates unconventional uranium resources from phosphate deposits to amount to roughly 22 million tonnes of uranium worldwide; the *Red Book* from 2006 (NEA/IAEA, 2006a), however, reports much lower figures, ranging from 7 to 7.2 million tonnes of uranium. In addition, various unconventional uranium resources considered in the past are not included, as it is likely that these are mineral inventories rather than rigorous resource estimates.

Another potentially vast resource is seawater. Uranium resources associated with the oceans are estimated at around 4000 million tonnes; however, the uranium concentration in seawater is only around 0.003 ppm. The recovery of uranium from seawater is still subject to basic research. Considerable technological developments as well as significant improvements of economics (or drastic increases in uranium prices) are crucial for the commercial use of this resource, which is unlikely in the foreseeable future. As the energy demand for uranium extraction increases with lower concentrations, the net energy balance of the entire fuel cycle is also critical.

4.1.4 Thorium

Thorium is a radioactive metal that occurs naturally in several minerals and rocks usually associated with uranium. However, it is approximately three times more abundant in nature than uranium. On average, soil contains 6 to 10 ppm of thorium. Thorium is most commonly found in the rare-earth thorium-phosphate mineral, monazite, which contains 8%–10% thorium. Current production of thorium is, therefore, linked to the production of monazite, which varies between 5500 and 6500 tonnes per year, with approximately 300 to 600 tonnes of thorium recovered (NEA/IAEA, 2006a).

Thorium, as well as uranium, can be used as a nuclear fuel. Although not fissile itself, thorium–232 (^{232}Th) can be used as a nuclear fuel through breeding to ^{233}U, which is fissile. Hence, like ^{238}U, it is fertile.

Thorium resources have been reported by the Nuclear Energy Agency until the late 1980s and after that some correlations have been made with uranium studies, as these are likely to be found in the ground together. Owing to a lack of exploration, detailed information on currently known deposits is limited and available resource estimates show large variations. Worldwide thorium resources listed in Table 4.3 are estimated to total about 6000 kt thorium, though no economic potential is implied for these resources (NEA/IAEA, 2006a). The NEA/IAEA 2005 *Red Book* reports reserves and additional resources of 4500 kt thorium, but this does not include a wide range of possible resources worldwide. The WNA (2007) reports the estimated world thorium

Table 4.3. *Worldwide thorium resources, 2005*

Country	Resources (kt Th)	Percentage (%)
CIS States	1 650	27.1
Brazil	1 306	21.5
Turkey	880	14.5
United States	432	7.1
Australia	340	5.6
India	319	5.2
Egypt	295	4.9
Norway	180	3.0
Canada	173	2.8
South Africa	115	1.9
Other	388	6.4
World total	**6 078**	**100**

Source: (NEA/IAEA, 2006a).

resources for the categories RAR+IR below $80 at 2500 kt thorium. Thorium has the same energy content per tonne as uranium (BGR, 2007).

Research and development activities for thorium fuel cycles have been conducted in Germany, the USA, India, Japan, Russia and the UK during the last 30 years at a much smaller scale than uranium and uranium–plutonium cycles. Nowadays, India, in particular, has made the utilisation of thorium a major goal in its nuclear power programme, as it has ambitious nuclear expansion plans and significant indigenous thorium resources.

The development of thorium-based nuclear power cycles still faces various problems and requires much more R&D to be commercialised. As a nuclear fuel, thorium could play a more important role in the coming decades, partly as it is more abundant on Earth than uranium and also because mined thorium has the potential to be used completely in nuclear reactors, compared with the 0.7% of natural uranium. Its future use as a nuclear source of energy will, however, depend greatly on the technological developments currently investigated in various parts of the world and the availability of and access to conventional uranium resources.

4.1.5 Nuclear fusion

Nuclear fusion is a physical phenomenon observed in the energy from the Sun, which is obtained from nuclear fusion reactions of small nuclei, primarily hydrogen isotopes to produce larger ones, mostly the transformation of hydrogen into helium. The resulting mass of the two atoms is smaller than the original one and the difference is represented in the form of energy in very large quantities. In the Sun and stars, owing to high gravitational forces, this fusion takes place in high-temperature plasma (15 million °C) in a sustained manner. However, these conditions are not the same on Earth, where natural gravitational confinement is impossible. Considerable energy is needed before fusion can occur, as the electrostatic repulsion between the positively charged protons has to be overcome.

For this reason, several technologies have been developed or are currently being investigated, to be able to produce electricity and heat in a controlled environment manner, preventing energetic particles from escaping before creating a suitable nuclear reaction. Among the most important technologies are *magnetic confinement*, in which very-high-temperature plasma is contained by a strong magnetic field for suitable periods of time, and *inertial confinement*, where fusion is realised in a small concentrated volume of plasma heated and compressed very rapidly with high energy lasers (Olah *et al.*, 2006).

For electricity production, the magnetic confinement option is preferred and widely used. It has been researched since the 1950s and major advances occurred at the end of 1960s, when the Russians obtained a very-high-temperature plasma in a so-called *Tokamak*, which has served as a basis for nuclear-energy research and fusion experiments in the 1980s. The temperature needed to maintain a fusion reaction in such systems is close to the order of 100 million °C. The plasma is injected with highly energetic neutral particles to reach such temperatures and is kept apart with magnetic fields that allow fusion to occur. The heat generation by this reaction is transferred to heat exchangers placed on the reactor walls. This heat produces steam and electricity, with the help of turbines and generator systems.

Nuclear fusion is still considered by many experts as a technology in development and for the future; although many advances have already been obtained, there is still the need for extensive research and development, especially on nuclear fuels combinations such as deuterium–tritium and deuterium–helium fusion, which are expected to reduce radiation impacts, and on power-generation systems. The fusion process offers various advantages if used as a process for energy production. Among the most important is the fact that the fuels (deuterium and lithium, isotopes of hydrogen) are widespread and abundant resources, playing an important role for energy security. Fusion could also be applied in large-scale facilities with the potential to generate electricity, heat and even hydrogen with very low emissions of air pollutants and greenhouse gases. There would be no danger of a runaway fusion reaction, as this is intrinsically impossible and any malfunction would result in a rapid shutdown of the plant (WNA, 2007).

The ITER R&D project is the latest international research initiative on nuclear fusion for power generation. The ITER initiative has partners from the European Union, including Switzerland, Japan, Russia, China, South Korea, India and the United States. The project is expected to run for the next 30 years, of which 10 years will be needed for construction of the reactor in Cadarache, France, and 20 years for operation. The cost is approximately €10 billion and the first plasma operation is expected to be in 2016 (ITER, 2007). The main objective of the ITER project is to demonstrate scientifically and technologically that fusion is feasible and can be used for energy production and for peaceful purposes.

Fusion power offers the prospect of an almost inexhaustible source of energy for future generations, but so far it has also presented insurmountable scientific and engineering challenges. Nuclear fusion is unlikely to play any role before 2050.

4.2 Summary

There is currently a 'renaissance' worldwide with respect to nuclear power to cover an increasing electricity demand, especially in emerging economies (largely China and India) as well as to contribute significantly to the reduction of CO_2 emissions. In the long term, if nuclear energy is to contribute a major part to global energy supply, the development of breeder reactors for commercial operation is crucial; otherwise, there are likely to be resource constraints.

However, future challenges faced by nuclear power also relate to operating security concerns, nuclear weapons proliferation issues and final waste management, which are reflected in a mixed public acceptance. These are crucial areas to be addressed and developed, if nuclear power should expand its share in electricity and hydrogen generation in the future.

References

Abram, T. and Ion, S. (2008). Generation-IV nuclear power: a review of the state of the science. *Energy Policy*, **36** (12), 4299–4668.

BGR (Bundesanstalt für Geowissenschaften und Rohstoffe) (Federal Institute for Geosciences and Natural Resources) (2003). *Reserven, Ressourcen und Verfügbarkeit von Energierohstoffen 2002*. Rohstoffwirtschaftliche Länderstudien, Heft XXVIII. Hannover.

BGR (Bundesanstalt für Geowissenschaften und Rohstoffe) (Federal Institute for Geosciences and Natural Resources) (2007). *Reserves, Resources and Availability of Energy Resources 2005*. www.bgr.bund.de.

Deutch, J. M. and Moniz, E. J. (2006). The nuclear option. *Scientific American*, **295** (3).

ESA (Euratom Supply Agency) (2007). *Annual Review 2006*. Luxembourg: Office for Official Publications of the European Communities.

EWG (Energy Watch Group) (2006). *Uranium Resources and Nuclear Energy*. EWG Series 1. www.energywatchgroup.org.

IEA (International Energy Association) (2006a). *World Energy Outlook 2006*. Paris: OECD/IEA.

IEA (International Energy Association) (2006b). *Energy Technology Perspectives 2006. Scenarios and Strategies to 2050*. Paris: OECD/IEA.

IEA (International Energy Association) (2007). *Nuclear power*. IEA Energy Technology Essentials No. 4. www.iea.org/Textbase/techno/essentials.htm.

ITER (2007). www.iter.org.

Lake, J. A., Bennett, R. G. and Kotek, J. F. (2002). Next-generation nuclear power. *Scientific American*, **286** (1).

MIT (Massachusetts Institute of Technology) (2003). *The Future of Nuclear Power. An Interdisciplinary MIT Study*. Cambridge: MIT. http://web.mit.edu/nuclearpower.

NEA (Nuclear Energy Agency) and International Atomic Energy Agency (NEA/IAEA) (2006a). *Forty Years of Uranium Resources, Production and Demand in Perspective. The Red Book Retrospective*. Paris: OECD.

NEA (Nuclear Energy Agency) and International Atomic Energy Agency (NEA/IAEA) (2006b). *Uranium 2005: Resources, Production and Demand*. Paris: OECD.

Olah, G. A., Goeppert, A. and Prakash, G. K. S. (2006). *Beyond Oil and Gas: The Methanol Economy*. Weinheim: Wiley.

Sokolov, Y. and McDonald, A. (2006). *Nuclear Power – Global Status and Trends*. Nuclear Energy 2006, Touch Briefings, 15–19. www.iaea.org/OurWork/ST/NE/Pess/publications.shtml.

WEC (World Energy Council) (2007). *Survey of Energy Resources 2007*. www.worldenergy.org/publications.

WNA (World Nuclear Association) (2007). www.world-nuclear.org.

5

Assessment of the potentials for renewable energy sources

Gustav Resch, Anne Held, Felipe Andrés Toro and Mario Ragwitz

In this chapter, an assessment of the global and European potentials of renewable energy sources is given. With respect to hydrogen as a promising future energy carrier, a clear focus is put on electricity generation from renewable energy sources (RES), serving as a sustainable solution for producing hydrogen based on electrolysis (see also Chapter 16). Biomass gasification and solar-thermal technology provide further options to produce hydrogen using RES. In the following, first, different categories of renewable energy potentials are defined, followed by a general overview of the potentials on a global level and a more detailed picture of the European potentials of renewable energy sources. Finally, the global potential for biofuels is shortly addressed.

5.1 Potential categories

The possible use of RES depends in particular on the available resources and the associated costs. In this context, the term 'available resources' or RES potential has to be clarified. The RES potential might be represented by the overall theoretically available resources or might take into account different aspects that restrict the theoretically available resources. Subsequently, the RES-potential categories used within this book are described.

- *Theoretical potential* The highest potential of an energy source is the theoretical potential. To derive the theoretical potential, general physical parameters have to be taken into account, based on the determination of the energy flow resulting from a certain energy resource within the investigated region. Examples are the kinetic energy of the wind, the total energy content of the existing biomass or the amount of solar energy radiated to a defined territory during one year. The theoretical potential consists of the overall physical energy supply available and represents a theoretically upper limit for the use of RES.
- *Technical potential* The technical potential comprises the share of the theoretical potential that is technically feasible. That means that the technical potential considers technical and

The Hydrogen Economy: Opportunities and Challenges, ed. Michael Ball and Martin Wietschel. Published by Cambridge University Press.

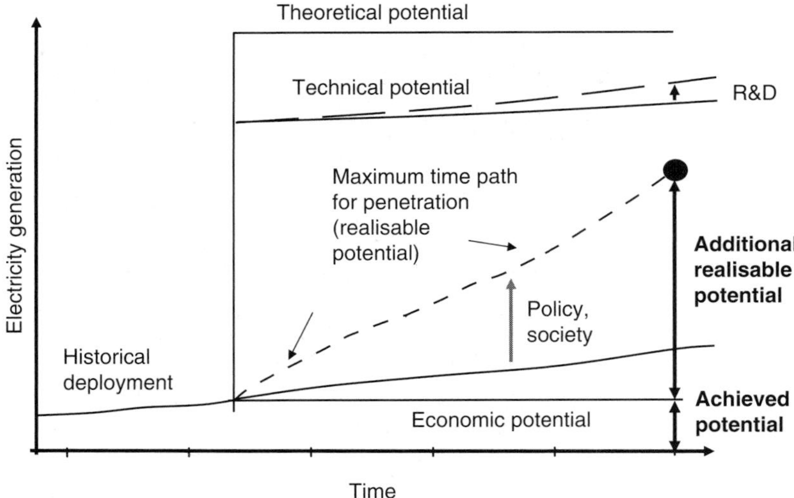

Figure 5.1. Methodology for the definition of potentials.

geographical restrictions, such as available conversion technologies, their efficiencies, availability of locations (e.g., possible locations to install wind turbines), competitive use of RES (e.g., biomass resources) or other limitations. The technical potential changes as technology develops over time. The theoretical potential is an upper limit of the technical potential.

- *Realisable potential* The realisable potential represents the maximum potential that can be exploited up to a certain year considering dynamic realisation restrictions, such as maximum market growth rates and planning constraints, as well as political and societal drivers and assuming that sufficient economic incentives exist. The technical potential is an upper limit of the realisable potential.

Figure 5.1 illustrates graphically the general concept of the realisable potential up to 2020, as well as the technical and the theoretical potential.

Renewable energy technologies for electricity generation (RES-E), as considered in the following, include biogas, biomass, biowaste, onshore wind, offshore wind, small-scale hydropower, large-scale hydropower, solar thermal electricity, photovoltaics and tidal and wave energy, as well as geothermal electricity. Primary biomass potentials include forestry products and residues and agricultural products and residues as well as biodegradable waste. Detailed definitions of these renewable energy sources are given in the glossary at the end of this chapter.

5.2 The global potential for renewable energy sources (RES)

Solar energy is the most abundant permanent energy resource on Earth and it is available for use in its direct (solar radiation) and indirect (wind, biomass, hydro, ocean, etc.) forms. The total annual solar radiation falling on the Earth is more than

Table 5.1. *Annual global primary technical and theoretical energy potentials for various renewable energy sources in 2004 (physical energy content method)*

Resource	Current use (2004) (EJ)	Technical potential (EJ)	Theoretical potential (EJ)
Biomass energy	50.0	250	2 900
Geothermal energy	2.0	5 000	140 000 000
Hydropower	10.0	50	150
Ocean energy	–	–	7 400
Solar energy	0.2	1 600	3 900 000
Wind energy	0.2	600	6 000
Total	**62.4**	**7 500**	**143 916 450**

Sources: (IEA, 2007; Johansson *et al.*, 2004; Rogner *et al.*, 2004).

7500 times the world's total annual primary energy use (WEC, 2007). The analysis of global RES potential shows that there is an important and enormous potential (technical and theoretical), especially from solar and geothermal resources as well as from ocean, wind and biomass resources (see Table 5.1). However, this enormous potential is constrained by serious economic barriers, which hinder its current and future contribution to the world energy supply. Further developments on technologies are required, but policy support is also necessary to guarantee conditions for sustained investments in this emerging market.

Renewable energy sources (RES) worldwide, including combustible renewables and waste (CRW), hydro, geothermal, solar, wind, tide and wave energy, amount approximately to 13% of the world's *total primary energy supply (TPES = 468 EJ)*, following oil, coal and gas, as shown in Fig. 5.2 (IEA, 2006a) (see also Section 3.1).[1] Currently, the mostly used RES in the world is solid biomass, owing to its widespread non-commercial use in developing countries, representing almost 11% of the world's TPES or 79% of the global renewable energy supply. Hydropower has the second-largest share, with about 2% of world's TPES or roughly 17% of global renewables. Looking at the remaining RES, geothermal energy is identified as the third-largest, with approximately 0.4% of the world's TPES and 3.2% of renewable supply. The other emerging technologies, such as wind, solar, tide and wave represent less than 0.1% of the current TPES.

In the period between 1971 and 2004, RES have experienced an average annual growth of 2.3% per year, as illustrated in Fig. 5.3. This is slightly above the growth in total primary energy supply in the same timeframe worldwide. The categories hydro and CRW show similar growth patterns, while emerging 'new' renewable energy technologies, such as wind, solar and geothermal technologies, exhibit a much higher

[1] Total primary energy supply is calculated using the IEA conventions, among others the physical energy content method for renewable sources (IEA, 2006a; 2007) (see also Section 3.1).

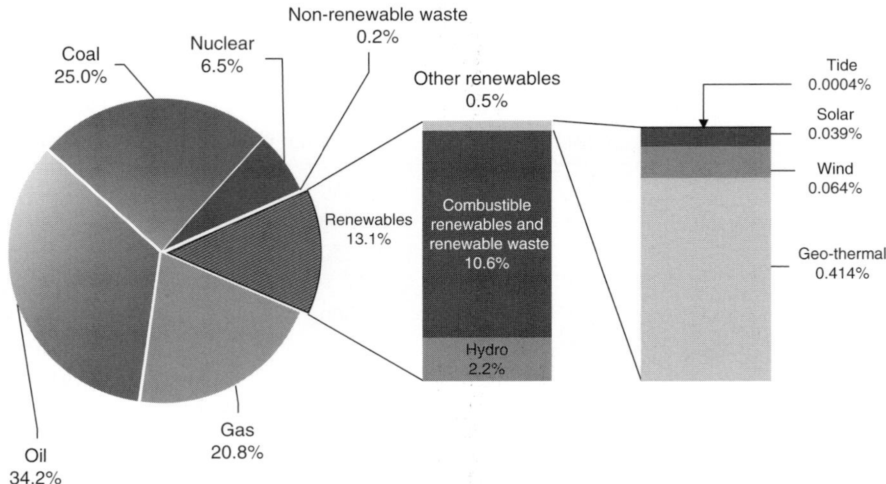

Figure 5.2. Fuel shares including renewable energy sources (RES) of world total primary energy supply in 2004 (IEA, 2006b; 2007).

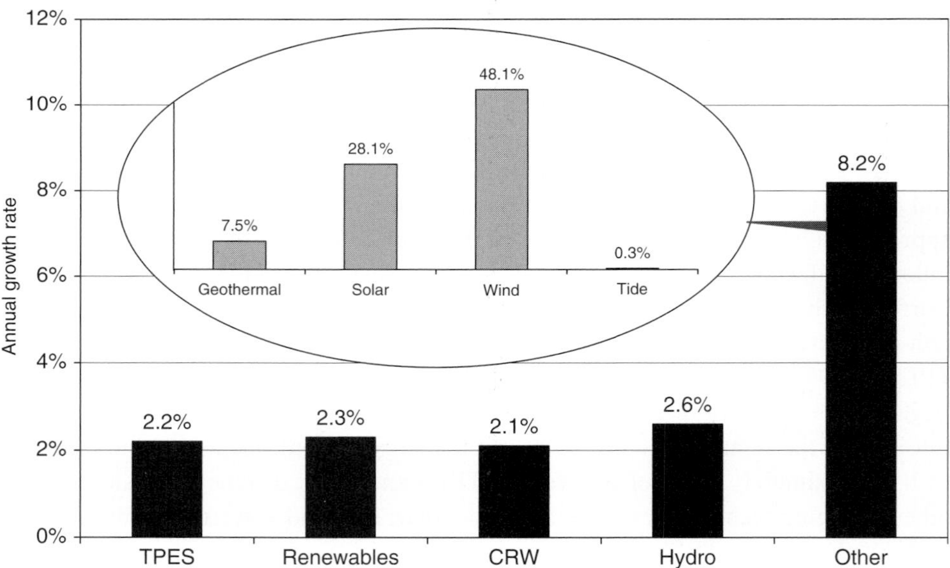

Figure 5.3. Annual growth of renewable energy sources (RES) from 1971 to 2004 (IEA, 2006b; 2007).

annual growth (on average 8.2% per year), mostly in the last decade (IEA, 2007; Martinot *et al.*, 2006). As emerging technologies, these are likely to record higher annual growth rates in the first phase of penetration to the markets, as the initial values are low or zero.

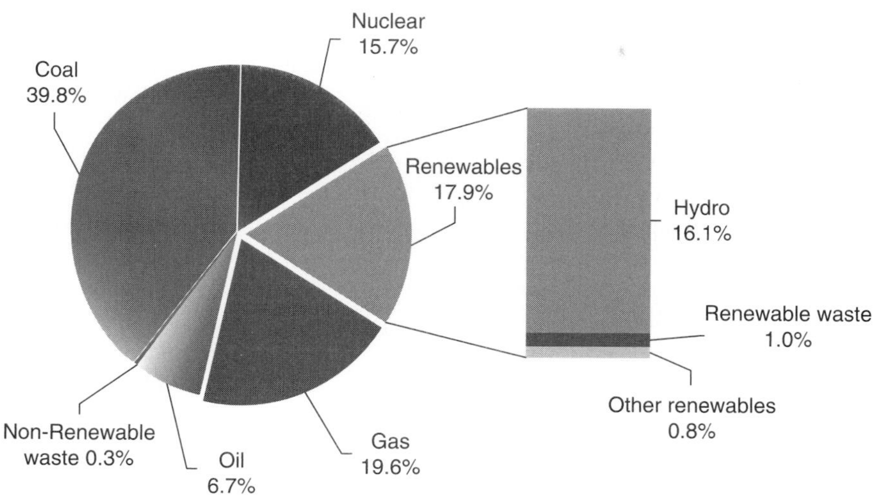

Figure 5.4. Renewable Energy Sources (RES) in electricity production in 2004 (IEA, 2007; IEA, 2006b).

With respect to electricity generation, RES amounted to 3257 TWh in 2004 on the global scale, equal to a share of almost 18% of total electricity production; following coal and gas, but with a larger share in electricity generation than that of nuclear and oil (see Fig. 5.4). Compared with 1971, the RES share shows a decrease of 5%, meaning that electricity demand grew faster than RES generation. As shown in Fig. 5.4, hydropower represents the dominating RES-E technology, holding a share of almost 90% of total RES electricity generation in 2004. Furthermore, biomass (CRW) exhibits a high increase, reaching almost a 6% share while the remaining 'new' renewable technologies, in particular wind power, represent roughly 4.5% of the total electricity production from renewable sources (IEA, 2006b; 2007; Martinot *et al.*, 2006).

Table 5.2 displays the installed power generation and heating capacities of renewable energies in 2006 worldwide. The total renewable power capacity amounted to some 930 GW, of which large hydropower had a share of 80%. In particular, the wind energy sector has experienced a tremendous growth: the worldwide generating capacity of wind turbines has increased by more than 25% a year, on average, for the last decade, reaching 75 GW at the end of 2006; offshore capacity currently only stands at close to 1 GW. Including off-grid applications, total photovoltaic power amounted to 7.8 GW; Germany's cumulative PV capacity exceeded Japan's for the first time in 2005. The installed capacities of solar-thermal and ocean power are negligible. Total investment in new renewable capacity in 2005 amounted to about US$38 billion (up from US$30 billion in 2004) (CERA, 2007; Martinot *et al.*, 2006; 2008).

The global potential for RES as an indication of long-term availability of the resource can be defined as a theoretical potential (e.g., theoretical maximum), which

Table 5.2. *Total installed renewables power generation capacity in 2006*

Power generation	
Large hydropower	770 GW
Small hydropower	73 GW
Wind power capacity	75 GW
Biomass power	45 GW
Geothermal power	9.5 GW
Solar PV capacity, grid connected	5.1 GW
Solar PV capacity, off grid	2.7 GW
Solar thermal power	0.4 GW
Ocean (tidal) power	0.3 GW
Total	**981 GW**
Hot water or heating	
Biomass heating	235 GW_{th}
Solar collectors (hot water, space heating)	105 GW_{th}
Geothermal heating	33 GW_{th}
Total	**373 GW_{th}**

Source: (Martinot *et al.*, 2008).

is compared with what is known as resources for other types of fuel. This theoretical potential is shown for various RES types in Table 5.1. These renewable resources, when demanded for energy purposes, should, however, be analysed taking into account appropriate technological and economic possibilities for exploiting the resources available. The potential of renewable resources based on technological and engineering criteria is illustrated here as the technical potential (see definitions above).

The concept of technical potential can be used in a similar fashion to the concept of energy reserves (see also Chapter 3). The fundamental difference, of course, is that renewable potentials represent flows available, in principle, on an indefinitely sustainable basis, while fossil energy reserves and resources, although expanding in time, are fundamentally finite quantities. Life-cycle analyses remain important, because although the energy flows are sustainable, they still require materials like concrete and copper and the commitment of land and other resources. The renewable energy potentials identified in Table 5.1 are theoretically large enough to provide the current primary energy needs for the world and the technical potentials are large enough to cover most of the conceivable future growth in global energy demand (Johansson *et al.*, 2004).

Hydropower is the mechanical conversion of the potential and kinetic energy of water into electricity in hydroelectric plants. The geographical conditions of the regions, as well as detailed information on water conditions, such as available head or flow volume per unit of time, play a direct and important role in assessing the potential of hydropower. According to potential analysis in this respect, hydropower has a theoretical potential of some 150 EJ per year while the technical potential, taking into account technical, structural, legal and ecological restrictions for energy generation, amounts to 50 EJ or a third of the theoretical potential. Dynamic changes in this potential are possible, as rainfall variations across world regions could increase or decrease the annual power output. In this respect, it is expected that climate change influences the hydropower potentials in the long term (Lehner *et al.*, 2005).

Four general categories of *biomass* energy resource are used as energy fuels, including forestry biomass, agricultural biomass, waste and energy plantations. Biomass waste originates from farm crops, animals, forestry waste, wood processing by-products, and municipal waste and sewage. The potential of biomass energy crops and plantations depends on the land area available, the harvestable yield, its energy content and the conversion efficiency. Theoretical biomass energy potentials, as illustrated in Table 5.1, amount to 2900 EJ. The technical potential of biomass energy crops and plantations is especially difficult to estimate, as land-use patterns have a very dynamic character and there is competition between crops for different uses, such as food, material or energy. Estimates of biomass energy technical potentials are likely to be higher than 250 EJ, which corresponds to approximately 10% of the theoretical potential.

As observed in Table 5.1, *solar energy* has an immense theoretical potential, over almost 4 million EJ, as it reflects the vast areas intercepting solar radiation across the globe. However, large scale availability of solar energy depends greatly on a region's geographical position as well as weather conditions and primarily on the assumptions regarding land availability (the average solar irradiation is between 900 and 1700 $(kWh/m^2)/year$). Table 5.3 shows the results of the maximum and minimum potential assessment in primary energy terms for different regions of the world: the final energy will depend on the efficiencies of the solar technologies used in these regions.

The theoretical potential of *wind energy*, as illustrated in Table 5.1, amounts to 6000 EJ, which seems to be enormously high when compared with its current use. A technical potential is estimated to be 6–10% of the theoretical one, as given above (Hoogwijk *et al.*, 2004; Johansson *et al.*, 2004). The ultimate potential of wind-generated electricity worldwide could indeed be very large: other estimates state that the contribution of wind in 2030 could be from 5% to 30% of the world's electricity; for 2050, this figure varies from 7% to 35% (Grubb and Meyer, 1993; GWEC, 2006; Häfele, 1981; Rogner *et al.*, 2004). The height limitations of wind converters, the distance of offshore sites, insufficient wind velocities, grid constraints and land use all limit the practical potential. The average power density estimated to derive the technical potentials given above is 300–400 W/m^2.

Geothermal energy is the RES with the highest theoretical and technical potential worldwide, as shown in Table 5.1. The theoretical potential amounts to 140 million EJ and a much more limited technical potential of 5000 EJ. There are four types

Table 5.3. *Annual solar primary energy potentials*

Region	Minimum potential (EJ)	Maximum potential (EJ)
North America	181	7 410
Latin America and the Caribbean	112	3 385
Western Europe	25	914
Central and Eastern Europe	4	154
Former Soviet Union	199	8 655
Middle East and North Africa	412	11 060
Sub-Saharan Africa	371	9 528
Pacific Asia	41	994
South Asia	38	1 339
Central Asia	115	4 135
Pacific OECD	72	2 263
Total	**1 570**	**49 837**

Sources: (Nakicenovic *et al.*, 1998; Rogner *et al.*, 2004).

Table 5.4. *Annual technical geothermal primary energy potentials*

Region	Million (EJ)	Percentage
North America	26	18.9
Latin America and the Caribbean	26	18.6
Western Europe	7	5.0
Eastern Europe and the former Soviet Union	23	16.7
Middle East and North Africa	6	4.5
Sub-Saharan Africa	17	11.9
Pacific Asia	11	8.1
China	11	7.8
Central and South Asia	13	9.4
Total	**140**	**100**

Sources: (Rogner *et al.*, 2004; WEC, 1994).

of geothermal occurrence: hydrothermal sources, hot dry rock, magma, and geo-pressurised sources. As illustrated in Table 5.4, geothermal energy is widely and almost evenly dispersed across the globe. High-temperature fields used for conventional power production (with temperatures above 150 °C) are largely confined to areas with young volcanism, and seismic and magmatic activity; geographic conditions will limit or expand its technical potential. Low-temperature resources suitable for direct use can be found in most countries.

However, the technical potential given in Table 5.1 is significantly reduced when only considering the easily accessible layers of the crust and a limitation of the

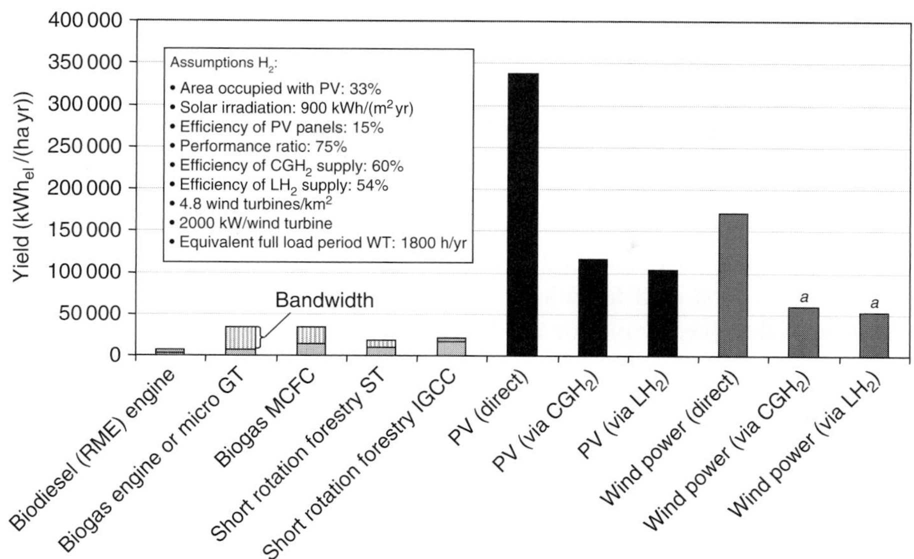

Figure 5.5. Electricity yield per ha and year for different renewable energy sources (Ludwig Bölkow Systemtechnik GmbH, personal communication, 2007). *a*) more than 99% of the land area can still be used for other purposes, e.g., agriculture.

drilling depth to 7000–8000 metres. The long-term technical potential based on these limitations would be of the order of 140 EJ per year, especially if deep drilling costs can be reduced, as these are a major limitation to this energy source. Thus, the technological availability rather than the availability of geothermal resources will determine its future share (WEC, 1994; Rogner *et al.*, 2004; Sørensen, 1979; 1991; Palmerini, 1993).

Ocean energy flows include thermal energy, waves, tides and the sea–freshwater interfaces as rivers flow into oceans. The low temperature gradients and low wave heights lead to an annual flow up to 7400 EJ per year of electricity. The technical potential is about 10 to 100 times smaller (see, e.g., Rogner *et al.*, 2004; Sørensen, 1979; 1991). The ocean, tidal and wave energy resources are rather diffuse, posing a very difficult challenge for commercial use and, like geothermal resources, technology advances will determine its use in the future.

Generally, renewable energy scenarios cover a wide spectrum, depending on the degree of future policy action, fuel prices, carbon prices or technology cost reductions. Global, European, and country-specific scenarios show a 10% to 50% share of primary energy from renewables by 2050; by 2020, many targets and scenarios show a 20% to 35% share of electricity from renewables, increasing to the range of 50% to 80% by 2050 under the most optimistic scenarios (Martinot *et al.*, 2007); resource constraints for renewables are mainly expected for biomass and biofuels. The bioenergy potential is large, but equally large are the uncertainties surrounding the future share of bioenergy in the global energy mix. The possible contributions to the global energy supply in 2050 found in the literature range on average from about 100 EJ to 400 EJ, with extreme cases covering the spectrum from as low as 40 EJ to as much as 1100 EJ (Martinot *et al.*, 2007; Worldwatch Institute, 2007).

Figure 5.5 shows a comparison of electricity yields per ha for different renewable energy sources. An interesting fact to note is that the highest electricity yields per ha can be achieved by photovoltaic and wind, significantly higher than for power generation from biomass. In addition, in the case of wind, more than 99% of the land area can still be used for other purposes, such as agriculture. The illustration further shows that even if solar or wind-generated electricity were used for hydrogen production and then the hydrogen converted again to electricity, the area yield would be higher than for bioenergy.

5.3 The European potential for RES

5.3.1 Comparison of technical and realisable potentials for RES

The aim of this section is to compare the technical potentials of RES with the realisable mid-term potentials for selected RES in Europe before the realisable mid-term potential is presented at country level in Section 5.3.2. The comparison aims to indicate the magnitude of the different potential categories rather than show exact potential figures, for the following reasons:

- The regional coverage of both categories differs slightly. Technical potentials were mostly taken from global potential studies and are shown for Western Europe, including EU15, Turkey, Switzerland and Norway for most of the RES, whereas data about the realisable mid-term potential comprises mainly the European Union as of 2006 (EU25).
- The potential definitions are of wide scope and potential determination depends strongly on the methodology applied and assumptions made.
- Different feedstocks were included in the analyses in the case of the biomass potentials or data about certain biomass categories were missing on a regional disaggregated level.

A comparison of the technical potential with the realisable mid-term potential for wind energy is shown in Fig. 5.6.

Grubb and Meyer (1993) estimate the technical *wind potential* for Western Europe to be 17 280 PJ/year, corresponding to 15% of the gross electric or theoretical potential (113 040 PJ/year). They exclude areas unsuitable for wind energy production, such as cities, forests and inaccessible mountains, as well as social, environmental and land-use constraints from the theoretical potential and estimate the technical potential. Only sites with an average wind speed above 6 m/s are included, assuming an efficiency factor of 0.3.

Hoogwijk *et al.* (2004) calculate a technical wind onshore potential of about 14 400 PJ/year including wind speeds above 4 m/s at 10 m. Assuming that only sites

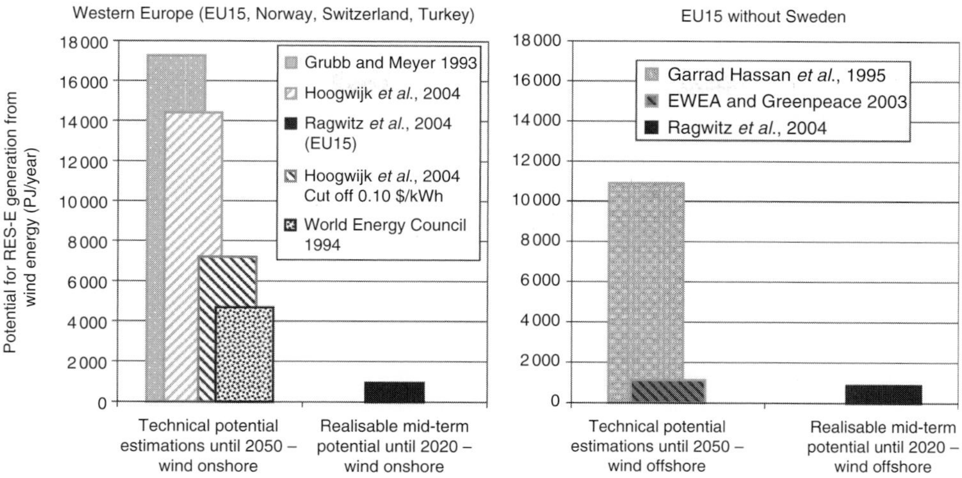

Figure 5.6. Comparison of the technical potential with the realisable mid-term potential for wind onshore and offshore energy in Europe.

where electricity generation costs amount to less than $0.1/kWh are included, the technical potential is reduced by 50%.

Based on the assumption that 4% of the area with a wind speed higher than 5.1 m/s at 10 m can be used for wind energy, WEC (1994) assesses the technical wind onshore energy potential to be 4680 PJ/year. A further restriction within this study is that areas with a distance of more than 50 km from the existing grid were excluded.

As shown in Figure 5.6, the realisable mid-term potential until 2020 shows significantly lower values. Only 964 PJ/year are expected to be realised until 2020. This is only partly because of the differing geographical coverage. In particular, it can be explained by assuming that additional barriers, such as grid restrictions and planning constraints, restrict the growth rates of wind energy.

In the area of wind offshore energy, Garrad Hassan *et al.* (1995) place the electricity production potential at 10 904 PJ/year including areas with a distance up to 30 km from the coast and a water depth of less than 40 m. The EWEA (2003) and Greenpeace (2001) apply further constraints leading to a significantly lower value for the offshore electricity potential (see Fig. 5.6). In this way they restrict the area available for offshore production to a water depth of 20 m and reduce the capacity density.

The potential comparison for wind energy shows a wide range of estimations. This arises from differing assumptions, in particular with respect to the methodology of site exclusion and the assumed capacity density. The assumption of maximal market growth rates and planning constraints leads to a significantly reduced potential which might realistically be achieved by 2020. In particular, surplus wind electricity might be used for hydrogen production as a means of energy storage (see Section 16.1).

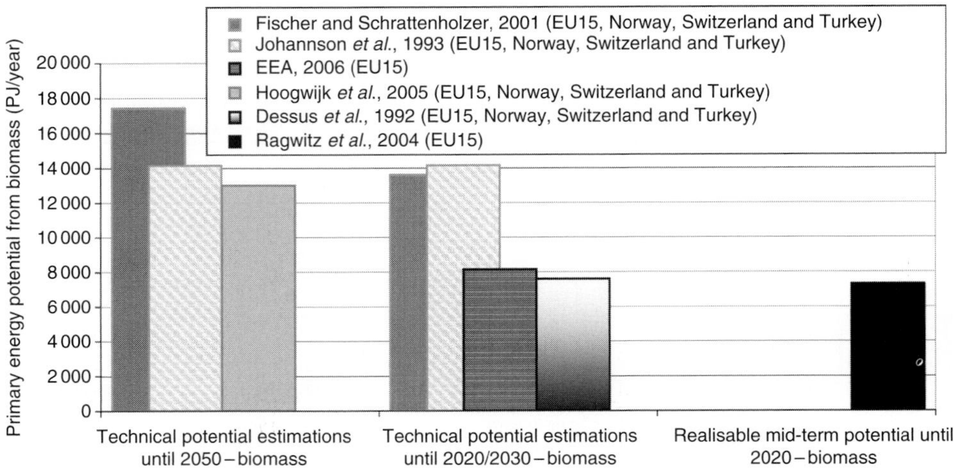

Figure 5.7. Comparison of the technical potential with the realisable mid-term potential for biomass primary energy in Europe.

The potential of *biomass*, as illustrated in Fig. 5.7, is shown in primary figures, since there are different possibilities for converting bioenergy into final energy, such as transportation fuel, electricity or heat. Resources in biomass include agricultural products and residues, wood and wood waste, animal waste and the biogenic fraction of municipal solid waste. The most important input factors for the determination of the bioenergy potential are the availability of land that can be used for biomass resources, the land productivity and competition of energetic biomass use with material use of bioenergy and food demand.

Fischer and Schrattenholzer (2001) estimate the Western European biomass potential, consisting of energy crops and residues, wood and forest residues, to be 17 435 PJ/year by 2050 and 13 635 PJ/year by 2020. The potential figures do not include bioenergy from animal waste and municipal waste, since no regional disaggregation was undertaken for either of those bioenergy categories. Potential calculations are based on a land-use model of IIASA and are supplemented with data from Dessus *et al.* (1992) for the bioenergy from wood products and residues. Assumptions about future food demand and supply are considered within the study.

The *Renewables-Intensive Global Energy Scenario* (*RIGES*) predicts a primary energy potential from biomass resources for Western Europe to be 14 160 PJ/year by 2025 and 14 170 PJ/year by 2050 (Johansson *et al.*, 1993). Thereby the biomass potential comprises resources from wood, energy crops, agricultural residues and industrial biomass residues. The estimates are based on the biomass production at that time in combination with assumptions of future growth rates.

The study initiated by the EEA (2006) aims at determining the environmentally compatible bioenergy potential. That means that a number of environment criteria

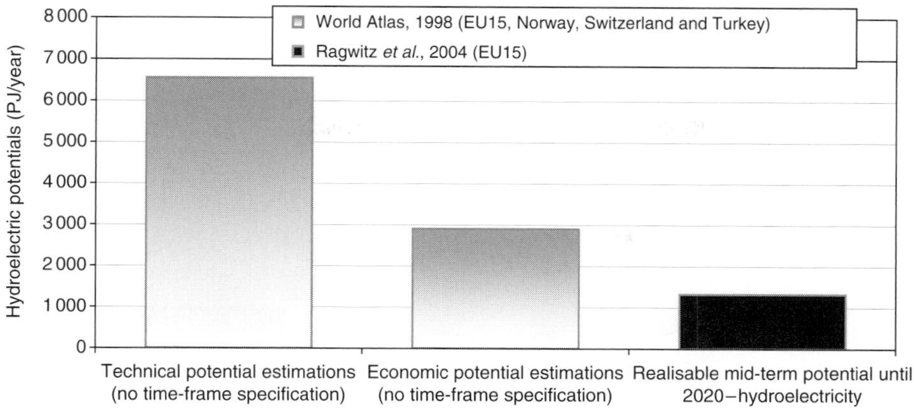

Figure 5.8. Comparison of the technical potential with the realisable mid-term potential for hydroelectric energy in Europe.

were selected and used as assumptions and restrictions for the potential calculations. Bioenergy crops, agricultural residues, forest products, forest residues and wastes of biological origin from agriculture, industry and households were included for the potential estimations. The estimated potential amounts to 7394 PJ/year in 2020 and 8918 PJ/year in 2030. This value corresponds to about 50% to 60% of the biomass potential calculated by Fischer and Schrattenholzer (2001) and Johansson *et al.* (1993) for the period between 2020 and 2030.

Hoogwijk *et al.* (2005) assume the biomass energy potential in Western Europe from energy crops, agricultural residues, forest residues and industrial biogenic residues to be of the order of 10 000 PJ/year and 16 000 PJ/year by 2050. The analysis is based on the IMAGE 2.2 model using the four scenarios from the *Special Report on Emissions Scenarios (SRES)*, (Nakicenovic, 2000) as main assumptions for the included food demand and supply.

Although Dessus *et al.* (1992) do not consider competition of energy crops with food production, the estimated bioenergy potential (wood, energy crops and waste) of 7620 PJ/year by 2020 is lower than in the other studies considered for the respective timeframe.

Figure 5.7 shows that the realisable mid-term potential is of similar size as the potential determined by EEA (2006) and Dessus *et al.* (1992).

The technical, economic and realisable mid-term potential of *hydroelectric energy* is shown in Fig. 5.8. A large part of the potential, especially for large-scale hydropower, is already exploited. A total of 1895 PJ of hydroelectricity was produced in 2005 in OECD-Europe (IEA, 2006b), amounting to 65% of the economic potential shown in Fig. 5.8. Here, the realisable potential is significantly reduced, owing to the different country coverage.

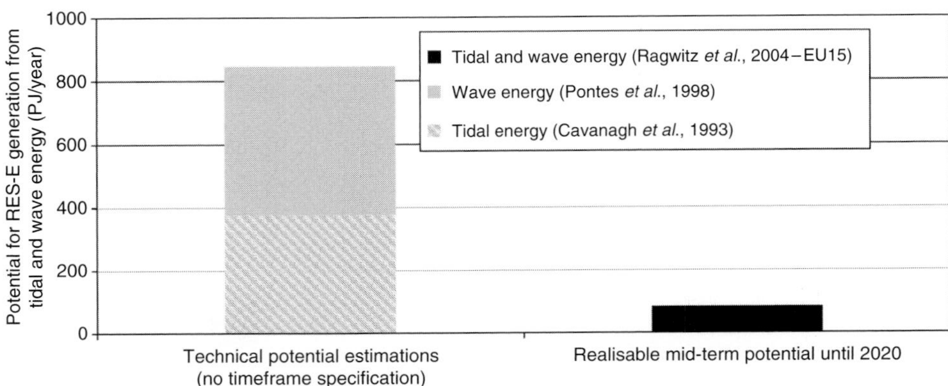

Figure 5.9. Comparison of the technical electricity generation potential with the realisable mid-term potential for tidal and wave energy in Europe. For the estimate of the technical tide potential, exploitable sites with a mean tidal range above 3 m are included.

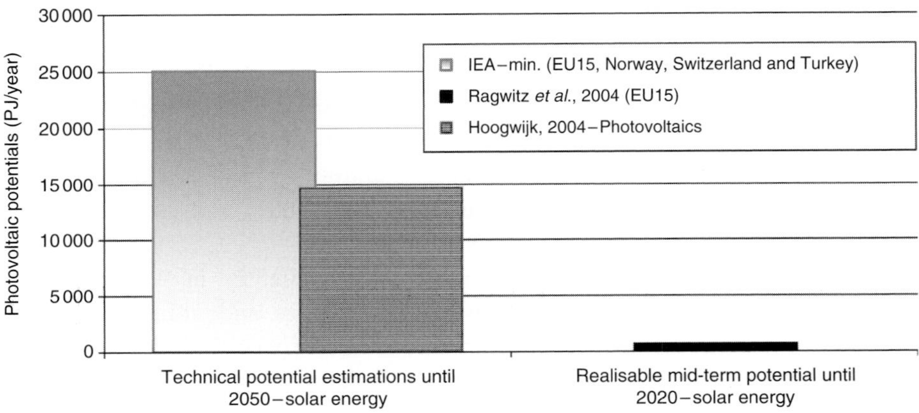

Figure 5.10. Comparison of the technical electricity generation potential with the realisable mid-term potential for photovoltaics in Europe.

Figure 5.9 provides a comparison of technical potential estimations for *tidal and wave* energy with the realisable mid-term potentials of both ocean technologies until 2020.

Compared with the potential of wind or hydroenergy, the energetic potential from the oceans is estimated to be low. According to Cavanagh *et al.* (1993), the tidal energy potential amounts to 378 PJ/year; 90% of the tidal energy potential is located in the United Kingdom and France. Pontes *et al.* (1998) place the wave potential between 468 PJ/year and 684 PJ/year. Presently the use of ocean energy is limited to some demonstration projects and the realisable mid-term potential is below 80 PJ/year.

The technical potential for photovoltaics as shown in Fig. 5.10 is of the same order as the potential of wind energy and biomass. According to the IEA minimum estimate, the PV potential amounts to 25 100 PJ/year. Hoogwijk *et al.* (2004) place the PV potential at about 15 000 PJ/year. Generally, the potential of photovoltaics is rather evenly distributed within Europe. In contrast, the use of PV energy is presently concentrated to some countries with the major share of the total capacity installed in Germany and the realisable mid-term potential is below 700 PJ/year.

5.3.2 Overview of realisable mid-term potential for RES in Europe

The following depiction aims to illustrate to what extend RES may contribute in the electricity sector within the European Union (EU25 plus selected candidates) up to the mid-term (i.e., the year 2020) by considering the specific resource conditions in the investigated countries. As explained before, *realisable mid-term potentials* are derived.

A broad set of different renewable energy technologies exists today. Obviously, for a comprehensive investigation of the future development of RES it is of crucial importance to provide a detailed investigation of the country-specific situation – e.g., with respect to the potential of the certain RES in general as well as their regional distribution and the corresponding generation cost. Major efforts have been taken within the FORRES 2020 study to assess Europe's RES resource base in a comprehensive manner. Consequently, this survey builds directly on these consolidated outcomes as presented in the Commission's communication, *The Share of Renewable Energy* (European Commission, 2004).

Renewable energy sources for electricity generation, such as hydropower or wind energy, represent energy sources characterised by a natural volatility. Therefore, to provide an accurate depiction of the future development of RES-E, historical data for RES-E are translated into electricity generation potentials[2] – the *achieved potential* at the end of 2004 – taking into account the recent development of this rapidly growing market. The historical record was derived in a comprehensive data collection – based on Eurostat (2006) and the IEA (2005) and statistical information gained at a national level. In addition, *future* potentials – the *additional realisable mid-term potentials* up to 2020 – were assessed taking into account the country-specific situation as well as overall realisation constraints. A brief description of the potential assessment is given by Resch *et al.* (2006).

Figure 5.11 depicts the achieved and additional mid-term potential for RES-E in the EU15 by country as well as by RES-E category. A similar picture is shown for the new member states as of 2006 (EU10) as well as Bulgaria and Romania, in Fig. 5.12. For EU15 countries, the already achieved potential for RES-E equals 441 TWh, whereas the additional realisable potential up to 2020 amounts to 1056 TWh

[2] The *electricity-generation potential* with respect to existing plants represents the output potential of all plants installed up to the end of 2004. Of course, figures for actual generation and generation potentials differ in most cases – because, in contrast to the actual data, potential figures represent, e.g., in the case of hydropower, the normal hydrological conditions and, furthermore, not all plants are installed at the beginning of each year.

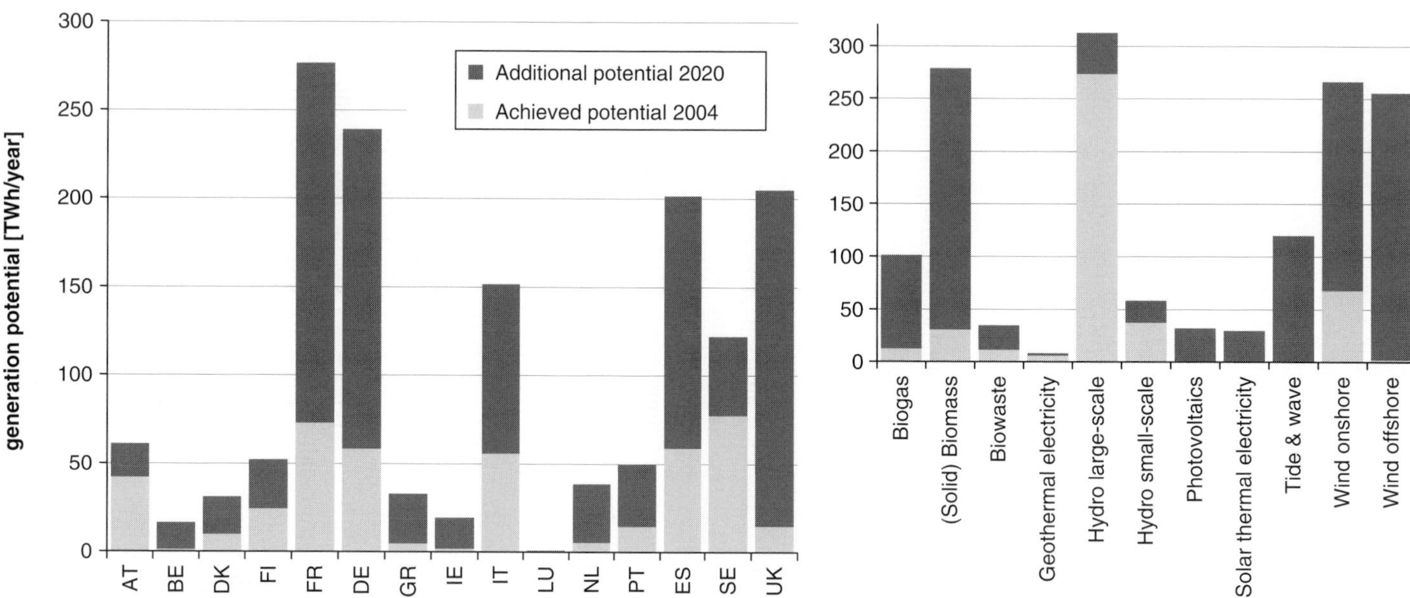

Figure 5.11. Achieved (2004) and additional mid-term potential 2020 for electricity from RES in the EU15 – by country (left) and by RES-E category (right).

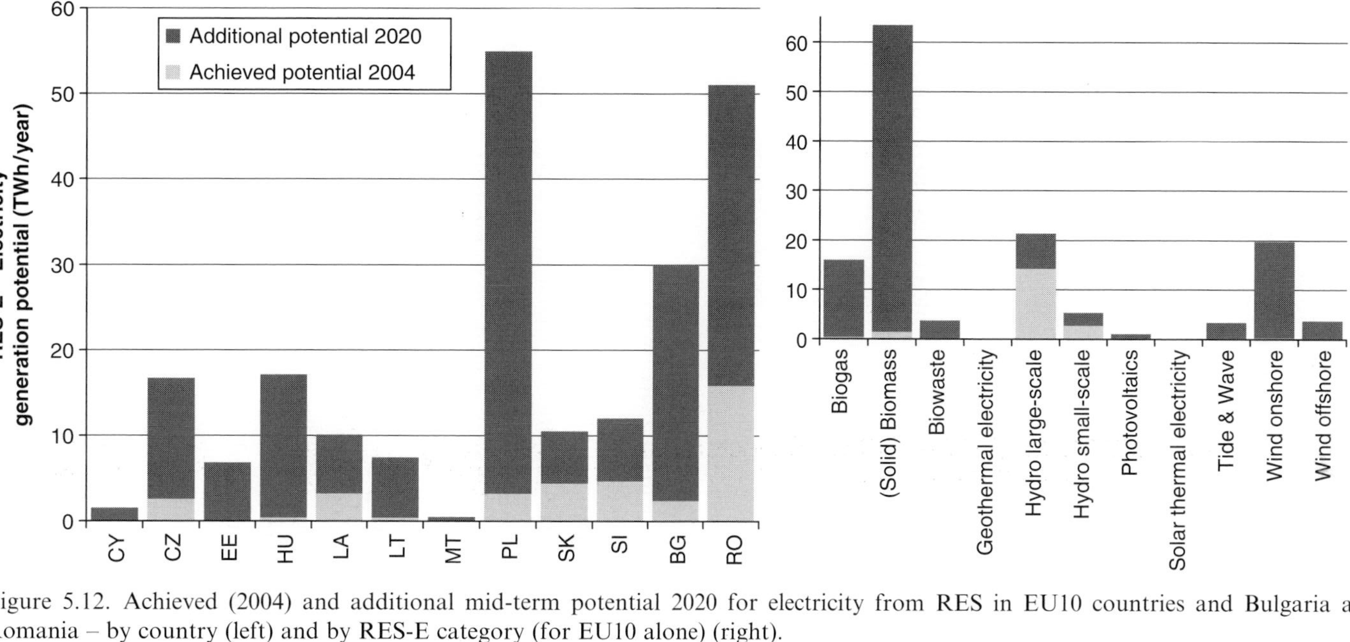

Figure 5.12. Achieved (2004) and additional mid-term potential 2020 for electricity from RES in EU10 countries and Bulgaria and Romania – by country (left) and by RES-E category (for EU10 alone) (right).

(about 38% of current gross electricity consumption). Corresponding figures for the EU10 are 19 TWh for the achieved potential and 119 TWh for the additional mid-term potential (about 36.1% of current gross electricity consumption).

As already mentioned, hydropower dominates current RES-E generation in most EU countries, followed by wind, biomass, biogas and biowaste.

Next, future perspectives are indicated at the country level. Figure 5.13 shows the share of different energy sources in the *additional* RES-E mid-term potential for the EU15 for 2020. The largest potential is found in the sector of wind energy (43%), followed by solid biomass (23%) and biogas (8%), as well as promising future options, such as tidal and wave (11%) or solar thermal energy (3%).

Figure 5.14 illustrates the share of different energy sources in the *additional* RES-E mid-term potential of the EU10 countries as well as Bulgaria, Romania and Hungary for 2020. In contrast to the EU15, the largest potentials for these countries exist in the sectors of solid biomass (52%) and wind energy (19%) followed by biogas (13%). Unlike the situation in the EU15, the refurbishment and construction of large hydro plants holds significant potential (6%).

Finally, Fig. 5.15 relates derived potentials to gross electricity demand. More precisely, it depicts the total realisable mid-term potentials (up to 2020) for RES-E as the share of gross electricity demand in 2004 and 2020 – for all EU25 countries as well as the EU25 in total. (The total realisable mid-term potential comprises the already achieved (as of 2004) and the additional realisable potential up to 2020.) The impact of the expected demand increase is crucial; if the indicated realisable mid-term potential for RES-E, covering all RES-E options, was fully exploited up to 2020, only 41% of gross consumption could be covered, if the demand increases as expected under 'business as usual' conditions. (Demand figures for 2020 are taken from DG TREN's *BAU* forecast (Mantzos *et al.*, 2003).) In contrast, if a stabilisation in demand is achieved, RES-E may contribute to meet about 53% of total demand.

The availability of biomass and the allocation of biomass resources across energy sectors are crucial, as this energy source is faced with high expectations with regard to its future potentials. Although the potential analysis undertaken here is focused on the electricity sector, with regard to biomass, all energy sectors have been considered. The total domestic availability of solid biomass is approximately 221 Mtoe/yr (9.2 PJ/yr).[3] To indicate the European perspective in a broader context, it is assumed that biomass can be imported to the European market. Specifically:

- Solid biomass in the form of wood products and wood residues can be imported to a maximum of 30% of the total additional primary input of forestry biomass, which represents about 9.7 Mtoe.
- Liquid biofuels in the form of ethanol and biodiesel products can be imported to a maximum of 30%, corresponding to a default case based on solely domestic biofuel supply.

[3] For example, the EEA (2006) report, *How Much Bioenergy Can Europe Produce Without Harming the Environment?*, gives 235 Mtoe in 2020 for total biomass under the assumption of significant ecological constraints on biomass use.

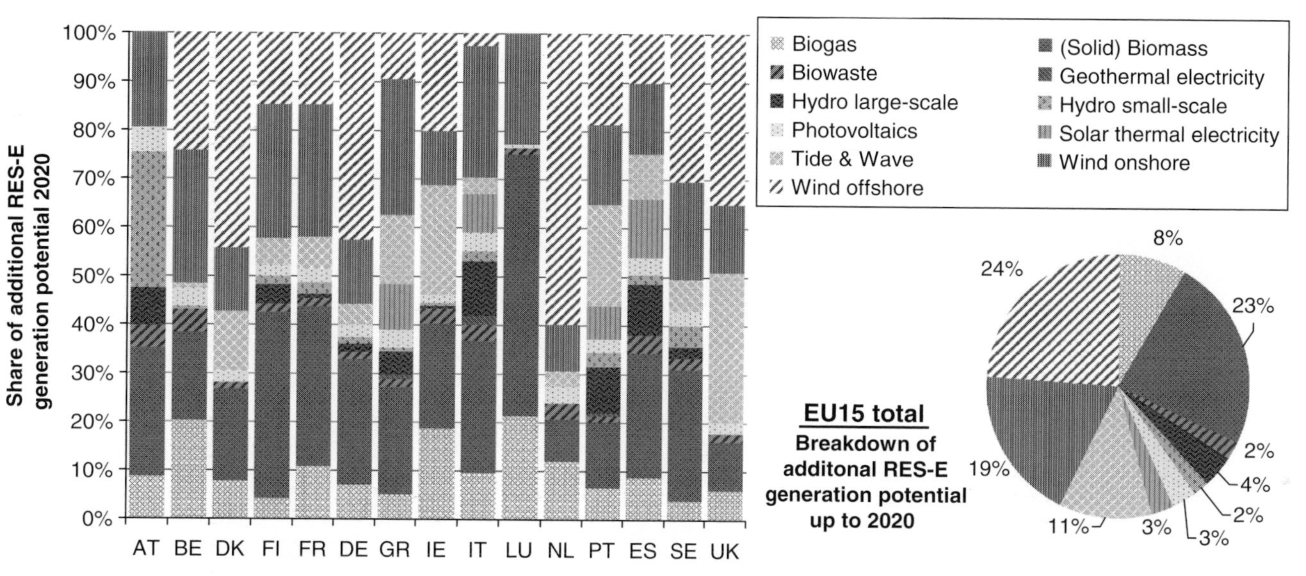

Figure 5.13. RES-E as a share of the total additional realisable potential in 2020 for the EU15 – by country (left) as well as for total EU15 (right).

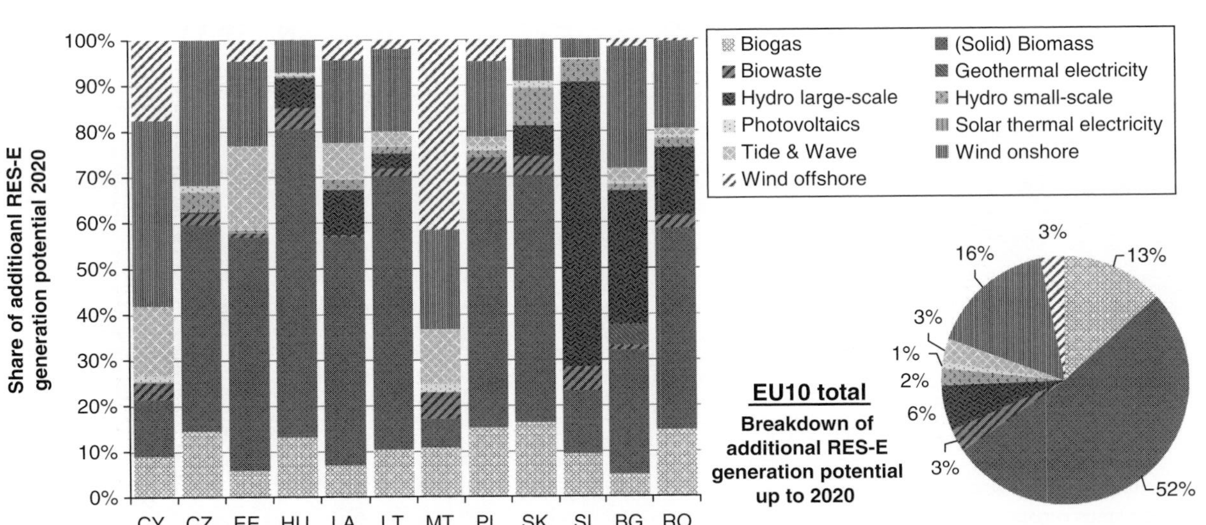

Figure 5.14. RES-E as a share of the total additional realisable potential in 2020 for the EU10 and Bulgaria and Romania – by country (left) as well as for total EU10 (right).

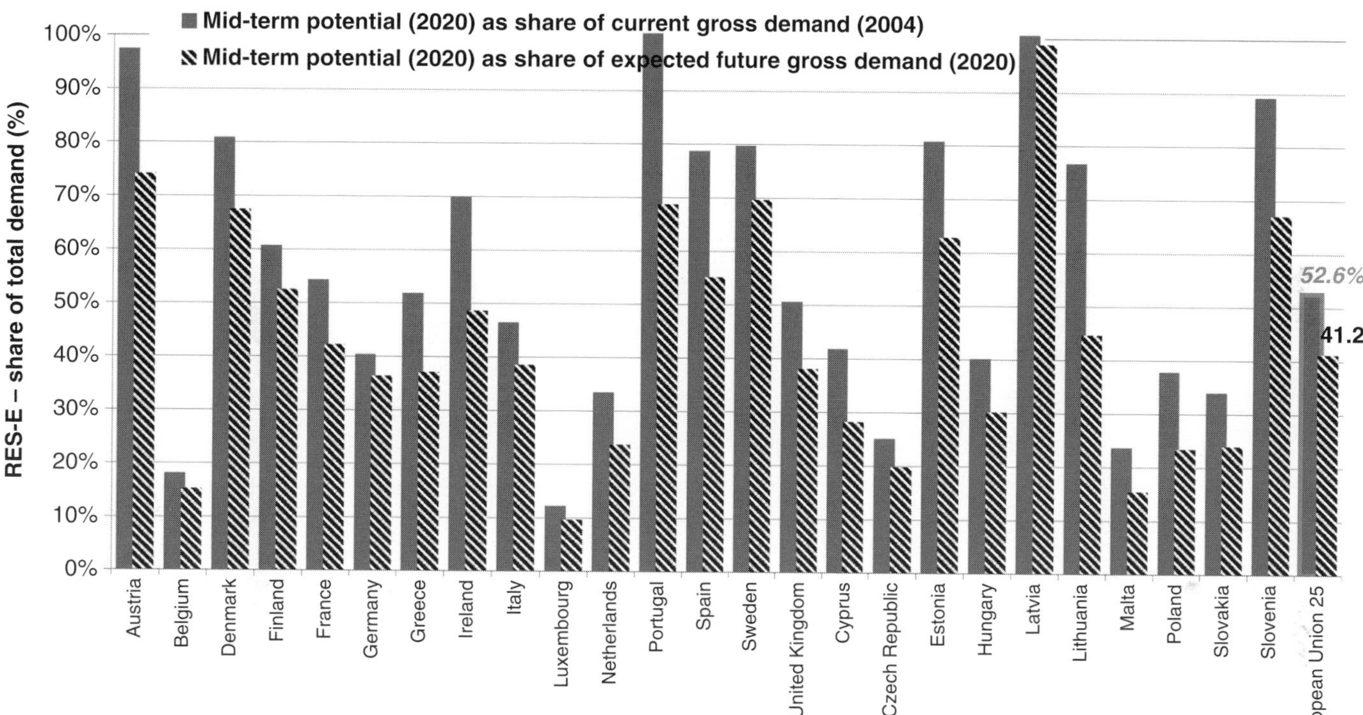

Figure 5.15. Total realisable mid-term potentials (2020) for RES-E in EU25 countries as a share of gross electricity demand (2004 and 2020).

In this context, Fig. 5.16 indicates the dynamic evolution of the identified biomass primary potentials at the EU25 level, whilst Table 5.5 shows a detailed breakdown of corresponding fuel costs for the considered biomass options, including agricultural products or energy crops (e.g., rapeseed and sunflower, miscanthus), agricultural residues (straw), forestry products (e.g., wood chips), forestry residues and biowaste.

Table 5.5. *Breakdown of fuel cost and corresponding primary potentials by fuel category*

Solid biomass – primary potentials and corresponding fuel cost by 2020	Realisable mid-term potential for 2020 in terms of primary energy	Fuel cost ranges (2005)		
		Minimum	Maximum	Weighted average
	(Mtoe/year)	(€/MWh-p)	(€/MWh-p)	(€/MWh-p)
AP1 – rape and sunflower		32.3	40.4	37.2
AP2 – maize, wheat (corn)		26.6	33.2	30.6
AP3 – maize, wheat (whole plant)		29.8	29.8	0.0
AP4 – SRC willow		27.4	32.9	29.2
AP5 – miscanthus		27.1	34.1	30.0
AP6 – switch grass		17.9	31.9	25.9
AP7 – sweet sorghum		31.0	40.9	40.9
Agricultural products – total	**75.8**	**17.9**	**40.9**	**31.9**
AR1 – straw		12.2	14.7	13.4
AR2 – other agricultural residues		12.2	14.7	13.5
Agricultural residues – total	**27.9**	**12.2**	**14.7**	**13.4**
FP1 – forestry products (current use; wood chips, log wood)		17.8	22.3	20.6
FP2 – forestry products (complementary fellings, moderate)		19.1	23.8	21.7
FP3 – forestry products (complementary fellings, expensive)		25.8	32.3	29.4
Forestry products – total	**51.9**	**17.8**	**32.3**	**23.0**
FR1 – black liquor		5.6	7.7	6.0

Table 5.5. (cont.)

Solid biomass – primary potentials and corresponding fuel cost by 2020	Realisable mid-term potential for 2020 in terms of primary energy (Mtoe/year)	Fuel cost ranges (2005) Minimum (€/MWh-p)	Maximum (€/MWh-p)	Weighted average (€/MWh-p)
FR2 – forestry residues (current use)		6.3	8.6	7.0
FR3 – forestry residues (additional)		12.5	17.1	13.9
FR4 – demolition wood, industrial residues		5.0	6.8	5.9
FR5 – additional wood processing residues (sawmill, bark)		6.3	8.6	6.9
Forestry residues – total	**47.8**	**5.0**	**17.1**	**6.9**
BW1 – biodegradable fraction of municipal waste		−3.8	−3.8	−3.8
Biowaste – total	**17.2**	**−3.8**	**−3.8**	**−3.8**
FR6 – forestry imports from abroad	9.7	16.0	16.8	16.8
Solid biomass – total	**230.3**	**−3.8**	**40.9**	**16.2**
of which domestic biomass	**220.6**	**−3.8**	**40.9**	**16.4**

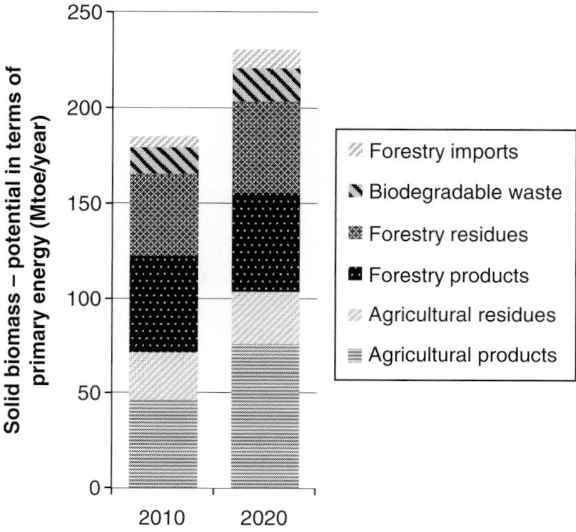

Figure 5.16. Biomass potentials in terms of primary energy for the years 2010 and 2020.

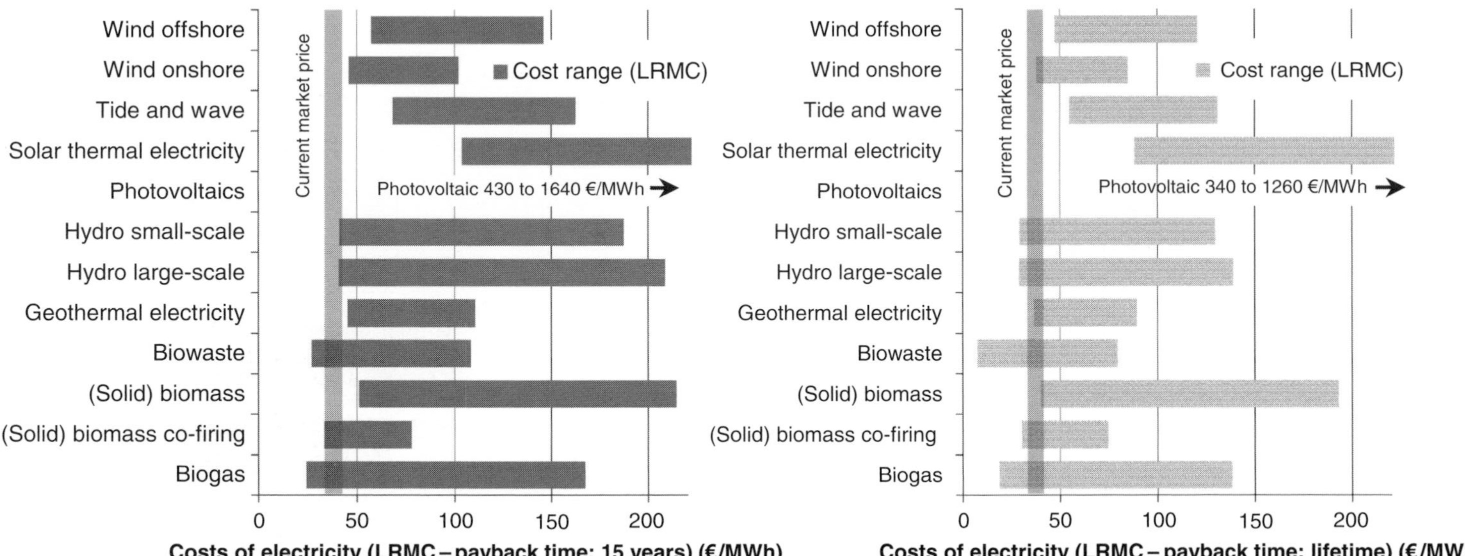

Figure 5.17. Long-run marginal generation costs (for the year 2005) for various RES-E options in EU countries – based on a default payback time of 15 years (left) and by setting payback time equal to lifetime (right).

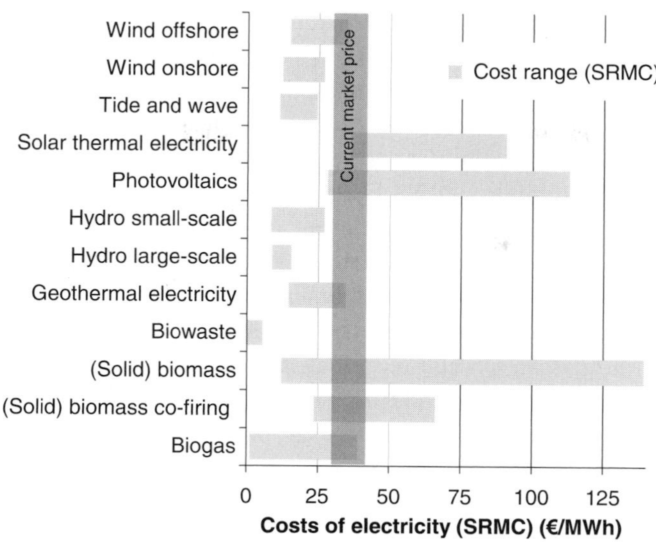

Figure 5.18. Short-run marginal generation costs (for the year 2005) for various RES-E options in EU countries.

5.3.3 Economic data for RES

To determine the potential use of RES for hydrogen, the economic performance has to be considered in addition to the theoretical availability of RES. For this reason, ranges of electricity generation costs concerning the various RES technologies are shown next.

First, Fig. 5.17 depicts the *long-run marginal generation costs* by RES-E category as calculated by Ragwitz and Resch (2006).[4] Thereby, to calculate the capital recovery factor, two different settings are applied with respect to the payback time: on the one hand, a default setting, i.e., a payback time of 15 years, is used for all RES-E options – see Fig. 5.17 (left), and on the other hand, the payback is set equal to the technology-specific lifetime – see Fig. 5.17 (right). For both cases, a default weighted average cost of capital (WACC) of 6.5% is used. The broad range of costs for several RES-E represents, on the one hand, resource-specific conditions, as are relevant, e.g., in the case of photovoltaics or wind energy, which appear between and also within countries. On the other hand, costs also depend on the technological options available – compare, e.g., co-firing and small-scale CHP plants for biomass.

Figure 5.18 illustrates *short-run marginal generation costs* by RES-E category. Short-run marginal costs are relevant for the economic decision of whether or not

[4] Sources for technical and economic data: DLR/WI/ZSW/IWR/Forum, 1999; BTM Consult ApS, 1999 to 2005; Beurskens and de Noord, 2003; Binnie Black and Veatch, 2001; BMU, 2002; DTI/ETSU, 1999; DLR *et al.*, 2004; Enquete, 2002; EUBIONET, 2003; Fischer & Schrattenholzer, 2001; Greenpeace, 2001; Haas *et al.*, 2001; Kaltschmitt *et al.*, 2003; Lorenzoni, 2001; Michael, 2002; Neubarth and Kaltschmitt, 2000; Nowak *et al.*, 2002; Neij *et al.*, 2003; Quaschning and Ortmanns, 2003; Resch *et al.*, 2001; Schäffer *et al.*, 2004; Thorpe, 1999).

to operate an existing plant. It is evident that for most RES-E options these short-run generation costs, i.e., the running costs, are low compared with conventional power generation based on fossil fuels. One exception in this context is biomass, where fuel costs and conversion efficiencies have a huge impact on the resulting running costs.

The current situation, without consideration of expected technological change, may be described as follows: RES-E options, such as landfill and sewage gas, biowaste, geothermal electricity, (upgrading of) large-scale hydropower plant or co-firing of biomass, are characterised from an economic point of view by comparatively low cost and by, in contrast, rather limited future potentials in most countries. Wind energy and, in some countries, also small-scale hydropower or biomass combustion (in large-scale plants) represent RES-E options with economic attractiveness accompanied by a high additional realisable potential. A broad set of other RES-E technologies are less competitive at present, compare, e.g., agricultural biogas and biomass (both if utilised in small-scale plants), photovoltaics, solar thermal electricity, tidal energy or wave power, although future potentials are in most cases huge.

5.4 Global biofuel potential

A wide range of biomass feedstock is currently available worldwide to produce liquid biofuels with several transformation technologies, including chemical processes, such as transesterification of oils and fats (e.g., biodiesel from palm oil, rapeseed oil, sunflower oil, soybean, tallow, used cooking oil, fish oil) to produce fatty acid methyl ester, commonly known as biodiesel. Furthermore, and widely available since the 1980s in Brazil and later in the USA and other countries worldwide, there is ethyl alcohol, or ethanol, which is obtained from the fermentation of sugar of starch crops (e.g., sugar cane, sugar beet, maize (corn), wheat, barley, sorghum and rye). At present, the predominant biofuels, ethanol and biodiesel, account for about 1.3% of total road-transport fuel consumption according to the IEA (2006a); around 85% of this comes from ethanol and the remainder from biodiesel (Doornbosch and Steenblik, 2007).

Second-generation biofuel technologies make use of a much wider range of biomass feedstock (e.g., forest residues, biomass waste, wood, woodchips, grasses and short rotation crops, etc.) for the production of ethanol biofuels based on the fermentation of lignocellulosic material, while other routes include thermo-chemical processes such as biomass gasification followed by a transformation from gas to liquid (e.g., synthesis) to obtain synthetic fuels similar to diesel. The conversion processes for these routes have been available for decades, but none of them have yet reached a high scale commercial level.

The global potential of conventional and emerging biofuels is limited by land availability for energy crops as well as the high cost of most existing and emerging conversion technologies (see also Chapter 7). In terms of land availability, estimates from FAO and IIASA (Fischer *et al.*, 2000; 2006) indicate that the maximum worldwide 'gross' available area that could be used for dedicated energy crops would amount to roughly 0.7 Gha after taking into consideration various areas for growing food and other types of crop. This translates into possible primary energy potentials for bioenergy (e.g., heat, electricity and transport) that amount to 245 EJ according to Hoogwijk *et al.* (2005). This corresponds to a lower end of the wide range of potentials (125–760 EJ) reported in the IPCC Fourth Assessment report (Doornbosch and Steenblik, 2007). Reasonable assumptions estimate that approximately a quarter to, at most, half of the biomass surplus for dedicated energy crops could be made available for biofuel production until 2050. This is translated in a global biofuel potential ranging from 22 EJ to roughly 43 EJ, if an average efficiency of 35% for biofuel conversion is used.

Estimates of the future contribution of biofuels to global transport fuel supply vary considerably and are very sensitive to the underlying assumptions. At a global level, it is estimated that biofuels could substitute up to 30% of today's total vehicle fuel consumption. For instance, the IEA projects biofuels to meet 4% of world road-transport fuel demand by 2030 in the Reference Scenario, and 7% in the Alternative Policy Scenario (IEA, 2006a); the IEA's Energy Technology Perspectives scenarios show a range of biofuel shares of transport energy demand by 2050, from a 3% Reference Scenario to 25% in the most optimistic scenario (IEA, 2006c). These shares are plausible, if feedstock and conversion costs are managed to be reduced substantially in the decades ahead. But it is clear that biofuels alone will not solve all of the world's transport-related energy problems.

5.5 Summary

Although a high theoretical potential of RES exists, this potential decreases significantly when further restrictions, such as the technical availability, market constraints, environmental constraints and the economic feasibility of renewable technology options for energy conversion, are taken into account. For these reasons, it is also hard to predict how renewable capacities will develop in the future.

Currently, electricity generation costs of RES-technologies tend to be higher than costs for conventional electricity generation and need financial incentives to be economically feasible. Given the restricted realisable potential of RES and the electricity generation costs, the option of producing hydrogen based on RES has to be evaluated carefully. As the realisable renewable potentials in the short- to mid-term until 2020 are only moderate, the energy produced can be directly consumed as electricity rather than for hydrogen production. Only in the longer timeframe, beyond 2020, might surplus potentials be available. In this context, in particular, the future technology progress of RES-technologies determines the feasibility of hydrogen-production based on RES. Energy storage will be a key enabling technology for the enhanced integration of renewable energies.

Glossary

Biofuel Fuel derived from organic sources, e.g., biogas, biomass and the biodegradable fraction of waste. The use of biofuel is neutral in terms of carbon dioxide emissions.

Biogas The combustible mix of methane (50–75%), carbon dioxide (25–50%), oxygen and nitrogen derived from the anaerobic digestion of organic material, especially wastes. Agricultural, sewage, landfill and organic waste produce biogas by anaerobic digestion that can be collected and combusted for electricity generation.

Biomass Forestry and agricultural crops and residues used as fuel. Energy crops are grown specifically as a biomass fuel.

Geothermal electricity The geothermal heat derived from the hot underground environment is used to generate electricity or may be used to supply heat for hot water and for heating buildings.

Hydropower for electricity can use either a dam or the natural flow of water in a 'run of the river' system. Large hydropower (larger than 10 MW) and small hydropower (smaller than 10 MW) are differentiated.

Municipal waste Municipal waste can be used as a fuel to produce electricity and heat.

Renewable electricity (RES-E) Electricity generated from renewable non-fossil energy sources, i.e., wind, solar, geothermal, wave, tidal, hydropower, biomass, landfill gas, sewage treatment plant gas and biogas (this corresponds to the definition in Directive 2001/77/EC on renewables, Article 2).

Renewable energy sources (RES) In general, all energy sources 'obtained from persistent and continuing flows of energy occurring in the environment'. European Union countries have historically taken differing approaches to defining which technologies are classified as renewable. This particularly applies to sources linked to waste and to large hydro plants. Likewise, categorisation of the many forms of agricultural 'biomass' and 'biofuel' may vary between countries. These decisions have partly been dependent on government policy objectives and public perceptions in each country. Directive 2001/77/EC on renewables, Article 2, defines RES as 'non-fossil energy sources (including wind, solar, geothermal, wave, tidal, hydropower, biomass, landfill gas, sewage treatment plant gas and biogas).'

Solar energy Energy initially absorbed from sunshine. If the *solar radiation* is absorbed in a device providing a controlled energy supply, e.g., hot water or electricity, this is an *active solar* system. A *solar thermal* device uses heat, e.g., a solar water heater or a heat engine for electricity generation. Alternatively, the solar radiation may be absorbed as light in a *solar photovoltaic* (PV) device for immediate electricity generation based on the photoelectric effect.

Tidal energy There are two different technologies: tidal barriers and tidal currents.

Tidal barriers utilise the rise and fall of the tide (the tidal range) to trap seawater at high tide in a reservoir behind a barrage. As the water leaves or enters the reservoir in a constrained duct, submerged hydroturbines generate electricity, as in conventional hydropower. There is only one significant tidal barrier power plant, which is at La Rance, Brittany, in France.

Tidal-current (or stream) power is derived from water turbines submerged in the wide expanse of a tidal flow or current; there is no constructed barrier. Such a turbine is, therefore, the water-equivalent of a wind turbine. As yet, there are no commercial tidal-current power plants.

Wave power The energy in waves can be captured in a number of ways. One method is to funnel the waves into a partially filled vertical tube, to form an oscillating water column. The motion of the water forces air back and forth through an air turbine to produce electricity. Power from such devices is already sold commercially to the grid in Scotland. Several other types of wave energy device are under development.

Wind energy Wind turbines capture the energy from the wind to produce electricity. They have been developed for various purposes, from large groups of grid-connected wind turbines, wind farms, both onshore and offshore, to very small autonomous turbines used for battery charging or in combined wind–diesel projects for off-grid application.

Wind onshore Wind turbines that are installed on land, instead of being installed offshore (in the sea). The term onshore is not limited to coastal areas.

Wind offshore Wind turbines that are installed in the sea.

References

Beurskens, L. and de Noord, M. (2003). *Offshore Wind Power Developments. Report.* Energy Research Centre of the Netherlands (ECN).

Binnie Black and Veatch (2001). *The Commercial Prospects for Tidal Power. Report in Association with IT Power.* United Kingdom: DTI.

BMU (Bundesministerium für Umwelt, Naturschutz und Reaktorsicherheit) (2002). *Fachtagung geothermische Stromerzeugung – eine Investition in die Zukunft*, 20–21 June 2002, Landau/Pfalz, Germany: German Ministry for Environment.

BTM Consult ApS (1999 to 2005). *International Wind Energy Development – World Market Update.* Yearly updated reports. Denmark: BTM Consult ApS.

Cavanagh, J. E., Clarke, J. H. and Price, R. (1993). Ocean energy systems. In Renewable Energy. Sources for Fuels and Electricity, ed. Johansson, T. B., Kelly, H., Reddy, A. K. N. and Williams, R. H. Washington, DC: Island Press.

CERA (Cambridge Energy Research Associates) (2007). *Crossing the Divide? The Future of Clean Energy.* Cambridge, MA: CERA.

Dessus, B., Devin, B. and Pharabod, F. (1992). World potential of renewable energies. *La Houille Blanche*, **47** (1), 21–70.

DLR/WI/ZSW/IWR/Forum (1999). *Klimaschutz durch Nutzung erneuerbarer Energien.* Germany: German Ministry for Environment.

DLR, ifeu and WI (2004). *Ökologisch optimierter Ausbau der Nutzung erneuerbarer Energien in Deutschland.* Germany: German Ministry for Environment.

Doornbosch, R. and Steenblik, R. (2007). *Biofuels: Is the Cure Worse Than the Disease?* Paris: OECD, General Secretariat. OECD Round Table on Sustainable Development SG/SD/RT(2007) 3.

DTI/ETSU (Department of trade and industry/Energy technology support office) (1999). *New and Renewable Energy. Prospects in the UK for the 21st century.* United Kingdom: DTI, ETSU.

EEA (European Environment Agency) (2006). *How Much Bioenergy Can Europe Produce Without Harming the Environment*? EEA Report No. 7/2006.

Enquete (2002). *Analysis Grid.* The German Enquete Commission on Sustainable Energy Policy.

EUBIONET (2003). Various reports, e.g., *Fuel Prices in 2002/2003*. www.eubionet.net.

European Commission (2004). *The Share of Renewable Energy in the EU*. COM(2004) 366 final. Brussels: European Commission.

Eurostat (2006). *Energy: Yearly Statistics. Data 2004*. Eurostat.

EWEA (European Wind Energy Association) and Greenpeace (2003). *Wind Force 12*. www.ewea.org/fileadmin/ewea_documents/documents/publications/reports/wf12-2005.pdf.

Fischer, G., van Velthuizen, H. and Nachtergaele, F. O. (2000). *Global Agro-ecological Zones Assessment: Methodology and Results*. Interim report IR-00–064. Rome: International Institute for Applied Systems Analysis (IIASA) and Food and Agricultural Organisation (FAO) of the United Nations.

Fischer, G. and Schrattenholzer, L. (2001). Global bioenergy potentials through 2050. *Biomass and Bioenergy*, **20** (3), 151–159.

Fischer, G., Shah, M., van Velthuizen, H. and Nachtergaele, F. (2006). *Agro-Ecological Zones Assessment*. Interim Report RP-06–003. Luxembourg: International Institute for Applied Systems Analysis (IIASA).

Garrad Hassan, Germanischer Lloyd and Windtest (1995). *Study of the Offshore Wind Energy in the EC*.

Greenpeace (2001). *North Sea Offshore Wind – A Powerhouse for Europe*. Germany: Greenpeace.

Grubb, M. and Meyer, N. (1993). Wind energy: resources, systems and regional strategies. In *Renewable energy. Sources for Fuels and Electricity*, ed. Johansson, T. B., Kelly, H., Reddy, A. K. N. and Williams, R. H. Washington, DC: Island Press, pp. 157–212.

GWEC (Global Wind Energy Council) and Greenpeace (2006). *Global Wind Energy Outlook 2006*. www.gwec.net/fileadmin/documents/Publications/Global_Wind_Energy_Outlook_2006.pdf.

Haas, R., Berger, M. and Kranzl, L. (2001). *Erneuerbare Strategien - Strategien zur weiteren Forcierung erneuerbarer Energieträger in Österreich unter besonderer Berücksichtigung des EU-Weissbuches für Erneuerbare Energien und der Campaign for Take-off*. Vienna: Energy Economics Group, Vienna University of Technology.

Häfele, W. (1981). *Energy in a Finite World*. IIASA, Report, Cambridge, MA: Ballinger, vol 2., p. 220.

Hoogwijk, M., de Vries, B. and Turkenburg, W. (2004). Assessment of the global and regional geographical, technical and economic potential of onshore wind energy. *Energy Economics*, **26** (5), 889–919.

Hoogwijk, M., Faaij, A., Eickhout, B., de Vries, B. and Turkenburg, W. (2005). Potential of biomass energy out to 2100, for four IPCC SRES land-use scenarios. *Biomass and Bioenergy*, **29** (4), 225–257.

IEA (International Energy Association) (2005). *Renewables Information 2005*. Paris: OECD/IEA.

IEA (International Energy Association) (2006a). *World Energy Outlook 2006*. Paris: OECD/IEA.

IEA (International Energy Association) (2006b). *Renewables Information 2006*. Paris: OECD/IEA.

IEA (International Energy Association) (2006c). *Energy Technology Perspectives 2006. Scenarios and Strategies to 2050*. Paris: OECD/IEA.

IEA (International Energy Association) (2007). *Renewables in Global Energy Supply*. IEA Factsheet. Paris: OECD/IEA.

Johansson, T. B., Kelly, H., Reddy, A. K. N. and Williams, R. H. (1993). A renewables-intensive global energy scenario (Appendix to Chapter 1). In *Renewable Energy. Sources for Fuels and Electricity*, ed. Johansson, T. B., Kelly, H., Reddy, A. K. N. and Williams, R. H. Washington, DC: Island Press, pp. 1071–1142.

Johansson, T. B., McCormick, K., Neij, L. and Turkenburg, W. (2004). The potentials of renewable energy: thematic background paper. *International Conference for Renewable Energies*. Bonn (January 2004).

Kaltschmitt, M., Wiese, A. and Streicher, W. (2003). *Erneuerbare Energien*. 3rd edn. Germany: Springer-Verlag.

Lehner, B., Czisch, G. and Vassolo, S. (2005). The impact of global change on the hydropower potential of Europe: a model-based analysis. *Energy Policy*, **33** (7), 839–855.

Lorenzoni, A. (2001). *Blue Energy for a Green Europe. Final report, EU Altener II Programme*. Milano: IEFE – Universitá Commerciale L. Bocconi.

Mantzos, L., Capros, P., Kouvarikatis, N. *et al.* (2003). *European Energy and Transport – Trends to 2030. On behalf of the European Commission, DG TREN*. Greece: University of Athens.

Martinot, E. *et al.* (2006). *Renewables, Global Status Report, 2006 Update*. Worldwatch Institute and Tsinghua University, Renewable Energy Policy Network for the 21st Century (REN21). www.ren21.net.

Martinot, E., Dienst, C., Weiliang, L. and Qimin, C. (2007). Renewable energy futures: targets, scenarios, and pathways. *Annual Review of Environment and Resources*, **32**, 205–239.

Martinot, E. *et al.* (2008). *Renewables, Global Status Report, 2007*. Worldwatch Institute and Tsinghua University, Renewable Energy Policy Network for the 21st Century (REN21). www.ren21.net.

Michael, P. (2002). *DTI Wave Report 2002*. United Kingdom: Department of Trade and Industry.

Nakicenovic, N. (2000). *Special Report on Emission Scenarios*. Cambridge University Press.

Nakicenovic, N., Grübler, A. and McDonald, A. (eds.) (1998). *Global Energy Perspectives*. Cambridge: Cambridge University Press.

Neij, L., Andersen, P. D., Durstewitz, M. *et al.* (2003). *Experience Curves: A tool for Energy Policy Assessment (EXTOOL)*. Final Report of the Project EXTOOL – A European Research Project Funded by the EC, DG RESEARCH. Sweden: Lund University.

Neubarth, J. and Kaltschmitt, M. (2000). *Erneuerbare Energien in Österreich*. Vienna: Springer-Verlag.

Nowak, St., Gutschner, M. and Favaro, G. (2002). *Impact of Technology Developments and Cost Reductions on Renewable Energy Market Growth*. Six Technology-Specific Reports Prepared Within the Project REMAC2000 – in Co-operation With the International Energy Agency. Switzerland: NET, Ltd.

Palmerini, C. G. (1993). Geothermal Energy. In *Renewable Energy: Sources for Fuels and Electricity*, ed. Johansson, T. B., Kelly, H., Reddy, A. K. N. and Williams, R. H. Washington, DC: Island Press.

Pontes, M. T., Athanassoulis, G. A., Barstow, S. *et al.* (1998). The European wave energy resource. *Proceedings of the 3rd European Wave Energy Conference*. Patras, Greece.

Quaschning, V. and Ortmanns, W. (2003). Specific cost development of photovoltaic and concentrated solar thermal systems depending on the global irradiation – a study performed with the simulation environment GREENIUS. *The ISES Solar World Congress 2003, 14–19 June 2003 Gothenburg, Sweden*. Almeria, Spain: DLR e.V.

Ragwitz, M., Schleich, J., Huber, C. *et al.* (2004). *FORRES 2020 – Analysis of the Renewable Energy's Evolution up to 2020. Draft Final Report of the Project FORRES 2020–on Behalf of the European Commission, DG TREN; Co-ordinated by FhG-ISI with Contribution from EEG, Ecofys, Kema and REC*. Karlsruhe, Germany.

Ragwitz, M. and Resch, G. (2006). *Economic Analysis of Reaching a 20% Share of Renewable Energy Sources in 2020*. Karlsruhe, Germany.

Resch, G., Berger, M. and Kranzl, L. (2001). *Database for the Model ElGreen. Report of the European Research Project ElGreen–Funded by the European Commission, DG TREN*. Vienna: Energy Economics Group (EEG), Vienna University of Technology.

Resch, G., Ragwitz, M., Held, A. *et al.* (2006). *Potentials and Cost for Renewable Electricity in Europe. A Report of the OPTRES-project: Assessment and Optimisation of Renewable Energy Support Measures in the European Electricity Market. A Research Project Supported by the European Commission, DG TREN, Intelligent Energy for Europe – Programme (Contract No. EIE/04/073/S07.38567)*. Vienna.

Rogner, H. H. *et al.* (2004). Energy resources. In *World Energy Assessment – 2004 update,* Chapter 5. United Nations Development Programme, United Nations Department of Economic and Social Affairs, World Energy Council.

Schäffer *et al.* (2004). *Learning From the Sun. Final report of the Photex project*. Netherlands: ECN (Energy Research Centre of the Netherlands).

Sørensen, B. (1979). *Renewable Energy*. London: Academic Press.

Sørensen, B. (1991). Renewable energy – a technical overview. *Energy Policy*, **19** (4), 386–391.

Thorpe, M. (1999). *A Brief Review of Wave Energy*. Report ETSU-R-120 for the DTI, AEA Technology.

World Atlas (1998). 1998 World Atlas and Industry Guide. *International Journal on Hydropower and Dams*. Surrey, UK: Aqua-Média International.

WEC (World Energy Council) (1994). *New Renewable Energy Resources. A Guide to the Future*. London: Kogan Page.

WEC (World Energy Council) (2007). *Survey of Energy Resources 2007*. www.worldenergy.org/publications.

Worldwatch Institute (2007). *Biofuels for Transport: Global Potential and Implications for Sustainable Agriculture and Energy in the 21st Century*. Earthscan Publications, Ltd.

Further reading

Martinot, E. (2007). *Renewable Energy Futures*. www.martinot.info/futures.htm.

Martinot, E., Dienst, C., Weiliang, L. and Qimin, C. (2007). Renewable energy futures: targets, scenarios, and pathways. *Annual Review of Environment and Resources*, **32** (2007), 205–239.

Trans-Mediterranean Renewable Energy Cooperation (TREC) (2008). *Clean Power from Deserts. The DESERTEC Concept for Energy, Water and Climate Security*. Hamburg: The Club of Rome. www.trec-eumena.org.

WEC (World Energy Council) (2007). *Survey of Energy Resources 2007*. www.worldenergy.org/publications.

Worldwatch Institute (2007). *Biofuels for Transport: Global Potential and Implications for Sustainable Agriculture and Energy in the 21st Century*. Earthscan Publications Ltd.

Zweibel, K., Mason, J. and Fthenakis, V. (2008). A solar grand plan. *Scientific American*, Jan. 2008.

6

Carbon capture and storage

Clemens Cremer

This chapter describes the option of energy conversion of carbon-containing fuels with the capture and storage of the associated CO_2. The main capture technologies in the field of hydrogen production are compared with capture technologies for electricity generation. Following the process chain of carbon capture and storage (CCS), transport and storage options for CO_2 are discussed. Further societal issues, such as legal and regulatory aspects, as well as public perception are examined.

6.1 Why carbon capture and storage?

The world energy supply is still strongly dependent on fossil fuels. According to the IEA (2006), in 2004 some 80% of the world total primary energy supply originated from fossil fuels. As described in Section 2.1.2, the use of fossil fuels and the associated greenhouse-gas emissions are the major source for human-induced climate change; nevertheless, there is a good probability that fossil fuels will also play an important role for energy supply in the coming decades. This holds true not only for the conventional applications, such as electricity, but possibly also for the generation of hydrogen. Reasons for the continued use of fossil fuels are, amongst others, the favourable economics and the physical properties, such as high energy density for use in the transport sector.

Increasing energy efficiency and energy production from renewable sources have the potential to reduce GHG emissions in the long term. However, implementing energy-efficiency measures and adopting an energy-source switch from fossil fuel to renewable energy at a realistic pace will alone not be sufficient to meet the reduction in CO_2 emissions required over the next half century.

Another approach to reducing CO_2 emissions from fossil fuels use is the development of CO_2 capture and storage technologies (CCS).[1] With these, fossil fuels should be converted into electricity, hydrogen or other secondary energy carriers while

[1] The capture and storage of CO_2 has been investigated for several decades already. However, it was only during the 1990s that this approach of avoiding GHG emissions reached an intensive discussion in the science community.

The Hydrogen Economy: Opportunities and Challenges, ed. Michael Ball and Martin Wietschel. Published by Cambridge University Press.

keeping CO_2 emissions into the atmosphere low. This should be reached by separating the bulk of the CO_2 produced (capture) and storing it in a location outside the atmosphere (storage). The underlying principle could be applied to any energy conversion process involving carbon-based energy carriers. Carbon capture and storage plays an important role in strategies for future electricity generation systems, as today, this sector contributes roughly 40% of world CO_2 emissions, of which coal-fired power plants account for some 70% and natural gas fired plants for some 20%. Consequently, the electricity industry and policy makers for this sector are the main drivers for these technologies. Nevertheless, CCS is also important for hydrogen production, as it allows the production of hydrogen from fossil fuels without the negative impact of significant greenhouse-gas emissions. Moreover, with CCS coming into play, the integration or co-production of hydrogen and electricity becomes an interesting option (see also Chapter 16).[2]

6.2 Capture of CO_2

The capture of CO_2 from process streams has been performed already for some decades in various industries. Examples of the capture of CO_2 from industrial applications are the production of hydrogen-containing synthesis gas for the production of ammonia or synthetic fuels or the purification of natural gas (from contaminated gas fields). Since in these processes, CO_2 is usually considered an undesired by-product, it is then released into the atmosphere.

In the field of electricity generation, in particular, intensive research is carried out on capture processes, as this industrial sector contributes heavily to worldwide greenhouse-gas emissions. Basically, three concepts for CO_2 capture in electricity generation can be distinguished: post-combustion capture, pre-combustion capture and oxyfuel (see Fig. 6.1; for a detailed description of the various capture processes, see, for example, BMWA (2003); IPCC (2005)).

The capture of CO_2 involves the introduction of a separation process somewhere along the conversion chain. There are different separation techniques that could be applied, depending on the specific circumstances. First of all, CO_2 can be separated from a gas stream by *absorption* with a solvent, where one can differentiate between chemical and physical solvents. In a *chemical solvent*, the gas undergoes a chemical bonding with the solving agent. In chemical scrubbers, the gas is brought into contact with the scrubbing solution in absorber columns to remove the CO_2. The scrubbing solution is then recycled and the solvent regenerated in a desorber (stripper tower); to

[2] There are several initiatives for the development and demonstration of CCS projects in combination with the co-production of electricity and hydrogen, particularly from coal in IGCC power plants. Among the earliest announced was the US FutureGen Initiative (see www.futuregenalliance.org); however, at the time of writing it was not yet clear whether the project would still go forward or be cancelled. Comparable in the general concept is the EU's DYNAMIS/HYPOGEN programme, aiming at the large-scale demonstration of electricity and hydrogen production from decarbonised fossil fuels (see www.dynamis-hypogen.com). Initiatives are not limited to the EU or the USA; in China, for instance, the GreenGen initiative pursues similar goals as well (see www.greengen.com.cn/en).

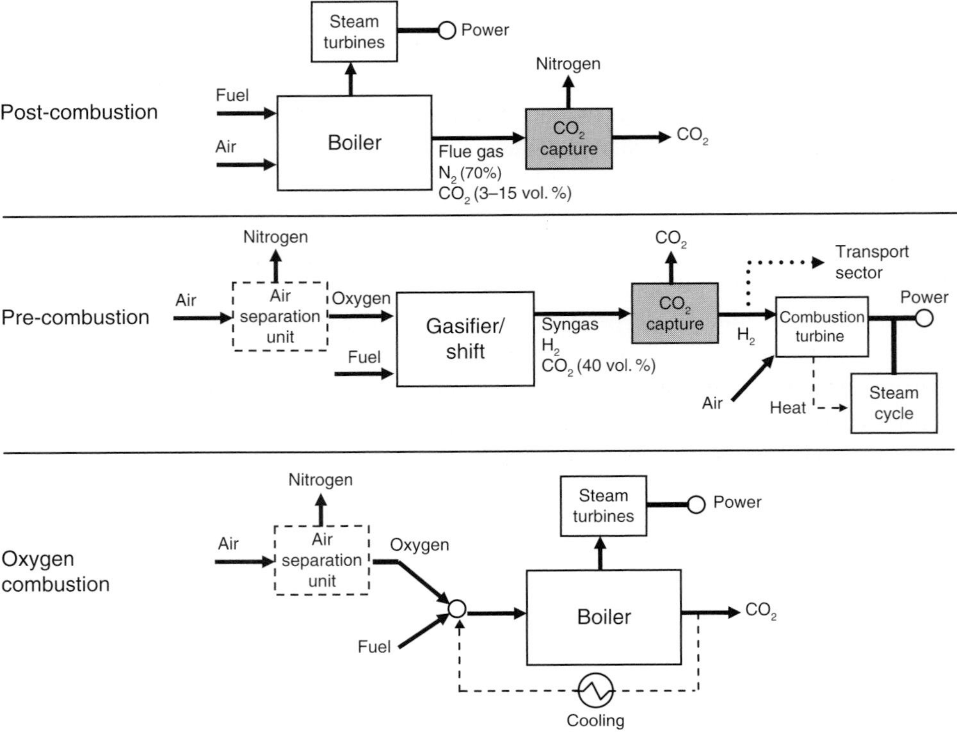

Figure 6.1. CO_2 capture configurations.

desorb the gas, the temperature has to be increased by applying thermal energy (steam) to the complex of solvent and absorbed gas molecules; at best, the heat requirement is between 2.7 and 3.3 GJ/t CO_2 (IPCC, 2005). Chemical absorption is more effective than physical absorption when the gas that has to be separated is under low to medium partial pressure. (Partial pressure is defined as the product of the total pressure of the gas stream and the CO_2 mole fraction.) When the gas is under high pressure and in high concentration, *physical absorption* is more appropriate. In this process, the gas undergoes physical solution in a liquid; the physical absorbent is regenerated by pressure reduction (release of the dissolved CO_2). Generally, chemical absorption uses an aqueous alkaline solvent, usually an amine (e.g., mono-ethanolamine (MEA)) or potash; for physical absorption, e.g., methanol-based solvents (Rectisol) or solvents based on polyethylene glycol (Selexol) can be applied.

Besides absorption processes, *adsorption* processes, cryogenic separation or membrane separation can be applied. Adsorption processes are based on the physical attachment and bonding of components from the gas mixture on the surface of solid sorbents. As with absorption a distinction can be made between physical and chemical adsorption; the first one is referred to as pressure-swing adsorption (PSA), where the desorbtion of the CO_2 is achieved by pressure-swing operations, the second one as temperature-swing adsorption (TSA), with the desorbtion taking place by temperature-swing operations. Adsorption processes are mainly employed for CO_2 removal from synthesis gas for hydrogen production (see also Section 10.9). All capture processes can reach capture rates above 90 vol.%.

In post-combustion capture, the currently most favourable option for separating the CO_2 from the flue gas, is with the help of a scrubber using a chemical solvent. The pre-combustion capture is based on the generation of a hydrogen-rich syngas, either by gasification of solid fuels or by steam reformation of gaseous fuels. From this syngas, the CO_2 can be separated by a physical solvent or by the use of pressure-swing adsorption; for low partial pressures, chemical solvents also are used. The oxyfuel concept follows a different approach: here the carbon-containing fuel is combusted with pure oxygen instead of air, producing a flue gas containing mainly CO_2 and H_2O. The latter can be condensed, leaving CO_2 in high concentration as a product. Different from the post-combustion and the pre-combustion capture, where CO_2 is separated from a gas stream, the main separation task in the oxyfuel concept is the preparation of pure oxygen. For the production of hydrogen, only the option of decarbonising the syngas is applicable. Compared with the decarbonisation of the flue gas, pre-combustion capture takes place in a smaller reaction volume with lower volumetric flow rates, which results in smaller absorber and stripper towers (and potentially lower costs of the capture plant).

The application of physical or chemical absorption processes strongly depends on the partial pressure of the CO_2 in the considered gas stream. Principally, the energy demand for the separation decreases with partial pressure, making processes dealing with a syngas more efficient. The flue gases of electricity generation processes are usually at ambient pressure and contain CO_2 in the range of 3 to 15 vol.% (2.5%–4% for natural gas fired power plants and 12%–15% for coal-fired power plants), making chemical absorption more efficient. Unlike flue gases, the syngases of gasification processes or steam reforming are mostly generated at a total pressure of 40 to 70 bar and have a CO_2 concentration typically above 30 vol.%; this makes processes of physical absorption more efficient. Also advantageous for physical scrubbers is the easy regeneration of the solvent by pressure reduction, compared with the much more energy-intensive regeneration process of chemical scrubbers. As the generation of a syngas is a prerequisite for hydrogen production from fossil fuels, physical-absorption processes and pressure-swing adsorption will play a greater role for a hydrogen economy than absorption with chemical solvents.

Reducing the energy demand for the separation of CO_2 is a main parameter for optimisation in the development of CO_2-capture processes (particularly for post-combustion processes, where thermal energy is needed for the solvent regeneration). This is the case, irrespective of whether the separation is performed in connection with hydrogen production (where the separation is inevitable even without CO_2 storage) or whether it is performed in connection with electricity generation. However, the ultimate appraisal of the energy efficiency of these processes can only be made by taking into account the complete system design with all preceding and following steps and their energy requirements. This also includes the energy needed to compress the CO_2 recovered to the final pressure required for pipeline transport (or, likewise, the liquefaction of the CO_2). Compression power can be significantly reduced if the CO_2 can be captured under pressure.

The principal processes for CCS plants, such as gasification of solid fuels or gas-separation processes, have been employed in large industrial scale for a long time already. Nevertheless, there are still some key points that have to be resolved for a successful implementation of CCS. On the one hand, the reliable and, at the same time, highly energy-efficient operation of capture processes has to be achieved and proven in real long-term operation. On the other hand, specific detailed technology elements are missing. In connection with the use of hydrogen for electricity generation, adapted gas-turbine burners have to be developed. Presently, existing high-efficiency gas turbines are not designed for the use of pure hydrogen instead of natural gas; only hydrogen-rich mixtures (up to 45 vol.% hydrogen) are currently used. As this change in fuel mainly concerns the burner, the entire turbine does not have to be redesigned. When looking at electricity generation with CCS only, the further development of combustion technologies suitable for pure oxygen use is another critical point for the advancement of CO_2 capture.

6.3 Transport of CO_2

6.3.1 Pipeline transportation

Large-scale pipeline transportation of CO_2 has been performed in the United States since the 1980s, supplying CO_2 for enhanced oil recovery (EOR) projects.[3] The demand for CO_2 for EOR, especially in Western Texas, has led to the construction of the largest network of CO_2 pipelines worldwide (see Figure 6.2). In total, there are about 5 800 km of CO_2 pipelines in the United States (CRS, 2007). Worldwide, there are CO_2 pipelines with a transport capacity of about 45 million tonnes of CO_2 per year (Gale and Davison, 2002). Compared with the pipeline networks for natural gas and other gases (e.g., 800 000 km in the USA), the existing CO_2 pipeline networks are of minor extent.

The technical requirements of a CO_2 pipeline can be derived largely from the example of natural gas pipelines (Skovholt, 1993) with the main elements:

- Piping with high-quality coated carbon steel, protected against exterior corrosion and mechanical damaging,
- Initial compressor station,

[3] Enhanced oil recovery is a method used for oil extraction in partly depleted oil fields. One variant of EOR is based on injecting CO_2 into the reservoir. This has a twofold effect: first, it increases the pressure in the reservoir and, second, the CO_2 can reduce the viscosity of the oil. Both effects help improve the flow of oil to the production wells and, thus, increase the oil production, compared with recovery without EOR.

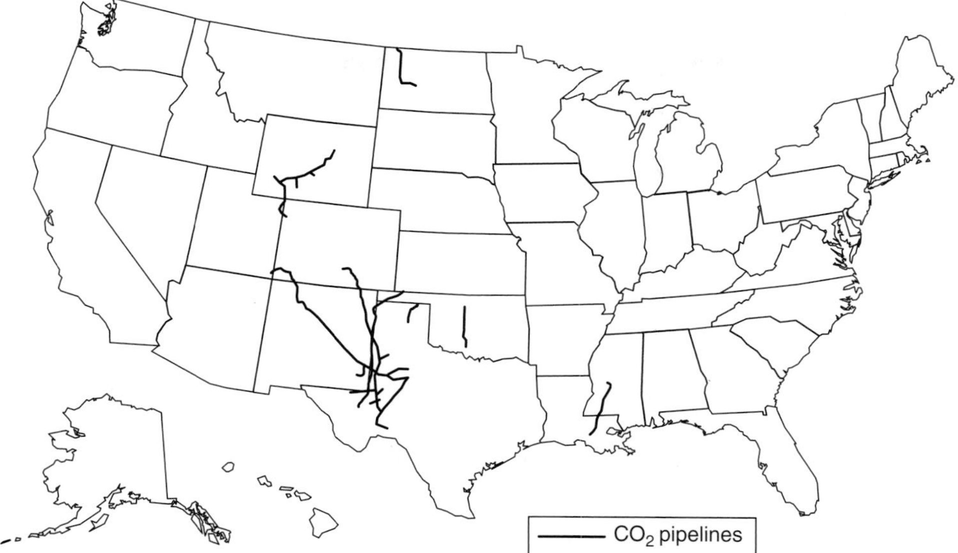

Figure 6.2. Map of the CO_2 pipeline network of the United States (CRS, 2007).

- Pumping or recompression stations,
- Section valves and security valves,
- Cathodic corrosion protection,
- Stations for corrosion monitoring.

To reach the highest possible mass transfer in a given pipeline diameter, the CO_2 should be conditioned into a state of high density. The most suitable solution for this requirement is to compress the CO_2 up to the supercritical or dense phase. This operation mode allows a high density of the transported goods, whilst avoiding the risk of two-phase flow, during which cavitation and damage to the infrastructure could occur (Hendriks *et al.*, 2003). Consequently, a pipeline should be designed such that the pressure in the pipeline stays above the critical pressure of 7.38 MPa along its entire length. With a safety margin, the pressure demand is also stated to be as high as 8 MPa (Egberts *et al.*, 2003).

Owing to the flow friction in the pipeline, a pressure drop occurs, making recompression of CO_2 necessary along the way. Doctor *et al.* (2000) state a pressure drop of 0.15 bar/km calculated in a baseline study for the Weyburn project. According to Heddle *et al.* (2003), recompression will be needed for distances larger than 150 km. There are cases in the United States of longer distances without recompression, realised by making use of an overall downward slope of the pipeline. Another alternative to avoid recompression is to compress CO_2 to a higher pressure at the beginning of the pipeline, by taking into account the ratio of pressure drop and pipeline length.

Doctor *et al.* (2000) point out the technical problems in the transport infrastructure that could arise from impurities in the CO_2. Any transport system requires the CO_2 to be dried to prevent the formation of CO_2 hydrates. Considerable problems with the formation of iron sulphide in natural gas pipelines indicate that CO_2 also has to be cleaned of hydrogen sulphide content.

6.3.2 Risks and safety with pipeline transportation

During transportation, substantial amounts of CO_2 could escape from pipelines if major leaks or breaches occur. Such failures can be provoked by corrosion or external damage. Major causes for externally induced failures are construction works with associated excavations. Carbon dioxide itself is not toxic but can affect the human body when concentrations reach 6% to 7%. Concentrations of 10% and more are usually lethal (Gale and Davison, 2002). Because the density of CO_2 is larger than the density of air, it can accumulate in depressions, imposing a risk for human and animal life. The selection of the pathways of a CO_2 pipeline outside of depressions and valleys, but along topographically exposed positions with higher ventilation can reduce the risk of CO_2 accumulation. Like methane, CO_2 is colourless and odourless and, thus, cannot be sensed in time by human beings. The addition of strongly scented trace gases in the gas stream would increase the safety, as is done in natural gas pipelines.

Because CO_2 is not flammable, the risks arising from CO_2 leaking from a pipeline would be much lower than the risk from natural gas leakages. Simulations have shown that leaking CO_2 from a buried pipeline moves mainly vertically upward and is dispersed quickly (Hendriks *et al.*, 2003).

Besides the risk of external damage, another implementation problem for CO_2 pipelines is corrosion prevention. Corrosion can be caused by the presence of water in the CO_2, which generates the corrosive acid H_2CO_3. Therefore, it is important to transport CO_2 in a dry condition, which is achieved by the dehydration after the initial compression.

Measures for the minimisation of risks could be:

- Safety zones along both sides of the pipelines (distances to buildings),
- Increased wall thickness of pipelines in inhabited areas,
- Reduced distance of safety valves in inhabited areas,
- Suitable above-ground marking of the pipeline to prevent damages resulting from construction work,
- Monitoring of the pipeline.

Altogether, the risks of failure of CO_2 pipelines are considered lower than the risks associated with pipelines for hazardous liquids. Compared with natural gas, the occurrence of failures is considered to be in the same level but with significantly fewer harmful consequences (Gale and Davison, 2002).

6.3.3 CO_2 transportation in ships

Transportation of CO_2 by ships is suitable for offshore geological CO_2 storage and is favourable for longer distances (Heddle *et al.*, 2003). One of the options where the use of ships could be favourable is the transportation of CO_2 from the onshore harbour with an intermediate storage facility to the underground geological storage site located offshore. There, CO_2 could be transferred into an injection well via a vertical pipeline. Ships would most probably come into play to supply offshore enhanced oil recovery projects with CO_2 (compare, e.g., Barrio *et al.*, 2004).

At the time of writing, there are a few specialised shipping companies in Europe transporting CO_2 on a small scale as well as companies providing industrial gases, which also offer this kind of transport service. The tank vessels used for this purpose have capacities of the order of 850–1400 tonnes. The tankers are designed for transporting CO_2 at 1.4 to 1.7 MPa and at temperatures of −25 to −30 °C (Odenberger and Svensson, 2003).

When transported in ships, CO_2 is usually stored in cold liquid phase. This state is preferred to the supercritical phase because the wall thickness required for maintaining a pressure of about 8 MPa would become unacceptably high. Depending on the size of the individual tanks, the necessary wall thickness could even exceed the material quality that could be manufactured with existing technologies.

Given the large amounts of CO_2 generated at sensible energy-conversion plants, special ships designed for the transport of these quantities of CO_2 would have to be built. The design could be similar to the design of ships transporting LNG. Initial studies for the design and operation of ships for CO_2 transport have been published by the IEA-GHG (2004).

The transportation of CO_2 by sea necessitates the construction of temporary storage facilities at the points of loading and at the injection points, depending on the rate of injection into the storage well.

6.4 Storage of CO_2

6.4.1 General storage options

Any storage solution for CO_2 has to fulfil two first-order requirements. The storage has to retain the CO_2 for a sufficient duration outside the atmosphere and it has to have acceptably low ecologic impacts. Further requirements are sufficiently large storage potential and acceptable costs.

There are several options that principally could be used for the storage of CO_2 while fulfilling these requirements. Generally, one could distinguish between storage in the water column of the oceans, storage in geological formations, storage by mineralisation and storage by fixation in organic matter. Each of the four approaches can again be divided into a number of variants. This analysis will be limited to the variants of storage in geological formations. (For a more detailed discussion of the various storage options, see Cremer (2005).) The reasons for the exclusion of the other options are as follows.

Although storage in the ocean has attracted a lot of attention in several countries (particularly the USA and Japan) and there have been numerous research projects exploring this option, it still seems hard to realise. The opposition of stakeholders, especially of environmental groups against storage in the ocean waters is very high. There are strong misgivings that the introduction of large amounts of CO_2 could have adverse effects to the biosphere in the oceans. Further, the durability of storage in the oceans is doubted. Owing to these strong objections against storage in the open ocean, many industrial stakeholders of CO_2 storage do not pursue this option any further; in the EU CCS Directive, the storage of CO_2 in the water column is prohibited (EP, 2008a). The options for mineralisation and storage by fixation in organic matter are still at the level of basic research and not yet ready for implementation. Nevertheless, these limitations should in no way imply that these approaches are not feasible or could not contribute to the storage of CO_2 in the future.

6.4.2 Storage of CO_2 in geological formations

The storage of CO_2 in geological formations principally would mimic natural occurring processes where CO_2 is withheld in geological formations. Natural CO_2 accumulations can be a result of biological activity; of the decay of organic matter. It can also be derived from volcanic activity or from fluid–rock interaction. The storage mechanism can be trapping by impermeable structures in gaseous or supercritical form or solution in fluids. Carbon dioxide is, e.g., often found as a by-product in natural gas reservoirs. The existence of CO_2 accumulations with ages up to millions of years gives clear indication that natural storage mechanisms exist that exceed the duration required to address climate change. This does, however, not prove a priori that it will be feasible to copy these mechanisms with means of technology. These reservoirs are investigated to gain an understanding of the transferability of natural CO_2-storage mechanisms to storage reservoirs created by human activity (see, e.g., Baines and Worden (2000) or Stevens *et al.* (2000)).

Probably the most important options for geological storage are storage in oil and gas fields, storage in saline aquifers and storage in coal seams. Further, there are options like the storage in abandoned mines or in salt caverns (see Table 6.1).

6.4.2.1 Storage in coal seams

Storage in coal seams is a technically feasible option because CO_2 has a high affinity for adsorbtion to coal. Normally, coal is at least partly adsorbed with methane. Of these two molecules, CO_2 has a greater affinity to the coal. The ratio of affinity varies greatly, depending on the type of coal. It may range from close to one in anthracite to ten in younger coal types (Stanton *et al.*, 2001). In consequence, if entered into a coal

Table 6.1. *Storage of CO_2 in geological formations*

Main variant	Options
Storage in oil and gas fields	Depleted oil and gas fields
	Enhanced oil or gas recovery
Storage in saline aquifers	Low-temperature aquifers
	Geothermal aquifers
Storage in coal seams	Enhanced coal-bed methane production
	Storage in unminable coal seams
Storage in salt caverns	
Storage in abandoned mines	

seam, CO_2 replaces the present methane. In molecular terms, coal seams can adsorb at least twice as much CO_2 as methane. So, if any methane released from the coal seams was burned, there would still be a reduction of overall CO_2 emissions into the atmosphere of at least 50%. It should be noted, though, that the reduction of greenhouse-gas emissions depends drastically on the capture efficiency of the released methane. The net emission balance might change drastically towards a lower emission reduction or even a net increase of emissions, if the complete amount of the released methane was not burned. This change in balance is caused by the higher global warming potential of methane compared with CO_2. A very important prerequisite for the feasibility of CO_2 storage in coal seams is a sufficient *in-situ* permeability of the coal seams in question. According to Christensen and Holloway (2003), this technology is still at an early stage of development.

6.4.2.2 Storage in oil and gas fields

The storage of CO_2 in oil and gas fields has generally to be separated into storage in depleted reservoirs and storage in connection with enhanced recovery of hydrocarbons. Both options principally use the pore volume that previously had been filled with hydrocarbons or is to be depleted from hydrocarbons at the very moment of storage.

Storage in oil and gas fields makes use of generally well explored geologic structures, since the production of oil and gas usually requires intensive and systematic exploration. Furthermore, the presence of oil and gas that had remained in the reservoir over geological time scales indicates that the structures are fully confined by sealing layers of low permeability. Hydrocarbon reservoirs offer especially favourable storage sites when located in structural or stratigraphic traps (such traps are bounded by unconformities, anticlines, non-transmissive faults or facies changes). Owing to the hydrocarbon production activity, there is usually a fully developed infrastructure existing, that could theoretically be used at least in part for the CO_2 storage operations.

6.4.2.3 CO_2 storage in the context of enhanced hydrocarbon recovery

Carbon dioxide could not only be stored in depleted hydrocarbon reservoirs but can also be used for the improvement of production of operational oil fields today. The background for these so-called enhanced oil recovery (EOR) activities is the objective of increasing the overall production and the production rate of existing oil fields. Usually only a fraction of the overall oil in a field is produced, with the remainder staying in place. For this remaining oil it would not be economically attractive to undertake further measures to allow its production. The actual fraction recovered depends on the specific reservoir geology and behaviour, as well as on the oil market situation during production. As a rule of thumb, it could be said that on average one-third of the oil in place is produced, with two-thirds staying in the field.

With decreasing production rates of a field, the producer can undertake measures to increase productivity or, at least, to slow down the production decrease. Next to secondary production (which usually means an increase of reservoir pressure by water or gas injection), measures for tertiary production can be undertaken, which aim at reducing the viscosity of the oil in the reservoir, with a simultaneous pressure increase. One way of achieving this is the injection of CO_2, which is easily dissolved in oil with medium to low density. The dissolution of CO_2 in the crude oil causes it to swell and reduces its viscosity. Together with the pressure increase resulting from the mass injection of CO_2, an increase of production rate can be achieved. Based on data from 25 case studies in the United States, Holt *et al.* (1995) calculated an average of 13% of the original oil in place that could be extracted by CO_2 injection. Hendriks *et al.* (2004) calculated a value of 12% of the original oil in place as a best estimate of what could be produced by CO_2 injection. (As a low estimate, Hendriks *et al.* (2004) calculated 5% and, as a high estimate, 20%.) Enhanced oil recovery with CO_2 injection is widely used in oil fields in Texas, which are supplied with CO_2 from natural and industrial sources. The CO_2 is transported in a long-distance pipeline network (see Section 6.3).

The economic viability of enhanced oil recovery with CO_2 flooding is limited to the cases where CO_2 is available at low cost. This is the case for the West-Texas fields, where CO_2 from natural sources is available, or at the Weyburn field in Canada, where CO_2 can be made available from the North Dakota Gasification Plant. Unlike the situation in the United States, almost the entire European oil production is located offshore. There, enhanced oil recovery activities would be more costly, simply resulting from the larger spatial extent of production units and the entire higher costs of offshore operations. Furthermore, there are no low-cost industrial CO_2 sources available in the closer vicinity of the North Sea oil fields, nor is there an infrastructure for CO_2 transport.

Another important issue for the operation of EOR is the temporal restriction in connection to the field development in the North Sea. A large number of fields whose development had been started in the 1970s are approaching the end of economic

Table 6.2. *Estimated storage potential of the most promising geological storage options*

Reservoir type	Lower estimate of storage capacity (Gt CO_2)	Upper estimate of storage capacity (Gt CO_2)
Oil and gas fields	675	900
Unminable coal seams (ECBM)	3–15	200
Deep saline formations	1000	Uncertain, but possibly 10 000

Source: (IPCC, 2005).

production with primary and secondary production methods. As the oil industry does not maintain the production installations after the end of (economic) production, the offshore installations will be dismantled and the wells will be closed if there is no prospect for further economic production at a field within a short time period. For the chances for realisation of CO_2 EOR, this indicates that there is a rather narrow time window of possibly ten years in the future. After that time, a larger number of fields might be closed. Of course, the production could be started over from scratch. For a new start, however, it has to be assumed that the costs would be higher than in the case of a transition with the use of existing infrastructure and wells.

6.4.2.4 Saline-aquifer CO_2 storage

From the theoretical potential, saline aquifers are the prospectively largest sink for CO_2, apart from storage in the water column of the oceans (see Table 6.2). The principle is comparatively simple. Storage in saline aquifers is performed by introducing supercritical CO_2 into a well that reaches into a deep saline aquifer with a sufficiently high permeability. The well is designed with an appropriate filter length, to allow the CO_2 to flow into the aquifer at the required rate. The storage in saline aquifers should be done with the CO_2 in a dense, supercritical phase (i.e., at a pressure above 7.38 MPa), in order to use the available pore space most efficiently and at low costs. In consequence, the storage reservoir should be located at a depth of at least 1000 m to guarantee a sufficient reservoir pressure. On the other hand, the reservoir formation should not be located at too large depths as then the drilling costs would be prohibitively high.

A series of processes will control the behaviour of CO_2 in saline aquifer formations. First, the CO_2 will displace the formation water (brine) originally in place and will lead to a local increase in pore fluid pressure (van der Meer, 1992). The injected CO_2 will not be distributed evenly, but will finger out, owing to the lower density than the pore waters and the heterogeneities of the aquifer. Doughty *et al.* (2001) point out that the shape of the CO_2 plume in the aquifer will be highly site- and case-specific. Carbon dioxide will rise to the top of the aquifer and migrate at the bottom of the confining layer up to the locally highest point in the aquifer. Simultaneously, dissolution will take place into the pore waters. The water with dissolved CO_2 has a higher density than the original brine and will have a tendency to flow downward in the aquifer. Lindeberg and Wessel-Berg (1997) have postulated that this behaviour could lead to convective cycles in the aquifer. The results of simulations on the dissolution processes vary from a 30% dissolution of the total amount of CO_2 in the water of the aquifer up to complete dissolution in the formation water being reached ultimately (Ennis-King *et al.*, 2003; Law and Bachu, 1996; McPherson and Cole, 2000). Beyond simple trapping and dissolution in the brine, chemical interactions of the CO_2 with the rock matrix can lead to mineralisation and provide an even more secure storage mechanism.

Mineralisation leads to the strongest fixation of the CO_2 in the aquifer. Whether, and at what reaction rate, it occurs depends on the specific geochemical situation. It should be noted that mineralisation of CO_2 is considered as desirable, since it leads to a strong fixation but could have adverse effects too. The mineralisation processes usually go hand in hand with dissolution of other parts of the rock matrix. So mineralisation could lead to an increased concentration of hazardous trace elements in the brine that were formerly bound to the rock matrix.

6.4.3 Global CO_2 storage potential

A key question related to the storage of CO_2 is the available storage potential in the reservoirs under discussion. When putting a preference to geological storage of CO_2, sedimentary basins are the structures of primary interest, as these are most likely to offer sufficient pore space for the storage of CO_2. Figure 6.3 shows the sedimentary basins with geological storage prospectivity, including saline formations, oil or gas fields, or coal beds.[4] There is also a potentially good correlation between major CO_2 sources and prospective sedimentary basins, with many sources lying either directly above, or within reasonable distances (less than 300 km) of areas with potential for geological storage (IPCC, 2005).

Even though there is a broad distribution of sedimentary basins, it is not possible to draw direct conclusions with respect to the storage potential. For any single geological formation that principally could be suitable for storage, the specific properties have to be examined, to determine the feasibility of safe storage. The extent of sedimentary basins does not give a direct indication of the global potential for geological CO_2, as several more criteria, such as the existence of a confining structure or the availability of pore space have to be favourable. Existing estimations are hence of a very preliminary nature and show a considerable degree of variation (see Table 6.2).

[4] Prospectivity is a qualitative assessment of the likelihood that a suitable storage location is present in a given area based on the available information.

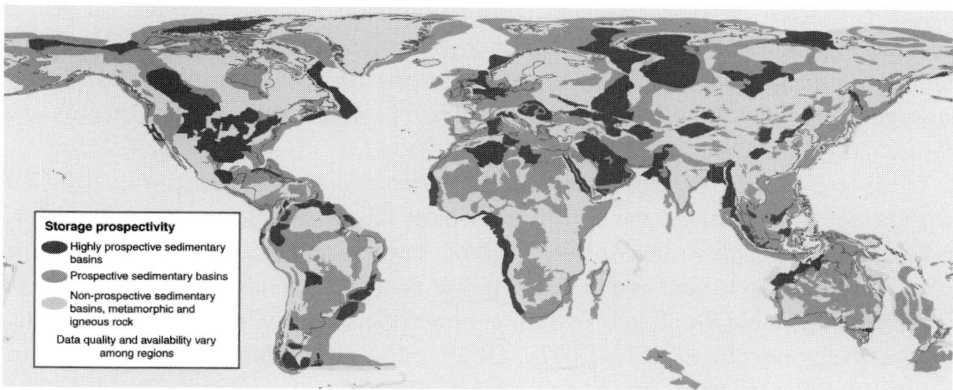

Figure 6.3. Prospective areas in sedimentary basins for CO_2 storage (IPCC, 2005).

6.4.4 Current CO_2 storage projects

At the time of writing, there were only three major CO_2 storage projects worldwide, all of them related to hydrocarbon recovery: Sleipner/SACS (Norway), In Salah (Algeria) and Weyburn (Canada/USA). For all three, the amount of CO_2 injected is in the range of 1 Mt CO_2 per year.

Large-scale storage of CO_2 in saline aquifers has been carried out in the SACS (Saline Aquifer CO_2 Storage) project at the Sleipner gas field in the North Sea since 1996. It involves the production of natural gas requiring gas treatment for the reduction of the CO_2 content. The main economic driver for this early project is the tax on CO_2 emissions originating from offshore operations on the Norwegian continental shelf (around €50/t). Additional to conventional gas treatment, the SACS project involves the storage of the CO_2 in an aquifer below the reservoir structure (the Utsira sandstone, a formation with very good permeability, is used as the storage structure). The SACS project has been used for extensive research on the CO_2 storage in saline aquifers. Investigations have focused on the storage potential of the formation (Chadwick *et al.*, 2000), rock alteration caused by CO_2 injection, simulation and prediction of CO_2 behaviour in the aquifer (Lindeberg *et al.*, 2000; van der Meer *et al.*, 2000; Zweigel *et al.*, 2000) and also on the monitoring of the CO_2 behaviour (Arts *et al.*, 2000).

The In Salah gas project took up operation in Algeria during summer 2004. There as well, a CO_2-rich natural gas is treated to reach commercial concentrations of CO_2. After separation in an amine plant, the CO_2 is stored in the gas-containing reservoir structure. Using the same structure for production and storage of course involves the risk of a breakthrough of CO_2 into the production wells.

Unlike the SACS project and the In Salah project, the Weyburn project has combined the storage of CO_2 with EOR operations in the Weyburn oil field in Canada since 2000. The CO_2 for storage originates from a gasification plant in North Dakota (USA), where synthetic natural gas is produced from a lignite resource. The distance between the North Dakota gasification plant and the Weyburn oil field is covered by a pipeline of some 300 km long. With the combination of a CO_2 capture unit in an energy conversion plant, of a pipeline for CO_2 transport and of CO_2 storage in combination with EOR, the Weyburn project combines key elements of a future CO_2 value chain.

There are several initiatives for the development and demonstration of CCS worldwide. For instance, the European Union has an ambition to deploy 10–12 full-scale CCS demonstration projects within Europe by 2015, testing various ways of integrating CCS in coal and gas-fired power generation; it also aims for CCS to be commercially viable for all new fossil fuel power plants by 2020, with existing plants progressively retrofitted (EC, 2007). Other pilot and demonstration plants are planned in the United States, Australia and China.[5]

6.5 Costs of CCS

The costs of CCS mainly arise from its three constituting components:

- *Capturing* the CO_2 from fossil-fuel conversion processes. The costs for capturing CO_2 result from (1) higher investments for the system with CO_2 capture compared to a system without, (2) the lower efficiency of the process as a result of the operation of the capture plant (e.g., from the additional energy demand for solvent recovery, CO_2 compression, etc.) and (3) the operational cost for the capture process (e.g., for absorbent replacement).
- *Transportation* of CO_2 to the point of storage.
- *Storage* of CO_2. These costs are strongly determined by the number and depth of wells required for underground storage and the operation of the necessary surface facilities.

Although costs for CCS are very case- and location-specific and estimates change rapidly in response to technological developments, Fischedick *et al.* (2007a; b) indicate the average distribution of costs for the total CCS chain as follows: 63% capture, 15% compression, 10% transport and 12% storage. The actual capture costs are highly volatile and difficult to compare among different studies, as they depend very much on technical, economic and financial factors specific to the design and operation of the production process or power plant, as well as the design and operation of the CO_2-capture technology. Uncertainty in the cost of CO_2 capture and storage has been analysed by the IEA (Gielen, 2003). In this study, the reference system is identified as the largest uncertainty for future cost of separation; other major influences on the economics of capture and storage are fuel prices, technology learning and possible leakage of CO_2 from reservoirs. Moreover, the degree of integration of syngas production, electricity production (as a by-product), hydrogen separation and CO_2 separation can vary strongly, making evaluation of the cost of

[5] For more information about proposed CCS projects refer to CERA (2007), IEA (2006), or the IEA GHG R&D Programme (www.ieagreen.org.uk); general CO_2 projects database is also available at www.co2captureandstorage.info.

CO_2 capture difficult. It is, therefore, necessary to specify separate cost and efficiency penalties for the different processes.

6.5.1 Capture costs for hydrogen production

The (additional) costs of CO_2 capture in connection with hydrogen production from natural gas or coal are mainly the costs for CO_2 drying and compression, as the hydrogen production process necessitates a separation of CO_2 and hydrogen anyway (even if the CO_2 is not captured). Total investments increase by about 5%–10% for coal gasification plants and 20%–35% for large steam-methane reformers (see also Chapter 10).

6.5.2 Transportation costs

A key factor in the CO_2 capture and storage system is the need for a functioning transportation system for the CO_2 captured at energy conversion units. The transport options suitable for the quantities produced by large power stations or industrial plants would mainly be pipeline transport for onshore distances and ship transport for the offshore area.[6]

Transport costs are strongly influenced by the infrastructure costs for pipelines or by the costs for capital of ships. For the case of pipeline transportation, the mass flow rate also has strong influences on the costs, since the transport capacity of pipelines is related to the square of the diameter. Main parameters governing the unit transport costs are, hence, the transport distance, the CO_2 mass flow rate and the nature of the environment (onshore or offshore; mountains, rivers or populated areas); in addition, there are legal costs. Additional costs for recompression (booster pump stations) that may be necessary for longer pipelines are rather negligible. Lastly, steel costs account for a significant share of pipeline costs and, thus, volatility in steel prices can markedly affect overall pipeline economics. According to the IPCC (2005), the pipeline transport costs for a 250 km distance vary from US$1–8 per tonne of CO_2, depending on the mass flow rates.

6.5.3 Storage costs

Storage costs basically include screening and exploration costs, well-drilling costs and injection equipment and flowline costs. On top of that come costs for measurement, monitoring and verification (MMV) of CO_2 storage. There is a large variation for the storage cost estimations of CO_2. Some indications can be found, for example,

[6] Large railway trains could be used as a means of transport for onshore distances as well (compare Odenberger and Svensson, 2003). However, the specific costs estimated by these authors for railway transportation were considerably higher than for pipeline transportation.

in the summary report to the European GESTCO project (Christensen and Holloway, 2003). Assuming a euro–dollar parity, the costs obtained in this work for storage range from a minimum value of US\$0.3 per tonne of CO_2 up to a maximum value of \$37.7 per tonne, with a mean of \$3.1 per tonne. The IPCC (2005) states costs for geological storage in the range of US\$0.5–8.0 per tonne CO_2 stored, with, additionally, US\$0.1–0.3 per tonne for monitoring and verification. The cost ranges indicated in this report for storage in the ocean or storage by mineralisation are higher than those for geological storage.

The high degree of variation in cost estimates for storage operations comes from several factors. First of all, the large differences between the geological conditions of possible reservoirs create a variation in costs. Reservoirs may differ in depth or in permeability, which determines the need to drill additional wells or to extend the inflow distance by horizontal drilling. Further differences can be related to the cap rock quality or to the location of the reservoir. Cost estimates are also dependent on the applied costs for drilling campaigns and material. These vary greatly with the economic conditions in oil and gas industries.

As CO_2 is principally a marketable product to the hydrocarbon industries, the economically most attractive solution for storage would certainly be an EOR project. Within such a project, the storage process of CO_2 could generate a stream of income for this mitigation option. Within Europe, the core problem standing against this option is to provide for a transport solution at lower specific cost than the revenue obtained for EOR. Storage in connection with EOR offers the possibility of improving the economics of the carbon capture and storage. Although the value of CO_2 for EOR is still limited, this option should not be underestimated, as a future rise in oil price would increase the value significantly.

On average, the costs for CO_2 transport and storage are between US\$5 and \$8 per tonne CO_2. Assuming 280 g CO_2/kWh_{H_2} for hydrogen production from natural gas and 570 g/kWh_{H_2} for hydrogen from coal, CO_2 transport and storage costs translate to 0.14–0.22 ct/kWh_{H_2} and 0.28–0.46 ct/kWh_{H_2} for natural gas and coal respectively. This increases total hydrogen production costs by 3%–5% in the case of natural gas and 10%–15% in the case of coal.

6.6 Legal and regulatory framework for CO_2 capture, transport and storage

Carbon dioxide capture and storage represents a novel technology concept. To be successful and become an integral part of the energy system, the entire concept, as well as the individual activities, have to be embedded into a multidimensional societal framework. First of all, this means that the entire value chain has to become an accepted part of greenhouse-gas emission mitigation strategies. Second, the activities have to be embedded into a legal and regulatory framework under which allowances and liabilities are defined. Third, they have to find acceptance in the public, as large-scale industrial activities usually become hardly practicable when there is strong and organised resistance arising from the general public. Fourth, CO_2 capture and storage projects, with their specific profitability prospects and risk structure, have to find capital on the equity and capital markets. The critical requirement of establishing a legal and regulatory framework to govern the deployment of CCS is addressed in more detail in the following.

6.6.1 General legal and regulatory aspects of CCS

The analysis of the legal aspects of CO_2 capture and storage could be divided into three fields, paralleling the three technical domains of the process and leading, ultimately, to storage. There is the regulatory environment for the construction, operation and the dismantling of an industrial facility. Then there is the regulatory environment for the construction and operation of a transport system for CO_2 and finally there is the legal and regulatory environment for the storage of CO_2.

When investigating the entire regulatory framework of carbon capture and storage, it becomes quickly evident that the extent of legislation and regulation decreases drastically along the chain from capture to transport to storage.

Principally, there is no fundamental difference between an industrial plant such as a chemical plant or a power plant releasing exhaust gases including CO_2 through a stack and a plant with similar purpose but equipped with CO_2 capture. Notably, the extent and requirements of regulations for the installation and operation of such industrial facilities varies across the world. Nevertheless, regulation for this kind of activity is in place and should not impose major barriers to CO_2 capture. (The EU CCS Directive regulates CO_2 capture under an existing Directive (EC, 2008).)

The transportation of CO_2 in significant amounts on land would have to be performed in pipelines (compare, e.g., Radgen *et al.*, 2005). So far, in the United States only, there are examples of industrial-size CO_2 pipelines. On the other hand, there is an abundance of pipelines for the transportation of high-pressured gases, especially natural gas, and for the transportation of liquids, such as crude oil or naphtha, in many parts of the world. (According to Eurostat (2004), there are around 25 000 km of oil pipelines operating in Europe.) The ex-ante evaluation of the regulations for the construction and operation of pipelines for CO_2 transportation could thus draw on the legal framework of existing pipelines in the specific regions considered and on experiences of operating CO_2 pipelines in the USA. Furthermore, there are also pipelines transporting industrial gases, amongst them hydrogen (e.g., pipelines operated by *Air Liquide* in Northern France, Belgium and the Netherlands, see Chapter 12). There might be some differences in the regulatory process for CO_2 pipelines, especially concerning safety regulations, compared with pipelines transporting hazardous industrial gases. But altogether, the fact that natural gas pipelines and even pipelines transporting industrial gases can obtain permission in densely populated areas should justify the conclusion that the permission of CO_2 transporting pipelines will not impose insurmountable obstacles to CO_2 capture and storage activities either. (The EU CCS Directive only regulates third-party access to transport infrastructures, not technical issues of transportation of CO_2.)

For the last part of the process chain, the geological storage of CO_2, an important step has been achieved in the EU by a Directive on the geological storage of carbon dioxide (EC, 2008), closing a gap that has been discussed in the scientific community and by enterprise representatives of CO_2 intensive industries for a long time.[7]

Nevertheless, existing laws and regulations have to be analysed and interpreted as to how they could be applied to specific CO_2 storage cases. The situation is aggravated, as there are hardly any examples for CO_2 storage worldwide. Up to now, there are only the SACS CO_2-injection project at the Sleipner field in the North Sea and the In Salah gas project with a comparable activity in Algeria. These two activities could be judged as more-or-less true commercial CO_2 storage projects, as they are not associated with enhanced gas and oil recovery activities.[8] But even though they are not connected with enhanced production measures, they are connected with hydrocarbon production. As a consequence, their regulation falls under the regime of the respective regulations for oil and gas production. These regulations, however, do not cover CO_2 storage operations that have no connection with oil and gas production.

There is further the Weyburn field in Canada, where CO_2 from the Dakota gasification plant is used for enhanced oil recovery. The Weyburn case stands for an example of the use of CO_2 from a fossil-fuel plant for enhanced oil recovery and for underground storage. Nevertheless this activity is also performed within the context of hydrocarbon recovery, where special regulations apply in many countries worldwide.

Besides the Sleipner project, the In Salah project and the Weyburn project, there are many sites, especially in the United States, where CO_2 from natural and industrial sources is used commercially for enhanced oil recovery. Even if those projects allow a lot of technical experience to be gained, the appropriateness of their regulative element as an example for CO_2 storage in general is not given.

Apart from activities where the gases are intended to remain ultimately in a geological formation, there are many subsurface installations for the storage of gas and petroleum products. Generally speaking, there are storage reservoirs in porous media and in caverns. In Germany, for example, there is an extractable volume of 12.47 billion m^3 of natural gas storage in porous media and an extractable storage volume of 6.13 billion m^3 in caverns (Sedlacek, 2004). This storage volume is distributed amongst 22 storage reservoirs in porous media and 145 caverns, of which many are operated in clusters. Both types are used to match intertemporal variations of demand and supply and to maintain strategic reserves.

[7] The Greenhouse Gas Control Technologies Conferences since 1990 is a good indicator of the intensity of scientific discussion.

[8] There is a difference between the two with respect to the drivers involved. The Sleipner case is mainly motivated by the Norwegian CO_2 tax for the offshore industry, whereas the In Salah project has no direct financial driver but is mainly driven by the objective of demonstrating CO_2 storage on a commercial scale and of gaining experience.

The operators of underground gas storage reservoirs have a strong interest in keeping the gas in a small confined area, to be able to recover it with minimum losses. The long-term storage safety on geological time scales is not an issue, as the reservoir content is not meant to remain in place. As a consequence, the regulations for underground gas storage will probably not be directly applicable to CO_2 storage. However, the existing safety regulations for daily operations and for the risk of catastrophic failure for these reservoirs might stand as an example for long-term CO_2 storage.

6.6.2 Implications of the quality of the stored CO_2

The principle evaluation of the legal aspects of CO_2 usually refers to the gas as being a homogeneous, pure substance. In reality, the CO_2 stream delivered by industrial power-conversion activities for storage will be a technical gas with at least minor impurities. The energy and cost requirements for purification rise exponentially with higher degrees of purity. As a consequence, the operator of the capture units will be motivated to minimise the efforts for purification. On the other hand, the existence of trace gases, especially those derived from the sulphur content of many fossil fuels, may create an unacceptable toxicity of the gas stream. Furthermore, any water content in the gas may provoke a strong increase in the risk of corrosion.

The CO_2 transported and stored is generally assumed to be of such purity that it can be judged as a non-toxic and non- or low-corrosive gas. Given these classifications, the properties of the CO_2 should not impose a risk to prevent transport and storage activities. The processes required to reach these properties will, however, influence the economics of the facility.

6.6.3 Legal aspects of geological CO_2 storage

6.6.3.1 CO_2 storage at an on-land location

At first, any CO_2-storage project on land falls under the sovereignty of the state in which the project is located. Whereas this simple statement is true for any country worldwide, it has to be kept in mind though, that for cases in Europe the supranational law of the European Union would also be applicable to CO_2-storage activities in the European Union as well.

One example where the legal aspects of CO_2 storage have been analysed quite early are the Netherlands. There, within a research project, the legal situation of CO_2 storage has been investigated by a task force of lawyers from different government ministries (CRUST legal task force, 2002).

In general, it can be imagined that three fields of legislation will be touched by a commercial large-scale CO_2 storage activity:

- Waste legislation,
- Water legislation,
- Mining legislation.

Waste legislation could principally be applied to CO_2 storage as, under the circumstances, the CO_2 is an undesired material that the owner wants to dispose of. Water legislation could be applicable, as geological storage would take place in water-containing structures (aquifers), which should usually be subject to water legislation. Mining legislation, finally, could be applicable, as possible storage reservoirs in aquifers or oil and gas fields can be interpreted as a natural geological reserve. The regulation for the use of such natural geological reserves is usually part of the mining legislation. Further, the respective administrative procedure legislation will come into play for a permitting procedure. Different national authorities usually represent these three fields of legislation. Because of the lack of precedence, it is not clear which authorities will have to be addressed for a CO_2-storage project.

The specific legal framework for CO_2 storage will be different from country to country, based on different kinds of law. Irrespective of these differences, certain aspects of CO_2 storage will have to be regulated in order to allow for its feasibility.

- Exploration and right of use of a reservoir for storage. Searching for and identifying a suitable reservoir structure for storage is an activity comparable to seeking for mineral resources. Here, aspects of competition, licensing and conflicts with the use of other resources play an important role.
- Environmental and hazard safety. For any technical operation, provisions have to be made to avoid or at least minimise effects on other environmental media and to manage safety hazards.
- Long-term custody. The integrity of the CO_2-storage reservoir and of the CO_2 stored has to be maintained for a period of time longer than the business cycles of enterprises of today. Hence, the long-term custody and implied activities have to be regulated.

6.6.3.2 CO_2 storage in or beneath the marine water column

The storage of CO_2 in the marine area can principally be divided into two distinct cases: storage in the open waters of the ocean and storage in a geological formation beneath the seabed. The first case is generally not considered, owing to negative environmental impacts. The EU CCS Directive, for instance, forbids the storage of CO_2 in the water column.

In the second case, two international conventions are seen as the most important sources of regulations for the storage of CO_2 in the marine environment:

- London Convention, with the 1996 Protocol being applicable worldwide,
- OSPAR Convention, being applicable in the North-East Atlantic region.

Originally, neither of the two conventions made explicit provisions for the regulation of CO_2 storage. At the time they were negotiated, this technology option was simply not yet part of the scientific or political agenda. As a result, there was an ambiguity as to how to interpret these regulations.

However, a new amendment in 2006 to the 1996 London Protocol[9] on waste dumping at sea now allows for CO_2 to be stored in rocks below the sea, removing a significant legal hurdle to the implementation of large-scale CO_2-capture and storage projects, which aim at storage in geological formations under the ocean. This represented the first international law explicitly addressing carbon sequestration in international waters and a step towards creating a positive international legal framework for CCS activities.

The changes to the London Protocol also led to a revision of the OSPAR Convention for the Protection of the Marine Environment of the North-East Atlantic, which is the North-East Atlantic's version of the London Protocol, and which combines the 1972 Oslo Convention on dumping waste at sea and the 1974 Paris Convention on land-based sources of marine pollution. In 2007, the OSPAR Convention also adopted amendments to the Annexes to the Convention to allow the storage of CO_2 in geological formations under the seabed (www.ospar.org).

6.6.3.3 Liability for CO_2 storage

The long-term liability for CO_2 storage and possible leaks is one of the most essential regulatory issues facing CCS projects (both on land and offshore) and still needs to be largely sorted out (see Solomon *et al.* (2007)). It will affect the costs of CCS projects and will be crucial in advancing public acceptance of the technologies and processes involved. A first milestone was achieved by the EU CCS Directive. According to this Directive, the operator should remain responsible for maintenance, monitoring and control and reporting, once the storage site has been closed. Only if and when all available evidence indicates that the stored CO_2 will be completely contained for the indefinite future, should the responsibility for the storage site, including all ensuing legal obligations, be transferred to the competent national authority (EC, 2008).

6.6.4 Implications of the Kyoto Protocol for CO_2 storage

The United Nations Framework Convention on Climate Change (UNFCCC) and the Kyoto Protocol, as such, have not foreseen CO_2 capture and storage as a means of emissions reduction. The UNFCCC defined emissions as, 'The release of greenhouse gases and/or their precursors into the atmosphere,' (Article 1(4), UNFCCC, 1992). Consequently, CO_2 captured at source and stored outside the atmosphere is not an emission according to the definition in the Convention. Since industrial activity with CCS (and a theoretical 100% capture rate) does not create emissions according to the UNFCCC definition, one could interpret the action of CO_2 capture and storage as an emission reduction. Purdy and Macrory (2004) point out that this

[9] A modernised version of the international 1972 London Convention on the Prevention of Marine Pollution by Dumping of Wastes and Other Matter (London Convention, 1972; London Protocol, 1996).

distinction is of importance, as Parties to the Convention were more restricted in how to deal with emissions.

The Convention and the Kyoto Protocol (1997) encourage the protection and increase of sinks and reservoirs, meaning activities that remove greenhouse gases from the atmosphere and components of the climate system where greenhouse gases are stored. The definition of emissions as well as the definitions of sinks and reservoirs do not cover the activity of CO_2 capture and storage. However, the Protocol calls the Parties to promote the development of, 'new and renewable forms of energy, of carbon dioxide sequestration technologies and of advanced and innovative environmentally sound technologies,' (Kyoto Protocol, Art. 2(1)(a)(iv)). So in principle, activities such as CO_2 capture and storage should be in agreement with the purpose of the Convention and of the Protocol.

For any CO_2 capture and storage activity to be carried out in the first commitment period (2008 to 2012) there is still an obstacle to the credibility of this activity as a means of emission reduction under the terms of the UNFCCC. The Annex I Parties to the UNFCCC have to report their emissions according to the IPCC Guidelines (IPCC, 1997; 2000). These Guidelines stipulate that the inventories are calculated by the use of fuel-specific emission factors. A reduction of emissions by capture and storage of greenhouse gases prior to release into the atmosphere is not foreseen. So, under the Guidelines applicable until 2012, capture activities would not lead to a reduction in emissions accounted in the national inventory. With the adoption of new Guidelines for the reporting of emissions by the IPCC in 2006 (IPCC, 2006) the situation has changed, as these make special provisions for CO_2 capture and storage. Apart from the accountability of capture activities, general regulations have been developed to deal with leakage during CO_2 transport and the storage process. Further detailed regulations will have to be developed on the monitoring and verification of storage integrity and duration.

Beyond the existence of a regulatory environment that principally allows CCS operations, regulations have also to be implemented that make CCS economically attractive, if this option for CO_2 emissions reduction were to be pursued. The most important existing example of such a regulation is the European Emissions Trading Scheme (EU ETS implemented by the Directive 2003/87/EC). With this regulation, CO_2 emissions have been attributed a value for the industrial enterprises covered by the Directive. Given a stable market for the tradable emissions allowances based on stable policies, operators of energy conversion plants may find sufficient incentives to invest into CCS technologies. According to the revised Directive on the European Emissions Trading Scheme, CO_2 captured and stored will be credited as not emitted under the ETS (EP, 2008b).

6.7 The public perception of CO_2 capture and storage

As the storage of CO_2 in geological formations will be a key element of CO_2-capture and storage strategies, the public perception of this activity will be of outmost importance. Even though the direct effect of any CO_2-storage activity on the wider public will probably be very limited, many people may still feel affected. The strong influence of public opinion on siting of facilities, on legislation development and on licensing processes could consequently have a significant effect on the realisation of CO_2-capture and storage activities.

The public opinion on CO_2 capture and storage has not yet been explored thoroughly in the European Union. So far, only investigations limited in regional coverage and in statistical terms of sample size have been undertaken (Shackley *et al.*, 2004). In the study of Shackley *et al.* (2004), undertaken in the United Kingdom, several important conclusions were drawn: at first glance, the majority of the people interviewed for the study showed a slight rejection against carbon capture and storage or stated that they were not able to give an opinion. This was true for the case when the topic was presented without any other information. When put in the context of climate change and the need for emissions reductions, CO_2-capture and storage was reported to be seen as a 'potentially important carbon mitigation option for the UK'. It can be observed that after being provided with some information, people spread into a group of 'supporters' and a – in this survey – smaller one of 'opponents'. Shackley *et al.* (2004) found a higher support for carbon capture and storage when people were given a choice of mitigation options, such as renewable energies, energy efficiency or nuclear energy. In contrast to these results, Christensen (2004) reported that, when given a choice of mitigation options, only a small part of the people interviewed for their study would choose carbon capture and storage. Here, even more people would be favourable to the use of nuclear energy to address global warming. In both studies, renewable energies were chosen by most of the interviewed people as an option to use to address climate change.

It is interesting to note that in the work of Shackley *et al.* (2004) more than half of the respondents would categorically oppose higher energy bills to address climate change. This alone does not give a clear picture about the willingness to pay for the mitigation of climate change. The opposition to higher energy bills indicates that the preferences towards mitigation options should be investigated in relation to the mitigation costs. Possibly, the support for options such as solar energy will be less prominent when viewed in light of the economic implications. When judging the results concerning energy bills, the authors of the study put it into the framework of equity matters. In the UK, energy costs can play a major role in the budgets of private households with low incomes. The still-occurring 'fuel poverty' is seen as a serious problem leading to a widespread objection of higher energy prices.

When judging the public impact of carbon-capture and storage activities, it should be taken in mind, that apparently only very few people are aware of the extraordinary challenge imposed by the need to reduce greenhouse-gas emissions by 60% to 80% within the next five decades. This finding of Shackley *et al.* (2004) clearly reveals the need for a 'bottom-up' information policy to prepare the grounds for any new mitigation option.

When investigating the processes and policies that could raise the potential acceptance of carbon capture and storage, Shackley *et al.* (2004) found that more certainty about risks would be helpful. The main concerns about risks connected with carbon capture and storage were possible leakage, ecosystems and environmental impacts, the untested nature of the technology and human health impacts.

The acceptance within the research panels could be raised significantly when putting carbon capture and storage as an explicit bridging strategy to a zero-emissions energy system (e.g., a hydrogen-based energy system). Enhanced oil recovery or enhanced gas recovery options led to a more positive attitude of a part of the respondents, whilst the rest would keep the same opinion.

6.8 Challenges for implementation of CCS

The capture and storage of CO_2 from energy conversion processes is a promising option for the reduction of greenhouse-gas emissions, but definitely does not constitute an exhaustive approach to tackle climate change. Carbon capture and storage may become a viable technology route for point sources of CO_2 – meaning large power plants or large industrial plants, such as for hydrogen production. For medium-scaled plants, distributed sources and the transport sector, CCS will, however, not constitute a solution for emission reduction. Besides the limitations imposed by the type of source, the availability of suitable storage volume within an acceptable distance will also impose limitations on the feasibility of CCS.

Even though climate change is perceived more and more as an urgent challenge, CCS has not been implemented apart from the few demonstration projects mentioned previously. Obviously, there are barriers to the application of CO_2 capture and storage present (at least at the time of writing this assessment). Important barriers are:

- *Maturity of technology* Although the principal technical elements for CCS are available, the entire system within a plant is not yet to be considered as mature. Stakeholders are concentrating on research and development of the components and of system design. In particular, safe and permanent CO_2 underground storage needs to be proven.
- *Economic incentives* There is no sufficiently strong economic incentive for undertaking CO_2-capture and storage operations. For the deployment of CCS, economic incentives provided by existing climate protection agreements or through existing international mechanisms, such as the flexible Kyoto mechanisms, are essential. Principally, market instruments such as the European Emissions Trading Scheme could incentivise investment in emission-reduction technologies like CCS, if they are to be considered a technology for CO_2 reduction. But also including CCS under the Clean Development Mechanism (CDM) is of central importance, as it would provide economic incentives to deploy CCS in countries with steeply rising CO_2 emissions, such as China and India. To be effective, market-based instruments implemented by policy makers must create long-term reliable and credible framework conditions.
- *Legal and regulatory environment* The operation of CO_2 capture, and of CO_2 storage in particular, on a large industrial scale is to be considered as a new industrial activity. So far,

there is no distinct legal framework for CO_2 storage in any country worldwide. The absence of regulation is a barrier to the realisation of CCS. A regulatory framework (long-term liability, licensing, royalties, leakage cap) is needed for private investment and public acceptance.

- *Competing technologies* CCS competes with other energy conversion technologies with low greenhouse-gas emissions, such as renewable energy sources and nuclear energy, that principally could supply not only electricity but also hydrogen.
- *Public perception* Literally (in all OECD countries), CCS is a novel concept to the general public. Consequently, a low degree of knowledge about the technological concepts and implications of CCS has to be assumed and, hence, it is difficult – if not impossible – to research the view of the general public. Experts working in the field of CCS, however, estimate that the general public will have a rather negative attitude towards CCS. If these estimations were to prove true, the public perception could become a significant barrier to the implementation of adequate regulation and to the implementation of CO_2-capture plants and CO_2-storage facilities. In particular questions about the long-term liability of CO_2 storage are of considerable significance for public acceptance.

The IPCC estimates global CO_2 storage capacity at between 1678 and 11 100 Gt CO_2, with 2000 Gt CO_2 classed as technically viable, more than 70 times current global CO_2 emissions (IPCC, 2005). However, the task of scaling up CCS to make a significant contribution to the reduction of global CO_2 emissions is daunting. To provide some orders of magnitude: global CO_2 emissions are about 27 Gt per year; the three worldwide current CCS projects taken together store about 3–4 Mt of CO_2 per year, i.e., 0.15%. Storing all the CO_2 from fossil-based power generation, roughly 10 Gt CO_2 per year, would equate to around 240 million barrels per day of CO_2 to be stored (assuming a CO_2 density of 700 kg/m^3 in dense phase), roughly three times today's global oil production. If CCS is to play a significant role in the coming decades, demonstration must be accelerated. In Europe, for instance, there is a strong political push to support CCS to become a viable option for CO_2 reduction, particularly for fossil-fired power plants. Nevertheless, the deployment of CCS is also a prerequisite for fossil hydrogen production, in particular from coal, if an overall CO_2 reduction is to be achieved.

6.9 Summary

The principle idea of CO_2 capture and storage is to redesign energy-conversion processes in such a manner that the generated CO_2 can be captured in a high concentration, compressed or liquefied and taken to a reservoir outside the atmosphere. In this reservoir, the CO_2 should remain long enough so that it cannot contribute to climate change.

Carbon dioxide capture and storage is especially studied in connection with electricity generation, where different technology routes – post-combustion capture, pre-combustion capture and oxyfuel combustion – are under research and development. Carbon dioxide capture is also crucial in connection with hydrogen production from carbon-based fuels, and, for the case of the pre-combustion capture, there is a link to electricity generation.

Implementing CCS would create a whole new value chain of plants with CO_2 capture, of CO_2 transport and of CO_2 storage. Carbon dioxide transport could be performed by pipelines on land or in the marine environment. For marine transport, ships could also be used. Creating a new CO_2 infrastructure is a challenging task, similar to the build-up of a hydrogen infrastructure; that's why a combined build-up should be envisaged, where possible.

Carbon dioxide can be stored in geological structures, in the water column of the oceans or possibly by mineralisation with suitable rock materials. Among these options, storage in connection with EOR activity, storage in depleted gas and oil fields and storage in saline aquifers are most promising. Research on storage in the water column of the ocean is only followed in parts of the world, with stronger activities in Japan or the United States, because there are strong concerns about the environmental effects to the ocean biosphere. Mineralisation is still in the development stage and not yet ready for implementation.

The application of CCS will need a set of enabling framework conditions, consisting of appropriate economic incentives for CO_2 emissions reductions and a legal and regulatory environment providing the means to obtain permission for the activities of CO_2 capture, transport and storage. Also, the public perception of CCS needs to be positive enough to allow policy makers and administrations to create the appropriate regulations and take favourable decisions on licensing. The principal advantages and disadvantages of CCS are summarised.

Advantages of the technology option CO_2 capture and storage

- Carbon dioxide-capture and storage is potentially the least expensive method for supplying 'low-carbon' hydrogen. The same might apply for the supply of low-carbon electricity.
- Fossil-fuel reserves can be utilised while limiting the contribution to global warming.
- Large CO_2 reductions can be achieved in the short to medium term.

Disadvantages of CCS

- Many issues need to be resolved with respect to underground storage and other forms of storage.
- Social acceptance of CO_2 storage is seen as a critical barrier by many stakeholders.
- Additional consumption of fossil fuels is required in the separation process.

To conclude, CCS should not be considered a permanent or long-term solution for emissions reduction, but rather to fulfil a bridging function for the transition to an energy system based largely on renewable energies.

References

Arts, R. J., Brevik, I., Eiken, O. *et al.* (2000). Geophysical methods for monitoring marine aquifer CO_2 storage – Sleipner experiences. *Proceedings of the 5th International Conference on Greenhouse Gas Control Technologies (GHGT-5)*, ed. Williams, D., Durie, B., McMullan, P., Paulson, C. and Smith A. Sydney, Australia: CSIRO, pp. 366–371.

Baines, S. J. and Worden, R. H. (2000). Geological CO_2 disposal: understanding the long term fate of CO_2 in natural occurring accumulations. *GHGT5*, pp. 311–316.

Barrio, M., Aspelund, A., Weydahl, T. *et al.* (2004). Ship based transport of CO_2. *Seventh International Conference on Greenhouse Gas Control Technologies*, Vancouver, Canada.

BMWA (Bundesministerium für Wirtschaft und Arbeit) (2003). *Forschungs- und Entwicklungskonzept für emissionsarme fossil befeuerte Kraftwerke*. Bericht der COORETEC-Arbeitsgruppen. Berlin: BMWA.

CERA (Cambridge Energy Research Associates) (2007). *Crossing the Divide? The Future of Clean Energy*. Cambridge, MA: CERA.

Chadwick, R. A., Holloway, S., Kirby, G. A., Gregersen, U. and Johannessen, P. N. (2000). The Utsira Sand, Central North Sea – an assessment of its potential for regional CO_2 storage. *Proceedings of the 5th International Conference on Greenhouse Gas Control Technologies (GHGT-5)*, ed. Williams, D., Durie, B., McMullan, P., Paulson, C. and Smith, A. Sydney, Australia: CSIRO, pp. 349–354.

Christensen, D. (2004). *CO_2 Capture Project's Policies and Incentives Study*. Presentation at the CO_2 Capture Projects Phase 1 Results Workshop. Brussels, June 2nd 2003.

Christensen, N. P. and Holloway, S. (2003). *Geological Storage of CO_2 from Combustion of Fossil Fuels*. Summary report, European Union Fifth Framework Programme for Research and Development, Project No. ENK6-CT-1999–00010.

Cremer, C. (2005). *Integrating Regional Aspects in Modelling of Electricity Generation – The Example of CO_2 Capture and Storage*. Dissertation No. 16119. ETH Zurich.

CRS (Congressional Research Service) (2007). *CRS Report for Congress. Carbon Dioxide (CO_2) Pipelines for Carbon Sequestration: Emerging Policy Issues*. Prepared by Parfomak, P. W. and Folger, P., CRS Report RL33971. http://ncseonline.org.

CRUST Legal Task Force (2002). *Legal Aspects of CO_2-Underground Storage*. Translation of the Dutch report *Juridische Aspecten van Ondergrondse CO_2-Bufferopslag*, developed within the CO_2 Re-use Through Underground Storage project. www.CO2reductie.nl.

Doctor, R. D., Molburg, J. C. and Brockmeier, N. F. (2000). Carbon dioxide recovered from fossil-energy cycles. *GHGT5*, pp. 567–571.

Doughty, C., Pruess, K. S., Benson, S. M. *et al.* (2001). Capacity investigation of brine-bearing sands of the Frio Formation for geologic sequestration of CO_2. *Proceedings of First National Conference on Carbon Sequestration*, 14–17 May 2001, Washington, DC.

EC (European Commission) (2007). *Sustainable Power Generation From Fossil Fuels: Aiming for Near-Zero Emissions from Coal after 2020*. Communication from the

Commission to the Council and the European Parliament, COM(2007), 843 final.

EP (European Parliament) (2008a). *Position of the European Parliament adopted at First Reading on 17 December 2008 with a View to the Adoption of Directive 2009/.../EC of the European Parliament and of the Council on the Geological Storage of Carbon Dioxide and Amending Council Directives 85/337/EC, 96/61/EC, Directives 2000/60/EC, 2001/80/EC, 2004/35/EC, 2006/12/EC and Regulation (EC) No 1013/2006.*

EP (European Parliament) (2008b). *Position of the European Parliament Adopted at First Reading on 17 December 2008 with a View to the Adoption of Directive 2009/.../EC of the European Parliament and of the Council Amending Directive 2003/87/EC so as to Improve and Extend the Greenhouse Gas Emission Allowance Trading System of the Community.*

Egberts, P., Keppel, F., Wilednborg, T. *et al.* (2003). *GESTCO-DSS – A Decision Support System for Underground Carbon Dioxide Sequestration.* Utrecht, Netherlands: TNO/Ecofys.

Ennis-King J., Gibson-Poole, C. M., Lang, S. C. and Paterson L. (2003). Long term numerical simulation of geological storage of CO_2 in the Petrel sub-basin, North West Australia. *Proceedings of the 6th International Conference on Greenhouse Gas Control Technologies (GHGT-6)*, ed. Gale, J. and Kaya Y., 1–4 October 2002. Kyoto, Japan: Pergamon, pp. 507–511.

Eurostat (2004). *Length of Pipelines Operated – 2001.* Eurostat. http://epp.eurostat.ec.europaeu.

Fischedick, M., Esken, A., Luhmann, H. J., Schüwer, D. and Supersberger, N. (2007a). *CO_2 Capture and Geological Storage as a Climate Policy Option. Technologies, Concepts, Perspectives.* Wuppertal: Wuppertal Institute for Climate, Environment and Energy, Wuppertal Spezial 35e.

Fischedick, M., Esken, A., Pastowski, A. *et al.* (2007b). *RECCS – Strukturell-ökonomisch-ökologischer Vergleich regenerativer Energietechnologien mit Carbon Capture and Storage.* Final report of a research project on behalf of the German Federal Ministry for the Environment, Wuppertal Institute for Climate, Environment and Energy; DLR - Institut für Technische Thermodynamik; Zentrum für Sonnenergie und Wasserstoff-Forschung and Potsdam-Institut für Klimafolgenforschung.

Gale, J. and Davison, J. (2002). Transmission of CO_2 – safety and economic considerations. In *Proceedings of the 6th International Conference on Greenhouse Gas Control Technologies*, October 1–4, 2002, Kyoto, Japan, pp. 517–522.

Gielen, D. (2003). *Uncertainties in Relation to CO_2 Capture and Sequestration: Preliminary Results.* Paris: IEA. www.iea.org/textbase/papers/2003/gielen.pdf.

Heddle, G., Herzog, H. and Klett, M. (2003). *The Economics of CO_2 Storage.* MIT-LFEE 2001–003RP. Cambridge, MA: Laboratory For Energy and Environment, Massachusetts Institute of Technology (MIT).

Hendriks, C., Wildenborg, T., Feron, P., Graus, W. and Brandsma, R. (2003). *EC-Case – Carbon Dioxide Sequestration.* Ecofys, TNO, M70066.

Hendriks, C., Graus, W. and von Bergen, F. (2004). *Global Carbon Dioxide Storage Potential and Costs.* Ecofys-Report EEP-02001 in co-operation with TNO-NITG. Utrecht, Netherlands.

Holt, T., Jensen, J. L. and Lindeberg, E. (1995). Underground storage of CO_2 in aquifers and oil reservoirs. *Energy Conversion and Management*, **36** (6–9), 535–538.

IEA-GHG (International Energy Agency Greenhouse Gas Research and Development Programme) (2004). *Ship Transport of CO_2*. Report No. PH 4/30.

IEA (International Energy Agency) (2006). *CO_2 Capture & Storage: IEA Energy Technology Essentials*. Paris: OECD/IEA.

IPCC (Intergovernmental Panel on Climate Change) (1997). *Revised 1996 IPCC Guidelines for National Greenhouse Gas Inventories*. www.ipcc-nggip.iges.or.jp/public/gl/invs1.htm.

IPCC (Intergovernmental Panel on Climate Change) (2000). *Good Practice Guidance and Uncertainty Management in National Greenhouse Gas Inventories*. www.ipcc-nggip.iges.or.jp/public/gp/english.

IPCC (2005). *IPCC Special Report on Carbon Dioxide Capture and Storage. Prepared by Working Group III of the Intergovernmental Panel on Climate Change*, ed. Metz, B., Davidson, O., de Coninck, H. C., Loos, M. and Meyer, L. A. Cambridge: Cambridge University Press.

IPCC (2006). *2006 IPCC Guidelines for National Greenhouse Gas Inventories, Prepared by the National Greenhouse Gas Inventories Programme*, ed. Eggleston, H. S., Buendia, L., Miwa, K., Ngara, T. and Tanabe, K. Hayama, Japan: Institute for Global Environmental Strategies (IGES).

Law, D. H-S. and Bachu, S. (1996). Hydrogeological and numerical analysis of CO_2 disposal in deep aquifers in the Alberta sedimentary basin. *Energy Conversion and Management*, **37** (6), 1167–1174.

Lindeberg, E. and Wessel-Berg, D. (1997). Vertical convection in an aquifer column under a gas cap of CO_2. *Energy Conversion and Management*, **38** (suppl.), 229–234.

Lindeberg, R., Zweigel, P., Bergmo, P., Ghaderi, A. and Lothe, A. (2000). Prediction of CO_2 distribution pattern improved by geology and reservoir simulation and verified by time lapse seismic. *Proceedings of the 5th International Conference on Greenhouse Gas Control Technologies (GHGT-5)*, ed. Williams, D., Durie, B., McMullan, P., Paulson, C. and Smith, A. Sydney, Australia: CSIRO publ., pp. 372–377.

London Convention (1972). *Convention on the Prevention of Marine Pollution by Dumping of Wastes and Other Matter*. www.imo.org/home.asp/?topic_id=1488.

London Protocol (1996). *1996 Protocol to the Convention on the Prevention of Marine Pollution by Dumping of Wastes and Other Matter, 1972*. London: International Maritime Organisation.

McPherson, B. J. O. L. and Cole, B. S. (2000). Multiphase CO_2 flow, transport and sequestration in the Powder River basin, Wyoming, USA. *Journal of Geochemical Exploration*, **69–70** (June), 65–70.

Odenberger, M. and Svensson, R. (2003). *Transportation Systems for CO_2 Application to Carbon Sequestration*. Gothenburg: Department of Energy Conversion, Chalmers University of Technology. www.entek.chalmers.se/~klon/msc.

Purdy, R. and Macrory, R. (2004). *Geological Carbon Sequestration: Critical Legal Issues*. Tyndall Centre for Climate Change Research.

Radgen, P., Cremer, C., Warkentin, S. *et al.* (2005). *Bewertung von Verfahren zur CO_2-Abscheidung und -Deponierung*. Report of the German Federal Environmental Agency (UBA) prepared by the Fraunhofer Institute for Systems and Innovation Research and the Federal Institute for Geosciences and Natural Resources (BGR). UBA Forschungsbericht 203 41 110. Dessau.

Sedlacek, R. (2004). Untertage-Erdgasspeicherung in Deutschland – Underground Gas Storage in Germany. *Erdöl, Erdgas Kohle*, **120** (11), 368–378.

Shackley, S., McLachlan, C. and Bough, C. (2004). *The Public Perceptions of Carbon Capture and Storage*. Tyndall Centre for Climate Change Research, Working Paper 44. www.tyndall.ac.uk/publications/working_papers/working_papers.shtml#wp44.

Skovholt, O. (1993). CO_2 transportation system. *Energy Conversion and Management*, **34**, 1095–1103.

Solomon, S., Kristiansen, B., Stangeland, A., Torp, T. A. and Kårstad, O. (2007). *A Proposal of Regulatory Framework for Carbon Dioxide Storage in Geological Formations*. Prepared for International Risk Governance Council Workshop, March 15–16, 2007. Washington, DC. www.irgc.org/IMG/pdf/IRGC_CCS_BellonaStatoil07.pdf.

Stanton, R., Flores, R., Warwick, P. D., Gluskoter, H. and Stricker, G. D. (2001). Coal bed sequestration of carbon dioxide. *Proceedings of the First National Conference on Carbon Sequestration*, 14–17 May 2001. Washington, DC.

Stevens, S. H., Fox, C. E. and Melzer, L. R. (2000). McElmo Dome and St. Johns natural CO_2 deposits: analogs for geologic sequestration. *GHGT5*, pp. 317–321.

UNFCCC (1992). United Nations Convention on Climate Change. http://unfccc.int/resource/convkp.html.

van der Meer, L. (1992). Investigation regarding the storage of carbon dioxide in aquifers in the Netherlands. *Energy Conversion and Management*, **33** (5–8), 611–618.

van der Meer, L. G. H., Arts, R. J. and Peterson, L. (2000). Prediction of CO_2 after injected in a saline aquifer: reservoir history matching of a 4D seismic image with a compositional gas/water model. *Proceedings of the 5th International Conference on Greenhouse Gas Control Technologies (GHGT-5)*, ed. Williams, D., Durie, B., McMullan, P., Paulson, C. and Smith, A. Sydney, Australia: CSIRO, pp. 378–384.

Zweigel, P., Hamborg, M., Arts, R. *et al.* (2000). Prediction of migration of CO_2 injected into an underground depository: reservoir geology and migration modelling in the Sleipner Case (North Sea). *Proceedings of the 5th International Conference on Greenhouse Gas Control Technologies (GHGT-5)*, ed. Williams, D., Durie, B., McMullan, P., Paulson, C. and Smith, A. Sydney, Australia: CSIRO, pp. 360–365.

Further reading

Freund, P. and Kårstad, O. (2007). *Keeping the Light On: Fossil Fuels in the Century of Climate Change*. Oslo: Universitetsforlaget.

International Energy Association (IEA) (2008). *CO_2 Capture and Storage – A Key Carbon Abatement Option*. IEA Energy Technology Analysis Series. Paris: OECD/IEA.

IPCC (2005). *IPCC Special Report on Carbon Dioxide Capture and Storage*. Prepared by Working Group III of the Intergovernmental Panel on Climate Change, ed. Metz, B., Davidson, O., de Coninck, H. C., Loos, M. and Meyer, L. A. Cambridge: Cambridge University Press.

Marchetli, C. (1977). On geoengineering and the CO_2 Problem. *Climate Change*, **1** (1), 59–68.

7

Energy-chain analysis of hydrogen and its competing alternative fuels for transport

Werner Weindorf and Ulrich Bünger

The driving forces for the development of alternative fuels are, on the one hand, anxiety about security of supply with oil, on which the transport sector still depends almost entirely, and, on the other hand, a reduction of transport-related emissions of greenhouse gases and air pollutants. In this respect, hydrogen and fuel cells are in competition with a number of other energy carriers and transformation technologies. For instance, hydrogen has to compete with improved gasoline and diesel engines, but also with synthetic fuels, biofuels or natural gas. With regard to drive trains, petrol and diesel engines still dominate. Besides an improvement of the efficiencies of these conventional combustion engines, there are also vehicle concepts under development, which are based on electric drives and which rely to varying degrees on batteries as a source for motion energy. Hence, this chapter briefly discusses the major alternatives to hydrogen and fuel cells in the transport sector and their characteristics.

7.1 Overview of alternative fuel options

Two thirds of today's oil use of more than 81 million barrel per day is for transportation, of which land transport for people accounts for some 55%, land transport for freight for some 35% and air transport for people and freight for around 10%. Almost 97% of road transport is fuelled by oil. The three most important targets with respect to transportation energy use, which are also increasingly favoured by policy makers around the world, are reduction of local air pollution, greenhouse gas-emissions reduction and energy security.[1] As a consequence, there is an enforced search for alternative transport fuels.

The choice of possible alternatives is manifold, not only with respect to the fuels themselves, but also with respect to the primary energy sources used and the production processes. Figure 7.1 shows for the currently most discussed fuels the major

[1] The focus in this chapter is on CO_2 reduction and feedstock availability, rather than on fuel quality and composition and related local emissions.

The Hydrogen Economy: Opportunities and Challenges, ed. Michael Ball and Martin Wietschel. Published by Cambridge University Press.

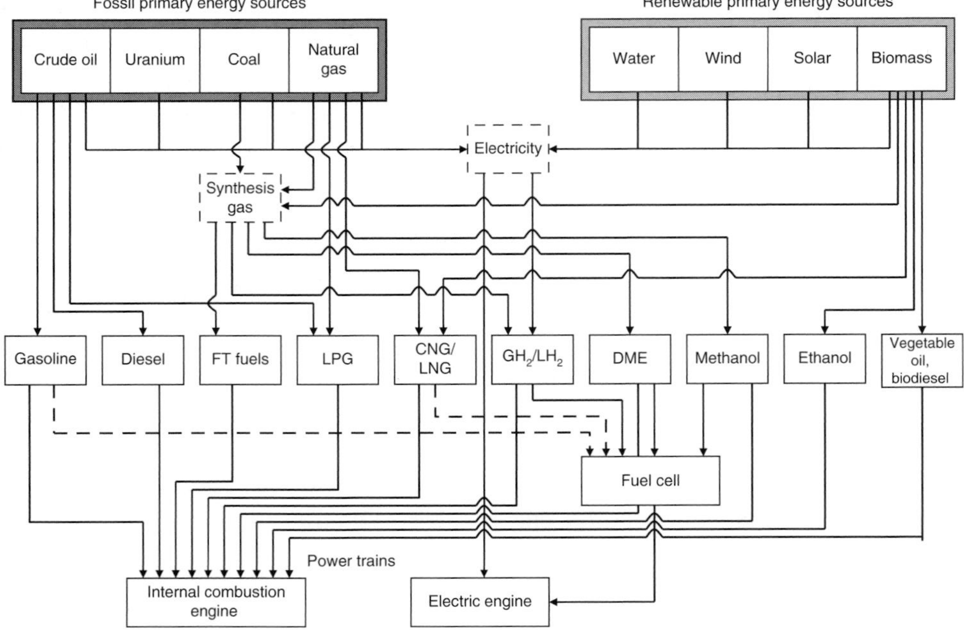

Figure 7.1. Alternative fuel supply options (Krüger, 2002, amended).

supply paths, split into fossil and renewable primary energy sources, as well as their application in the motor industry. The major alternative fuels in use today are liquefied petroleum gas (LPG) (also referred to as autogas), natural gas (CNG), (bio)ethanol – all of them best suited for petrol engines – and biodiesel (FAME). In the future, synthetic fuels on the basis of coal, natural gas or biomass, as well as hydrogen and electricity, are likely to gain greater market shares.

The typical hydrocarbons used today as fuels can also be produced synthetically (synthesised) from other organic carbon-containing raw materials. Incentives are the 'direct' formulation of the synthetic fuel's properties with regard to an optimal combustion in the engine (e.g., low sulphur and aromatic content, low particle emissions), the use of cheap raw materials such as coal or stranded gas (compared with crude oil) and the use of biogenic feedstock. The manufacture of these synthetic fuels is through gasification from synthesis gas (syngas), a mixture of hydrogen and carbon monoxide. As feedstock, natural gas, coal and biomass can be used. The most important synthetic fuels are Fischer–Tropsch (FT) fuels, with Fischer–Tropsch synthesis subsequent to the gasification step (see also Section 7.3.4), methanol and dimethyl ether (DME).[2]

[2] In the broader sense, fuels on the basis of oil sands or oil shale can also be considered synfuels. From the latter, synthetic fuels can be produced more easily than from coal or biomass, as they have a higher H:C ratio and thus less hydrogen needs to be added (see also Section 7.3.4). However, large mass flows of inorganic material need to be moved and heated, which significantly reduces the process efficiency.

In the case of FT fuels a distinction is made between gas-to-liquids (GTL), coal-to-liquids (CTL; synthetic fuels were first produced from coal using Fischer–Tropsch synthesis in Germany during World War II) and biomass-to-liquids (BTL); they together are also referred to as XTL. Both CTL and GTL are already commercially applied processes in South Africa and Malaysia, by Sasol and Shell, respectively; new plants are planned or under construction, for example in Qatar and China (see also Sections 3.4.3.2 and 3.5.4); a BTL pilot plant is operated by Choren in Germany and a larger one is currently being planned. Between 65% and 75% of the final product state of the FT synthesis are typically middle distillates, mainly diesel, but gasoline fractions can also be produced, as for example in the case of Sasol in South Africa. While the interest for methanol as an alternative fuel for conventional internal-combustion engines (ICE) has largely declined, a potential use of methanol was, for some time, seen in methanol fuel cells (DMFC) for vehicle propulsion (see also Chapter 13). The discussion about DME as an alternative fuel is still relatively new but, owing to its high cetane number and its synthetic properties, it has received a lot of attention as an ultraclean diesel replacement.

Another alternative fuel already in use in many countries is (compressed) natural gas (CNG). Compressed natural gas is stored on board the vehicle at a pressure of around 200 bar and the range of CNG cars is comparable to gasoline cars. Compressed natural gas requires primarily the implementation of new refuelling stations, as a natural gas distribution infrastructure is already largely in place in many countries. Certain infrastructure components (e.g., pipelines or fuelling components) may possibly advance the introduction of hydrogen.

In the last couple of years, the interest in fuels from renewable raw materials as a means to respond to growing concerns about energy security, environmental impacts and, not least, tightening supplies of conventional oil is soaring. Figure 7.2 illustrates the multitude of possible production routes for the supply of fuels from renewable energies. These can be broadly divided into extractive, fermentative and thermo-chemical processes. The most important renewable fuels are vegetable oils and their esters (biodiesel), the alcohols methanol and ethanol, synthetic liquid hydrocarbons (FT fuels), synthetic methane and biogas, as well as hydrogen.

Bioethanol is the largest biofuel today and is used in low 5%–10% blends with gasoline (E5, E10), but also as E85 in flexible-fuel vehicles. Conventional production is a well known process, based on the enzymatic conversion of starchy biomass (cereals) into sugars, and fermentation of 6-carbon sugars with final distillation of ethanol to fuel grade.

Ethanol can be produced from many feedstocks, the most important being cereal crops, corn (maize), sugarcane and sugarbeet. While conventional processes use only the sugar and starch biomass components, advanced processes are under development that can utilise ligno-cellulosic materials to extend the feedstock base. A wide range of ligno-cellulosic biomass wastes can be considered from agriculture (e.g., straw, corn stover, bagasse), forestry, the wood industry and the pulp and

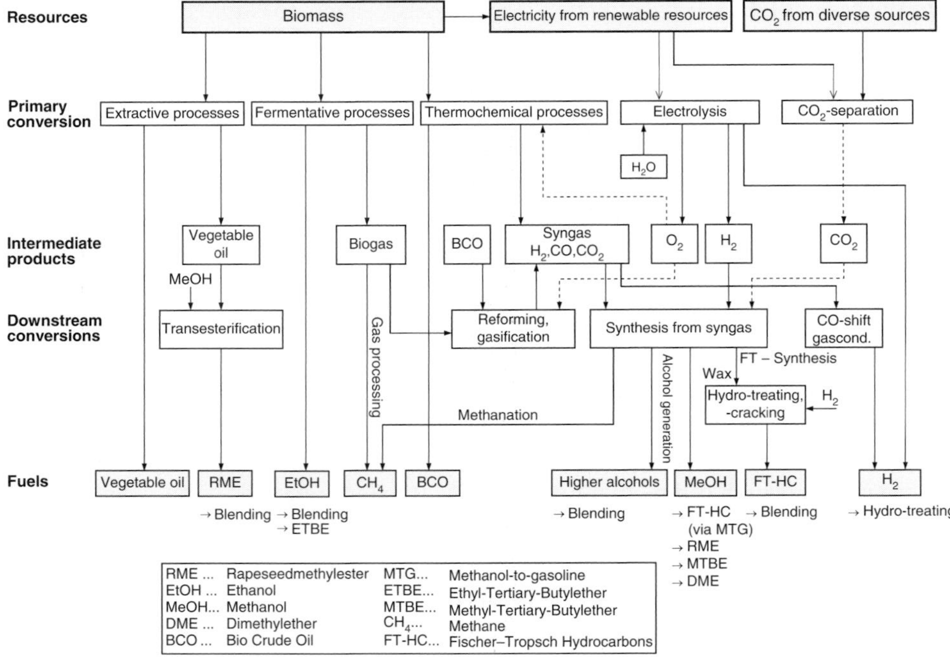

Figure 7.2. Supply paths for renewable fuels (FVS, 2003; Specht *et al.*, 2001).

paper industry, as well as dedicated fast-growing plants, such as poplar trees and switch grass. Cellulosic feedstock could be grown on non-arable land, thus avoiding interference with the food chain, or produced from integrated crops, which could considerably increase land availability (IEA, 2007).

Biodiesel (fatty acid methyl ester (FAME)) production is based on transesterification of vegetable oils and fats through the addition of methanol (or other alcohols) and a catalyst, giving glycerol as a by-product (which can be used for cosmetics, medicines and food). Oil-seed crops include rapeseeds, sunflower seeds, soy beans and palm oil seeds, from which the oil is extracted chemically or mechanically. Biodiesel can be used in 5%–20% blends with conventional diesel, or even in pure form, which requires slight modifications in the vehicle.

Another option to extend the ligno-cellulosic feedstock base is the development of BTL through biomass gasification and subsequent Fischer–Tropsch synthesis. Although BTL is fully compatible with diesel fuel, ligno-cellulosic BTL has not yet been commercialised.

Irrespective of the fuel supply chain, alternative fuels have generally lower tailpipe emissions in terms of local pollutants (such as NO_x, CO, SO_2, VOC and particle emissions) than conventional gasoline and diesel engines: for instance, natural gas completely eliminates particle emissions; synfuels are manufactured with very low sulphur and aromatic contents; alcohol-based fuels have high octane numbers, which improve engine efficiency and result in significantly lower NO_x, CO and SO_2 emissions; because of its high cetane number, biodiesel results in very low sulphur and particulate emissions.

Apart from hydrogen and electricity, of all the alternative fuels, natural gas (CNG) achieves the greatest CO_2 reduction in the vehicle, of 20%–25%, compared with gasoline (while no significant advantage exists with respect to the more efficient diesel engine). Tailpipe CO_2 emissions from biofuels are not much different from those for gasoline and diesel, but as the CO_2 released has previously been fixed by photosynthesis in the plants, biofuel combustion is generally considered CO_2 neutral (i.e., ideally, the carbon emitted during the combustion of the fuel is equal to the carbon absorbed by the biomass during growth). Synfuels generally have slightly lower tailpipe CO_2 emissions, owing to their slightly higher H:C ratio. However, an assessment of the full CO_2-equivalent footprint of alternative fuels needs to take into account the entire fuel supply chain, including feedstock extraction, fuel production and distribution, as this can be quite significant (see Section 7.2).

Besides the development of alternative fuels and hydrogen- or electricity-fuelled vehicles, in the field of conventional drive trains numerous efforts are undertaken to reduce fuel consumption and emissions of both CO_2 and local pollutants. Today's consumption of new gasoline passenger vehicles in the EU is on average 5.9 l/100 km, for diesel vehicles around 4.9 l/km; in the USA, for comparison, the consumption of gasoline cars is around 9.2 l/100 km.[3] For the reduction of fuel consumption by purely technical measures, a wide range of options is available, which can generally be divided into *vehicle-related measures, engine-related measures* and *measures affecting transmission* (Krüger, 2002).

Vehicle-related measures comprise, among others, a reduction in the weight of the vehicle, as well as in the various drive resistances (rolling resistance, air resistance and acceleration resistance). As for the reduction in vehicle weight, however, it has to be noted that the average empty weight in recent years has significantly increased, despite the use of lighter materials such as aluminium, plastics or so-called tailored blanks for steel, owing to the ever-increasing number of gadgetries in vehicles and more demand for comfort and passive safety. (For instance, the empty weight of the Golf 1 in 1975 was about 700 kg and of the Golf 5 in 2007 around 1250 kg.) *Engine-related measures* concern improvements of the thermodynamic engine efficiency, as in ICE only a small fraction of the fuel energy is used for propulsion (generally between 20% and 35% for spark ignition engines and between 30% and 40% for compression ignition engines).[4] *Transmission-related measures* concern the improvement of the transmission efficiency of

[3] While fuel-efficient diesel vehicles represent roughly 50% of passenger cars in the EU (with a rapidly growing share), their shares in the USA are still less than 10%, and in China even more negligible with less than 5%.

[4] However, the optimisation of an ICE is of high complexity and often characterised by counteracting effects, as a reduction in fuel consumption does not necessarily lead to an optimisation of pollutant emissions: for instance, high thermodynamic efficiencies can only be achieved with high combustion temperatures, which in turn favours the formation of NO_x.

Table 7.1. *Global warming potential of different greenhouse gases*

	CO_2 equivalents
CO_2	1
CH_4	23
N_2O	296

the drive train as well as the shift of the engine operating points into areas of lower consumption.[5] To reduce emissions of air pollutants, a change in the composition of today's petrol and diesel fuels also plays an important role, as contrary to technical emission reduction measures, which require the implementation of new vehicle technologies, the introduction of improved fuels directly affects the entire vehicle fleet, for example, in the case of synthetic fuels.

7.2 Introduction to energy-chain analysis

In recent years, full energy-chain analysis has developed to become a well accepted method to support consensus processes on the advantages and disadvantages of energy supply alternatives specifically when environmental constraints have been in focus. Earlier calculations had revealed that a rigorous and transparent analysis of full energy supply chains from 'cradle to grave' needs to replace the comparison of individual processes along energy supply chains, such as, e.g., fuel cells and internal combustion engines.

Although in principle stationary and transport-specific energy chains can be analysed, here the assessment of the latter is explained in more detail, and is then referred to as well-to-wheel (WTW) analysis. The primary focus of WTW analysis in Europe is on global environmental impact, i.e., greenhouse-gas emissions expressed as CO_2-equivalents. Other issues of interest are (a) primary energy demand (which equals resource utilisation), (b) local pollutant emissions and (c) full energy or fuel supply costs. Well-to-wheel analysis covers the entire fuel supply chain from feedstock extraction, feedstock transportation, fuel manufacturing and fuel distribution to fuel use in a vehicle.

The greenhouse gases (GHGs) considered are carbon dioxide (CO_2), methane (CH_4) and nitrous oxide (N_2O). Other greenhouse gases are CFCs, HFCs and SF_6, but they are not relevant in this context. Their global warming potential is expressed in CO_2 equivalents. Table 7.1 shows the global warming potential for a time period of 100 years, according to the Intergovernmental Panel on Climate Change (IPCC).

In this evaluation only CO_2 from the combustion of fossil fuels is considered. The combustion of biomass is assumed to be CO_2 neutral because the amount of CO_2

[5] Examples of novel engine and transmission technologies are homogeneous combustion compression ignition (HCCI), combined combustion system (CCS), combined autoignition (CAI) and continuously variable transmission (CVT).

emitted during the combustion of the biomass is the same as the amount of CO_2 that has been removed from the atmosphere during the growth of the plants, although, in reality, substantive CO_2-equivalent emissions may be produced throughout the biofuel supply chain (see Section 7.12).

To calculate the energy requirements, the so-called 'efficiency method' (also called the 'physical energy content method') has been used, like the procedure adopted by international organisations (IEA, EUROSTAT, ECE). For nuclear electricity in this method the efficiency of electricity generation is based on the energy lost as heat by nuclear fission, which reduces total electrical efficiency to about 33%. For electricity generation from hydropower and other renewable energy sources, the efficiency cannot be measured in terms of a calorific value (wind, solar energy). Hence, the energy input is assumed to be equivalent to the electricity generated, which leads to total efficiencies of 100%.

A combined comparative WTW analysis of specific global emissions and fuel supply costs is typically presented in a pathway portfolio analysis. Portfolio analysis helps to identify rapidly those alternative fuels and drive trains, or combinations of these, which can lead to the highest specific GHG emission savings.

Well-to-wheel analysis needs to be applied for all relevant time steps to understand the evolution of environmental effects and possibly costs in the short to long term. This is of specific importance when innovative processes are considered, as these are characterised by technology development and cost curves with high gradients.

Well-to-wheel analysis is a specific form of life-cycle analysis (LCA). In contrast to WTW analysis, LCA typically also takes factors other than global GHG emissions of a product or an energy carrier into consideration (such as air pollutants), including provision of all construction materials for the necessary processing plants and, furthermore, plant decommissioning. The full detail of a general LCA analysis is not needed at the level of policy discussion to reach a broad consensus on alternative fuels or drive systems. As a subset of WTW analysis, well-to-tank (WTT) analysis is often used to separate environmental or economic effects of fuel supplies and drive systems.

The overall WTW energy use is calculated by

$$\mathrm{Energy}_{\mathrm{WTW}} = \mathrm{Energy}_{\mathrm{TTW}}[\mathrm{MJ}_{\mathrm{final\ fuel}}/\mathrm{km}] \cdot \Sigma \mathrm{Energy}_i[\mathrm{MJ}_i/\mathrm{MJ}_{\mathrm{final\ fuel}}],$$

where:

$\mathrm{Energy}_{\mathrm{TTW}}$ = fuel consumption of the vehicle,
Energy_i = input of primary energy source i to generate the final transportation fuel.

The input of primary energy includes the energy content of the final fuel. The overall WTW GHG emissions expressed in g CO_2 equivalents per km are calculated by

$$\begin{aligned}\mathrm{GHG}_{\mathrm{WTW}} = {} & \mathrm{GHG}_{\mathrm{WTT}}[\mathrm{g/MJ}] \cdot \mathrm{Energy}_{\mathrm{TTW}}[\mathrm{MJ/km}] \\ & + \mathrm{GHG}_{\mathrm{TTW}}[\mathrm{g/MJ}] \cdot \mathrm{Energy}_{\mathrm{TTW}}[\mathrm{MJ/km}].\end{aligned}$$

Well-to-wheel and WTT discussions are often accompanied by an assessment of further closely related issues, such as:

- Analysis of technology and cost learning, typically by applying standard development curves or other specific background information from the literature,
- Analysis of primary fossil- or nuclear-energy availability, (regional) renewable energy potentials, identification of competing resource requirements, and
- Availability of regional potentials to apply CO_2 capture and storage (CCS).

Monte-Carlo analysis is typically applied to handle uncertainty if several data sources are available to identify most probable costs.

Well-to-wheel and WTT analysis have been applied recently to a number of important high-level consensus finding processes among industry and its relevant automobile and energy branches and public representatives throughout the world. Although applying different software-based tools in Europe, Japan and the USA, international information exchange in 2006 has proven that the WTW or WTT methodology produces converging results.

In Europe, WTW and WTT analysis for the road transport sector emerged from the German Transport Energy Strategy (TES), a group of initially seven automobile and energy companies that successfully developed a consensus on a phased introduction of alternative fuels and drive-systems for Germany, with hydrogen and fuel cells being the most relevant (medium- and) long-term options. Using the General Motors WTW study as a stepping stone (GM, 2001), the automotive and energy industries then joined in a European partnership to develop a successful broad European consensus (CONCAWE/EUCAR/JRC study (JEC, 2007)).

In Section 7.3, process-specific technical information on alternative fuels, which is needed for the WTW analysis is presented, and in Section 7.4 drive-system-specific data are provided, which are then merged in a WTW analysis of complete energy chains in comparison in Section 7.5. In reality, the potential number of realistic alternative fuel chains and drive system combinations is much larger. Owing to limited space, a set of most relevant processes is presented. A separate section (7.6) discusses the resource utilisation of the energy chains presented in Section 7.5. Section 7.7 finally combines specific GHG emissions for relevant alternative fuel supply chains with specific costs in a portfolio analysis.

7.3 Characteristics of alternative fuel supply

Gasoline, ethanol and the gasoline fraction of BTL, CTL and GTL generated via the MtSynfuel trademark by Lurgi process are used in petrol engines. Diesel, fatty acid methyl ester (FAME) (also referred to as biodiesel) and the diesel fuel fraction of BTL, CTL, GTL are used in diesel engines. Compressed methane gas (CMG) from natural gas and biogas are used in adapted petrol engines. Ethanol can be used up to an ethanol content of 85% by volume (E85) in dedicated petrol engines; without

Table 7.2. *Fuel properties*

Fuel	LHV (MJ/l)	LHV (MJ/kg)	CO_2 (g/MJ)
Gasoline	32.2	43.2	73.3
Diesel	36.0	43.1	73.3
LPG	23.4[a]	46.4	64.7
Ethanol	21.2	26.8	71.3
FAME	32.8	36.8	76.1
BTL, CTL, GTL	34.3	44.0	70.8
DME	19.0[a]	28.4	67.2
Methanol	15.8	19.9	68.9
Hydrogen	0.0108[b]	120.0	0
CMG from biogas	0.036[b]	45.1	56.4
CMG from natural gas	0.036[b]	49.2	55.0

Notes:
[a] Liquid state at 20 °C, 0.5–1.0 MPa (LPG); ~1.6 MPa (DME).
[b] 0.1013 MPa, 0 °C.

modification of the engine, ethanol-blended gasoline with an ethanol content of up to 10% by volume can be used. The European standard EN DIN 228 for conventional gasoline allows an ethanol content of up to 5% by volume. Flexible-fuel vehicles (FFV) can be fuelled with any ethanol and gasoline mixture from 0% to 85% by volume (E85).

For the calculation of WTW energy requirements and GHG emissions we have made the simplification that the fuel consumption of a vehicle fuelled with ethanol (e.g., E85) is the same as that of a vehicle fuelled with pure gasoline. Methanol is used in fuel cell vehicles with on-board fuel processors. Table 7.2 shows the properties of different transportation fuels.

In the case of biomass-derived fuels, the CO_2 emitted by carbon-containing fuels is absorbed from the atmosphere during the growth of the plants. As a result, the combustion of biomass-derived fuels is assumed to be CO_2 neutral at a global scale.

Dimethyl ether (DME) is stored at a slightly elevated pressure (~1.6 MPa), at which it becomes liquid at ambient temperature. The physical properties of DME are similar to those of liquid petroleum gas (LPG). Liquid petroleum gas is a mixture of propane and butane, and is derived from crude oil refining and natural gas processing. The fraction of LPG generated in refineries and in natural-gas processing plants is relatively low. Therefore, only a small fraction of all vehicles can be operated on LPG. Liquid petroleum gas is also used for heat generation, e.g., for residential buildings in remote locations. In Germany and the United Kingdom, LPG mainly consists of propane (~95%).

Table 7.3. *Refinery data for gasoline and diesel production (JEC, 2007)*

Feedstock	Unit	Gasoline	Diesel
Crude oil	MJ/MJ	1.08	1.10
Final fuel	MJ	1.00	1.00
CO_2 equivalent	g/MJ	25.2	31.0

The use of oil sands and tars to produce liquid fossil fuels (gasoline and diesel) is part of today's non-conventional oil business and, hence, has not been considered here as a fuel alternative for a future mostly renewable-based energy system (for details see Chapter 3).

The alternative fuels discussed in the following are benchmarked against gasoline and diesel from conventional crude oil. The energy requirement and GHG emissions for the supply of gasoline and diesel include the extraction of crude oil, transport to a refinery in the EU and distribution of the final fuels. The energy requirement for crude oil extraction is assumed with 1.025 MJ/MJ crude oil, and for related CO_2 emissions with 3.3 g/MJ.

In the CONCAWE/EUCAR/JRC study (JEC, 2007) a marginal approach has been used to allocate the GHG emissions and energy use to the final products, gasoline and diesel. In recent years, Europe has seen an unprecedented growth in diesel-fuel demand while demand for gasoline has been stagnating or even dropping. According to all forecasts, this trend will continue in future years, driven by increased dieselisation of the personal car and the growth of freight transport. At the same time, jet-fuel demand also steadily increases as air transport develops. The ratio of an increasing demand for 'middle distillates' and a constant demand for gasoline exceeds the 'natural' capabilities of a refining system that was designed with a focus on gasoline production. Reducing diesel demand, therefore, 'de-constrains' the system, whereas decreasing gasoline demand makes imbalances worse. Therefore, in Europe, marginal diesel fuel is more energy-intensive than marginal gasoline (Table 7.3).

The gasoline and diesel are transported to a depot via ship, train and pipeline. From the depot, the gasoline and diesel are transported to the filling station via truck.

7.3.1 Liquefied petroleum gas (LPG)

Liquefied petroleum gas (LPG) was used as fuel for the first time in the USA in 1912. Under the general term natural gas liquids (NGL), 60% of global LPG originates as a fraction separated from methane during the production of oil and gas; the remaining 40% are generated as a by-product from the fractionated distillation of crude oil in refineries. Liquefied petroleum gas is a mixture of propane and butane, with the mixing ratio dependent on the country and season.

For the calculations, it has been assumed that LPG is derived from natural gas processing. Natural gas is extracted from natural gas fields in the North Sea. The separated LPG is liquefied and transported via ship across a distance of 1000 km to the coast and distributed to filling stations via truck across an average distance of 500 km.

7.3.2 Compressed methane gas (CMG)

7.3.2.1 CNG from natural gas

For WTW analysis, it is a sufficiently accurate assumption, that natural gas mainly consists of methane (CH_4). Compressed natural gas is also referred to as 'CNG'. Natural gas is extracted, processed, transported and distributed via pipeline to the filling stations, where it is compressed to about 25 MPa. Natural gas sources may vary for different countries. Depending on the source (natural gas quality) and the transport distance (e.g., 4000 km or even 7000 km from Russia, depending on the relevant gas fields) the auxiliary energy needs or energy losses, and hence the GHG-relevant emissions can vary. For the calculation of the energy requirement and GHG emissions for the supply of natural gas, a transport distance of 4000 km is assumed.

7.3.2.2 CMG from biogas

Biogas is generated by fermentation of wet biomass, such as organic municipal waste, sludge from sewage plants, manure and from plantation of fermentable crops such as corn or grasses. Biogas mainly consists of 50% to 75% CH_4 and 25% to 50% CO_2. For the use of biogas as transportation fuel, upgrading is required. As biogas also mainly consists of methane gas it is often referred to as compressed methane gas (CMG) if it is processed to become an alternative vehicle fuel.

Storing large amounts of wet manure results in high emissions of methane (CH_4) because of uncontrolled anaerobic fermentation in the storage tanks. According to Boisen (personal communication, European Natural Gas Vehicle Association, 22nd February, 2005), about 1.6 kg of CH_4 emissions per ton of manure will be avoided if the manure is sent to a biogas plant. At a dry matter content of 8%, about 560 MJ biogas are formed per ton of manure (dry matter content is defined as dry biomass contained in biomass/water mixture, in mass. %). As a result, and especially with wet manure, the installation of biogas plants would avoid significant amounts of CH_4 emissions.

Storing dry manure instead of wet manure by admixing straw, CH_4 emissions can decisively be further reduced. For dry manure, only about 10% of the CH_4 emissions for wet manure are produced.

In energy chains related to the provision of CMG, several GHG-emission relevant credits need to be taken into account. Biogas from municipal organic waste is credited for its fertilising effect, accounting for savings of 0.54 g of synthetic nitrogen

Table 7.4. *Electricity and heat demand for the production of biogas, CH_4 emissions and credits*

Feedstock	Unit	Municipal organic waste	Wet manure	Dry manure
Electricity	MJ/MJ	0.062	0.043	0.043
Heat	MJ/MJ	0.087	0.150	0.150
Credit for N fertiliser	g/MJ	0.54	0.34	0.34
CH_4 emissions from biogas plant	g/MJ	0.20	0.20	0.20
Avoided CH_4 emissions	g/MJ	–	2.86	0.268
Net CH_4 emissions	**g/MJ**	**0.20**	**−2.66**[a]	**−0.068**[a]

Note:
[a] Negative values from credits by avoided CH_4 emissions from open storage of manure.

(N) fertiliser (calculated as N) per MJ biogas (Boisen, 2005). A credit is also assumed for wet and dry manure processed in biogas plants for fertiliser saved, as it increases the fertilising effect of residue from biogas plants. Finally, when installing a biogas plant, the reduction of nitrogen losses from the complete energy chain section 'manure storage and distribution' need to be taken into account (Möller, 2003). According to Boisen (2005), about 0.19 kg nitrogen fertiliser are replaced per ton of manure fermented in biogas plants. Assuming the generation of about 560 MJ of biogas per ton of manure, biogas plants then save fertiliser (N) of the order of 0.34 g per MJ of biogas.

The electricity requirement for upgrading biogas to pure methane (CH_4) using a pressurised water scrubber is about 0.03 MJ per MJ of methane (Schulz, 2004; W. Tentscher, personal communication, Eco Naturgas Handels GmbH, 14th July, 2004). In addition, small amounts of CH_4 are released to the atmosphere (0.2 g per MJ of biogas). The electricity and heat demand of a biogas plant are assumed to be provided by a biogas-fuelled combined heat and power (CHP) plant using a gas engine, excess electricity being fed into the electricity grid. The upgraded biogas, which mainly consists of methane (>96% CH_4), is transported to the filling station via the natural gas grid where it is compressed to 25 MPa.

The data are given in Table 7.4.

7.3.3 *Vegetable-oil-based fuels*

In Europe, vegetable-oil-based fuels are mainly produced from rapeseed. In the USA, vegetable-oil-based fuels are mainly derived from soybeans. Another feedstock used in Europe and North America is sunflower seed. Most of the vegetable oil that is used as energy source for the generation of transportation fuel is converted to fatty acid methyl ester (FAME), often called 'biodiesel'.

Table 7.5. *Input and output data for the plantation of rapeseed (Dreier and Geiger, 1998; Edwards, personal communication, Joint Research Centre (JRC), Ispra, August 2005; Kaltschmitt and Reinhardt, 1997; Kraus* et al., *2000)*

	I/O	Unit	Amount
Diesel	Input	MJ/(ha yr)	2963
N fertiliser	Input	kg N/(ha yr)	146
CaO fertiliser	Input	kg CaO/(ha yr)	19
K_2O fertiliser	Input	kg K_2O/(ha yr)	30
P_2O_5 fertiliser	Input	kg P_2O_5/(ha yr)	53
Pesticides	Input	kg/(ha yr)	1.23
Seeding material	Input	kg/(ha yr)	2
Yield	Output	$GJ_{rapeseed}$/(ha yr); t/(ha yr)	71.5; 3.0
CO_2 emissions	–	kg/(ha yr)	217
CH_4 emissions	–	kg/(ha yr)	0.0
N_2O emissions	–	kg/(ha yr)	3.1

The Malaysian government has encouraged the construction of biodiesel plants for the conversion of palm oil to FAME, owing to the increasing demand for biodiesel in the world. In Indonesia, peatlands are converted to oil-palm plantations because of the increased biodiesel demand. Whenever peatlands are converted to oil-palm plantations, the CO_2 emissions of land-use change are extremely high. Drainage of peatlands leads to CO_2 emissions of 70 to 100 t per ha and year from the decomposition of dried peat. As a result, the production of palm oil typically results in CO_2 emissions that are up to ten times higher than the emissions from the production and use of crude-oil based diesel fuel (Hooijer *et al.*, 2006).

Another possible feedstock is jatropha oil. Deforestation is, in general, no problem with jatropha, as it is a typical crop for arid and degraded land. However, the use of jatropha is still under investigation.

Alternatively, vegetable oil can be upgraded to a fuel with similar properties as BTL by hydro-treating. For optimum use of vegetable oil, dedicated diesel engines are required.

A large portion of the overall GHG emissions from the supply of biomass based fuels results from the formation of N_2O in fertilised soils. To calculate N_2O emissions, the 'European Soil Model' is typically applied, developed by the European Joint Research Centre in Ispra (Italy). The data in Table 7.5 represent the average N_2O emissions from the plantation of rapeseed in EU25 (R. Edwards, personal communication, Joint Research Centre (JRC), Ispra, August 2005).

The different inputs are connected with upstream processes, i.e., the provision of diesel fuel, pesticides and fertilisers, which can affect overall GHG emissions decisively (the supply of, e.g., nitrogen fertiliser generates large amounts of N_2O; see Table 7.6).

Table 7.6. *Energy demand and GHG emissions for the supply of nitrogen-based fertilisers (Kaltschmitt and Reinhardt, 1997)*

	I/O	Unit	Amount
Hard coal	Input	MJ/kg_N	3.95
Diesel oil	Input	MJ/kg_N	0.86
Electricity	Input	MJ/kg_N	0.626
Heavy fuel oil	Input	MJ/kg_N	4.38
Natural gas	Input	MJ/kg_N	33.0
CO_2 emissions	–	g/kg_N	2468
CH_4 emissions	–	g/kg_N	0.45
N_2O emissions	–	g/kg_N	9.63

Table 7.7. *Input and output data for vegetable oil extraction (Dreier and Geiger, 1998; Kraus* et al., *2000)*

	I/O	Unit	Amount
Rapeseed	Input	MJ/MJ	1.633
Electricity	Input	MJ/MJ	0.008
Steam	Input	MJ/MJ	0.040
n-hexane	Input	MJ/MJ	0.003
Crude vegetable oil	Output	MJ	1.000
Rapeseed meal	Output	kg/MJ	0.041
CO_2 emissions	–	g/MJ	0.2[a]

Note:
[a] Originate from the input of fossil fuel derived *n*-hexane.

Table 7.8. *Input and output data for refining of vegetable oil (Dreier and Geiger, 1998; Kraus* et al., *2000)*

	I/O	Unit	Amount
Crude vegetable oil	Input	MJ/MJ	1.0417
Electricity	Input	MJ/MJ	0.0006
Steam	Input	MJ/MJ	0.0082
Fuller's earth	Input	kg/MJ	0.00017
Pure vegetable oil	Ouput	MJ	1.000

Once harvested, rapeseed is transported to an oil mill where vegetable oil is extracted. The crude vegetable oil is then refined. The refined vegetable oil is converted to FAME by esterification, which generates glycerol as a by-product (see Tables 7.7 to 7.9).

Table 7.9. *Input and output data for esterification of vegetable oil (Dreier and Geiger, 1998; Dreier, 2000; Kraus* et al., *2000; Borken* et al., *1999)*

	I/O	Unit	Amount
Pure vegetable oil	Input	MJ/MJ	1.0065
Electricity	Input	MJ/MJ	0.0029
Steam	Input	MJ/MJ	0.0718
H_3PO_4	Input	kg/MJ	0.00005
HCl	Input	kg/MJ	0.00054
Methanol	Input	MJ/MJ	0.0585
Na_2CO_3	Input	kg/MJ	0.00007
NaOH	Input	kg/MJ	0.00018
FAME	Ouput	MJ	1.000
Glycerol	Output	kg/MJ	0.00283

Alternatively, refined vegetable oil could be converted to a fuel similar to BTL by hydro-treating, e.g., in a refinery. Rapeseed cake, which is produced as a by-product, is assumed to replace animal fodder, i.e., from imported soybeans. Alternatively, the rapeseed cake can also be converted to biogas.

As FAME or hydro-treated vegetable oil are typically blended with conventional diesel fuel from crude oil, they do not require a separate fuel distribution infrastructure to the filling station.

7.3.4 Synthetic gasoline and diesel

7.3.4.1 Fischer–Tropsch synthesis

Fischer–Tropsch (FT) synthesis is a catalysed chemical reaction in which carbon monoxide (CO) and hydrogen (H_2) are converted into liquid hydrocarbons of various forms. Typical catalysts used are based on iron (Fe) and cobalt (Co). The production of liquid hydrocarbons using FT synthesis is a well known process. It was invented by Franz Fischer and Hans Tropsch in the 1920s in Germany. It follows the reaction:

$$n\ \mathrm{CO} + 2n\ \mathrm{H_2} \rightarrow (-\mathrm{CH_2}-)_n + n\ \mathrm{H_2O} \quad \Delta H = -152\ \mathrm{kJ/(mole\ CO)}.$$

The process of producing synthetic hydrocarbons consists of the three steps of syngas generation, FT synthesis and product upgrading and processing. The actual FT synthesis can be understood as a chain growth reaction, in which long-chain hydrocarbons are generated from CO and H_2 over metal catalysts (Fe or Co). To achieve high fuel product yields and quality, the primary long-chain FT synthesis products are chemically processed (hydro-cracking and isomerisation for diesel and gasoline fractions, respectively). The FT process has substantial requirements for hydrogen,

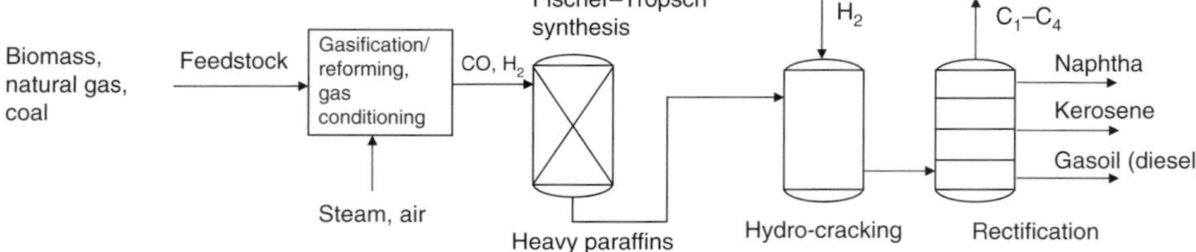

Figure 7.3. Production of hydrocarbon to liquid fuel (XTL) by FT synthesis.

especially for fuel synthesis on the basis of fuels with a low H:C ratio, such as coal and biomass; hydrogen needs to be added to adjust the required H_2:CO ratio and, thus, to increase the carbon conversion rate and the product yield;[6] hydrogen is also needed for hydro-cracking the FT products (FVS, 2003; Specht *et al.*, 2001).

The reaction being exothermal, the process is typically carried out at temperatures of 200 to 350 °C and a pressure of about 2.5 MPa. Applying temperatures of 300 to 350 °C leads rather to lighter hydrocarbons. Lower temperatures of 200 to 250 °C lead to heavier hydrocarbons instead.

For maximum yield of liquid hydrocarbons and minimum yield of gases, FT synthesis is optimised to produce predominantly heavy products (heavy paraffins),[7] i.e., producing hydrocarbon chains as long as possible at maximum hydrocarbon chain growth probability.

Consecutively, the heavy paraffins are cracked into lighter hydrocarbon fractions by hydro-cracking. For example, for the Shell Middle Distillate Synthesis (SMDS) process, the liquid product stream is composed of 60% gasoil (diesel), 25% kerosene and 15% naphtha. The gaseous product mainly consists of LPG (a mixture of propane and butane) (Eilers *et al.*, 1990). Figure 7.3 shows a simplified diagram comprising all process steps to produce synthetic hydrocarbons from biomass, natural gas and coal.

7.3.4.2 Methanol route

An alternative to FT synthesis is the production of synthetic transportation fuels from methanol. From the synthesised gas, methanol is produced in a first step following the reaction:

$$CO + 2H_2 \rightarrow CH_3OH \quad \Delta H = -91\,\text{kJ/(mole CO)}.$$

[6] Fischer–Tropsch synthesis requires a stochiometric H_2:CO ratio of 2.1:1. If coal or biomass are used as feedstock, the raw syngas contains much less hydrogen than needed. Hence, CO is reacted with water to form CO_2 and hydrogen in the shift reactor. As the CO_2 cannot be used in the Fischer–Tropsch synthesis, part of the carbon for fuel production is lost in this process. If external hydrogen is added to increase the H_2:CO ratio, the carbon of the coal or biomass is more effectively used and the hydrocarbon product yield is improved.

[7] A measure for the hydrocarbon growth is the 'chain growth probability'. For optimum liquid hydrocarbon yields, it is typically in the order of 0.85 to 0.90.

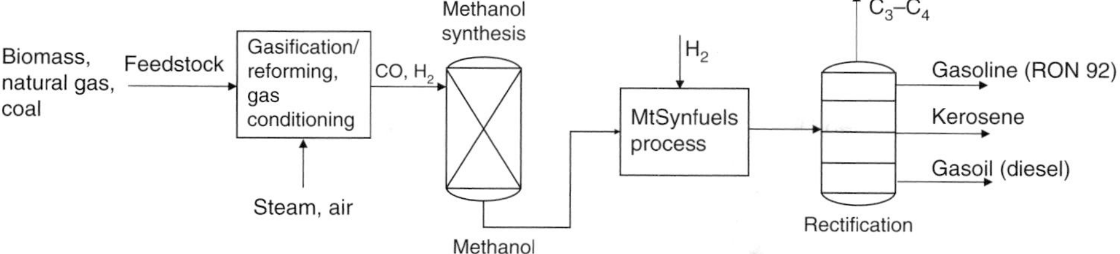

Figure 7.4. Generation of XTL via methanol synthesis and MtSynfuels process.

The reaction is exothermal at temperatures of 220 to 280 °C and pressures of 5 to 10 MPa. Methanol is then converted to synthetic transportation fuels (gasoline, diesel) by the MtSynfuels (trademark by Lurgi) process (see Fig. 7.4).

Apart from FT synthesis for naphtha, the MtSynfuels process has the advantage of sufficiently high octane numbers (RON 92) of the naphtha fraction. Therefore, this naphtha can be used directly as a fuel (gasoline) for petrol engines without further upgrading (Liebner *et al.*, 2004).

7.3.4.3 Biomass to liquids (BTL)

Ligno-cellulosic biomass, such as residual wood, residual straw or wood from the plantation of short-rotation forestry can be used as feedstock. Short-rotation forestry typically harvests fast-growing trees, which need small amounts of synthetic N fertiliser. According to Murach (personal communication, Fachhochschule Eberswalde (Germany), 27th August, 2003), 20 to 30 kg of N per ha and year need to be provided for a yield of about 10 tons of dry substance per ha and year. According to Kaltschmitt and Hartmann (2001), the lower heating value of wood from poplar is about 18.5 MJ per kg dry substance.

In a first step, biomass is converted to synthesis gas by gasification. If Fischer–Tropsch (FT) synthesis is applied, a H_2:CO ratio of about 2.1:1 is required for a maximum yield of liquid hydrocarbons. To adjust the H_2:CO ratio, CO shift reactors are used to convert a part of the CO to H_2 and CO_2 according to the following reaction:

$$CO + H_2O \rightarrow CO_2 + H_2.$$

In Hamelinck (2004), the production of liquid hydrocarbons via gasification of biomass and downstream FT synthesis and rectification have been investigated. For a hydrocarbon chain growth probability of 0.85, about 139.1 MW of liquid hydrocarbons are produced for a biomass input of 367 MW. The gaseous by-products are used for electricity generation, resulting in a net electricity output of 33.3 MW_{el}. As a result, about 0.38 MJ of liquid hydrocarbons are produced per MJ of biomass. If the hydrocarbon chain growth probability were assumed to be 0.90, about 154.8 MW of liquid hydrocarbons and 21.3 MW_{el} of electricity are generated with a liquid hydrocarbon yield of about 0.42 MJ per MJ of biomass (see Table 7.10).

Table 7.10. *Input and output data for a BTL plant with gasification and Fischer–Tropsch synthesis (chain growth probability 0.90)*

	I/O	Unit	Amount
Biomass	Input	MJ/MJ_{BTL}	2.371
Dolomite	Input	MJ/MJ_{BTL}	0.0055
NaOH	Input	MJ/MJ_{BTL}	0.00001
BTL	Output	MJ	1.000
Electricity	Output	MJ/MJ_{BTL}	0.138

Table 7.11. *Input and output data for a GTL plant with combined reforming and Fischer–Tropsch synthesis*

	I/O	Unit	Amount
Natural gas	Input	MJ/MJ_{GTL}	1.587
GTL	Output	MJ	1.000
CO_2 emissions	–	g/MJ_{GTL}	16

The liquid hydrocarbon yield from the BTL production via gasification and FT synthesis is about 42% based on the LHV, which is similar to the production of BTL via gasification, methanol synthesis and the MtSynfuel process (Dena, 2006).

7.3.4.4 Gas-to-liquids (GTL)

In the case of synthesis gasoline or diesel fuel from natural gas (GTL), synthesis gas is produced by a combination of steam reforming and partial oxidation processes (combined reforming) to achieve a H_2:CO ratio of generally 2.1:1. This means that the overall process energy demand can be reduced to its minimum. The individual reactions are:

$$CH_4 + H_2O \rightarrow CO + 3H_2 \quad \text{(steam reforming)},$$

$$CH_4 + 0.5O_2 \rightarrow CO + 2H_2 \quad \text{(partial oxidation)}.$$

Gas-to-liquids plants are generally located close to natural gas fields, as the transport costs for liquid fuels are less than those for gaseous fuels. The production of GTL is considered to be an alternative to liquefied natural gas (LNG), specifically when focusing on the end-product 'vehicle fuel' and not the long distance transport of energy. In 1993, a first large-scale GTL plant was erected by Shell in Bintulu, Sarawak in Malaysia, based on Fischer–Tropsch synthesis. The plant's total thermal process efficiency is about 63% (Shell, 1995) (see Table 7.11); a second plant is under construction in Qatar, with production expected to begin in 2010.

Table 7.12. *Input and output data for a CTL plant with gasification and Fischer–Tropsch synthesis (Gray and Tomlinson, 2001; D. Gray, personal communication, Mitretek Systems, 21st July, 2005)*

	I/O	Unit	Amount
Hard coal	Input	MJ/MJ_{CTL}	2.471
CTL	Output	MJ	1.000
Electricity	Output	MJ/MJ_{CTL}	0.330
CO_2 emissions	–	g/MJ_{CTL}	167

Table 7.13. *Input and output data for a methanol plant with gasification and methanol synthesis*

	I/O	Unit	Amount
Biomass	Input	$MJ/MJ_{methanol}$	1.959
Methanol	Output	MJ	1.000

7.3.4.5 Coal-to-liquids (CTL)

In this energy chain, coal is gasified to generate synthesis gas. The H_2:CO ratio required for an optimum efficiency is adjusted via the CO shift reaction of a part of the carbon monoxide (CO) contained in the synthesis gas. The remaining synthesis gas is converted to liquid hydrocarbons via Fischer–Tropsch synthesis or via methanol synthesis with a downstream MtSynfuels (trademark by Lurgi) process (see beginning of Section 7.3.4). The liquid hydrocarbon yield amounts to about 0.40 MJ per MJ of hard coal, which is of the same order of magnitude as in the case of BTL (~0.40 MJ/MJ); to calculate the thermal process efficiency, the electricity export must also be taken into account (see Table 7.12).

In South Africa, large-scale coal gasification and FT plants have been operated by Sasol since 1955, mainly producing gasoline.

7.3.5 Methanol and DME

Methanol has been considered as a fuel for fuel-cell vehicles with on-board fuel processors for some time. Dimethyl ether (DME) has been suggested as a fuel alternative for diesel engines in Japan and Sweden. The synthesis of DME is based on methanol synthesis followed by DME formation:

$$2CO + 4H_2 \rightarrow 2CH_3OH,$$

$$2CH_3OH \rightarrow CH_3OCH_3 + H_2O.$$

The formation of DME is a result of the selection of adequate catalysts. Methanol and DME plants achieve similar total process energy efficiencies.

7.3.5.1 Methanol from biomass

Methanol is produced from biomass by gasification and downstream methanol synthesis (see Table 7.13).

Table 7.14. *Input and output data for a methanol plant with combined reforming and methanol synthesis*

	I/O	Unit	Amount
Natural gas	Input	$MJ/MJ_{methanol}$	1.467
Methanol	Output	MJ	1.000
CO_2 emissions	–	$g/MJ_{methanol}$	12

Table 7.15. *Input and output data for a DME plant with combined reforming and DME synthesis*

	I/O	Unit	Amount
Natural gas	Input	MJ/MJ_{DME}	1.403
Electricity	Input	MJ/MJ_{DME}	0.0043
O_2	Input	kg/MJ_{DME}	0.0046
DME	Output	MJ	1.000
CO_2 emissions	–	g/MJ_{DME}	10

7.3.5.2 Methanol from natural gas

Methanol is produced from natural gas via combined reforming and downstream methanol synthesis. The technical data for the methanol plant have been derived from a methanol plant located in Tjeldbergodden in Norway (Larsen, 1998) (see Table 7.14).

7.3.5.3 DME from natural gas

Dimethyl ether is produced from natural gas via combined reforming and downstream DME synthesis. Technical data for a typical DME plant are based on information provided by Haldor Topsoe (personal communication, October, 2002) (see Table 7.15).

The required oxygen is produced in an air-separation plant. The electricity for the air-separation plant is provided by a natural-gas fuelled combined-cycle gas turbine (CCGT) power plant.

7.3.6 Ethanol

7.3.6.1 Conventional fermentation

Sugar-containing plants The conventional fermentation of sugar-containing plants involves micro-organisms that use the fermentable sugars for food, simultaneously producing ethanol and other by-products. The highest process efficiencies can be

Table 7.16. *Input and output data for the production of ethanol from wheat (Kaltschmitt and Hartmann, 2001; Punter* et al., *2004)*

	I/O	Unit	Amount
Wheat	Input	$MJ/MJ_{ethanol}$	1.864
Steam[a]	Input	$MJ/MJ_{ethanol}$	0.364
Electricity	Input	$MJ/MJ_{ethanol}$	0.054
Ethanol	Output	MJ	1.000
DDGS	Output	$kg/MJ_{ethanol}$	0.0425

Note:
[a] Supplied by a natural-gas fuelled CHP plant.

achieved when feeding the micro-organisms with C6 sugars (glucose). Therefore, biomass containing high levels of glucose or precursors to glucose are the preferred option for the lowest fuel conversion complexity to ethanol.

Biomass feedstocks with high sugar content (best known as saccharides) are sugarbeet, sugarcane, sweet sorghum and various fruits. To produce ethanol as transportation fuel from sugar-containing plants, sugarbeet is used in regions with moderate climates, such as parts of Europe and sugarcane is used in tropical regions, such as Brazil, as feedstock today.

Starch-containing plants Another potential ethanol feedstock is starch. Starch molecules are made up of long chains of glucose molecules. Hence, starch-containing materials can also be fermented after the starch molecules have been broken down into simple glucose molecules. Examples of starchy materials commonly used around the world for ethanol production include cereal grains, potatoes, sweet potatoes and cassava. Typical cereal grains commonly used for ethanol production in the EU are rye and wheat.

In most of today's ethanol plants for the conversion of wheat, rye and corn, the required thermal energy is provided by natural gas, heavy fuel oil or coal. The protein-rich by-products of ethanol plants are referred to as 'dried distillers' grains with solubles', abbreviated as DDGS, and are mostly used for animal fodder. Alternatively, they can be converted to biogas for heat and electricity production. The resulting residue can then be used as fertiliser (see Table 7.16).

The most energy-efficient production of ethanol is from sugarcane (Brazil), since the crop produces high yields per hectare and the sugar is relatively easy to extract; if bagasse (the crushed stalk of the plant) is used to provide heat and power for the process, net GHG emissions are significantly reduced. As starchy crops first have to be converted to sugar in a high-temperature enzymatic process, ethanol production from cereals can be very energy-intensive and debate exists on the net energy gain.

7.3.6.2 Hydrolysis and fermentation of ligno-cellulosic biomass

Only recently has ligno-cellulosic biomass, such as wood and straw, been discovered for ethanol production based on new processes. As a result, the potential of biomass for ethanol production has been increasing substantially. Ethanol production from ligno-cellulosic feedstock includes biomass pre-treatment to release cellulose and hemicellulose, hydrolysis to release fermentable 5- and 6-carbon sugars, sugar fermentation, separation of solid residues and non-hydrolysed cellulose, and distillation to fuel grade (IEA, 2007). To provide better conversion, new chemical and enzymatic processes (pre-treatment, hydrolysis, fermentation) are being examined. However, production of ethanol from ligno-cellulosic feedstock is not yet commercially viable and requires further research and development.

Several different plant layouts for ligno-cellulosic biomass conversion to ethanol are known using enzymatic hydrolysis:

- Enzymatic hydrolysis with glucose fermentation, but without the fermentation of pentoses.
- Separate hydrolysis and fermentation (SHF): hydrolysis and fermentation occur in separate process steps. The process is suitable for the fermentation of pentoses, also in a separate process step.
- Simultaneous saccharification and fermentation (SSF): one-stage enzymatic hydrolysis, but the fermentation of pentoses and hexoses takes place in separate process steps.
- Simultaneous saccharification and co-fermentation (SSCF): one-stage enzymatic hydrolysis of cellulose and fermentation of pentoses and hexoses all in one process step. The upstream hydrolysis of the hemicellulose takes place in a separate process step.

For ethanol from ligno-cellulosic feedstock, the total energy input for the production process may be even higher than for ethanol from corn, but the ethanol yield is greatly improved and conversion efficiencies of 60%–70% may ultimately be possible (IEA, 2006b). Moreover, much of the energy can be provided by the biomass itself (for example by burning the lignin, an unfermentable part of the organic material), at the same time reducing emissions.

A pilot plant using residual straw as feedstock to produce ethanol is in operation by Iogen in Ottawa, Canada. The plant can produce up to 3000 m^3 of ethanol annually (Iogen, 2005).

Also, a Spanish company (Abengoa Bioenergy) has developed a process for the conversion of ligno-cellulosic biomass to ethanol based on SSF. A demonstration plant on the basis of wheat and barley straw has been operating in Salamanca since 2006, with an annual production capacity of five million litres of ethanol (Abengoa, 2006).

Table 7.17 shows inputs and outputs for the conversion of wheat straw to ethanol based on data provided by Iogen.

Table 7.17. *Input and output data for the production of ethanol from wheat straw*

	I/O	Unit	Amount
Wheat straw	Input	$MJ/MJ_{ethanol}$	2.377
CaO	Input	$kg/MJ_{ethanol}$	0.0024
H_2SO_4	Input	$kg/MJ_{ethanol}$	0.0041
Ethanol	Output	MJ	1.000
Electricity	Output	$g/MJ_{ethanol}$	0.052

Table 7.18. *Input and output data for the production of hydrogen via onsite steam reforming of natural gas*

	I/O	Unit	Amount
Natural gas	Input	MJ/MJ_{H_2}	1.441
Electricity	Input	MJ/MJ_{H_2}	0.016
Hydrogen	Output	MJ	1.000
CO_2	–	g/MJ_{H_2}	286
CH_4	–	g/MJ_{H_2}	0.075

The plant is fully energy self-sufficient by using the lignin residue to provide the required thermal input energy. In addition, some excess electricity is co-generated. The technology is still in the stage of research and development.

7.3.7 Hydrogen

For a more detailed description of hydrogen-production technologies, see Chapter 10.

7.3.7.1 Steam reforming of natural gas

To calculate the specific energy demand and GHG emissions for the supply of natural gas, a transport distance of 4000 km is assumed (see Section 7.3.2). Hydrogen is then generated onsite at the filling station (see Table 7.18).

The pressure at the outlet of the pressure-swing adsorption (PSA) plant is 1.5 MPa. The specific electricity requirement for the compression from 1.5 to 88 MPa at the compressed gaseous hydrogen (CGH_2) filling station (required for 70 MPa on-board vehicle storage) amounts to about 0.077 MJ_{el}/MJ of CGH_2.

Alternatively, hydrogen can be generated in a large, central steam methane reformer (SMR) plant and transported or distributed to the filling stations via pipeline. The efficiency of the large plant is higher, leading to slightly lower overall GHG emissions. But an additional infrastructure (hydrogen pipelines) is required.

Table 7.19. *Input and output data for the production of hydrogen via biomass gasification*

	I/O	Unit	Amount
Biomass	Input	MJ/MJ_{H_2}	1.931
Hydrogen	Output	MJ	1.000
Electricity	Output	MJ/MJ_{H_2}	0.005

Table 7.20. *Input and output data for the production of hydrogen via coal gasification without CO_2 capture and storage*

	I/O	Unit	Amount
Coal	Input	MJ/MJ_{H_2}	1.967
Hydrogen	Output	MJ	1.000
CO_2	–	g/MJ_{H_2}	189

7.3.7.2 Gasification of biomass

A gasifier with a capacity of 10 MW_{th} (biomass input) based on the 'staged reforming' method suggested by the German company D.M.2 has been assumed for the calculations. The advantage of the gasifier from D.M.2 over other gasification concepts (e.g., that put forward by the Battelle Columbus Laboratories (BCL) or Shell) is that hydrogen can be generated in relatively small plants, which simplifies its introduction into the transportation fuel market. The plant could be installed near large filling stations, e.g., at motorways (see Table 7.19).

Hydrogen pressure at the outlet of the purification stage, which is typically a pressure-swing adsorption plant, is about 2 MPa. Hydrogen is distributed to the filling stations by means of a small pipeline (<5 km). The electricity requirement for compression from 2 to 88 MPa at the CGH_2 filling station amounts to about 0.070 MJ_{el}/MJ of CGH_2.

7.3.7.3 Gasification of coal

In hydrogen fuel supply schemes using coal as feedstock, the same gasification processes as for the production of CTL fuels are applied. The synthesis gas from the gasifier is then converted to hydrogen by CO shift and pressure-swing adsorption (see Table 7.20).

For the conversion of coal to pure hydrogen, data provided by a study from Foster Wheeler (1996) have been used. As the specific CO_2 emissions already surpass those from the operation of conventional gasoline cars, CO_2 capture and storage needs to be applied to dispose of harmful GHG emissions down to about 6 gCO_2/MJ of pure hydrogen. In turn, the specific coal demand increases to about 2.302 MJ/MJ of pure hydrogen.

Hydrogen has been assumed to be distributed by means of a hydrogen pipeline grid (main pipelines: 50 km, distribution pipelines: 5 km) to the filling stations. The electricity requirement for the compression from 2 to 88 MPa at the CGH_2 filling station amounts to about 0.070 MJ_{el}/MJ of CGH_2.

7.3.7.4 Water electrolysis

Pressurised alkaline water electrolysis technology has been state-of-the-art for many years. The system efficiency of real electrolysers ranges from 62% to 70%, including all auxiliaries (AC/DC converter, pumps, blowers, controls, etc.) based on the lower heating value of hydrogen.

The electricity requirement for large-scale hydrogen liquefaction amounts to about 0.3 MJ_{el} per MJ of LH_2 today (suction pressure typically 2 to 3 MPa, which corresponds to the outlet pressure of the PSA of a typical steam reforming plant), but can be reduced to about 0.16 MJ_{el} per MJ of LH_2, if the feed pressure can be increased (Quack, 2001).

The overall production of GHG emissions greatly depends on the electricity source, which feeds the electrolyser at an annual average. In the case of renewable electricity, the specific GHG emissions become almost negligible. Furthermore, if adding a liquefaction stage to provide liquid hydrogen (LH_2) to the filling station, small amounts of specific GHG emissions have to be taken into account, which result from the transport of LH_2 in diesel-operated trucks.

In the case of compressed hydrogen (CGH_2), the electrolysis is assumed to take place onsite at the filling station. The pressure at the outlet of the electrolyser is 3 MPa. The specific electricity requirement for the compression from 3 to 88 MPa at the CGH_2 filling station amounts to about 0.062 MJ_{el}/MJ of CGH_2.

7.4 Vehicle drive systems

Typically, internal combustion engines, fuel-cell and battery-electric drives are under consideration in today's WTW analysis. Hybridisation[8] can reduce the fuel consumption of passenger cars both with internal combustion engines and fuel cells significantly in European driving cycle mode (NEDC) and is typically assumed for all drive systems for the medium to long term. All vehicle data are based on the technology expected to be available by 2010 as agreed between the EUCAR (European Council of Automotive Research and Development) members for the CONCAWE, EUCAR, JRC study (JEC, 2007). Table 7.21 shows the fuel consumption of various non-hybrid and hybrid vehicles.

[8] Hybridisation of a vehicle means the combined use of an internal combustion engine (or a fuel cell), an electric motor and a battery. The battery is used to store energy from regenerative braking. During braking, the electric motor works as generator. Furthermore, the electric motor assists the ICE, e.g., during acceleration and can be the sole source of power in inner-city driving mode. The ICE can be operated more often at the point of maximum efficiency or is dimensioned for less power output.

Table 7.21. *Fuel consumption of various passenger vehicles (reference vehicle: Volkswagen Golf) (IFP, personal communication, 2005; JEC, 2007)*

Propulsion system	Non-hybrid (MJ/km)	Hybrid (MJ/km)
Gasoline/ethanol PISI	1.900	1.617
Gasoline/ethanol DISI	1.879	1.630
LPG PISI	1.900	1.617
CNG dedicated	1.872	1.394
Diesel DICI DPF	1.767	1.456
DME DICI	1.721	1.411
Methanol FPFC	–	1.480
CGH_2 ICE	1.675	1.485
CGH_2 FC	0.940	0.837
LH_2 ICE	1.675	1.414
LH_2 FC	0.940	0.837
Battery-electric vehicle	0.460	–

The lower heating value (LHV) of gasoline is assumed to be about 32 MJ/l, and that of diesel about 36 MJ/l.

All fuel cells for use in vehicles are based on proton-exchange-membrane fuel cell (PEMFC) technology. The methanol fuel-processor fuel cell (FPFC) vehicle comprises an on-board fuel processor with downstream PEMFC. On-board methanol reforming was a development focus of industry for a number of years until around 2002. Direct-methanol fuel cells (DMFC) are no longer considered for the propulsion of commercial vehicles in the industry (see also Chapter 13).

If the battery pack of hybrid vehicles can additionally be charged with electricity from the public electricity grid, the vehicle is called plug-in hybrid. Batteries of plug-in hybrids have to be larger than those of a conventional hybrid vehicle. For short distance trips, plug-in hybrid vehicles can be operated in the electricity-only mode. Plug-in hybrids are not explicitly considered here, but addressed in more detail in Section 7.9.

7.5 Well-to-wheel analysis (GHG emissions and costs)

Figure 7.5 shows the overall primary energy requirement and Figs. 7.6 and 7.7 show the overall GHG emissions for different WTW pathways. It has been assumed that all fuels are used in hybrid vehicles, except electricity in battery-electric vehicles. Methanol and hydrogen are used in fuel-cell hybrid vehicles.[9] Gasoline, diesel, CNG, CMG from biogas, BTL, GTL, CTL, DME and ethanol are used in hybrid vehicles with internal combustion engines. As a reference vehicle, a compact car, the Volkswagen Golf, was selected as most representative.

[9] If hydrogen was used in an ICE, WTW CO_2 emissions might be twice as high, because of the low efficiency of the ICE compared with a fuel cell.

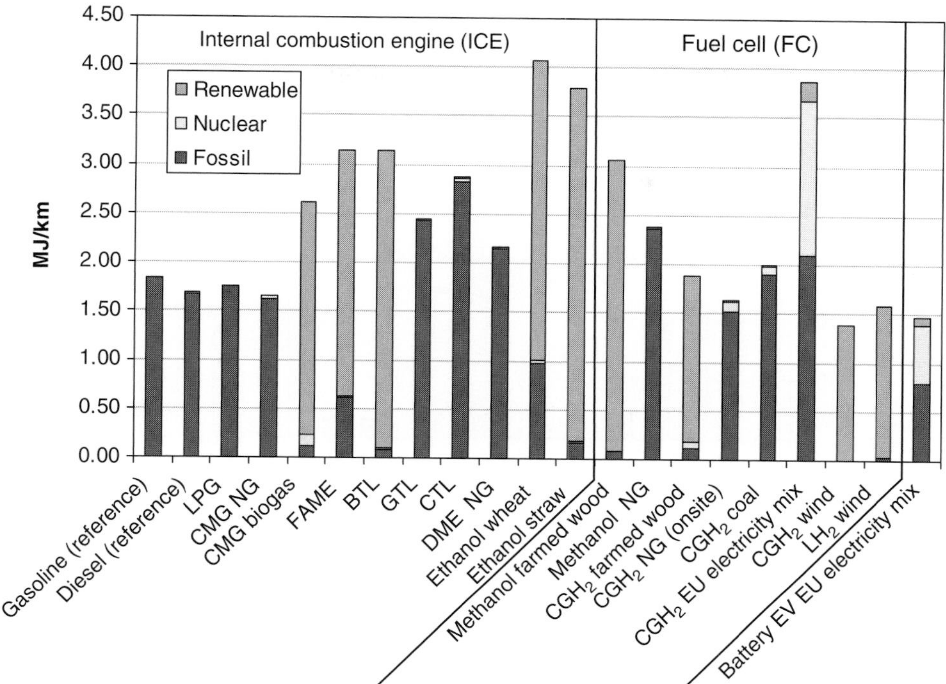

Figure 7.5. Overall well-to-wheel (WTW) primary energy requirement for different energy chains (hybrid ICE and FC vehicles, electric vehicle (EV)).

For reference, gasoline and diesel from conventional crude oil are used to benchmark the alternative fuels, leading to CO_2-equivalent emissions of about 140 g/km (gasoline) and 129 g/km (diesel). With the exceptions of CTL, hydrogen and electricity, roughly 75%–80% of WTW CO_2 emissions result from the combustion of the fuel in the vehicle (TTW), while the remaining 20%–25% are attributable to feedstock extraction, feedstock transportation, processing (refining) and product distribution (WTT).

For comparison, gasoline and diesel from non-conventional oil (here, oil sands) would result in 170 to 190 g/km (gasoline) and 150 to 170 g/km (diesel fuel) based on data from ACR (2004) and Söderbergh *et al.* (2006).

Furthermore, the extraction of non-conventional oil has other detrimental environmental impacts, such as water pollution and loss of biodiversity. Depending on the depth of the deposits, oil sands are either strip mined in open pits or heated so that the bitumen from which the non-conventional oil is extracted can flow to the surface (*in-situ* extraction). Both forms of oil-sands extraction require considerable amounts of energy (i.e., natural gas) and water, and lead to significant detrimental environmental impacts (Woynillowicz *et al.*, 2005; see also Chapter 3).

With carbon-containing biomass-derived fuels, well-to-tank (WTT) GHG emissions are negative because the carbon bound in the fuel is removed from the

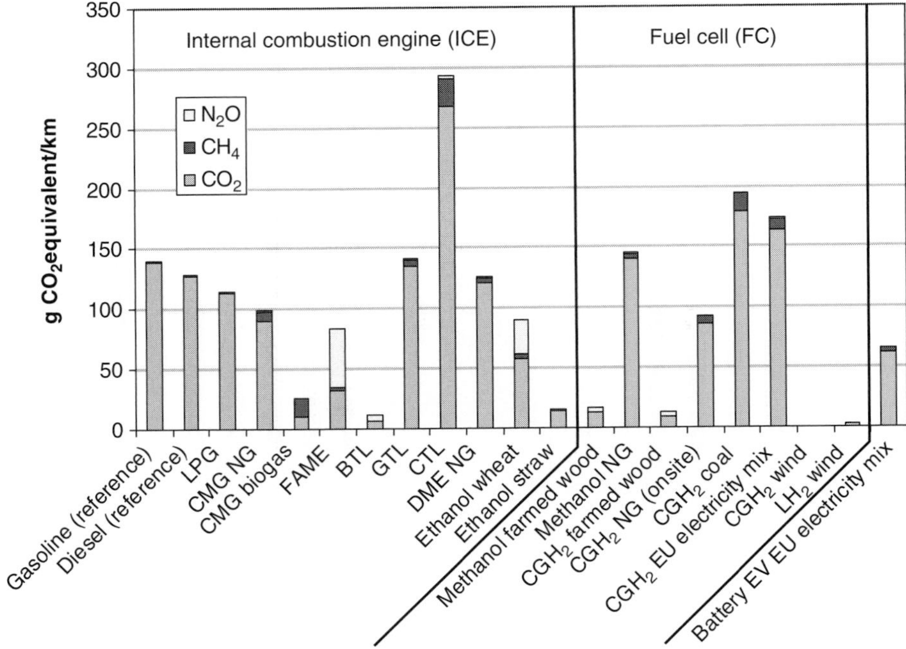

Figure 7.6. Overall well-to-wheel (WTW) emissions of greenhouse gases for different energy chains (hybrid ICE and FC vehicles, electric vehicle (EV)).

atmosphere during the growth of the plants. The carbon is then emitted during operation of the vehicle (TTW), leading to positive GHG emissions as shown in Fig. 7.6. The resulting net GHG emissions are the sum of the positive TTW emissions and the negative WTT emissions.

Fuel cells typically use hydrogen directly, mostly as compressed gas, possibly also as liquid hydrogen stored on board. Even though the CO_2-equivalent emissions were assessed to be low, methanol from farmed wood as a fuel for fuel cells turned out to be no optimum solution. The reason is not revealed by the WTW graph. Industry has decided against methanol as a fuel-cell fuel for two reasons:

- The need of complex on-board reforming systems, and
- Lower efficiencies as a future renewable fuel from sources other than (fairly limited) biomass.

Among liquid fuels (XTL), only biomass-derived hydrocarbons (BTL) are a relevant option from the perspective of lowering GHG emissions; not so other fossil-based liquids (CTL, GTL). Even if CTL fuel supply paths were upgraded by carbon capture and storage, the resulting specific CO_2-equivalent emissions would only be reduced to the level of conventional gasoline or diesel energy chains.

The relatively high N_2O emissions from the supply and use of FAME from rapeseed and ethanol from wheat are caused by soil N_2O emissions and N_2O

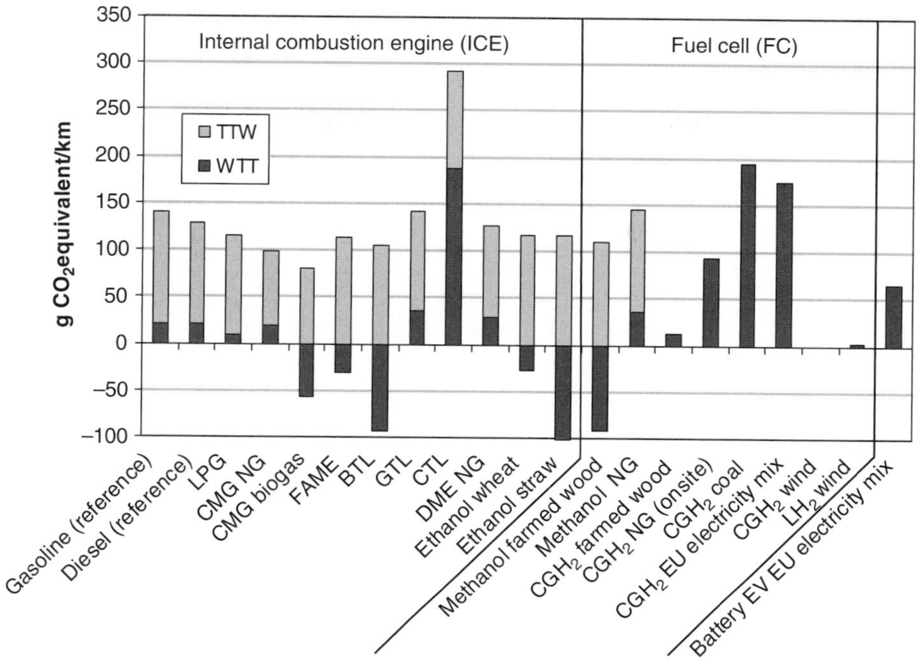

Figure 7.7. Overall well-to-wheel (WTW) emissions of greenhouse gases for different energy chains (hybrid ICE and FC vehicles, electric vehicle (EV)), with TTW and WTT contributions.

emissions from the supply of synthetic nitrogen fertiliser. It is assumed here that biogas is generated from organic municipal waste.

Battery-electric vehicles would lead to relatively low WTW GHG emissions, even if electricity from the EU electricity mix is used.

7.6 Impact of alternative fuels on resource availability

Today, large amounts of biomass are already used to generate heat and electricity (mainly wood) and are predicted to increase further (e.g., wood-pellet-fuelled boilers, wood-chip-fuelled CHP plants, electricity generation from biogas).

The additional and steadily increasing demand for biofuels could lead to a situation where production of biomass derived fuels finally compete with food production. People who can afford cars can pay more for biomass for fuels than people in non-industrialised countries can pay for food production. Fertile soil in non-industrial countries might then be used for energy crops instead of food. This may eventually lead to a situation where only bad soil is left for food crops and the poor, which in addition would eventually also lead to further deforestation of the World's rainforests.

The potential for renewable electricity production from wind and solar energy is by far higher than the potential for biomass production. Therefore, in the case of

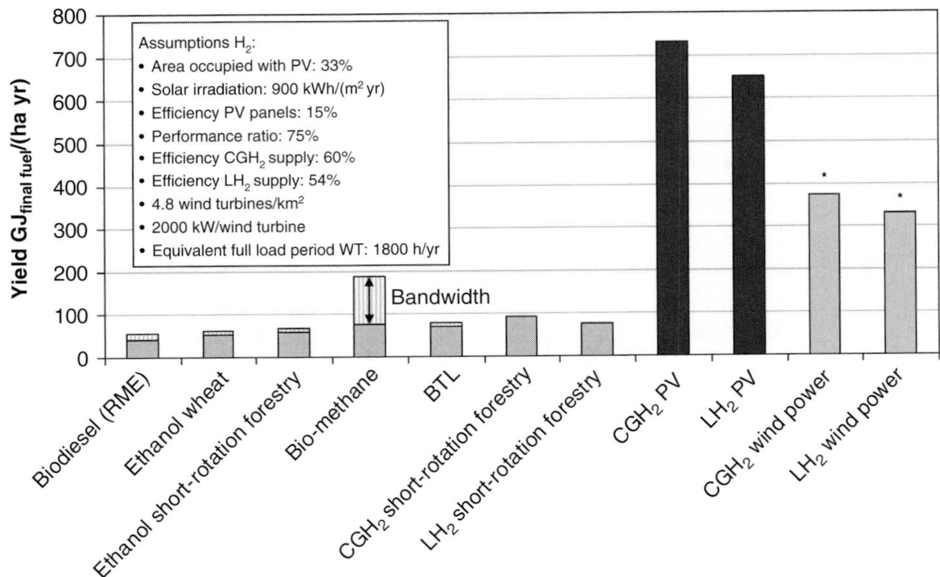

Figure 7.8. Yield of final fuel per ha and year for different transportation fuels.

electricity, competing uses are much easier to handle than for biomass. Figure 7.8 shows the yield of various biomass-derived transportation fuels compared with hydrogen from photovoltaic panels and from wind power.

Figure 7.8 is based on the following assumptions. The rapeseed yield ranges between 3 and 4 t per ha and year at a water content of 10%. The yield of wheat grain is assumed to be about 7 t per ha and year at a water content of 16%. For short-rotation forestry (plantation of fast growing trees, such as poplar and willow) the yield is assumed to be about 10 t of dry substance per ha and year (average yield per year, harvested at three- to five-year intervals). For energy crops used in biogas plants, the yield is assumed to be 10 to 24 t of dry substance per ha and year.

Biodiesel (RME) gives the lowest yield of all renewable transportation fuels per ha and year, whereas hydrogen from renewable electricity gives the highest yield. The yield for hydrogen from photovoltaic panels is higher than for the best biofuel (upper limit for compressed methane from biogas), even if it is assumed that only one-third of the land area is occupied by photovoltaic panels. In the case of hydrogen from wind power, most of the land can still be used for other purpose, e.g., for the plantation of food crops. Therefore, competing use of land need not be taken into account in the case of wind power.

7.7 Portfolio analysis

A combined presentation of specific greenhouse-gas (GHG) emissions and specific fuel-supply costs is dubbed portfolio analysis. It serves to identify rapidly the most

Table 7.22. *Assumptions for energy and feedstock prices, basis 2010 (GEMIS, 2005; JEC, 2007)*

	€/bbl	€/t	€/GJ
Crude oil	50		9.1
Natural gas remote			4.0
Natural gas at EU border			7.3
Hard coal			2.5
Rapeseed		248[a]	10.4
Wheat grain		100[b]	6.7
Wood chips from farmed wood		81[c]	4.5
Residual straw (bales)		60[c]	3.5
Electricity (0.4 kV)			18.1

Notes:
[a] Water content: 10%, 23.8 MJ/kg.
[b] Water content: 13%, 14.8 GJ/t.
[c] Dry substance: 18 GJ/t.

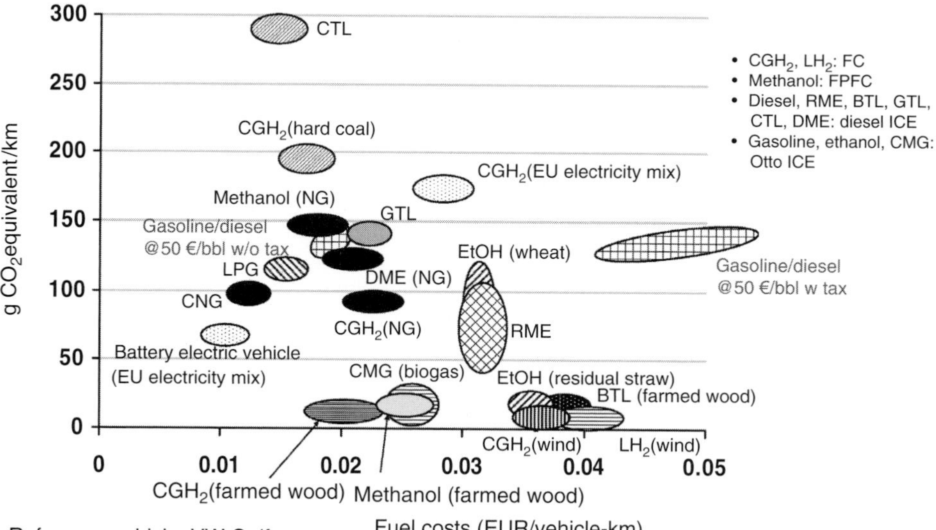

Figure 7.9. Fuel costs versus GHG emissions, 'Well-to-Wheel' (until 2010).

relevant fuel supply paths, drive system alternatives or combinations of both at a glance.

Table 7.22 shows the assumption for energy and feedstock prices underlying the portfolio analysis in Fig. 7.9. In general, energy costs are highly time dependent. The analysis also needs to be seen in the context of complete energy systems and assumptions about them. Specifically: (a) cost learning curves need to be considered for innovative technologies, (b) increasing costs are to be expected for exhaustible primary energies and those renewable energies which are more limited, e.g., biomass and (c) exchange rates will change for the world regions. As many of these parameters are constantly changing, a portfolio presentation is typically valid only for a short period and based on a set of well defined parameters.

Figure 7.9 shows a typical portfolio analysis of selected combinations of alternative fuel supply and vehicle drive systems for the period until 2010, based on the assumptions in Table 7.22.

The fuel costs typically do not include the vehicle costs, as these can vary widely depending on cost and technology development assumptions and vehicle equipment.

The analysis has to be understood as an example, because of the changing cost assumptions and system constraints at the time of the assessment. The following exemplary conclusions can be drawn from Fig. 7.9:

- Hybrid fuel-cell vehicles fuelled with CGH_2 from natural gas can achieve lower GHG emissions than hybrid CNG vehicles.
- For ethanol and FAME from rapeseed (RME), a large bandwidth for the overall WTW GHG emissions is observed.
- The use of fuel-cell vehicles does not always result in GHG emission reductions. For example, in the case of hybrid fuel-cell vehicles fuelled with CGH_2 from the EU's current electricity mix and from hard coal lead to higher GHG emissions than hybrid ICE vehicles fuelled with crude oil based gasoline and diesel.
- Although CTL is a cost-competitive alternative fuel for internal combustion engines, its GHG emissions are significantly higher than those from conventional gasoline and diesel cars. Gas-to-liquids fuel shows no advantage in emissions or costs for internal combustion engines, except that of extending the reach of oil-based fuels, if natural gas reserves should be larger.
- Compressed hydrogen from farmed wood (and other biomasses, such as residual wood) can, in combination with fuel-cell cars, achieve fuel costs that are close to today's costs of untaxed conventional gasoline and diesel, and simultaneously lower overall GHG emissions significantly. If the potential were not limited, farmed wood could be an optimal alternative fuel source.
- Compressed hydrogen from renewable electricity in combination with fuel-cell cars can achieve fuel costs close to today's costs of taxed conventional gasoline and diesel cars, less overall GHG emissions, but at considerably higher costs, at least in the short term.

Battery-electric vehicles would lead to both lower fuel costs and WTW GHG emissions, owing to the higher efficiency throughout the fuel supply chain and the lower fuel consumption of the battery-electric vehicle (battery-electric vehicle: ~0.46 MJ/km (0.13 kWh/km); FC hybrid: ~0.84 MJ/km. Since, as yet, no battery is available that meets the different requirements, such as long lifetime, short recharging time, high energy densities (kWh/kg, kWh/l) and durability concerning cold weather simultaneously, battery vehicles currently have not developed to become a competitive option (see also Section 7.9).

7.8 Critical reflection on WTW studies

The fundamental principles of WTW studies are straightforward, but their application to the analysis of different fuel or vehicle pathways is complex, due to the large amount of detailed information needed. The differences between different studies are thus generally not due to major differences in approach, but to the detailed assumptions made.

For instance, as results depend on system boundaries in time and space, it is essential to fix the boundaries with respect to both dimensions. Most processes to produce fuels generate co-products (such as gases, gasoil, glycerine, etc.), and the treatment of these co-products is critical to the results of WTW studies (the two approaches used are the 'allocation' method and the generally preferred 'substitution' method). Finally, biofuel analyses, particularly, are prone to large uncertainties, owing to the large variability and relatively poor scientific understanding of emissions associated with biofuel crop production, and in particular with the emissions impact of land use and land-use change. Biofuel pathways are a contentious area, where results are highly sensitive to the input assumptions. The two key areas to consider, apart from process energy inputs, are emissions from the land used for production of the biomass, i.e., farming, and the use of co-products.

7.9 Battery-electric technology

Issues of greenhouse-gas emissions and local air pollution, coupled with the increase in oil and gasoline prices, are triggering renewed interest in electric vehicles as a means to reduce emissions, improve the fuel economy of the present automobiles and reduce depletion of oil resources (CERA, 2008a; 2008b; Deutsche Bank, 2008).

Besides fuel-cell (electric) vehicles (FCV), there are other vehicle concepts under development, which are also based on electric drives: ranked by increasing battery involvement in the propulsion system, and thus extended battery driving range, these are hybrid-electric vehicles (HEV), plug-in hybrid-electric vehicles (PHEV) – which both incorporate an ICE – and, finally, pure battery-electric vehicles (BEV), without an ICE. While electric mobility in its broadest sense refers to all *electric-drive vehicles*, that is, vehicles with an electric-drive motor powered by batteries, a fuel cell, or a hybrid drive train, the focus in this chapter is on (primarily) battery-driven vehicles, i.e., BEV and PHEV, simply referred to as *electric vehicles* in the following.

7.9.1 Hybrid-electric vehicles

Conventional fuels and drive trains today show some system-inherent disadvantages in real operation, such as unfavourable fuel consumption at partial load (e.g., during urban driving) for ICE or limited driving range of electric vehicles with battery storage. Vehicles with a hybrid drive train try to overcome those shortcomings by combining the respective advantages of the individual drive trains. Generally speaking, hybrid vehicles are vehicles with two energy-conversion machines (contrary to bi-fuel vehicles with two different fuels). Today, most *hybrid-electric vehicles (HEV)*, also simply referred to as hybrids, are equipped with an ICE and an electric motor. They can have a *parallel* design, a *series* design or a combination of *both*.

In a *parallel hybrid*, the electric motor *or* the engine can drive the full hybrid vehicle independently or together. At low speeds, the electric motor powers the car and at high speeds, the ICE takes over. The ICE engine is used for highway driving; the electric motor is used during urban driving and provides added power during hill climbs or acceleration. The engine is also used to recharge the battery, when it produces more power than is needed to drive the wheels. An example of a parallel hybrid is the *Toyota Prius*. Parallel hybrids can realise fuel savings of 30%–40% in urban driving compared with non-hybrid vehicles (IEA, 2006c).

Series hybrids are electric cars with support from a small ICE, where *only* the electric motor propels the vehicle. There is no mechanical link between the combustion engine and the wheels. The ICE drives a generator that produces electricity that either flows to the electric motor that turns the wheels, or to a battery for storage. When the car is running solely on batteries, the engine turns on after the batteries have drained to a certain level and begins to recharge them.

It is also important to note that the *fuel-cell vehicles (FCV)* currently being developed are, in fact, advanced series fuel-cell hybrids. Instead of an ICE generating the power, it is the fuel-cell that generates the electricity, which either drives the motor momentarily or that is used to charge up the battery for later use. While for hybrid ICE vehicles, efficiency is the main driver for hybridisation, for the (hybrid) FCV, it is as much the operational control of the fuel cell and the sizing thereof that drives FCV makers towards hybridisation. Just as hybridisation allows an ICE-based HEV to be equipped with a smaller engine, the FCV can be equipped with a smaller fuel cell. Given the high cost of the fuel cell, this is the strongest driver. Additionally, it allows the fuel cell to be operated under more steady (more optimal) conditions, so as to improve the longevity of the fuel cell or lower the technical challenges on the way towards commercialisation of the fuel cell.

Hybrid-electric vehicles can also be built to use the series configuration at low speeds and the parallel configuration for highway driving and acceleration. Since electric motors have excellent torque for acceleration, a smaller ICE can be used with an electric-drive vehicle for improved gasoline mileage, and yet give the same performance on acceleration. Another advantage of HEV is that the batteries can be smaller than in a PHEV or BEV. All hybrids also incorporate regenerative braking to recover the energy that is normally lost during braking. In Europe, gasoline hybrids are likely to face strong competition from diesel engines, unlike in the USA.

7.9.2 Plug-in hybrid-electric vehicles

Plug-in hybrid-electric vehicles (PHEVs) combine operational aspects of both BEVs and HEVs. A PHEV, like a BEV, can be recharged from the electricity grid, stores significant energy in an on-board battery and depletes the battery during driving. Unlike a BEV, a PHEV has an internal combustion engine that is also used for propulsion.

As PHEVs focus on increasing the range on electric drive alone, they incorporate a considerably larger battery to enable a longer range on electric drive. Plug-in hybrid-electric vehicles are generally classified with respect to their utility factor, i.e., the fraction of driving in a PHEV that is performed by electricity. The vehicles' all-electric range can be variable, posing different requirements for battery performance, but is generally expected to be between 20 and 40 miles (32–64 km): PHEV-20, PHEV-40, etc., refer to vehicles that may be driven about 20 or 40 miles from their batteries. The battery of a PHEV may be recharged overnight by plugging into the regular household electric power, not by the ICE in the car. The ICE starts to power the car and recharge the battery only after the battery becomes depleted during driving operations. Depending on regional factors such as electricity–fuel mix, fuel economy or driving range, significantly lower emissions and higher fuel economy ratings result from battery-alone operation. The vehicle will still operate as an HEV when the range on battery alone is exceeded. Charging the battery overnight is particularly attractive for utilities, as they could increase plant loading during low-demand periods. In fact, increased off-peak charging would effectively improve the production efficiency of the existing power generation fleet by 'valley filling' the load profile during the night-time hours.

The current electric range of hybrids, such as the Toyota Prius, is only between 15 and 20 km, but it is to be expected that 'plug-in' hybrid vehicles with an electric range of 50–80 km will come on the market in the coming decade; the development goal in the USA is PHEV-40s. Such vehicles could 'fuel up' from the power grid for 60%–80% of their energy needs (as on average less than 20% of trips exceed 60 km in distance), thus drastically reducing the liquid-fuel demand for that same vehicle;[10] further, a 60 km all-electric range would cover up to two thirds of annual mileage (which is largely sufficient for daily commuting and short distance drives) (Graham, 2005; Duvall and Knipping, 2007; CERA, 2008a; Deutsche Bank, 2008). But also PHEVs with a 30 km range, which matches urban driving patterns well, could displace between one third and two thirds of liquid fuel. For instance, US oil consumption would decrease by up to 70% if the entire US LDV fleet were replaced by PHEV-40s, completely eliminating the need for petroleum imports (Denholm and Short, 2006; Kammen, 2006; CERA, 2008a).

[10] 50% of US cars drive less than 40 km daily, 80% less than 80 km. In the EU25, the average daily drive is about 27 km.

The relative CO_2 performance of PHEVs will depend mainly on two factors: the increase in conventional vehicle efficiency and related lower tailpipe CO_2 emissions of standard ICE vehicles due to very stringent emission targets, and the fuel mix in the power generation fleet during the charging period, as CO_2 emissions in electricity operation mode depend on the power source: for example, PHEVs offer no CO_2 savings if charged using the fossil fuel dominated electricity mix of most countries; if charged using electricity from coal, they would even be more CO_2 intensive than the average vehicle fleet (around 3 t CO_2 per car per year in Europe), all the more when compared against evolving efficiency standards for new cars. Ultimately, the CO_2 benefits of PHEVs would also depend on the marginal power plant during the charging period.

The large scale adoption of PHEVs (or BEVs) in Europe might create a regulatory issue: as electric vehicles will shift CO_2 emissions from the transport sector to the electric power sector, which – unlike road transport – is covered by the European Emissions Trading Scheme (ETS), a compensatory mechanism or the inclusion of road transportation within the scheme would be required.

The current high interest in PHEVs, especially in the USA, is fuelled by good overall energy efficiency and as they allow a smooth and flexible transition away from oil dependence in the transport sector. By using the electrical grid in the night, average electricity consumption is more constant, and more regenerative wind power could be fed into the grid. A critical point for the success of PHEVs will be economics, which are heavily influenced by the evolution of transport fuel and electricity prices, i.e., the relative difference between gasoline/diesel prices and electricity prices. PHEVs will incur higher up-front costs, as they always have two drive trains (electric motor and ICE) and two fuel supply systems (battery and tank), compared to single-drive train/single fuel system vehicles. Ultimately, in order for the economics to favour the PHEV, lower fuel costs would have to compensate for the higher up-front purchase price of the PHEV. In most parts of the world, the consumer price of a kilometre driven electrically is up to 70% cheaper than a kilometre driven on gasoline, due to the relatively high level of fuel taxes relative to electricity taxes, but an important question in this context is whether a change in taxation for electricity used by electric vehicles is to be expected, if transport electrification takes off, to make up for potentially significant petroleum tax losses incurred by the reduction of liquid fuels demand.[11]

7.9.3 Battery-electric vehicles

Finally, pure battery-electric vehicles (BEV) operate on batteries alone, with all normal operations powered by the battery. As a result, a BEV has a sizable battery pack. Recharging BEVs today still requires between three and eight hours and is

[11] For further reading about plug-in hybrids refer to Axsen *et al.* (2008); Duvall (2004); Duvall and Knipping (2007); Kempton and Tomic (2005a; b); Kempton (2007) and Tomic and Kempton (2007).

often done overnight from a household circuit. Alternative concepts under discussion are fast charging as well as charging stations, where empty batteries can simply be exchanged for fully charged batteries, so-called battery swapping stations.

From the electrochemical reaction, when a battery is discharged, about 90% of the reaction energy is given off in the form of electric energy (the remaining 10% is lost as heat). At this efficiency level, the battery is superior to any other form of chemical energy conversion (almost double the efficiency of a fuel cell). Further, as the electric energy stored in the battery can be far more efficiently converted into traction power (typically 85% 'tank-to-wheel' efficiency; electric motors are about 95% efficient in converting electric power to mechanical power) than the energy stored in gasoline (around 25%): a MJ of battery energy gets a vehicle about four times as far as a MJ of gasoline energy; in other words, an ICE consumes about four times the energy per km travelled than an electric motor. Compared with a 10 km/litre car ($\approx$1 kWh gasoline/km), a kWh of battery capacity will give the car a range of about 4 km. As a result, in terms of final energy demand, electric mobility will increase electricity demand, but to a lesser extent than it will reduce the demand for liquid fuels. The average electricity consumption of today's electric vehicles is about 20 kWh/100 km, with a longer term outlook of achieving 13–15 kWh/100 km (see also Section 7.7).

Electric vehicles have had a long and rather unsuccessful history. The first electric car was built in 1873 in England by R. Davidson. The BEV had clear advantages over the combustion engine at the beginning of car manufacturing and around 1890 in the USA almost twice as many BEVs as petrol vehicles were constructed (Krüger, 2002). However, they could not prevail, because of the enormous improvements of vehicles with ICE, and around 1930 the construction of electric cars was practically stopped. Even though the development of electric cars started again in the 1960s in the USA because of the air pollution resulting from gasoline cars, and even though this trend was also revived in Europe as a consequence of the energy crises of the 1970s, their appeal to motorists has been limited as no significant breakthroughs with regard to the disadvantages typical for 'electric-traction' vehicles, such as low storage capacity and hence relatively short driving range, limited power density, long recharging times and high costs could be achieved (Gerl, 2002).[12] The prevalence of cheap and plentiful petroleum-based fuels has also historically disadvantaged electric cars.

Hence, BEVs are still largely applied in niche markets, such as for inner urban traffic. In this context electric vehicles will further profit from the introduction of environmental zones and toll charges on ICE vehicles in cities.[13] Another promising application is the electric two-wheeler (scooter), which is becoming increasingly popular in Asian megacities.

[12] From 1992 to 1996, a large-scale trial was conducted with 60 battery-powered vehicles (minibuses, vans and private cars) on the German Baltic island of Rügen, to prove the technical feasibility of advanced battery technologies for electrical vehicles (Eden *et al.*, 1996). Technical, economic and ecological deficits revealed during this test period led to a halt of electric vehicle development and related battery research in Germany at that time.

[13] For example, Daimler is running a fleet of BEV with ZnS batteries at a car-rental agency in London; the cars are economical because they are exempted from the city congestion charge.

7.9.4 Battery types

There are several factors that need to be fulfilled for batteries to become an alternative option to power vehicles. These are energy density, power density, cycle life, safety and costs. Whereas full HEV applications pose the most demanding requirements for high power density of batteries, BEVs demand the highest possible energy density (see also Table 7.23).

There are various battery systems that have or will be involved in electric-vehicle propulsion. Lead–acid starter batteries are found in all ICE vehicles and power the present motive power applications. However, the lead–acid technology has with 30–40 Wh/kg insufficient energy storage capability to give satisfactory performance for propulsion of the modern automotive applications. The energy storage capability of the various rechargeable battery systems varies considerably. The nickel metal hydride (Ni-MH) battery is the current choice for HEV applications, such as the Toyota Prius and achieves an energy density of about 70 Wh/kg. Only lithium-ion (Li-ion) batteries show promise of meeting the requirements developed for advanced electric-vehicle applications.[14] But at 120 Wh/kg, the energy density of a present-day Li-ion battery is around a hundred times less than that of gasoline.[15] The low energy density is one of the major challenges for batteries, which also affects the weight of the battery pack: for a 10 km/litre car and a driving range of 500 km, the weight of the Li-ion battery pack with the current 120 Wh/kg energy density level would be about 1000 kg.

Today's best performing Li-ion batteries have an energy density of 170 Wh/kg (Axsen *et al.*, 2008), with a physical limit of about 400 Wh/kg. However, the long-term target of 200 Wh/kg as proposed for instance by the *US Advanced Battery Consortium* (USABC) still requires significant development efforts. But their biggest advantage, their high energy density, can also make them react like an explosive under certain conditions. There is currently a lot of R&D on designing larger, explosion-proof Li-ion batteries. Further advantageous properties are their high voltage and low tendency to self-discharge.

Car manufacturers have selected Li-ion batteries as the system of choice for their high volumetric and weight energy storage capability. The Li-sulphur and Li-air batteries, which have very high energy densities, are still very early in their development. Table 7.23 shows the performance requirements for energy-storage systems of electric vehicles (for more details see Axsen *et al.*, 2008).

[14] Li-ion technology continues to follow multiple paths of development, each using different electrode materials in efforts to optimise power, energy, safety, life and cost performance, for which the composition of the cathode is the single biggest determinant. Lithium battery technologies for automotive applications typically fall into four major categories, based on different chemistries: lithium nickel cobalt aluminium, lithium manganese spinel, lithium titanate and lithium iron phosphate (Axsen *et al.*, 2008; Deutsche Bank, 2008).

[15] Gasoline has an energy density of around 12 kWh/kg. A battery would have to be approximately 10× the size of a gasoline fuel tank in order to provide an equivalent driving range, assuming a drive-system of equal efficiency. For comparison, hydrogen has a mass-specific energy density 33.3 kWh/kg.

Table 7.23. *Electric vehicle energy storage system performance requirements*

Vehicle type	Weight (max. kg)	Peak power (min. kW)	Power density (min. W/kg)	Energy storage capacity (min. kWh)	Energy density (min. Wh/kg)
HEV	50	40–60	800–1200	1.5–3	30–60
PHEV	120	65; 50[a]	540; 400[a]	6; 12[a]	50; 75[a]
BEV	250	50–100	200–400	25–40	100–160

Note:
[a] Requirements for PHEV-20 and PHEV-40, respectively.
Source: (Kalhammer *et al.*, 2007).

Cycle life is another important factor because it determines the longevity of the battery in practical use. The number of lifetime cycles depends strongly on the so-called depth of discharge: if only some 10–20% of the full discharge capacity is used (as in the Toyota Prius for instance) the batteries can handle millions of 'shallow cycles'. However, for PHEVs or BEVs the number of 'deep cycles' (typically 80% discharge) is a relevant characteristic.

A key challenge for electric vehicles is that batteries will remain expensive, even if target costs are achieved, i.e., if battery costs can be reduced from €700–€800/kWh today to about €300–€400/kWh. With 20 kWh/100 km and a reasonable electric range of around 150 km (100 miles), a 30 kWh battery will be required which will still cost a minimum of €9000. For example, the Tesla Roadster which has an electric range of 400 km, costs around US$ 100 000 (Tesla Motors, 2008). Although the incremental costs may be justified if amortised over the vehicle's lifetime, there are questions about whether consumers will be willing to bear the high up-front cost.[16]

In the same way as fuel cells, batteries need to fulfil certain performance criteria: high energy, cost-effective, long lasting, and abuse tolerant batteries will be the key technical enablers for customer acceptance and user convenience, as they determine vehicle costs, driving range etc. But it must be understood that there will always be inherent trade-offs in battery development among power density, energy density, longevity, safety and cost, and some of these technical requirements will have to be compromised (Axsen *et al.*, 2008). For instance: increasing power density requires higher voltage that reduces longevity and safety and increases cost; increasing energy density tends to reduce power density; simultaneously optimising power, energy, longevity and safety will increase battery cost. Due to the low energy density and the high costs of batteries, increasing the electric range of vehicles will also be very costly.

[16] For this reason, leasing concepts have been proposed, which would involve providing the battery packs to the vehicle owner on an operating-cost basis with little or no up-front investments for the customer.

In large volume production, the availability of electrode materials for batteries must also be considered, such as cobalt, nickel, iron or manganese, and above all lithium (see also Deutsche Bank, 2008). The lithium demand is 0.3 kg lithium metal equivalent/kWh (Tahil, 2006);[17] for a 30 kWh battery (20 kWh/100 km and 150 km range) this results in 9 kg lithium/vehicle. To avoid stresses on lithium supply battery recycling will be crucial.

7.9.5 Prospects of battery-drive vehicles

Today, there is a general consensus that in the coming two decades electric vehicles, i.e., PHEVs and BEVs, are going to gain a material share of the vehicle fleet in many countries. However, the upsides as well as limitations of electric mobility need to be addressed realistically.

Although the electrification of transport is seen as a great business opportunity by utilities, the question arises as to what impact the large-scale supply of electricity to the transport sector would have on the existing energy system and how electric mobility could best be integrated with the electricity system in the future; in particular, there is the need to assess the electricity infrastructure reinforcements specifically at low-voltage level, taking a fully fledged electric-vehicle market into account.[18]

The main areas of the electricity system affected by large-scale electrification of the transport sector concern resulting electricity demand and load profile and impacts on generation capacity, as well as transmission and distribution capacity of the grid (and how they are affected by the type of charging), potentials of vehicle-to-grid (V2G) power[19] and opportunities for the integration of intermittent renewable energies (i.e., storage of surplus electricity). In addition, smart metering technology will be essential for billing of electricity consumption of electric vehicles. As it would be beyond the scope of this book to address these topics in more detail, only some key questions that need to be addressed in this respect from a utilities' perspective regarding the integration of electric road transport are listed:

- What will the expected *electricity demand* be, and depending on type and pattern of charging, what will the corresponding *load profile* look like, as this affects grid management?
- How will *demand peaks* be managed?
- What will the impact be on (current/future) *generation capacity*? With proper load levelling ('valley filling'), to what extent will the current (over)capacities (i.e., medium and peak load power plants) be able to cope with the additional demand? What is the critical penetration rate that would require adding new generation capacity?

[17] In lithium carbonate (Li_2CO_3) terms this corresponds to 1.4 to 1.5 kg Li_2CO_3 per kWh capacity.

[18] Interestingly, a total switch of the global LDV fleet of around 900 million vehicles to 100% electricity would lead to only a less than 5% increase of global electricity demand.

[19] The basic concept of vehicle-to-grid (V2G) power is that electric-drive vehicles (BEV, PHEV or FCV) provide power to the grid while parked. For instance, BEVs can charge during low demand times or serve as a flexible electricity sink by storing excess loads from fluctuating renewable sources and discharge when power is needed. The V2G concept requires three elements for each vehicle: a connection to the grid for electrical energy flow, a control or logical connection necessary for communication with the grid operator, and controls and metering on board the vehicle. See also Kempton and Tomic (2005a; b) and Tomic and Kempton (2007).

- Will the capacities of the *transmission and distribution grid* be sufficient?
- Will the standard *domestic electric sockets* and *local substations* be sufficient to cope with the different charging methods and patterns?
- What are the opportunities of *vehicle-to-grid (V2G)* in terms of providing control power, contributing to load levelling and improving grid stability with increasing shares of fluctuating renewable energies?
- Are there opportunities for improving the *integration of intermittent renewable energies*? To what extent could batteries of electric vehicles act as (large-scale) storage for renewable surplus electricity (see also V2G concept) and how does it compare with alternative methods of storage in terms of feasibility and economics, such as pumped storage, compressed air storage or hydrogen production?

A key question from a utilities' perspective is what the method and pattern of charging electric vehicles will be, as this will largely determine what impact the increased demand they will create will have on generation, networks and supply, as well as other areas such as energy services.[20] There are three recharging options: slow charging, fast charging and modular swapping stations. For *slow charging* there is a causal link between parking and charging, as it will last a minimum of 2 to 4 hours. Charging stations for slow charging could be installed at homes or as public access charging points in streets, or alternatively in public parking stalls or at work, where space constraints are an obstacle for home charging or street charging in densely populated urban areas. *Fast charging* (within 15–30 minutes) would not be suitable for homes, as current domestic systems would not be able to handle these voltage levels; but fast charging could be set up for refuelling stations with network support.[21] The battery replacement mechanism of *swapping stations* (like that e.g., proposed by Project Better Place (2008)), is based on swapping depleted batteries for recharged ones. But this concept is likely not to be feasible, because (1) charging standards are emerging, whereas the swapping stations solution assumes a single standard; (2) it will be impossible to shape a universal battery pack that fits each car; (3) the mechanics of each car are very different; and (4) because battery technology is emerging and the solution cannot be sustained for more than a few years. For any of the above options, availability, accessibility and user convenience of recharging stations will be crucial for customer acceptance. It will also be important to integrate the implementation of a recharging infrastructure early on into city and land-use planning.

If electric vehicle ownership should be viable and attractive, it is to be expected that electric vehicle owners would want to be able to charge their vehicles as frequently as they wish to the level they require, to ensure that the car is ready when it is next required. On the other hand, to avoid charging patterns that would risk the power

[20] To illustrate the impact of charging power and charging time: charging a 20 kWh battery in 10 hours requires a charging power of 2 kW, in 2 hours of 10 kW and in 1 hour of 20 kW.

[21] For instance, charging a 30 kWh battery in 10 minutes requires a minimum of 180 kW of power, equivalent to an office block. Fast charging, therefore, also poses a particular challenge to the battery-management system.

network (as the uptake of electric vehicles and thus electricity demand increases), such as 'after driving' charging that largely coincides with electricity demand peaks, utility solutions for peak demand management will be required. There are a number of different methods available, both regulatory and market led, which would essentially drive off-peak charging and apply the demand increase to current low-demand periods such as overnight: electric vehicle tariffs and time-of-use (ToU) pricing to differentiate peak and off-peak tariffs, smart metering control or regulatory restrictions, which would dictate times of charging. However, as these solutions are restrictive, the question remains how they will suit customers/vehicle owners.

7.9.6 Impacts on hydrogen

Assessing the potential of hydrogen without taking into account competing options would result in misleading conclusions. Both PHEVs and BEVs are major competitors for hydrogen FCVs (see, for example Dixon (2007)). In the same way as hydrogen, BEVs or PHEVs help to meet the three most important targets with respect to transportation energy use, which are also increasingly favoured by policy makers around the world: clean air, GHG-emission reductions and energy security. When driven in the electric mode, cars are zero-emission vehicles and thus the gasoline-to-electric switch helps to clean the air. As energy security is first and foremost an oil-dependency issue, the switch is helpful in reducing the 95% oil dependence of the transportation sector, by opening it up to the much wider portfolio of primary energy resources that fuel the power sector, notably coal and renewables.

Plug-in hybrids (and BEVs) could be attractive to both consumers and governments. For consumers in most parts of the world, the consumer price of a kilometre driven electrically is up to 70% cheaper than a kilometre driven on gasoline, because of the relatively high level of fuel taxes relative to electricity taxes (unless an 'electric' petroleum tax will be introduced to make up the tax losses incurred by the reduction of liquid fuels demand) and the superior tank-to-wheel energy use of the electric mode. Kilometres driven electrically are also zero-emission in terms of local air pollution, and lower in well-to-wheel CO_2 emissions (even with electricity made from coal from new coal power stations). On top of that, an energy switch from the fully oil-based transportation sector to the more flexible and thereby more 'energy-secure' power sector would alleviate energy security concerns. Finally, if 'green' or carbon-neutral electricity is used to charge up the plug-in hybrid, this is by far the cheapest option to reduce transport GHG emissions. If renewable electricity is used to charge the batteries, BEVs represent another zero-carbon solution.

At present, the size and weight of existing batteries compared with the amount of energy they store heavily constrains the range of battery-powered cars, limiting their suitability to largely urban operation; long recharging times, high cost and scarcity of some metals are further constraints on this option. The main attraction of this option is that the 'fuel supply' infrastructure (electricity) already exists. If battery performance was to improve markedly and cost was to reduce, BEVs could represent a complete solution to decarbonising transport, thus making the discussion about the introduction of hydrogen in the transport sector largely obsolete. However, at this time, both are a significant challenge. Given these constraints, it follows that the impact of batteries on the transportation sector will probably come through vehicles that are only partly dependent on battery power, such as HEVs or PHEVs.

What would kill the prospects for hydrogen are the 'ideal battery' offering 'unlimited range' (as hydrogen is less efficient than electricity) and/or 'unlimited' supply of liquid 'low-carbon fuels' (i.e. in principle 2nd generation biofuels), because hydrogen is more cumbersome to distribute and use than liquid fuels. However, it is wise to assume that neither of these will come true. While the fraction of driving performed by electricity will undoubtedly grow, there is unlikely to be a 'silver bullet' in the coming decades and the transport sector will witness a much more diversified portfolio of fuels in the future. In the short to medium term, hydrogen will be additional to what biofuels and electricity can offer for energy security and CO_2 emissions reduction. In the long run, however, hydrogen holds promise to overcome some of the limitations of biofuels and electricity, allowing for further decarbonisation of transport. Particularly in this respect, when compared with electricity, hydrogen is more promising, as fuel cell vehicles cover the entire driving spectrum, allow fast refuelling and have the same potential for reduction of CO_2 and local pollution as electric vehicles. In addition, longer term, batteries could act as potential range extenders for fuel cell vehicles, thus benefiting from the development of PHEVs (or vice versa: hydrogen/fuel cells as range extenders of battery vehicles). Recognising the limitations of electric vehicles, hydrogen fuel cell vehicles are also generally seen as the long-term solution by major car manufacturers.

7.10 Current production of alternative transportation fuels

Today ethanol and biodiesel (FAME) are the most common biofuels. Alternative fuels from fossil energy sources are mainly LPG and CNG. Synthetic gasoline and diesel from coal (CTL) and natural gas (GTL) are produced mainly in South Africa. Electricity used in battery-electric vehicles plays a minor role today. The fuel consumption for road transport in the world today amounts to about 65 700 PJ per year (IEA, 2006a); in total, the share of alternative fuels for transport at the time of writing was about 2.7% (Table 7.24).

In 2005, global production of biofuels amounted to some 850 PJ or 37 billion litres (643 kb/day), equal to about 1.3% of total road-transport fuel consumption and about half of total global alternative fuel production in energy terms (IEA, 2006b; see Table 7.24). Bioethanol is the most common biofuel, amounting to 33 billion litres; global production of biodiesel is relatively small, reaching only about four billion litres. Brazil and the United States together account for more than 80% of global supply. Global production of ethanol has more than doubled between 2000 and 2005,

Table 7.24. *Current production of alternative transportation fuels (Bensaid, 2004; Earth Policy Institute, 2006; EC, 2007; IEA, 2006a; REN21, 2006)*

	Ethanol (PJ/yr)	FAME (PJ/yr)	CNG (PJ/yr)	LPG (PJ/yr)
OECD Europe	20	138	19	195
OECD North America	323	13	23	90
Latin America	323	3	169	1
Transition economies	0	0	10	34
Middle East	0	0	0	17
East Asia, South Asia, China	29	0	4	25
OECD Pacific	1	2	11	301
Africa	0	0	11	13
World	**697**	**156**	**247**	**675**

and biodiesel expanded nearly fourfold. About 1% of the world's available arable land (about 14 million hectares) is currently used for the production of biofuels. Biodiesel and ethanol are mainly used blended with diesel or gasoline, respectively.

Brazil has been the world's leader (and primary user) of fuel ethanol for more than 25 years, to reduce the oil dependence of the country, producing slightly less than the world's total. In Brazil, ethanol from sugarcane constitutes some 28% by energy content (40% by mass) of transport fuel (non-diesel) and is available at all fuelling stations; 'flexible fuel' vehicles attained a 70% share of the (non-diesel) vehicle-sales market (Martinot *et al.*, 2006). Fuel ethanol production (corn ethanol) and consumption in the United States caught up with that of Brazil for the first time in 2005. The growth of the US market is a relatively recent trend: by the end of 2005, there were 95 operating ethanol plants in the USA, with a total capacity of 16.4 billion litres/year (compared to four billion litres in 1996; Martinot *et al.*, 2006). Ethanol constitutes 99% of all biofuels in the USA, and currently amounts to about 2% of all gasoline sold by volume (Farrell *et al.*, 2006). In 2006, about 20% of the US corn crop was used to make ethanol; the increased demand for ethanol biofuel was also the main driver behind the drastic price increase of almost 70% for corn in the USA experienced during 2006 (MIT, 2007). The annual subsidy for corn-based ethanol in the USA amounts to about $2 billion. The EU also increased fuel-ethanol production, although still at low levels relative to Brazil and the USA; here, ethanol is mainly produced from wheat and, to a lesser extent, sugarbeet.

As for biodiesel, more than 90% of global production comes from the EU. Germany alone accounts for about half of global biodiesel production (with about 1500 fuelling stations selling biodiesel). Biodiesel in the EU is mainly produced from rapeseed.

The share of LPG of today's fuel consumption for road transport is roughly equal to ethanol and amounts to about 1.0%, and the share of CNG to about 0.4%.

Worldwide, there are more than 10 million LPG vehicles, with the majority in Italy, Poland, the Netherlands and France. Of around seven million CNG vehicles, about two-thirds are found in Argentina, Pakistan and Brazil, while in Bangladesh, Armenia, Pakistan, Iran and Argentina, CNG vehicles account for more than 20% of the total vehicle fleet; in Europe, Italy has most CNG vehicles (IANGV, 2008; NGV, 2008). The South African company Sasol produces about 7.5 million t (~320 PJ) CTL and GTL (Sasol, 2007), which is about 0.5% of today's consumption of transportation fuel in the world. Sasol produces the major fraction of CTL and GTL worldwide. Adding up, the contribution of all alternative fuels today means that about 97% of the fuel demand for road transport is met by crude-oil-based gasoline and diesel.

7.11 Policy measures to promote biofuels and GHG-emissions reduction in the transport sector

Many countries are enacting policy measures in support of biofuels and for greenhouse-gas emissions reduction in the transport sector. The most important policy measures at the time of writing for the EU and the USA are given below. For an overview of worldwide policy support measures for biofuels, see the EC (2006); the IEA (2006b) and Martinot *et al.* (2006).

7.11.1 The European Union

In 2003, the *European Biofuels Directive* (CEU, 2007; EC, 2003) was enacted. It sets an indicative target of a 2% market share for biofuels in 2005, a 5.75% share in 2010 and a 10% share in 2020 for road transport related to the energy content of the fuel. It has to be noted, however, that biofuel mandates are increasingly being scrutinised from a sustainability perspective and the targets in the EU are currently being reconsidered.

In addition, the European Commission aims to set a binding *target for CO_2 emissions for new cars* of 120 g/km by 2012. The Commission proposed to bring down average emissions to 130 g/km through vehicle-technology improvements alone: the remaining cuts (10 g/km) are to be achieved by complementary measures, such as the further use of biofuels, fuel-efficient tyres and air conditioning, traffic and road-safety management and changes in driver behaviour (eco-driving).

In the course of the review of the *EU Fuel Quality Directive* (2003/17/EC), petrol and diesel specifications are being reviewed, to lower their environmental and health impact, as well as to take into account new EU-wide targets on biofuels and greenhouse-gas emissions reduction.

7.11.2 United States

In 1991, the governor of Nebraska – a leading farming state (or cornhuskers, as they call themselves) – initiated the Governors' Ethanol Coalition (GEC, see www.ethanol-gec.org). The Coalition's goal is to increase the use of ethanol-based fuels, to decrease the nation's dependence on imported energy resources, improve the environment and stimulate the national economy. As of January 2007, 36 US states plus international representatives from Brazil, Canada, Mexico, Queensland, Australia, Sweden and Thailand have become members.

In 2005, a *Renewable Fuel Standard (RFS)* was enacted in the United States, which requires the annual volume of biofuels to be blended into gasoline to increase to 7.5 billion gallons (28 billion litres) by 2012. The mandatory *Alternative Fuel Standard (AFS)* from 2007 further requires 35 billion gallons (132 billion litres) of renewable and alternative fuels in 2017.

The *Low Carbon Fuel Standard (LCFS)* in California requires fuel providers to ensure that the mix of fuel they sell to the California market meets, on average, a declining standard for GHG emissions measured in CO_2-equivalent gram per unit of fuel energy sold. By 2020, the LCFS will produce a 10% reduction in the carbon content of all passenger vehicle fuels sold in California.

7.12 Advantages and disadvantages of different transportation fuels

Table 7.25 shows a comparison of the major alternative fuels with respect to feedstock availability, supply security, handling and volumetric energy density. Gasoline, diesel, FAME and synthesis gasoline and diesel (XTL) are simple to handle, have a high volumetric energy density and are simple to store on-board a vehicle. In addition, a supply infrastructure is already largely in place. Both CGH_2 and LH_2 have a low volumetric energy density but are characterised by a high feedstock flexibility, i.e., hydrogen can be supplied from a large variety of primary energy sources, such as biomass and renewable electricity, as well as fossil fuels. Compared with XTL, the production of hydrogen from synthesis gas shows a higher thermal process efficiency, as the Fischer–Tropsch synthesis step is eliminated.

Equally, electricity can be generated from a wide variety of energy sources, and battery-electric vehicles have a far higher efficiency than fuel-cell vehicles, as the high discharge rate of the battery is almost double the efficiency of a fuel cell. Battery-electric vehicles or PHEVs are also advantageous, as they can rely on an existing supply infrastructure.

The potential of biomass to make a large contribution towards replacing conventional fuels is constrained by land availability and competition with other end-use sectors. In particular, the potential for oil seeds to generate FAME is limited. Generally, yields of biofuels from purpose-grown crops depend on the species, soil type and climate.[22] At a global level, it is estimated that biofuels could substitute up

[22] Cereals can yield around 1500–3000 litres of gasoline equivalent (lge)/ha; sugarcane, 3000–6000 lge/ha; sugarbeet, 2000–4000 lge/ha; vegetable oil crops, 700–1300 litres of diesel equivalent (lde)/ha and palm oil, 2500–3000 lde/ha (IEA, 2007). In addition, there are novel biofuel production processes under development, for example biodiesel from marine algae, which are claimed to have a 15 times higher yield per ha than rapeseed.

Table 7.25. *Advantages and disadvantages of different transportation fuels*

Fuel	Feedstock availability	Supply security/ supply potential	Handling	Volumetric energy density
Gasoline	−−	−−	++	++
Diesel	−−	−−	++	++
LPG	−−	−−	0	+
CNG, CMG	0	0	−	0
FAME	−−	−	++	++
XTL	0	0	++	++
Methanol	0	0	+	+
DME	0	0	0	+
Ethanol	−	0	++	+
CGH_2	++	++	−	−− (35 MPa) − (70 MPa)
LH_2	++	++	−−	0
Electricity	++	++	+	−−

to 30% of today's total vehicle-fuel consumption. The costs of biofuels are very location specific, as they highly depend on feedstock, process, land type, crop yield, labour costs and agricultural subsidies; co-products (e.g., animal feed) help reduce production cost.

For alternative fuels requiring a new dedicated fuel-supply infrastructure (i.e., refuelling stations), such as natural gas or hydrogen, mono- and bi-fuel vehicle types are to be distinguished. It can generally be stated that mono-fuel vehicles, where the engine is optimised for one specific fuel, have a better fuel and engine efficiency and thus a better emissions performance than bi-fuel vehicles. Bi-fuel vehicles, however, greatly increase the driving range and thus offer clear advantages during the introduction phase of a new fuel, when the supply infrastructure is still being implemented.

Biomass is not an inexhaustible resource. In addition, there are various critical issues associated with the supply of biofuels, which may constrain large-scale production and challenge their overall sustainability: competition for water resources, use of pesticides and fertilisers (energy embedded in fertilisers, N_2O emissions from fertiliser use), land use (increased deforestation and biodiversity loss) and impact on soil quality, as well as competition with food production for arable land availability, which may drive up food and fodder prices. In particular, conventional agricultural crops for 'first-generation' biofuels, such as rapeseed and cereals, generally require high-quality farm land and substantial amounts of fertiliser and chemical pesticides. Moreover, biomass use for transportation fuels is increasingly in competition with stationary heat and power generation for feedstock availability. Local supply, as well as transport of the low-calorific biomass, might further limit the size of biofuel production plants. Novel biofuel technologies currently under development, such as enzymatic hydrolysis and gasification of ligno-cellulosic feedstock (such as biomass waste or dedicated plants grown on poorer-quality land), could potentially extend the feedstock base and avoid interference with the food chain. However, more R&D is needed before these 'second-generation' biofuels become commercially available.

The net impact on GHG emissions of replacing conventional fuels with biofuels depends on several factors (IEA, 2006b). These include the type of crop, the amount and type of energy embedded in the fertiliser production and related emissions, the water use and the resulting crop yield. Moreover, the (fossil) energy used in gathering and transporting the feedstock to the biorefinery, the energy intensity of pre-treating the biomass and of the subsequent conversion process, as well as alternative land uses are to be considered. Calculating the energy and emissions balance of biofuel production requires estimates of and assumptions on all these factors as well as the energy or emission credits or penalties that should be attributed to the by-products. Accordingly, the reported GHG reductions achievable by biofuel use show large variations.

The overall balance of GHG emissions over the entire supply chain of certain biofuels is questionable, especially if emissions from land-use change are considered, as, for example, in the case of palm oil from Indonesia (see Section 7.3.3) (Fargione *et al.*, 2008; Searchinger *et al.*, 2008); and there is also an increasing debate about whether manufacturing of ethanol takes more non-renewable energy than the resulting fuel provides (Farrell *et al.*, 2006). Besides this, from the overall perspective of CO_2 reduction in the energy sector, the massive extension of biofuel production is to be reflected critically, as biomass can be used up to three times more efficiently in heating and combined heat and power than in producing the currently used biodiesel and bioethanol (SRU, 2007).[23] However, it would be beyond the scope of this publication to assess the sustainability of biofuels in greater detail, as it is very sensitive to processes and feedstocks, as well as local conditions (such as CO_2 emissions from land-use change).[24]

7.13 Stationary hydrogen pathways

A group of stakeholders from industry propose to produce pure hydrogen via biomass gasification and to distribute it for use in stationary fuel cells for combined heat and electricity generation. Two cases have been considered: co-generation of heat and electricity, and electricity generation only.

[23] The same holds true for hydrogen; however, biomass yields more kilometres when used via hydrogen in fuel-cell cars than liquid biofuels in ICE cars (see Fig. 7.5). Moreover, as hydrogen is produced via gasification, it is equivalent to second-generation biofuels, as it can use feedstock that does not interfere with the food chain.

[24] For further reading on the various options of biofuel production processes and their characteristics see, for example, BMELV (2005); IEA (2004); Schaub *et al.* (2003; 2004); van Thuijl *et al.* (2003a; b); Worldwatch Institute (2007) or Further reading section.

7.13.1 Co-generation of heat and electricity

In contrast to the operation of vehicles, electricity and heat for stationary applications can be generated by the combustion of solid biomass without upstream biomass conversion to pure hydrogen (or methanol, BTL or DME). The efficiency of the direct use of solid biomass is generally higher. The overall efficiency of a solid-biomass-fuelled heat and power (CHP) plant is typically about 70% to 80%; direct combustion of solid biomass (e.g., wood chips, wood pellets) in suitable boilers for heat generation only can reach an efficiency of more than 90%.

In contrast, if hydrogen is the energy carrier for electricity and heat generation, then the maximum efficiency to be reached is 60% (based on LHV). This result assumes a two step approach:

- Production of pure hydrogen from solid biomass via gasification: total efficiency about 50 to 60%;
- CHP plant: condensing heat exchanger, assuming an overall efficiency of about 100% (heat and electricity).

For natural-gas-fuelled CHP plants, the same line of argumentation holds as for the stationary use of hydrogen from biomass. It is more reasonable to use natural gas directly than to convert it to hydrogen first and then to heat and electricity. High electrical efficiencies can be reached in the stationary sector by feeding natural gas to molten-carbonate fuel cells (MCFC) and solid-oxide fuel cells (SOFC). Molten-carbonate fuel cells have the added advantage of using CO_2 for the electrolyte (see also Chapter 13).

7.13.2 Electricity generation only

In the electricity-only mode without hydrogen a modern solid-biomass-fuelled steam turbine power plant with a capacity of 10 MW_{el} has an efficiency of more than 30%. Alternatively, the biomass can be converted to synthesis gas and fed into a gas engine, gas turbine or high-temperature fuel cell (MCFC, SOFC) without upgrading the synthesis gas to pure hydrogen, leading to a higher efficiency ($>40\%$) than conventional steam turbine power stations ($>30\%$). Even relatively small plants ($\sim$1 MW biomass input) based on biomass gasification with downstream gas engines can achieve electrical efficiencies of about 25% (total: 80%).

In the hydrogen case, the efficiency of a stationary hydrogen-fuelled fuel-cell power plant can be assumed to be about 50%, leading to an overall electricity generation efficiency of about 25% to 30%, if the hydrogen is generated by biomass gasification with downstream CO shift and pressure-swing adsorption (PSA).

To conclude, the generation of pure hydrogen from solid biomass is only reasonable if transportation fuel should be produced. But it is not reasonable to produce pure hydrogen from biomass for stationary heat and electricity generation.

7.13.3 Other considerations for the stationary end use of hydrogen

Electricity from solar and wind energy is preferably used directly for economic reasons. The conversion of renewable electricity to hydrogen for electricity generation may be reasonable to compensate fluctuating supply of wind and solar energy and fluctuating electricity demand. Hydrogen filling stations could be used for load levelling to compensate for fluctuating electricity supply from solar and wind energy.

The use of ethanol and FAME for heat and power generation is not reasonable. By combustion of solid biomass or of biogas, more heat and electricity can be produced per ha of land area.

If FAME is used for electricity generation, between 3500 and 7000 kWh electricity can be generated per ha and year (efficiency: 30% to 45%) and in the case of biogas, between 7500 and 33 500 kWh per ha and year. For biogas the variation results from the broad range of assumptions for biomass yields of between 10 and 25 t dry substance per ha and year and electricity generation efficiencies of between 26% (micro gas turbine) and 46% (large gas engine).

In contrast, the yield of electricity from photovoltaics (PV) is about 337 500 kWh per ha and year, even if it is assumed that the area of the PV panels cover about one-third of the total plant area. If PV electricity were converted to liquid hydrogen (LH_2), stored and then converted back to electricity by a combined cycle gas turbine (CCGT, efficiency: 57.5%) about 104 000 kWh electricity could be generated per ha and year. This yield is still more than three times the yield of the best biomass pathway (upper end of bandwidth: electricity from biogas via large gas engine).

7.14 Summary

The ever-growing demand for transportation fuels and the related growth in CO_2 emissions in the transport sector require new solutions for transport energy use. In the near and medium term, smaller vehicles, lightweight construction, improved conventional internal combustion engines, hybridisation and dieselisation can improve the fuel economy of vehicles and help reduce fuel consumption and emissions. However, longer term strategies must focus on developing alternative fuels and propulsion systems.

Among the various choices, hydrogen seems especially promising, as it can contribute to each of the three most important targets with respect to transportation energy use, which are increasingly favoured by policy makers around the world: GHG-emissions reduction, energy security, and reduction of local air pollution. As a secondary energy carrier that can be produced from any primary energy source, it can contribute to a diversification of automotive fuel sources. Hydrogen offers significant advantages in combination with fuel-cell vehicles on account of their high conversion efficiencies – as compared with conventional gasoline or diesel vehicles – particularly at partial load, such as in urban driving. Moreover, hydrogen is nearly emission-free at final use and thus reduces transport-related emissions of both CO_2 and air pollutants. In the case of hydrogen produced from fossil fuels in centralised plants, CCS (assuming it will eventually be realised at a large scale) offers an advantage by allowing the capture of CO_2 from a single point source, instead of from the highly dispersed CO_2 emission sources of vehicles, from which it is challenging to capture the CO_2. The major competitor to hydrogen in the long term is electricity, which, however, hinges on a breakthrough in battery technology.

References

Abengoa (2006). *Abengoa Biomass: BCyL Biomass Plant – The First Commercial Demonstration of Abengoa Biomass Ethanol Technology in the World*. www.abengoabioenergy.com.

Alberta Chamber of Resources (ACR) (2004). *Oil Sands Technology Roadmap*. www.acr-alberta.com.

Axsen, J., Burke, A. and Kurani, K. (2008). *Batteries for Plug-in Hybrid Electric Vehicles (PHEVs): Goals and the State of Technology Circa 2008*. Report UCD-ITS-RR-08–14. Davis: Institute of Transportation Studies, University of California.

Bensaid, B. (2004). *Alternative Motor Fuels Today and Tomorrow*. Rueil-Malmaison Cedex, France: IFP.

BMELV (Bundesministerium für Ernährung, Landwirtschaft und Verbraucherschutz) (2005). *Synthetische Biokraftstoffe. Techniken – Potenziale – Perspektiven*. Schriftenreihe 'Nachwachsende Rohstoffe' Band 25. Münster: Landwirtschaftsverlag GmbH.

Borken, J., Patyk, A. and Reinhardt, G. A. (1999). *Basisdaten für ökologische Bilanzierungen*. Vieweg.

CERA (Cambridge Energy Research Associates) (2008a). *From the Pump to the Plug – What is the Potential of Plug-in Hybrid Electric Vehicles?* Cambridge MA: CERA.

CERA (Cambridge Energy Research Associates) (2008b). *Taking Up the Charge – Plug-in Hybrid Electric Vehicles in Europe*. Cambridge MA: CERA.

CEU (Council of the European Union) (2007). *Presidency Conclusions – Brussels, 8–9 March 2007*. European Commission. http://europa.eu/european_council/conclusions/index_en.htm.

Dena (2006). *Biomass to Liquid – BTL Realisierungsstudie: Zusammenfassung*. Deutsche Energie-Agentur GmbH (dena), unter Beteiligung des Verbandes der Automobilindustrie (VDA) sowie Adam Opel GmbH, Audi AG, BMW Group, DaimlerChrysler AG, Ford-Werke GmbH, MAN Nutzfahrzeuge, AG Volkswagen AG, BASF AG, Deutsche BP AG, TOTAL Deutschland GmbH, Lurgi AG, Choren Industries GmbH.

Denholm, P. and Short, W. (2006). *An Evaluation of Utility System Impacts and Benefits of Optimally Dispatched Plug-in Hybrid Electric Vehicles*. Technical Report NREL/TP-620-40293, National Renewable Energy Laboratory (NREL), Colorado.

Deutsche Bank (2008). *Electric Cars: Plugged In*. Deutsche Bank Securities Inc., FITT Research.

Dixon, R. K. (2007). *Building the Hydrogen Economy: Enabling Infrastructure Development*. International Energy Agency (IEA), presentation UNIDO–ICHET Workshop, Istanbul, Turkey. www.iea.org/textbase/speech/2007/Dixon_UNIDO.pdf.

Dreier, T. (2000). *Ganzheitliche Systemanalyse und Potenziale biogener Kraftstoffe*. IfE Schriftenreihe, **42**, Lehrstuhl für Energiewirtschaft und Anwendungstechnik (IfE). Munich: Technische Universität München.

Dreier, T. and Geiger, B. (1998). *Ganzheitliche Prozeßkettenanalyse für die Erzeugung und Anwendung von biogenen Kraftstoffen*. Studie im Auftrag der Daimler Benz AG, Stuttgart und des Bayerischen Zentrums für Angewandte Energieforschung e.V. (ZAE), Lehrstuhl für Energiewirtschaft und Kraftwerkstechnik. Munich: TU München (IfE), Forschungsstelle für Energiewirtschaft (FfE).

Duvall, M. (2004). *Advanced Batteries for Electric-Drive Vehicles: A Technology and Cost-effectiveness Assessment for Battery Electric Vehicles, Power Assist Hybrid Electric Vehicles, and Plug-in Hybrid Electric Vehicles*. Report 1009299. Palo Alto, CA: Electric Power Research Institute (EPRI).

Duvall, M. and Knipping, E. (2007). *Environmental Assessment of Plug-in Hybrid Electric Vehicles: Volume 1 Nationwide Greenhouse Gas Emissions*. Report 1015325. Palo Alto, CA: Electric Power Research Institute (EPRI).

EC (European Commission) (2003). *The Promotion of the Use of Biofuels or Other Renewable Fuels for Transport*. Directive 2003/30/EC of 8 May 2003. OJ L 123/42.

EC (European Commission) (2006). *An EU Strategy for Biofuels*. Communication from the Commission, Commission of the European Communities, COM(2006), 34 final.

EC (European Commission) (2007). *Fact Sheet: Biofuels in the European Union: An Agricultural Perspective*. European Commission, Directorate-General for Agriculture and Rural Development.

Eden, U. *et al.* (1996). *Vergleichende Ökobilanz: Elektrofahrzeuge und konventionelle Fahrzeuge – Bilanz der Emission von Luftschadstoffen und Lärm sowie des Energieverbrauchs im Rahmen des BMBF-Vorhabens 'Erprobung von Elektrofahrzeugen der neuesten Generation auf der Insel Rügen'*. Heidelberg: Institut für Energie-und Umweltforschung (IFEU).

Eilers, J., Posthuma, S. A. and Sie, S. T. (1990). The Shell Middle Distillate Synthesis Process (SMDS). *Catalysis Letters*, **7** (1–4), 253–269.

EPI (Earth Policy Institute) (2006). *Supermarkets and Service Stations Now Competing for Grain*. www.earth-policy.org/Updates/2006/Update55_data.htm.

Fargione, J., Hill, J., Tilman, D., Polasky, S. and Hawthorne, P. (2008). Land clearing and the biofuel carbon dept. *Science Express*, published online, 7th February, 2008.

Farrell, A. E., Plevin, R. J., Turner, B. T. *et al.* (2006). Ethanol can contribute to energy and environmental goals. *Science*, **311** (27th January 2006), 506–508.

Foster Wheeler (1996). *Decarbonisation of Fossil Fuels*. Report No. PH2/2, prepared for the Executive Committee of the IEA Greenhouse Gas R&D Programme.

FVS (ForschungsVerbund Sonnenenergie) (2003). *Workshopband 2003: Regenerative Kraftstoffe*. November 13–14, ZSW, Stuttgart. www.fv-sonnenenergie.de/publikationen/publikation/download/workshop-2003-regenerative-kraftstoffe.

GEMIS (2005). *Globales Emissions-Modell Integrierter Systeme (GEMIS)*. Version 4.3.0.0. www.oeko-institut.org/service/gemis/index.htm.

Gerl, B. (2002). *Innovative Automobilantriebe*. Landsberg/Lech: Verlag Moderne Industrie.
GM (General Motors) (2001). *Well-to-Wheel Energy Use and Greenhouse Gas Emissions of Advanced Fuel/Vehicle Systems*. Argonne National Laboratory.
Graham, R. (2005). *Plug-in Hybrid Electric Vehicles: Changing the Energy Landscape*. Palo Alto, California: Electric Power Research Institute (EPRI).
Gray, D. and Tomlinson, G. (2001). *Coproduction: A Green Coal Technology*. Mitretek Systems (MTS), MP 2001–28. Technical report for the US Department of Energy. www.angtl.com/pdfs/GREENCOAL.pdf.
Hamelinck, C. N. (2004). *Outlook for Advanced Biofuels*. Dissertation. Utrecht: Utrecht University.
Hooijer, A., Silvius, M., Wösten, H. and Page, S. (2006). *Peat CO_2: Assessment of CO_2 Emissions From Drained Peatlands in South-East Asia*. Delft Hydraulics Report Q3943. www.wetlands.org.
IANGV (International Association for Natural Gas Vehicles) (2008). www.iangv.org.
IEA (International Energy Agency) (2004). *Biofuels for Transport: An International Perspective*. Paris: OECD/IEA.
IEA (International Energy Agency) (2006a). *IEA Statistics: Energy Statistics of Non-OECD Countries 2003–2004*. Paris: OECD/IEA.
IEA (International Energy Agency) (2006b). *World Energy Outlook 2006*. Paris: OECD/IEA.
IEA (International Energy Agency) (2006c). *Energy Technology Perspectives 2006. Scenarios and Strategies to 2050*. Paris: OECD/IEA.
IEA (International Energy Agency) (2007). Biofuel production. *IEA Energy Technology Essentials*, **1** (2007). www.iea.org/Textbase/techno/essentials.htm.
Iogen (2005). *Cellulose Ethanol Demonstration Facility*. www.iogen.ca/company/facilities/index.html.
Joint Research Centre, EUCAR, CONCAWE (JEC) (2007). *Well-to-Wheels Analysis of Future Automotive Fuels and Powertrains in the European Context*: Well-to-Wheels Report. Version 2c, March 2007. http://ies.jrc.ec.europa.eu/wtw.html.
Kalhammer, F. R., Kopf, B. M., Swan, D. H., Roan, V. P. and Walsh, M. P. (2007). *Status and Prospects for Zero Emissions Vehicle Technology*. Report of the ARB Independent Expert Panel 2007. California: State of California Air Resources Board Sacramento. www.arb.ca.gov/msprog/zevprog/zevreview/zev_panel_report.pdf.
Kaltschmitt, M. and Reinhardt, G. A. (1997). *Nachwachsende Energieträger: Grundlagen, Verfahren, ökologische Bilanzierung*. Vieweg.
Kaltschmitt, M. and Hartmann, H. (eds.) (2001). *Energie aus Biomasse – Grundlagen, Techniken und Verfahren*. Berlin: Springer.
Kammen, M. K. (2006). The rise of renewable energy. *Scientific American*, **295** (3).
Kempton, W. (2007). Vehicle to grid power. *IEEE Conference Plug-in Hybrids: Accelerating Progress*, Washington, DC. www.ieeeusa.org/policy/phev/presentations/Tutorial%20Kempton.pdf.
Kempton, W. and Tomic, J. (2005a). Vehicle-to-grid power fundamentals: calculating capacity and net revenue. *Journal of Power Sources*, **144** (1), 268–279.
Kempton, W. and Tomic, J. (2005b). Vehicle-to-grid power implementation: from stabilizing the grid to supporting large-scale renewable energy. *Journal of Power Sources*, **144** (1), 280–294.

Kraus, K., Niklas, G. and Tappe, M. (2000). *Aktuelle Bewertung des Einsatzes von Rapsöl/RME im Vergleich zu Dieselkraftstoff*. Texte 79/99. Berlin: Umweltbundesamt (UBA) (Federal Environment Agency).

Krüger, R. (2002). *Systemanalytischer Vergleich alternativer Kraftstoff - und Antriebskonzepte in der Bundesrepublik Deutschland*. Dissertation. VDI Fortschritt-Berichte 12, No. 499, Düsseldorf: VDI Verlag.

Larsen, H. H. (1998). The 2400 MTPD Methanol Plant at Tjeldbergodden. Haldor Topsoe A/S. *1998 World Methanol Conference*. Frankfurt, Germany.

Liebner, W., Koempel, H. and Wagner, M. (2004). *Gas-To-Chemicals-Technologien von Lurgi: von Ergas/Synthesegas zu hochwertigen Produkten*. 55. Berg-und Hüttenmännischer Tag, Freiberg: Lurgi AG.

Martinot, E. *et al.* (2006). *Renewables, Global Status Report, 2006 Update. Renewable Energy Policy Network for the 21st Century (REN21)*. Worldwatch Institute and Tsinghua University. www.ren21.net.

MIT (2007). www.technologyreview.com/Energy/18173.

Möller, K. (2003). Systemwirkungen der 'Biogaswirtschaft' im ökologischen Landbau: Pflanzliche Aspekte. Auswirkungen auf den N-Haushalt und auf die Spurengasemissionen. *Biogas Journal*, **1** (May 2003), 20–29.

NGV (Natural Gas Vehicles Group) (2008). *Gas Vehicles Report, June 2008*. www.ngvgroup.com.

Project Better Place (2008). www.betterplace.com.

Punter, G., Rickeard, D., Larivé, J-F. *et al.* (2004). *Well-to-Wheel Evaluation for Production of Ethanol from Wheat*. Report by the LowCVP Fuels Working Group, WTW Sub-Group; FWG-P-04-024.

Quack, H. (2001). Die Schlüsselrolle der Kryogentechnik in der Wasserstoff-Energiewirtschaft. *Wissenschaftliche Zeitschrift der Technischen Universität Dresden*, **50** (5/6).

REN21 (2006). *Renewable Global Status Report 2006 Update*. Paris: REN21 Secretariat and Washington, DC: Worldwatch Institute.

Sasol (2007) *Sasol Facts: Introducing Sasol*. Sasol. http://sasol.investoreports.com/sasol_sf_2007/html/sasol_sf_2007_1.php.

Schaub, G., Unruh, D. and Rohde, M. (2003). Kraftstoff-Bereitstellung über die Biomassevergasung – Herausforderungen und Perspektiven. *Proceedings Internationale Tagung 'Biomasse-Vergasung'*. Leipzig.

Schaub, G., Unruh, D. and Rohde, M. (2004). Synthetische Kraftstoffe aus Biomasse über die Fischer-Tropsch-Synthese – Grundlagen und Perspektiven. *Erdöl, Erdgas, Kohle*, **10** (120).

Schulz, W. (2004). *Untersuchung zur Aufbereitung von Biogas zur Erweiterung der Nutzungsmöglichkeiten, Aktualisierung einer im Juni 2003 vorgelegten gleichnamigen von Wolfgang Schulz, Maren Hille unter Mitarbeit von Wolfgang Tentscher durchgeführten Untersuchung*. On behalf of Bremer Energie-Konsens GmbH. Bremen: Bremer Energieinstitut.

Searchinger, T., Heimlich, R., Houghton, R. A. *et al.* (2008). Use of US croplands for biofuels increases greenhouse gases through emissions from land-use change. *Science Express*, **319** (4th January), 1238–1240.

Shell (1995). *Shell MDS Malaysia*. Corporate Affairs Shell Malaysia Ltd.

Söderbergh, B., Robelius, F. and Aleklett, K. (2006). *A Crash Program Scenario for the Canadian Oil Sands Industry*. Uppsala, Sweden: Uppsala University.

Specht, M., Bandi, A. and Pehnt, M. (2001). *Regenerative Kraftstoffe – Bereitstellung und Perspektiven*. Berlin: FVS Themen 2001, Forschungsverbund Sonnenenergie. www.fv-sonnenenergie.de.

SRU (Sachverständigenrat für Umweltfragen) (2007). *Klimaschutz durch Biomasse*. Berlin: Erich Schmidt Verlag.

Tahil, W. (2006). *The Trouble with Lithium. Implications of Future PHEV Production for Lithium Demand*. Meridian International Research. www.meridian-int-res.com/Projects/Lithium.htm.

Tesla Motors (2008). www.teslamotors.com.

Tomic, J. and Kempton, W. (2007). Using fleets of electric-drive vehicles for grid support. *Journal of Power Sources*, **168** (2), 459–468.

van Thuijl, E., Ree, R. and Lange, T. J. (2003a). *Biofuel Production Chains. Background Document for Modelling the EU Biofuel Market Using the BIOTRANS Model*. ECN Report C–03–088.

van Thuijl, E., Roos, C. J. and Beurskens, L. (2003b). *An overview of Biofuel Technologies, Markets and Policies in Europe*. ECN Report C–03–008.

Worldwatch Institute (2007). *Biofuels for Transport: Global Potential and Implications for Sustainable Agriculture and Energy in the 21st Century*. Worldwatch Institute.

Woynillowicz, D., Severson-Baker, C. and Raynolds, M. (2005). *Oil sands Fever – The Environmental Implications of Canada's Oil Sands Rush*. The Pembina Institute. www.pembina.org/pubs.

Further reading

IEA (International Energy Association) (1998). *Automotive Fuels for the Future. The Search for Alternatives*. Paris: OECD/IEA.

IEA (International Energy Association) (2008). *Energy Technology Perspectives 2008. Scenarios and Strategies to 2050*. Paris: OECD/IEA.

Joint Research Centre, EUCAR, CONCAWE (JEC) (2007). *Well-to-Wheels Analysis of Future Automotive Fuels and Powertrains in the European Context*. Well-to-Wheels Report, Version 2c. http://ies.jrc.ec.europa.eu/wtw.html.

Kamm, B., Gruber, P. R. and Kamm, M. (eds.) (2006). *Biorefineries – Industrial Processes and Products: Status Quo and Future Directions*. Two Volumes. WILEY-VCH.

Marland, G., Obersteiner, M., Schlamadinger, B., Righelato, R. and Spracklen D. V. (2007). The carbon benefits of fuels and forests. *Science*, **318** (16th November), 1066–1068.

Righelato, R. and Spracklen, D. V. (2007). Carbon mitigation by biofuels or by saving and restoring forests? *Science*, **317** (17th August), 902.

Robertson, G. P., Dale, V. H., Doering, O. C. *et al.* (2008). Sustainable biofuels redux. *Science*, **319** (3rd October), 49–50.

Scharlemann, J. P. W. and Laurance, W. F. (2008). How green are biofuels? *Science*, **319** (4th January), 43–44.

Searchinger, T., Heimlich, R., Houghton, R. A. *et al.* (2008). Use of US croplands for biofuels increases greenhouse gases through emissions from land-use change. *Science*, **319** (4th January), 1238–1240.

8

Hydrogen today

Martin Wietschel, Michael Ball and Philipp Seydel

Worldwide, the number of attempts and efforts to develop and test hydrogen-related technology in vehicles and implement the necessary hydrogen supply infrastructure has been increasing tremendously in recent years, resulting in numerous hydrogen demonstration and lighthouse projects around the globe. In this chapter, first a brief summary of this development is presented, while recognising that this can only be a snapshot, as development is very quick and the information provided here can rapidly become outdated. Next, international roadmapping activities that show possible developments towards the introduction of hydrogen are described. Finally, the issue of social acceptance of hydrogen technology and the need for regulations and standards, are briefly discussed, since these are important factors for the hydrogen penetration and infrastructure transition process.

8.1 Hydrogen in the transport sector

With a share of more than 80% in total energy use in the transport sector, the automotive sector is the driving force for the introduction of hydrogen as fuel.[1] Hence, the focus of hydrogen-vehicle manufacturing is on passenger cars and buses.[2] Heavy-goods vehicles are not in the spotlight, as neither fuel cells nor hydrogen combustion engines are likely to manage a breakthrough in this market segment any time soon, because of the dominance and high performance of the diesel engines for long-transport applications.

[1] While prototype aeroplanes fuelled by hydrogen have also been successfully demonstrated (IEA, 2005), an inherent disadvantage of hydrogen for use in aviation is its low energy density. Hydrogen has also been used as a fuel for scooters and boats.

[2] See also Section 8.2 for roadmapping. For a discussion of scenarios and strategies relating to the introduction of hydrogen vehicles, see Chapter 14.

The Hydrogen Economy: Opportunities and Challenges, ed. Michael Ball and Martin Wietschel. Published by Cambridge University Press.

Table 8.1. *Comparison of hydrogen and conventional drive trains*

Passenger car	Otto-motor[a]	Diesel-motor[a]	PEM FC – today	PEM FC – target	H_2 ICE – today	H_2 ICE – target
Efficiency (NEDC) (%)	20–35[b]	30–40[b]	>37%	40–45	≈26	35
Investment ($/kW)	32–38	40–46	≈2000	≪100	40–46[a]	≪40
Lifetime (h)	>8000	>8000	>2000	>5000[c]	≈5000	>5000

Notes:
[a] On the basis of a 70 kW engine.
[b] Lower values in urban driving, higher values at best operating point.
[c] Up to 20 000 hours for buses.
Sources: (BMWA, 2005; HFP, 2007; IEA, 2005, amended).

8.1.1 The use of hydrogen in vehicles

Principally, hydrogen can be used as fuel both in modified internal combustion engines (ICE)[3] and in fuel cells with electric drive trains.[4] For passenger cars, fuel-cell systems are almost exclusively developed on the basis of the PEMFC, the reason being their capability to respond dynamically and quickly to load changes (as, e.g., during urban driving), their immediate start-up ability compared with other fuel cell systems, their ideal partial load behaviour, and, not least, their advanced development stage. Hydrogen-fuelled ICE vehicles offer the advantage that they can be designed in bi-fuel mode, i.e., they can run on hydrogen and gasoline, a capability which may be essential for the introduction of hydrogen during the transition period. While hydrogen-fuelled fuel cell vehicles are so-called zero-emission vehicles, hydrogen vehicles on the basis of ICEs have very low NO_x emissions only.[5]

Table 8.1 shows a comparison of conventional and hydrogen-based drive trains for passenger cars, with respect to efficiency (related to the entire drive train), investments and lifetimes. Current fuel-cell cars have efficiencies of around 37% in NEDC (new European driving cycle), with the potential of reaching up to 45% in the medium term. (To improve the efficiency of fuel-cell drive trains, the concept of hybridisation with a conventional battery is also applied.) The efficiency of hydrogen ICE vehicles is around 26%. The requirement for the lifetime of fuel cells is about 5000 hours in passenger cars and up to 20 000 hours in buses. The greatest challenge

[3] Hydrogen is a suitable fuel for spark ignition engines. Hydrogen cannot be used directly in a diesel (or 'compression ignition') engine, since hydrogen's autoignition temperature is too high. Thus, diesel engines must be fitted with spark plugs or use a small amount of diesel fuel to ignite the gas.

[4] In addition, fuel cells can be used in vehicles as auxiliary power units (APUs) for on-board power supply. The increasing number of electronic functions in vehicles requires new strategies to supply the growing power demand. When aiming at reducing the fuel consumption of vehicles, however, providing the required electrical energy by means of larger generators and batteries is not advisable. Here, the fuel cell as APU (PEMFC for hydrogen, SOFC for conventional gasoline or diesel fuel) can make use of its advantages.

[5] Nitrogen oxide emissions from hydrogen ICEs are around 90% lower than for a gasoline ICE, because the engine can operate in 'lean-burn' mode with an excess of air, which leads to lower engine temperatures and less NO_x production.

is to reduce the fuel cell costs drastically from more than \$2000/kW to less than \$100/kW for passenger cars and to between \$135 to \$200/kW for buses and light duty vehicles (IEA, 2005) (see also Chapter 13). Thereby, the reduction of costs is in a continuous interplay with the requirements for efficiency and lifetime. Internal combustion engines with hydrogen as fuel can be manufactured today at costs similar to diesel engines.

The storage of hydrogen on board the vehicles can be realised in liquid or compressed gaseous form. Most of the hydrogen vehicles today use pressurised hydrogen storage. With the currently used 350 bar pressure tanks, driving ranges of 200–300 km are possible. With the gradual introduction of 700 bar pressure tanks, driving ranges of 400–600 km (as of today's conventional passenger cars) are achievable, which are necessary for customer acceptance. Today's liquid-hydrogen tanks contain between 70 and 100 l LH_2 (4.5 to 6.5 kg LH_2), which corresponds to a gasoline equivalent of about 20 to 30 l. With this, ranges of more than 400 km are realised (1 kg H_2 $\approx$ 80 km). Besides the stored hydrogen volumes, the driving ranges of the vehicles depend largely on the drive train – fuel cell or internal combustion engine – with the ICE in bi-fuel configuration yielding a greater vehicle mileage, despite the higher hydrogen consumption.

8.1.2 The evolution of hydrogen vehicles

Car producers and research institutes have been working on hydrogen vehicles for many years. While the first experiments with a hydrogen gas engine date back as far as 1807, it was only in the second half of the twentieth century that vehicle prototypes started to emerge. In 1967, General Motors designed a first operational fuel-cell-powered electric vehicle, which was a six-passenger electrovan, using liquid hydrogen. It had a top speed of 105 km/h and a driving range of 200 km.

The most recent wave for hydrogen-vehicle market introduction arose with announcements by Daimler and Toyota in 1997 of the production of fuel-cell cars in 2004 (see Fig. 8.1). After this, other car manufacturers joined the hydrogen race.

Today, almost every car maker possesses its own prototypes and development experience. Different car manufacturers follow different design concepts, with regard to drive train, hydrogen storage or market segment to be addressed. Most of the prototypes developed after 2000 were fuel-cell vehicles rather than vehicles with internal combustion engines (an exception being, for example, BMW). The preferred storage option is for compressed gaseous hydrogen tanks, although liquid hydrogen storage can also be found.[6]

According to Butler (2008), the number of fuel-cell cars in the world at the end of 2007 was around one thousand. Considering the geographic distribution of *vehicle*

[6] Liquid hydrogen is preferred in combination with internal combustion engines, as the low temperature of the hydrogen yields a higher efficiency (Carnot efficiency).

Figure 8.1. Daimler F-Cell car (LBST, 2009a).

development and construction, Europe is the predominant region of fuel-cell vehicle manufacturing, with more than half of the market share, followed by North America with a fifth and Asia (largely Japan) with around a quarter. Europe's high market share is mainly accounted for by Daimler, who have also announced the start of small series production of fuel-cell cars from the middle of 2009, looking to reach annual production of 100 000 vehicles within five years at cost levels competitive with conventional vehicles. Considering the *deployment of fuel-cell vehicles*, at the time of writing, North America and Europe had the lion's share, close to 50% each, with Asia taking up the remainder (Butler, 2008). Most car manufacturers currently see commercial production of fuel-cell cars around 2015.

Hydrogen buses (see Fig. 8.2) have their own development history, since in some cities authorities showed demand for them. They favoured the buses for their low pollution as well as for social reasons, such as raising public hydrogen awareness and promotion of further research. The first bus was the Ballard p1 and was released in 1993. In the 2000s, the development of buses accelerated and many manufacturers got involved. In 2007, a little over one hundred hydrogen buses were used in various cities around the world. One important hydrogen bus demonstration project worth mentioning in Europe is the HyFLEET:CUTE initiative (HyFLEET:CUTE, 2007). Further information about hydrogen buses and related demonstration projects is provided by Jerram (2008) and FuelCells2000 (2009a).

Figure 8.2. Daimler fuel cell bus (LBST, 2009a).

It would be outside the scope of this book to address the various hydrogen vehicles developed so far and their characteristics in more detail, and the information given here would be rapidly outdated. Regular market surveys of the development and use of hydrogen cars and buses, as well as the strategies of different car manufacturers, are carried out for instance by *FuelCellToday* or *FuelCells2000* (see Butler, 2008; FuelCells2000, 2009a; 2009b), and also www.netinform.net/h2/h2mobility. See also www.fuelcelltoday.com and www.fuelcells.org. Further websites are provided at the end of this chapter.

8.1.3 Hydrogen-supply infrastructures

There are numerous demonstration projects for the use of hydrogen in the transport sector, with the aims of gaining first experiences with the operation of hydrogen vehicles, testing a hydrogen infrastructure (i.e., hydrogen supply and operation of refuelling stations) under real-world conditions and promoting public perception and acceptance. Hydrogen refuelling stations can be separated into stationary and mobile ones. Mobile stations demand less capital investment, allow flexible refuelling and are ideal for fuel-cell vehicle demonstrations. They supply compressed hydrogen to hydrogen refuelling stations, thus being suitable for mother–daughter stations.

Figure 8.3 shows the global distribution of existing hydrogen refuelling stations. Black spots represent operational stations, grey spots, planned stations and white ones indicate expired stations.

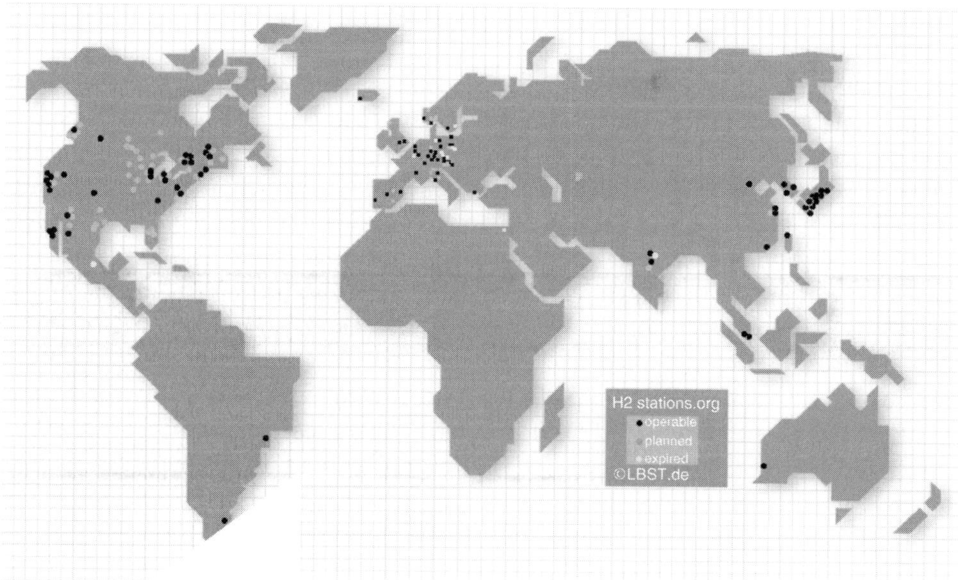

Figure 8.3. Global distribution of existing hydrogen refuelling stations (LBST, 2009b).

Worldwide, there were close to 200 hydrogen refuelling stations in operation at the time of writing (see www.netinform.net/h2/h2stations and Huleatt-James, 2008). There are three leading geographical regions where hydrogen refuelling stations are found: North America, Europe and Eastern Asia, especially Japan. In the USA, there were about 65 operating hydrogen stations (compared with 170 000 gasoline stations) concentrated in the north-eastern parts of the country (Chicago, New York) as well as in the metropolitan areas of California. In Japan, there were about 20 operating hydrogen refuelling stations at the time of writing, with the highest density in the area of Tokyo. In Germany, there were 21 operating hydrogen refuelling stations, which is the highest number in Europe; this compares with about 16 000 conventional refuelling stations in Germany (and about 1500 fuelling stations selling biodiesel and close to 800 natural gas stations). There are also some interesting projects in China around Beijing and Shanghai, as well as in Singapore and near New Delhi in India.

Important demonstration projects for testing a hydrogen infrastructure are the California Fuel Cell Partnership (CaFCP), the Clean Energy Partnership (CEP) in Berlin and Hamburg and the already-mentioned HyFLEET:CUTE initiative for hydrogen buses. (See also www.cafcp.org and www.cleanenergypartnership.de. Further websites are provided at the end of the chapter.) More information about other infrastructure projects can be found in Huleatt-James (2008) and Roads2HyCom (2007b).

8.2 Hydrogen roadmaps

The implementation of advanced, highly innovative technologies, such as hydrogen applications, is not just a matter of achieving the right payback time. A transition towards a sustainable energy system involves changes at various levels of the economy and society. To facilitate a smooth and beneficial introduction of hydrogen into their energy systems, besides the global roadmapping activities by the IEA (*Prospects for Hydrogen and Fuel Cells* (IEA, 2005), many industrial countries and regions are currently working on their own hydrogen roadmaps (for an overview of roadmaps, see Roads2Hycom (2007a)).

In the EU, the development of a hydrogen and fuel-cell roadmap is ongoing. Progress in the formulation of a pan-European strategy was initiated by the setting up of the High Level Group (HLG) in 2002 and subsequent consultations (see HLG (2003) and HyNet (2004)). Indeed, the European Commission has made the hydrogen economy one of its long-term priorities for its energy system. To this end, it has created the Hydrogen Technology Platform (HFP), to devise an action plan aimed at creating a completely integrated hydrogen economy based on renewable energy sources and nuclear power by the middle of this century. This initiative was launched on the 10th September, 2003 in a Commission communication entitled *A European Partnership for a Sustainable Hydrogen Economy*. Under the HFP, a *Deployment Strategy* (HFP, 2005a), a *Strategy Research Agenda* (HFP, 2005b) and an *Implementation plan* (HFP, 2007) were developed. This led to the establishment of the Fuel Cell and Hydrogen Joint Technology Initiative (JTI) in 2008. Two comparatively large multiregional roadmapping activities in Europe have been carried out under the HyWays (*A Roadmap for Hydrogen Energy in Europe* (HyWays, 2007)) and the WETO-H2 (*World Energy Technology Outlook 2050* (EC, 2006a)) projects. HyWays was an integrated project, co-funded by research institutes, industry, national agencies and by the European Commission under the 6th Framework Programme. Table 8.2 summarises the HyWays roadmap and action plan.

The EU's main competitors in the hydrogen and fuel-cell field are the USA, Japan and, to a lesser extent, Canada. Each of these countries has established H_2 and FC RTD&D support frameworks and developed long-term technology roadmaps describing technical milestones over the coming decades. Worthy of mention are the *National Hydrogen Energy Roadmap* (DOE, 2002) and *Hydrogen Posture Plan* (DOE, 2004) of the United States Department of Energy (DOE) (see Fig. 8.4) and the *Strategic Technology Roadmap* (see Fig. 8.5) of Japan's Ministry of Economy, Trade, and Industry (METI, 2006). However, other countries assessed (Australia, China, India and South Korea) are significantly less well developed and are just beginning to develop H_2 and FC RTD&D activities and programmes.

In all these roadmaps, the main application for hydrogen is seen in the transport sector. Most studies see an initial market penetration for the transport sector after 2010 and the development of a mass market after 2020. In 2050, the majority of

Table 8.2. *Summary of the deployment phases, targets and main actions outlined in the Roadmap and Action Plan for the EU*

	2010	2015	2020	2030	2050
Phases	➡ **Technology development with focus on cost reduction**	➡ **Pre-commercial technology refinement and market preparation** ➡ **Start of commercialisation**	HFP snapshot 2020 ◆ **Materialisation of first impacts** • New hydrogen supply capacities partially based on low-carbon sources • Improvements in local air quality • More than 5% of new car sales H_2 and FC	HyWays snapshot 2030 ◆ **Hydrogen and FC are competitive** • Creation of new jobs and safeguarding existing jobs (net employment effect of 200 000–300 000 labour years) • Shift towards carbon-free hydrogen supply • More than 20% of new car sales H_2 and FC	**H_2 and FC dominant technologies high impact** • 80% of light duty vehicles and city buses fuelled with CO_2-free hydrogen • Reaching more than 80% CO_2 reduction in passenger car transport • In stationary end-use applications, hydrogen is used in remote locations and island grids
Targets	**LHPs facilitate initial fleet of a few thousand vehicles by 2015** • PPP 'Lighthouse Projects' • Increase R&D budgets to M€80/year • Financial support for large-scale demonstration projects		Vehicles: **2.5 million of fleet** Cost: **H_2: €4/kg (€50/barrel)** **FC: €100/kW** **Tank: €10/kWh**	Vehicles: **25 million of fleet** Cost: **H_2: €3/kg (€50/barrel)** **FC: €50/kW** **Tank: €5/kWh**	

Table 8.2 (cont.)

	2010	2015	2020	2030	2050
Required policy support actions	**Development of H_2-specific support framework** • Create and support early markets • Implement performance-monitoring framework • Long-term security for investing stakeholders • Education and training programmes • Harmonisation of regulations, codes and standards		**H_2-specific support framework** • In place before 2015 at MS level • Deployment supports, e.g., tax incentives of M€180/year • Public procurement • Planning and execution of strategic development of hydrogen infrastructure	**Gradual switch from hydrogen-specific support to generic support of sustainability (2020→)**	**Incentives provided through general support schemes for sustainability**
	2010	**2015**	**2020**	**2030**	**2050**

Source: (HyWays, 2007).

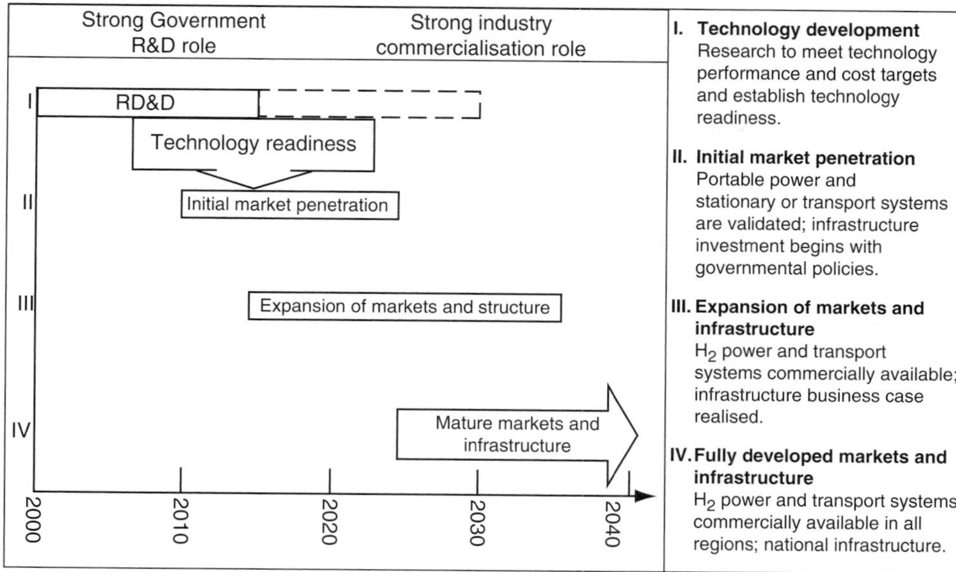

Figure 8.4. Roadmap from United States Department of Energy (information taken from DOE (2006)).

vehicles (without trucks) could be hydrogen vehicles (however, the range between the studies varies between 30% and 80% of all vehicles). Hydrogen internal combustion engines could play a role only in the beginning of the market development; later on, fuel-cell vehicles will clearly dominate. The target cost for hydrogen at the pump of a filling station (without tax) varies between \$2/kg and \$4/kg (6 and 12 ct/kWh). Only in roadmaps before 2004 could lower cost targets be found. However, these cost targets were adopted among others because of the cost increase of conventional fuels. Normally, the cost targets use the cost of conventional fuels as a benchmark.

At a long-term perspective, the cost targets of the propulsion system are often oriented on the cost of highly efficient conventional combustion engines (e.g., modern diesel drive systems). Most roadmaps point out that the cost reduction of the propulsion system is a major challenge and technical breakthrough has to be reached here for a successful market introduction of hydrogen. Other stimulating elements for hydrogen are high prices of oil and gas and ambitious climate policies.

Stationary use of fuel cells for the industrial and residential sectors is often foreseen in combined heat and power applications, but these are mainly run on syngas and/or natural gas instead of hydrogen.

For the first decades, all roadmaps show a focus on fossil-based hydrogen production options, mainly onsite and decentral steam methane reformers (SMR), electrolysers and hydrogen as a by-product from the chemical industry. In some regions, hydrogen is also produced to a certain extend by nuclear, electrolysis, biomass and waste gasification. Later on, with a significant increase of hydrogen, the production

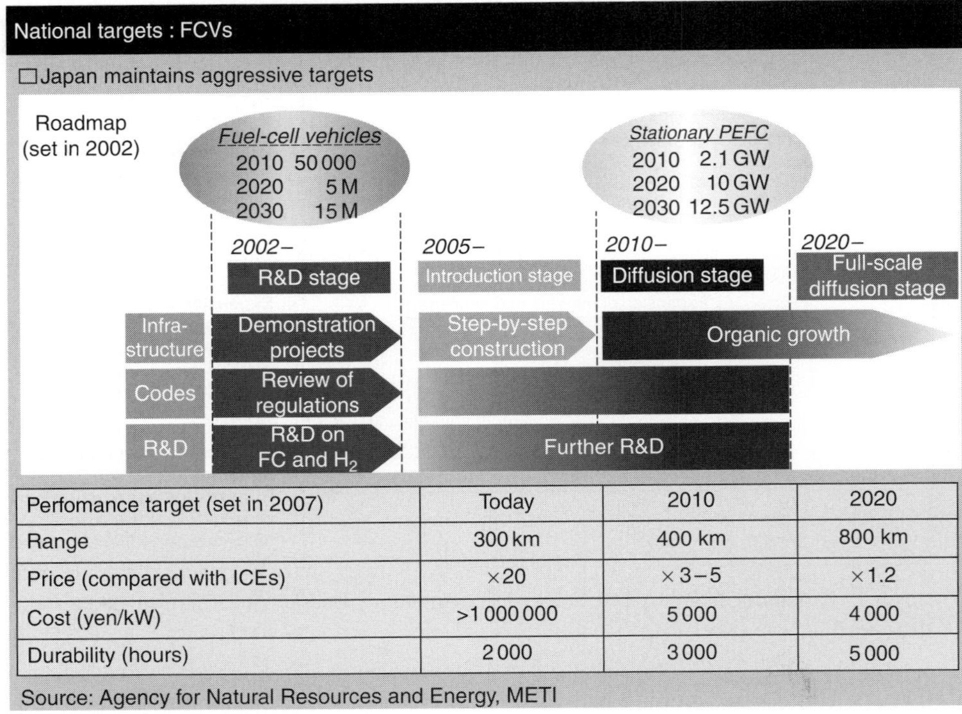

Perfomance target (set in 2007)	Today	2010	2020
Range	300 km	400 km	800 km
Price (compared with ICEs)	×20	×3–5	×1.2
Cost (yen/kW)	>1 000 000	5 000	4 000
Durability (hours)	2 000	3 000	5 000

Figure 8.5. Roadmap from Japanese Ministry of Trade and Industry (information taken from Maruta, (2008)) (1 Yen = US\$0.007).

will become more centralised. Central coal gasification with carbon capture and sequestration or SMR could play a significant role. In most roadmaps, a relevant increase of the renewable share of hydrogen production is seen after 2030. The possible role of nuclear after 2030 varies in the roadmaps. The WETO-H_2 study forecasts a share of 40% of nuclear energy for hydrogen production next to 50% for renewable energies in 2050, whereas the IEA assumes that 80% of the hydrogen will be produced centrally by SMR and coal gasification, both with CCS, in this timeframe. However, all roadmaps forecast a CO_2-lean production for hydrogen at a long-term perspective.

8.3 Social acceptance

8.3.1 Importance of social acceptance

Growing public expenditures for developing hydrogen technology and applications give rise to concerns over whether the public will accept the change in its energy system. All parts of a hydrogen economy, including production, distribution and consumption, are affected by public acceptance. Because of that, the public acceptance of hydrogen technology has to be analysed in more detail.

Public knowledge of hydrogen technology is still relatively low (Fuhrmann and Bleischwitz, 2007). Many people still do not know the difference or have difficulties with separating the usage of hydrogen as a fuel from the fuel cell, which is a device to turn hydrogen into useful power.

To improve the public acceptance of hydrogen, it is important to understand the current perception of hydrogen by the broad public so that conclusions can be made by producers. Knowledge about the current state of social acceptance is a basis for detecting improvement points to increase it in future.

8.3.2 Surveys

So far, only a few studies exist to analyse the acceptance of hydrogen by the general public, most of which were conducted in European countries. Most of them focus on the transport sector, especially on public transport, such as buses and taxis. Studies that take into account the whole production line of hydrogen do not yet exist.

A study that was conducted in Canada consisted of a survey of passengers of a test hydrogen bus (Hickson *et al.*, 2007). The results showed a wide-scale preference for hydrogen buses over conventional buses and a strong overall support for hydrogen as a motive fuel. The positive attitude was higher amongst males and more frequent bus users.

Two other surveys were held in Iceland (Maak and Skulason, 2006): one before hydrogen buses started to operate in Iceland, and another among actual passengers of both conventional and hydrogen buses after the latter were launched. The first survey showed a very positive attitude towards using hydrogen, as well as large curiosity towards this technology. The second survey pointed out a readiness to pay 10%–20% more for hydrogen transport during the introduction phase. Hydrogen was considered a clean and safe fuel; however, many admitted that they were not very well informed. Contact with the technology (among passengers) seemed to increase the acceptance.

The AcceptH2 project compared public attitudes in London, Luxemburg, Munich, Perth and Oakland, enabling international comparisons of perception to be made (for more information, see www.accepth2.com). Its objective was to assess economic preferences towards the potential and actual use of hydrogen buses by conducting 'before' and 'after' economic valuation studies. In addition, the project assessed the level of influence of the hydrogen-bus demonstration projects in these cities. The results were made on the basis of surveys in each of the cities before and after the introduction of hydrogen-transportation projects. The detailed results showed that people strongly support hydrogen and fuel cells. They showed no objections to hydrogen technology – neither in general, nor when these technologies are applied in vehicles. However, most people indicated that they need more information and would make a decision for a hydrogen-vehicle dependent on existing infrastructure.

Many people showed acceptance towards an extra charge for using hydrogen buses (Altmann *et al.*, 2004).[7]

No explicit studies concerning the perception of the security aspect of hydrogen have been made so far and it is only possible to draw conclusions from the overall attitude towards hydrogen technologies. As long as public knowledge of the use of hydrogen remains low, it is difficult to expect adequate and sensible data on the perception of hydrogen security. However, this aspect is of particular importance, as security concerns seriously have the potential to slow down the diffusion of hydrogen applications.

To conclude, surveys generally show that there is a great acceptance, but a low knowledge level for hydrogen technologies. Males and people with a higher education level seem to have a greater acceptance. There is practically no opposition to the introduction of hydrogen as a fuel. However, more educational activities and possibilities for practical experience with hydrogen vehicles are critical measures to increase public acceptance.

8.3.3 General consumer behaviour

Besides studies and surveys, which are available in limited numbers, general publications on factors that influence consumer decision can be considered as well. Most of the authors agree that customer satisfaction is the key factor for a positive attitude towards a new technology, and this can only be achieved if the services offered by this technology exceed the old ones. However, many agree that customer satisfaction is a very subjective matter and depends on a number of personal factors, such as personality, education and interests, so that universal selling arguments are hard to find.

An OECD study identifies several barriers for a new technology to penetrate the market (OECD, 2002). These are lack of awareness, cognitive dissonance, lack of concern for future generations, fear of and resistance to change and, finally, lack of adequate professional advice. Each of these barriers must be countered with special measures if the acceptance for a new technology, in this case hydrogen, is to be increased.

According to Spitzley *et al.* (2000), there are six major factors to influence consumers to switch to alternative fuel vehicles: performance, fuel consumption, noise and vibration, consumer costs, durability and safety. A new technology must be superior in some of these points without being significantly worse in others. Other authors point out the necessity of broader media coverage of hydrogen issues, since greater knowledge positively correlates with the acceptance of hydrogen. Also, companies that develop new sensitive technology must try to embed public anticipation and do this regularly and frequently from an early phase of the technology design process.

[7] Other studies had comparable results: see Roads2HyCom (2007c); Schulte *et al.*, (2004), or, particularly for Germany, Altmann and Gräsel (1998) and Dinse (2000).

According to Steinberger-Wilckens (2003), there is a perception of hydrogen as a 'green technology' regarding its zero-emission performance in fuel-cell vehicles. Although the real environmental added value of hydrogen use will depend on the production technologies, the visible clean part of its usage is contributing to its green image.

8.4 Regulations, codes and standards

Hydrogen is in common use as a feedstock in a range of industries, where its use is tightly regulated. However, its use in energy systems is still very novel, and appropriate regulations and standards have not yet been developed. If hydrogen is to become a significant energy carrier and fuel, the development and promulgation of regulations, codes and standards (RCS) are essential to establish a market-receptive environment for commercialising hydrogen-based products and systems.

Regulations, codes and standards must be internationally harmonised and consistent with the maximum extent. Significant efforts are underway to develop regulations and harmonise these across countries. The development of internationally recognised regulations, codes and standards is vital not to impede the development of new hydrogen and fuel cell products and projects. An adequate level of standardisation and regulation is equally required in ensuring the safe deployment of hydrogen technologies in the market. Regulation and standards have to be put in place for the entire hydrogen supply chain, i.e., production, transportation and storage, hydrogen refuelling infrastructure, fuel-cell technologies and hydrogen-fuelled vehicles. If these aspects are not appropriately considered, RCS may become a barrier to the early introduction into the market and hydrogen systems may encounter resistance from insurers.[8]

8.5 Summary

Today's rapidly increasing activities on hydrogen focus mostly on vehicle applications and less on stationary applications. For fuel cells, stationary applications are also relevant, but natural gas will be the dominant fuel here. The dominance of the transport sector is also reflected in the hydrogen roadmaps developed, among others, in the EU, the USA, Japan, or at an international level. Whereas in the beginning, onsite or decentralised production options based on fossil fuels or electricity are seen as the major option for hydrogen production, later on central production options will dominate the market. Here, several options could play a role, from coal, with carbon capture and sequestration, through natural gas and renewables (wind, biomass) to nuclear. A CO_2-free or lean vision can be identified in every roadmap. The cost

[8] For an overview of ongoing activities and developments see, for instance, the EC (2006b); the EIHP (2004); HarmonHy (2006), Joseck and Davis (2006) and http://hcsp.ansi.org.

targets for hydrogen as a fuel, as well as for the whole hydrogen propulsion, are oriented on conventional fuel prices, and on the conventional drive system. Hydrogen internal combustion engines are only seen in the introduction phase of hydrogen: this introduction phase could start around 2015. In the end, fuel-cell vehicles will dominate the market with high shares of the total vehicle population in 2050. The development of a mass market is seen only after 2020.

The consumer acceptance in the transport sector seems to be no barrier to the introduction of hydrogen. All analysed studies show that there is a great acceptance, but a low knowledge level for hydrogen technologies. Males and people with a higher education level seem to have a greater acceptance.

The establishment of internationally harmonised and consistent regulations, codes and standards is essential for the commercialisation of hydrogen-based products and systems.

Websites

Hydrogen vehicles

www.netinform.net/h2/h2mobility,
www.fuelcells.org.

Hydrogen refuelling stations

www.netinform.net/h2/h2stations.

Hydrogen and fuel-cell demonstration projects

www.hylights.eu,
www.h2moves.eu,
www.roads2hy.com,
www.ieahia.org,
www.iphe.net,
www.hfpeurope.org.

References

Altmann, M. and Gräsel, C. (1998). *The Acceptance of Hydrogen Technologies.* www.HyWeb.de/accepth2.

Altmann, M., Schmidt, P. *et al.* (2004). AcceptH2: public perception of hydrogen buses in five countries. In *Proceedings of the International German Hydrogen Energy Congress 2004*, Essen.

BMWA (Bundesministerium für Wirtschaft und Arbeit) (2005). *Strategiepapier zum Forschungsbedarf in der Wasserstoff-Energietechnologie*. Berlin: BMWA.

Butler, J. (2008). *2008 Light Duty Vehicle Survey.* www.fuelcelltoday.com/media/pdf/surveys/2008-Light-Duty-Vehicle.pdf.

Dinse, G. (2000). *Hydrogen – A New and Yet Unfamiliar Fuel*. Berlin: Institute for Mobility Research, BMW. www.ifmo.de.

DOE (2002). *National Hydrogen Energy Roadmap*. US Department of Energy (DOE).
DOE (2004). *Hydrogen Posture Plan*. US Department of Energy (DOE).
DOE (2006). *Hydrogen Posture Plan*. US Department of Energy (DOE).
EC (2006a). *World Energy Technology Outlook – 2050 (WETO-H2)*. Brussels: Directorate-General for Research, Directorate Energy, European Commission.
EC (2006b). *Introducing Hydrogen as an Energy Carrier – Safety, Regulatory and Public Acceptance Issues*. EUR 22002. Brussels: Directorate-General for Research, European Commission.
EIHP (2004). *European Integrated Hydrogen Project – Phase II*. Final Technical Report. www.eihp.org.
FuelCells2000 (2009a). *Fuel Cell Buses*. www.fuelcells.org/info/charts/buses.pdf.
FuelCells2000 (2009b). *Fuel Cell Vehicles*. www.fuelcells.org/info/charts/carchart.pdf.
Fuhrmann, K. and Bleischwitz, R. (2007). *Existing Acceptance Analysis in the Field of Hydrogen Technologies*. www.roads2hy.com.
HarmonHy (2006). *'HarmonHy': Harmonization of Regulations, Codes and Standards for a Sustainable Hydrogen and Fuel Cell Technology*. Support Action funded by the European Commission under the 6th Framework Programme. www.harmonhy.com.
HFP (European Hydrogen and Fuel Cell Technology Platform) (2005a). *Strategic Research Agenda*. July 2005, Brussels. www.HFPeurope.org.
HFP (European Hydrogen and Fuel Cell Technology Platform) (2005b). *Deployment Strategy*. August 2005, Brussels. www.HFPeurope.org.
HFP (European Hydrogen and Fuel Cell Technology Platform) (2007). *Implementation Plan – Status 2006*. March 2007, Brussels. www.HFPeurope.org.
Hickson, A., Phillips A. and Morales G. (2007). Public perception related to a hydrogen hybrid internal combustion engine transit bus demonstration and hydrogen fuel. *Energy Policy*, **35** (4), 2249–2255.
HLG (2003). *High Level Group for Hydrogen and Fuel Cell Technologies: Hydrogen Energy and Fuel Cells – A Vision for Our Future*. Final Report of the High Level Group. EUR 20719 EN. Brussels: European Commission – Directorate-General for Research & Directorate-General for Energy and Transport.
Huleatt-James, N. (2008). *2008 Infrastructure Survey*. www.fuelcelltoday.com/media/pdf/surveys/2008-Infrastructure.pdf.
HyFLEET:CUTE (2007). www.global-hydrogen-bus-platform.com.
HyNet (2004). *Towards a European Hydrogen Energy Roadmap*. www.hyways.de/hynet.
HyWays (2007). *Hydrogen Energy in Europe*. Integrated Project under the 6th Framework Programme of the European Commission to develop the European Hydrogen Energy Roadmap (Contract No. 502596), 2004–2007. www.hyways.de.
IEA (International Energy Association) (2005). *Prospects for Hydrogen and Fuel Cells*. IEA Energy Technology Analysis Series. Paris: OECD/IEA.
Jerram, L. C. (2008). *2008 Bus Survey*. www.fuelcelltoday.com/media/pdf/surveys/2008-Bus.pdf.
Joseck, F. and Davis, P. (2006). Hydrogen safety, codes and standards: overview of US DOE program. *HarmonHy Final Conference*. Brussels. www.harmonhy.com/harmonhydocs/HarmonHyFin-005.pdf.
LBST (2009a). *H2Mobility: Hydrogen Vehicles Worldwide*. www.netinform/net/h2/h2mobility.

LBST (2009b). *Hydrogen Filling Stations Worldwide*. www.netinform/net/h2/h2stations.

Maak, M. and Skulason J. (2006). Implementing the hydrogen economy. *Journal of Cleaner Production*, **14** (1), 52–64.

Maruta, A. (2008). Fuel cell vehicles on the line? Challenges and opportunities for the next decades. *F-cell 7th Forum for Producers and Users*, September 24–25, 2008, Stuttgart. http://www.f-cell.de/de/programm-2007.php.

Ministry of Economy, Trade and Industry (METI) (2006). *Strategic Technology Roadmap (Energy Sector) – Energy Technology Vision 2100*. Tokyo. www.iae.or.jp/2100.html.

Organisation for Economic Co-operation and Development (OECD) (2002). *Policy Instruments for Achieving Environmentally Sustainable Transport*. Paris: OECD.

Roads2HyCom (2007a). *Snapshot of Hydrogen Uptake in the Future – A Comparison Study*. ECN (Ros, M. E., Weeda, M. and Jeeninga, H. (eds.)), Roads2HyCom. www.roads2hy.com.

Roads2HyCom (2007b). *European Hydrogen Infrastructure Atlas and Industrial Excess Hydrogen Analysis*. Steinberger-Wilckens, R. and Trümper, S. C. (eds.). Roads2HyCom.

Roads2HyCom (2007c). *Existing Acceptance Analysis in the Field of Hydrogen Technologies*. College of Europe (Fuhrmann, K. and Bleischwitz, R. (eds.)), Roads2HyCom.

Schulte, I., Hart, D. and van der Vorst, R. (2004). Issues affecting the acceptance of hydrogen fuel. *International Journal of Hydrogen Energy*, **29** (7), 677–685.

Spitzley, D. V., Brunetti, T. A. and Vigon, B. W. (2000). *Assessing Fuel Cell Power Sustainability*. SAE Technical Papers, No. 2000–01–1490. SAE International.

Steinberger-Wilckens, R. (2003). Not cost minimisation but added value maximization. *International Journal of Hydrogen Energy*, **28** (7), 763–770.

9

Fundamental properties of hydrogen

Maximilian Fichtner and Farikha Idrissova

With the expected increasing significance of hydrogen as a universal chemical and as an energy vector, its physical and thermodynamic properties are undergoing extensive investigation. To provide a basis of understanding for the themes covered in the remainder of the book, this chapter briefly describes the fundamental properties of hydrogen.

9.1 Discovery and occurrence

Named by a French chemist, Lavoisier, hydrogen (H) is the first chemical element of the periodic table of elements with an atomic number of one. At standard temperature and pressure, hydrogen is a colourless, tasteless, odourless and easily flammable gas. With its atomic mass of 1.00797 g/mol, hydrogen is the lightest element. The British scientist, Henry Cavendish, was the first to identify H as a distinct element in 1766, publishing precise values for its specific weight and density (NHA, 2007).

Hydrogen is also one of the most abundant chemical elements in the Universe (70–80 wt.% H_2 content); more than 50 wt.% of the Sun consists of hydrogen. However, on Earth it mostly occurrs naturally in the form of chemical compounds, most frequently water and hydrocarbons. As a gas in its free state, hydrogen is very rare (1 ppm by volume in the Earth's atmosphere), owing to its light weight, and it can only be found in natural gas and some volcanic gases, as well as trapped in small quantities in some minerals and rocks (Ullmann, 2003).

9.2 Atomic structure and isotopes

The hydrogen atom has the simplest atomic structure of all elements and consists of a nucleus and one electron. A neutral H atom can join a second electron, which forms the negative ion, H^-. Atomic hydrogen is formed as a result of different chemical reactions, but its lifetime is extremely short, as the atoms join each other to form a

The Hydrogen Economy: Opportunities and Challenges, ed. Michael Ball and Martin Wietschel. Published by Cambridge University Press.

Figure 9.1. Schematic representation of a H_2 molecule formed by two hydrogen atoms and sharing their electrons.

hydrogen molecule (H_2) (Fig. 9.1). The role of hydrogen and its compounds is very important in chemistry, as the so-called hydrogen bond influences the properties of many organic and inorganic compounds (Zittel and Wurster, 1996).

Other characteristics of hydrogen are its isotopes, which are occur naturally in the following forms: ^1H, *protium*; ^2H, *deuterium* (D) and ^3H, *tritium* (T). Protium consists of a single proton and is the most common of hydrogen isotopes (99.9% abundance). Deuterium is a stable isotope and contains one proton and one neutron in its nucleus. It is used for so-called 'heavy water', when water is enriched with molecules of D_2O instead of normal H_2O. Heavy water is used as a neutron moderator and cooling agent for nuclear reactors. Tritium has one proton and two neutrons in its nucleus and emits low energy beta radiation, disintegrating into helium-3 (^3He), which is a light non-radioactive isotope of helium. Its half-life is 12.26 years. Tritium occurs naturally in small amounts, owing to the interaction of cosmic rays with atmospheric gases (Ullmann, 2003).[1]

9.3 Chemical and physical properties

The physical and chemical properties of hydrogen impose technical barriers on standard methods of storing H_2 in pure form, such as a pressurised gas or cryoliquid (see also Chapter 11). Moreover, other H_2 properties, such as minimal ignition

[1] The fusion of hydrogen atoms is also the energy source of the Sun. The fusion reaction in the Sun is a multistep process, in which hydrogen is burned into helium. The cycle starts with the thermal collision of two protons ($^1H + {}^1H$) to form a deuteron (^2H), with the simultaneous creation of a positron and a neutrino. The positron encounters a free electron and both particles annihilate, their mass energy appearing as two gamma-ray photons. The deuterium nucleus collides with another proton and forms a ^3He nucleus and a gamma ray. Two such ^3He nuclei may eventually fuse to form ^4He and two hydrogen nuclei. In the centre of the Sun, at a temperature of about 15 million °C, 600 million tonnes of hydrogen fuse every second, forming helium and radiating energy at the rate of 3.9×10^{26} W.

Table 9.1. *Physical properties of hydrogen, methane, and* n-*heptane at the triple point, the boiling point and the critical point, and under standard conditions*

		Hydrogen H_2	Methane CH_4	*n*-Heptane C_7H_{16}
Molar weight	g/mol	**2.016**	16.043	100.204
Heating value	kJ/g	**120.0**	50.0	44.7
Triple point				
Temperature	K	**13.8**	90.7	182.6
Pressure	mbar	**70.4**	117.2	0.0
Density of liquid	g/l	**77.0**	451.2	771.6
Density of gas	g/l	**0.125**	0.251	–
Boiling point				
Temperature	K	**20.3**	111.6	371.6
Density of liquid	g/l	**70.8**	422.5	614.6
Density of gas	g/l	**1.338**	1.82	3.47
Viscosity of liquid	μPa s	**11.90**	19.30	–
Heat of vaporisation	J/g	**445.5**	510.4	317.7
Heat of vaporisation	kJ/l	**31.5**	215.7	195.3
Heating value of liquid	MJ/l	**8.5**	21.1	27.5
Heating value of gas	kJ/l	**160.5**	90.9	155.1
Critical point				
Temperature	K	**33.0**	190.6	540.2
Pressure	bar	**12.9**	46.0	27.4
Density	g/l	**31.4**	162.2	234.1
Standard properties (273 K, 1 bar)				
Density of liquid	g/l	**–**	–	702.3
Density of gas	g/l	**0.090**	0.718	4.48
Vapour pressure	Mbar	**–**	–	15.3
Viscosity of gas	μPa s	**8.9**	10.9	–
Specific heat capacity	KJ/(kg K)	**Cp = 14.199** **Cv = 10.074**	Cp = 2.22	Cp = 2.24
Explosion limits in air	vol.%	**4.0–77.0**	4.4–17.0	1.1–6.7
Detonation limits in air	vol.%	**18.3–59.0**	6.3–17.0	–
Minimum ignition energy	mJ	**0.017**	0.29	0.24
Spontaneous combustion temperature	K	**833**	868	488

Sources: (DWV, 2007; LBST, 2007).

energy and explosion limits, should also be considered for storage safety requirements. Table 9.1 lists the physical properties of H_2, in comparison with methane and *n*-heptane, which were chosen as representatives of natural gas and gasoline, respectively (for more details, see Ullmann (2003)).

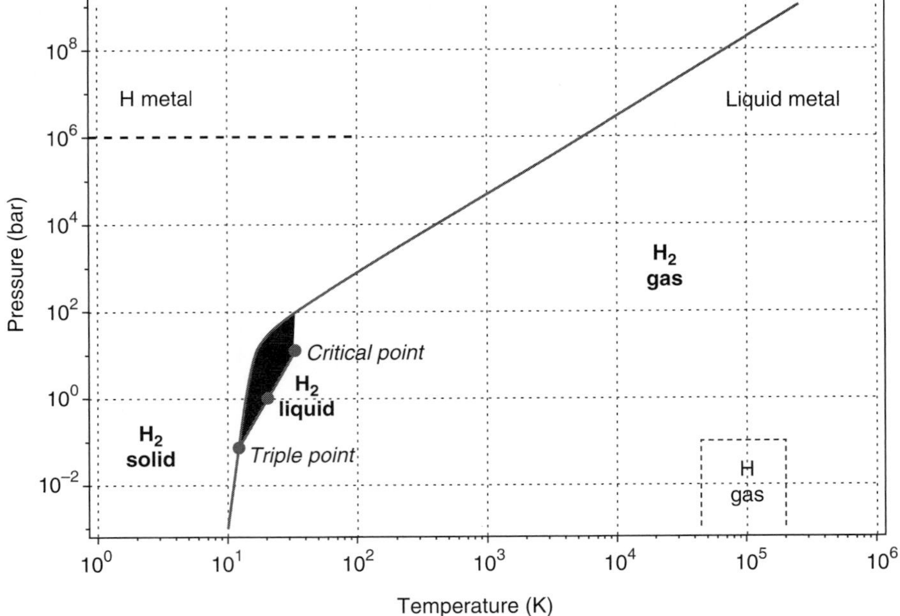

Figure 9.2. Primitive phase diagram of hydrogen (Leung *et al.*, 1976) reproduced with permission from (Züttel, 2003).

At standard conditions, hydrogen has a density of 0.09 g/l, in the liquid state, however, its density is 70.8 g/l and its boiling point is 20.3 K (−252.77 °C).

The phase diagram in Fig. 9.2 indicates that liquid hydrogen exists only in a small region between the solid line and the line from the triple point at 21.2 K and the critical point at 32 K. This implies that once hydrogen is evaporated from liquid and stays at higher temperature, it is not possible to reliquefy it by applying elevated pressure, a method which works with many other gases.

As seen from Table 9.1, hydrogen burns in air at concentrations in the range of 4%–77% by volume. The highest burning temperature of hydrogen, of 2318 °C is reached at 29% concentration by volume. While burning in an oxygen atmosphere, hydrogen can reach up to 3000 °C. As little as 0.02 mJ is the minimum ignition energy required for a stoichiometric fuel:oxygen mixture, whereas for methane and *n*-heptane, this value is 0.29 and 0.24 mJ respectively. For instance, the energy of a static electric discharge from the arcing of a spark is enough to ignite natural gas, so it is quite significant that hydrogen requires only a tenth of this energy to be ignited. The explosive range of hydrogen is much greater than that of methane; the latter is able to explode at a much lower concentration (Zittel and Wurster, 1996).

9.4 Adiabatic expansion

When a real gas like hydrogen expands freely, its temperature may either decrease or increase, depending on the initial temperature and pressure. The change of temperature in relation to a change of pressure is described by the Joule–Thomson coefficient. For hydrogen, the Joule–Thomson coefficient is negative at room temperature, unlike for most other gases, which means that the gas temperature rises when hydrogen is released from a pressurised vessel. This has to be taken into account in safety considerations, because the effect may lead to self-ignition of the released hydrogen under certain conditions.

9.5 Energy content

Current technologies that use hydrogen as an energy carrier are based exclusively on the well known strongly exothermic reaction of hydrogen with oxygen, thus forming low energetic water. On the basis of the respective combustion reaction, hydrogen has nearly three times the energy content of gasoline (120 MJ/kg (33.3kWh/kg) for hydrogen versus around 44 MJ/kg for gasoline). In other words, 1 kg of H_2 has the same energy content as 2.4 kg of methane or 2.8 kg of gasoline. Thus, with such properties, hydrogen has the highest energy to weight ratio of all fuels. On a volume basis, however, the situation is reversed and hydrogen has only about a quarter of the energy content of gasoline (8.5 MJ/l for liquid hydrogen versus 32.6 MJ/l for gasoline, Fig. 9.3). The latter is the main reason why hydrogen-storage applications need to be developed further to reach high storage densities for the lightweight gas, to avoid voluminous storage tanks for the fuel transportation purposes (see also Chapter 11).

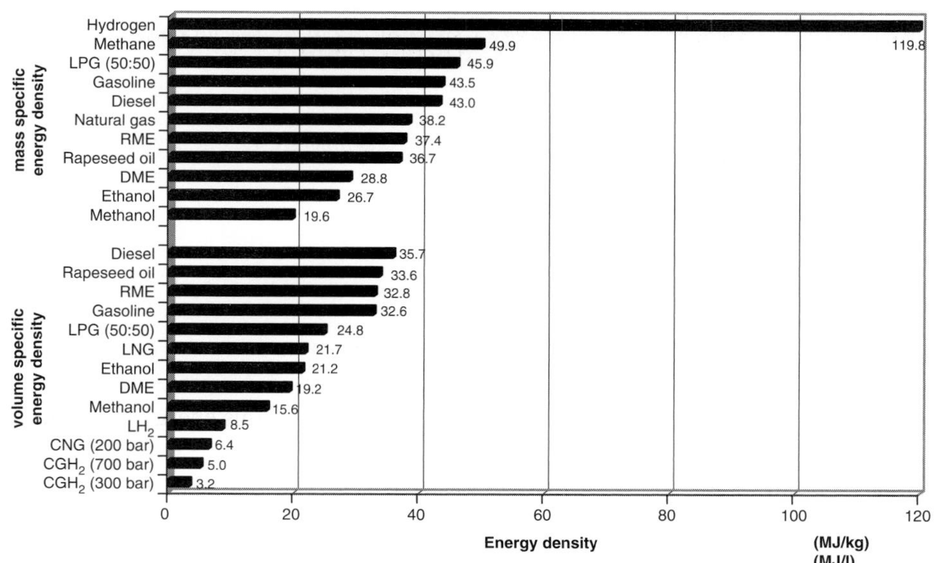

Figure 9.3. Energy densities of various alternative fuels (LHV) (LPG: 50% propane, 50% butane; natural gas: 83% methane).

Figure 9.3 illustrates the volume- and mass-specific energy densities of hydrogen compared with other fuels. It can be seen that hydrogen has the highest gravimetric energy density and the lowest volumetric energy density of all the fuels.

9.6 Summary

Hydrogen is expected to gain increasing significance as a universal chemical and as an energy vector. Knowledge of its particular physical and thermodynamic properties is an important prerequisite for application development. An understanding of these properties is particularly essential for the development of hydrogen-storage systems as well as for safety considerations.

References

DWV (Deutscher Wasserstoff- und Brennstoffzellen-Verband e.V.; German Hydrogen and Fuel Cell Association) (2007). *Vergleichstabelle für physikalische und chemische Eigenschaften von Wasserstoff und anderen Stoffen (Gasen, Energieträgern).* www.dwv-info.de/wissen/tabellen/wiss_vgl.html.

LBST (2007). *Hydrogen Data.* Ottobrunn: HyWeb, Ludwig-Bölkow-Systemtechnik GmbH. www.h2data.de.

Leung, W. B., March, N. H. and Motz, H. (1976). Primitive phase diagram for hydrogen. *Physics Letters*, **56**A (6), 425–426.

NHA (2007). *The History of Hydrogen.* Fact Sheet Series. Washington, DC: US National Hydrogen Association. www.hydrogenassociation.org/general/factSheets.asp.

Ullmann (2003). Hydrogen. In *Ullmann's Encyclopedia of Industrial Chemistry.* 6th edn. vol. **17**. Weinheim: WILEY-VCH, pp. 85–240.

Zittel, W. and Wurster, R. (1996). *Hydrogen in the Energy Sector.* Issue 8.7.1996. www.hyweb.de/index-e.html.

Züttel, A. (2003). Materials for hydrogen storage. *Materials Today*, **6** (9), 24–33.

10

Hydrogen production

Michael Ball, Werner Weindorf and Ulrich Bünger

This chapter provides an overview of the various hydrogen production methods. In this respect, the chapter aims especially at outlining the technical fundamentals of the most important commercial processes of hydrogen production and quantifying their technical and economic parameters, which are used in the context of modelling the build-up of a hydrogen infrastructure in Chapter 14. Novel hydrogen-production technologies that still require basic research are also briefly addressed. The chapter finishes with an assessment of the availability of industrial surplus hydrogen as a potential hydrogen source for the transition phase towards its widespread use as vehicle fuel.

10.1 Overview of production processes

Since hydrogen only occurs naturally in a bonded form, it first has to be released from its various compounds by using energy. Hydrogen can be produced from all primary energy sources. Figure 10.1 shows an overview of the various relevant hydrogen-production processes and the respective primary energy sources used, differentiated into renewable and non-renewable sources.

Hydrogen can be produced directly from primary as well as from secondary energy sources. Today's commercially applied methods based on fossil raw materials include natural gas reforming and the partial oxidation of feeds with lower quality, such as petroleum coke or other refinery residues. The gasification of coal to produce hydrogen has undergone further development in the last decade and is now also a commercially available process. Apart from this, there are other methods still at the research and development stage, particularly those based on biomass, but also biological hydrogen production. However, commercialisation in the near future is only expected for biomass gasification. Hydrogen is produced from secondary energy sources almost exclusively by water electrolysis using electricity. Nuclear energy could produce hydrogen in one of three ways: (1) through electrolysis with electricity

The Hydrogen Economy: Opportunities and Challenges, ed. Michael Ball and Martin Wietschel. Published by Cambridge University Press.

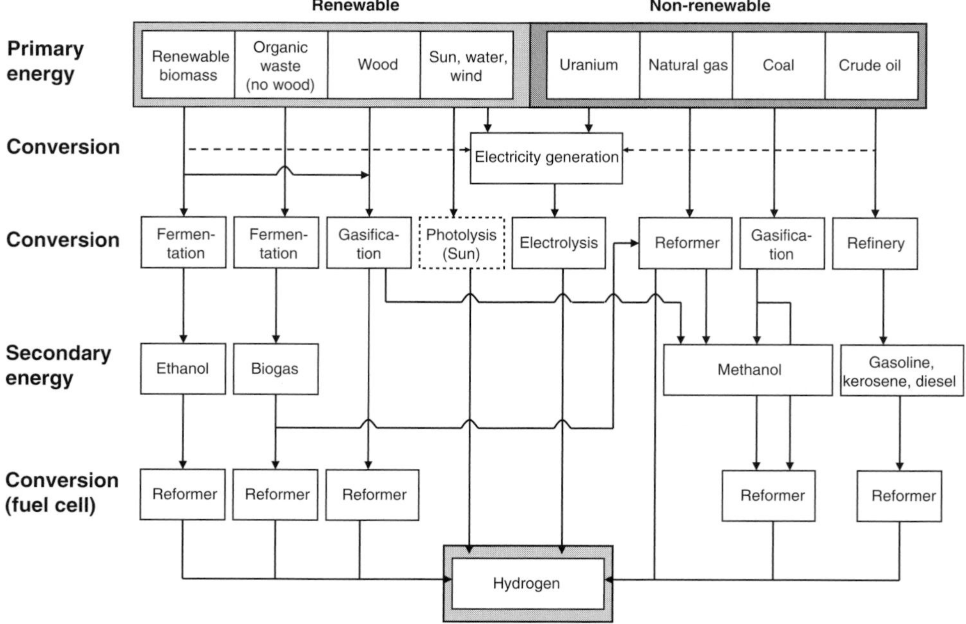

Figure 10.1. Hydrogen production options by energy source (Nitsch, 2002; Pehnt, 2001).

generated by dedicated nuclear power plants; (2) through process heat provided by advanced high-temperature reactors for the steam reforming of methane; or (3) through a thermochemical cycle, such as the sulphur–iodine process. As expressed in the vision of a 'solar' hydrogen economy, in the long term, electricity produced from renewables is seen as the main energy source for producing hydrogen, both to offer a real alternative to dwindling oil and gas reserves and to reduce CO_2 emissions. In addition, hydrogen can also be produced from various fossil-based secondary energy sources or biogas. Here, methanol reforming especially could play a role in fuel cells for mobile applications.

Available statistics or market surveys about hydrogen capture the real production volumes only partially, as they usually consider only captive production, i.e., the *directly* produced hydrogen (e.g., in steam reformers), as for instance in refineries or fertiliser plants.[1] Besides this, hydrogen is produced in significant amounts as a *by-product* from the manufacture of various chemical products, such as chlorine or ethylene, as well as from refinery processes (see also Section 10.9). Where this hydrogen cannot be internally utilised further, for instance for hydrogenation in

[1] Within the series of the *Chemical Economics Handbook* published by SRI Consulting, nearly all known *direct* hydrogen producers worldwide are cited (see www.sriconsulting.com). Another possibility to estimate the produced hydrogen volumes is from the respective hydrogen demand of the final products (e.g., from ammonia, methanol or refinery products) (see LBST (1998)).

refineries, it is mostly energetically used as fuel for power and heat generation or vented or flared. Such data are, however, often not accessible or not published, and consequently the real production volumes can only be estimated, both at national level, and even more at global level.[2]

Global hydrogen production has been growing for several years and is indicated today with 600 to 720 billion Nm^3 (54–65 Mt) per year (IEA, 2007; Linde, 2003), enough to fuel between 600 and 720 million fuel cell vehicles (80%–95% of the world's vehicle fleet); however, this is less than 2% of the global primary energy use. Total annual hydrogen production in the European Union is estimated at about 80 billion Nm^3, led by Germany with 22 billion Nm^3 and the Netherlands with 10 billion Nm^3; the production in the USA amounts to about 84 billion Nm^3 (Roads2HyCom, 2007). About 96% of the total hydrogen is produced from fossil fuels; split into primary energy sources, about 48% comes from natural gas (steam reforming), 30% from crude oil fractions in refineries (partial oxidation of heavy refinery residues) and recovered from refinery or chemical industry off-gases, and 18% from coal (gasification); the remaining 4% is accounted for by electrolysis, mainly in countries where cheap electricity from hydropower is available, such as in Canada. In Germany, for instance, almost half of the hydrogen is generated as a by-product in refineries, where it is largely used for hydrogenation (BMWA, 2005).[3]

Hydrogen for industrial facilities is mainly produced where it is also immediately used (so-called 'captive hydrogen'). Only around 5% of total hydrogen production is sold on the free market and transported in liquid or gaseous form in trailers or pipelines (so-called 'merchant hydrogen'). Hydrogen pipelines have already been operated by the chemical industry in the United States and in Europe (particularly Germany, France and the Netherlands) for decades (see also Chapter 12).

In the following, the most important (commercial) hydrogen production processes available today are briefly described and analysed from the perspective of technology and economics, including their parameterisation for the MOREHyS infrastructure model described in Chapter 14. They include the reforming of natural gas, the gasification of coal and biomass and electrolysis. Other methods of hydrogen production, some of which are still in the research and development stage, are briefly outlined in Section 10.6. (For more detailed coverage of the technical aspects of hydrogen production technologies, see the section on Further reading at the end of this chapter.)

Cost assessments of technologies and, even more, projections into the future inherently bear high uncertainties, with assumptions on feedstock prices having a major impact on economics (also closely related to efficiencies via the feedstock volumes). The economics presented here should, therefore, not be interpreted as

[2] In 1997, LBST carried out a statistical survey among hydrogen producers and chemical companies in the EU15 to capture the actual volumes of hydrogen production better (LBST, 1998). Although deepened insights were gained, the problem of the incomplete data basis could not be fully resolved (see also Section 10.9).

[3] The most important hydrogen source in a refinery is the catalytic reforming of naphtha to produce a light gasoline with a higher octane number. Hydrogen is also generated in smaller amounts by the different cracking processes.

precise predictions, but rather indicative values for a specific set of assumptions. As prices for natural gas and electricity are subject to the effective quantities specified in the supply contracts and generally decrease with increasing quantities, it is assumed that natural gas and electricity prices for onsite technologies (onsite steam reformers and electrolysers) are 30% higher than the respective prices for large, central production plants. The energy price assumptions used to calculate specific production costs are detailed in Chapter 14.

Economies of scale of production are taken into account according to the following formula, with I_1 and I_2 and C_1 and C_2, respectively, denoting the (absolute) investments and capacities of the reference plants; investments typically scale according to a so-called extrapolation coefficient, which from experience is between 0.5 and 0.7 (Chauvel *et al.*, 1976). Further, to project investments into the future, learning curve effects are considered.[4]

$$\frac{I_1}{I_2} = \left(\frac{C_1}{C_2}\right)^{0.5-0.7}. \tag{10.1}$$

10.2 Steam reforming of natural gas

Natural gas can be converted to hydrogen by involving the reaction with either steam (steam reforming), oxygen (partial oxidation), or both in sequence (autothermal reforming). Today, steam reforming is globally the most commonly used method to produce hydrogen. Steam reforming is generally understood to be the endothermic catalytic conversion of light hydrocarbons (methane or naphtha, but also biogas or methanol) using steam. This is usually done by applying heat over nickel catalysts. Because of the catalysts used, it is vital to use clean, residue-free, vaporisable feed material. The main catalyst poison is sulphur; concentrations as low as 0.1 ppm form a deactivating layer on the catalyst. The focus in the following is on steam methane reforming (SMR).

The equilibrium of the reactions imposes several conditions to achieve maximum hydrogen yields: a high temperature at the end of the reforming reactor, a high excess of steam (a molar steam-to-carbon ratio of about 2.5–3), and moderate pressure. Since the reaction is endothermic, the thermodynamic formation of hydrogen and CO is aided by increasing temperature and hindered by increasing pressure, in accordance with Le Chatelier's principle. Steam reforming is generally conducted on an industrial scale at temperatures of 800–900 °C and pressures of approximately

[4] Learning curves (also called experience curves) describe technological progress as a function of accumulating experience with that technology. They provide a simple, quantitative relationship between the cost and the cumulative production or use of a technology, by quantifying how costs decline with cumulative production (which is used as an approximation for the accumulated experience). The learning (cost reduction) is expressed for a doubling of the cumulative volume; the corresponding change in cost is referred to as *progress ratio*: e.g., a progress ratio of 0.8 means that the cost is reduced to 0.8 of its previous level after a doubling of cumulative sales.

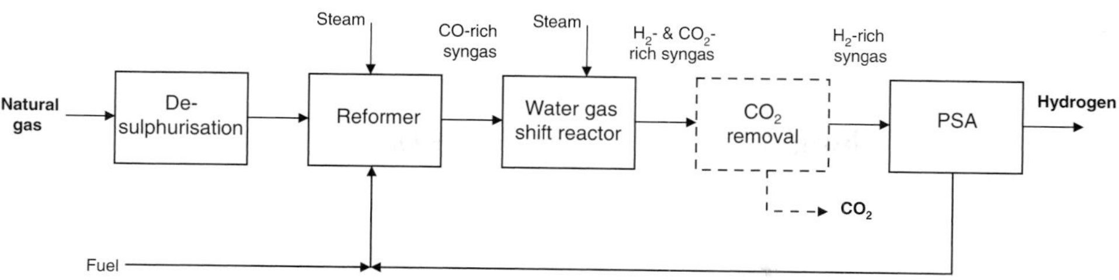

Figure 10.2. Block diagram of natural gas reforming.

20 to 40 bar (if the pressure is too low, additional energy is needed to compress outlet gases) and takes place according to the following basic reaction:

$$CH_4 + H_2O \rightarrow CO + 3H_2 \quad \Delta H = 206\,\text{kJ/mol},$$

or, more generally,

$$C_xH_y + H_2O \rightarrow xCO + (x + y/2)H_2.$$

After the reforming reaction, the gas is quickly cooled down to about 350–450 °C before it enters the (high-temperature) water-gas shift reaction (CO shift). Here, the exothermic catalytic conversion takes place of the carbon monoxide formed with steam to hydrogen (H_2) and carbon dioxide (CO_2) in the following reaction:

$$CO + H_2O \rightarrow CO_2 + H_2 \quad \Delta H = -41\,\text{kJ/mol}.$$

Figure 10.2 shows a schematic diagram of hydrogen production based on the steam reforming of natural gas. First, the natural gas fed into the system is desulphurised. Then the gas is mixed with steam and preheated before it is channelled into the reformer. This consists of reactor pipes containing a nickel catalyst. In this way, the natural gas–steam mix is transformed into a synthesis gas of hydrogen, carbon monoxide, carbon dioxide and water. The heat required for this process is produced by combusting some of the inlet gas (up to 25%) and the tail gas of the pressure-swing adsorption (PSA) plant. Considerable excess steam is used to avoid carbon deposition on the catalyst. The synthesis gas exits the reformer and enters a shift converter. The carbon monoxide contained in the gas is then converted into hydrogen and carbon dioxide using steam in a ferric-oxide catalytic converter. After the shift conversion, the gas is cooled down to ambient temperature and subsequently purified or treated. The gas produced then undergoes pressure-swing adsorption to remove the carbon dioxide and any remaining residues, such as CO; this produces hydrogen with a purity of at least 99.9 vol.%. In larger production facilities, the hydrogen is usually available at an outlet pressure of approximately 30 bar. Depending on the plant design, part of the excess steam could be exported to generate electricity; the water demand would then increase accordingly.

The gas purity demanded determines the extent of the gas treatment necessary. In the case of deliberate CO_2 separation with subsequent storage, the CO_2 can be removed after the shift reaction by washing (e.g., with Rectisol), achieving capture rates of about 90%; otherwise it is emitted after the PSA.[5] The (additional) costs of CO_2 capture in connection with hydrogen production from natural gas are mainly the costs of CO_2 drying and compression, as the hydrogen production process necessitates separation of CO_2 and hydrogen anyway (even if the CO_2 is not captured). Total investments increase by about 20%–35% for large steam methane reformers (4%–6% of total costs; IEA, 2005). On average, the costs for CO_2 transport and storage are between €5 and €8 per tonne CO_2 (IPCC, 2005). Assuming 280 g CO_2/kWh_{H_2} for hydrogen production from natural gas, CO_2 transport and storage costs translate to at least 0.14–0.22 ct/kWh_{H_2} (see Chapter 6). This increases total hydrogen production costs by 3%–5%. In the case of CCS, these costs have to be added to the production costs (which is done in the MOREHyS model, see Chapter 14).

Table 10.1 shows technoeconomic data of natural gas reforming for various plant sizes, used in the MOREHyS model in Chapter 14. The systems were scaled up according to an extrapolation factor of 0.6. Large steam reformers achieve thermal efficiencies of 71%–76%, with CO_2 compression in the case of CCS slightly lowering the efficiency. The process's operating efficiency is slightly reduced when taking other energy demands into account, such as the electrical energy for pumps or fans. As can be seen, the specific costs of supplying hydrogen are clearly dominated by the variable fuel costs. The economic efficiency of natural gas reforming is thus significantly dependent on the price of natural gas (it should be pointed out that no distinction is made in the consumer prices between production in centralised and in onsite systems). The regular changeover of catalysts or the water demand, which can amount to 0.6 to 2.1 l/Nm^3 hydrogen, depending on the system design, were not included in the costs, because of their low overall share. The specific investments of small reformer units, which are still comparatively high today for onsite production, were reduced for 2020, taking into account learning-curve effects based on a progress ratio of 0.96 and related assumptions about the production figures expected in the future.

The centralised reforming of natural gas, especially steam reforming, is a fully developed commercial technology. Refineries and chemical industries (ammonia, methanol) have a lot of experience in this technology. The large volumes of hydrogen needed by these industries are generally produced onsite. Other consumers can be supplied with hydrogen in gas form by a network of pipelines or with hydrogen in liquid form by road transport.

Decentralised hydrogen production from natural gas for onsite applications (fuel cells, refuelling stations for hydrogen vehicles) eliminates or reduces the problems of distribution and storage. Nevertheless, current technology has high costs because it lacks economy of scale. Lower pressure and temperature and lower-cost materials are

[5] To reach a better CO conversion, it is possible to add a low-temperature shift reactor, which increases the CO_2 capture rate (see also Fig. 10.3). If both clean CO_2 for storage and clean hydrogen for fuel cell applications are required, a combination of a CO_2-capture plant (e.g., absorption with Rectisol) and a PSA plant is necessary. If only pure hydrogen is required, a PSA unit would be sufficient (and is standard practice), but the CO_2 stream would be contaminated by impurities, such as H_2, N_2 or CO, which have to be removed for geological storage.

Table 10.1. *Techno-economic data of steam reforming*

		Central				Onsite	
Technical data							
Capacity	MW_{H_2}	10	50	100	300	2.4	2.4
Capacity	Nm^3/h	3300	16 700	33 300	100 000	800	800
Capacity	t_{H_2}/day	7.2	36	72	216	1.7	1.7
Annual full load hours	h/year	8000	8000	8000	8000	6500	6500
Outlet pressure	bar	30	30	30	30	15	15
Natural gas demand	kWh_{gas}/kWh_{H_2}	1.44	1.42	1.40	1.39	1.5	1.5
CO_2 emissions	g/kWh_{H_2}	285	281	277	275	297	297
Thermal efficiency	%	69.4	70.4	71.4	72.0	66.7	66.7
Lifetime	years	30	30	30	30	20	20
Economic data							
Specific investment	$€/kW_{H_2}$	1000	540	400	260	1250	750
Fixed costs [a]	% Investment/year	3	3	3	3	2	2
Year of availability		Today	Today	Today	Today	2010	2020

Note:
[a] Maintenance, labour etc.
Sources: (Valentin, 2001; Wagner *et al.*, 2000; Zittel *et al.*, 1996).

needed to make decentralised reforming competitive. Carbon capture and storage is not economically viable for onsite steam reforming units, due to the specifically high CO_2 separation and transport costs.

In industry, large steam reformers generally produce between 20 000 and 100 000 Nm^3/h of hydrogen. These reformers can be scaled down to 1000 Nm^3/h. Their disadvantages are their large size and a high cost for materials, imposed by the conditions of pressure and temperature. Compact steam reformers have been developed for use with fuel cells. These reformers operate at a lower pressure and temperature (3 bar, 700 °C); the requirements for materials are thus less. For these units, energy conversion efficiency can reach 70%–80%.

10.3 Coal gasification

A combination of pyrolysis and gasification is applied to produce hydrogen from solid fuels. In the past, a variety of methods has been used to gasify solid fuels, to

Table 10.2. *Characteristics of gasifier types*

	Entrained flow	Fluidised bed	Moving bed
Pressure (bar)	20–85	20–30	20–25
Temperature (°C)	1400–1600	800–1000	370–600
Moderator	Steam–water	Steam	Steam
Amount of moderator	Low	Medium	High
Oxidant	Oxygen	Oxygen–air	Oxygen, air
Amount of oxidant	High	Medium	Low
Fuel mesh size (mm)	0.05–0.1	3–4	5–50
Fuel feeding type	Dry, wet	Dry	Dry
Syngas LHV	High	Medium–high	High
Syngas type	Mostly H_2 and CO	Low CH_4%	High CH_4%
Slag	Molten	Dry-caking	Dry, molten
Carbon conversion	>95%	80–95%	80–90%
Process	GE/Texaco, Shell, Prenflo	HTW, KRW	Lurgi, BGL

produce gases industrially. The large number of possible reactor configurations derives from the large number of influencing factors, which give wide scope to the process design. The gasification technologies can be differentiated using the following criteria:

- External or internal heat generation (allothermic or autothermic gasifier),
- Contact between oxidation agent and fuel (fixed bed, fluidised bed and entrained flow),
- Direction of the material flows of the fuel and the gasification agent (co-current and counter-current),
- Gasification agent (air, oxygen, steam or a mix).

Different gasification technologies are present on the market. The most common way of classifying them is by flow regime, i.e., the way in which the fuel and oxidant flow through. Three main groups can be distinguished: entrained flow (co-current), fluidised bed (counter-current) and moving bed. The feed can be supplied to the gasifier either dry or as a coal–water mixture slurry. Furthermore, gasification can take place either with air or with oxygen. The most important characteristics of the different types of gasifiers are presented in Table 10.2.[6]

The gasification of solid fuel occurs at temperatures between 300 and, at maximum, 2000 °C and can be divided into the substages of drying, pyrolysis, gasification and combustion (see Treviño (2002)). During the actual gasification, the fuel is pyrolysed into a combustible gas by adding a gasification agent; air, oxygen, hydrogen, steam, or a mix of these may be used. After the gasifier, the raw syngas is cooled and ash particles are removed, as well as sulphur components, before the gas is supplied to the shift section. The resulting synthesis gas is treated to produce hydrogen in analogy with natural gas reforming.

[6] The process-related details of the various types of gasifier are not discussed here in depth. For more information, refer to the relevant literature (see, e.g., Treviño (2002); Chiesa *et al.* (2005); IE/IPTS (2005); BMELV (2005)).

The ideal process for producing hydrogen using coal gasification is the *entrained-flow gasifier*, for process-related reasons (high degree of carbon conversion, >95%; production of a synthesis gas rich in hydrogen and carbon monoxide on account of the gasification using oxygen). The entrained-flow gasifier combines a high efficiency with low methane content in the product gas. For hydrogen production, oxygen-blown is preferred over air-blown, because it avoids having to separate N_2 and H_2. In entrained-flow gasifiers, pulverised coal flows co-currently with the oxidiser (typically O_2 and steam). The key characteristics are their very high temperatures (usually more than 1000 °C, and up to 1400 °C) and the short residence time of the fuel in the gasifier. Solids fed into the gasifier must be ground very fine and be of homogeneous quality, making the gasifier technology not suitable for biomass or waste without pre-treatment (e.g., pyrolysis). The ash is removed as a molten slag. An important characteristic of the entrained flow process, as compared with other types of gasifier, is the versatility with regard to the type of coal that can be used. Entrained-flow gasifiers can process all coal ranks, depending on their ash and moisture contents (10% or less is preferred). They are designed to process coals with ash fusion temperatures lower than 1400 °C. The most important commercial processes here are the *GE/Texaco* and *Shell* gasifiers.

Besides conventional coal gasification, the possible co-production of electricity and hydrogen is seen as having great potential in the medium to long term in power stations with integrated coal gasification (IGCC; integrated gasification combined cycle) (see IE/IPTS (2005); see also Section 16.2). Figure 10.3 shows a block diagram of an IGCC plant for producing hydrogen and electricity. The process corresponds to conventional coal gasification, with the exception that part of the hydrogen produced in an IGCC plant can also be used to generate electricity in a gas turbine. In contrast, in IGCC power stations designed purely for power generation, there is no shift reaction and the synthesis gas produced is converted directly into electricity in a gas turbine after purification.

Like natural gas reforming, if the intention is to capture and store the CO_2, this has to be removed from the shifted synthesis gas by washing before the hydrogen enters the PSA or the gas turbine. Depending on the size of the gas turbine, the synthesis gas flow can either be used to produce more hydrogen within a certain system-specific load range or to generate more electricity. Because here the focus is to analyse the production of hydrogen, an IGCC plant is considered that is designed to produce the maximum amount of hydrogen. By analogy, however, the plant could also be designed to produce the maximum amount of electricity; unlike systems with priority hydrogen production, the gas turbine would then have to be correspondingly larger. With regard to the gas turbine's degree of electrical efficiency, for modelling reasons it was assumed that this is constant within the possible load range.

Table 10.3 shows the assumed capacity ranges and efficiencies of an IGCC plant designed to produce hydrogen with the possibility of co-producing electricity (using hard coal). According to Chiesa *et al.* (2005), the net electrical output (efficiency) of

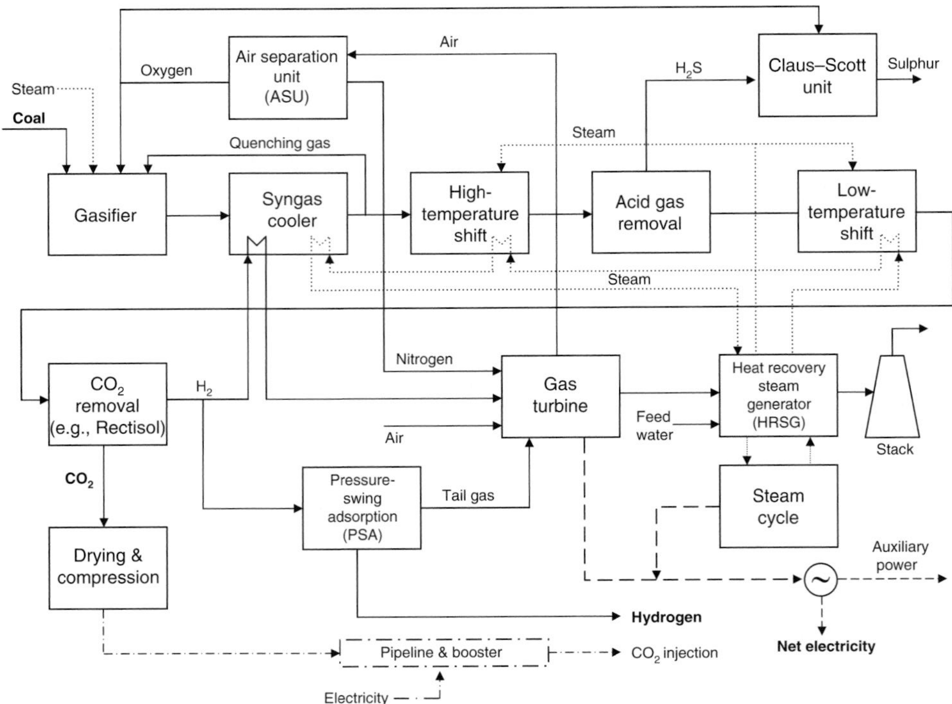

Figure 10.3. Block diagram of an IGCC plant with CCS to produce hydrogen (IE/IPTS, 2005).

the plant can fluctuate between 4% and 15% based on the coal input if the plant is designed for hydrogen production, but the maximum output is also determined by the size of the gas turbine. Even if the plant is operated exclusively to produce hydrogen, the residual gas from the PSA (separation rate 85%) is used to generate electricity. For the combined cycle, an efficiency of 52% has been assumed (IE/IPTS, 2005).

Table 10.4 shows technical data for both conventional coal gasification (without CO_2 capture) and for the IGCC plant (with CO_2 capture), each based on the use of hard coal. Typical thermal-process efficiencies for coal gasification are in the range of 51%–63%. As in the case of steam reformers, the (additional) costs of CO_2 capture for hydrogen production from coal are only the costs for CO_2 drying and compression to about 100 bar for subsequent pipeline transport. Therefore, the specific investment for coal gasification plants designed for CO_2 capture are only 5% to 10% higher (9%–12% of total costs) (IEA, 2005; Kreutz *et al.*, 2005). Including the same generic costs for CO_2 transport and storage as in the case of SMRs, 0.28–0.46 ct/kWh_{H_2} have to be added (see also Chapter 6). This increases total hydrogen production costs from coal by 10%–15%.

In IGCC plants used only to generate power, in which the synthesis gas is directly converted into electricity in the gas turbine and thus neither a shift reaction nor CO_2 separation are necessary, the specific investment for CCS is 30% to 40% higher

Table 10.3. *IGCC co-production of power and hydrogen*

IGCC (H_2)		H_2 production, min.	H_2 production, average	H_2 production, max. (reference)
η_{H_2} [a]	%	40	50	57
Coal input hydrogen	kWh_{coal}/kWh_{H_2}	2.5	**2.0**	1.74
Max. hydrogen output	MW_{H_2}	**211**	**263**	**300**
$\eta_{net,el}$ [a]	%	**15**	10	**4**
Coal input electricity	kWh_{coal}/kWh_{el}	6.7	**10.0**	25.0
Max. power output	$MW_{net,el}$	**79**	**53**	**22**
Power output ratio	$MW_{net,el}/MW_{H_2}$	0.37	0.2	0.07
Specific investment [b]	€/kW_{H_2}	**1140**	**912**	**800**

Notes:
[a] Based on the lower heating value of hard coal input to supply 300 MW_{H_2}.
[b] Based on the reference plant with 300 MW_{H_2}; only covers CO_2 separation and compression.
Source: calculations based on Chiesa *et al.* (2005).

Table 10.4. *Coal gasification (conventional and IGCC)*

		Conventional coal gasification (without CO_2 capture)	IGCC (H_2) (with CO_2 capture)
Technical data			
Capacity	MW_{H_2}	300	300
Max. hydrogen capacity	MW_{H_2}	300	263
Max. electricity capacity	MW_{el}	0	53
Annual full load hours	h/year	7000	7000
Outlet pressure	bar	50	50
Coal input hydrogen	kWh_{coal}/kWh_{H_2}	1.74	2.0
Coal input electricity (net)	kWh_{coal}/kWh_{el}	–	10.0
CO_2 emissions	g/kWh	570	60
Lifetime	year	30	30
Economic data			
Specific investment	€/kW_{H_2}	800	912
Fixed costs	% Invest/year	3.0	3.0
Year of availability	–	Today	2020

Sources: (Chiesa *et al.*, 2005; IE/IPTS, 2005; Kreutz *et al.*, 2005; Ogden, 1999; Yamashita and Barreto, 2003).

(BMWA, 2003; IE/IPTS, 2005); while the electrical efficiency in these power stations is decreased by 6% to 8%, owing to CCS. Carbon capture and storage in hydrogen production results in about 2% less electrical efficiency (mainly caused by the compression of the CO_2; Chiesa *et al.*, 2005).

Coal gasification is a commercially available technology. Large gasification plants can be found worldwide. The front runner is South Africa, using coal gasification for Fischer-Tropsch synthesis. Large gasifiers are also in operation in Russia and North America. At the time of writing, only four commercial IGCC power plants for power generation are operating worldwide: one in Spain (Puertollano), one in the Netherlands (Buggenum) and two in the USA (Tampa and Wabash).

10.4 Biomass gasification

Resources such as biomass could provide a clean and sustainable resource for hydrogen production. As with fossil fuels, the processes that produce hydrogen gas from biomass all create carbon dioxide, but because the biomass acts as a carbon sink during the growing phase, the net carbon emission of the whole cycle is neutral.

Despite the fact that numerous processes are being developed to use biomass for hydrogen production, at present there is no commercially available process to produce hydrogen from biomass. The following observation is limited to the gasification of solid biomass, since this process is closest to commercialisation owing to the extensive experience with coal gasification and also because the estimated potential of solid biomass far exceeds that of gaseous or liquid biomass. The biomass gasification procedure then corresponds largely to that of coal gasification.

A promising method for biomass gasification is the so-called 'staged reforming' method developed by the German company D.M.2 (see www.blauer-turm-im-c-port.-de) and the indirectly heated circulating-bed gasifier developed by the Battelle Columbus Laboratory (BCL) in the USA, whose technical and economic data are shown in Table 10.5.[7] The advantage of the gasifer from D.M.2 over other gasification concepts (e.g., by BCL or Shell) is that hydrogen can be generated in relatively small plants, which simplifies its introduction into the transportation fuel market; the plant could be installed near large filling stations. Another indirectly heated circulating-bed gasification process has been developed by AE Energietechnik together with the Technical University of Vienna and is operated in Güssing in Austria. Indirect gasification refers to systems that use external heating to drive the gasification reactions.

The D.M.2 gasifier is also based on an indirectly heated gasification process. The heat needed for the gasification stage is introduced into the process by a heat carrier (e.g., spheres of corundum). One possible process layout for hydrogen production based on the 'staged-reforming process' is shown in Fig. 10.4.

To produce pure hydrogen, the product gas, which mainly consists of hydrogen and CO, has to be converted to pure hydrogen via CO shift and pressure-swing adsorption (PSA). The tail gas from the PSA plant is used for the generation of electricity and heat (e.g., in a gas engine). To simplify the process, the CO shift stage can be left out, which would result in a smaller hydrogen yield, simultaneously

[7] A review of various biomass gasification processes can be found in BMELV, 2005; Boukis *et al.*, 2005; Ni *et al.*, 2006; Williams *et al.*, 2007; Zittel *et al.*, 1996.

Table 10.5. *Biomass gasification*

		Decentral (D.M.2)		Central (BCL)	
Technical data					
Output	MW_{H_2}	4.0	5.2	216	255
	MW_{el}	1.0	–	–	–
	MW_{th}	1.6	1.6	–	–
H_2 production capacity	Nm^3/h	1330	1750	72 100	84 900
Annual full load hours	h/year	6500	6500	8000	8000
Outlet pressure	bar	20	20	30	75
Biomass input	$kWh_{biomass}/kWh_{H_2}$	2.5	1.9	2.0	1.5
Electricity input	kWh_{el}/kWh_{H_2}	–	–	0.024	0.082
Thermal efficiency	%	0.40 (0.66)[a]	0.52	0.50	0.67
Lifetime	years	20	20	25	25
Economic data					
Special investment	€/kW_{H_2}	2900	1700	700	600
Fixed costs[b]	% Invest/year	8.1	8.9	6.7	7.0
Year of availability		2010	2010	2020	2020

Notes:
[a] Including electricity and useable heat (by-product).
[b] Maintenance, labour, etc.
Sources: (D.M.2, 2001; Katofsky, 1993; Spath *et al.*, 2005).

increasing electricity and heat output. In the early transition phase to a hydrogen-fuel infrastructure, this can be a reasonable approach at lower hydrogen demand. The capacity of the suggested plant is about 10 MW_{th} (biomass input), which can supply about 4 MW of hydrogen, 1 MW of electricity and 1.6 MW of useable heat (D.M.2, 2001).

Another indirectly heated gasification process has been developed by the Battelle Columbus Laboratory (BCL). The heat carrier applied for the BCL gasification process is sand. In contrast to the D.M.2 proprietary process, the BCL gasifier is a circulating-bed gasifier (see Fig. 10.5). The coke formed in the gasifier is also burnt in the circulating-bed combustion furnace. Simultaneously, the heat carrier is heated up, which, together with the coke, has been separated from the product gas stream beforehand. Like the 'staged-reforming process', the product gas leaving the scrubbing plant has to be converted to pure hydrogen through a CO-shift reactor and a pressure-swing adsorption plant.

Table 10.5 shows the major technoeconomic parameters of the aforementioned gasification processes.

The gasification process can use a variety of biomass resources, such as agricultural residues and wastes, or specifically grown energy crops. The technologies for gasifying

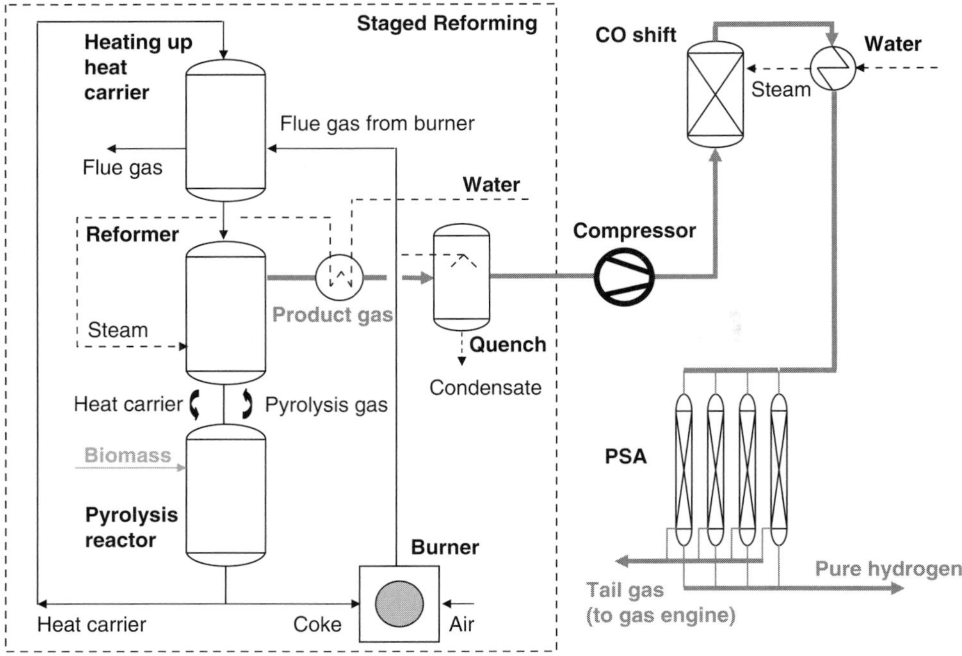

Figure 10.4. Layout of a plant for the production of high purity hydrogen based on the 'staged-reforming' process from D.M.2.

biomass in IGCC plants for power generation are subject to intensive R&D. A number of demonstration units are in operation, but no concept has so far reached the technical maturity required for hydrogen production. Improved process economics and biomass are needed if this option is to become attractive. The commercial production of hydrogen from biomass gasification is assumed to be available after 2010.

10.5 Electrolysis

In electrolysers, the reaction occurring in fuel cells is reversed; that is, water is decomposed using electricity. Electrolysis means splitting water to hydrogen (the desired product) and oxygen by supplying direct current electricity to the process. To produce a standard cubic metre of hydrogen, theoretically 3.54 kWh_{el} energy has to be expended, which corresponds to the upper heating value of hydrogen. The water decomposition taking place in an electrolyser involves two reactions that occur at the two electrodes: hydrogen is formed at the cathode, oxygen at the anode. The charge equalisation necessary to do this occurs in the form of ionic conduction through an electrolyte. There is a separator between the reactions at the electrodes, which ensures that the hydrogen and oxygen gases produced remain isolated. Electrolysers are always operated with direct current. Upscaling electrolysers is very straightforward because of their modular construction.

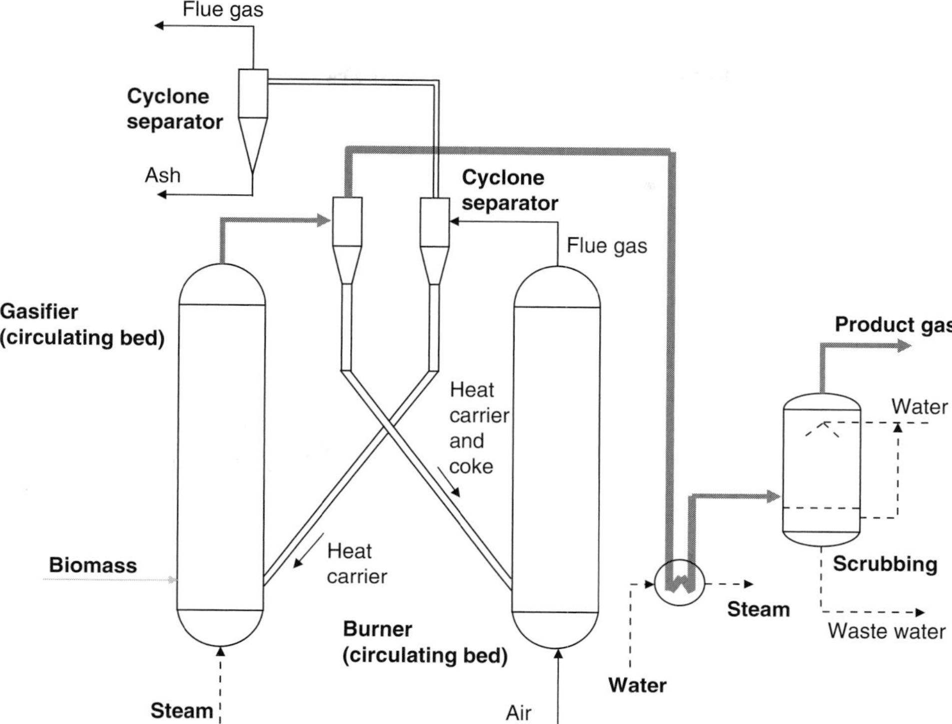

Figure 10.5. Production of synthesised gas via gasification of biomass according to the BCL process.

$$\text{Anode:}\quad 2OH^- \rightarrow \tfrac{1}{2}O_2 + H_2O + 2e^-,$$

$$\text{Cathode:}\quad 2H_2O + 2e^- \rightarrow H_2 + 2OH^-.$$

Electrolysers can be classified into three basic kinds, based on the type of electrolyte used: alkaline water electrolysers, membrane electrolysers (PEM electrolysers) and high-temperature electrolysers. Alkaline water electrolysis, the oldest and, therefore, most widely used technology, is described in more detail below. Figure 10.6 shows a diagram of an alkaline water electrolyser.

The core of the electrolyser is the cell block, which is made up of a large number of usually bipolar cells in a modular structure.[8] Typical sizes of a cell block range from 1 to 800 Nm^3/h. (Most electrolysers sold today to laboratories, the semiconductor industry, etc., have a capacity less than 60 Nm^3/h.) The biggest capacities realised are about 150 MW. An alkali solution, usually 20% to 40% potassium hydroxide (KOH), is used as the electrolyte that flows between the electrodes. In alkaline solutions, the electrodes must be resistant to corrosion, have good electronic

[8] Two distinct classes of cell design exist: the monopolar and the bipolar. Most commercial stacks have the bipolar design, which means that the single cells are connected in series both electrically and geometrically. The bipolar cell design has the advantages of compactness and shorter current paths with lower voltage losses.

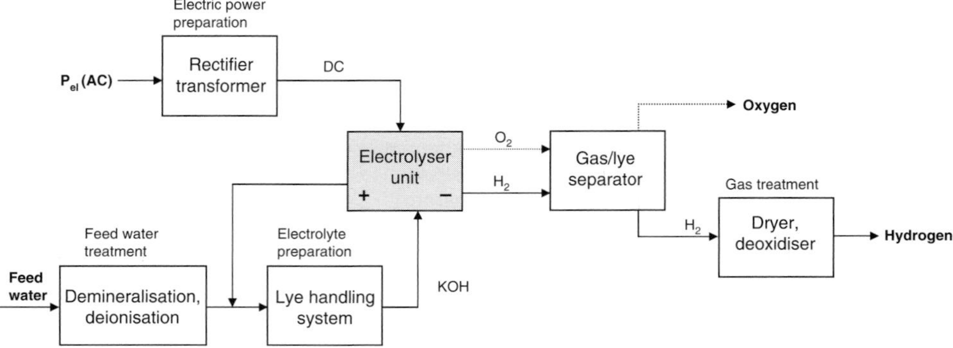

Figure 10.6. Block diagram of alkaline water electrolyser.

conductivity and good catalytic properties; the electrodes are usually made of nickel or chromium-nickel steel. To prevent the reaction products (hydrogen and oxygen gases) mixing, the individual cells are isolated by a separator, a so-called diaphragm, which should have a low electrical resistance. Apart from the actual cell block, other parts of an electrolyser include the adjustment of the electricity supply's current and voltage, the deionisation of the feed water supply and the treatment of the operating medium (potassium hydroxide). On the discharge side, separators ensure the separation of the gas and the operating agent: gas purification and drying then follow. The theoretical maximum efficiency of electrolysers is about 85%; under real operating conditions, efficiencies are between 65% and 75% (based on the lower heating value). Typical operating conditions are electrolyte temperatures of 70–90 °C, a cell voltage of 1.8–2.2 V, a current density of 2–3 kA/m^2 and a power consumption of 4–5 kWh/Nm^3 hydrogen (Ullmann, 2003). Hydrogen purities >99.8% are achieved.

The electrolysers for alkaline water electrolysis are usually distinguished with regard to their operating pressure into atmospheric pressure electrolysers (1 bar absolute) and pressurised electrolysers (10–30 bar). The basic functional procedure is the same in both types, but the high-pressure technique has the main advantage that the hydrogen produced is supplied at a higher pressure level and, therefore, less compression is necessary for its later distribution in pipelines or during storage. The drawbacks of high-pressure electrolysis are problems with gas purity, arising from the fact that the diaphragms show increasing gas permeability at higher pressures and higher temperatures. An important feature of electrolysers is that they cope very well with load shifts and thus when combined with wind power systems, for example, have no problems in matching the fluctuations of the intermittent supply.

Table 10.6 shows the technoeconomic characteristics for alkaline pressurised electrolysis. The production costs are clearly dominated by the variable electricity costs, which make up a share of at least 80%. In addition, the specific investments today are, at around €1000/kW, high when compared with steam reformers or coal gasification plants. The investments for future electrolysers are scaled in analogy to steam reformers, taking learning curve effects into account to get down to about €625/kW from around 2020, bringing hydrogen costs at the fuelling station down to less than 9 ct/kWh (€3 /kg).

Table 10.6. *Electrolysis (alkaline pressurised electrolysers)*

Technical data			
Capacity	MW_{H_2}	2.4	2.4
Capacity	Nm^3/h	800	800
Annual full load hours	h/year	8000	8000
Outlet pressure	bar	30	30
Electricity demand	kWh_{el}/kWh_{H_2}	1.43	1.43
Electrical efficiency	%	70.0	70.0
Water demand	l/Nm^3	0.8	0.8
Lifetime	years	20	20
Economic data			
Specific investment	€/kW	1000	625
Electrolyser	%	63	63
Transformer and Rectifier	%	20	20
DeOxo dryer	%	8	8
Electrical control unit	%	6	6
Deioniser	%	3	3
Fixed costs[a]	% Invest/year	1.5	1.5
Year of availability	–	Today	2020

Note:

[a] Maintenance, labour etc.

Sources: (Ullmann, 2003; Valentin, 2001; Wagner *et al.*, 2000).

Another class of electrolysers is the PEM (polymer electrolyte membrane) electrolyser, which is principally a PEM fuel cell operating in reverse mode. At present, PEM technology has a higher electricity consumption than the alkaline-based technology; however, the potential for increased energy efficiency in the long term is better. High-temperature electrolysis also offers efficiency advantages. Although the heat energy consumption of the electrolysis process slightly increases when operating at a higher temperature, the electricity requirement decreases. Therefore, high-temperature electrolysis (800–1000 °C) may offer a favourable energy balance, if high-temperature residual heat is available from other processes (IEA, 2005).

10.6 Other hydrogen-production methods

As well as the previously described methods of hydrogen production, there are other commercial processes whose application is restricted to specialised production conditions. These include the partial oxidation of heavy hydrocarbons, autothermal reforming and the *Kværner* process. In addition, there are numerous production processes that are still at the basic research stage, but show promising potential. These primarily include thermochemical hydrogen production, photochemical and biological processes. The main characteristics of these methods are outlined below. For a more detailed discussion, please refer to the relevant specialist literature.

10.6.1 Partial oxidation

The catalytic conversion of heavy hydrocarbons, such as heavy oil or sulphurous organic residues, from the oil industry via steam reforming is not feasible because solid carbon starts to be deposited at temperatures above 800 °C, which renders the catalyst inactive in a short period of time and, furthermore, blocks the gas flow in the reactor. Heavy hydrocarbons are, therefore, converted to hydrogen using partial oxidation (POX). Note that in refineries the term 'gasification' is more commonly used: 'partial oxidation' is the scientific terminology.

Partial oxidation is the exothermic conversion of hydrocarbons using oxygen with the addition of steam. Because the reaction is exothermic, no additional heat has to be supplied. The chemical reactions occur at temperatures between 1300 and 1500 °C and a pressure of 30 to 100 bar, which renders the use of a catalytic converter unnecessary. In contrast to steam reforming, partial oxidation does not make any specific demands of the quality of the raw materials and thus has the advantage of being applicable to lower-quality feeds, such as petroleum coke or other refinery residues. Coal can also serve as a feedstock if it is pulverised and mixed to a suspension with a solid fraction of 50% to 70%. The drawbacks of this process over natural gas reforming are the high CO content of the produced gas and the use of pure oxygen. The partial oxidation of crude oil has been put into practice in two commercial processes by Texaco/GE and Shell.

Partial oxidation is mainly used in refineries, since the raw materials, i.e., refining residues, are available here at low cost. As far as the hydrogen produced is concerned, it can be assumed that this is primarily used by the refineries themselves, since the availability of lighter crudes is decreasing and hydrogen is being increasingly used to process heavier oils.[9]

10.6.2 Autothermal reforming

Autothermal reforming (ATR) combines catalytic partial oxidation and steam reforming to convert both lighter and heavier hydrocarbons, where the exothermic oxidation supplies the necessary reaction heat for the subsequent endothermic steam reforming. Autothermal reforming has a greater process efficiency, since the heat required is generated as part of the process (by partial oxidation of part of the feedstock). An advantage of ATR is that the H_2:CO ratio can be adjusted according to the requirements of, for instance, the Fischer-Tropsch process or methanol synthesis. Autothermal reforming has not yet been widely applied and is primarily used today for very large conversion units, where it has an economic advantage over steam methane reforming.

[9] Since EU Directive 2005/33/EC bans high-sulphur heavy oil (bunker fuel) as fuel on ships from 2010, this 'source' may become relevant as a cheap feedstock for hydrogen production in the future.

10.6.3 Kværner process

The *Kværner-Carbon Black & Hydrogen* process was developed by the *Kværner Oil & Gas Company* in Norway as a more environmentally friendly variant of natural-gas steam reforming. The core idea of this process is the separation and commercial use of the carbon contained in the natural gas (but also in heavy oil) as soot. This process only makes sense, therefore, if there is sufficient demand for soot, which is primarily used at present in the manufacture of rubber and paints. Natural gas is split in a plasma arc process at temperatures over 1500 °C into carbon and hydrogen; the plasma is generated using a high-voltage source. Oxygen is not required in this process. As a result, there are no significant emissions, so that the CO_2 balance depends only on the upstream electricity supply. However, the electricity demand, at approximately 0.41 kWh_{el}/kWh_{H_2} is only about one-third of that required for electrolysis (Pehnt, 2001). So far, there are only two such plants worldwide: a pilot plant commenced operation in Sweden in 1992 and the first commercial plant started production in Canada in 1999.

10.6.4 Thermochemical methods

The direct thermal splitting of water is technically challenging, since it occurs at a very high temperature. If thermally activated at temperatures above 2200 °C, water begins to break down into hydrogen and oxygen. By injecting heat at this temperature, hydrogen can, in principle, be formed directly from steam. The fundamental and so far unresolved problem, apart from technical control of the necessary operating temperature, is the separation of hydrogen at these high temperatures. To be economic, thermochemical water splitting needs the development of new materials for the process equipment, cost reductions and access to cheap supplies of high-temperature heat (e.g., from nuclear or solar thermal energy).

The temperature of thermal water splitting can be considerably lowered to below 900 °C by applying coupled chemical reactions, namely the use of two parallel thermal cycles, in which hydrogen and oxygen are produced separately. Various thermochemical cycles were proposed as early as the 1970s to integrate the heat from high-temperature reactors, some of which are also suitable for the use of concentrated solar radiation (see also NRC, 2004). Today's view is that an improved sulphur–iodine process (S–I cycle) has the highest system efficiencies and the biggest potential for improvements: iodine and sulphur dioxide react with water at 120 °C to form hydrogen iodide and sulphuric acid. After the reaction products have been separated, the sulphuric acid is split into oxygen and sulphur dioxide at 850 °C; hydrogen and the base product iodine are formed from the hydrogen iodide at 300 °C. The high thermal efficiencies of the thermochemical cycles (up to 50%) have to be set against the currently still unresolved material and process engineering difficulties. Thermochemical hydrogen production is still the subject of basic R&D. Processes that use the waste heat of high-temperature reactors are being investigated, as is the use of concentrated solar radiation (Verfondern, 2005).

10.6.5 Photoprocesses for hydrogen production

As well as using solar energy in thermochemical processes, there is research on the possibility of producing hydrogen by photochemical reactions (photo-electrolysis, photolysis). The basic idea here is to make direct use of solar radiation by energy-rich photons being absorbed by reactants. This requires the use of semiconductor materials whose energy gap is so large that electrons are extracted from the water by the uptake of photons, which results in water splitting. The conversion processes thus triggered are facilitated or made possible using photocatalysts. The main problem is that the photoactive materials have to be both highly active catalysts and at the same time stable over long periods in contact with water. In the long term, combining photo- and thermochemical processes also seems promising.

10.6.6 Biological hydrogen production

Another technique, which is still at an early stage of research, is biological hydrogen production, where hydrogen is generated by micro-organisms in biological processes. Even if this is not yet competitive with the established methods of hydrogen production, the expertise already attained allows biological hydrogen to be produced on a laboratory scale and optimisation strategies to be explored experimentally. The central step of all biological hydrogen production processes is the enzymatic conversion of protons and electrons into molecular hydrogen. The possible metabolic pathways to achieve this can be divided into three processes: biophytolytic hydrogen production by green algae, photoproduction of bacteria and fermentation of biomass (see Wagner *et al.*, 2000).

10.7 Hydrogen purification

Hydrogen has to be cleaned according to the demands made by the final use, to guarantee the necessary quality. Different methods are applied, depending on the type of contamination and the purity required. If, for example, hydrogen is produced by reforming or pyrolysis, then the unwanted components of the reactant gas can be removed in advance. This mainly involves dust removal for coal and biogas, desulphurisation of natural gas and the removal of CO_2. Fine cleaning is only carried out on the final product hydrogen.

The highest demands made on the purity of the produced hydrogen are for its use in fuel cells, since even the slightest trace of carbon monoxide impedes the functioning of the precious metal catalyst in the fuel cell.[10] For the platinum used in hydrogen fuel

[10] There are differing degrees of purity for gases. The classification for hydrogen is based on the number of nines appearing as percentages: e.g., hydrogen with a hydrogen share of 99.9999 vol.% is referred to as 'Hydrogen 6.0'.

cells, a volume share of 10 ppm CO is regarded as being the upper acceptable limit. Such values can only be achieved with a downstream purification unit, in which all the undesired substances contained in the hydrogen gas are removed. This mainly affects products resulting from incomplete reforming, such as CO, H_2O, O_2, NH_3 and CO_2.

The most important purification processes can be differentiated into catalytic, membrane and adsorption processes. While catalytic processes are used only to remove CO, the other processes can also remove other substances depending on the material involved.

In catalytic processes, such as *CO conversion* ($CO + H_2O \rightarrow CO_2 + H_2$), *selective methanisation* ($CO + 3\ H_2 \rightarrow CH_4 + H_2O$) or *selective CO oxidation* ($CO + \frac{1}{2}\ O_2 \rightarrow CO_2$), the achievable efficiencies depend on reaction parameters, such as temperature, pressure, volume flow, raw gas concentration and catalyst material. These are capable of achieving contamination levels from 1% down to a few ppm. The selection of different reaction paths is based on the use of different types of catalyst.

Membrane processes are based on the selective transmission characteristics of the membrane material for different molecules, whereby the most effective membranes are usually also the most expensive. For example, the purest hydrogen can be captured by palladium membranes with suitable additives, but their low permeability make it necessary to use large membrane surfaces and high pressures, which result in high costs.

Adsorption methods are used in large numbers to separate and extract gases selectively from gas mixtures and to clean gases. The pressure-swing adsorption (PSA) method is also the most widely used process to purify hydrogen; the related process of temperature-swing adsorption (TSA) is carried out more rarely. The highest purities, of up to 99.999% hydrogen content, are obtained using pressure-swing adsorption (Zittel *et al.*, 1996). In this process, the raw gas is forced under pressure (approximately 20 bar) through an activated carbon filter or a carbon molecular sieve.[11] The adsorbent can be loaded until a state of equilibrium is reached, at which point its adsorption capacity is exhausted. To be able to repeat the gas separation process, the loaded adsorbent has to be regenerated (desorption) by lowering the pressure. The pressure-swing adsorption is thus a discontinuous process.

The investments for catalytic processes are mostly incorporated in the investments for a reformer, since these are linked. Pressure swing, in contrast, is a downstream ancillary process and its specific investment cost ranges from €120 to €150/kW_{H_2} (0.3 to 0.5 ct/kWh_{H_2}) (Kreutz *et al.*, 2005; Zittel *et al.*, 1996).

10.8 Industrial use of hydrogen

Hydrogen has been used as industrial gas for decades. Hence, besides its potential use as a future automotive fuel, the conventional use of hydrogen as an important

[11] Adsorption processes make use of the fact that the adsorbent's loadability depends on pressure and temperature. The process is conducted such that the gas mixture to be separated is fed through an adsorbent packed bed whereby easily adsorbable gas components accumulate on the surface and those components that are harder or impossible to adsorb pass through the packed bed.

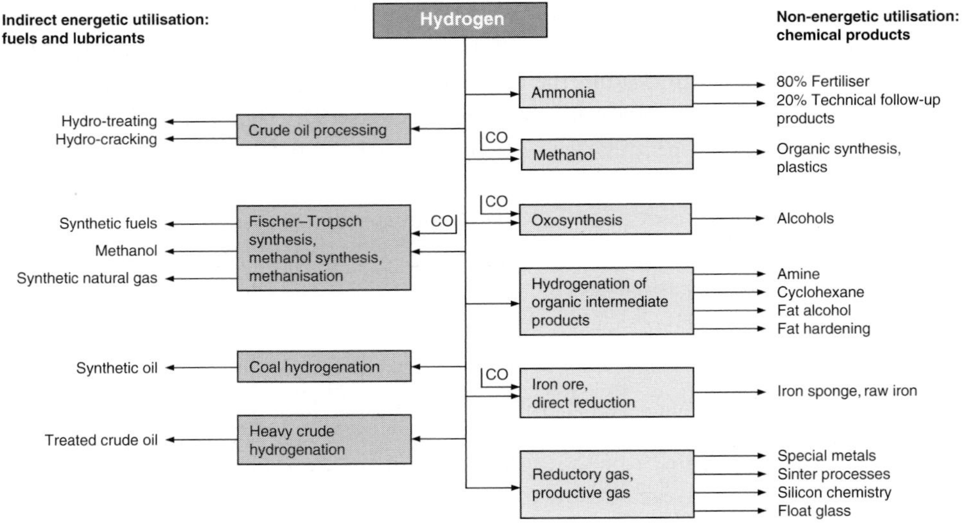

Figure 10.7. Industrial use of hydrogen (NRW, 2006).

feedstock for the chemical and petrochemical industry also has to be addressed: mainly for the production of ammonia – with the subsequent products, fertilisers and plastics – as well as for the processing of crude oil into fuels and high-quality chemical products (hydro-cracking and hydro-treating). In addition, hydrogen is used for reduction processes in metallurgy and steel production, as a cooling agent for electrical engineering, as a protective gas for electronics, for welding and cutting in mechanical engineering and for fat hardening in the food industry as well as for rocket fuel. Figure 10.7 shows the different industrial areas in which hydrogen is used. A comprehensive overview of the industrial uses of hydrogen is found in Ramachandran and Menon (1998). The future demand for industrial hydrogen has been excluded in the remainder of the book, as it is assumed to be supplied by industry anyway.

The main three players in the hydrogen market are captive producers (which produce hydrogen for their direct customer or their own use), by-product hydrogen producers (which provide hydrogen resulting from chemical processes) and merchant companies (which trade hydrogen). Only about 5% of total hydrogen production is merchant hydrogen. The majority of hydrogen is used as a reactant in the chemical and petroleum industries. Of the total non-energetic hydrogen use worldwide, ammonia production has a share of around 50%, followed by crude-oil processing (37%) and methanol production (8%) (Ramachandran and Menon, 1998); in Western Europe, refineries consume about 50% and ammonia production 32% of the hydrogen (Roads2HyCom, 2007).

It is expected that the hydrogen demand for crude-oil processing will increase further, the reason for this being twofold: on the one hand, through an increasing fertiliser demand (ammonia production) due to a growing world population. On the other hand, the availability of light crude oils is rapidly decreasing, and heavier crudes contain more sulphur and nitrogen. To supply 'sulphur-free' fuel for compliance with tightening environmental standards, hydrogen for hydrogenation (hydro-treating) is as indispensable as for the processing of heavy crudes or oil sands (bitumen) through hydro-cracking.[12] These trends, together with a growing demand for lighter oil products as well as tighter fuel-quality standards, are likely to increase the need for investments in additional upgrading facilities in refineries.

10.9 Availability of industrial surplus hydrogen

As mentioned in the previous section, many industrial processes require hydrogen as an ingredient. But hydrogen is also formed as a by-product of many processes, for instance, in the chemical industry: these include, among others, the production of chlorine, ethylene and acetylene. Varying amounts of hydrogen are also formed during cracking and catalytic reforming in refineries; however, since hydrogen is also required for various refinery processes, e.g., hydrogenation, the net balance of a refinery depends heavily on the processes used and the products manufactured.[13] Moreover, hydrogen is produced as a by-product from coke oven gas.

If the hydrogen occurring as a by-product of these industrial processes cannot be directly utilised internally as chemical feedstock, as for hydrogenation in refineries, for example, at present there may not be a suitable use for this by-product hydrogen at reasonable added value, since there are often no suitable consumers nearby or its transportation to external users is not economically viable. For these reasons, this hydrogen is mostly directly used as an energy source for power and heat generation onsite by mixing it with other combustible gases, or it is blown off or burnt off.[14] This amount would theoretically be available for new hydrogen markets, for example, as a fuel for the transportation sector.

The amount of hydrogen occurring as a by-product is very difficult to estimate because of the limited data availability. The result of interviews conducted with hydrogen producers and chemical companies in the EU15 was that between 5% and 10% of the hydrogen produced is vented, burnt off or used as a substitute for fossil fuels (LBST, 1998). Roads2HyCom (2007) estimates by-product hydrogen production, i.e., hydrogen produced inadvertently as a by-product of a chemical process, at 23 billion Nm^3 per year in the EU, with Germany (6.8 billion Nm^3) and the UK (3.6 billionNm^3) being the largest producers. Of this total, 10%–20% (roughly

[12] The process of *hydro-cracking* was already developed in 1927 by the German company *I. G. Farben Industrie*, to transform lignite into gasoline. As a support for thermal or catalytic cracking, hydro-cracking is used nowadays to crack hydrocarbons that are more difficult to crack.

[13] During thermal cracking, catalytic cracking and catalytic reformulation, hydrogen is produced at a rate of about 3 m^3/t crude oil, 100 m^3/t and 200 m^3/t, respectively. Hydrogen consumption amounts to around 300 m^3/t of product for hydro-cracking, 80 m^3/t for catalytic cracking, 50 m^3/t for hydration of cokers, 20 m^3/t for gasoline hydro-treating and 35 m^3/t for distillates hydro-treating (Roads2HyCom, 2007).

[14] According to Nitsch (2002) and the BMWA (2005), the direct energetic use of hydrogen worldwide is estimated at 40% to 50% of total hydrogen production.

2–5 billion Nm^3 per year) is assumed to be available to fuel transport applications during the hydrogen transition phase.

However, this by-product hydrogen varies in its purity, depending on the different processes employed. Whereas hydrogen from chlorine–alkali electrolysis is of a very high purity, the hydrogen purity from refineries or ethylene production is only between 70% and 80% (LBST, 1998). Therefore, for further use of hydrogen (other than generation of process heat) a purification process by PSA or membrane is needed. The relevant purification of hydrogen is ultimately purely a question of cost, since, technically, every level of purity can be provided in principle (see Section 10.7).

By-product hydrogen is potentially one of the cheapest sources of hydrogen. The essential question, however, is not whether hydrogen will be produced per se and may be available as a by-product, but to what extent it can be supplied to an external (new) market and can be substituted by other energy sources (e.g., natural gas) in its use as a fuel. If the producer is prepared to supply by-product hydrogen, the final decision about whether to do so will depend on the price that can be obtained for the hydrogen or the price of the substitute energy source.

Figure 10.8 shows the geographical distribution of industrial hydrogen production sites in Western Europe. As can be seen, there are clusters of hydrogen production, mainly in the Benelux and Rhine–Main area (Germany) as well as in the Midlands (UK) and Northern Italy.

The by-product hydrogen from chlorine production seems best suited and most promising to supplying a new market, not least because of its already high purity.[15] For the manufacture of chlorine, electrochemical processes are almost exclusively used – i.e., chlorine–alkali electrolysis. The basic principle of alkaline–chloride electrolysis is the electrochemical splitting of alkaline chloride solutions into the by-products chlorine gas, sodium hydroxide solution (caustic soda) and hydrogen. Without going into the differences between the individual manufacturing processes in detail, the membrane cell process, the diaphragm cell process and the mercury cell process can be distinguished. Approximately 315 Nm^3 hydrogen are generated in all three processes per tonne of chlorine (VCI, 2005).

With respect to the German case study used in Chapter 14 to discuss the build-up of a hydrogen infrastructure, Fig. 10.9 shows where surplus hydrogen capacities (from chlorine–alkali electrolysis) exist in Germany. If these capacities are added up, the resulting total amount is about 1 billion Nm^3 per year (around 4% of total German hydrogen production).

Of the chlorine production capacity installed in Germany, which totalled 4.4 million tonnes in 2003, 50% were from the membrane cell process, 27% from the mercury cell process and 23% from the diaphragm cell process. The mercury cell process has been the subject of environmental policy criticism for years because of its use of mercury cathodes and resulting pollutant emissions. Hence, no new mercury plants will be

[15] For example, *Bayer* in Germany is building a plant in Leverkusen to clean and compress by-product hydrogen from Dormagen and from chlorine–alkali electrolysis in Leverkusen, to feed this into the hydrogen pipeline network of the Ruhr region (Bayer, 2005).

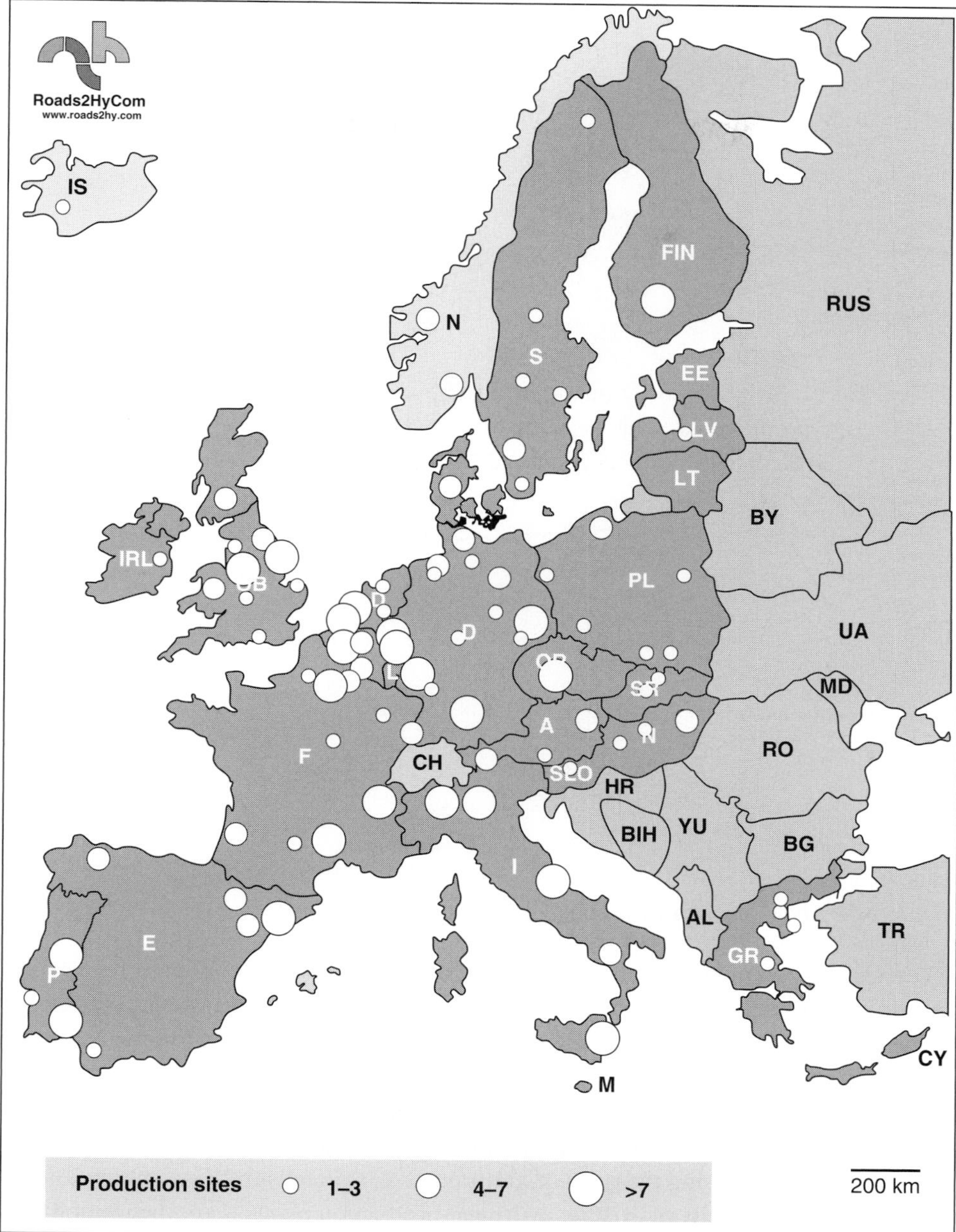

Figure 10.8. Geographical distribution of industrial hydrogen-production facilities (Roads2HyCom, 2007).

constructed in Europe and the existing facilities are to be phased out (VCI, 2005), resulting in a reduction of hydrogen available from these processes. Considering a further development of the manufacturing processes, it can be expected that the 'surplus' hydrogen from chlorine production will drop significantly in the future.

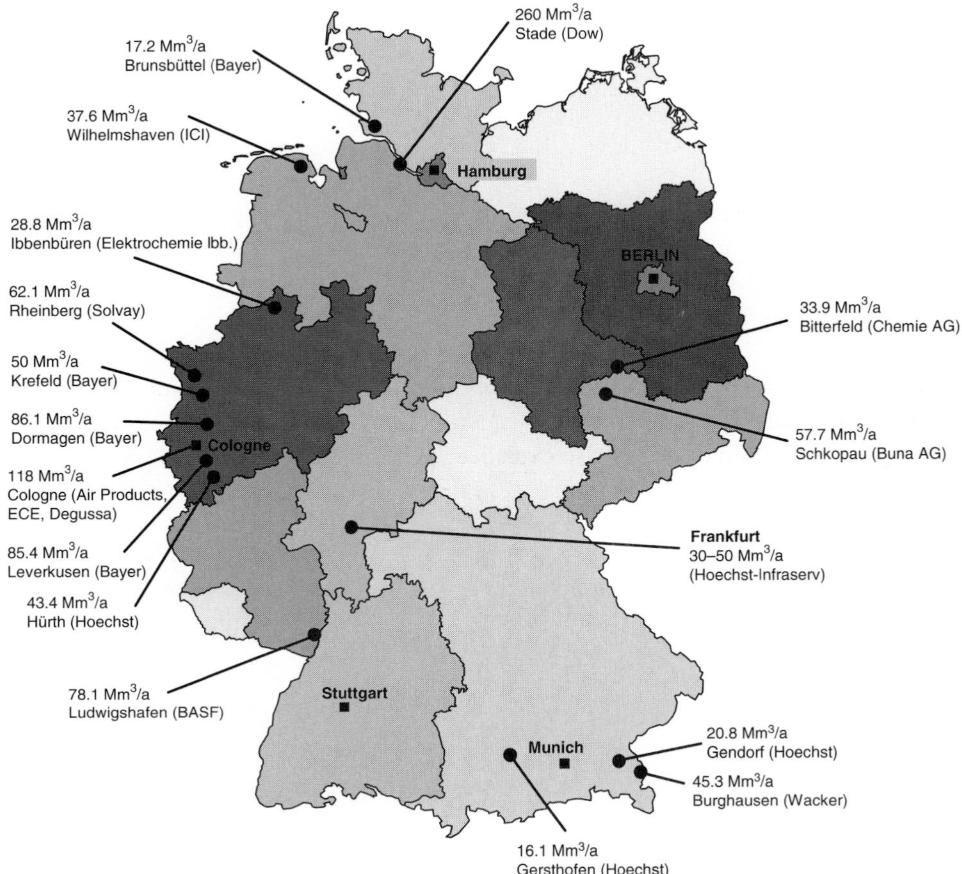

Figure 10.9. Hydrogen as a by-product in Germany (DWV, 2005).

It can also be assumed that the total volume will not be available. Therefore, a rather conservative total estimate of 500 million Nm^3 of by-product hydrogen is assumed for modelling purposes, which can be used annually until 2030 to provide hydrogen for vehicles. This amount would be sufficient to power around 610 000 fuel-cell cars or 340 000 cars with hydrogen internal combustion engines. The available amounts are distributed in the model according to their real geographical location. The price of the hydrogen is derived using the price of natural gas as a direct substitute energy source.

Extending the capacity of existing hydrogen-production plants is another option that should be considered when looking at the most cost-efficient hydrogen supply possible. In certain cases, depending on the additional volume required, this may be the more economic alternative to building new plants. A principal difficulty with estimating the increase in production that can be achieved in this way is the availability of the technoeconomic data required to assess the different retrofit measures. Furthermore, the possible technical options and the related production increases cannot be generalised, but are plant-specific and depend on the respective process management and the primary energy sources used (see, for example, Cromarty and Hooper (1997)). Roads2HyCom (2007) assumes a 10% margin of excess capacity from captive users, amounting to up to 5 billion Nm^3 of hydrogen available in the EU. Owing to the uncertainties around these numbers, the option of capacity extension of existing plants is excluded in the MOREHyS model. This approach also seems justified, since the achievable expansion is limited, hydrogen producers increasingly adapt their production to their needs and, in the long term, it is inevitable that a production infrastructure expressly for supplying hydrogen as a fuel will have to be constructed.

As for refineries, it can be assumed that in a business-as-usual scenario hydrogen will not be available for external utilisation on account of the increasing internal demands there for hydrogenation of heavier crudes or for the manufacturing of sulphur-free fuels. Therefore, the partial oxidation of heavy refinery residues as a hydrogen production option is also excluded, as it is assumed that the hydrogen thus produced remains in the refineries and is not available for external use as a fuel. However, despite the growing demand for hydrogen, refineries have a significant potential for an extension of their hydrogen production capacity by 25%–50% to supply hydrogen for the transport sector in the future.

Potential *surplus hydrogen* that could be used as vehicle fuel in the early phase of building up a hydrogen infrastructure can come from two sources: *excess capacity* (from captive industries) and *by-product* hydrogen. In total, a potential of between two and ten billion Nm^3 (0.18–0.9 Mt) of hydrogen might be available as surplus hydrogen in Europe: up to five billion Nm^3 in the form of excess capacity and two to five billion Nm^3 as by-product hydrogen. While these figures should be taken with care, as they are largely based on statistical assumptions and not on a site-by-site assessment, this surplus hydrogen volume is not negligible, as it could supply between two and ten million fuel-cell cars, between 1% and 5% of all vehicles in the EU (estimated at 190 million). Surplus hydrogen can hence be considered as a potential hydrogen source during the transition phase, most probably in locations closest to the surplus. Consideration has to be given to the purity of the surplus hydrogen, especially when intended for use in fuel-cell vehicles. Because of a lack of information, only the situation in Europe has been addressed, but it can be assumed that a similar situation exists in the USA.

10.10 Summary

Natural gas reforming, coal gasification and water electrolysis are proven technologies for hydrogen production today and are applied at an industrial scale all over the world. These are also the most likely hydrogen production technologies to be employed until 2030 and beyond.

It is evident that hydrogen needs to be produced in the long term from processes that avoid or minimise CO_2 emissions. Renewable hydrogen (made via electrolysis

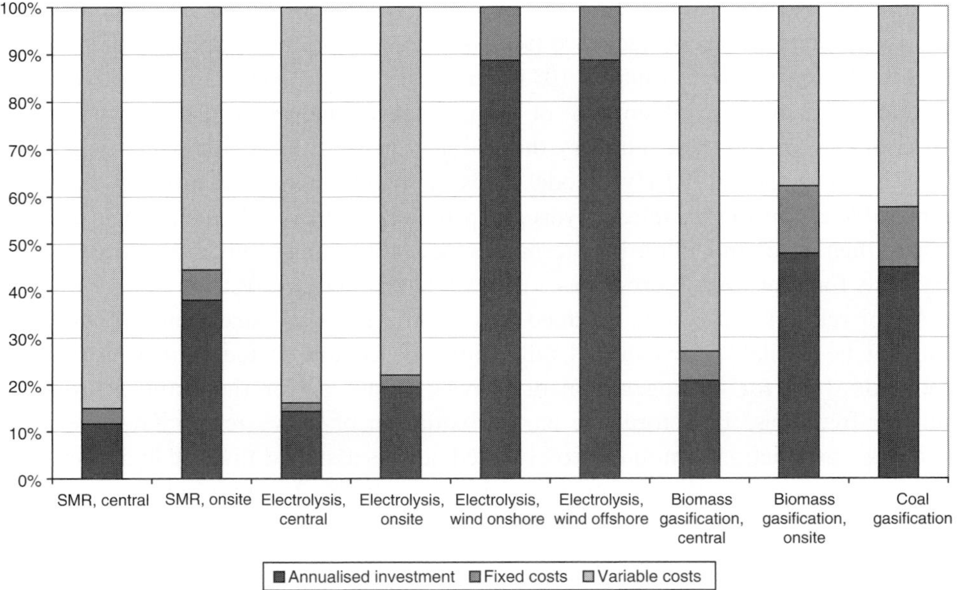

Figure 10.10. Composition of hydrogen-production costs.

from wind- or solar-generated electricity or gasification of biomass) is surely the ultimate vision (particularly from the point of view of mitigating climate change), but not the precondition for introducing hydrogen as an energy vector. As long as this goal is not yet reached, hydrogen from fossil fuels will prevail, for which, however, the capture and storage of the produced CO_2 is an indispensable condition, if hydrogen is to contribute to an overall CO_2 reduction in the transport sector.

The economics and CO_2 emissions of the different hydrogen production technologies are summarised in Figs. 10.10 and 10.11, which illustrate the major differences of specific hydrogen-production costs for different technologies and feedstocks.

Figure 10.10 shows the relative shares of feedstock and capital costs. It becomes clear that for most technologies the total costs are dominated by variable costs, i.e., feedstock prices, particularly so for steam reformers, biomass gasifiers and electrolysers. This also stresses the fact that any projection of future hydrogen-production costs depends critically on underlying assumptions about the development of feedstock prices (as well as CO_2 certificate prices) and has to be taken with care. Via the feedstock quantities required, the efficiency of the processes is also indirectly linked to the variable costs. While for steam reformers total costs are dominated by feedstock prices, owing to the high natural gas prices, capital costs are dominant for coal gasification. Onsite generation has generally much higher capital costs than large centralised plants, as it cannot benefit from economies of scale.

Figure 10.11 indicates cost ranges for various hydrogen production technologies for the time until around 2030 on the basis of the energy price scenarios outlined in

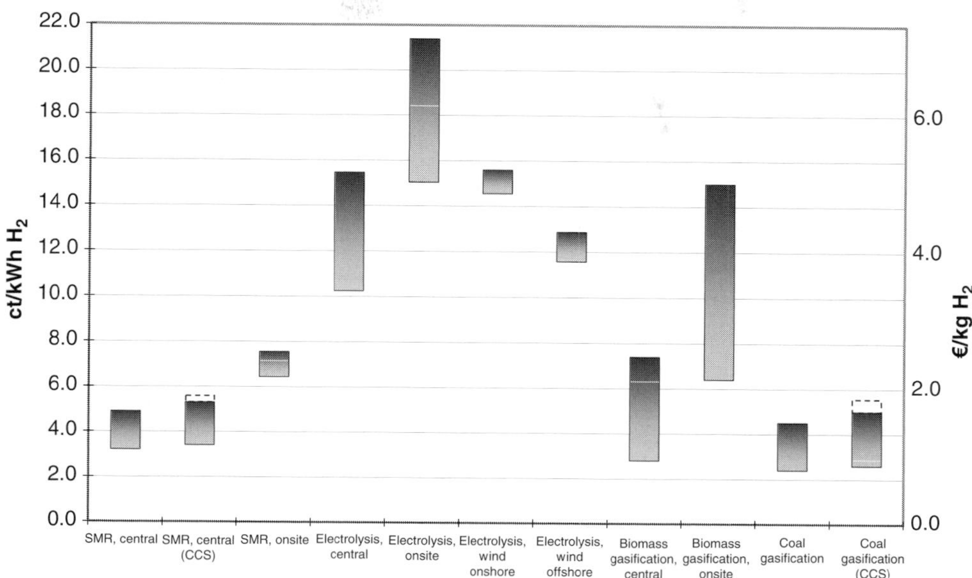

Figure 10.11. Hydrogen-production costs.

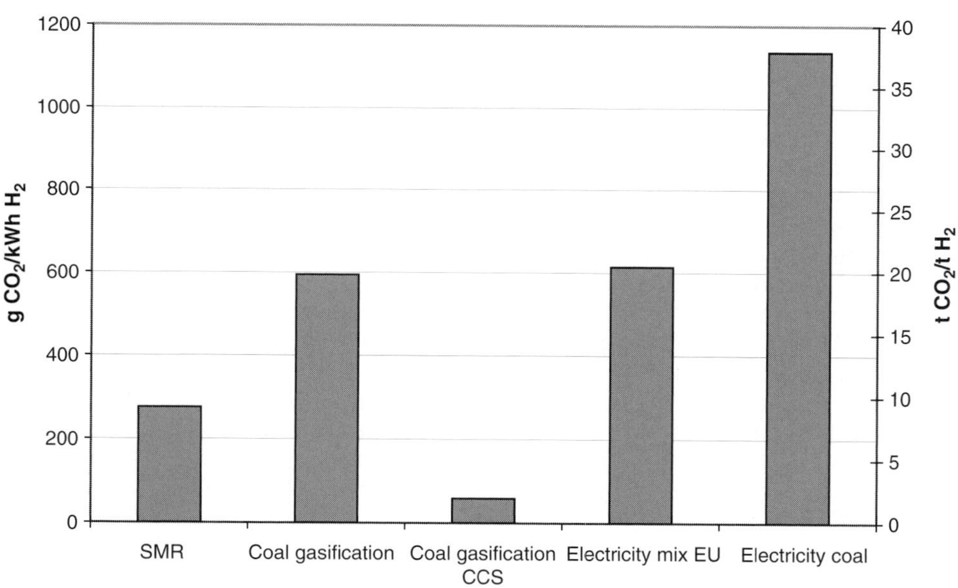

Figure 10.12 CO_2 emissions from hydrogen production.

Section 14.4.2 (but *excluding* CO_2 prices for fossil fuels, unlike in the MOREHyS model); the dotted lines for the CCS cases indicate the additional costs for CO_2 transport and storage. Increases in feedstock prices could significantly increase hydrogen-production costs, owing to their high shares of total costs for some technologies (see Fig. 10.10). The most expensive technology in this timeframe is electrolysis from grid-mix electricity, followed by wind electricity; for wind electrolysis, however, this picture might change with further depletion and price increases of fossil fuels. Costs for hydrogen from biomass show a large bandwidth because of the large spread in biomass costs. The switch between natural gas and coal-based hydrogen production is very sensitive to slight changes in the natural gas to coal price ratio.

Figure 10.12 shows the specific CO_2 emissions for various hydrogen production technologies. With the exception of electrolysis from EU grid-mix electricity, the highest CO_2 emissions are incurred for hydrogen from coal. This underlines again that CCS will be necessary, to avoid an overall increase of CO_2 emissions through fossil hydrogen production, primarily from coal.

References

Bayer (2005). *Bayer to Invest in Hydrogen Network*. www.bayertechnology.com.

BMELV (Bundesministerium für Ernährung, Landwirtschaft und Verbraucherschutz) (2005). *Synthetische Biokraftstoffe. Techniken – Potenziale – Perspektiven.* Schriftenreihe 'Nachwachsende Rohstoffe' Band 25, Münster: Landwirtschaftsverlag GmbH.

BMWA (Bundesministerium für Wirtschaft und Arbeit) (2003). *Forschungs- und Entwicklungskonzept für emissionsarme fossil befeuerte Kraftwerke*. Berlin: Bericht der COORETEC-Arbeitsgruppen. BMWA.

BMWA (Bundesministerium für Wirtschaft und Arbeit) (2005). *Strategiepapier zum Forschungsbedarf in der Wasserstoff-Energietechnologie*. Berlin: BMWA.

Boukis, N., Diem, V., Galla, U. *et al.* (2005). Wasserstofferzeugung aus Biomasse. Forschungszentrum Karlsruhe. *Nachrichten*: *Energieträger Wasserstoff*, **37** (3/2005).

Chauvel, A., Leprince, P., Barthel, Y., Raimbault, C. and Arlie, J. P. (1976). *Manuel d' Evaluation Economique des Procédés. Avant-projets en Raffinage et Pétrochimie*. Institut Français du Pétrole (IFP), Collection pratique du pétrole, Editions Technip.

Chiesa, P., Consonni, S., Kreutz, T. and Williams, R. (2005). Co-production of hydrogen, electricity and CO_2 from coal with commercially ready technology: part A: performance and emissions. *International Journal of Hydrogen Energy*, **30** (7), 747–767.

Cromarty, B. J. and Hooper, C. W. (1997). Increasing the throughput of an existing hydrogen plant. *International Journal of Hydrogen Energy*, **22** (1), 17–22.

D.M.2 (2001). *Wirtschaftlichkeitsbetrachtung zur Erzeugung von Strom und Wasserstoff in Anlagen zur gestuften Reformierung mit 10 MW thermischer Inputleistung*. (June 30, 2001).

DWV (Deutscher Wasserstoff-und Brennstoffzellen-Verband German Hydrogen and Fuel Cell Association) (2005). *H_2-Roadmap*. Berlin: DWV. www.dwv-info.de.

IEA (International Energy Agency) (2005). *Prospects for Hydrogen and Fuel Cells*. IEA Energy Technology Analysis Series. Paris: OECD/IEA.

IEA (International Energy Agency) (2007). *Hydrogen Production & Distribution*. IEA Energy Technology Essentials. Paris: OECD/IEA.

IE/IPTS (Institute for Energy and Institute for Prospective Technological Studies) (2005). *Hypogen Pre-feasibility Study*. Report EUR 21512 EN. European Commission, Directorate-General, Joint Research Centre.

IPCC (2005). *IPCC Special Report on Carbon Dioxide Capture and Storage. Prepared by Working Group III of the Intergovernmental Panel on Climate Change*, ed. Metz, B., Davidson, O., de Coninck, H. C., Loos, M. and Meyer, L. A. Cambridge: Cambridge University Press.

Katofsky, R. E. (1993). *The production of Fluid Fuels from Biomass*. PU/CEES Report No. 279. Princeton: Centre for Energy and Environmental Studies, Princeton University.

Kreutz, T., Williams, R., Consonni, S. and Chiesa, P. (2005). Co-production of hydrogen, electricity and CO_2 from coal with commercially ready technology: part B: economic analysis. *International Journal of Hydrogen Energy*, **30** (7), 769–784.

LBST (Ludwig-Bölkow-Systemtechnik GmbH) (1998). *Identification of Hydrogen By-product Sources in the European Union*, ed. Zittel, W. and Niebauer, P. (LBST), study funded by the European Commission under Contract No. 5076–92 11 EO ISP D Amendment No. 1.

Linde (2003). *Linde Technology Report 2/2003*. Wiesbaden, Germany: Linde AG.

Ni, M., Leung, D. Y. C., Leung, M. K. H. and Sumathy, K. (2006). An overview of hydrogen production from biomass. *Fuel Processing Technology*, **87** (5), 461–472.

Nitsch, J. (2002). *Potenziale der Wasserstoffwirtschaft. Gutachten für den Wissenschaftlichen Beirat der Bundesregierung Globale Umweltveränderungen (WBGU)*. Stuttgart: DLR – Institut für Technische Thermodynamik.

NRC (National Research Council) (2004). *The Hydrogen Economy: Opportunities, Costs, Barriers, and R&D Needs*. Washington, DC: The National Academies Press.

NRW (Energy Agency State of Nordrhein-Westfalen) (2006). *Hydrogen – Sustainable Energy for Transport and Energy Utility Markets*. Düsseldorf: Energy Agency NRW.

Ogden, J. M. (1999). Prospects for building a hydrogen energy infrastructure. *Annual Review of Energy and the Environment*, **24**, 227–279.

Pehnt, M. (2001). *Ganzheitliche Bilanzierung von Brennstoffzellen in der Energie- und Verkehrstechnik*. Dissertation. VDI-Verlag Fortschrittsberichte Reihe 6, No. 476.

Ramachandran, R. and Menon, R. K. (1998). An overview of industrial uses of hydrogen. *International Journal of Hydrogen Energy*, **23** (7), 593–598.

Roads2HyCom (2007). *European Hydrogen Infrastructure Atlas and Industrial Excess Hydrogen Analysis*, ed. Steinberger-Wilckens, R. and Trümper, S. C. Roads2HyCom. www.roads2hy.com.

Spath, P., Aden, A., Eggeman, T. *et al.* (2005). *Biomass to Hydrogen Production – Detailed Design and Economics Utilizing the Battelle Columbus Laboratory Indirectly-heated Gasifier*. Battelle: National Renewable Energy Laboratory (NREL). www.nrel.gov/docs/fy05osti/37408.pdf.

Treviño, M. C. (2002). *Integrated Gasification Combined Cycle Technology: IGCC. Its Actual Application in Spain*. ELCOGAS, Puertollano Club Español de la Energía.

Ullmann (2003). Hydrogen. In *Ullmann's Encyclopedia of Industrial Chemistry*, 6th edn. vol. 17 Weinheim: WILEY–VCH, pp. 85–240.

Valentin, B. (2001).*Wirtschaftlichkeitsbetrachtung einer Wasserstoffinfrastruktur für Kraftfahrzeuge*. Diploma thesis. Münster, Germany: University of Applied Sciences Münster and Linde Gas AG.

VCI (Verband der Chemischen Industrie (German Chemical Industry Association)) (2005). *Positionen zur Chemie mit Chlor*. www.vci.de/default2~rub~0~tma~0~cmd~shd~docnr~64356~nd~~ond~pb.htm.

Verfondern, K. (2005). Nukleare Wasserstoffproduktion. Forschungszentrum Karlsruhe. *Nachrichten: Energieträger Wasserstoff*, No. **37** (3/2005).

Wagner, U., Angloher, J. and Dreier, T. (2000). *Techniken und Systeme zur Wasserstoffbereitstellung. Perspektiven einer Wasserstoff-Energiewirtschaft (Teil 1)*. Munich: Wiba (Koordinationsstelle der Wasserstoff-Initiative Bayern) and Lehrstuhl für Energiewirtschaft und Anwendungstechnik of the Technical University München.

Williams, R., Parker, N., Yang, C., Ogden J. and Jenkins, B. (2007). *Hydrogen Production via Biomass Gasification*. Advanced Energy Pathways (AEP) Project, Task 4.1 Technology Assessments of Vehicle Fuels and Technologies, Public Interest Energy Research (PIER) Program, California Energy Commission, UC Davis, Institute of Transportation Studies (ITS-Davis).

Yamashita, K. and Barreto, L. (2003). *Integrated Energy Systems for the 21st Century: Coal Gasification for Co-producing Hydrogen, Electricity and Liquid Fuels*. Interim Report IR-03-039. Laxenburg, Austria: International Institute for Applied Systems Analysis (IIASA).

Zittel, W., Wurster, R. and Weindorf, W. (1996). *Wasserstoff in der Energiewirtschaft*. Ludwig-Bölkow-Systemtechnik (LBST). www.hyweb.de.

Further reading

Cormos, C. C., Starr, F., Tzimas, E. and Peteves, S. (2008). Innovative concepts for hydrogen production processes based on coal gasification with CO_2 capture. *International Journal of Hydrogen Energy*, **33** (4), 1286–1294.

Dynamis. www.dynamis-hypogen.com.

Ewan, B. C. R. and Allen, R. W. K. (2005). A figure of merit assessment of the routes to hydrogen. *International Journal of Hydrogen Energy*, **30** (8), 809–819.

IEA (International Energy Association) and Hydrogen Implementing Agreement (HIA) (2006). *Hydrogen Production and Storage. R&D Priorities and Gaps*. Paris: OECD/IEA.

IEA (International Energy Association) (2006). *Energy Technology Perspectives 2006. Scenarios and Strategies to 2050*. Paris: OECD/IEA.

NRC (National Research Council), Committee on Alternatives and Strategies for Future Hydrogen Production and Use (2004). *The Hydrogen Economy: Opportunities, Costs, Barriers, and R&D Needs*. Washington, DC: The National Academies Press.

Sørensen, B. (2005). *Hydrogen and Fuel Cells. Emerging Technologies and Applications*. Elsevier Academic Press.

Starr, F., Tzimas, E. and Peteves, S. (2006). *Near-Term IGCC and Steam Reforming Processes for the Hydrogen Economy: the Development Issues*. Report EUR 22340 EN, European Commission (DG JRC), Institute for Energy, Petten (The Netherlands).

Starr, F., Tzimas, E. and Peteves, S. (2007). Critical factors in the design, operation and economics of coal gasification plants: the case of the flexible co-production of hydrogen and electricity. *International Journal of Hydrogen Energy*, **32** (10–11), 1477–1485.

Tetzlaff, K.-H. (2005). *Bio-Wasserstoff*. Norderstedt: Books on Demand.

Ullmann (2003). Hydrogen. In *Ullmann's Encyclopedia of Industrial Chemistry*, 6th edn. vol. 17. Weinheim: WILEY-VCH, pp. 85–240.

11

Hydrogen storage

Maximilian Fichtner

The need for new and sustainable energy technologies is particularly urgent in the transport sector, where energy demands keep growing and give rise to significant global and local pollution. Hydrogen is expected to play a key role in this development (Satyapal *et al.*, 2006). Hydrogen storage is regarded as one of the most critical issues that has to be solved before a technically and economically viable hydrogen economy can be established. In fact, without effective storage systems, a hydrogen economy will be difficult to achieve. One of the most challenging applications in this field is hydrogen storage for mobile applications. This chapter addresses the current state of the various on-board hydrogen-storage systems.

11.1 Requirements for hydrogen storage

In hydrogen-fuelled passenger cars, 4–5 kg (130–160 kWh) H_2 must be stored in a small, preferably lightweight, tank in order to achieve a driving range of 500 km (i.e., 80–125 km/kg H_2). However, whereas the gravimetric energy density of hydrogen is extremely high, the volumetric storage density of the lightweight gas is low. At ambient temperature and pressure, 5 kg H_2 would fill a ball 5 m in diameter, which is roughly comparable to the volume of an inflated hot-air balloon. Consequently, the most important technical and economic challenges to be overcome in a practical hydrogen-storage system are the storage density related to the system (including tank, heat management, and valves), the costs of the system, its safety, a short refuelling time, and the ability to deliver enough hydrogen during the driving cycle.

Technical and economical targets have been established by the US Department of Energy (US DOE) in collaboration with car manufacturers and are summarised in Table 11.1. It should be mentioned that there may be variations regarding the assessment of specific targets between car manufacturers from Europe, the USA and Japan.

The Hydrogen Economy: Opportunities and Challenges, ed. Michael Ball and Martin Wietschel. Published by Cambridge University Press.

Table 11.1. *US DOE Technical targets for on-board hydrogen-storage systems*

Storage parameter	Units	2007	2010	2015
System *gravimetric* capacity usable energy density from H_2 (net useful energy/max system mass)	kWh/kg (kg H_2/kg system)	1.5 (0.045)	2 (0.06)	3 (0.09)
System *volumetric* capacity usable energy density from H_2 (net useful energy/max system volume)	kWh/l (kg H_2/l system)	1.2 (0.036)	1.5 (0.045)	2.7 (0.081)
Storage-system cost (and fuel cost)	\$/kWh net (\$/kg H_2)	6 (200)	4 (133)	2 (67)
	\$/gge at pump	–	2–3	2–3
Durability, operability				
Operating ambient temperature	°C	−20/50 (sun)	−30/50 (sun)	−40/60 (sun)
Minimum/maximum delivery temperature	°C	−30/85	−40/85	−40/85
Cycle life variation	% of mean (min) at % confidence	N/A	90/90	99/90
Cycle life (1/4 tank to full)	Cycles	500	1000	1500
Minimum delivery pressure from tank	atm (abs)	8 (FC)/ 10 (ICE)	4 (FC)/ 35 (ICE)	3 (FC)/ 35 (ICE)
Maximum delivery pressure	atm (abs)	100	100	100
Charging and discharging rates				
System fill time (for 5 kg)	min	10	3	2.5
Minimum full flow rate	(g/s)/kW	0.02	0.02	0.02
Start time to full flow (20 °C)	s	15	5	5
Start time to full flow (−20 °C)	s	30	15	15
Transient response 10–90% and 90–0%	s	1.75	0.75	0.75
Fuel purity (H_2 from storage)	% H_2	99.99 (dry basis)		
Environmental health and safety				
Permeation and leakage	scc/h	Meets or exceeds applicable standards		
Toxicity	–			
Safety	–			
Loss of useable H_2	(g/h)/kg H_2 stored	1	0.1	0.05

Source: (Satyapal *et al.*, 2006).

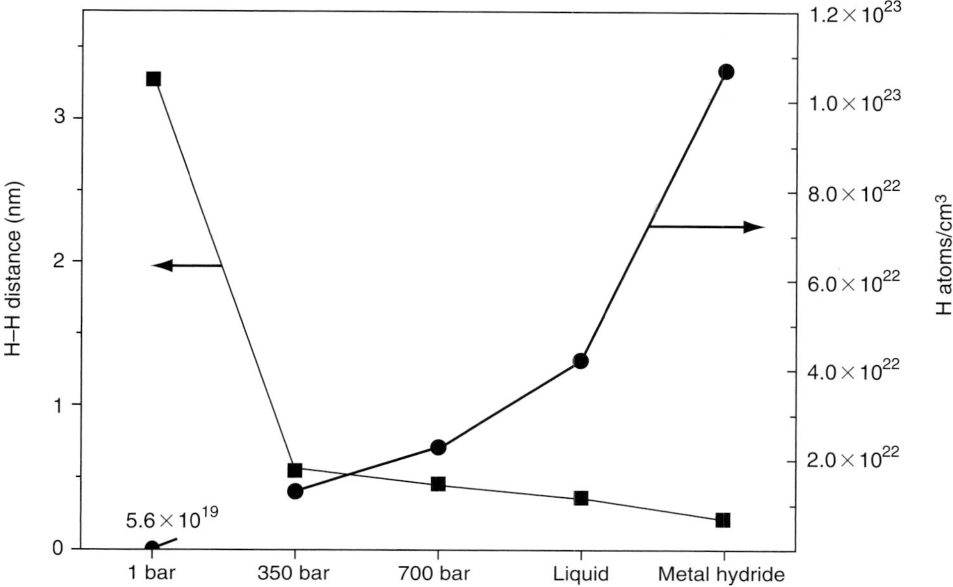

Figure 11.1. Average H–H distances and volumetric storage densities of gaseous (pressurised) hydrogen, liquid hydrogen and the physical limit of hydrogen in a metal lattice.

11.2 Overview of hydrogen storage options

The physical and chemical properties of hydrogen impose technical boundary conditions on standard methods of storing H_2 in pure form, such as a pressurised gas or cryoliquid. Table 9.1 lists the physical properties of H_2, in comparison with methane and *n*-heptane, which were chosen as representatives of natural gas and gasoline, respectively.

According to the current state of the art, five ways of storing hydrogen on-board vehicles have been proposed: pressurised hydrogen, liquid hydrogen, storage in solids, hybrid storage systems and regenerative off-board systems. Of the various options, the two conventional and technically most advanced storage systems are based on the storage of pure hydrogen in pressurised or liquid form. Both methods exhibit principal drawbacks or limitations, however, and optional methods, such as storage in solids, in so-called hybrid systems and in the form of regenerative off-board systems, have been proposed as alternatives and are under development at the moment. The following sections shall outline some of the advantages, drawbacks and limitations of the various methods currently considered as storage options in vehicles. As a fundamental comparison, Fig. 11.1 depicts the physical limits of different storage options as compressed gas, as a liquid, or chemically absorbed in a metal hydride.

11.2.1 Pressurised hydrogen

Pressurised hydrogen (CGH_2) can be stored in containers made of composite materials that have to withstand high pressures in order to carry enough fuel for an envisaged driving cycle of 400–500 km. Such a system's major drawback is that the already limited volumetric density does not increase proportionally to the operating pressure at high values because of the real gas behaviour of the hydrogen. Recent studies revealed that the gravimetric capacity of a 700 bar system (approximately 4.5 wt.% H_2) is less than in a 350 bar system with approximately 6 wt.% H_2. However, 700 bar tanks are nevertheless considered more attractive, owing to their slightly better volumetric capacity and, thus, the smaller size of the tank.

High-pressure vessels consist of an outer shell, which is reinforced by lightweight and highly stable carbon fibres and a polymer liner inside that acts as a permeation barrier for hydrogen.[1] The outer shell is enclosed by an additional protective shell to prevent mechanical damage. The tanks were tested for their safety properties and are regarded as being safe at the moment. However, there are still safety concerns related to a potential tank rupture in an accident. Severe damage is expected in certain scenarios, if the tank ruptures and hydrogen is released instantaneously, owing to the high mechanical energy which is stored in the compressed gas and the chemical energy released if the hydrogen is ignited in a mixture with air. Materials research has been performed to understand degradation effects better in pressurised storage systems and to reduce the risk of failure from fatigue and corrosion. Fatigue may be induced by mechanical stress resulting from repeatedly changing the load of the vessel and the other components of the storage system. Hydrogen corrosion may occur by physicochemical interaction of the pressurised gas with materials of the pipes and valves, or, to a smaller extent, with the carbon of the vessel.

As shown in Fig. 11.2, current cost estimates based on 500 000 units are in the range of US$ thousands, which may be still too high to be economically viable.[2] Hence, lowering the manufacturing costs will be a key aspect of future work and efforts are underway to identify lower-cost carbon fibre that can meet the required stress, strain and safety specifications for high-pressure hydrogen gas tanks. However, lower-cost carbon fibre must not interfere with the specific storage capacity and must still meet tank thickness constraints, to help meet volumetric capacity targets.

11.2.2 Liquefied hydrogen

As an alternative, liquefied hydrogen (LH_2) with a density of 70.8 kg/m^3 (compared with 39 kg/m^3 for H_2 at 700 bar) is particularly attractive for attaining higher storage densities. The hydrogen has to be cooled down to 21 K for liquefaction which,

[1] The gravimetric energy density of a pressure vessel is very much dependent on the material of the container, which in the case of carbon fibres accounts for around 60% of total storage-system costs.

[2] Current tank costs are still in the range of $2500–3000/kg hydrogen ($75–90/kWh). Costs of gasoline tanks are around $50 per piece or around $1/kWh.

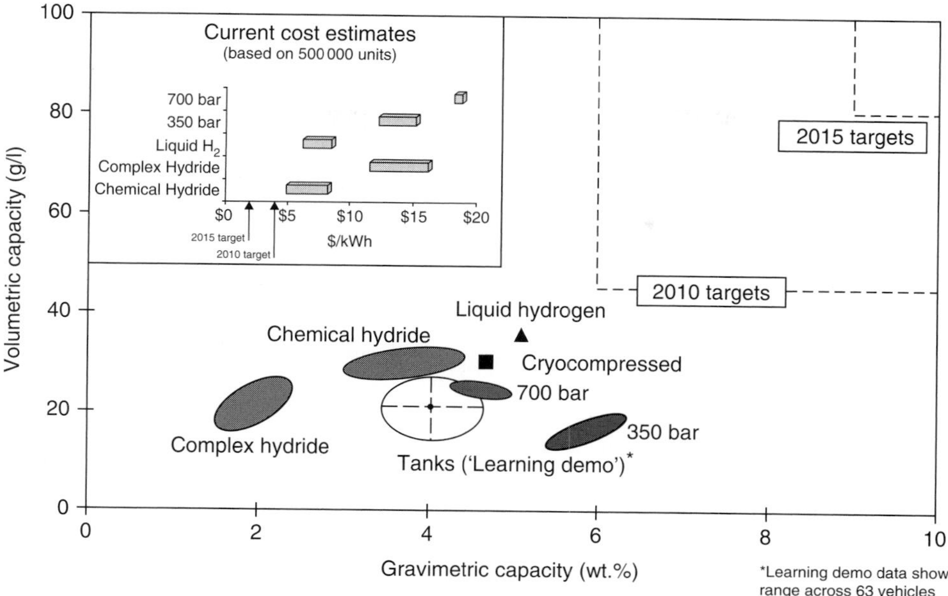

Figure 11.2 Status of current technologies relative to key system performance and cost targets (Satyapal *et al.*, 2006). Costs exclude regeneration and processing. Data are based on R&D projections and independent analysis (FY05–FY06), and are updated periodically.

however, needs over 30% of the lower heating value of hydrogen. This remains a key issue and influences fuel costs as well as fuel cycle efficiency. Overall efficiency is further reduced by the so-called boil-off phenomenon. The stored cryogenic liquid starts to evaporate after a certain period of time, owing to unavoidable heat input into the storage vessel, leading to a loss of 2%–3% of evaporated hydrogen per day (Eberle *et al.*, 2006). This cannot be prevented, even with a very effective vacuum insulation and heat-radiation shield in place. To prevent a high pressure build-up (the critical temperature of hydrogen is 32 K), the overpressure must be released from the tank, e.g., via a catalytic converter. Hydrogen boil-off is considered an issue in terms of refuelling frequency, cost, energy efficiency and safety, particularly for vehicles parked in confined spaces, such as parking garages. Recently manufactured hydrogen cars with liquid hydrogen tanks are only allowed to be parked in open spaces.

A system developed by *Linde* has minimised the evaporation losses. The system draws in the surrounding air, which is dried and then liquefied by the energy released as the hydrogen increases in temperature. The cryogenically liquefied air (−191 °C) flows through a water cooling jacket surrounding the inner tank and, thus, acts as a refrigerator causing a significant delay in the temperature increase of the hydrogen (Bossel *et al.*, 2005).

11.2.3 Hydrogen storage in solids

Several fuel storage alternatives were proposed in recent years, which are mostly based on storage in a solid that can readily take up and release large amounts of hydrogen. The thermal properties of such a system should match the operation conditions of the fuel cell, however, which means that the temperature necessary to release the hydrogen from storage should not exceed the temperature and heat content of the exhaust gas of the fuel cell (in the case of fuel-cell-driven vehicles). In this case, the waste heat of the fuel-cell system is used to release the hydrogen from the solid and no additional energy is required.

In practice, two basic bonding mechanisms are considered for hydrogen storage in solid-state materials:

1. *Chemisorption* (i.e., absorption of hydrogen), which involves dissociation of hydrogen molecules into hydrogen atoms and chemical bonding of the atoms to a host matrix. Thus, the hydrogen is integrated in the lattice of a metal, an alloy or a chemical compound.
2. *Physisorption* (i.e., adsorption of hydrogen) of molecular hydrogen by weak van der Waals forces to the inner surface of a highly porous material. Adsorption has been studied on various nanomaterials, e.g., nanocarbons, metal organic frameworks and polymers.

A principle advantage of storing hydrogen in *chemisorbed* form, e.g., as an atom in a metal hydride, is the high volumetric storage density that can be achieved by this method (see also Fig. 11.1). Hydrogen in the gaseous (70 MPa, 300 K) or liquid state consists of H_2 molecules at a mean distance of approximately 0.45 nm or 0.36 nm, respectively. These distances result from repulsive molecular interactions. The minimum H–H separation in ordered binary metal hydrides is 0.21 nm, owing to the repulsive interaction generated by the partially charged hydrogen atoms. A principal drawback of storing hydrogen as an atom in a metal matrix is the necessity to split or recombine the hydrogen molecule and form chemical bonds with the material. This requires a thermal management of the storage in order to supply or remove the heat of reaction. During the refuelling of the hydride, heat is released from the material that has to be removed. Heat recovery at the fuel station would increase the overall efficiency of the technique. During the driving cycle, heat has to be supplied to release the hydrogen from the material.

Storing hydrogen by *physisorption* needs insulated cryovessels but less heat management, because the hydrogen stays in its molecular form and the enthalpy change between the loaded and the empty form of the storage is low. The problem rather is to provide light carrier materials with a sufficient amount of bonding sites for the hydrogen per volume.[3] Moreover, physisorption interaction between the H_2 molecule and the surface is in the lower kJ/mol range (5–8 kJ/mol H_2), as a result of which it

[3] It should be mentioned that carbon nanotubes do not store more than 0.5 wt.% H_2 at room temperature, although it was claimed earlier that much higher capacities can be obtained. The initial results of carbon nanotubes with 30–60 wt.% of stored hydrogen are now considered to have been an experimental error.

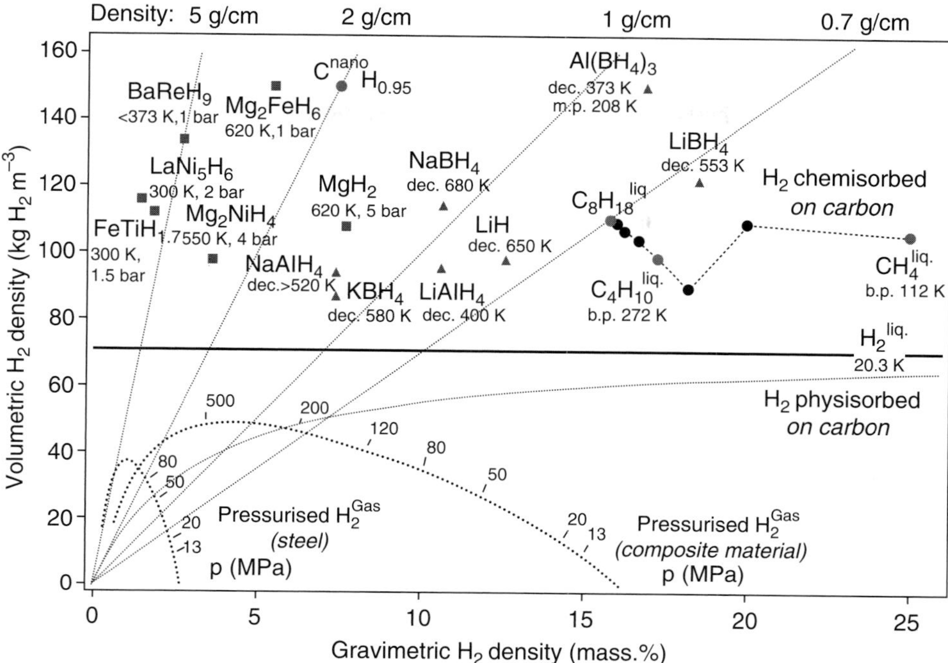

Figure 11.3. Volumetric and gravimetric hydrogen density of selected hydrides (Züttel, 2003).

may be necessary to work at very low temperatures. As a consequence, most systems have been studied at liquid nitrogen temperature (77 K).

Figure 11.3 summarises volumetric and gravimetric storage capacities of various hydrides. Physisorption materials are represented by a curve depicting the relationship between gravimetric and volumetric storage capacity for physisorbed hydrogen on carbon materials. It should be mentioned that the values in the diagram are material properties and not system values, i.e., for instance, the storage tank is excluded.[4] Hence, a trade-off of 20%–50% capacity has to be taken into account when the materials are to be integrated in a system that includes a high-pressure vessel and heat management by a heat exchanger structure.

The temperatures and pressures indicated for chemisorbed hydrogen are necessary for the release of the hydrogen (temperature) or are experimental data of equilibrium pressures for the respective temperature. The values for natural gas, diesel and gasoline are expressed as hydrogen equivalents. Mg_2FeH_6 shows the highest known volumetric hydrogen density of 150 kg/m^3, which is more than twice the value of liquid hydrogen. $BaReH_9$ has the largest H:M ratio, of 4.5, i.e. 4.5 hydrogen atoms per metal atom. $LiBH_4$ exhibits the highest gravimetric hydrogen density of 18 wt.%.

[4] This explains, for instance, the higher volumetric density of LH_2 as compared with Fig. 11.2 (which shows system performance targets), as the tank is not included.

Pressurised gas storage is shown for steel (tensile strength $\sigma_v = 460$ MPa, density 6500 kg/m^3) and a hypothetical composite material ($\sigma_v = 1500$ MPa, density 3000 kg/m^3). Chemisorbed hydrogen shows better storage characteristics than physisorbed hydrogen, both for volumetric and gravimetric density. The storage density of chemical hydrides is higher than for liquid and pressurised hydrogen.

11.2.4 Hybrid storage systems

Hybrid in this context means that a metal hydride storage tank is kept under elevated hydrogen pressure to increase both the gravimetric capacity and operational features of the system. As a pressurised storage vessel can hold a larger volume of hydrogen when using metal hydrides than without them, a greater driving range is provided. Present-day fuel-cell vehicles typically offer an inadequate driving range because of the limited amount of fuel carried by their first-generation high-pressure hydrogen-storage cylinders. Storing hydrogen in a metallic alloy at moderate pressures represents a safer and more practical method. A Japanese car manufacturer presented a first system in 2005, with 1.7 wt.% H_2 system capacity, the tank being filled with a conventional AB_2 alloy which has 1.9 wt.% storage capacity. The tank volume was 180 l, the weight was 420 kg (Mori *et al.*, 2006). Second-generation systems, which were presented recently, contain an optimised AB_2 alloy based on Mg and Ti and yield 2.2 wt.% H_2 in the system (>50 kg/m^3 H_2), which is higher than the volumetric capacity of hydrogen in a 700 bar vessel by 50%. With some 200 kg, the weight of such a system still is comparably high. However, the volumetric density is the more important parameter in this respect, according to Japanese car manufacturers (Hirose and Mori, 2006).

11.2.5 Regenerative off-board systems

Hydrogen storage in chemical compounds, combined with hydrogen released by hydrolysis or thermal annealing, may give rise to options with high energy densities and potential ease of use. This may be particularly interesting, if systems involve liquids that may be handled easily using infrastructure similar to today's gasoline refuelling stations. The chemical reactions for releasing the hydrogen are irreversible on board. Hence, the spent storage material would have to be regenerated off-board the vehicle (Satyapal *et al.*, 2007). The regeneration of the material could be accomplished in chemical plants, where the materials could be regenerated by an input of energy and hydrogen. A number of chemical systems with both exothermic and endothermic hydrogen release are currently under investigation. Some of them seem to have severe drawbacks in terms of efficiency, owing to the formation of very stable dehydrogenated products, such as oxides or hydroxides, which are produced in a hydrolysis reaction, for example.

AlH_3 might be a viable system, because it can deliver hydrogen timely by annealing at moderate temperatures. However, there is still no simple and cost-effective process to regenerate the Al, which is produced during the driving cycle and further research is needed to develop efficient regeneration methods.

11.3 Hydrogen compression and liquefaction

The energy required for the *compression* of atmospheric hydrogen depends on the final pressure as well as on the chosen compression technology. For a state-of-the-art multistage process, around 10 and 17 MJ are needed to compress 1 kg of hydrogen from atmospheric pressure to 20 and 80 MPa (200 and 800 bar), respectively, the latter needed for storage at 700 bar. This is between 9% and 15% of the LHV energy content of hydrogen (8% and 13% of the HHV). Including other losses in the process, between 15 and 20 MJ of electrical energy are needed for the compression of 1 kg hydrogen to 200 or 800 bar (Bossel *et al.*, 2005).

For *liquefaction*, the gas has to be cooled down to 20 K or −253 °C using a combination of compressors, heat exchangers, expansion engines and throttle valves. The simplest liquefaction process is the Linde or Joule–Thomson cycle. In this process, the pre-cooled gas is compressed at ambient pressure and then further cooled down in a heat exchanger. As part of the process, the hydrogen passes through 'ortho–para' conversion catalyst beds that convert most of the 'ortho' hydrogen into the 'para' form. These two types of hydrogen have different energy states. 'Ortho' hydrogen is less stable than 'para' at liquid hydrogen temperatures. It spontaneously changes to the 'para' form, releasing energy, which vaporises a portion of the liquid. By using a catalyst, such as platinum or hydrous ferric oxide, most of the hydrogen is converted into the more stable form during the liquefaction process.

Theoretically, only about 4 MJ/kg have to be removed to cool hydrogen down sufficiently and to condense the gas at atmospheric pressure. However, with a Carnot efficiency of 7%, the cooling process is very energy intensive. Current liquefaction processes are complex and efficiency analyses are mostly based on the operating data of existing hydrogen liquefaction plants. Typically, small plants are less energy-efficient than large facilities. Smaller plants produce, e.g., 182 kg H_2/h at a specific energy consumption of about 54 MJ/kg, while larger plants require only 36 MJ/kg to liquefy hydrogen. A plant for 12 500 kg/h would consume some 30.3 MJ of electricity for the liquefaction of 1 kg hydrogen, according to a recent study. This value corresponds to 21% of the HHV energy content of the liquefied hydrogen.

11.4 Hydrogen storage in stationary applications and fuel stations

Achieving a high gravimetric storage capacity is one of the greatest challenges in automotive and mobile applications. However, for stationary applications, the volumetric storage density is the more important parameter, because weight does not necessarily play a role and does not reduce the overall efficiency of the stationary energy system. Instead, there can be restrictions in space that makes systems with high volumetric storage capacities more attractive.

For large-scale underground storage of gaseous hydrogen, large underground cavities are used, similar to those for natural gas storage. Typical storage capacities for pressures of up to 50 bar (salt caverns) range from several million to several hundred million Nm^3 of hydrogen. The quantities of energy involved have the potential to meet the needs of large communities for extended periods, such as might be needed to ensure security of supply or to meet seasonal variations in energy production. This gives underground storage a special importance. Two methods of underground storage that are suitable for both hydrogen and natural gas are the use of cavities left after the mining of salt, and the use of empty aquifers (Larsen *et al.*, 2004).

At present, all the different storage options of hydrogen as pressurised gas, as a liquid or in a solid storage material have been realised in various demonstration systems. For fuel stations, CGH_2 and LH_2 seem to be the most attractive methods at the moment. Liquefied hydrogen offers the opportunity for fuelling both LH_2- and CGH_2-driven cars and its higher density can make delivery and transport easier when larger amounts of hydrogen are to be transported. However, as already mentioned in Section 11.3, liquefaction is an energy-intensive process and may not be a sustainable option on a long-term basis.

A potential future application of stationary systems is intermediate chemical storage in association with non-continuous energy sources such as wind and solar power. Systems based on transition metals, which are not considered for hydrogen storage in cars because of their heavy weight, may be suitable for chemical energy storage in stationary applications. Moreover, a hydrogen storage based on AB_5 and AB_2 alloys exhibits very high volumetric densities, and is safe and robust. Several systems have already been tested successfully in demonstration projects, to store hydrogen produced from electricity, for example generated in wind parks.

11.5 Hydrogen safety

Table 11.2 presents some of the most important safety parameters of hydrogen in comparison to the data of methane and *n*-heptane. The data show that hydrogen has by far the lowest minimum ignition energy of the three energy carriers. Moreover, the explosion or detonation range in mixtures with air exhibits the widest spread for hydrogen. Hence, the formation of hydrogen and air mixtures has to be strictly avoided in uncontrolled environments, because there is a high risk of severe incidents, mostly because of the low ignition energy and the wide detonation range. These properties, together with the fact that hydrogen is 15 times lighter than air, are the reasons why an adapted safety strategy is needed to be able to benefit from hydrogen without exposing persons to unnecessary risks.

Table 11.2. *Safety-relevant physical and chemical properties of hydrogen, methane and* n*-heptane*

		Hydrogen H_2	Methane CH_4	*n*-Heptane C_7H_{16}
Lower detonation limit	vol.%	18.3	6.3	–
Stoichiometric mixture	vol.%	29.6	9.5	1.9
Upper detonation limit	vol.%	59.0	13.5	–
Upper explosion limit	vol.%	77.0	17.0	6.7
Minimum ignition energy	mJ	0.017	0.290	0.24
Self-ignition temperature	K	833	868	488

Source: (DWV, 2007).

Research has been performed by fluid dynamics modelling and experimental validation to assess the behaviour of hydrogen and air mixtures in various spatial environments, such as private and public garages, tunnels, etc. The H_2-release situation, mixing, ignition, flame propagation and pressure wave expansion can be modelled in three dimensions for various scenarios and mechanical and thermal loads at various places are obtained in the calculations. As a result of the studies, suggestions can be made for optimised geometries, where the light gas cannot accumulate and is released at low concentrations into the environment.

First studies have shown that critical situations can be avoided if the construction of private and public garages is adapted. It was demonstrated that natural convection may be sufficient to guide rising hydrogen away from a leaking tank to the outer environment, where it is diluted in air. A slightly tilted roof may channel the hydrogen to an area where it can leave the room, for example. Other combined theoretical and experimental studies have resulted in the recommendation of minimum distances between hydrogen fuel stations and surrounding buildings. Of particular importance is the modelling of the interior in cars. Detailed knowledge of the behaviour of hydrogen in voids and the passenger cabin may reduce the number of necessary hydrogen sensors on board and may give hints for safer construction of the car.

11.6 Summary

On-board hydrogen storage for vehicles is challenging and may have significant impact on hydrogen infrastructure and standards. A great deal of progress was achieved during recent years concerning H_2-propelled vehicles. Most of the development effort concentrated on the propulsion system and its vehicle integration. Nowadays, it is generally agreed in the automotive industry that on-board storage of hydrogen is one of the critical bottleneck technologies for future car fleets.

The target is to store 4–5 kg of hydrogen while minimising volume, weight (gravimetric density >5–6 wt.%), storage energy, refuelling time, costs and hydrogen on-demand release time. Still, no approach exists to comply with the technical requirements for a range greater than 500 km, while meeting all performance parameters, regardless of costs. The physical limits for the storage density of compressed and liquid hydrogen have more or less been reached, while there is still potential in the development of solid materials for hydrogen storage (see Fig. 11.2). Storage in solid materials may offer decisive advantages (smaller volume, low pressure and energy input), but development is still in progress, with a number of materials under investigation, of which metal hydrides are the most developed. The development of on-board reformers to produce hydrogen from fossil fuels also proved to be very challenging and expensive and is no longer a major option for car manufacturers.

A CGH_2 system's major disadvantage is that the already limited volumetric density does not increase proportionally with the operating pressure at high values. Furthermore, the fabrication costs of a 700 bar vessel are still too high, mostly because of the high costs of carbon fibres. A technological breakthrough is needed in this case. Otherwise, the costs will be too high for the consumers, especially if the lifetime of the vessel is less than the lifetime of the car, so that it has to be exchanged in a service interval.

For LH_2 storage systems, the problems of cooling-down losses during refuelling at the filling station and the so-called boil-off phenomenon during parking have to be addressed. However, it is not possible to reduce the losses to zero and further insulation efforts or active cooling will increase the overall costs of the system.

As far as alternative storage systems are concerned, hybrid systems of a metal hydride and pressurised hydrogen seem to be most promising at the moment. It is a much safer option than the other methods, and systems of this kind can absorb or deliver large amounts of hydrogen even at temperatures below freezing point. However, the weight of such a system is still too high and has to be further reduced by using optimised storage materials.

As far as the economics of hydrogen storage are concerned, it should be noted that currently (in 2007), any hydrogen solution will be more expensive than oil, owing to the inherent characteristics of hydrogen handling compared with liquid gasoline or diesel. However, a drastic rise in oil and gas prices is foreseen for the time after 2010 by the International Energy Agency (IEA, 2007) and it is expected that the situation will change in the mid-term.

References

Bossel, U., Eliasson, B. and Taylor, G. (2005). *The Future of the Hydrogen Economy: Bright or Bleak?* Report E08, 26 February 2005, European Fuel Cell Forum, www.efcf.com/reports.

DWV (2007). *Vergleichstabelle für physikalische und chemische Eigenschaften von Wasserstoff und anderen Stoffen (Gasen, Energieträgern).* Deutscher Wasserstoff- und Brennstoffzellen-Verband e.V. (German Hydrogen and Fuel Cell Association). www.dwv-info.de/wissen/tabellen/wiss_vgl.html.

Eberle, U., Arnold, G. and von Helmolt, R. (2006). Hydrogen storage in metal–hydrogen systems and their derivatives. *Journal of Power Sources*, **154** (2), 456–460.
Hirose, K. and Mori, D. (2006). Toyota's vision of the development of hydrogen storage materials for vehicular applications. *MH2006 International Symposium on Metal–Hydrogen Systems Fundamentals and Applications*, Lahaina, Maui, Hawaii, 1–6 October, 2006.
International Energy Agency (IEA) (2007). *Medium Term Oil Market Report – July 2007*. Paris: OECD/IEA.
Larsen, H., Feidenhans'l, R. and Sønderberg P. L. (2004). *Hydrogen and its Competitors*. Risø Energy Report 3. Roskilde: Risø National Laboratory.
Mori, D., Kobayashi, N., Shinozawa, T. *et al.* (2005). Hydrogen storage materials for fuel cell vehicles high-pressure MH system. *Journal of Japan Institute of Metals*, **69** (3), 308–311.
Satyapal, S., Read, C., Ordaz, G. and Thomas, G. (2006). *Hydrogen Storage*. US DOE Hydrogen Program, 2006 Annual Merit Review Proceedings. www.hydrogen.energy.gov/annual_review06_plenary.html.
Satyapal, S., Petrovic, J., Read, C., Thomas, G. and Ordaz G. (2007). The US Department of Energy's National Hydrogen Storage Project: progress towards meeting hydrogen-powered vehicle requirements. *Catalysis Today*, **120** (3–4), 246–256.
Züttel, A. (2003). Materials for hydrogen storage. *Materials Today*, **6** (9), 24–33.

Further reading

Bossel, U. (2003). The physics of the hydrogen economy. *European Fuel Cell News*, **10** (2). www.efcf.com/reports.
Fichtner, M. (2005). Nanotechnological aspects in materials for hydrogen storage: review. *Advanced Engineering Materials*, **7** (6), 443–455.
International Energy Association and Hydrogen Implementing Agreement (IEA/HIA) (2006). *Hydrogen Production and Storage. R&D Priorities and Gaps*. Paris: OECD/IEA.
Satyapal, S., Read, C., Ordaz, G. and Thomas, G. (2006). *Hydrogen Storage*. US DOE Hydrogen Program, 2006 Annual Merit Review Proceedings. www.hydrogen.energy.gov/annual_review06_plenary.html.
Sørensen, B. (2005). *Hydrogen and Fuel Cells. Emerging Technologies and Applications*. Elsevier Academic Press.

12

Hydrogen distribution

Michael Ball, Werner Weindorf and Ulrich Bünger

An area-wide supply of hydrogen will, in the medium to long term, require the implementation of an extensive transport and distribution infrastructure. In addition, a dense network of refuelling stations will have to be put in place. This chapter first addresses the various options for hydrogen transport and their characteristics. Subsequently, different fuelling station concepts will be discussed.

12.1 Transport options for hydrogen

Three main options are used today for hydrogen transport: delivery of compressed gaseous and liquid hydrogen by trailers and of gaseous hydrogen by pipelines. The technical and economic competitiveness of each transport option depends on transport volumes and delivery distances. As hydrogen transport costs could be considerably reduced if the existing natural gas pipeline infrastructure could be used, further possibilities under consideration are the adaptation of natural gas pipelines for hydrogen transport or the transport of hydrogen and natural gas mixtures. As for hydrogen transport by ship, so far only different concept studies have been developed.

12.1.1 Gaseous-hydrogen transport

12.1.1.1 Hydrogen compression

Hydrogen compression is a prerequisite for the transport of hydrogen either by pipeline or in gaseous form by trailers. The compression of hydrogen is less energy intensive than liquefaction. While today's liquefaction plants need about 0.33 kWh_{el}/kWh_{H_2} ($11 kWh_{el}/kg_{H_2}$), for the compression of hydrogen from ambient pressure to 800 bar via a five-stage compressor only about 0.13 kWh_{el}/kWh_{H_2} (21 MJ/kg) is required (0.10 kWh_{el}/kWh_{H_2} (16 MJ/kg) for 200 bar). This is between 10% and 13% of the LHV energy content of hydrogen. But the hydrogen pressure at the outlet

The Hydrogen Economy: Opportunities and Challenges, ed. Michael Ball and Martin Wietschel. Published by Cambridge University Press.

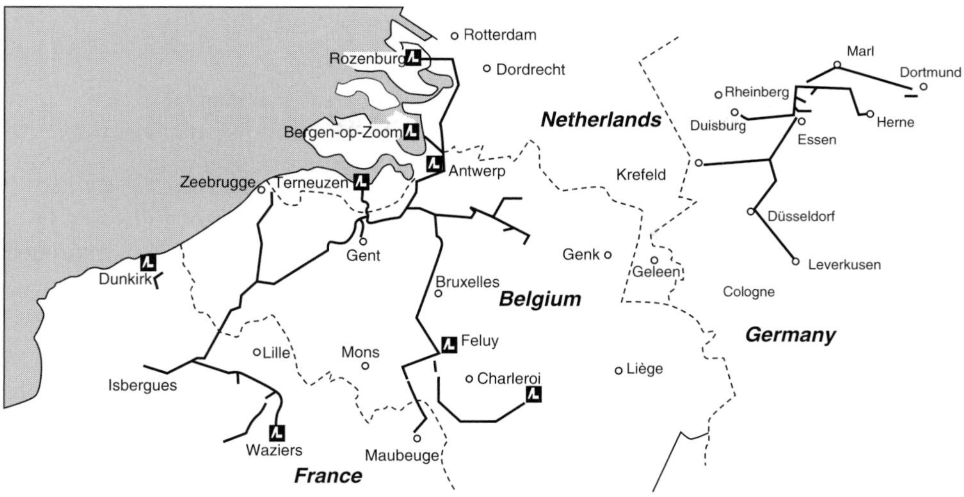

Figure 12.1. Hydrogen pipeline network in North-western Europe (Roads2HyCom, 2007).

of electrolysers or pressure-swing adsorption plants, such as from steam methane reformers, is at least 20 bar, which would significantly reduce the compression energy, owing to the logarithmic relation between compression power and pressure. (Compressing hydrogen from ambient pressure to 10 bar takes almost as much energy as compressing hydrogen from 10 to 100 bar.) For hydrogen from a pipeline grid, the pressure should be at least 5 bar. In the longer term, hydrogen filling stations will probably be connected on a pipeline grid, which delivers hydrogen at a pressure of 20 bar. If the pressure at the outlet of the electrolyser was 30 bar, the electricity requirement for the compression to 800 bar would only amount to about 0.06 kWh per kWh of hydrogen or 6% of the LHV of the delivered compressed gaseous hydrogen.

For this reason, electrolysers or steam reformers, which supply hydrogen at ambient pressure, would not be used. Compression costs have been calculated based on the assumption that hydrogen is available at a minimum pressure of 30 bar from any production technology and that the pipeline outlet pressure is also 30 bar.

12.1.1.2 Pipeline transport

Pipelines have been used to transport hydrogen for more than 50 years and, today, there are about 16 000 km of hydrogen pipelines around the world that supply hydrogen to refineries and chemical plants; dense networks exist for example between Belgium, France and the Netherlands, in the Ruhr area in Germany and along the Gulf coast in the United States (see Fig. 12.1). Existing hydrogen pipelines are about 25–30 cm in diameter and usually operate at a pressure of 10–20 bar, but pressures up to 100 bar can also be used (IEA, 2005).

For transporting hydrogen by pipeline, three principal options exist:

1. build new, dedicated hydrogen distribution networks,
2. adapt existing natural gas pipelines for hydrogen transport (if possible), or
3. blend hydrogen with natural gas up to a certain extent and either separate the two at the delivery point, or use the mixture, e.g., in stationary combustion applications.

In the following, the characteristics of pure hydrogen pipelines are addressed first. The pressure drop occurring in the pipe due to friction when transporting hydrogen (or any other gas) depends on the pipe diameter, the gas throughput, the surface properties of the pipe material, the pressure level in the pipe and the density of the gas. Generally, the pressure drop needs to be compensated by recompression (booster stations) every 200–300 km. The pressure drop in high-pressure gas pipelines can be calculated according to the Darcy–Weisbach equation as follows:

$$p_{\text{in}}^2 - p_{\text{out}}^2 = \lambda \cdot \frac{16}{\pi^2} \cdot \rho_0 \cdot p_0 \cdot \frac{T}{T_0} \cdot l \cdot K \cdot q^2 \cdot \frac{1}{d^5}, \qquad (12.1)$$

where:

p_{in}, p_{out}	Inlet and outlet pressure of pipeline (Pa),
λ	Pipe friction factor,
ρ_0	Gas density under normal conditions (kg/m^3),
p_0	Normal pressure (101 300 Pa),
T	Gas temperature (K),
T_0	Normal temperature (273.2 K),
l	Pipeline length (m),
K	Compressibility factor of the gas,
q	Volume flow under normal conditions (m^3/s),
d	Pipeline diameter (m).

For a given pipeline length, transmission capacity, pipeline diameter and outlet pressure, the required inlet pressure to the pipeline can be calculated according to Eq. (12.1). It is assumed that the hydrogen is available from the production plants at a pressure of at least 30 bar, so that an energy-intensive primary compression from ambient pressure is not needed. The compressor power P to bring the hydrogen to the required pipeline inlet pressure can be calculated according to the equation for isentropic compression, as follows:

$$P = \frac{\kappa}{\kappa - 1} \cdot R_S \cdot T \cdot \dot{V}_0 \cdot \rho_0 \cdot \left[\left(\frac{p_1}{p_0} \right)^{\frac{\kappa-1}{\kappa}} - 1 \right], \qquad (12.2)$$

where:

κ	Isentropic exponent of gas (hydrogen: 1.41)
R_S	Specific gas constant (J/(kg K))
ρ_0	Gas density under normal conditions (kg/m^3)
T	Gas temperature (K)
$\dot{V}_0$	Normal temperature (273.2 K)
p_0, p_1	Inlet and outlet pressure of the compressor (Pa).

Table 12.1. *Pipeline transport*

		Transport		Distribution	
Technical data					
Diameter	m	0.25	0.25	0.25	0.10
Capacity	MW_{H_2}	100	600	1500	2.4
Inlet pressure	bar	30	30	30	30
Outlet pressure	bar	30	30	30	30
Lifetime	years	30	30	30	30
Economic data					
Specific investment pipeline: min.	*k€/km*	*500*	*500*	*500*	*180*
Specific investment pipeline: Reference	k€/km	560	560	560	250
Specific investment pipeline: max.	*k€/km*	*620*	*620*	*620*	*350*
Specific investment compressor	k€/km	10	60	140	0
Fixed costs[a]	% Investment/year	1	1	1	1
Variable costs, compressor[b]	$ct/MWh_{H_2}/km$	0.093	0.42	0.64	0.0001
Average distance	km	300	300	300	50
Total costs	$\mathbf{ct/kWh_{H_2}}$	**2.51**	**0.57**	**0.40**	**9.30**
Share annualised investment	%	90	71	47	91
Share fixed costs	%	9	7	4	9
Share variable costs	%	1	22	49	0

Notes:
[a] Maintenance, etc.
[b] Electricity price: 4.4 ct/kWh.
Sources: Own calculations according to (Amos, 1998; Mintz *et al.*, 2002; Parker, 2004).

Table 12.1 displays the technoeconomic data for hydrogen pipeline transport used in the MOREhyS model. In this model, a distinction is made between transport pipelines and distribution pipelines (to the fuelling stations), which is reflected in different pipeline diameters. Pipeline capacities and diameters are oriented at today's industrial hydrogen pipelines. The delivery pressure at the fuelling station is set at 30 bar. Capital costs for hydrogen pipelines are derived from natural gas pipelines; owing to the higher material requirements for hydrogen pipelines (e.g., to avoid embrittlement of steel) the specific investments are estimated to be up to twice that of comparable natural gas pipelines (see also later). A detailed assessment of pipeline capital costs, however, is difficult as it depends very much on local conditions such as topography, land use or rights of way. In addition, steel costs account for a significant share of pipeline costs and thus volatility in steel prices can markedly affect overall pipeline economics (as experienced in recent years).[1] The variable costs for pipeline transport are largely due to compression: the specific compression costs have been calculated for an average distance of 300 km for transport pipelines and 50 km for distribution pipelines. For the compression, an electric drive is further assumed; alternatively, a part of the transported hydrogen could be used to drive the compressor (as in the case of natural gas pipelines). It can be taken from Table 12.1 that pipeline transport is very capital-intensive.

Because of the specific physical and chemical properties of hydrogen, the pipelines must be made of (non-porous) materials of high quality, such as stainless steel. Therefore, the investments for a hydrogen pipeline of a given diameter are 1.4 to 2 times that of natural gas pipelines (Castello *et al.*, 2005; Mintz *et al.*, 2002). Because of its lower molecular weight and viscosity, hydrogen flows 2–2.5 times faster than natural gas in a pipeline under the same conditions of pipe diameter and pressure drop. However, because of the lower heating value of hydrogen,[2] such a hydrogen pipeline carries about 30%–40% less energy than a natural-gas pipeline. In consequence, hydrogen pipelines need to operate at higher pressures to supply the same amount of energy, or have to be of larger diameter.

The pressure drop, and as a result the energy requirement to compensate the pressure drop, strongly depends on the layout of the pipeline. The energy required to move hydrogen through a pipeline is between 3.8 and 4.6 times higher per unit of energy than for natural gas, if the same pipeline diameter and the same energy throughput is used. But increasing the pipeline diameter decreases the pressure drop significantly.

There is a compromise between the capital costs of the pipeline and the pressure drop and, as a result, the energy requirement for the hydrogen transport. For economic reasons, the layout of a hydrogen pipeline would never be the same as for a natural-gas pipeline. Since hydrogen is more expensive per unit of energy transported than natural gas, the layout of the pipeline would be probably more towards a larger diameter (as this has the biggest impact on the transport volume) and, as a result, a lower pressure drop and a lower energy requirement. Therefore, the energy requirement for the transport of hydrogen in a future hydrogen pipeline need not be higher than that for the transport of natural gas today. In Wagner *et al.* (2000), the energy loss from hydrogen-fuelled compressors for the transport of hydrogen over a distance of 2500 km would amount to about 8% of the transported hydrogen if a reasonable layout of the pipeline were selected.

If existing pipeline configurations were used, the energy transport capacity would generally be lower than for natural gas. But it can be assumed that the introduction

[1] In this respect, the specific investments for pipelines assumed here are rather conservative. Owing to the high steel prices, for typical long-distance natural gas pipelines (with a diameter of 10 inches or 25 cm) capital costs are currently (2008) estimated at about k€1000/km.

[2] The lower heating value of hydrogen in the pipeline is about one-third that of natural gas until at a pressure of 100 bar. Long-distance pipelines typically are designed for a pressure of 30 to 100 bar.

of hydrogen is likely to come along with a reduction in energy demand, especially in the transport sector, and hence also a lower demand of hydrogen. Therefore, the energy throughput would be lower too, leading to a lower pressure drop, even if existing natural-gas pipelines were used for hydrogen transport. Pipelines also have an additional advantage in providing a store for hydrogen; excess hydrogen can be stored in pipelines by allowing the pressure to rise.

In the event of a leak in confined situations, hydrogen can be ignited more easily than natural gas and has much wider flammability and detonability ranges, which makes it potentially more hazardous than natural gas. Outdoors, however, hydrogen will disperse rapidly upon leakage, owing to high buoyancy, thereby reducing hazards to safe levels in a much shorter time than for other fuels.

12.1.1.3 Hydrogen transport in natural-gas pipelines

Several studies discuss the use of existing natural-gas pipelines to transmit either pure hydrogen with modifications or hydrogen and natural gas blends, because the hydrogen transport costs could be considerably reduced if the natural-gas infrastructure could be adapted to hydrogen.

The feasibility of adapting natural-gas pipelines for hydrogen transport depends on the materials of the gas pipes (notably the carbon content of steel) and, in principle must be investigated on a case-by-case basis. Hydrogen pipelines need to be made of non-porous materials, such as stainless steel,[3] as hydrogen can diffuse quickly through most materials and seals and can cause severe degradation of steels, mainly by hydrogen embrittlement (Castello *et al.*, 2005). (Hydrogen embrittlement is characterised by a loss of ductility of a steel, leading to eventual fracture of and cracks in the material.)

The oil and gas industry has always been troubled by internal and external hydrogen attack on steel pipelines, described variously as hydrogen-induced cracking (or corrosion), hydrogen-corrosion cracking, stress-corrosion cracking, hydrogen embrittlement and delayed failure (Leighty *et al.*, 2003). Elements that form metal hydrides, such as Ti, Zr, Nb, Ta and V, should be avoided in alloys for hydrogen pipelines (Müller and Henning, 1989). Thermomechanical-treated steels of the StE 480.7 TM (X 70) type are usually used for the construction of natural-gas pipelines. From former investigations, it can be deduced that hydrogen embrittlement also occurs in the case of thermomechanical-treated steels of StE 480.7 TM (X 70) type. Through admixture of 150 ppm O_2, its influence (that of embrittlement) on elongation at fracture and contraction at fracture can practically be eliminated, if an appropriate rate of elongation can be maintained (i.e., at continuous haul-off speed,

[3] For the transport of hydrogen via pipeline, alternative materials such as fibre-reinforced polymer (FRP) are also being discussed. The polymer consists of polyethylene terephthalate (PET). Reinforced-polymer-made pipelines need a liner to avoid hydrogen permeation. The hydrogen losses due to leakage and permeation must not exceed 0.5% of the hydrogen throughput of the pipeline. The investment for a 4-inch (~100 mm), 1000 psi-rated (~7 MPa) fibre-reinforced-polymer-made pipeline is expected to be US$ 50 000 to US$ 100 000 per mile including installation (Smith *et al.*, 2005).

a constant elongation rate can be maintained, so long as no contraction of the sample occurs) (Kußmaul and Deimel, 1998). Therefore, no hydrogen embrittlement occurs, if the pipeline is not deformed beyond a certain extent (e.g., after an earthquake). As a result, a natural-gas pipeline could be used for the transport of hydrogen even for high pressures, if minor amounts of oxygen are added. Internal coating and lining of pipelines could be another solution for transmission pipelines made from steel (Castello *et al.*, 2005).

Next to the potential problem of embrittlement with high-pressure transmission of hydrogen in steel pipes, valves, manifolds and, in particular, compressors of existing natural-gas pipelines would need to be adapted to hydrogen use, as they are optimised to work under a certain range of conditions, such as gas composition (Castello *et al.*, 2005). As for the natural gas distribution network, pressure is low (around 4 bar), so cheaper plastic pipes are usually used, such as PVC (polyvinyl chloride) and HDPE (high density polyethylene). Because these materials are too porous, hydrogen diffusion would prohibit the transport of hydrogen in low-pressure natural-gas distribution pipelines.

Hydrogen could potentially be used in all stationary applications where natural gas is also being used, such as in domestic boilers, gas engines or gas turbines. To do so, hydrogen could be mixed with natural gas and distributed through the existing natural-gas pipeline network. This leads to the question of how much hydrogen can be tolerated in the different parts of the pipeline infrastructure (high- and medium-pressure steel pipelines, low-pressure plastic pipelines and associated components) as well as in the end-use applications. According to the IEA (2005), up to 3 vol.% hydrogen can be added to natural gas without any changes to pipelines or devices being needed.

As for the transport of hydrogen and natural gas mixtures, it is not yet possible to say conclusively what is the maximum amount of hydrogen that can safely be added to the existing natural-gas network. Current assessments indicate that it may be possible to add up to 30 vol.% (10% in energy terms) of hydrogen to the existing high-pressure natural-gas transmission lines with existing steel materials, requiring no modifications (NATURALHY, 2007). A special problem with transporting hydrogen and natural gas mixtures is that variations in the consumption of gas over the year would lead to variations in the composition of the gas mixture, if the supply of hydrogen is constant over the year, a likely pre-condition for the economic production of hydrogen. Alternatively, either the load factor of the hydrogen plant would need to be lower, raising the cost of hydrogen or hydrogen gas storage would be needed (IEA, 2005).

The maximum tolerable percentage of hydrogen in boilers, for instance, is still a matter of research; depending on the addition, a modification of the burners would be required. Current estimates indicate a feasible range from less than 10% to up to 50% hydrogen (by volume) in natural gas, depending on the type of boiler (NATURALHY, 2007). Adding the maximum amount of hydrogen to natural-gas pipelines is, therefore, likely to require new end-use devices. Another difficulty is that of possible fluctuations of the hydrogen to natural gas ratio.

Separating the hydrogen from the natural gas and hydrogen mixture to supply, for example, pure hydrogen for transport applications would be cost-effective only if there is a cheap technology to separate the pure hydrogen from the natural gas at the point of use. Technically it could be achieved by means of suitable membranes, but owing to the high energy penalty and relatively high hydrogen loss for separating hydrogen at low concentrations (below 30–50 vol.%), membrane separation is not a real option.

Given that hydrogen is an expensive and valuable commodity, it is still a matter of debate as to what extent it is sensible to transport or use natural gas and hydrogen mixtures. From the above, it can be concluded that the widespread introduction of hydrogen would largely require a new dedicated pipeline transportation and distribution infrastructure.

12.1.1.4 Gaseous trailer transport

Seamless steel pressure vessels are the most common method in use today for hydrogen transportation at short distances (<200 km) and when small quantities are involved (up to about 500 kg). The different vessel options include cylinders, manifolded cylinder pallets and tube trailers. While single cylinders or manifolded pallets are trucked to the destination and off loaded, tube trailers, which consist of several steel cylinders mounted to a protective framework, are often left in place and replaced when empty. Transporting hydrogen in liquefied form is seven times more efficient in terms of actual hydrogen weight transported than using compressed gas cylinders.

The trailer transport of compressed hydrogen is – because of the low energy density of gaseous hydrogen – only an economic alternative for the distribution of small quantities and short distances (such as during the early phase of hydrogen penetration) and can, therefore, be excluded as an option for the area-wide implementation of a supply infrastructure in a later phase.

12.1.2 Liquid-hydrogen transport

12.1.2.1 Hydrogen liquefaction

Compared with hydrogen as a gas, liquid hydrogen has a much higher volume-related energy density and is, therefore, principally better suited for storage, for example, on-board vehicles and for transport applications (see also Chapter 11).[4] However, hydrogen liquefaction is an energy-intensive process.

To liquefy hydrogen, it has to be cooled down to about –253 °C. The thermodynamically ideal liquefaction method consists of isothermal compression and subsequent

[4] In energy terms, a 50 litre gasoline tank is equivalent to a 560 litre tank of compressed hydrogen at 350 bar, or a 330 litre tank of hydrogen at 700 bar, or a 190 litre tank of liquid hydrogen.

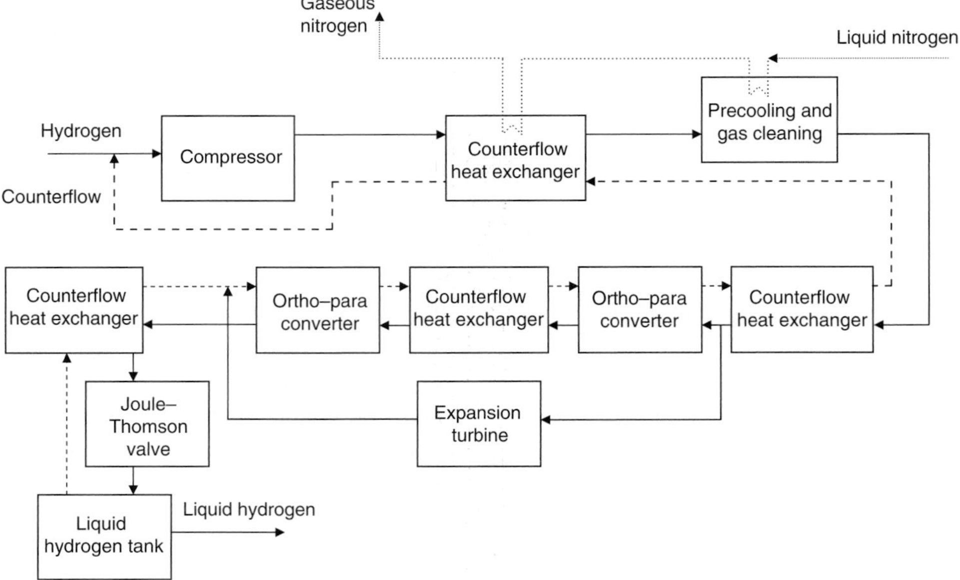

Figure 12.2. Hydrogen liquefaction.

adiabatic expansion. The expansion cools down the gas by the Joule–Thomson effect (the Joule – Thomson effect describes the property of real gases to cool under throttled expansion conditions) but this is only true for hydrogen at low temperatures. Hydrogen's thermodynamic behaviour, however, means that, in practice, multistage precooling of the gas, e.g., using liquid nitrogen, is necessary until a temperature is reached at which the Joule–Thomson effect is triggered.

Today, the liquefaction of hydrogen on an industrial scale is performed almost exclusively using the Claude process. Figure 12.2 shows a diagram of hydrogen liquefaction based on a single-stage Claude process. To start with, the hydrogen is compressed and then cooled to about –190 °C using liquid nitrogen. The coldness for further cooling is usually supplied by hydrogen expansion turbines which are operated in a closed refrigeration cycle. The closed-loop hydrogen cools as a result of expansion in the turbines and then removes the heat from the product hydrogen to be liquefied in counterflow heat exchangers. The final cooling and partial liquefaction then takes place under expansion in a throttle valve, the Joule–Thomson valve; the remaining gas is used to cool the hydrogen to be liquefied again in a counterflow.

At ambient temperature, hydrogen is composed of 75% ortho- and 25% parahydrogen, which are distinguished by whether the hydrogen molecule's atomic nuclei spin in the same direction or the opposite one. At lower temperatures, the equilibrium shifts in favour of para-hydrogen until in liquid hydrogen the para- form exists almost exclusively. During cooling, therefore, a transformation of ortho-hydrogen into parahydrogen occurs. (The 'boil-off effect' when storing liquid hydrogen is also a problem for liquid-hydrogen storage.) The point of equilibrium is reached autonomously, but without catalytic acceleration this occurs very slowly. The conversion is exothermic and hence heat is released that needs to be removed, as the slow conversion of cold ortho- to para-hydrogen in a closed tank would result in a shorter dormancy period owing to the internal heat release (i.e., re-evaporation of the liquefied hydrogen) (see the ortho–para converter in Fig. 12.2). If hydrogen were liquefied without catalytic conversion, the liquid hydrogen could consist of considerable shares of ortho-hydrogen, which would gradually transform into para-hydrogen by releasing heat, which would result in the unwanted evaporation of the liquid phase. Therefore, when liquefying hydrogen from ambient temperature, its heat, the enthalpy of condensation and the energy released by the ortho–para transformation all have to be extracted.

The theoretical minimum work for hydrogen liquefaction depends on the pressure of the hydrogen feed, the rate of ortho–para conversion and the temperature difference between ambient temperature and the temperature of the liquid hydrogen. The following formula is valid for ambient input and output pressures:

$$W_{\text{theoretical}} = \int_{T_0}^{T_{\text{evap}}} \frac{T_0 - T}{T} c_p \mathrm{d}T + \frac{T_0 - T_{\text{evap}}}{T_{\text{evap}}} \cdot Q_{\text{liquefaction}} + \int_{c_1}^{c_2} \frac{T_0 - T_{\text{evap}}}{T_{\text{evap}}} Q_{\text{op}} \mathrm{d}c, \quad (12.3)$$

where:

T_0	ambient temperature,
T_{evap}	temperature of the cold hydrogen at boiling point,
c_p	specific heat capacity,
$Q_{\text{liquefaction}}$	heat of evaporation or liquefaction,
c_1, c_2	ortho-hydrogen content before and after o–p conversion, respectively,
Q_{op}	reaction enthalpy of ortho–para (o–p) conversion.

The first term represents the sensible heat of cooling down from ambient to boiling point temperature, the second term represents the latent heat of the phase change, and the third term represents the energy contained in the ortho–para (o–p) conversion.

The theoretical minimum demand of work for hydrogen liquefaction depends on the pressure (Fig. 12.3). At a hydrogen feed-gas pressure of 0.1 MPa, the theoretical minimum demand of work of a liquefaction is 3.92 kWh/kg of LH_2 (Peschka, 1992).

Table 12.2 shows the technoeconomic parameters of hydrogen liquefaction plants of varying sizes. For the calculation of the investment, the plants were scaled up based on an extrapolation factor of 0.6 (see Chapter 10). The Carnot efficiency of hydrogen liquefaction was calculated from the power demand of the compressors in relation to a theoretical minimum energy demand. The consumption of other operating resources, such as cooling water and nitrogen, is not explicitly recorded because of their negligible share in the total costs of liquefaction, but these are taken into account in the fixed operating costs. The CO_2 emissions of hydrogen liquefaction are calculated (within the MOREHyS model) based on the average CO_2 emissions of the energy mix used to generate electricity; assuming a CO_2 emission factor of power

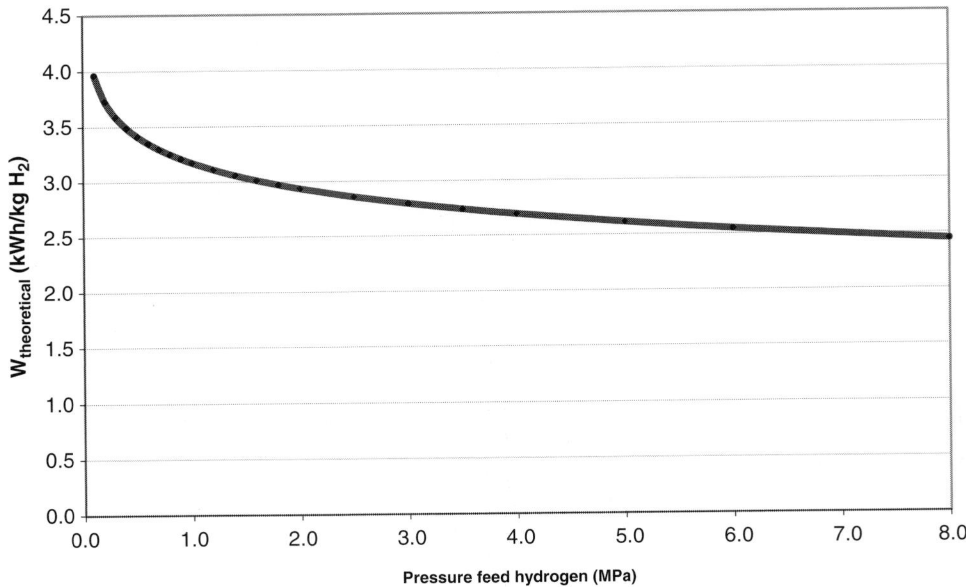

Figure 12.3. Theoretical minimum liquefaction work as a function of hydrogen pressure.

generation of 470 g CO_2/kWh (EU average), this would result in about 150 g CO_2/kWh_{H_2}.

As can be seen in Table 12.2, there are clear economies of scale in the specific supply costs of large plants. The variable costs related to electricity make up 30% to 60% of the annual total costs, depending on plant size. The electricity demand results from compressing the hydrogen to the pressure level required by the process design, as well as compressing the cooling agent (usually also hydrogen) in the refrigeration cycle.[5] The specific electricity use amounts to 10–13 kWh_{el}/kg_{H_2} (0.3–0.4 kWh_{el}/kWh_{H_2}) in today's hydrogen liquefaction plants. Typical operating pressures lie between 25 and 40 bar; the supply of liquid hydrogen usually takes place at ambient pressure (HySociety, 2004; Ullmann, 2003). The hydrogen from steam reforming or electrolysis is usually supplied at a primary pressure of 10 to 30 bar.

Figure 12.4 shows the composition of liquid hydrogen delivery costs. It can be seen that the electricity prices and costs clearly have the highest impacts on the delivery costs, while the influence of delivery distance is much smaller.

Increasing the operating pressure of the plants is one way to increase efficiency, as less energy then has to be applied later for cooling (Quack, 2001). Moreover, if the output hydrogen is available at increased pressure, as is the case, for example, in high-pressure electrolysis or coal gasification, less energy is also required for compression. The specific electricity demand could be lowered even further in the future

[5] It is assumed that the compressor is powered by electric motors. When using alternative cooling agents, such as a neon–helium mix, more efficient turbo compressors could be used for the cooling cycle, which can be powered using gas or steam turbines (Quack, 2001).

Table 12.2. *Technoeconomic data of hydrogen liquefaction*

Technical data					
Capacity	MW_{H_2}	10	50	100	300
Capacity	t LH_2/day	7	36	72	216
Annual full load, hours	h/yr	8000	8000	8000	8000
Inlet pressure	bar	30	30	30	80
Specific electricity demand	kWh_{el}/kWh_{H_2}	0.40	0.33	0.31	0.22
Electrical nominal power	MW_{el}	3.9	16.7	31.2	66.6
Theoretical demand of work	kWh/kWh_{H_2}	0.084	0.084	0.084	0.073
Carnot efficiency	%	21	25	27	33
Lifetime	years	30	30	30	30
Economic data					
Specific investment	€/kW_{H_2}	2800	1500	1000	733
Fixed costs[a]	% Invest/year	2.5	2.5	2.5	2.5
Variable costs[b]	ct/kWh_{H_2}	1.76	1.45	1.36	0.97
Year of availability		Today	Today	Today	2020
Total costs	**ct/kWh_{H_2}**	**6.35**	**3.91**	**3.00**	**2.17**
Share annualised investment	%	58	51	45	45
Share fixed costs	%	14	12	10	10
Share variable costs	%	28	37	45	45

Notes:
[a]Maintenance, labour etc.
[b]Electricity price: 4.4 ct/kWh.
Sources: own calculations based on (Bossel and Eliasson, 2003; Bossel *et al.*, 2005; Syed *et al.*, 1998; Valentin, 2001; Zittel *et al.*, 1996).

by optimising the cooling cycles (e.g., by using a helium–neon gas mixture as a cooling agent to improve the efficiencies of compressors and turbines), using magnetocaloric refrigeration processes or shifting the ortho–para conversion to higher temperature levels using catalysts. Different authors assume that an electricity demand of about 7 kWh_{el}/kg_{LH_2} is feasible at a pressure of 0.1 MPa for both feed-gas hydrogen and product and with the corresponding plant sizes. At a feed-gas hydrogen pressure of 6 MPa and a pressure of the product (LH_2) of 0.1 MPa, an electricity requirement of less than 5 kWh/kg_{LH_2} is feasible (Quack, 2001). However, whether large liquefaction plants are built will depend decisively on the future development of hydrogen demand. Onsite liquefaction at filling stations can be ruled out for both technical and economic reasons, because of the complexity of the process.

Up to now, there are only about ten commercial-scale hydrogen liquefaction plants worldwide, of which the largest production capacities are located in the USA, which

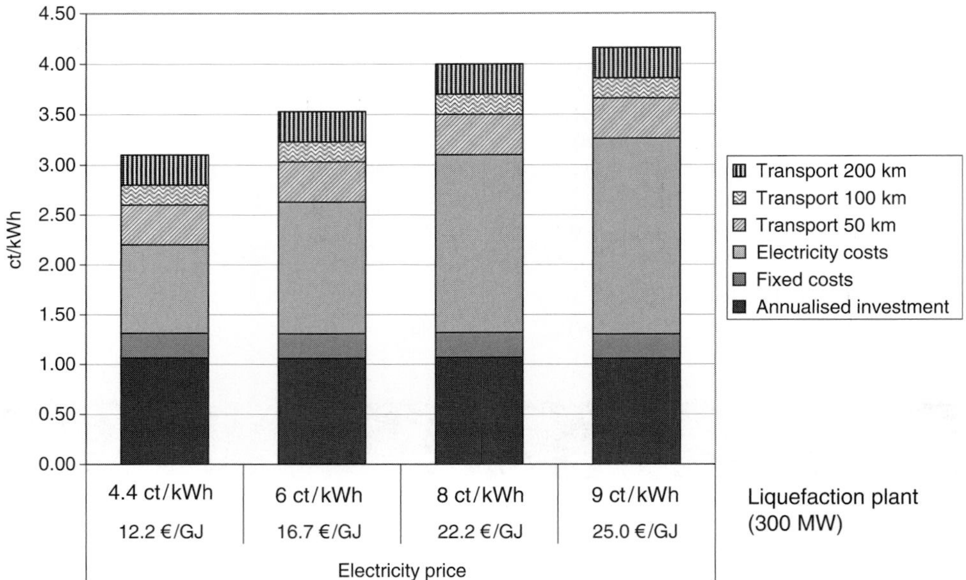

Figure 12.4. Composition of LH_2 delivery cost.

supply 25 to 60 tonnes of LH_2 per day. The semiconductor industry is one of the main consumers of liquid hydrogen, owing to its ultra-high purity requirements. There are only four hydrogen liquefaction plants in Europe: one operated by Air Liquide in Waziers (France) with a capacity of 10.5 t LH_2/d, one by Air Products in Rozenburg (the Netherlands) with 5.4 t/d, and two by Linde in Ingolstadt and Leuna (both Germany), with a capacity of 4.4 t/d and 5 t/d, respectively (Roads2HyCom, 2007). The total hydrogen liquefaction capacity in Europe is presently 25 t/day (equivalent to about 300 GWh per year at an equivalent full load period of 8000 hours).

12.1.2.2 Liquid trailer transport

For road transport, liquid hydrogen (LH_2) is transported in cylindrical super insulated cryogenic vessels in a semitrailer (see Fig. 12.5). The gross weight of a truck capable of carrying the LH_2 container is typically about 40 t. The investment for a LH_2 semitrailer amounts to about €500 000. The investment for a tractor capable to haul a semitrailer is about €160 000. Table 12.3 displays the technical characteristics of hydrogen trailers.

Table 12.4 shows the technoeconomic assumptions for trailer transport of liquid hydrogen. A large part of the trailer costs are the wages of the driver. The variable costs increase with increasing transport distance, because of the fuel consumption of the trailers themselves. To be able to make a direct cost comparison between trailer and pipeline transport, the liquefaction costs (of which electricity costs account for some 30%–60%) need to be taken into account as well.

Table 12.3. *Technical data of a 53 100 l liquid hydrogen tank used for LH_2 semitrailers*

Warm gross volume	53 100 m^3
Cold volume	52 570 m^3
Maximum pressure	160 Psi (~1.1 MPa)
Transport capacity	3463 t LH_2
Length	45 ft (~13.7 m)
Gross weight of the container	25.5 t

Source: Gardner Cryogenics, Personal communication, 1994.

Figure 12.5. LH_2 semitrailer with tractor (Linde).

12.1.2.3 Maritime-hydrogen transport

Maritime-hydrogen transport is a novel concept and has not yet been realised. But detailed design studies for large seagoing containers for capacities of 3600, 24 000, 50 000 and 100 000 Nm^3 have been performed in Europe (EQHHPP, Germanischer Lloyd/HDW/LGA) and Japan (IHI in WE-NET). The autonomy of these transport vessels (i.e., the time before the first evaporated LH_2 has to be blown off in gaseous form) depending on layout criteria for design is between 30 and 60 days.

In a study undertaken by Germanischer Lloyd, Howaldtswerke Deutsche Werft (HDW) and Noell-LGA during 1988 and 1990, a number of different designs for LH_2 carriers for maritime LH_2 transport were examined. The study was supported by the German Ministry for Research and Technology (BMFT). One concept for

Table 12.4. *Trailer transport, liquid hydrogen*

		300 km (one way)	50 km (one way)
Technical data			
Gross weight truck (including H_2 load)	t	40	40
Transport capacity	t	3.5	3.5
Max. energy delivered per trip	MWh	117	117
Average driving speed	km/h	50	50
Time for loading and unloading	h	1.5	1.5
Number of trips per day		1	4
Capacity	MW_{H_2}	8.6	33.3
Annual full load hours	h/year	3240	3360
Fuel consumption (diesel)[a]	l/100 km	35	35
Lifetime	years	10	10
Economic data			
Investment	k€	500	500
Number of drivers		1	1
Wage	€/h	50	50
Working days per year		240	240
Fixed costs[a]	% Investment/a	2	2
Variable costs[b]	ct/(MWh_{H_2} km)	0.24	0.24
Total costs	**ct/kWh_{H_2}**	**1.05**	**0.26**
Share annualised investment	%	28	29
Share fixed costs	%	59	62
Share variable costs	%	13	9

Notes:
[a]Maintenance, etc.
[b]Diesel price: 0.8 €/l.
Sources: Own calculations according to (DWV, 2005; HySociety, 2004).

maritime transport of hydrogen was the SWATH (Small waterplane area twin hull) carrier concept. The transport capacity amounts to about 8150 t LH_2. For loading and unloading, the LH_2 carrier navigates between two stationary supports.

For unloading (loading), at first the draught is decreased by pumping out ballast water until the platform of the ship is higher than the surface of the two stationary supports. Then the LH_2 carrier navigates between the two stationary supports. Subsequently, the draught is increased by pumping ballast water into the ship until the platform of the LH_2 carrier with the full LH_2 tanks (empty LH_2 tanks) rests on the stationary supports. Then the ship navigates to the other pair of stationary supports with the empty LH_2 tanks (full LH_2 tanks) and the procedure is reversed.

Figure 12.6. SWATH carrier (HDW).

Therefore, two pairs of bases are required for a LH_2 terminal. Figure 12.6 shows the LH_2 carrier based on the SWATH concept.

The maximum speed of the LH_2 carrier is about 17.5 kn (32.4 km/h). The propulsion power is 41 MW. The propulsion of the SWATH carrier is based on a steam-injection gas turbine with an efficiency of 50% (Würsig, 1996). The investment for the SWATH carrier is indicated at about € 440 million, including five LH_2 tanks. The investment for one LH_2 tank is € 40 million. The investment for the LH_2 terminals (both export and import terminal) is indicated at about € 820 million (G. M. Würsig, Germanischer Lloyd, Hamburg, personal communication, January 1999).

12.1.3 Comparison of pipeline and trailer transport

In the following, the advantages and disadvantages of the various hydrogen transport options are summarised. In principle, the choice of transport means depends on the volume and the transport distance.

- Generally, distance and volume are decisive factors. For a short distance, a pipeline can be very economic because the capital expense of a short pipeline may be close to the capital cost of trailers, and there are no transportation or liquefaction costs. As the distance increases, the capital cost of a pipeline increases rapidly, and the economics will depend on the quantity of hydrogen: pipelines will be favoured for larger quantities of hydrogen. For small quantities of hydrogen, at some point the capital cost of the pipeline will be higher than the operational costs associated with delivering and liquefying the hydrogen.

- For large quantities of hydrogen, pipeline delivery is cheaper than any other option. Pipelines are characterised by very low operating costs, mainly for compressor power, but high capital costs. The investment for pipelines is in proportion to the delivery distance, while the influence of capacity is lower.
- Liquid hydrogen has a high operating cost, owing to the electricity needs for liquefaction (which account for 30%–60% of the total liquefaction costs and may also represent a significant CO_2 footprint), but lower capital cost, depending on the quantity of hydrogen and the delivery distance. The break-even point between liquid hydrogen and a pipeline will vary depending on the distance and quantity.
- Distance is a deciding factor between liquid and compressed gaseous hydrogen transport, because it may be possible to use the same tube trailer for several trips per day for a short distance. Compared with liquid hydrogen, compressed gas has lower power requirements and slightly lower capital costs for the tube trailers, but many more tube trailers are required to deliver the same quantity of hydrogen. At long distances, the capital and transport costs of the number of trucks required to deliver a given quantity of compressed hydrogen will be greater than the increased energy costs associated with liquefaction and fewer trucks. If the distance is relatively short and the quantity of hydrogen transported is small, compressed gas is likely to be the preferred option.
- Tube trailer trucks delivering compressed gaseous hydrogen have high variable costs (also per distance), owing to the small volume of hydrogen, but they are flexible and have comparatively low fixed costs. Their flexibility is also advantageous where fuelling station utilisation is too low to apply onsite production in a technically and economically reasonable way, given that the delivery distance is smaller than about 100 km. Truck delivery of liquid hydrogen results in lower variable costs per distance, owing to considerably higher hydrogen capacity and, therefore, longer distances can be covered between production and use location.
- In short: pipelines are the preferred option for large quantities and long distances; liquid hydrogen is for smaller volumes and transport over long distances, and compressed gas is for small quantities over short distances.

Figure 12.7 illustrates the costs of the various transport options and capacities as a function of transport volume and distance. For LH_2 trailer transport, the liquefaction costs are also included, and have been calculated for an electricity price of 4.5 and 8 ct/kWh$_{el}$, respectively.

12.2 Hydrogen refuelling stations

Depending on the physical condition required for the storage of hydrogen on-board the vehicle, compressed hydrogen (CGH_2) and liquid hydrogen (LH_2) refuelling stations can be distinguished. (A detailed discussion of the design and layout of different hydrogen fuelling-station concepts can be found in DWV (2005).) Typically, small, medium and large fuelling stations with capacities of 50 kg H_2/day, 500 kg H_2/day and

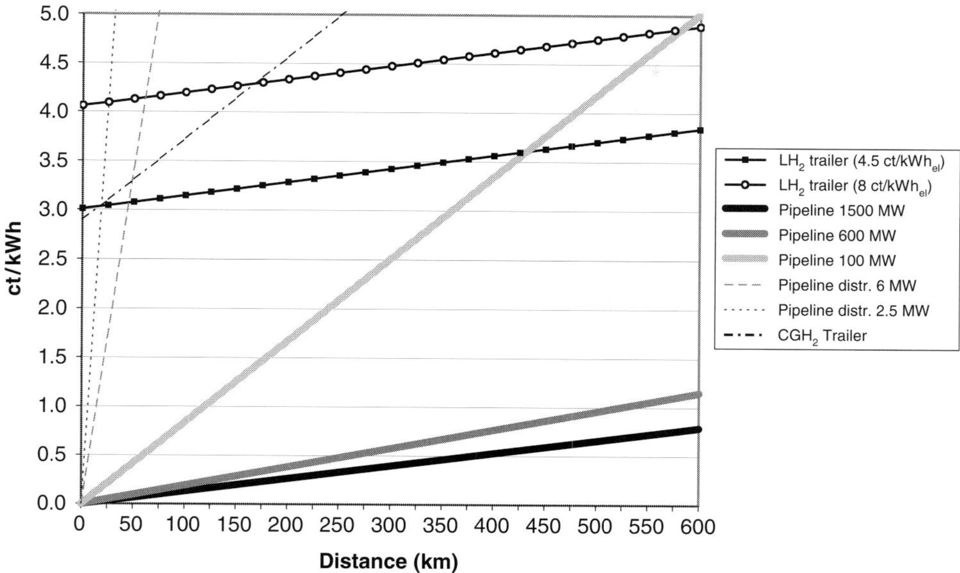

Figure 12.7. Comparison of various hydrogen transport options.

1300 kg H_2/day are distinguished. Assuming an average hydrogen consumption of fuel cell cars of 0.8 kg/100 km and a driving range of 500–600 km (4–5 kg H_2 per car), the small station can supply hydrogen for at least ten cars per day, the medium for 100 cars per day and the large station for 300–400 cars per day (which is comparable to a large conventional station). Worldwide, there are currently around 300 hydrogen fuelling stations, especially in the USA and Japan; in Europe most hydrogen stations are found in Germany. (An overview of hydrogen refuelling stations worldwide is at www.h2stations.org.) The development of codes and standards will be necessary for the widespread commissioning of fuelling stations.

12.2.1 Compressed-hydrogen (CGH_2) stations

The CGH_2 refuelling station concept is based on gaseous hydrogen being stored in compressed form, so that it only needs to be filled in the pressure tank of the vehicle during refuelling. The main components of a refuelling station for compressed gaseous hydrogen storage and dispensing are the compressors, (high-pressure) storage vessels and the dispenser with filling nozzle. Additionally, filter and gas-cleaning plants for gas conditioning might be important, as well as hydrogen sensors and other safety equipment. The compressor normally has several stages and needs to fulfil high requirements for impermeability, owing to the low density of hydrogen. The storage vessels have several storage banks, so that the compressor can feed continuously into the banks, while gaseous hydrogen is taken from another bank.

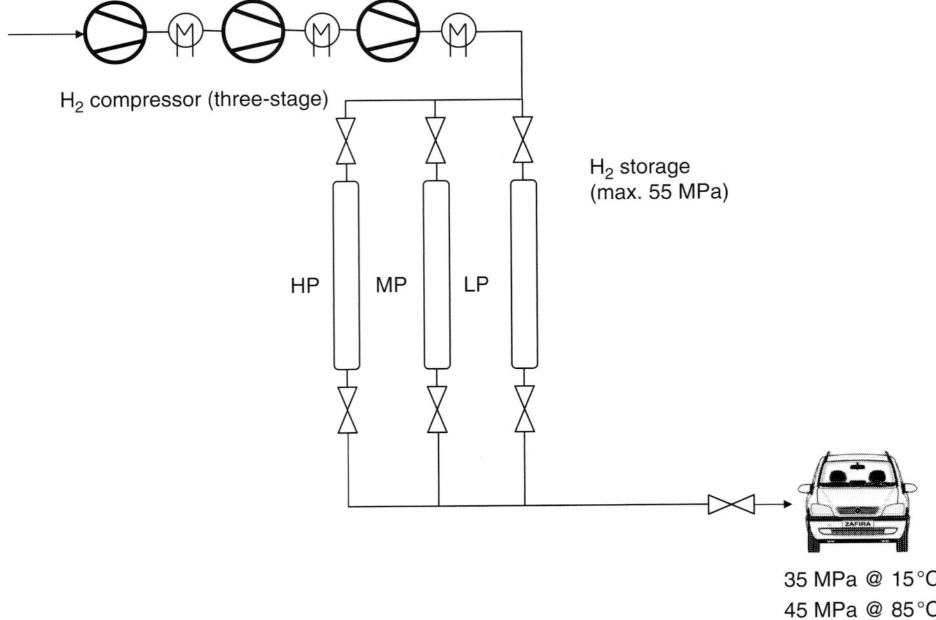

Figure 12.8. Three-bank system for vehicles with 35 MPa vehicle tanks.

For instance, in the case of 70 MPa (700 bar) storage tanks in the vehicle, hydrogen is compressed to more than 80 MPa to compensate the temperature increase during filling the tank and to allow for a short refuelling time. The design of a CGH_2 refuelling station is, in principle, comparable to a natural gas refuelling station, with differences in the technical layout resulting from the different physical properties between hydrogen and natural gas.

There exist two principle concepts for a filling station, the multibank system and the booster dispensing system.

12.2.1.1 Multibank dispensing concept

The multibank dispensing concept is similar to the one used in most natural-gas filling stations today (see Fig. 12.8).

The filling procedure automatically follows three steps (for the example of a filling station for vehicle tanks with 35 MPa):

1. Hydrogen flows from the low pressure bank to the vehicle tank until a pressure of about 10 to 15 MPa is reached in the vehicle tank (or the pressure in the low pressure bank is reduced to 16 MPa). In fill mode, the low and medium pressure banks are filled to the same pressure of the high pressure bank, to reduce the filling time and to make the maximum use of the storage vessels.
2. Hydrogen flows from the medium pressure bank to the vehicle up to 20 to 25 MPa.
3. Hydrogen flows from the high pressure bank to the vehicle up to 35 MPa.

This procedure allows filling pressure in the vehicle tank of 35 MPa to be attained even if the pressure in the low- and medium-pressure storage bank is well below this value. These data are suitable, if the temperature of the gas can be held at 15 °C. Because of the temperature increase during fast fill operation, an overpressurisation is needed to achieve the maximum fill at nominal conditions (35 MPa at 15 °C).

A temperature increase to about 85 °C would require a pressure of about 43.8 MPa in the vehicle tank (or a pressure of about 44.8 MPa in the high pressure bank). If the vehicle tank should be filled to the upper level at all conditions, the worst case (85 °C) has to be considered. In this case, the maximum pressure of the storage banks must be well above 45 MPa.

Praxair (Halvorson *et al.*, 1996) investigated two filling station concepts, a three-bank system and a booster system. For the three-bank system, Praxair laid out the storage banks to a maximum pressure of 48.3 MPa and the high-pressure bank to a minimum pressure of about 34.7 MPa for vehicle tanks with 35 MPa. They did not, however, consider the temperature increase. The pressure difference between the pressure of the full vehicle tank (34.5 MPa) and the minimum pressure of the high pressure bank (34.7 MPa) seems to be very low (only 0.2 MPa). If a temperature increase up to 85 °C and a pressure difference of 1.0 MPa is considered, then a pressure difference of only about 3.5 MPa is available for storage.

In the hydrogen filling station layout by Ferrel *et al.* (1996), a maximum pressure of 56.9 MPa is selected, although BOC did not take into account the effects of heat of compression. Usually the maximum pressure has to be about 25% above the pressure level of the vehicle tank. For a pressure of 43.8 MPa at 85 °C, the pressure in the storage banks must be about 54.8 MPa, which is a good fit for the 56.9 MPa assumed.

In a hydrogen filling station from Hydro, which has been in operation in Iceland since 2003, the pressure in the storage tanks is 44 MPa (ECTOS, 2003).

A cascade storage system is divided into three or more units, operating at different minimum pressure levels. The filling process of a vehicle starts with a hydrogen flow out of the low pressure cascade. If the pressure difference between this cascade and the vehicle tank is too low, the system switches to the next stage, consecutively. The last stage completes the filling process. The big advantage of a cascade storage system over single-stage storage is that the storage capacity of the stationary storage tanks can be better utilised. The compressor is required only to recharge the cascade stages.

The car industry is developing vehicle tanks with a pressure of 70 MPa. For 70 MPa, extremely high pressures of more than 100 MPa for the stationary hydrogen storage would be required. A booster dispensing concept (see below) avoids the need of such high pressure vessels for stationary hydrogen storage.

12.2.1.2 Booster dispensing concept

At first, the hydrogen is compressed to about 30 MPa and stored in stationary pressure vessels. During the refuelling procedure, the hydrogen is compressed to about 45 MPa for 35 MPa vehicle tanks and to about 88 MPa for 70 MPa vehicle

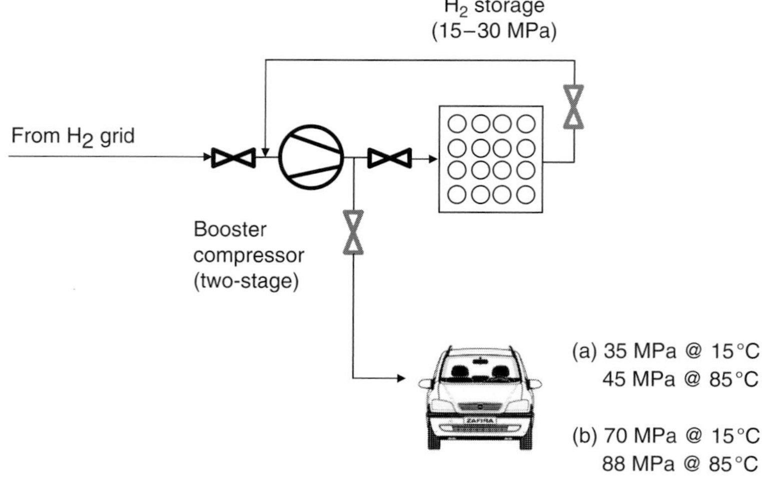

Figure 12.9. Layout of a filling station according to the booster compression dispensing concept; the booster compressor is also operated in recharging mode.

tanks. The higher pressure (e.g., 88 MPa instead of 70 MPa) is required to compensate for the temperature increase during the fast filling procedure and to provide a full 70 MPa vehicle tank at 15 °C in any case. Furthermore, a pressure difference of about 0.5–1.0 MPa is required to ensure a short refuelling time. The stationary CGH_2 tank is emptied to a minimum pressure of 15 MPa. The booster compressor is designed for a refuelling time of three minutes.

There exist two different concepts for booster compression dispensing at the filling station. One possible concept is that the booster compressor is also used for recharging the stationary hydrogen storage system. For booster operation, two-stage hydraulic compressors are used. Therefore, for hydrogen compression from 1.5 to 30 MPa (primary compression), a two-stage compression is also assumed. If a vehicle has to be refuelled in the 'one compressor case', the valves are switched in such a way that the booster compressor operates in the booster compression mode (see Fig. 12.9).

Another concept is to use a separate primary compressor (see Fig. 12.10). The second concept is preferred in larger filling stations or in filling stations with onsite hydrogen generation, to avoid an intermediate low-pressure hydrogen storage.

According to one manufacturer, hydraulic compressors capable of delivering hydrogen at a pressure level of about 45 MPa and even at 85 MPa are commercially available. These compressors also consist of two stages. Therefore, in both cases (35 and 70 MPa), a two-stage compression can be used in the calculation of the

Table 12.5. *Electric power consumption for hydrogen compression at the filling station at a suction pressure of 2.0 MPa.*

	35 MPa vehicle tanks (kWh/kWh_{CGH_2})	70 MPa vehicle tanks (kWh/kWh_{CGH_2})
Primary compression[a]	0.038	0.038
Booster compression[a]	0.019	0.032
Total	0.057	0.070

Note:
[a] Isentropic compression; η (compressor) = 75%; η (electric motor) = 90%.

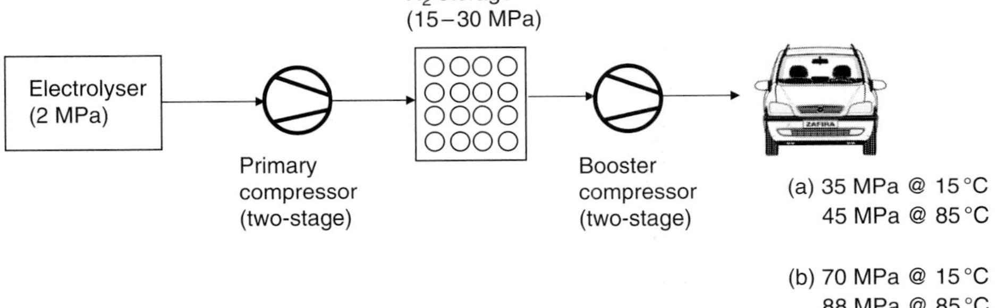

Figure 12.10. Layout of a filling station according to the booster compression dispensing concept, with separate compressors for loading and boosting.

electric power consumption required for booster compression. Precooling of the hydrogen can reduce the required pressure.

Table 12.5 shows the electric power consumption data used to calculate the energy efficiency.

A mixture of multibank and booster dispensing concepts is also possible; e.g., a small high pressure CGH_2 buffer storage (85 MPa) can be employed to assist the booster compressor.

The supply of a hydrogen refuelling station can be realised either by onsite production (via small-scale electrolysers or steam reformers) or in the future, in analogy to today's natural gas fuelling stations, by connection to a hydrogen pipeline grid. There is further the possibility of supplying gaseous hydrogen via the evaporation of liquid hydrogen.

Table 12.6 shows a breakdown of the investments for compressed hydrogen fuelling stations, for a capacity of 50, 500 and 1300 kg H_2/day. It can clearly be seen that large fuelling stations show substantial economies of scale; the investments per capacity for small fuelling stations are around three times higher than for larger stations. With increasing station size, compressors account for an increasing share of total investments (up to 50%), which makes cost reductions for compressors essential in the future.

Table 12.6. *Investment breakdown for compressed hydrogen fuelling stations*

		35 MPa, multibanking			35 MPa, booster concept			70 MPa, booster concept		
Daily sale	**(kg H_2/day)**	**50**	**500**	**1300**	**50**	**500**	**1300**	**50**	**500**	**1300**
Compressor	(€1000)	30	150	352	196	392	588	285	712	855
Storage	(€1000)	23	139	361	19	96	252	19	96	252
Refuelling system	(€1000)	44	88	176	44	88	176	63	126	252
Installation cost	(€1000)	24	75	107	65	144	196	92	187	204
Total	**(€1000)**	**121**	**452**	**996**	**324**	**720**	**1212**	**459**	**1121**	**1563**

Note:
This assumes that hydrogen is supplied at pipeline pressure of 7 MPa (70 bar).
Sources: (DWV, 2005; IEA, 2005).

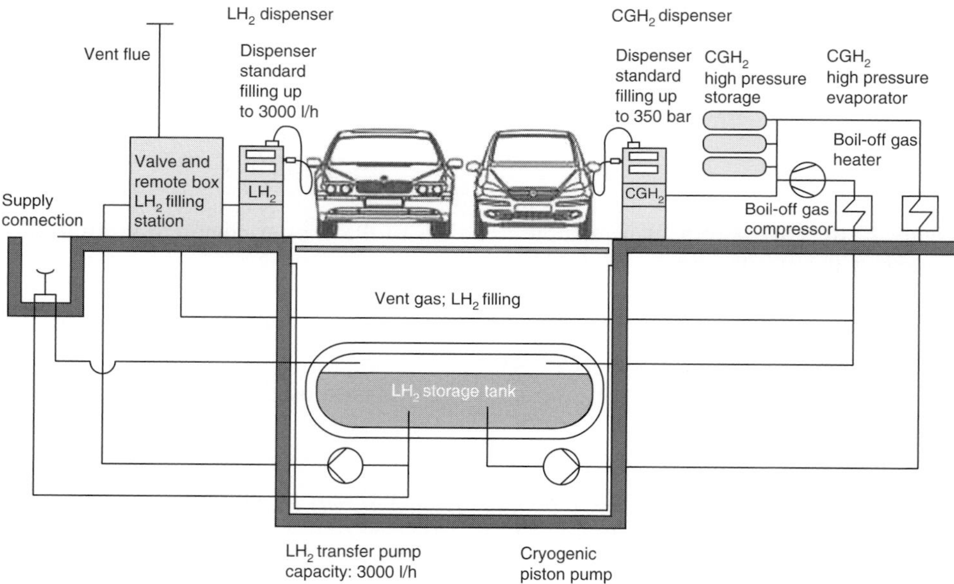

Figure 12.11. Filling station for refuelling of CGH_2 and LH_2 vehicle tanks (Linde).

12.2.2 Liquid-hydrogen (LH_2) stations

A liquid-hydrogen (LH_2) fuelling station has basically the same layout as a conventional refuelling station. The main components are the LH_2 storage tank, the LH_2 dispenser and controls, instruments and pipes (see Fig. 12.11). All components of a LH_2 fuelling station must have a special heat insulation. Today's LH_2 fuelling stations allow fuelling of 100 l LH_2 in about two minutes and also allow for several

Table 12.7. *Investment breakdown for liquid hydrogen fuelling stations supplying gaseous hydrogen*

Daily sale	**(kg H_2/day)**	**50**		**500**		**1300**	
On-board gaseous storage pressure	**(MPa)**	**35**	**70**	**35**	**70**	**35**	**70**
Liquid H_2 storage tank	(€1000)	100	100	484	484	484	484
Liquid H_2 high-pressure pump	(€1000)	78	99	156	197	312	394
Air-heated high-pressure vaporiser	(€1000)	36	36	72	72	143	143
Gaseous H_2 storage buffer	(€1000)	10	95	20	190	41	380
Refuelling system	(€1000)	44	63	88	126	176	252
Installation cost	(€1000)	67	98	164	214	173	248
Total	**(€1000)**	**335**	**491**	**984**	**1283**	**1329**	**1901**

Sources: (DWV, 2005; IEA, 2005).

fast subsequent fillings. The LH_2 is supplied in trailers; liquefaction at the fuelling stations is, because of the complexity of liquefaction plants, neither technically nor economically viable.

The use of LH_2 for CGH_2 supply is also possible. The hydrogen is transported to the filling station as LH_2. During refuelling the LH_2 is vaporised to supply CGH_2. Furthermore, both LH_2 and CGH_2 can be dispensed.

According to Linde (Reijerkerk, 2001), the electricity consumption for LH_2 dispensing amounts to about 0.0003 kWh/kWh of LH_2 (2.4 kW, 50 l LH_2/min) and the electricity consumption for CGH_2 dispensing in a filling station with ambient air vaporiser amounts to about 0.021 kWh/kWh of CGH_2 for 70 MPa vehicle tanks (20 kW, 315 Nm^3 H_2/h). The data indicated in Ferrel *et al.* (1996) lead to an electricity consumption of about 0.020 kWh/kWh of CGH_2 for 35 MPa vehicle tanks.

Table 12.7 shows a breakdown of the investments for liquid hydrogen fuelling stations supplying gaseous hydrogen, for a capacity of 50, 500 and 1300 kg H_2/day and on-board storage pressure of 35 and 70 MPa. If hydrogen was to be supplied in liquid form, the investments for the fuelling station would potentially be lower.

12.3 Summary

The principal options for hydrogen transport and distribution include pipelines, gaseous and liquid trailers. The choice for the most economic option depends on transport volumes and transport distances. For the transport of liquid hydrogen, additionally the costs of the liquefaction plant need to be taken into account. Another possibility could be to blend hydrogen with natural gas up to a certain extent and either separate the two at the delivery point, or use the mixture, e.g., in stationary combustion applications. To what extent this is feasible and reasonable (given that hydrogen is an expensive and valuable commodity), is still a matter of debate. Depending on the physical condition required for the storage of hydrogen on-board the vehicle, compressed hydrogen and liquid hydrogen refuelling stations can be distinguished.

References

Amos, W. (1998). *Costs of Storing and Transporting Hydrogen*. Report NREL/TP–570–25106. National Renewable Energy Laboratory Colorado/US Department of Energy.

Bossel, U. and Eliasson, B. (2003). Energy and the hydrogen economy. *European Fuel Cell News*, January 2003. www.efcf.com/reports.

Bossel, U., Eliasson, B. and Taylor, G. (2005). The future of the hydrogen economy: bright or bleak? *European Fuel Cell Forum*, February 2005. www.efcf.com/reports.

Castello, P., Tzimas, E., Moretto, P. and Peteves, S. D. (2005). *Techno-Economic Assessment of Hydrogen Transmission & Distribution Systems in Europe in the Medium and Long Term*. Joint Research Centre (JRC), Report EUR 21586 EN. Petten, The Netherlands: The Institute for Energy.

DWV (Deutscher Wasserstoff- und Brennstoffzellen-Verband) (2005). *H_2 Roadmap*. Study of the DWV, Berlin, December 2003 (published 2005). www.dwv-info.de.

ECTOS (2003). *Icelandic New Energy: ECTOS Ecological City Transport System*. Flyer, April 2003. www.newenergy.is.

Ferrel, J., Kotar, A. and Stern, S. (1996). *Direct Hydrogen Fuelled Proton Exchange Membrane (PEM) Fuel Cell System for Transportation Applications*. Final report, Section 3; Hydrogen Infrastructure Report. Prepared for the Ford Motor Company and the Department of Energy.

Halvorson, T. G., Terbot, C. E. and Wisz, M. W. (1996). *Hydrogen Production and Fuelling System Infrastructure for PEM Fuel Cell-Powered Vehicles*. Final report, prepared for the Ford Motor Company, Dearborn, MI, USA, under Ford Subcontract No. 47–2-R31157 'Direct Hydrogen Fuelled Proton Exchange Membrane (PEM) Fuel Cell System for Transportation Applications'.

HySociety (The European hydrogen based society) (2004). *Technology Database for Hydrogen Pathways, Analysis of Hydrogen Pathways*. European Commission (DG TREN), FhG-ISI. www.eu.fhg.de/h2database/index.html.

IEA (International Energy Agency) (2005). *Prospects for Hydrogen and Fuel Cells*. IEA Energy Technology Analysis Series. Paris: OECD/IEA.

Kußmaul, K. and Deimel, P. (1998). *Materialverhalten in H_2-Hochdrucksystemen*. Wasserstoff-Energietechnik IV, Munich, 17/18 October 1998, VDI Berichte 1201. Düsseldorf: VDI Verlag.

Leighty, W., Hirata, M., O'Hashi, K. *et al.* (2003). Large renewables – hydrogen energy systems: gathering and transmission pipelines for windpower and other diffuse, dispersed sources. *World Gas Conference 2003*. Tokyo, Japan, 1–5 June.

Mintz, M., Molburg, J., Folga, S. and Gilette, J. (2002). *Hydrogen Distribution Infrastructure*. Research paper. Argonne: Argonne National Laboratory, Transportation Technology R&D Centre.

Müller, E-I. and Henning, A. (1989). *Systemtechnische Untersuchung des stationären und instationären Gasferntransports von solar erzeugtem Wasserstoff*. Diploma thesis. Munich: Technical University of Munich and Ludwig-Bölkow-Systemtechnik (LBST).

NATURALHY (2007). *Preparing for the Hydrogen Economy by Using the Existing Natural Gas System as Catalyst*. Integrated Project under the 6th Framework Programme of the European Commission, Contract No. SES6/CT/2004/502661, Newsletter, Issue 6, November 2007. www.naturalhy.net.

Parker, N. (2004). *Using Natural Gas Transmission Pipeline Costs to Estimate Hydrogen Pipeline Costs*. Technical Report No. UCD-ITS-RR–04–35. University of California Davis, Institute of Transportation Studies.

Peschka, W. (1992). *Liquid Hydrogen – Fuel of the Future*. Wien: Springer-Verlag.

Quack, H. (2001). Die Schlüsselrolle der Kryogentechnik in der Wasserstoff-Energiewirtschaft. *Wissenschaftliche Zeitschrift der Technischen Universität Dresden*, **50** (5/6).

Reijerkerk, C. J. (2001). *Hydrogen Filling Stations Commercialisation*. Project carried out within the frame of an integrated International Master's Programme in Technology and Management at the University of Hertfordshire in conjunction with Fachhochschule Hamburg. Linde Gas AG, Market Development & Global Key Accounts.

Roads2HyCom (2007). *European Hydrogen Infrastructure Atlas and Industrial Excess Hydrogen Analysis*, ed. Steinberger-Wilckens, R. and Trümper, S. C. Roads2HyCom. www.roads2hy.com.

Smith, B., Frame, B., Eberle, C. *et al.* (2005). New materials for hydrogen pipelines. *2005 DOE Hydrogen Programme Annual Review*. Oak Ridge National Laboratory, US Department of Energy.

Syed, M. T., Sherif, S. A., Veziroglu, T. N. and Sheffield, J. W. (1998). An economic analysis of three hydrogen liquefaction systems. *International Journal of Hydrogen Energy*, **23** (7), 565–576.

Ullmann (2003). Hydrogen. In *Ullmann's Encyclopedia of Industrial Chemistry*. 6th edn. vol. 17. Weinheim: WILEY–VCH, pp. 85–240.

Valentin, B. (2001). *Wirtschaftlichkeitsbetrachtung einer Wasserstoffinfrastruktur für Kraftfahrzeuge*. Diploma thesis. Münster, Germany: University of Applied Sciences Münster and Linde Gas AG.

Wagner, U., Angloher, J. and Dreier, T. (2000). *Techniken und Systeme zur Wasserstoffbereitstellung. Perspektiven einer Wasserstoff-Energiewirtschaft (Teil 1)*. Munich: Wiba (Koordinationsstelle der Wasserstoff-Initiative Bayern) and Lehrstuhl für Energiewirtschaft und Anwendungstechnik of the Technical University München.

Würsig, G. M. (1996). *Germanischer Lloyd, Hamburg: Beitrag zur Auslegung von mit Wasserstoff betriebenen Hauptantriebsanlagen für Flüssig-Wasserstoff-Tankschiffe*. Dissertation, University of Hannover, Verlag Mainz – Wissenschaftsverlag Aachen.

Zittel, W., Wurster, R. and Weindorf, W. (1996). *Wasserstoff in der Energiewirtschaft*. Ludwig-Bölkow-Systemtechnik (LBST). www.hyweb.de.

13

Key role of fuel cells

Frank Marscheider-Weidemann, Elna Schirrmeister and Annette Roser

Some of the most important benefits of hydrogen can only be realised if hydrogen is used in fuel cells; for instance, the high overall conversion efficiency compared with the internal combustion engine, as well as the reduction of local pollution and noise. Therefore, the market success of fuel cells plays a key role in a hydrogen economy. The following chapter gives a brief introduction to the fuel cell as a technology and describes the various types of fuel cells and their potential uses in mobile, stationary and portable applications. However, preparing for the structural changes in industry is just as important as the technical optimisation of fuel cells, and the remainder of the chapter is devoted to this aspect.

13.1 Historical development of fuel cells

Fuel-cell technology first took off more than 170 years ago. In 1839, the Welsh judge, Sir William Grove, presented the first fuel-cell battery, in which he was able to generate an electrical current from hydrogen and oxygen by reversing the process of electrolysis (Grove, 1839). The electrodes were platinum and sulphuric acid was used as the electrolyte. Since the invention of the fuel cell, expectations of their broad market introduction have built up into waves several times, but have then crashed each time. One such wave is demonstrated by the speech of Wilhelm Ostwald, the famous electrochemist, to the Bunsengesellschaft in 1894, in which he stated that fuel cells are superior to steam engines and all other kinds of incineration technique (Ostwald, 1894). At this time R&D used coal directly as the anode in fuel cells for 'cold combustion'.

Despite this, the principle of the fuel cell was not able to be developed into a technically mature process for a long time. The main reasons, apart from insufficient knowledge of the electrochemical processes involved, were material problems. Around the turn of the century, the dynamo generator (1866, Siemens), combustion

The Hydrogen Economy: Opportunities and Challenges, ed. Michael Ball and Martin Wietschel. Published by Cambridge University Press.

engines (1876 Otto; 1897, Diesel) and the gas turbine (1900, Stolze) were successfully introduced to the market, so there was little interest on the part of industry in the development of an electrochemical generator. More intensive work on the basic principles of fuel cells was only begun around 1950 in England, then Germany and the USA. These research results led to an extensive NASA programme of development, which peaked in the equipping of the Apollo moon mission in 1968 with an alkaline fuel cell system. Since then, there has been continuous further development of various fuel-cell systems. (A description of the history of fuel-cell technology can be found in Hoogers (2003) and Ullmann (2003).)

The most recent wave for fuel-cell market introduction arose after announcements by DaimlerChrysler and Toyota in 1997 that they were going to produce fuel-cell cars in 2004. For certain mass applications of fuel cells, very fast market penetration has been projected several times over the last ten years (Frost and Sullivan, 2001; VDMA, 2002; Wengel and Schirrmeister, 2000; Wurster, 1999). These projections did not materialise, however, owing to unresolved technical issues and high risks for the investments in mass-produced elements of fuel cells. Today, scenarios developed by the Brennstoffzellen-Bündnis Deutschland (Fuel Cell Association of Germany) indicate that fuel cells will be ready for portable broad mass markets in 2010 and for mobile applications in 2015 (see also Chapter 14). Similar targets are given by the European Union (https://www.hfpeurope.org/uploads/2097/HFP_IP06_FINAL_20APR2007.pdf), Japan (www.fuelcelltoday.com/events/archive/2006-11/2006-Fuel-Cell-Hydrogen-Technolo), Canada (www.fuelcells.org/info/library/canada.pdf) and the USA (www.uscar.org/commands/files_download.php?files_id=81).

13.2 Fuel cells in a hydrogen economy

It seems obvious that fuel cells can work without hydrogen, but a hydrogen society without fuel cells does not make much sense in the long run because of the high electricity-to-heat ratio and high overall conversion efficiency of fuel cells powered by hydrogen. In addition, the fuel cell's exhaust produces zero emissions when fuelled by hydrogen, which could be generated by electricity from renewables in the long term or from bio-based methane. As the demand for heat and fuels is expected to fall more steeply in stationary applications than the demand for electricity in industrial countries (e.g., due to better insulation), fuel-cell technology is considered a better option to meet the increasing share of electricity in total energy demand. The main criticism of its technological rivals, i.e., the engine-driven co-generation plant, the microturbine or the Stirling engine, is the low electricity share in their total energy output (Krewitt *et al.*, 2004; 2006). On the other hand, the energy demand in transport applications is still increasing, so it may make sense to use hydrogen in combustion engines as an interim solution.

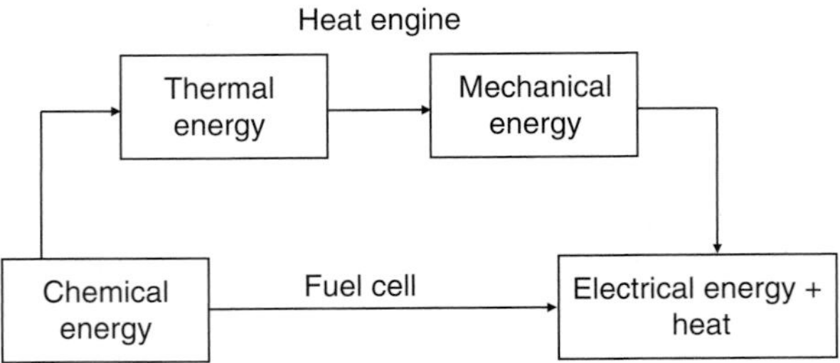

Figure 13.1. Comparison of fuel cell and conventional electricity generation.

13.3 Principles of fuel cells

Fuel cells belong to the so-called galvanic elements. Three classes of these electrochemical energy converters are distinguished: primary cells, also called batteries, which consume the reaction substances contained within them when discharging. Secondary cells, often called storage or rechargeable batteries, have the property of being able to be recharged by supplying electrical energy. Tertiary cells are systems in which the reactants are supplied continuously and externally during operation and the reaction products are also continuously discharged. These include fuel cells. In contrast to conventional electricity generation, which usually takes place in a three-stage conversion process, in a fuel cell, chemical energy is directly converted into electrical energy (see Fig. 13.1).

In fuel cells, electricity and water are produced from hydrogen or hydrogen-rich fuel and oxygen in an electrochemical reaction, which also releases heat. This is illustrated, based on the example of a PEM fuel cell, in Fig. 13.2. A single cell consists of an anode, electrolyte and cathode, and is separated from the adjacent cell in the stack by a bipolar separator plate. All electrochemical reactions in a fuel cell consist of two separate reactions: an oxidation half-reaction at the anode and a reduction half-reaction at the cathode. Hydrogen or a hydrogen-rich gas is introduced at the anode, which results in the formation of positively charged hydrogen ions (H^+ ions) by oxidation of the fuel. The released electrons are directed to the cathode with the help of an external circuit, where oxygen is reduced to form oxide ions (O^{2-} ions) by reaction with the electrons. (In contrast to the usual convention, in fuel cells the *anode* is the *negative* electrode and the *cathode* is the *positive* one (Larminie and Dicks, 2003).) To close the electrical circuit, H^+ ions have to migrate from the anode to the cathode. This is achieved using an ion-conducting electrolyte, which separates both gas chambers from each other and is impermeable to electrons. Owing to the chemical potential difference between the two sides of the electrolyte, the H^+ ions diffuse through it and react on the other side with the corresponding reaction partner. This process is known as 'cold combustion', since oxidation and reduction take place separately, so there is no direct combustion. The electrons can provide the electrical power for an external consumer integrated into the external circuit on their

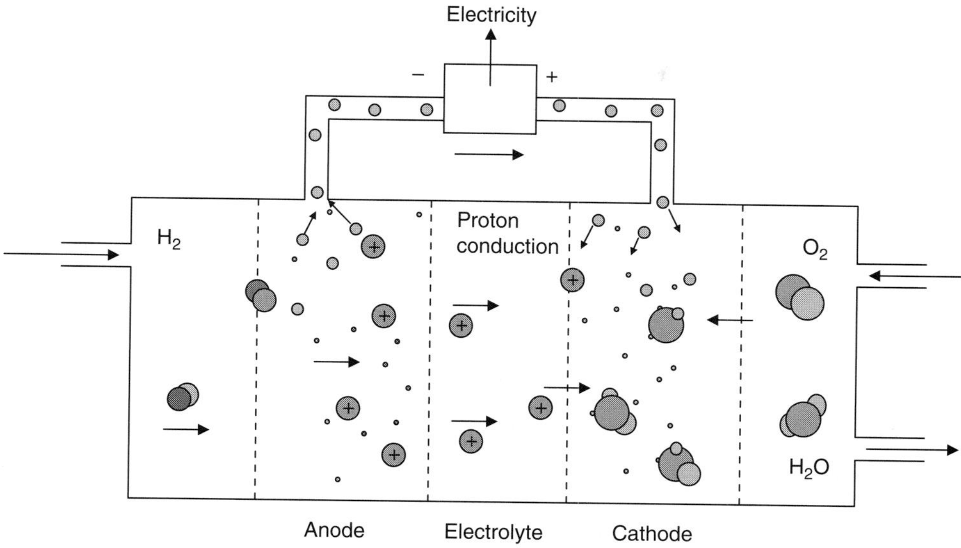

Figure 13.2. Principle of a Proton-Exchange-Membrane (PEM) fuel cell.

way from the anode to the cathode. Normally, the two half-reactions occur very slowly at the low operating temperature of the PEM fuel cell. To speed up the reaction of oxygen and hydrogen, each of the electrodes is coated on one side with a precious metal catalyst (e.g., platinum). The precious metal catalysts and the electrolytes require a high degree of fuel-gas purity.

A fascinating point, especially to physical chemists, is the potential theoretical efficiency of fuel cells. Conventional combustion machines principally transfer energy from hot parts to cold parts of the machine and, thus, convert some of the energy to mechanical work. The theoretical efficiency is given by the so-called Carnot cycle and depends strongly on the temperature difference, see Fig. 13.3. In fuel cells, the maximum efficiency is given by the relation of the useable free reaction enthalpy G to the enthalpy H ($\Delta G = \Delta H - T \cdot \Delta S$). For hydrogen-fuelled cells the reaction takes place as shown in Eq. (13.1a). With $\Delta H_R = 241.8$ kJ/mol and $\Delta G_R = 228.5$ under standard conditions (0 °C and $p = 100$ kPa) there is a theoretical efficiency of 95%. If the reaction results in condensed H_2O, the thermodynamic values are $\Delta H_R = 285.8$ kJ/mol and $\Delta G_R = 237.1$ and the efficiency can then be calculated as 83%.

$$H_2^g + \frac{1}{2} O_2^g \rightarrow H_2O^g \qquad (13.1a)$$

$$H_2^g + \frac{1}{2} O_2^g \rightarrow H_2O^{fl} \qquad (13.1b)$$

Like all galvanic elements, fuel cells produce direct current. The voltages of a cell are typically in the range of 0.7 to 0.75 V (the theoretical voltage being 1.23 V), the

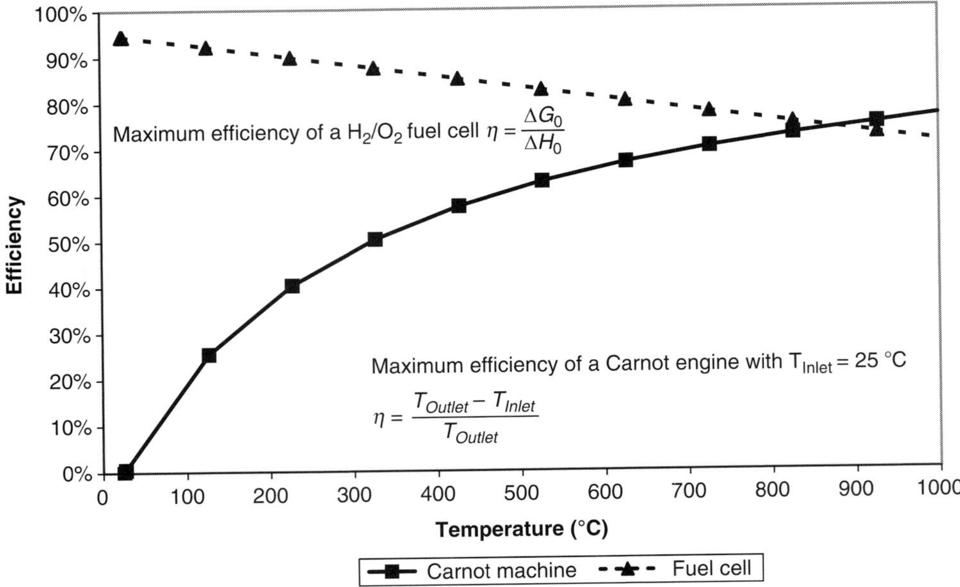

Figure 13.3. Theoretical thermodynamic efficiency of fuel cells and Carnot machines as a function of temperature.

theoretical efficiency thus being around 60%. To achieve higher voltages (and power), several cells must be stacked together (in serial or parallel connection) (see Fig. 13.4). In this 'stack', the individual fuel cells are connected through bipolar plates, which also serve as separators between adjacent cells. Gas ducts on the surface of the bipolar plates assure the fuel supply and the drainage of the resulting reaction product water.

As shown in Fig. 13.3, fuel cells theoretically have higher efficiencies at lower temperatures, whereas engines are the best choice for higher temperatures. In reality, however, the efficiency in existing systems is much lower due to resistance losses in the system.

13.4 Types of fuel cell

This section aims to give a brief review of the different types of fuel cell and their most important properties. With regard to hydrogen production, the focus is on the fuel gases used and the requirements made of their purity. Within the scope of this publication, it is not possible to describe the different fuel cell systems in any detail. References are made to the relevant specialist literature (see also, the Further reading section at the end of this chapter).

There is no such thing as 'the fuel cell', but many different types of fuel cell, which differ mainly with regard to the electrolyte, the chemical reaction and the working

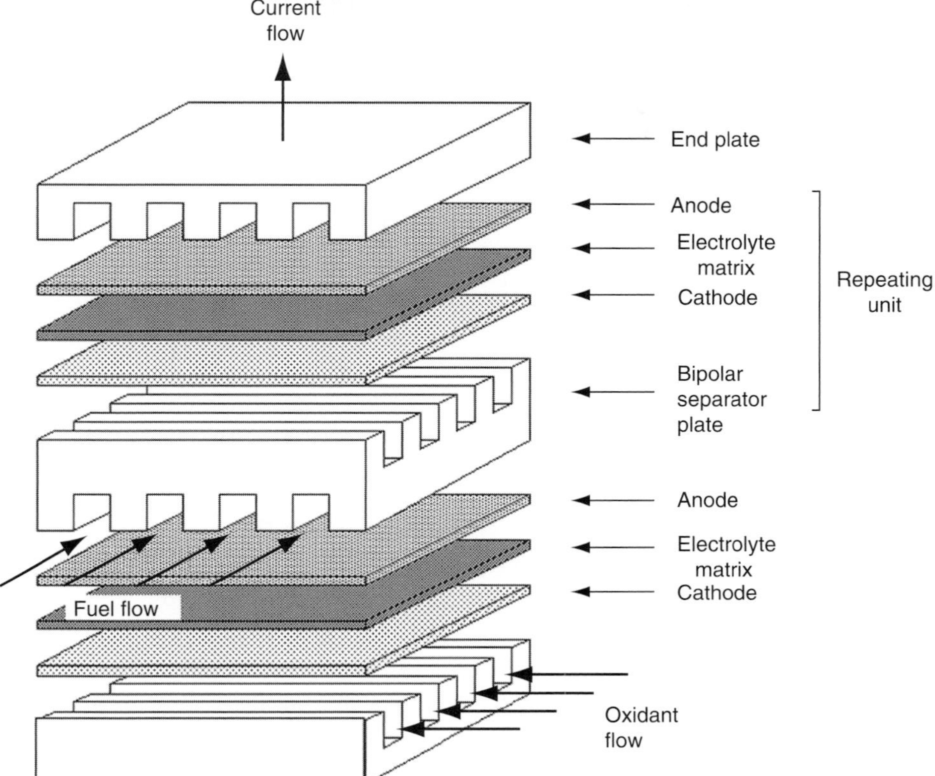

Figure 13.4. Schematic layout of a fuel-cell stack.

temperature involved. Their names are derived from the electrolyte used, see Table 13.1. A division can be made into low-temperature and high-temperature fuel cells. Membrane and alkali fuel cells have operating temperatures of approximately 80 °C, phosphoric-acid cells work at approximately 200 °C and high-temperature fuel cells, like MCFC and SOFC, work at 650 °C and 800–1000 °C, respectively.

Because hydrogen does not exist in molecular form in nature, it must be produced by reforming from fossil fuels such as natural gas or biogas, or by electrolysis from non-renewable or renewable energies, such as wind or solar energy. For low-temperature fuel cells, the reforming process takes place in external reformers, where hydrocarbons or alcohols are reacted to form a hydrogen-rich gas. After reforming, there are further stages of cleaning the hydrogen to reduce the CO content. In high-temperature fuel cells, the reforming process takes place as an intermediary stage in the fuel cell itself, so the efficiency losses of the entire system are lower than for low temperature cells.

The technically relevant fuel cell types are shown in Fig. 13.5; a newer development of the PEM fuel cell is the direct-methanol fuel cell (DMFC), which uses a diluted

Table 13.1. *Main characteristics of technically relevant fuel cells today*

Type	Name	Electrolyte	Charge carrier	Temperature range (°C)	Power range (electricity)	Electric efficiency (system)	Start-up time	Field of application
PEM	Proton-exchange-membrane fuel cell (Polymer-electrolyte-membrane fuel cell)	Proton-conducting polymer membrane (e.g., Nafion®)	H^+ (proton)	50–80	mW (Laptop); 50 kW (Ballard); modular up to 200 kW	25–≅45%	Immediate	Road vehicles, stationary electricity generation, heat and electricity co-generation, submarines, space travel
AFC	Alkaline fuel cell	30–50% KOH	OH^- (hydroxide)	60–90	7 kW (Apollo)	37–≅42%	Immediate	Space travel, road vehicles, submarines
PAFC	Phosphoric-acid fuel cell	Concentrated phosphoric acid	H^+ (proton)	160–220	50 kW (Fuji) to 200 kW (UTC)	37–≅42%	30 minutes from 'hot standby'	Stationary electricity generation, heat and electricity co-generation, road vehicles

MCFC	Molten-carbonate fuel cell	Molten carbonate (Li_2CO_3, K_2CO_3)	CO_3^{2-} (carbonate)	620–660	250 kW (MTU, Ansaldo)	40–≅47% (>50% combined with gas turbine; overall efficiency in co-generation >85%)	Several hours after cold start	Stationary electricity generation, heat and electricity co-generation
SOFC	Solid-oxide fuel cell	Ion-conducting ceramic (yttrium stabilised zircon oxide)	O^{2-} (oxide)	800–1000	1 kW (IKTS) to 250 kW (Siemens)	44–≅50% (>50% combined with gas turbine; overall efficiency in co-generation >85%)	Several hours after cold start	Stationary electricity generation, heat and electricity co-generation
DMFC	Direct-methanol fuel cell	Proton-conducting polymer membrane	H^+ (proton)	80–100	mW to kW (scooters)	15–≅30%	Immediate	Portable, mobile

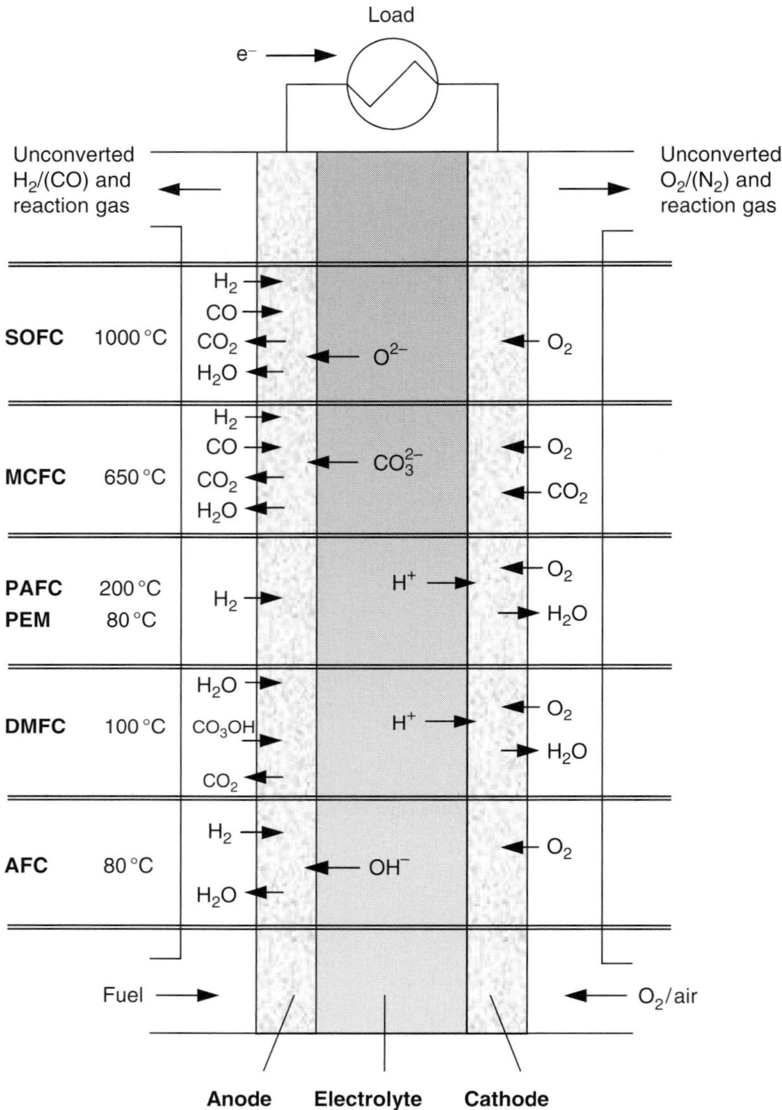

Figure 13.5. Different types of fuel cell (Jörissen and Garche, 2000).

methanol solution. Its technical development is still ongoing. Some organisations, including the Fraunhofer Society in Germany, are developing a PEMFC working with ethanol, the so-called direct-ethanol fuel cell (DEFC, see www.defc.de). On the other hand, companies like UTC, which announced a shift from the PAFC technology to PEM in 2002 for stationary applications, are now reverting to producing PAFC because they consider the PEMFC's lifetime to be insufficient.

Table 13.2. *Fuel gas composition requirements of different fuel-cell types*

Fuel cell	Fuel	Gas purity required	Tolerated inert compounds	Internally reformable compounds
AFC	pure H_2 and O_2	No CO_2, H_2S	–	–
PEMFC	pure H_2[a]; O_2/air	CO < 10–100 ppm	N_2, CO_2, CH_4	–
DMFC	MeOH; O_2/air	Little CO	No data	–
PAFC	H_2[a]; O_2/air	CO < 1–2 vol.% S < 50 ppm	N_2, CO_2, CH_4	–
MCFC	H_2, CO, natural gas, coal or biogas, etc.[b]; O_2/air	S < 1 ppm Cl < 1 ppm	N_2, CO_2	CH_4, higher HC
SOFC	H_2, CO, natural gas, coal or biogas, etc.[b]; O_2/air	S < 1 ppm Cl < 1 ppm	N_2, CO_2	CH_4, higher HC

Notes:
[a]With a reformer, also natural gas, MeOH, biogas, coal gas etc.; CO precision cleaning may be necessary.
[b]Internal reforming.
Sources: (Hoogers, 2003; Oertel and Fleischer, 2003; Pehnt, 2001).

In principle, any energy source containing hydrogen can be used in fuel cells. In high-temperature fuel cells, the electrochemical reaction can also be aided by the oxidation of the carbon monoxide. The highest efficiencies of fuel cells are achieved using pure hydrogen and pure oxygen but, in theory, any chemically bonded hydrogen in the form of natural gas, coal gas, sewage gas, biogas or methanol can be converted. However, these energy carriers first have to be transformed into a hydrogen-rich gas (by reforming) and may have to be purified; these reforming reactions can take place within the fuel cell or externally. The required purity of the fuel varies widely, depending on the type of fuel cell. Only sulphur compounds have to be removed to the greatest extent possible in all fuel-cell types (although research is currently being done on developing sulphur-resistant fuel cells). Table 13.2 summarises the demands made of the fuel gas composition by the various fuel cell systems.

The required purity of fuels differs greatly, according to the type of fuel cell. Among others, the purity requirements for the fuels employed are a function of the electrode materials used. These, in turn, are equipped with chemically sensitive catalysts, depending on the working temperature of the fuel-cell system. Therefore, fuel cells with electrode coatings made of precious metal catalysts require the highest degrees of purity. Low-temperature fuel cells such as AFC and PEMFC can only tolerate small amounts of carbon monoxide (CO), as this is a catalyst poison. The PAFC allows the use of gases containing CO_2 because of its acidic electrolyte and is

Table 13.3. *Fuels for fuel cells*

	Usable in fuel cells	
Fuel	Directly	After reforming
Natural gas	x[a]	x[b]
Biogas	x[a,c]	x[b,c]
Synthesis gas	x[a,c]	x[b,c]
Gasoline	–	o
Diesel	–	o
Hydrogen	x	not applicable
Methanol	x[d]	x
Ethanol	x[e]	x
Dimethylether	–	x

Notes:
x: usable or used; o: theoretically usable, but realisation not yet technically mastered or very complex; –: not usable.
[a]in MCFC/SOFC;
[b]in PAFC/PEMFC;
[c]advanced gas purification necessary;
[d]in DMFC;
[e]in DEFC.
Source: (Oertel and Fleischer, 2003; amended).

thus suited to hydrocarbons, since the CO_2 produced during reforming does not have to be removed. High-temperature fuel cells do not make the same demands of fuel purity. This leads to a grading of the required fuel purity depending on the working temperature in the fuel cell: *the higher the operating temperature, the lower the required fuel gas purity*. When using low-temperature fuel cells, therefore, energy sources such as natural gas, biogas or synthesis gas have to be externally reformed outside the fuel cell to obtain a hydrogen-rich gas. Table 13.3 shows various fuels and their usability in fuel cells.

The MCFC and the SOFC operate at temperatures around 650 °C and above. These working temperatures correspond to the temperature during the reforming of the various carbonaceous fuels, so that the heat losses of the exothermic fuel cell reaction can be used directly for endothermic steam reforming in the stack to produce hydrogen-rich gases. This is advantageous for the system's overall efficiency and complexity because a separate, upstream reforming reactor is not required. This case is referred to as *internal reforming*. Because of the high operating temperatures, gas mixtures, especially containing CO, can be directly transformed using internal reforming. This is why the MCFC and the SOFC are particularly well suited for generating electricity with various carbonaceous fuel gases (e.g., natural gas, coal gas or biogas) and carbon monoxide.

13.5 Status of fuel-cell development and application

13.5.1 Current status of fuel-cell development

As mentioned above, fuel cells may be used for mobile, stationary and portable applications. Table 13.4 shows the currents status of fuel cells for the three respective fields of application in terms of specific investment, lifetime and system efficiency as well as target values for the future.

Several thousand FC systems are produced per year worldwide, about 80% for stationary and portable uses, the rest for FCV demonstration projects (IEA, 2007). Total installed FC power capacity is some 50 MW. Stationary systems in

Table 13.4. *Fuel cells – current status (2004) and future targets*

Parameters	Unit	Today	Target
MCFC (*stationary*)			
Investment (250 kW system)	$/kW	8 500–13 000	1 500
Lifetime	h	20 000	40 000–60 000
Efficiency (system)	%	47	50
SOFC (planar design, *stationary*)			
Investment (system)	$/kW	16 000–20 000	1 500
Lifetime	h	2 000–5 000	40 000–60 000
Efficiency (system)	%	40–45	50
SOFC (tubular design, *stationary*)			
Investment (100 kW system)	$/kW	>10 000	1 500
Lifetime	h	20 000	40 000–60 000
Efficiency (system)	%	45	45–55
PEMFC (*mobile*, passenger car)			
Investment	$/kW	2 000–4 000	50–60
Lifetime	h	<2 000	>5 000
Efficiency (system)	%	38	>45
DMFC (*portable*)			
Investment	$/kW	10 000–100 000	3 000–5 000
Lifetime	h	<1 000	1 000–5 000
Efficiency (system)	%	20–30	30–35

Notes:
SOFC are produced with either tubular or planar stack configurations; investments for planar design are a rough estimate, as no prototypes exist. Specific investments for PAFC are in the range $4000–$4500/kW (IEA, 2007). For further fuel-cell R&D needs see IEA (2005).
Sources: (Hasenauer *et al.*, 2005; HFPE, 2005; IEA, 2005).

operation worldwide number roughly 3000, including more than 2000 small units (0.5 kW–10 kW) and some 1000 large units (>10 kW). A number of additional small units are being installed for remote applications and telecommunication. Proton-exchange-membrane fuel cells are the choice technology for fuel-cell vehicles, but they also represent 70%–80% of the current small-scale stationary FC market. While PAFC have been pioneering for the large-scale stationary market, MCFC and SOFC are expected to dominate this sector in the immediate future. They are used in niche markets (back-up, highly reliable or remote power generation). Solid-oxide fuel cells currently represent 15%–20% of the stationary market, but their share is expected to increase. Direct-methanol fuel cells appear to be close to entering the market for portable devices.

13.5.2 Mobile applications

From the viewpoint of fuel-cell producers and suppliers, the automobile market is most promising, although the targets for cost (for combustion engines, approximately $50/kW, see Table 13.4) and dynamics (load shifting and spreading) in this field are stricter than in other markets. Other important points are volume and weight constraints (for combustion engines, currently approximately 1 kW/kg; doubling the power of fuel-cell cars means doubling the weight), cold-starting (start-up time well under 1 minute), freezing resistance, operation in ambient temperatures of –30 to +40 °C, safety and range (Demuss, 2000). Proton-exchange-membrane fuel cells are at present the best candidates for powering fuel-cell vehicles, as they operate at low temperature (80 °C), offer short start-up times, high efficiency, good power density and good power-to-weight ratios; they also have a very good load change behaviour, i.e., they can rapidly change their power output as a function of demand. Another clear advantage for the application of PEMFC in the transport sector is that they can also operate with air. Because of their high operating temperature and long start-up time, MCFC and SOFC are not suitable for use in vehicles.

The biggest attraction of fuel-cell-powered vehicles for car manufacturers is the fact that they no longer emit nitrogen oxides or hydrocarbons (or carbon dioxide if they are fuelled with pure hydrogen). (Burning hydrogen in internal combustion engines results in NO_x emissions; fuel-cell vehicles emit only water.) This effectively does away with one of the main environmental discussion points about traffic. In California, these zero-emission cars have been demanded since the foundation of the California Fuel Cell Partnership in 1999.

The costs of a PEMFC stack are composed of the costs of the membrane, electrode, bipolar plates, platinum catalysts, peripheral materials and the costs of assembly. For the fuel-cell vehicle, the costs of the electric drive (converter, electric motor, inverter, hydrogen and air pressurisation, control electronics, cooling systems, etc.) and the hydrogen storage system have to be added. Current costs of PEM fuel-cell stacks are around $2000/kW, largely dominated by the costs of the bipolar plates and the electrodes (around 40% each), which are currently manufactured manually (IEA, 2005).

Consumers will only buy fuel-cell cars if these are not more expensive than combustion-engine cars. This means great efforts will have to be made by the automobile industry. Technology and cost projections of fuel-cell components and, hence, fuel-cell vehicles vary over a wide range, largely as a result of differing assumptions about the transformation from small-scale manufacturing to mass production. Industrial production is expected to drive costs down significantly; but to reach the costs of today's internal combustion engines of \$50–\$70/kW, fundamental advances in materials used in PEMFC and higher fuel-cell power densities are required. (For a detailed analysis of how the costs can be reduced for the different fuel-cell components, see the IEA (2005).)

Cost targets exist for all parts of the fuel cell: for bipolar plates, from \$10/kW (2004) to \$3/kW in 2015; for electrocatalysts, from \$40/kW (2005) to \$3/kW in 2015 and for membrane electrode assemblies (MEA), from \$50/kW (2005) to \$5/kW in 2015 (Freedom Car, 2005; these cost targets are somewhat different from those mentioned by the IEA (2005)). Since 2004, the number of fuel-cell cars has been growing and at the time of writing they numbered approximately 1000 worldwide; there are also around 100 fuel-cell buses in use worldwide in several demonstration projects. But these cars are produced as individual (hand-built) models and are extremely expensive, with production costs per vehicle currently estimated at around \$ one million; large-scale production is not expected before 2015, see Section 13.1.

Some of the first models obtained hydrogen from methanol using a reforming process. As methanol developed a bad reputation due to its toxicity, the focus turned to on-board reforming of gasoline, which also proved impractical (IEA, 2005). On-board reforming of hydrogen still poses significant technical and economic problems. While the on-board reforming concept for cars has not been completely abandoned, the current focus is on on-board storage of hydrogen. Nowadays, most fuel-cell cars store hydrogen under high pressure or as a liquid to increase the driving range to 500 km (see Chapter 11). This seems to be the required driving range for normal consumers.

Assuming a theoretical efficiency of the fuel-cell system of around 60% and an electric-drive-train efficiency of 90%, the overall fuel-cell system efficiency is about 55%. The theoretical efficiencies for a fuel cell cannot be realised in practice. The efficiency of the system (including fuel treatment, air supply and others) is already lower than that of the pure fuel-cell stack on its own; the overall efficiency of the FC drive train falls to less than 40% as a result of additional components, such as compressors, control electronics and others, see Fig. 13.6.

Still, the efficiency of hydrogen fuel-cell vehicles is about twice that of current internal combustion engines on the highway, and about three times as high in urban traffic (and between 1–1.5 times more than hybrid electric vehicles (IEA, 2005)). A clear advantage of fuel cells is that at part loads, fuel-cell drives have a higher

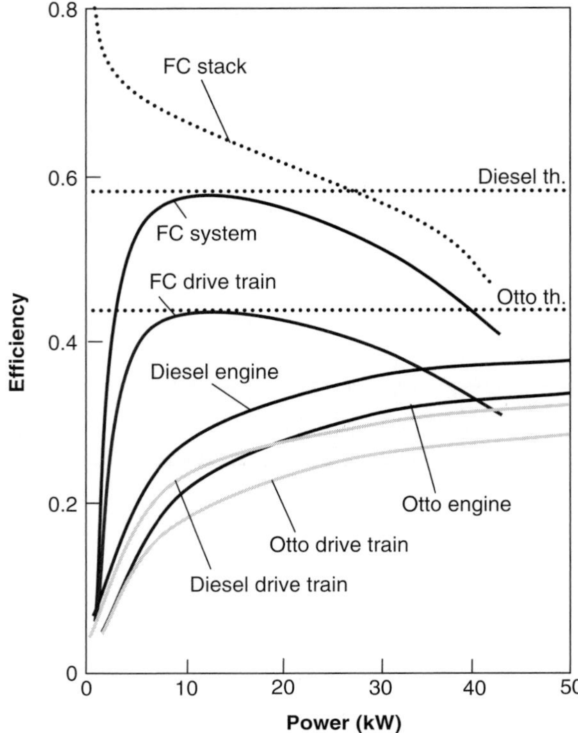

Figure 13.6. Efficiencies of automobile drive trains with combustion engines and fuel cells (DLR, 1997).

efficiency than at full loads; this suggests their application in motor vehicles, which are usually operated at partial load, such as during urban driving. Combustion motors are normally optimised for the maximum necessary propulsion, e.g., for steep inclines. Car manufacturers use charging and reduced cylinder capacity to give ICE better efficiencies at partial loads (Dauensteiner, 2001).

Today, test cars consume hydrogen in the range of high-class conventional cars. Evidently, fuel-cell vehicles have a clear advantage over internal combustion engines (ICE) vehicles in terms of hydrogen consumption. But the ICE offers the possibility for operation in a bi-fuel mode, which is a clear advantage regarding the implementation of an initial network of refuelling stations.

Besides the PEMFC being developed for vehicle propulsion, SOFC are being considered for APU applications in vehicles, since they operate at very high temperatures and therefore require long start-up times (an hour or more). In APU applications, the fuel cell can be left running most of the time, or could be started far in advance of an anticipated stop. The principal attraction of SOFCs is their high tolerance to hydrocarbon fuels. The heat of the SOFC can be used in the air-conditioning unit, either as heat or as cold.

Figure 13.7. Selection of fuel cell systems, 1 and 5 kW_{el}, in Japan.

13.5.3 Stationary applications

Stationary fuel cells (today largely PEMFCs and in the future most likely MCFCs and SOFCs) can be used for both distributed and centralised electricity generation. The high temperatures of the exhaust gases produced from MCFCs and SOFCs are ideal for co-generation and combined-cycle power plants, reaching an overall system efficiency of up to 90%. Another advantage of MCFCs and SOFCs is that they can be fuelled directly by hydrocarbons (e.g., natural gas) and do not require external reformers because of their high operating temperatures. Stationary fuel cells (and hence distributed heat and power generation) are not necessarily a market for hydrogen, because natural gas from the gas mains can easily be used directly. Conversion of natural gas or biogas to hydrogen would only reduce the overall efficiency.

Research and development activities regarding fuel-cell units for the residential sector can be located all over the world. In Japan, 400 1kW fuel-cell systems were tested in 2005 (see Fig. 13.7). But intense attention is also being paid to these applications in the United States, China and Korea.

In Europe, there are also several developers of fuel-cell heating appliances (see Table 13.5). They are mainly suppliers of traditional heating systems, like Vaillant, Viessmann or the Baxi Group. Hexis and Elco are new players on the market and do not have long-standing contacts with the heating sector.

In stationary applications, the cost targets for fuel cells, at approximately \$3000/kW, are not as stringent as in fuel-cell cars. For instance, the costs of MCFCs and SOFCs are currently in the range of \$8500–\$20 000/kW. The start-up time and the load

Table 13.5. *European companies manufacturing fuel-cell products for residential heating*

Producer	Hexis	Vaillant	Viessmann	Baxi Innotech	Elco
Technology	SOFC	PEM/SOFC	PEM	PEM	SOFC
Performance	1 kW_{el}	1.5 – 4.6 kW_{el}	2 kW_{el}	1.5 kW_{el}	1 kW_{el}
Electrical efficiency	25–30% target: >30%	29% target: ≥35 %	28% target: >32%	25% target: ≥30%	target: ≥35%
Total efficiency (CHP)	ca. 85%	85%	>87%	>80%	>85%
Brand name	Galileo 1000 N	Euro 2	HEVA II	Beta 1.5	n.a.
Already installed	110 units	60 units	–	15 units	–
Status quo of development	Field test at the beginning of 2007	Co-operation with Webasto	2006–2009 HEVA III field test	In 2007 Beta 1.5 plus systems	In 2007 prototypes (SOFC tubes)

changes are not the determining factors; the determining factor here is the required lifetime for household use, which should reach about 40 000 hours.

It is clear that a broad mass market is not expected until after 2010 (Gummert and Suttor, 2006). But there are already a number of pilot and demonstration systems installed worldwide. It is very difficult to obtain a complete overview of installations because, on the one hand, the data are published by different players, such as utilities, manufacturers or users for 'their' own fuel cells and, on the other hand, if an installation does not work, no data are published or sometimes the system is shut down. Nevertheless, Fuel Cells 2000 have set up a fuel cell database for stationary applications. Most entries are from the USA, but it should be pointed out that Japan installed 480 stationary plants in 2005 alone (Fuel Cell Development Information Centre, personal communication, 2006).

One can distinguish between different time phases for fuel cells in stationary applications: for example, in the 1990s, mainly PAFC systems were built (see Fig. 13.8; the numbers for 2005 may not be complete). The AFC had its field day during the nineties, only a dozen of them were installed in 2001, but since then they have appeared more often again up to the present (there were four installations in the 1980s and 150 in the 1990s). Then the PAFC re-entered the market, UTC being the largest manufacturer of PAFC (with 260 global installations).[1] The PEM entered the market in 1988 as the most promising technology and the number of installations

[1] Although UTC originally wanted to make a complete switch to the PEM technology, it is now once again backing the proven PAFC technology since the targets for PEM have not been achieved. Press Info., 3rd April, 2007.

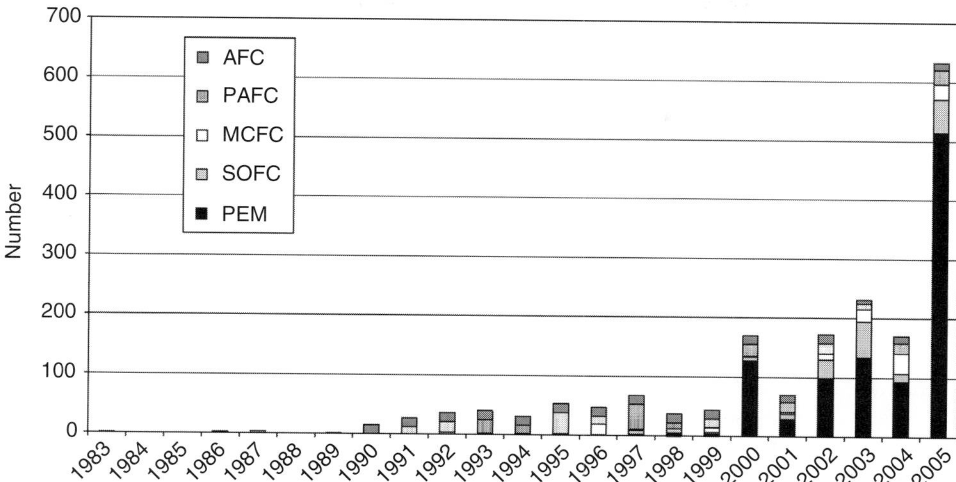

Figure 13.8. Worldwide new installations of stationary fuel cells 1990 to 2005 (www.fuelcells.org/db as of January 2005; FCDIC, personal communication 2006; Jochem *et al.*, 2007).

continued to grow up to 2004, while MCFCs and SOFCs are nowadays seen as the most promising candidates for stationary applications (heat and power generation). In terms of electrical and total efficiency the difference between MCFCs and SOFCs is small, but MCFCs are designed as big units, with 250 kW.

Stationary fuel cells help exergise the energy system.[2] For instance, (condensing) boilers of the central heating system of a building are highly energy efficient, as they convert almost 100% of the chemical energy content of the fuel (natural gas or light oil) to heat; but as they generate heat at a temperature for which no users exist (boiler flame temperatures of up to 1000 °C for supplying room radiator temperatures of 60–70 °C), their exergetic efficiency is miserable, only a few per cent (Winter, 2007). If a hydrogen-fuelled (pure hydrogen or natural gas-reformate) low-temperature fuel cell (<200 °C) is installed instead of a boiler, it generates electricity (i.e., pure exergy) from 35%–40% of the fuel's energy, while the remaining heat available at this temperature regime is sufficient to warm the building over most of the year.

Because of their capacity to use heat at higher temperatures, high-temperature fuel cells, like SOFCs and MCFCs, are better suited for stationary CHP applications. But whereas there are developments aiming to produce SOFC in a 1 kW_{el} scale for domestic heating, the highest efficiency results from bigger plants using a combined cycle gas turbine. Another problem of high-temperature fuel cells is their start-up time of several hours, because the ceramics have to be gradually pre-heated, increasing the temperature by only several degrees per minute.

[2] Thermodynamically, each conversion step converts energy into exergy and anergy: energy = exergy + anergy. Exergy is, by definition, the availability to perform technical work, it is the maximum work to be extracted from energy. Exergy can be converted into any other energy form, anergy cannot.

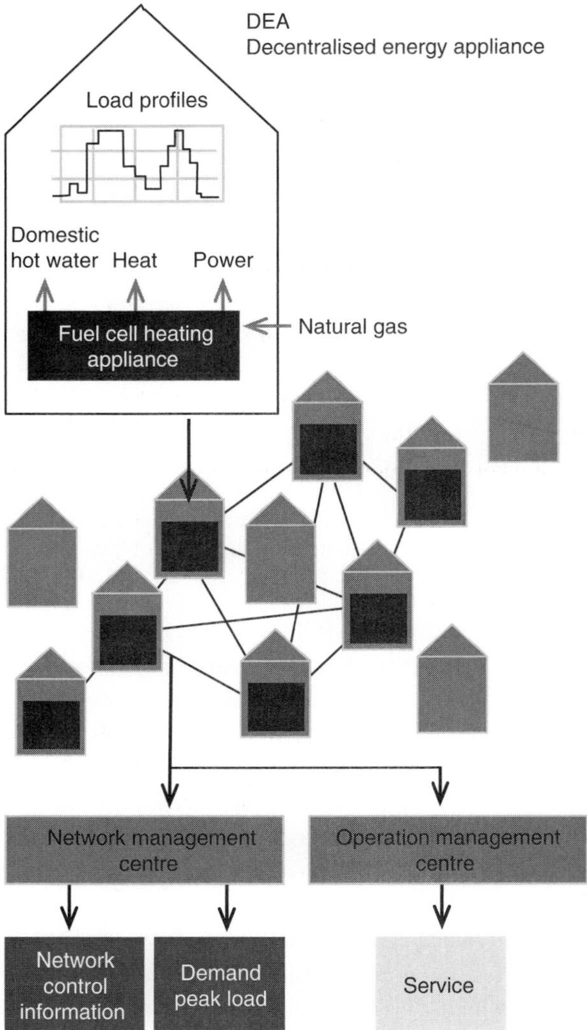

Figure 13.9. Schematic diagram of a virtual power plant (Vaillant, 2007).

Fuel cells are well suited to being applied as part of a *virtual power plant*, see Fig. 13.9. This is the term used for a centrally controlled network of small, decentralised CHP installations. Each installation produces heat and electricity independently for its host (e.g., enterprise, private house, public building, etc.), but if needed, all the installations together can also feed electricity into the public grid. This is controlled by a network management centre. The fuel-cell systems are monitored in an operation management centre and a technician is informed when maintenance is necessary. In this way virtual power plants can provide electricity during peak loads or help stabilise the grid by peak shaving.

13.5.4 Portable applications

Portable fuel-cell systems are systems that produce electricity for devices with a performance ranging from several watts to 10 kilowatts. The heat produced in the process is a by-product that is normally not used. The system has, therefore, to be cooled down by fans or cooling surfaces, etc. A wide range of applications is possible for fuel cells from small electronic devices like camcorders, mobile phones, laptops, etc. to electric tools, back-up systems, or power generation on boats or caravans.

Currently, direct-methanol fuel cells (DMFC) are the best candidates for portable applications, given that PEMFCs face hydrogen-storage problems and that the operating temperatures of MCFCs and SOFCs are too high for portable devices. Because methanol is easily transportable, DMFCs represent an option for replacing batteries in portable devices. With low efficiency (15%–30%) and low power density, they are not suitable for mobile or stationary use. Although the technology of DMFCs is close to market introduction and presently has a niche market in caravans and boats, R&D is still required to lower costs and improve the periphery systems and reliability. Proton-exchange-membrane fuel cells offer limited benefits for use in portable devices, as hydrogen storage offers no energy density advantage over batteries.

13.5.5 Development and potentials of fuel cells

The various types of fuel cells are at somewhat different stages in the technology cycle: the MCFC is ready for market introduction but faces the typical problem of a new technology, i.e., is expensive because of the lack of economies of scale for its production and the lower cost of its technical rivals (engine-driven co-generation and microturbines). From the technical point of view, the phase of euphoria has almost passed for PEMs and SOFCs and further R&D activities are necessary for these two types to match the technical performance of their competitors or the necessary cost level for fuel cells to be technologically and economically ripe for the market. Only the DMFC has reached a standard that allows its use in niche markets, like caravans or yachts, despite poor efficiency levels.

An analysis of the individual PEM components offers evidence of almost unbroken R&D; see Fig. 13.10 (Jochem *et al.*, 2007). The overall importance of the membrane is striking. Furthermore, the numbers of annual applications for bipolar plates (BPP) and the gas-diffusion layer (GDL) decrease after 2002, while the increase in membrane applications flattens out. This correlates with the equally lower number of fuel cell patents in the field of mobile applications.

The key question for mobilising further market potential is whether a steady decrease in fuel-cell costs and an increase in lifetime can be achieved. Many energy technologists believe the future of the different types of fuel cell to be very bright because of anticipated efficiency improvements and the low emissions associated

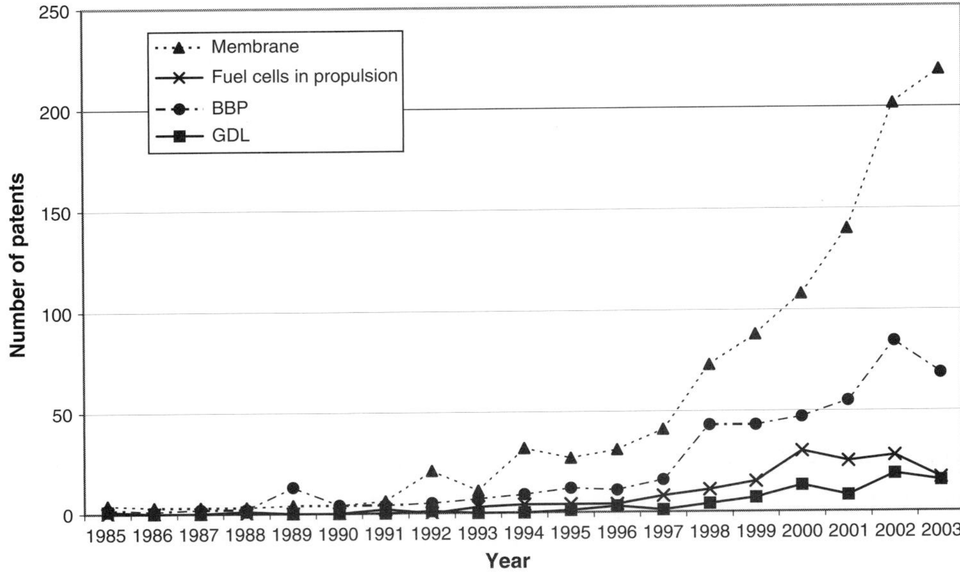

Figure 13.10. Patent applications for components of the PEM fuel cell, 1985 to 2003.

with them (zero emissions at the point of use in the case of hydrogen). The application potential is considered to be high in both stationary uses (decentralised heat and electricity production) and mobile uses, owing to the very large market of road transportation. Optimistic visions assume a 10% to 20% share of electricity generation in a future economy powered by fuel cells and even higher shares in road transport. However, these visions tend to overlook the further possible improvements of competing technologies (internal combustion engine, Stirling engine, microturbines), the necessary infrastructure innovations (e.g., hydrogen) and the substantial cost-reduction requirements of the different types of fuel cell necessary for them to compete successfully in open markets with alternative technologies.

The so-called 'microbial fuel cells' (MFC) are a completely different type of fuel cell. Here, bacteria are used to convert a bio-usable substrate directly into electricity. In the future, it might be possible to run an MFC for medical purposes by using glucose directly from the patient's bloodstream (Logan *et al.*, 2006).

For the near future, with a growing share of fluctuating regenerative electricity like wind energy, 'regenerative fuel cells' are being developed, which can operate in reverse to act as electrolysers. Regenerative fuel cells can thus play a role in energy-storage systems. The produced hydrogen can be stored and used for the fuel cell during wind free periods.

Nevertheless, it must be taken into account that the development of hydrogen fuel cells also depends on how competitive technologies develop (for details see Chapter 7). If batteries for electric cars are invented that, e.g., have twice the energy density and can be recharged in ten minutes, electricity would probably be the energy carrier of the mobile future. Perhaps a catalyst will be discovered that can convert methane directly and easily to methanol (without the energy-consuming intermediate step of synthesis gas), which could be used as fuel for DMFCs. In this case, a methanol society would be possible, e.g., as described by Olah *et al.* (2006). From today's viewpoint, hydrogen can be used in fuel cells as a secondary energy carrier without emitting any CO_2. It therefore does not make sense to convert hydrogen to methanol, which does emit CO_2 when used as a secondary energy carrier.

13.6 Sectoral changes induced by fuel cells

In recent years, the fuel cell has been discussed as an alternative to the internal combustion engine for use as a propulsion system in motor vehicles or small co-generation plants, which are powered today either by combustion engines or micro gas turbines. Many automobile and boiler manufacturers, as well as energy-supply companies, have reinforced their research activities in the area of fuel-cell technology and plan to develop mass-produced fuel-cell driven cars, co-generation units and other smaller fuel-cell-based energy packages. For example, many of the leading car producers announced the market entry of the first fuel-cell models of their fleets shortly after 2005 (Maruo, 1998; Nitsch and Dienhart, 1999). And heating boiler manufacturers wanted to introduce their new co-generation units into the market, but due to the higher investment cost per kW and lifetime limitations (see above), their stationary units cannot yet compete with boilers and delivered electricity.

In many regions worldwide, automobile manufacturing, together with its outfitters and suppliers, is a central pillar of the economy. For example, in 2005 every sixth job in the manufacturing industry of the German federal state of Baden-Württemberg was directly or indirectly dependent on automobile production (www.wm.baden-wuerttemberg.de). The outfitters and suppliers of boilers or car producers also contribute greatly to the number of indirect jobs, particularly in the area of machine tools, and metal and plastic production. The outfitter and supplier industries in these regions are particularly oriented towards conventional propulsion technology with spark-injection or diesel engines. A changeover to fuel-cell propulsion technology in cars and small co-generation would mean that certain production areas, particularly those of mechanical processing, such as turning, milling, grinding and cutting, could lose their current importance. On the other hand, suppliers of the components needed for fuel cells (such as electric motors, power electronics and small reformers) may not be available today in these regions, but may be an important industrial activity in the next 20 or 30 years.

Because of the high degree of dependence of the economy on the automobile or boiler industry in these regions, it seems worthwhile to investigate the impacts on the supplier and outfitter industries arising from the technological restructuring of changing the propulsion system of cars or small co-generation, i.e., the technology-induced structural change in industry. Furthermore, the innovation process has its own momentum with 'first-mover' benefits as a car-producing region, but may also involve risks because of the strong orientation towards traditional propulsion or boiler technology, both domains of mechanical engineering. Thus, today's leading position of an industrial region producing cars or boilers with traditional technologies may adversely affect its chances of taking a leading role in the new technologies for vehicle propulsion and co-generation, which are, to a large extent, based on process engineering.

13.6.1 Methodological approach

The cost of the new propulsion system has usually been estimated by engineering analyses, target costing and assumptions applied to experience curves to account for learning and economy-of-scale effects. The structural change within industry and the energy supply caused by diffusion of the new technology is calculated on this basis (see Erdmann and Grahl, 2000; Grahl, 2000; Walz *et al.*, 2001; Wengel and Schirrmeister, 2000).

13.6.2 Cost aspects of the fuel-cell application

The economic criteria for successful market diffusion of a new technology or product are cost competitiveness at similar or even better performance of the new technology, a reduced environmental burden and other advantages (Höhlein and Stolten, 1998; Kolke, 1999). So the first analysis step has to be to compare the future cost of both technological options from the engineering and business points of view. Distinctions among the different components of a fuel-cell powered product (e.g., vehicle, co-generation unit) have to be made: (1) unchanged components of the system, (2) adapted components, such as motor electrics, transmission, exhaust system, cooling, and fuel storage and (3) components substituted by the new technology (see Fig. 13.11).

These distinctions permit a detailed engineering and cost analysis to be made, and this has been conducted by various authors to different degrees. Erdmann and Grahl (2000), for instance, conclude in their cost analysis of an upper-middle-class car that the investment cost of the power train will range between \$4000 and \$10000, i.e., \$50 to \$130/kW$_{el}$, depending on the fuel used. The cost target derived from the target costing method suggests that the cost should not be more than that of the present turbo-diesel drive systems, i.e., below \$50/kW$_{el}$. Whether and when these specific costs become feasible is highly debated and heavily dependent on several factors (Friedrich and Noreikat, 1996).

A detailed analysis published for fuel-cell drive systems by Schirrmeister *et al.* (2002) showed a specific cost breakdown of the added value of a fuel-cell drive system powered by methanol (a methanol reformer was a reasonable alternative at that time)

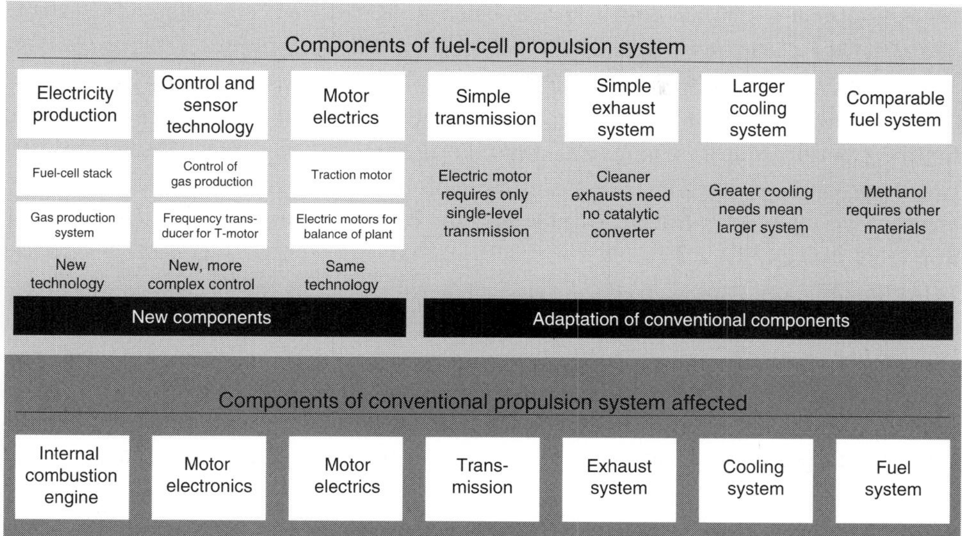

Figure 13.11. Overview of the differences between the drive system of a fuel-cell powered vehicle and a conventional vehicle (Wengel and Schirrmeister, 2000).

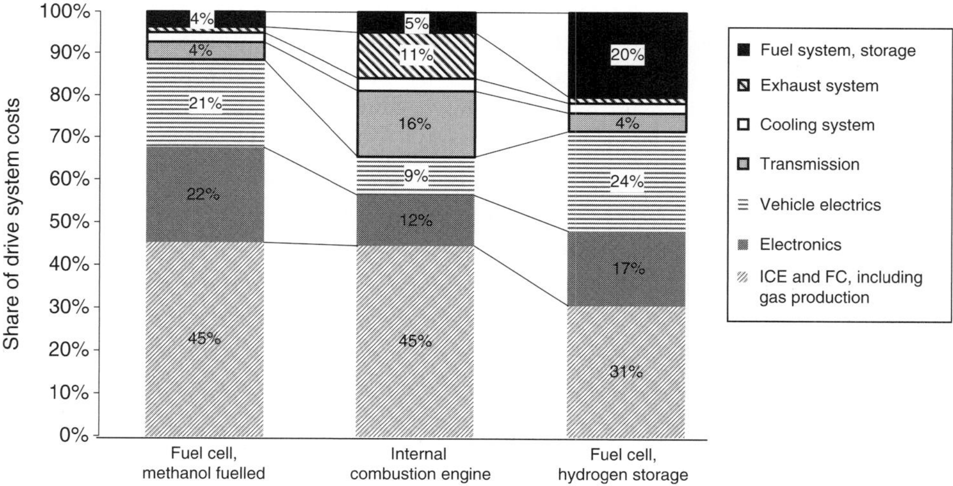

Figure 13.12. Shares of added value of the individual components of a fuel cell (by the target costing method) and a combustion engine drive system (Schirrmeister *et al.*, 2002).

with on-board storage of hydrogen (high-pressure storage). This was also used to identify the differences of added value in the conventional combustion engine drive system of a similar car, see Fig. 13.12.

The cost of the exhaust system is lowered by omitting the catalytic converter and the lambda sensor. The transmission is considerably simpler than the multigear transmission of a gasoline motor, so that its costs are reduced. The value-added share of the motor electrics will increase considerably, chiefly because of the large cost share of the traction motor. Similarly, the control and sensor technology will become more expensive, essentially because of the higher costs of the frequency transducer for the traction motor. Comparing the two alternatives of a fuel-cell vehicle, i.e., hydrogen on-board storage versus methanol fuelling, there will be a trade-off between the costs of hydrogen storage and the costs of on-board hydrogen production from methanol.

The greatest uncertainty today is the cost development of the electrochemical part of the system, the fuel-cell stack. Experts agree that it could be mass-produced at reasonable cost. The target-costing approach states that, in the long term, the specific costs of the fuel-cell stack must be in the order of €10/kW_{el} (based on a 75 kW_{el} power for each of the two fuel-cell stacks).

In contrast to the fuel-cell stack, there is very little information available about the costs of the gas production system, the compressor–expander unit and the balance of the system (pumps, separators, etc.). Cost objectives between €15/kW_{el} and €100/kW_{el} have been given for the gas production system (reforming). In the next few years, it will become clear whether it is possible to reach these low-cost objectives for the gas production system, the compressor–expander unit and the balance of system.

The costs of the gas production system are mainly determined by the manufacturing costs of the components, which have to be integrated thermally and physically into an efficient and compact system, and by the efficient use of cost-intensive catalysts. The material costs for catalyst materials may range from 20% of the catalytic converter of a methanol steam reformer to 80% of a catalytic burner. The costs of the hydrogen storage depend mainly on the technology used (liquid storage versus high-pressure gas storage) and the applied production technology, which has not yet been adapted to high volume mass production.

These cost considerations have to rely on some risk analysis: if the risks of producing smaller series for stationary markets, which can absorb fuel-cell systems at a specific cost of some €500/kW_{el} are smaller, it seems likely that stationary fuel cells will enter the market earlier than mobile fuel cells in cars at a specific cost of €50 to €80/kW_{el}, but with much larger production series.

13.6.3 Structural changes in industry

The type and extent of the structural effects of a change in the propulsion system on industry are determined by several factors of influence, such as the regional supplier and outfitter structure of a traditional propulsion system or a traditional boiler, the regional changes (due to changing net imports or net exports of components of the new propulsion system), and the overall development of car or boiler production in the economy considered (Feige and Goes, 1999).

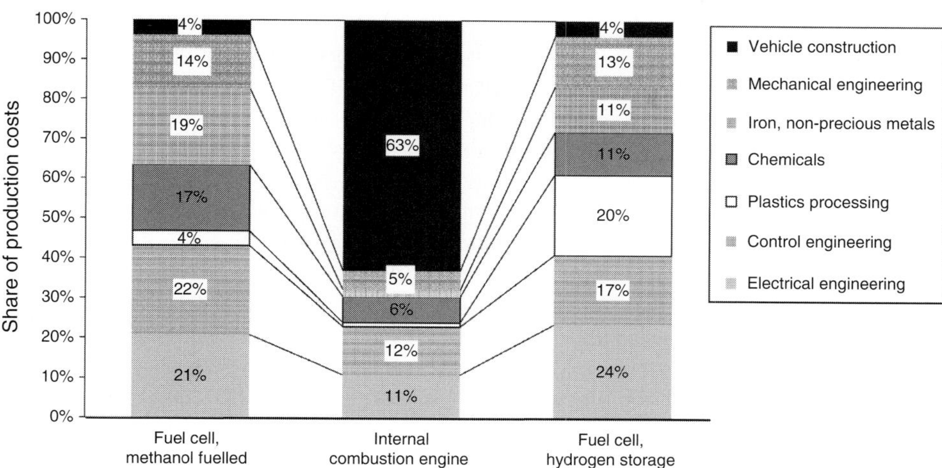

Figure 13.13. Structural changes in industry due to a complete replacement of internal combustion engines with fuel-cell drives (Schirrmeister *et al.*, 2002).

Assuming constant definition of industrial sectors and no changes in net foreign trade, there is a substantial structural effect of mobile fuel cells on industrial branches (see Fig. 13.13).

- The *electrical industry covering electrical and control engineering,* in particular, stands to gain considerably, from 23% to approximately 41%. The electric motor of the fuel-cell system alone, including the steering, is expected to claim over 35% of the total costs of the new drive train and thus causes substantial growth in the electrical-engineering sector. The omission of many smaller components of the standard drive train, such as the oil-level sensor, the pedal-travel sensor, the spark plugs and the ignition module are over-compensated for by the electric motor and the control system. Thus a shift in demand within the sector will take place.
- Even though the usual catalytic converter of the exhaust of the standard drive train is omitted, there are also net increases in the chemical industry. The increased demand is for the expensive coating of the electrode-membrane unit (MEA) and the catalysts needed for gas preparation (reformer).
- There is a substantial shift of production in favour of the industrial sectors for 'iron and non-precious metals', such as 'steel and light metal work products' and 'products of the drawing mill, cold-rolling mill, etc.'. Components with high shares of added value in these sectors include the vaporiser and probably the bipolar plates, as well as various containers needed for the reformer. Supplies from these sectors to the electric motor (plates in the centre of the electric motor) are not identified here, since only the first supply stage was considered and it is assumed that the electric motor is obtained as a system component from a supplier in the electrical-engineering industry. The production technology of the bipolar plates is still uncertain, since this is a new product and has, as yet, only been manufactured in small numbers.
- The demands on the *mechanical engineering* sector will increase slightly. However, there will be changes within this industrial sector. The simpler transmission of the fuel-cell propulsion

system, compared with standard drive trains, leads to a reduction in supplies from the mechanical engineering sector. However, this is essentially compensated for by the supplies of new components such as the compressor–expander unit, the catalytic burner and other pumps. The production methods of these new components are expected to be very different from those used for components of the combustion engine drive train.

- The reduction of the *vehicle industry*'s share is mainly a result of the omission of engine components with large added value (e.g., crankshafts, pistons, cylinder heads and bearings).
- While '*plastics processing*' has been of minor importance for today's production of the internal combustion engine and its respective drive system, this sector will become highly relevant if hydrogen is stored on board. A thin-walled aluminium storage sheeted by a carbon-fibre-reinforced plastic has been assumed for the above-shown structural changes in industry.

In standard internal combustion engine drive trains, about 60% of the added value results from the *vehicle industry*. This share may be reduced to only 10% for fuel-cell drive trains if the outsourcing potential is fully exploited. This shift is because the components of the fuel-cell propulsion system are not suited to current production structures in the automobile industry. Therefore, it can initially be assumed that they will be manufactured by other sectors. However, if there is a breakthrough of fuel cells, it is possible that the automobile industry will start to manufacture many of the components that are assigned to other sectors in Figure 13.13.

Although a comparable analysis does not exist for co-generation units, the changes in shares of added value of the different industrial branches may be similar with respect to the steel and light metal work and the chemical industry, as these are the branches that stand to gain. In contrast, the mechanical-engineering sector may face internal shifts and the electrical-engineering industry may suffer, as the omitted generators, whose added value may not be fully compensated for by additional control systems of the stationary fuel-cell system.

However, not only are the supplier industries of importance for a region or country producing cars or co-generation units; the outfitter industry is also affected by changes in the supplier structure. In general, in the case of a breakthrough of fuel-cell technology, there will be growth opportunities for the corresponding sectors, particularly, of course, for mechanical engineering. However, within the outfitter sector, there will be a shift in emphasis because of the technological changes (new components) in the drive train. The expected increases for outfitters in the electrical-engineering and chemical-industry sectors will probably be counterbalanced by decreases for outfitters in the vehicle-manufacturing sector. In the area of automation and assembly technology, it is not expected that there will be extensive adaptation to the new production technology resulting in considerable changes for outfitters. In general, the outfitter sector seems to be far more independent of the share of added value within an industrial economy than the supplier industry, assuming, however, that the new components are domestically produced (and not imported).

The basic tendency assumed here, i.e., that new components of the fuel-cell propulsion system will be manufactured in technologically competent subsectors, may be obstructed by the car manufacturers' decisions of what, where and by whom production and delivery will be carried out. The industrial division of work in the automobile industry has been successively developed over decades. There have recently been great efforts made to bring the increasing complexity under control by establishing system suppliers and by arranging the supplier structure into pyramids. With regard to fuel-cell innovation, there is the opportunity from the outset to set up a concept with few system suppliers, which is supported by the features of the technology. This change in supply policy is expected to be less pronounced in the case of co-generation units or boiler producers, who obtain the major part of their added value in-house.

Whereas the drive train of the standard combustion engine comprises many individual, diverse components, these are reduced in fuel-cell propulsion systems to a few expensive components. The decision on the production location of the important system components (i.e., fuel-cell stack, hydrogen storage, reformer and electric motor) will, therefore, be vital for the regional supplier structure.

The speed of the structural change will also be determined by the competition between internal combustion engines and fuel-cell propulsion systems. Measures to reduce fuel consumption and emissions may cause additional development and investments in the technical performance of combustion engines, reduce the advantages of fuel-cell applications, and slow down the diffusion of the mobile or even the stationary fuel cell.

13.6.4 Concluding remarks

Looking at the enormous cost reductions that will be necessary for fuel-cell systems to compete with traditional technologies and the risks involved in exploiting the economies of scale effects, one would expect market diffusion of the mass-produced fuel cell in its stationary application as a co-generation unit, and only later as a mobile fuel cell in cars. Techno-economic analyses also suggest that substituting engine-driven co-generation units and boilers would induce less changes in added value among the different industrial branches than the mobile fuel cell would, simply because of the smaller market volumes involved.

So the greatest challenges are in the mobile sector, but the pressure to act is much greater here as well, owing to oil scarcity, pollutants from vehicles, noise nuisance, etc. Compared with stationary applications, the alternative technologies in the mobile sector are also much poorer. This is why fuel-cell vehicles remain a possibility, despite the enormous sectoral changes that accompany this alternative. The question is when will they achieve market penetration? One of the main obstacles that will have to be overcome is the attendant position of both the automobile industry and the infrastructure industry concerning the investment. Which one is prepared to make the first large-scale investment and thus enable the market penetration of the other? This is a classical example of the chicken-and-egg dilemma.

But when fuel cells arrive, there could be substantial structural changes in industry and energy supply within two decades, even if the production of the fuel-cell system remained in the country where traditional drive trains, co-generation units or boilers are produced. Process-oriented industries, such as steel and light metal works, cold-rolling mills, the chemical industry and the plastics-processing industry stand to gain from the additional net value added; the losers are the car industry, producing components of the internal combustion engine with large added value, and the electricity producing industry. The electrical industry will both profit from the mobile fuel cell because of additional sales of the electric motor and the more complex control system, and will produce equipment for co-generation units. As long as hydrogen is made from natural gas – possibly with an intermediate fuel such as methanol or synfuel for on-board reforming – the oil industries could be a substantial loser in the case of the mobile fuel cell. The electricity sector could also lose in the long term, if stationary fuel cells not only replace small boilers, but also start substituting mobile fuel cells when parking in the garage at home or at work.

13.7 Impacts of fuel cells on the service sector

The introduction of a new product or technology is always accompanied by different circumstances that promote or prevent its success on the market. If there is no public acceptance, for example, fuel-cell technology will be hard put to gain important market segments. But there are also other barriers to be overcome.

Fuel-cell systems have to be installed, maintained and repaired. During the introduction of condensing-boiler technology, for example, there were technical problems due to a different gas composition, and organisational problems, as the installers were not integrated enough into demonstration and vocational training projects. To avoid these market introduction barriers, education and training aspects have to be considered within field tests and demonstration periods.

To analyse these problems, the impact of fuel-cell technologies on the service sector was analysed in Germany in a project of the BERTA programme financed by the German Federal Ministry of Economics.[3] The possible impacts were examined with regard to job profiles, qualifications and the job market, to prepare the branches involved most optimally.

Indeed, trade and small businesses will be strongly affected by the introduction of fuel cells. Considering the number of skilled workers in the different sectors, fuel-cell technologies will affect three of the four groups that have the highest employment rate in Germany (Fig. 13.14): electricians are affected by portable and stationary

[3] BERTA stands for 'Brennstoffzellen: Entwicklung und Erprobung für stationäre, mobile und portable Anwendungen' (Fuel Cell Development and Validation for Stationary and Mobile Applications) and was the fuel-cell part of the German ZIP ('Zukunfts-Investitions-Programm') Programme on investment in the future.

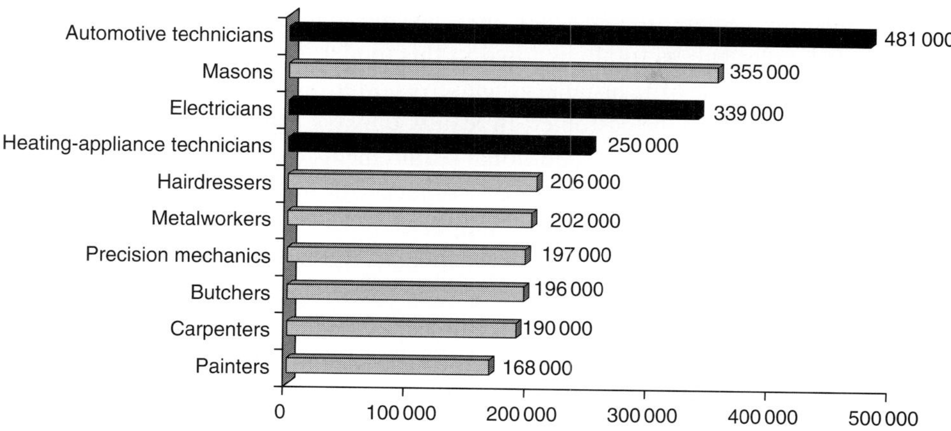

Figure 13.14. Number of employed workers by sector in Germany, 2006 (estimates; Central Association of the German Craft, personal communication, 2007).

applications, automotive technicians by transport, and heating-appliance technicians by stationary applications. Moreover, chimney sweepers, electronics technicians and other branches within trade and industry are involved.

Among these groups of professionals, the installers and plumbers, as well as the electricians, are the most affected by fuel-cell technology. As Hexis and Vaillant were already running field tests during the BERTA project, workers in the heating and electric sectors received training from the developers and were surveyed by the project team. The heating installers and electricians were asked about their experiences and training needs. The results confirm that the fuel-cell heating appliances are not just another technique that may replace the condensing boilers currently on the market. They require additional skills, for example, concerning the calculation of power requirement and operation efficiency. This know-how is comparable to that of micro combined heat and power systems (CHP).

Because the fuel-cell heating system is a CHP system, the installation requires both electricians and heating installers. So the vocational training should refer to both groups and requires not only new training content but also, at least in Germany, new organisational structures. These qualifications will also be required in other countries. The traditional division between heating systems and electronic systems will not be able to be upheld.

The scenarios developed in the project showed a wide range of possible impacts for trade. In the worst case, the installer will lose clients as the energy supplier may offer an all-inclusive service to the customer. This means that the energy supplier will install its own fuel-cell heating appliances in the cellar or basement of the customer's house, provide electricity and heat and be paid an all-inclusive price by the customer. The energy supplier employs its own technicians for installation, maintenance and repair. The customer does not know whether the electricity is generated by the fuel-cell heating system in his cellar or delivered by the energy supplier. The traditional installer will be totally ignored in this model.

In the opposite model, the installers themselves offer the same all-inclusive service as the energy supplier to the customer. In reality, only big installation companies will be able to do this, because of the additional requirements described above. However, it is usually the installers who have direct contact with the customer in most cases. This is a big advantage over the energy supplier.

There are two other possible scenarios. In one, the installer offers the fuel-cell heating appliance just as he does other heating systems; then he has to cope with additional training needs, because of the double qualification and the calculations for operation efficiency. In the other scenario, the installer works for the energy supplier, an alternative that is not profitable, as he loses the profit margin of the heating system.

To conclude, there is a high probability that the companies dealing with heating and power systems can only maintain or strengthen their market position if they are willing to engage in vocational training and if they are able to adapt to the changes in the power-market structures towards a decentralised energy supply. It is recommended that skilled workers from trade and small businesses deal with the CHP heating systems that are already on the market.

Concerning the skilled workers in the electric and electronic branches, they will probably have to repair bigger systems (e.g., power packs above 500 W) or replace defective modules, as they already do today, for example, in the case of technical problems with CD players. A new business opportunity may be the establishment of a hydrogen-cartridge infrastructure – selling, refilling or recycling the cartridges. So the craftsmen dealing with devices for potential portable fuel-cell applications will only be marginally affected by the introduction of fuel cells. The training requirements in industry consider engineers, technicians, craftsmen and field service personnel. The training will be closely linked to the applications and offered by the developer or supplier of the respective device.

Looking at the transport application and the concerns of automotive technicians, interest is still restricted. One reason is that fuel-cell vehicles are not yet ready for marketing. The market introduction scenarios are shown in Chapter 14. Another reason is that the biggest share of the automotive turnover comes from buying and selling vehicles (82%) and not from maintenance and repair. In Germany, maintenance only constitutes 6%, repairs 4% and accident repairs 8% (DAT-Veedol-Report, 2001).

Analysing the maintenance and repairs of vehicles shows that only a small part of the work will disappear, as there will be no more abrasion repairs on motors and motor electrics. Also, repairs on gears and the coupler will be reduced, as these parts will be simplified (see Fig. 13.15). On the other hand, it has not been estimated what kind of work the fuel-cell technology will bring in addition to the work done today. As the trend at present is towards more electronic components and control, fuel-cell

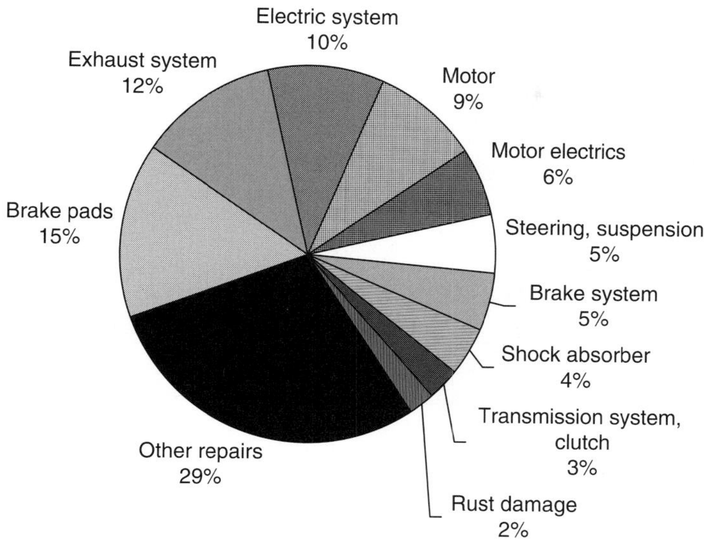

Figure 13.15. Frequency of replacements and repairs, 2000.

vehicles are likely to move in the same direction. At the end of the innovation process, according to Koschorke *et al.* (2005), there might be approximately 20% less maintenance and 26% less abrasion repairs. The overall business volume will be reduced by only 3.1%.

The education and training aspect is the most important for trade. So a master plan for education and training has been developed within the BERTA project. It calculates the requirements for each application in relation to market sales figures (see Fig. 13.16).

At the same time as training needs for trade were identified in the very early phase of demonstration projects in Germany, the first considerations for a concept of education and training were put forward at a European level. In 2002, the European Commission launched the High Level Group (HLG) on Hydrogen and Fuel Cells. For the first time, an integrated EU vision was formulated on the possible role of hydrogen and fuel cells in a sustainable-energy world. Recommended by the HLG, the European Hydrogen and Fuel Cell Technology Platform was established a year later (see also Chapter 8). Within the organisational structure of the platform, an initiative group on education and training made initial suggestions about how to train the different groups involved in the research, deployment and market introduction of fuel-cell systems. This group, as well as the group on public awareness, has since been integrated into the section on cross-cutting issues.

The importance of education and training was also noticed by the International Partnership for the Hydrogen Economy (IPHE), initiated in 2003 by the United States to co-ordinate research activities and to accelerate the transition to a hydrogen economy (see also Chapter 8). The IPHE Hydrogen Education Work Group (EWG)

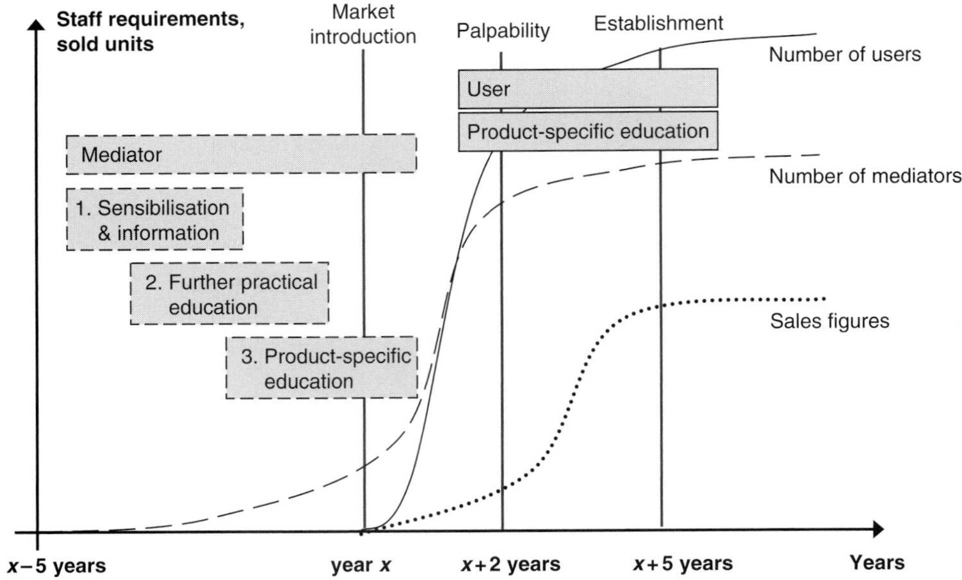

Figure 13.16. Needs for training in relation to sales figures.

is charged with examining the international dimensions of hydrogen education, including possibilities for collaboration among IPHE member countries. This also involves training technicians and other industry specialists through appropriate instruction at technical and vocational schools.

13.8 Summary

Fuel cells are electrochemical devices that use hydrogen, or hydrogen-rich fuels, and oxygen to produce electricity and heat. As shown in this chapter, fuel cells may be able to contribute a significant share to the success of the hydrogen economy. For them to be able to do so, it is essential that the targets set for costs, lifetime and reliability are achieved. These technology developments obviously always take longer than planned by industry. However, preparation for the structural changes in industry is just as important as the technical optimisation of fuel cells. Qualified service technicians and skilled workers must be available to ensure that the introduction of fuel-cell technology occurs as smoothly as possible.

Fuel-cell vehicles could gain significant market shares in the coming decades if fuel-cell costs are greatly reduced and if effective policies (incentives) are implemented to reduce CO_2 emissions in the transport sector. The main efficiency benefit of fuel-cell vehicles occurs at partial load during urban driving. But fuel-cell vehicles may also have to compete with alternative technologies and fuels currently under development, such as natural gas, biofuels, (plug-in) hybrids or battery-electric vehicles, which may play an equally important role in the future.

The potential of stationary fuel cells for distributed generation depends on feed-in tariff policies and electricity and gas prices, as well as on market competition from gas engines and small turbines. SOFCs and MCFCs, mostly fuelled by natural gas, are likely to play an important role for combined heat and power generation in buildings.

There is a global market for fuel cells. In the USA, the main R&D areas are in transportation fuel-cell systems, distributed or stationary fuel-cell systems and fuel-cell subsystems and components, all supported by the Department of Energy. Another important player is the Department of Defense, which is developing fuel cells for military purposes, such as 'portable soldier power'. Even though fuel-cell manufacturers in Japan are also experiencing too-high costs and insufficient cell lifetimes and have the same targets as elsewhere in the world, they are still investing much more in fuel-cell development and field tests. The European Union, which is not giving as much support to field tests, needs to be aware that it is running the risk of missing the boat here.

Portable DMFCs seem close to commercialisation. They are likely to be followed by stationary MCFC and SOFC systems for decentralised heat and power generation. More research is still needed before PEMFCs are ready for commercialisation in the transport sector.

References

DAT-Veedol-Report (2001). *Dossier Kfz-Betrieb*. Vogel Verlag und Druck GmbH & Co. KG.

Dauensteiner, A. (2001). *Der Weg zum Ein-Liter-Auto – Minimierung aller Fahrwiderstände mit neuen Konzepten*. Berlin: Springer Verlag.

Demuss, L. (2000). Technologische Veränderungen beim Übergang vom konventionellen Antriebsstrang zur Brennstoffzelle. In *Innovationsprozess vom Verbrennungsmotor zur Brennstoffzelle – Chancen und Risiken für die Baden-Württembergische Industrie*, ed. Wengel, J. and Schirrmeister, E. Karlsruhe: Fraunhofer ISI.

DLR (Deutsches Zentrum für Luft- und Raumfahrt) (1997). *Energie- und Schadstoffbilanzen von Elektrofahrzeugen mit Batterien und/oder Brennstoffzellen-Antrieben im Vergleich zu Kraftfahrzeugen mit Verbrennungsmotor*, ed. Carpetis, C. STB-Bericht Nr 16, DLR-97 44417 IB 404. Stuttgart: DLR, Institut für Technische Thermodynamik.

Erdmann, G. and Grahl, M. (2000). Competitiveness and economic impacts of fuel cell electric vehicles on the future German market. *Proceedings Hyforum 2000*, (September 11–15). Munich.

Feige, A. and Goes, F. (1999). Wandel in der Wertschöpfungskette. *Automobilproduktion*, **4** (1999).

Freedom Car: Freedom Car Fuel Partnership (2005). *Fuel cells Technologies Roadmap*. www1.eere.energy.gov/vehiclesandfuels/about/partnerships/freedomcar/fc_goals.html.

Friedrich, J. and Noreikat, K. E. (1996). State of the art and development trends for fuel cells vehicles. *Proceedings of the 11th World Hydrogen Energy Conference*. Stuttgart, pp. 1757–1766.

Frost & Sullivan (2001). *Stationary and Portable Fuel Cells – Developments, Markets and Opportunities*. Report D226. New York: Frost & Sullivan.

Grahl, M. K. (2000). *Ökonomische Systemanalyse zum Antrieb von Personenwagen mit Polymer-Elektrolyt-Brennstoffzellen unter Verwendung neuer Kraftstoffe*. Dissertation. Berlin: Technical University Berlin.

Grove, W. R. (1839). On voltaic series and the combination of gases by platinum. *Philosophical Magazine and Journal of Science*, **14** (86), 127–130.

Gummert, G. and Suttor, W. (2006). *Stationäre Brennstoffzellen – Technik und Markt*. Heidelberg: C. F. Müller Verlag.

Hasenauer, U., Ragwitz, M., Eichhammer, W. *et al.* (2005). *Energy Scientific & Technological Indicators and References (ESTIR). Lot 1: Fuel Cells and Hydrogen Technology*. Final Report for the Directorate General for Research. Karlsruhe: Ecofys (Utrecht, NL), Fraunhofer ISI.

HFPE (2005). *European Hydrogen & Fuel Cell Technology Platform. Strategic Research Agenda*. www.HFPeurope.org.

Höhlein, B. and Stolten, D. (1998). *Pkw-Antrieb mit Verbrennungsmotor und Brennstoffzellen im Vergleich*. 2. Euroforum-Fachtagung Brennstoffzellen. Stuttgart.

Hoogers, G. (2003). *Fuel Cell Technology Handbook*. Boca Raton, FA: CRC Press.

IEA (International Energy Agency) (2005). *Prospects for Hydrogen and Fuel Cells*. IEA Energy Technology Analysis Series. Paris: OECD/IEA.

IEA (International Energy Agency) (2007). *Fuel Cells*. IEA Technology Essentials. Paris: OECD/IEA.

Jochem, E., Bradke, H., Cremer, C. *et al.* (2007). *Developing an Assessment Framework to Improve the Efficiency of R&D and the Market Diffusion of Energy Technologies EduaR&D*. Report Contract No. 0327 287. Karlsruhe: Fraunhofer ISI.

Jörissen, L. and Garche, U. (2000). Brennstoffzellen für den Fahrzeugantrieb. In *Innovationsprozess vom Verbrennungsmotor zur Brennstoffzelle – Chancen und Risiken für die Baden-Württembergische Industrie*, ed. Wengel, J. and Schirrmeister, E. Karlsruhe: Fraunhofer ISI.

Kolke, R. (1999). *Technische Optionen zur Verminderung der Verkehrsbelastung – Brennstoffzellenfahrzeuge*. Berlin: Umweltbundesamt (Federal Environment Agency).

Koschorke, W., Bünger, U., Marscheider-Weidemann, F. *et al.* (2005). *Anforderungen an das Handwerk durch die Innovation Brennstoffzelle*. Fraunhofer IRB Verlag.

Krewitt, W., Pehnt, M., Fischedick, M. and Temming, H. V. (eds.) (2004). *Brennstoffzellen in der Kraft-Wärme-Kopplung – Ökobilanzen, Szenarien, Marktpotenziale*. Berlin: Erich Schmidt Verlag.

Krewitt, W., Nitsch, J., Fischedick, M., Pehnt, M. and Temming, H. (2006). Market perspectives of stationary fuel cells in a sustainable energy supply system – long-term scenarios for Germany. *Energy Policy*, **34** (2006), 793–803.

Larminie, J. and Dicks, A. (2003). *Fuel Cell Systems Explained*. West Sussex, England: John Wiley & Sons Ltd.

Logan, B., Hamelers, B., Rozendal, R. *et al.* (2006). Microbial fuel cells: methodology and technology. *Environmental Science and Technology*, **40** (17), 5181–5192.

Maruo, K. (1998). *Strategic Alliances for the Development of Fuel Cell Vehicles.* University of Gothenburg.

Nitsch, J. and Dienhart, H. (1999). Konkurrenzsituation und Marktchancen von Brennstoffzellen-Systemen. *Proceedings, Sechstes OTTI-Fachforum 'Einsatz von Brennstoffzellen'*: Leipzig.

Oertel, D. and Fleischer, T. (2003). *Fuel Cells. Impact and Consequences of Fuel Cells Technology on Sustainable Development.* Technical Report Series EUR 20681 EN. Seville: European Commission, Joint Research Centre (JRC), Institute for Prospective Technological Studies (IPTS).

Olah, G. A., Goeppert, A. and Prakash, G. K. S. (2006). *Beyond Oil and Gas: The Methanol Economy.* Weinheim: W-VCH.

Ostwald, W. (1894). Die Wissenschaftliche Elektrochemie der Gegenwart und die Technische der Zukunft. *Zeitschrift Elektrochemie*, **1** (4) (1894), 81–84 and 122–125.

Pehnt, M. (2001). *Ganzheitliche Bilanzierung von Brennstoffzellen in der Energie- und Verkehrstechnik.* Dissertation. VDI-Verlag, Fortschritt-Berichte Reihe 6, No. 476.

Schirrmeister, E., Marscheider-Weidemann, F. and Wengel, J. (2002). *Auswirkungen des Einsatzes der Brennstoffzelle im Kraftfahrzeug auf die Industrie in Nordrhein-Westfalen. Szenarien für die Einführung und spezielle Chance Nordrhein-Westfalens.* In co-operation with Agiplan ProjectManagement, Mülheim and Research Centre Jülich. Karlsruhe: Fraunhofer ISI.

Ullmann (2003). Hydrogen. In *Ullmann's Encyclopedia of Industrial Chemistry.* 6th edn. vol. 17. Weinheim: WILEY-VCH, pp. 85–240.

Vaillant (2007). www.initiative-brennstoffzelle.de.

VDMA (2002). *Markteinführung von Brennstoffzellen-Produkten: Auswirkungen auf den Maschinen- und Anlagenbau.* Frankfurt: VDMA.

Walz, R., Dreher, C., Marscheider-Weidemann, F. *et al.* (2001). *Arbeitswelt in einer nachhaltigen Wirtschaft – Analyse der Wirkungen von Umweltschutzstrategien auf Wirtschaft und Arbeitsstrukturen.* Texte, No. 44/01. Berlin: Umweltbundesamt.

Wengel, J. and Schirrmeister, E. (eds.) (2000). *Innovationsprozess vom Verbrennungsmotor zur Brennstoffzelle – Chancen und Risiken für die Baden-Württembergische Industrie.* Karlsruhe: Fraunhofer ISI.

Winter, C.-J. (2007). Energy efficiency, no: it's exergy efficiency! *International Journal of Hydrogen Energy*, **32** (17), 4109–4111.

Wurster, R. (1999). PEM fuel cells in stationary and mobile applications – Pathways to Commercialization. *Sixth International Technical Congress - BIEL'99.* Biel: Bienal de la Industria Eléctrica y Luminotécnica. CADIEM Cámara Argentina de Industrias Electromecánicas.

Further reading

Hoogers, G. (2003). *Fuel Cell Technology Handbook.* Boca Raton, FA: CRC Press.

Larminie, J. and Dicks, A. (2003). *Fuel Cell Systems Explained.* West Sussex, England: John Wiley & Sons Ltd.

Oertel, D. and Fleischer, T. (2003). *Fuel Cells. Impact and Consequences of Fuel Cells Technology on Sustainable Development.* Technical Report Series EUR 20681

EN. Seville: European Commission, Joint Research Centre (JRC), Institute for Prospective Technological Studies (IPTS).
Pehnt, M. (2001). *Ganzheitliche Bilanzierung von Brennstoffzellen in der Energie- und Verkehrstechnik*. Dissertation. VDI-Verlag, Fortschrittsbericht Reihe 6, Nr 476.
Pehnt, M. (2002). *Energierevolution Brennstoffzelle?* Weinheim: WILEY-VCH.
Sundmacher, K., Kienle, A., Pesch, H. J., Berndt, J. F. and Huppmann, G. (2007). *Molten Carbonate Fuel Cells*. Weinheim: WILEY-VCH.
Winkler, W. (2002). *Brennstoffzellenanlagen*. Berlin: Springer Verlag.

14

Hydrogen-infrastructure build-up in Europe

Michael Ball, Philipp Seydel, Martin Wietschel and Christoph Stiller

If a mass-market roll-out of hydrogen vehicles in the European Union takes place in the next 10 to 15 years, as promoted by the European Hydrogen and Fuel Cell Technology Platform (HFP), then infrastructure strategies will be crucial. At the core of any infrastructure analysis is the question of how the infrastructure should be developed over time and how the needs of both consumers and suppliers can be met. At the same time, such an analysis must also take into account the characteristics of different national energy systems (such as the availability of primary energy sources or competition for end uses), as well as national energy policies. What this infrastructure build-up could look like and what it might cost is shown in a case study for Germany as well as at the European level. On this basis, more general infrastructure strategies are derived with respect to roll-out strategies, production mix and distribution options, and their impacts on supply costs and CO_2 emissions. The chapter finishes with an outlook on global hydrogen scenarios.

14.1 The need for a hydrogen-infrastructure analysis

The potential benefits of a hydrogen economy are recognised to differing degrees by national governments and supranational institutions, although the pathways and timeframes to achieve such a transition remain highly contended. The development of hydrogen-powered fuel-cell vehicles that are economically and technologically competitive with conventional vehicles is a crucial prerequisite for the successful introduction of hydrogen as an automotive fuel. Besides this, there are various other factors that are vital for a successful transition to a hydrogen economy, in particular the build-up of an infrastructure for supplying hydrogen. Developing a hydrogen infrastructure involves selecting user centres, deciding on a mix of production technologies, siting and sizing production plants, selecting transport options and locating and sizing refuelling stations. Integrating all this

The Hydrogen Economy: Opportunities and Challenges, ed. Michael Ball and Martin Wietschel. Published by Cambridge University Press.

into an existing energy system constitutes a challenging task for the introduction of hydrogen as an energy carrier.

The implementation of an operational infrastructure will require considerable investments over several decades, involving a high investment risk regarding the future of hydrogen demand. In addition, the supply of hydrogen needs to be integrated into the energy system as a whole, as its production will affect the conventional energy system – especially the electricity sector – in various ways: some examples are the ensuing competition for renewable energies, the dispatch of electrolysers or the possible co-production of electricity and hydrogen in IGCC plants. Hydrogen can also be used as a storage medium for electricity from intermittent renewable energies, e.g., wind energy, thus facilitating load levelling. Moreover, favourable energy-policy framework conditions are important premises for the successful introduction of hydrogen.

Whether hydrogen can solve most of the energy issues in the long term needs to be evaluated through well defined deployment scenarios, which can provide quantitative information on the opportunities and risks related to large market introduction. In particular, the large investments required for hydrogen take-off must be known and accepted as affordable by all the stakeholders involved in such a critical transition. These issues, plus the fact that there is no clearly outstanding hydrogen pathway in terms of economics, primary energy use and CO_2 emissions, show that it is vital for all the stakeholders involved to start defining a strategic orientation as soon as possible.

Besides modelling energy chains and the energy system, hydrogen infrastructure analysis is a crucial task. The essence of this task is to create regional hydrogen demand and construct development scenarios over time by considering the available resources as well as national policies. The purpose is to evaluate different infrastructure options in economic terms and to derive recommendations for introducing hydrogen as a transport fuel in the next decades. An infrastructure analysis must be able to answer the following questions:

- How do different geographical demand scenarios and hydrogen introduction strategies affect the choice of production technology, production structure (centralised vs. decentralised production) and means of transportation (trailer or pipeline)?
- What could an optimal (geographical) infrastructure development look like and what are the system's expenses for supplying hydrogen?
- What impacts do price changes of primary energy sources have on the hydrogen production mix and related supply costs?
- What is the ratio of CO_2 emissions caused by hydrogen production based on fossil energy sources to emissions savings in the transport sector?
- What interactions between hydrogen production and electricity generation result from integrating a hydrogen energy economy into the energy supply system and what role can renewable energy sources play in supplying hydrogen?
- To what extent will energy supply security in the transport sector be improved as a result of introducing hydrogen as an alternative fuel?

14.2 Tools for the assessment of hydrogen-introduction strategies

The vision of a hydrogen energy economy and thus the use of hydrogen as an energy carrier is nothing new. However, significant advances in fuel-cell technology and increasing concern about future energy supplies have recently made hydrogen a serious alternative, especially with regard to meeting future fuel demand in the transport sector. Correspondingly, instruments have begun to be developed in recent years to support planning and decision-making in setting up a hydrogen infrastructure, its integration into the existing energy system and an estimation of the energy-economic consequences of a hydrogen economy.

(Model-based) instruments to support decision-making in the energy sector should contribute to a better understanding of complex energy systems and at the same time make it possible to design energy systems in such a way as to achieve set targets. It should also be possible to determine how the available resources can be used optimally within certain system boundaries, in order to be able to meet these targets. Hence, the development of analysis tools for planning and optimising the geographical and chronological development of a hydrogen infrastructure is vital.

How hydrogen is produced is influenced by various technical and energy policy developments and frame conditions, such as the further expansion of renewable energies, the development of clean coal technologies or the required reduction of CO_2 emissions, so that hydrogen production is simultaneously closely linked to the conventional energy supply system. To be an instrument capable of adequately supporting decision-making, the developed model has to be able to reflect these frame conditions and to handle the planning tasks of possible market actors, such as energy-supply companies (utilities, oil companies), plant manufacturers or gas producers, but also the demands of sector-specific policy consultation.

To be able to derive from the model results a sound basis for planning and reasonable recommendations for developing a hydrogen infrastructure and its integration in an existing energy system, the special characteristics of hydrogen supply and the energy supply system must be considered in a suitable degree of detail. The demands placed on such an instrument of analysis can be summarised based on the following points:

- The model should record the relevant techno-economic characteristics of the concrete system in sufficient detail so as to be able to understand the processes of change contained in the model results (e.g., fuel substitution, changes in the production programme, construction of hydrogen infrastructure).
- Country-specific features, such as different fuel supply options, expansion potentials for regenerative energy sources or political frame conditions have to be considered to cater sufficiently for their influence on the real decision options.
- Different strategies for introducing hydrogen have to be evaluated with regard to the timing and location of infrastructure development.
- System interactions between the introduction of a hydrogen energy economy and the electricity sector have to be covered.

The development of quantitative tools and methodologies for a hydrogen-infrastructure analysis and the derivation of introduction strategies has been increasing in recent years. Today, there are several models available to analyse the introduction of hydrogen that are supposed to display the transition processes. They use different modelling approaches, specialise in different aspects of the hydrogen transition, have different geographical scopes and significantly varying degrees of detail in modelling the introduction of hydrogen into the energy system in general and of hydrogen technologies in particular. The most simple model types are well-to-wheel (WTW) models (see also Chapter 7). Other model approaches are linear programming (LP), agent-based models (ABM), dynamic programming (DP) and system dynamics (SD).

Table 14.1 provides a non-exhaustive overview of existing models with respect to their methodological approach and respective fields of application.[1] Typical model input parameters include hydrogen demand, energy prices, technical and economic parameters of hydrogen technologies and geographical data (GIS data). Model outputs include hydrogen production mix, hydrogen supply costs, cumulative investments, CO_2 emissions and location of plants and fuelling stations.

Since the two oil crises in the 1970s, a growing number of models of energy systems have been developed and used in the energy sector. The conceptual structure of energy models can be found in Hafkamp (1984); however, in Lev (1983) and FORUM (1999) different energy models are presented. The investment planning of electric utilities in particular has received considerable attention and many formulations of the problem have been proposed over the last decades (see, for example, Anderson (1972); Caramanis (1983) and the IAEA (1995)). Therefore, the use of mathematical programming in the context of energy planning is a well known field of research (Gately, 1970), but is still of great interest, especially as a result of the new situation in liberalised energy markets (see, for example, Gonzalez-Monroy and Cordoba (2002); Song (1999) and Wietschel (2000)).

Optimising energy-system models are particularly suited to analysing the introduction of a hydrogen energy economy, since they permit optimal supply structures to be identified for alternative policy scenarios, which can be used later for aligning long-term investment strategies. Optimising energy-system models, which belong to the category of bottom-up models, have long been used in the corporate planning of energy-supply companies but also in policy consultation. They are primarily characterised by their strongly technology-based mapping. Usually these are energy and material flow models, which simulate real energy-supply systems in the form of directed graphs. The installation-oriented mapping that records the technoeconomic characteristics of the energy-supply processes in detail allows an in-depth analysis of

[1] Other studies (some model-based) that address particularly the build-up of a hydrogen infrastructure include Hart (2005); Karlsson and Meibom (2008); Ogden (1999); Ogden *et al.* (2005); Smit *et al.* (2007); Tseng *et al.* (2003) and Tzimas *et al.* (2007). For energy system models with a more aggregated representation of hydrogen technologies see WETO (2006); Uyterlinde *et al.* (2004); CASCADE MINTS (2005a; b) or the IEA (2006b).

Table 14.1. *Overview of existing hydrogen models and calculation tools for life-cycle analysis*

Model name	Focus	Type
E3 database (Schindler, 2008)	Life-cycle analysis (LCA) of conventional and alternative fuels and vehicle drives, LCA of the supply of stationary electricity and heat supply, LCA for the supply of other products and services	WTW
GREET (Wang, 2008)	Supply chain and life-cycle analysis of conventional and alternative fuels and vehicle drives	WTW
H2A model family: H2A Delivery (HDSAM), H2A Production (H2A, 2008; Mintz *et al.*, 2006)	Distribution infrastructure, modelling of the whole life cycle; production infrastructure	WTW
MSM (Ruth, 2007)	Integration of existing models of life cycles	WTW
Idealised City Model (ICM) (Yang *et al.*, 2006; Yang and Ogden, 2007)	GIS-based hydrogen infrastructure model for cities	Excel®-based
H2INVEST (Stiller, 2008)	Investment decision support for integrated regional infrastructure build-up	Heuristic
MARKAL (Krzyzanowski *et al.*, 2008; Joffe *et al.*, 2007)	Model for energy systems; contains all components of the energy sector;	LP
Almansoori (2006)	Production, transport and storage	LP, mixed integer
Forsberg and Karlström (2006)	Strategies for implementation of refuelling stations	LP
Hugo *et al.* (2005)	Strategic long-range investment planning and design of supply chains	LP, mixed integer
ETP (Gielen and Taylor, 2007)	Global model for energy systems; contains all components of the energy sector	LP
MOREHyS (Ball, 2006)	Spatial and temporal infrastructure construction	LP, mixed integer
HIT (Hydrogen Infrastructure Transition) (Lin *et al.*, 2006)	Dynamic programming optimisation of infrastructure development	LP, DP
HyDS (Hydrogen Deployment System) (Parks *et al.*, 2006)	Spatial and temporal construction planning for production infrastructure; focus on wind energy	LP, DP
HyTrans (Leiby *et al.*, 2006; Greene, 2005)	Transition to hydrogen in the transport sector; complete market for hydrogen and hydrogen vehicles	DP

Table 14.1 (cont.)

Model name	Focus	Type
HyDIVE (Welch, 2007)	Development of distribution infrastructure and demand	SD
Schwoon (2006)	Development of distribution infrastructure and demand	ABM
Stephan and Sullivan (2004)	Development of distribution infrastructure and demand	ABM
H2CAS (Tolley, 2005)	Demand development	ABM

technical modification processes in the system in the course of exogenous influences, such as policy measures or varying fuel prices.

The application of model-based system analyses to compare and evaluate different hydrogen pathways and their integration into national energy systems is only a recent development; this also holds true for the optimisation of hydrogen infrastructure construction, taking into account different energy-policy and country-specific framework conditions. Moreover, possible synergies with the electricity sector are often not explicitly modelled. In addition, modelling approaches generally lack an appropriate geographical representation of crucial infrastructure aspects, such as the location and distribution of hydrogen demand centres or production sites and related transport distances, modes and costs, or the regional distribution of renewable energies.

To overcome those deficits of existing instruments, the MOREHyS (Model for Optimisation of Regional Hydrogen Supply) model was developed as a novel tool to assess the introduction of hydrogen as a vehicle fuel by means of an energy-system analysis.[2] In the next section, the main features of the MOREHyS model are described.

14.3 A model-based approach for hydrogen-infrastructure analysis – the MOREHyS model

14.3.1 Scope of MOREHyS

MOREHyS was developed by the *German–French Institute for Environmental Research (DFIU/IFARE)*, in Karlsruhe (Germany), in co-operation with the *Fraunhofer Institute for System and Innovation Research (ISI)* (Karlsruhe) (Ball, 2006; Ball *et al.*, 2006; Kienzle, 2005). MOREHyS is based on the open-source BALMOREL model (Baltic model of regional energy market liberalisation), which was initially developed

[2] The MOREHyS model has been applied as a supporting tool for the hydrogen infrastructure analysis within the integrated EU project *Hyways* to develop the European Hydrogen Energy Roadmap (see www.hyways.de).

to support analyses of the energy sector in the Baltic Sea region, with emphasis on the electricity and combined heat and power sectors (see www.balmorel.com). The model is implemented in the algebraic modelling language GAMS.

MOREHyS, which allows for a high degree of regionalisation, identifies the cost-optimal way for constructing and implementing an (initial) hydrogen supply infrastructure as well as possible trade-offs and interactions between hydrogen production and electricity generation within a country-specific context. The model is generic – and for a given geographical distribution of hydrogen demand and other location-specific parameters – can be adopted to any geographical entity. The objective of the modelling approach is to optimise – for an exogenously given, regionally distributed hydrogen demand – the build-up of a hydrogen infrastructure over space and time, and to assess the corresponding economic and environmental effects as well as its integration into the existing energy system and the resulting implications. Applying an integrated system view is crucial for deriving realistic introduction scenarios for hydrogen (or any other alternative energy carrier), as otherwise only partial solutions are identified, which lack interdependencies with the energy infrastructures already in place (as, e.g., the case for WTW analyses).

Major outcomes of the model are the development of the hydrogen production mix, the shares of central and onsite production, the means of transportation (pipeline or trailer) and related supply costs and CO_2 emissions. By means of scenario analyses, the effects of varying country-specific framework conditions, such as geographical distribution of hydrogen demand, bounds on the use of renewable energies for hydrogen production, limitations of CO_2 emissions or price variations of primary energy carriers on the development of the infrastructure are being analysed. By simultaneously taking into account trade-offs with the electricity sector, the model approach guarantees that the introduction of hydrogen is assessed and optimised in the context of the energy system as a whole, as possible synergies between hydrogen and electricity generation, limited primary energy potentials and competitions of end uses (e.g., biomass for electricity and heat vs. hydrogen or other biofuels) are integrated. The hydrogen infrastructure analysis in this chapter focuses on the time horizon from 2010 to 2030, but also an outlook until 2050 is provided.

14.3.2 General modelling approach

Within MOREHyS, the complete electricity and hydrogen sector is modelled in a consistent approach, starting from the resources and progressing via several energy-conversion steps to the supply of final energy. This takes into consideration technical (conversion efficiency, installed capacity, etc.), economic (investments, fixed and variable costs) and ecological (emission factors, etc.) aspects (see Fig. 14.1). Each technology (technology class) is characterised by technoeconomic parameters and the technologies are linked via energy flows. Therefore, MOREHyS can be classified as a so-called technology-based (bottom-up) model of the energy system.

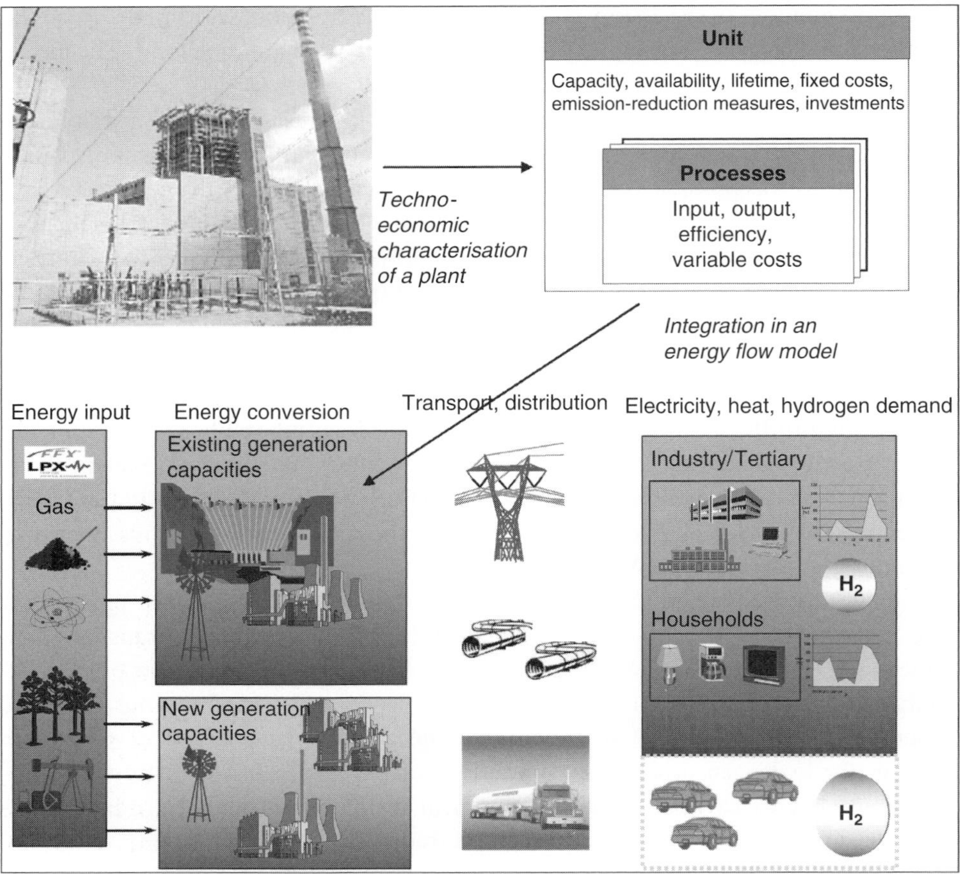

Figure 14.1. Steps of the MOREHyS modelling approach.

For an exogenously given demand for electricity and hydrogen, the model determines the minimal cost of production (as well as transport and transmission) and new investments given some constraints for each year, such as maximum emission levels or regional limitations for generation capacities and fuel availability, which might affect the current energy mix. Possible interconnections between hydrogen supply and the electricity sector concern CO_2 emissions, the competition for renewable energies regarding hydrogen or electricity or heat generation, the dispatch of electrolysers with regard to optimal load levelling and the co-production of hydrogen and electricity in IGCC plants (see Fig. 14.2).

The objective function used for the optimisation, which is sequentially carried out on a year-by-year basis, is cost minimisation for the whole region each year. In the optimisation procedure, existing technologies compete with their variable and fixed costs against new technologies with their additional annualised investments. Dynamics between the years are introduced by transferring the results of the optimisation

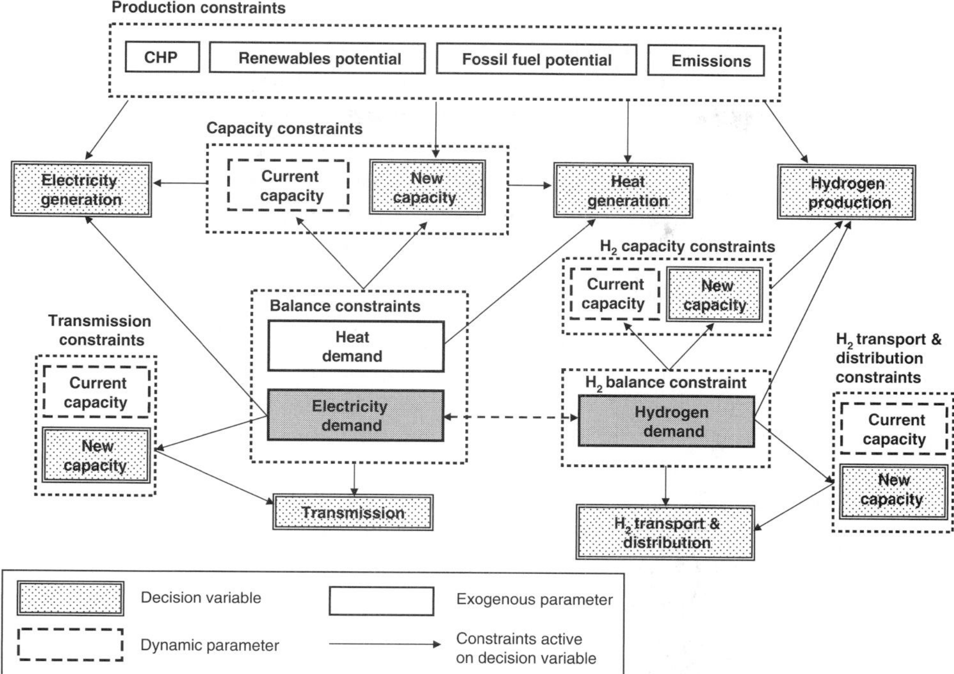

Figure 14.2. Simplified structure of MOREHyS.

(i.e., endogenously found capacities of production and transport or transmission) from the previous year to the beginning of the subsequent year. Thus, the single optimisation periods are interlinked through capacity accumulations and the yearly decommissioning of old capacities. Since the model has perfect foresight only within the year, but not beyond, decision-making can be described as myopic.[3]

The optimisation variables of the model include the level of electricity, heat and hydrogen production per time period, technology type and area; investment in new generation capacity per technology type and area; electricity transmission and new investments in transmission capacities; and transportation mode (pipeline, trailer) and amount of hydrogen transported between and within all hydrogen areas. At country level, the model further derives resulting CO_2 emissions, total primary energy use and total capital costs.

Integer variables are included in the model to ensure that new capacities can only be installed as integer multiples of a given plant capacity (if, e.g., hydrogen liquefaction plants in the model are only defined for a capacity of 100 MW_{H_2}, only liquefaction capacities of 100 MW_{H_2}, 200 MW_{H_2}, etc., can be installed). This approach allows the definition of different capacity classes of, e.g., production plants or transport

[3] Unlike the 'perfect foresight' approach used in most energy system models. For a discussion of the different methodological approaches of energy system models see Enzensberger (2003).

pipelines with different technoeconomic characteristics (such as steam methane reformers (SMR) for production onsite or in central plants with 50 MW_{H_2} or 300 MW_{H_2}) and thus to take economies of scale or learning effects of hydrogen technologies explicitly into account. This is necessary to create a realistic build-up of the hydrogen infrastructure, especially in the start-up phase. As the model also includes real variables, and as both the objective function and the constraints are mathematically implemented as linear equations only, MOREHyS can be described as a mixed-integer, linear optimisation model.

The principal novelties of the MOREHyS modelling approach can be summarised as follows:

- The focus on the start-up phase of a hydrogen infrastructure, by taking into account uncertainties of future market development over the step-by-step optimisation (myopic approach) and the influence of single production plants (using integer variables), thus allowing economies of scale to play out.
- The integration of geographical aspects in the energy system analysis (such as relations between geographical distribution of hydrogen demand, location of hydrogen production sites, location of refuelling stations, local availability of primary energy resources or potential sites for CO_2 storage), by linking a conventional energy system model with a geographical information system (GIS).
- Coupling the hydrogen-infrastructure model – in fact hydrogen production – with the power-sector model.

14.3.3 Modelling of a hydrogen infrastructure

MOREHyS permits the specification of geographically distinct entities: countries, regions and areas, where each country constitutes one or more regions, and each region has zero or more areas. The areas represent the building blocks with respect to the geographical dimension: all generation technologies and capacities, as well as hydrogen demand, are described at the level of areas. In this way, the model can be flexibly adjusted to any desired geographical partition within a given country-specific context. Since the modelling approach focuses on the introduction of hydrogen to the transport sector, only areas with an assumed potential for automotive hydrogen demand have been considered and areas of industrial hydrogen demand have been ignored; nevertheless, the geographical distribution and availability of hydrogen as an industrial by-product have been taken into account. A detailed mathematical description of the modelling approach is provided in Ball (2006).

Fig. 14.3 shows schematically how the hydrogen infrastructure options – comprising the whole supply chain of hydrogen from production (central or onsite), via transport and distribution to the (implementation of) refuelling stations – are modelled in MOREHyS. It has to be noted, that from the point of view of model implementation, transport refers to the transportation of hydrogen between different areas, while distribution is defined as the transportation of hydrogen within the

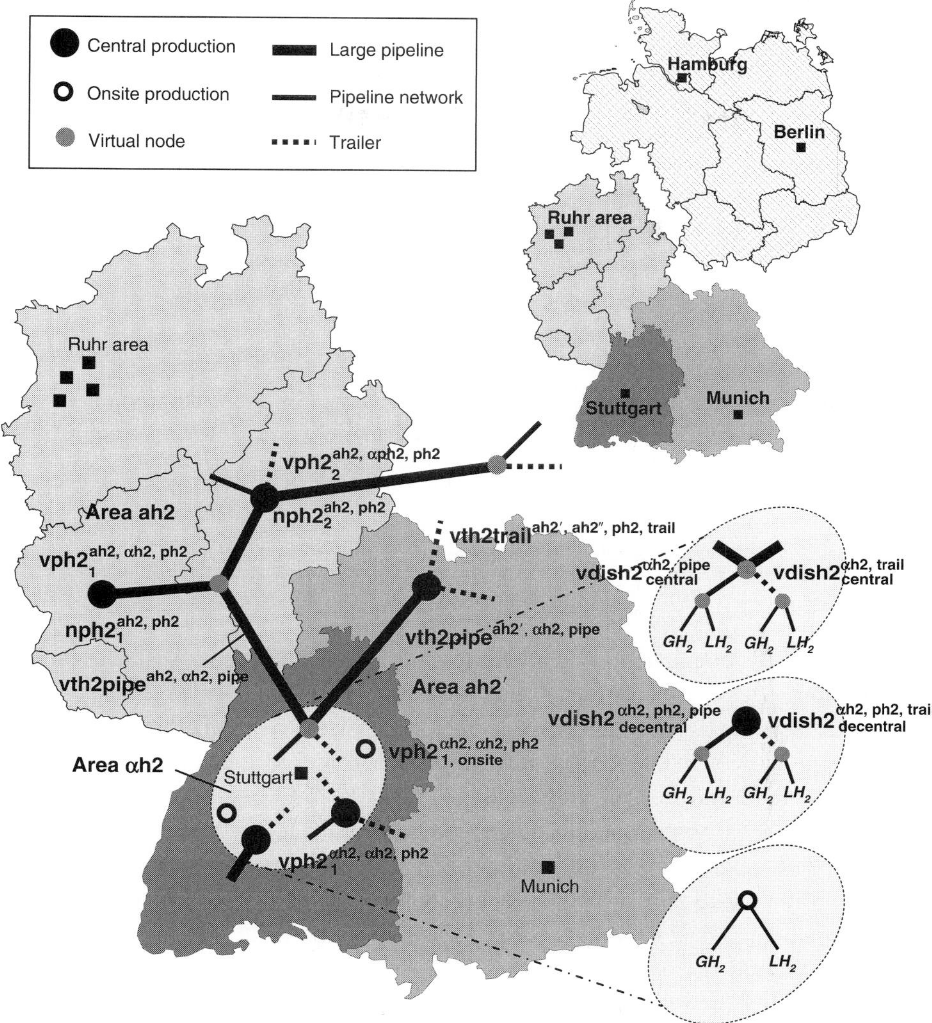

Figure 14.3. Modelling the hydrogen infrastructure in MOREHyS.

boundaries of an area. From a central production plant, the hydrogen can either be transported to another area or distributed within the same area to the fuelling stations (the latter also referred to as decentral production). For both transport and distribution pipelines, LH_2 as well as CGH_2 trailers can be used. Centrally produced hydrogen that is transported to the target area via pipeline can be distributed from the pipeline terminal via distribution pipelines or trailers.

Infrastructure costs for the transport and distribution of hydrogen by pipelines are largely determined by regional characteristics, such as the topology of the terrain. To take these 'real world' conditions into account for determining pipeline lengths and calculating related capital costs, a GIS (geographical information system) is applied,

Table 14.2. *Modelling the hydrogen infrastructure in MOREHyS*

$vph2^{ah2,\alpha h2,ph2}$	Variable H_2 production on technology ph2 in production area ah2 for demand area αh2
$nph2^{ah2,ph2}$	Integer number of production plants ph2 in production area ah2
$vth2pipe^{ah2,\alpha h2,pipe}$	Variable H_2 pipeline transport *from* production area ah2 *to* demand area αh2
$vth2trail^{ah2,\alpha h2,ph2,trail}$	Variable H_2 trailer transport *from* production plant ph2 in area ah2 *to* demand area αh2
$vdish2^{\alpha h2,pipe}_{central}$	Variable H_2 pipeline distribution from central transport node *within* demand area αh2
$vdish2^{\alpha h2,trail}_{central}$	Variable H_2 trailer distribution from central transport node *within* demand area αh2
$vdish2^{\alpha h2,ph2,pipe}_{decentral}$	Variable H_2 pipeline distribution from production ph2 *within* demand area αh2
$vdish2^{\alpha h2,ph2,trail}_{decentral}$	Variable H_2 trailer distribution from production ph2 *within* demand area αh2
GH_2, LH_2	Filling stations for gaseous and liquid H_2 demand

as outlined next. In particular, realistic estimations of transport distances are of major importance as they influence not only the optimal choice of transportation means, but also the associated transport costs. Besides pipeline routing (and derivation of pipeline costs) the GIS is also applied for the geographical allocation of hydrogen demand and for locating fuelling stations.

To integrate geographical details into the linear optimisation model MOREHyS, a GIS is applied, referred to in the following as a Hydrogen Infrastructure Model – H2GIS. In the H2GIS model, hydrogen trailer distances as well as hydrogen pipeline routes are calculated. The GIS is used to calculate pipeline lengths *between* and within areas based on real-world data. This means that the model considers land use, land costs, nature reserves and slope of the area to calculate the cheapest pipeline routes (Cremer, 2005). This procedure is followed for every single optimisation period, successively. In this way, decisions made in the previous optimisation period in MOREHyS influence the subsequent hydrogen pipeline routes in the H2GIS model.

The H2GIS model is implemented in a geographical information system to prepare geographical and topological input information for the MOREHyS model. The H2GIS model comprises different modules. One of the core modules is the *regional-demand module*, which estimates the hydrogen demand for each MOREHyS area based on an exogenous hydrogen penetration rate. Another module is the *filling-station-location module*, which approximates distribution distances to the average filling station *within* an area and locates fuelling stations. The *transportation module* calculates transport distances *between* the model areas for trailer and pipeline routes.

The geographical resolution is on the basis of NUTS (Nomenclature of Territorial Units for Statistics) areas, the regional classification of the European Union. For the German case study the NUTS3 classification (439 districts) has been used to analyse the demand allocation and fuelling station distribution as well as the regional infrastructure development.

The dynamic hydrogen pipeline module is an extension of the transportation module and determines the transport distances and investments to construct a pipeline between and within two areas. The pipeline routing is based on the shortest cost path algorithm, in which the least cost path is calculated based on investment assumptions for each 1000×1000 metre cell. The cost for each cell is calculated based on data such as slope, land-use information, streets, settlement areas and nature reserves. As for the calculation of trailer-transport distances, a real-world street network is used to estimate the distances between regions; within each region an amplifying factor of 1.2 is applied to transform straight-line distances to fuelling stations into road kilometres.

In the first calculation period of the H2GIS model, all the possible pipeline routes between the areas (starting from the geometrical centre points) are calculated and transferred to MOREHyS via a database. After receiving the information from H2GIS about pipeline investments, the MOREHyS model then optimises the infrastructure for this period on the basis of cost minimisation for the whole infrastructure system. This optimisation takes into account the different production technologies (central and onsite) as well as all the different transportation and distribution options (including LH_2 and CGH_2 trailer).

At the end of the period, the MOREHyS model transfers information about whether any pipeline should be constructed back to the H2GIS. In the second period, this information is used in the H2GIS model to calculate new pipeline routes. These routes might differ from the routes in the first period if some pipelines have already been built. If a pipeline is already existing, it might be cheaper to extend this instead of constructing a completely new one from scratch to connect another area. Therefore, possible pipeline routes are recalculated on the basis of established connections for the pipeline's further development. This new information is then passed back to the MOREHyS model, which recommences optimising the infrastructure for the second period, and so on.

The linear optimisation model, MOREHyS, and the GIS-based hydrogen infrastructure model, H2GIS, are interconnected in an iterative process. The H2GIS model calculates the best pipeline routes for a period, the MOREHyS model decides which of these pipelines should be built and returns this information to H2GIS, which then determines new pipeline routes for the next calculations in MOREHyS (see Fig. 14.4). This consecutive exchange takes place for all modelling periods. The high integration between MOREHyS and H2GIS allows analysis of the development of a pipeline grid instead of single point-to-point connection, as in an optimisation infrastructure model.

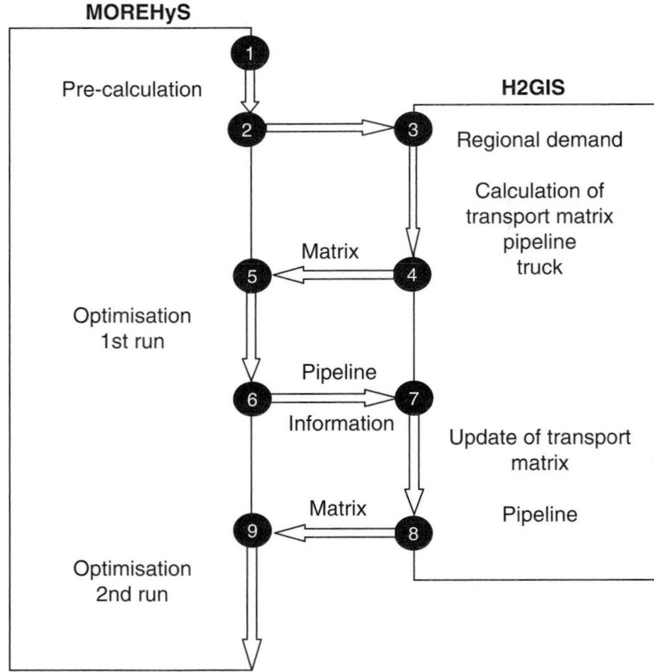

Figure 14.4. MOREHyS–H2GIS interaction.

14.3.4 Modelling the integration of hydrogen and electricity generation

Since the supply of hydrogen in the long term depends on different technical and energy policy developments and frame conditions, such as the expansion of renewable energy sources, the development of clean coal technologies or the required reduction of CO_2 emissions, hydrogen production is simultaneously closely linked to the conventional energy supply system.

The possible impacts of a hydrogen economy (i.e., hydrogen production) on the existing energy system, primarily electricity generation, can be examined by incorporating the hydrogen infrastructure model into an optimising energy-system model. Interfaces concern especially the economically optimally timed dispatch of electrolysers or IGCC power stations to (co-) produce hydrogen or competition for the use of renewable energy sources. The developed modelling approach thus integrates optimisation of hydrogen infrastructure build-up (in terms of time and space) with system planning and power-station management and takes into account interactions between the supply of hydrogen and the existing energy system.[4]

[4] The selected optimisation approach integrates system planning with power station management. System planning covers all the decisions for constructing new plants or for closing down old ones. Power-station management, in contrast, regulates when and to what extent which of the available plants contribute to meeting the total electricity demand. In addition to this, power transmission between the model's regions (e.g., at country level) is also subject to optimisation. The period regarded is flexible, but is usually between 20 and 30 years.

As expressed by the vision of a 'solar' hydrogen energy economy, in the long run, hydrogen must be produced using renewable energy sources to constitute a real alternative to dwindling oil and gas reserves. Here – alongside biomass-based production technologies – electrolysis based on electricity from renewable energy sources will play a leading role. With the progressive expansion of renewable energies, there will be increasing competition for their utilisation between electricity and heat generation on the one hand and hydrogen (or other biofuel) production on the other, particularly with respect to biomass. The selected model structure allows the user to determine an optimal allocation of the renewable energy sources with regard to hydrogen or conventional energy generation when alternative policy scenarios are prescribed such as, for example, reducing CO_2 emissions within the framework of energy-system modelling.

Whereas biomass represents a substitutional use for electricity and hydrogen, intermittent renewables, such as wind or photovoltaic, as well as being able to be used *directly* to produce hydrogen, also offer the additional possibility of storing surplus power in the form of hydrogen, for example, during off-peak periods. A significant increase in the use of wind energy in power generation, in particular, has important technical consequences for grid operation, which stem from the fluctuating and insufficiently predictable energy supply. In principle, the *excess capacities* resulting from time discrepancies between energy supply and demand could be used for the production of hydrogen, which in this case functions as an energy store (see also Chapter 16).

Finally, with the development of clean coal technologies and, not least from the viewpoint of resource economics, interest is increasingly being focused on the co-production of electricity and hydrogen in combined-cycle power stations with integrated coal gasification (IGCC). This technology has an important role to play, given the pending overhaul of the power system in such countries as Germany and Great Britain or a possible phase-out of nuclear power (such as in Germany), as well as an expected increase in natural gas prices, and will be even more important if these power stations manage to generate electricity and hydrogen with almost zero CO_2 emissions using carbon capture and storage. By recording the load characteristics, an optimised (i.e., cost-optimal) dispatch of IGCC power plants with regard to hydrogen and electricity production can be determined with the help of the model, besides the correctly timed dispatch of electrolysers (see Chapter 16).

14.3.5 Limitations of the modelling approach

One weakness of the developed model approach is the central, one-dimensional optimisation that assumes the same target function for all participants. The model identifies possible economic and environmental benefits of a hydrogen infrastructure build-up by determining the global optimum for the whole system instead of the optimum for each company. For this reason, the decisions reached are not necessarily the same as decisions made from the point of view of individual players. Therefore, it is the subject of further research to tackle this problem using game theoretic approaches, which have been successfully used for similar problems in other areas (Nouweland *et al.*, 1996). Besides this add-on solution, another future research topic is the transformation of the optimisation model to a multiagent system as a construct of software agents, representing each partner of the co-operation by a single software agent (Uyterlinde *et al.*, 2004). As a central idea in the paradigm of multiagent systems, the role concept will be used as the enabler for different players to pursue their individual objectives and strategies, in order to avoid an all-encompassing target function negating alternative goals for different participants of the network.

The scenario analysis has shown the influence of exogenous assumptions on the model results. To be able to consider uncertainties in hydrogen infrastructure build-up concepts adequately (e.g., share of gaseous vs. liquefied hydrogen, hydrogen demand figures, new environmental legislations) the chosen deterministic optimisation approach is to be enhanced by multiperiodic stochastic programming methods (Prekopa, 1995). With the help of these methods, it is possible to determine hedging strategies against the occurrence of events changing the assumed energy-economic framework assumptions. By assigning probabilities to possible scenarios, a strategy can be elaborated which is not necessarily 'optimal' for any specific case, but which can be adjusted to different outcomes of the political discussion without sacrificing competitiveness. With the help of this procedure, robust solutions are to be identified, to accommodate future internal and external uncertainties.

14.4 General input data for the MOREHyS model

14.4.1 Development of hydrogen demand

Regarding the use of hydrogen in the transport sector, three types of vehicle are distinguished in the model: passenger cars, light-duty vehicles (LDV, <3.5 t) and buses.[5] The (annual) demand for hydrogen transportation fuel is a function of the hydrogen vehicle stock in a certain year, the specific vehicle consumption and the driving range. Assumptions around these factors are described in the following with respect to the case study for Germany which is analysed in Section 14.5.

14.4.1.1 Transport scene setting and fuel consumption of hydrogen vehicles

In 2005, in total there were 45.4 million passenger cars, 1.8 million LDVs and 87 000 buses on the roads in Germany (Prognos/EWI, 2005). It is assumed that the number

[5] Heavy-goods vehicles are not considered, as it can be assumed that neither fuel cells nor hydrogen-combustion engines will manage a breakthrough in this market segment in the time period analysed up to 2030 because of the dominance and high performance of diesel engines, especially with regard to driving range and lifespan.

Table 14.3. *Consumption of hydrogen vehicles*

Year 2010	(kg/ 100 km)	(kWh/ 100 km)	(l oil equivalent/ 100 km)	kg/(year / vehicle)	Nm^3/(year / vehicle)	MWh/(year / vehicle)	Range (km/year)
Car, fuel cell	0.78	26	2.7	87	821	2.9	11400
Car, ICE	1.4	47	4.9	157	1474	5.2	11400
Light-duty vehicle, fuel cell	2.4	80	8.4	269	2527	8.9	25000
Light-duty vehicle, ICE	4.0	133	14.1	448	4211	14.9	25000
Bus, fuel cell	12.0	400	42.2	1344	12633	44.7	50000
Bus, ICE	16.0	533	56.2	1792	16844	59.6	50000

Sources: (DWV, 2005; JEC, 2007).

of passenger cars in 2030 will increase to 53.9 million and of LDVs to 3.3 million, while the number of buses will remain constant (TREMOVE, 2007).[6]

In line with recent trends, the average annual mileage in passenger cars will decrease from today's 11 400 km to 10 600 km in 2030, which is mainly caused by an above-average increase in older age groups who tend to have a lower driving range (Shell, 2004). Sometimes there are large differences in the available estimates of the mileages for LDV and buses, which also means a correspondingly large range in hydrogen demand. Based on a survey among local traffic companies and operators of vehicle fleets, Höllermann (2004) reported the average driving range of delivery vehicles to be approximately 25 000 km, and of buses approximately 50 000 km.[7] A constant mileage was assumed for the period as a whole since the future development in these sectors is very difficult to predict. It has to be noted that any projection of the future hydrogen demand is very sensitive to assumptions about driving ranges.

The specific and the average annual hydrogen consumption of the vehicles distinguished in MOREHyS is provided in Table 14.3. Consumption is differentiated between passenger cars, light-duty vehicles and buses, as well as between vehicles with fuel cells and internal combustion engines. The figures for both fuel-cell and ICE vehicles are averaged between hybrid and non-hybrid vehicles.

The specific consumptions shown are forecasts for the period 2010 to 2020 based on the prototypes available today. In the long term, an increasing share of hybrid-vehicle

[6] Other sources show lower estimates. For instance, Prognos/EWI (2005) projects the number of passenger vehicles in 2030 in Germany at 48.5 million and of LDV at 2.6 million.

[7] In comparison with other studies, these values are in the intermediate range. According to TREMOVE (2007), the driving ranges of LDV and buses equal 15 000 km and 33 500 km, respectively. The DWV (2005) in contrast, reports the annual mileage of delivery vehicles to be over 120 000 km and that of buses over 90 000 km. Prognos/EWI (2005) estimates the driving range of buses at about 42 000 km.

designs is assumed, which are expected to increase efficiency by between 10% and 20%. However, for want of available data, the simplification was made that the specific consumptions are constant throughout the entire modelling period in question; furthermore, the influence of possible reductions in hydrogen demand due to improvements in vehicle efficiency on the infrastructure build-up examined is negligible. As can be seen, the specific fuel consumption of fuel-cell vehicles is 60% to 80% of the consumption of vehicles with hydrogen-combustion engines. Since vehicles based on combustion motors can be designed in a bi-fuel way, the range of the vehicles is clearly increased, which is advantageous when implementing a hydrogen infrastructure, as long as the filling station network is still being developed.

14.4.1.2 Scenarios for penetration of hydrogen vehicles in Europe

The exogenously given hydrogen demand in the form of different penetration rates, which indicate the percentage of conventional vehicles substituted by hydrogen vehicles, is the driving factor of the model. Hydrogen deployment is developed on the basis of penetration rates of hydrogen vehicles in the European Union, which have been proposed by the HyWays project and which are the same for every country (HyWays, 2007).[8] The infrastructure build-up is divided into three phases:

- *Phase I* Early start-up phase with very low hydrogen penetration (demonstration phase). A few large-scale first-user centres are situated in European capitals (see also Roads2HyCom (2007)). Owing to its case-by-case selection of the technology options, this phase is not considered in the infrastructure analysis.
- *Phase II* Early commercialisation phase, with approximately two to five early-user centres per country.
- *Phase III* Full commercialisation phase, characterised by the extension of existing user centres, the development of new hydrogen regions and the installation of a dense local and corridor network (approximately 0.25%, 2% and 8% of all vehicles, respectively, and 25%, 50% and 85% of the population, respectively).

These phases are defined by the number of hydrogen cars on the roads rather than by calendar years. A connection to calendar years can be established through the hydrogen-vehicle market-penetration curves elaborated by the automotive industry (see Fig. 14.5). The infrastructure analysis focuses on the early phase of hydrogen deployment, with a relatively low penetration of hydrogen vehicles because regional aspects are crucial in this phase.

The hydrogen-demand scenarios differ in the following options:

- *Penetration rate* (very high, high and moderate policy support), coupled with fast or moderate technology learning for vehicles (i.e., cost reduction of hydrogen drive trains).

[8] Stationary hydrogen demand has not been considered. Stationary fuel cells are not necessarily a market for hydrogen, because natural gas from the gas mains can easily be used directly; a conversion of natural gas or biogas to hydrogen would only reduce the overall efficiency (see Chapter 13).

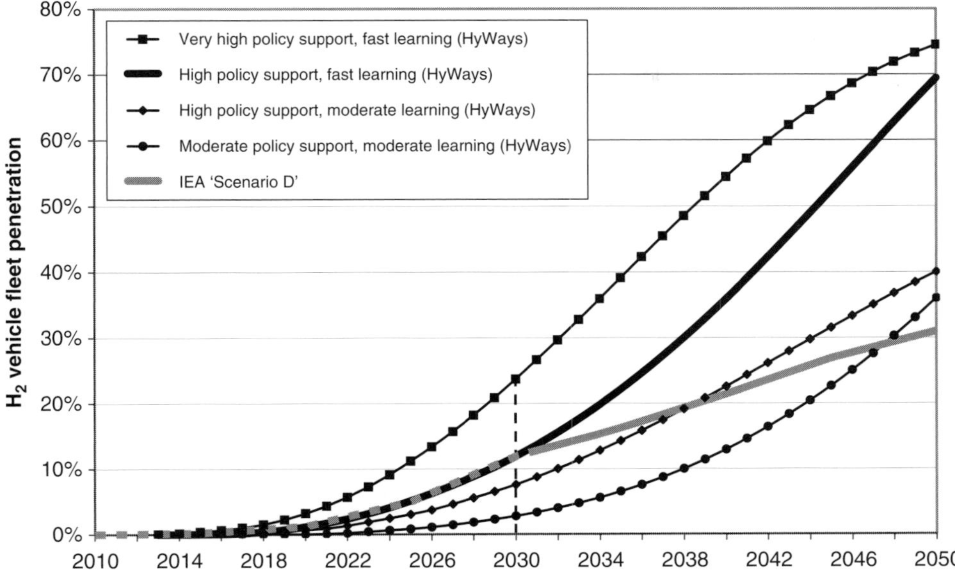

Figure 14.5. Hydrogen vehicle penetration rates (HyWays, 2007; IEA, 2005).

- *User access* (concentrated, distributed). In the 'concentrated users' scenario ('CON'), approximately 25% of the population have access to hydrogen by 2020, 50% by 2025 and 85% by 2030. In contrast, in the 'distributed users' scenario ('DIS'), 50% of the population already have access to hydrogen by 2020, 90% by 2025, and the whole population by 2030. The scenarios do not differ in the period until 2015, when only the early-user centres are covered.
- *Road networks* (early, late). In the 'early network' option, the early corridors are supplied from 2015 on, and the 'full highway network' is supplied from 2020 on. In contrast, in the 'late network' scenario, the early corridors are only supplied from 2020, and the full highway network from 2030.
- *Fuelling-station location strategy*. To supply the hydrogen demand, two scenarios have been developed with respect to the fuelling station distribution: one scenario, where number, capacity and location of fuelling stations are tailored to meeting the demand only ('FS demand') and one scenario where the number, size and location of fuelling stations are determined by customer convenience with respect to fuelling station access ('FS user'). Therefore, all existing stations in 2005 have been analysed with an indicator approach to identify the best-suited fuelling stations for a hydrogen introduction for each simulation year. The most important decision factors for the distribution of this network have been traffic counts on highways as well as the aim of reaching a full area network as soon as possible.

Figure 14.5 shows the different scenarios with respect to the development of the market penetration of hydrogen passenger cars until 2050, as developed by the HyWays project for the EU (HyWays, 2007); in addition, the most optimistic world hydrogen penetration scenario developed by the IEA is displayed (which also includes light/medium trucks) (IEA, 2005). For the development of the hydrogen vehicle stock, a lifetime of 10 years for passenger cars, 11 years for LDVs and 15 years for buses is assumed. Hence, it would take at least 10 to 15 years to switch over the entire fleet.

The beginning of the possible market introduction of hydrogen passenger cars in the most optimistic scenarios is around 2015, which is also in line with the implementation plan of the European Hydrogen and Fuel Cell Technology Platform (HFP, 2007). The *Reference scenario* features high policy support, fast learning, concentrated users, early network and no country-specific bounds.

Table 14.4 displays the initial market penetration of all hydrogen vehicles – passenger cars, light-duty vehicles and buses – for the Reference scenario relative to the estimated development of the total vehicle stock as well as the shares of the hydrogen vehicle drive trains, i.e., fuel cells and internal combustion engines.

A hydrogen refuelling infrastructure will not be fully developed when the market penetration of hydrogen vehicles begins. In fact, it can take several decades for hydrogen vehicles to penetrate a sizable fraction of the fleet.[9] According to the Reference scenario, by the year 2030, 20 years after the first vehicles have come on the market, passenger cars reach a market share of almost 12% of the total fleet, while LDVs and buses will reach around 18%; extending this scenario until 2050, all vehicle types are projected to reach market shares of 70%.

As a refuelling infrastructure will not yet be fully developed at the beginning of the market penetration of hydrogen vehicles, ICE vehicles with bi-fuel conversions can greatly extend driving ranges, so that an area-wide coverage of refuelling stations does not need to be immediately in place. As a consequence, the share of ICEs in passenger cars and LDVs is assumed to be the highest during the early introduction phase, with about 40% each. With time, FC vehicles prevail, as they have competitive advantages over ICE vehicles owing to a much better fuel economy. Because of minor construction-related restrictions in buses, in particular with respect to the storage of hydrogen, the advantages of fuel cells can be more immediately used, so that the share of ICEs in this market segment is rather insignificant.

There is a variety of barriers that hinder the introduction of hydrogen vehicles (or any new vehicle type) to the market, which can be technical, economic or institutional in nature: examples are the limited driving range of fuel-cell cars due to the challenges of hydrogen storage, the much higher costs of fuel cell vehicles or the lack of a refuelling infrastructure. Some of these barriers can be reduced by introducing the vehicles into niche applications for which they are particularly suited (Smith, 2001).

For example, hydrogen- (or other alternative) fuelled vehicles can be introduced into urban fleets where the refuelling infrastructure is located at a central depot.

[9] For example, hybrid vehicles (Ogden, 2006): although fundamental research on hybrid vehicles began in the 1970s, it was not until 1993 that Toyota began the development of the Prius hybrid. Initial sales began in 1997, and in 2005 hybrid models from several manufacturers still accounted for only 1.2% of new vehicle sales in the USA. As for electric vehicles, when introduced by major car manufacturers in California in the late 1990s, despite the enthusiasm of early adopters, the vehicles failed to reach more than a few hundred drivers for each model and in a few years electric-vehicle programmes were dropped (IEA, 2005).

Table 14.4. *Penetration of hydrogen vehicles in the Reference scenario until 2030*

Penetration rates	2010	2011	2012	2013	2014	2015	2020	2025	2030
Passenger cars	**0%**	**0%**	**0%**	**0.001%**	**0.01%**	**0.02%**	**1.2%**	**5.1%**	**11.9%**
Share FC	–	–	–	60.3%	60.4%	60.5%	61.6%	65.4%	73.9%
Share ICE	–	–	–	39.7%	39.6%	39.5%	38.4%	34.6%	26.1%
LDV	**0.001%**	**0.01%**	**0.02%**	**0.05%**	**0.17%**	**0.4%**	**3.2%**	**8.9%**	**17.6%**
Share FC	60.3%	60.3%	60.3%	60.3%	60.4%	60.5%	61.7%	65.4%	74.0%
Share ICE	39.7%	39.7%	39.7%	39.7%	39.6%	39.5%	38.3%	34.6%	26.0%
Buses	**0.001%**	**0.01%**	**0.02%**	**0.05%**	**0.17%**	**0.4%**	**3.2%**	**8.9%**	**17.6%**
Share FC	90.0%	90.1%	90.2%	90.2%	90.3%	90.4%	91.2%	93.4%	96.3%
Share ICE	10.0%	9.9%	9.8%	9.8%	9.7%	9.6%	8.8%	6.6%	3.7%

Sources: (Hart, 2005; HyWays, 2007).

Other applications include car rental, shared-car ownership or public transport – these transfer the risk of ownership of a vehicle with a new propulsion system away from the private person. Once established, these niche applications can help to raise the profile of the new technology, increase public acceptance and provide opportunities for feedback that can lead to technology improvement. Once their commercial viability in the niche market has been proved, the vehicles can expand into wider markets. Particularly suitable for introducing hydrogen (or other alternative fuels) are buses, fleet vehicles and rental vehicles (Smith, 2001).

Buses are suitable because:

- They run on short, regular routes and return to a central depot for refuelling, so the limited range is not usually a problem;
- The size and weight of the fuel tank are of less importance than for smaller cars;
- As the vehicles are expensive, the cost premium is less significant;
- It is relatively easy to set up and support clean bus projects because bus fleets are generally under public control and public support is seen as legitimate;
- Public ownership or funding often facilitates the payment of a cost premium in return for a better environment in the city centre;
- Buses operate in crowded city centres, where the benefits of reduced pollution (and noise) are particularly important;
- City-centre buses are highly visible and thus help to raise public awareness;
- Cleaner buses promote a better image of public transport and so can help to encourage a modal switch.

Fleet vehicles (such as taxis, couriers and delivery vans) are suitable because:

- Like buses, they share the advantage of returning to a central depot, so that the barriers of lack of refuelling and service infrastructure are reduced;

Table 14.5. *Annual hydrogen demand in the Reference scenario*

GWh H_2/a	2010	2015	2020	2025	2030	2040	2050
Germany	**1.1**	**380**	**5 119**	**18 186**	**39 870**	**105 971**	**184 357**
Share passenger car (%)	0	10	42	53	57	58	62
Share LDV (%)	80	72	46	38	35	34	31
Share bus (%)	20	18	12	9	8	8	7
EU27	**6.3**	**2 244**	**30 212**	**107 344**	**235 333**	**625 486**	**1 088 157**

- Fleets are easier for government programmes to target than private individuals, as they are under the control of a small number of people;
- Companies can benefit from the public relations advantage of being seen to use cleaner vehicles;
- Managers of large fleets should be able to negotiate a favourable deal with manufacturers and fuel suppliers for vehicle and fuel purchase.

Finally, *rental vehicles* are also suitable as they are depot-based, which simplifies maintenance, although refuelling will take place outside the depot. By renting a vehicle, the user eliminates the risks of ownership of a novel technology. As rental vehicles have many users, they are ideal for raising awareness of new technologies and for decreasing barriers for adopting hydrogen vehicles for private use.

In all scenarios, the first hydrogen vehicles penetrating the market are fleet vehicles (buses and LDV). From 2013 on, the first FC cars are assumed to appear. For FC passenger cars it is also assumed that the first ones will be commercially operated fleet vehicles.

Table 14.5 displays for the Reference scenario the resulting hydrogen demand in Germany, split into the different vehicle types. The demand has been calculated on the basis of the respective penetration rates and vehicle consumptions. The temporal distribution of hydrogen demand among the different geographical areas is described next. The hydrogen demand for the EU27 has been scaled up in proportion to the population, as the same vehicle penetration rates are assumed across all countries.

It is assumed that hydrogen's demand profile remains constant over the year. Admittedly, transport fuel demand is subject to seasonal fluctuations – for example, a rise in demand can be observed during the summer months – but these fluctuations are relatively marginal over the course of the year so that the above assumption adequately reflects the real situation. Hence, when hydrogen is being used as a transport fuel it is assumed that there is no need for large stationary storage outside the production plants (unlike for natural gas, where storage is necessary to compensate the difference between a steady supply and a more irregular demand) and the storage of hydrogen has not been explicitly modelled. (The storage of hydrogen at the fuelling stations is included in the capital costs of the fuelling stations, see Chapter 12.) By assuming that there are no major time differences between hydrogen supply and demand, the simplification is also made that hydrogen from the production plant can be continuously transported to the fuelling stations so that the production locations can also do without significant interim storage facilities.[10] Time-of-day demand fluctuations at the fuelling stations can be compensated for in principle by sufficiently large hydrogen storage facilities, which are calculated into the investments for the fuelling stations.

14.4.1.3 Geographical allocation of hydrogen demand

As previously mentioned, the regional resolution is based on the NUTS3 classification, and hence hydrogen-demand areas are defined based on NUTS areas. The areas used for modelling are aggregated NUTS3 areas. Owing to computing limitations, areas with similar indicators are combined. In total, 20–26 regions are distinguished per country. A distinction is made between urban and rural areas. Both types play an important role in the build-up of a hydrogen infrastructure.

On the basis of the hydrogen-demand scenarios, a *regional demand model* has been applied to analyse different scenarios of regional hydrogen-demand development. The H2Demand model is implemented in a geographical information system. The system includes detailed regional data for population density, vehicle distribution, traffic flows between regions, purchasing power and other important drivers for hydrogen-demand development. As input for the H2Demand model, the hydrogen penetration rate for a country as well as first hydrogen user centres are set upfront.

As a first step in the model sequence, it was determined in which order the different NUTS3 regions in a country will gain access to hydrogen. The 'first user centres', which are determined in each country by qualitative evaluation of a list of regional indicators, including local pollution, cars per household, size of cars, existing demonstration projects, favourable hydrogen production portfolio (renewable energy, by-product), customer base, political commitment and stakeholder consensus, are the starting point of the hydrogen-demand development. With an indicator approach, the next regions are selected that could be the most likely hydrogen areas because of the ranking of the indicator values for population density, purchasing power, cars per person and traffic connections to existing hydrogen areas, etc. New regions are connected for one year until the total hydrogen penetration rate for that year in the country is fulfilled (see also Section 14.5.1).

The long-distance road network is supplied by fuelling stations in two steps: the 'early corridors', which mainly serve to connect the early-user centres and allow for daily commuting in their vicinity in an early phase, and the 'full highway network', which includes all long-distance roads in later phases. For the allocation of long-distance traffic hydrogen demand, the simplifying assumption was made that the same amount of hydrogen is used on each kilometre of road segment.

[10] An exception to this is a hydrogen production plant using intermittent renewable energies such as, e.g., wind power, which requires storage facilities owing to the fluctuating hydrogen production. Since the investments for storage are not explicitly modelled, they are accounted for as part of the production plant investments – in this case, electrolysers.

14.4.1.4 Fuelling station network

Three different fuelling station capacities are considered, with 80, 320 and 800 refuellings per day, respectively (i.e., one, four and ten dispensers), assuming 4 kg hydrogen per refuelling. It has to be noted that these sizes are already dedicated to the early market phase; i.e., the 'small' station is already significantly larger than today's hydrogen fuelling stations. The number and size share of fuelling stations required in each region is not only a matter of the local traffic hydrogen demand. Accessibility must be guaranteed by having a certain minimum number of fuelling stations in a region with a certain overcapacity to compensate for fluctuations in fuelling station usage. A common assumption is that 10%–30% of all conventional fuelling stations must dispense an alternative fuel to achieve broad user acceptance (Stiller *et al.*, 2007). Based on assumptions about the ratio of hydrogen to conventional fuelling stations for user convenience, a minimum number of fuelling stations is calculated for an area. Together with the assumed refuelling overcapacity (decreasing over time) and the local traffic demand calculated for an area, the specific split of small, medium and large fuelling stations is determined for each area.

The hydrogen energy demand is split into local traffic and long-distance traffic and thus a distinction is made between fuelling stations for local traffic and for long-distance traffic. This is mainly because the latter implies a continuous fuelling-station network along main roads (also in areas where there is no local use), while the former will only be situated in the areas where hydrogen users reside. The regional allocation of both types of fuelling station is treated separately.

At the beginning of the infrastructure build-up, fuelling stations for local traffic hydrogen are only located in 'early-user centres'; for long-distance traffic, a few 'early corridors' are defined, which mainly serve to connect the early-user centres and to permit daily commuting in their vicinity.

The further regional roll-out of hydrogen fuelling stations for local traffic in later times is determined by a ranking of the areas, based on weighted socioeconomic indicators (catchment area population, purchasing power, cars per person). For the supply of long-distance traffic, all long-distance roads (motorways or E-roads, depending on the country) are equipped with hydrogen-fuelling stations. The required number of long-distance fuelling stations (e.g., on highways) is calculated, to reach a country-wide fuelling station network as soon as possible. To ensure convenient travelling, a distance between two highway fuelling stations of 80 km at the beginning is assumed, which is later reduced to 60 km (multilane roads have one fuelling station on each side). The most important decision factors for fuelling stations on an area level have been area coverage and accessibility for commuter and private households.

14.4.2 Energy-price scenarios

The approach described here combines infrastructure build-up and scenario analysis. However, its validity depends on the validity of the assumptions made on the energy

Table 14.6. *Projections of energy and CO_2 price development until 2050*

		2010	2015	2020	2025	2030	2040	2050
High-price scenario								
Oil price	2005€/GJ	6.5	7.4	8.1	9.1	10.0	13.0	16.7
Hard coal price	2005€/GJ	1.8	2.0	2.2	2.4	2.5	3.2	4.9
Natural gas price	2005€/GJ	3.9	5.1	6.8	7.6	8.9	11.0	15.3
Biomass price	2005€/GJ	0.9	5.0	5.4	5.5	8.1	8.43	10.2
Electricity price	2005€/GJ	14.0	16.7	22.1	24.8	28.2	29.4	30.6
Low-price scenario								
Oil price	2005€/GJ	6.3	7.1	7.9	8.6	9.4	11.0	14.3
Hard coal price	2005€/GJ	1.4	1.4	1.5	1.5	1.5	1.7	2.0
Natural gas price	2005€/GJ	3.3	3.6	3.8	4.0	4.4	4.9	5.6
Biomass price	2005€/GJ	0.9	3.8	4.0	4.2	4.4	4.7	5.1
Electricity price	2005€/GJ	12.8	15.0	19.7	21.7	24.4	26.2	28.4
Lignite price	2005€/GJ	1.14	1.18	1.21	1.21	1.21	1.21	1.21
CO_2 price	€/t CO_2	10	15	20	25	30	40	50

Source: (WETO, 2006; own calculations).

scene, in particular the development of energy prices. The authors acknowledge that there is an inherent unpredictability in energy prices (as for instance exemplified by the unexpected oil price peak of more than \$140/bbl during 2008 and subsequent price drop); hence, any projection of future energy prices bears a high uncertainty and has to be critically reflected; hydrogen supply costs, therefore, must always be seen in the context of the assumed feedstock prices. In this context, it also has to be noted that the oil, gas and coal prices seen during 2008 already exceeded the projections of the high-price scenario for 2030 (see Table 14.6).[11] Nevertheless, while the absolute energy prices are subject to a high uncertainty, the relative ratios of energy prices are more constant (e.g., gas-to-coal price), and eventually decisive for the choice of production technology. Table 14.6 shows two underlying energy-price scenarios until 2050; a high- and a low-price scenario.

The energy-price scenarios for fossil fuels used in MOREHyS also include CO_2 prices according to Table 14.6, assuming that market-based mechanisms for CO_2 reduction will continue to persist in the European Union. Electricity prices also include CO_2 prices; possible green certificates for renewable energies were not considered, to avoid a double penalty for fossil fuels. As prices for natural gas and electricity are subject to the effective quantities specified in the supply contracts and decrease with increasing quantities, it is assumed that natural gas and electricity prices for onsite technologies (onsite steam reformers and electrolysers) are 30% higher than the respective prices for large, central production plants.

[11] For instance, in March 2008, oil prices were above €16/GJ, natural gas prices at around €9/GJ and some hard coal import prices exceeded €4/GJ (e.g., FOB ARA coal prices from South Africa were close to €100/t).

14.4.3 Technology data

For the limitations of this publication, it is not possible to present a comprehensive set of the data used as input to the model. In principle, the model is based on the technoeconomic characteristics of hydrogen production and distribution technologies, as presented in Chapters 10 and 12, respectively, such as specific investments for certain plant sizes, full load hours, process efficiencies, maintenance and labour costs, etc.

The production technologies distinguished in the model are: steam reforming of natural gas, conventional coal gasification as well as integrated coal gasification (IGCC) with the possibility of co-producing hydrogen and electricity, biomass gasification and electrolysis (on the basis of different electricity sources, such as grid mix, wind power or nuclear power). The application of CCS (carbon capture and storage) to fossil-based production technologies is assumed to be commercially available from 2025 onwards; for the transport and storage of the CO_2, generic costs of 0.14–0.22 ct/kWh_{H_2} for hydrogen from natural gas and 0.28–0.46 ct/kWh_{H_2} for hydrogen from coal are assumed (see also Chapter 6). (The total CO_2 storage capacity in Germany is estimated at between 19 and 48 Gt (Fischedick *et al.*, 2007).) In addition, the supply of industrial by-product hydrogen (mainly from chlorine–alkaline electrolysis) is taken into account; for Germany, this amounts to around 500 million Nm^3/year (see Section 10.10). For hydrogen transport and distribution, liquid (LH_2) and gaseous (CGH_2) hydrogen trailers and hydrogen pipelines with different capacities are considered (see also Chapter 12).

To assess possible trade-offs with the conventional energy system, the model also contains a comprehensive data set of the German power sector, which may be extended to any other geographical scope (see also Chapter 16).

14.5 Case study: Germany

14.5.1 Geographical allocation of hydrogen demand

The regional demand distribution has been simulated with the H2GIS model (H_2 demand model) on a year-by-year basis from 2010 to 2030 for the Reference penetration scenario (see Fig. 14.5). Areas for the demand of automotive hydrogen have been defined on the basis of the distribution of population densities (being also closely correlated with the distribution of car densities), number of cars, local air pollution, purchasing power, etc. to take different demand densities into account, as they affect the infrastructure being built up. Car registration numbers for each area have been used to calculate the possible share of hydrogen vehicles per area. As starting points for the infrastructure development for Germany, Hamburg and Berlin have been chosen, because of the current activities in both cities (for instance the 'HafenCity Hamburg' project and 'Clean Energy Partnership Berlin'). For comparison reasons, two scenarios have been developed with the H2Demand model (as described in Section 14.4.1.2), one with a very concentrated demand pattern around the first-user centres (in the following, referred to as 'concentrated users scenario') and one with a more distributed demand pattern (referred to as 'distributed users scenario'). Nevertheless, for both scenarios an initial hydrogen penetration primarily in the first-user centres (large urban areas) is assumed, whereas rural areas follow later. The geographical allocation of hydrogen demand over time for these scenarios is displayed in Fig. 14.6.

Studies show that between 10% and 30% of all existing fuelling stations should offer hydrogen to achieve a sufficient coverage of fuelling stations and thus user acceptance (Greene, 1998; Melaina, 2003; van Benthem *et al.*, 2006). Therefore, it is assumed that the regional fuelling station number has to correspond to 15% of the available gasoline fuelling stations in 2015 and 20% in 2020; thereafter, the number of new fuelling stations is matched to the hydrogen demand. Figure 14.7 shows the distribution of all hydrogen fuelling stations for the 'distributed users' scenario in 2030 as well as for the emerging network of highway stations in 2020. The average number of hydrogen fuelling stations in Germany in 2030 is estimated at between 7000 and 9000, and projected to increase to 12 000 in 2050.

14.5.2 Scenario definition

A great number of scenarios can be defined for analysing the impact of various parameters on the build-up of a hydrogen infrastructure. These include, for instance, the geographical distribution of hydrogen demand (e.g., concentrated vs. distributed) or the fuelling-station location strategy. For hydrogen supply, country-specific feedstock and production technology bounds can be taken into account, such as minimum shares for renewable hydrogen, availability of CCS for fossil hydrogen or free, unbound choice of feedstock shares (the only decision criterion is then cost minimisation). Lastly, variations in energy prices have a significant impact on the production mix.

Among this multitude of possible scenarios, seven scenarios have been selected to be analysed in more detail (see Table 14.7; for scenarios assessing the impact of hydrogen production on the electricity sector, refer to Chapter 16). The hydrogen-penetration Reference scenario is the 'high policy support and fast learning' scenario from Fig. 14.5. The reference energy price scenario is the 'high-price' scenario from Table 14.6.[12]

The Reference scenario ('REF') is, among others, based on the 'distributed users' scenario for the geographical distribution of hydrogen demand, a fuelling-station strategy tailored to meeting demand only ('FS demand') and a late road network; it is further characterised by high natural gas prices and a stabilisation objective for CO_2

[12] Different hydrogen penetration rates would shift the build-up of a hydrogen infrastructure to earlier or later periods, which would be affected by the respective changes in energy prices.

(a)

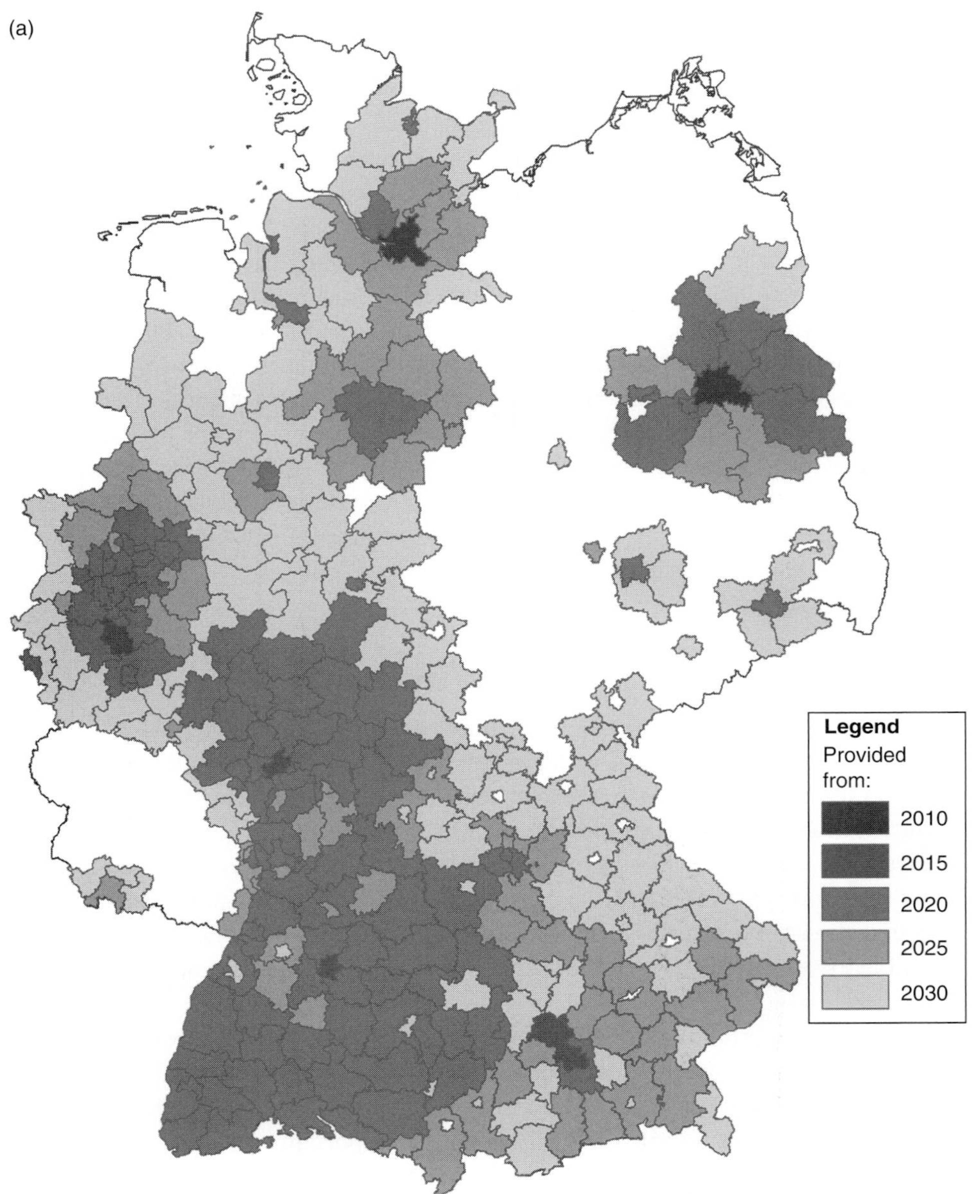

Figure 14.6(a). Geographical allocation of hydrogen demand in Germany until 2030: ‘concentrated users scenario’.

(b)

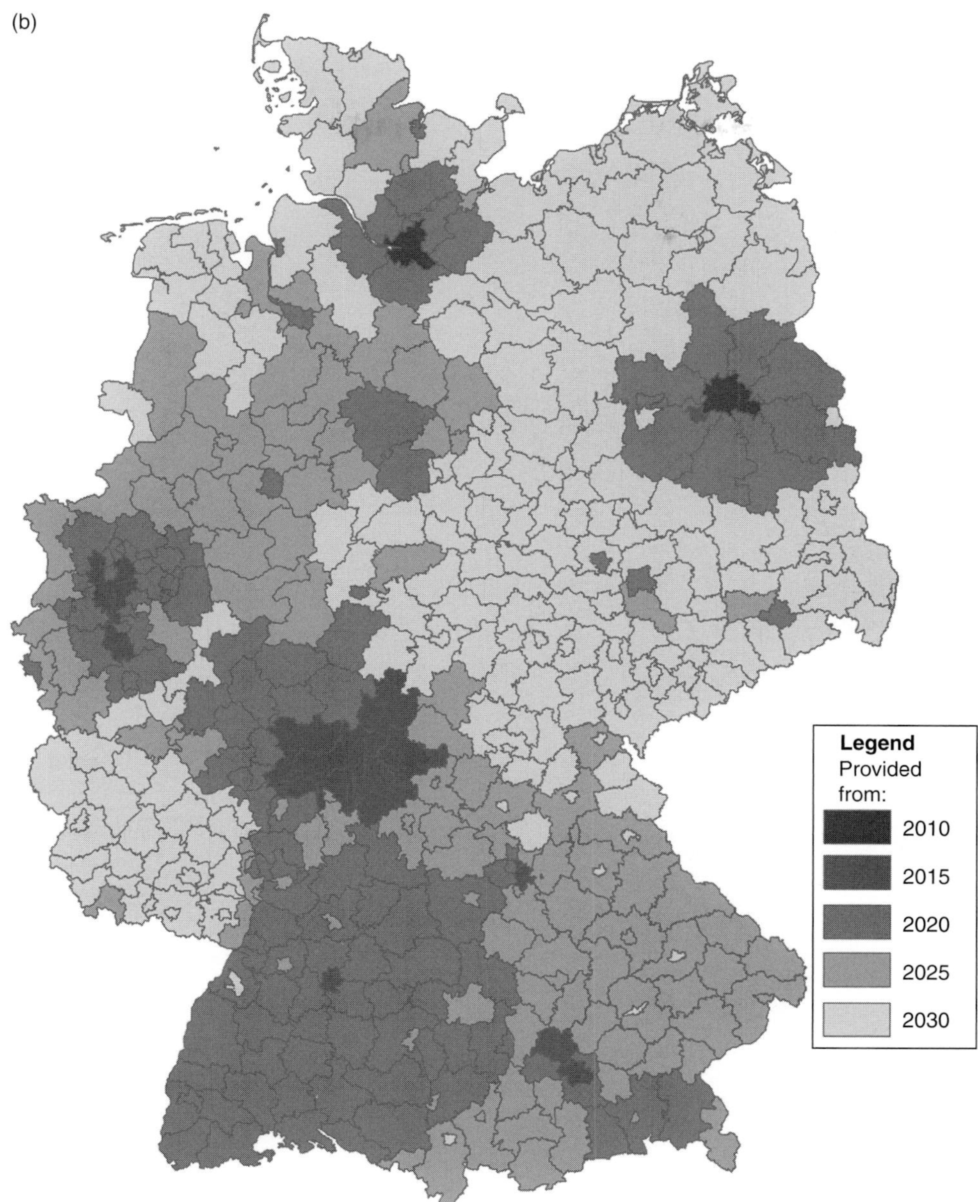

Figure 14.6(b). Geographical allocation of hydrogen demand in Germany until 2030: 'distributed users scenario'.

(a)

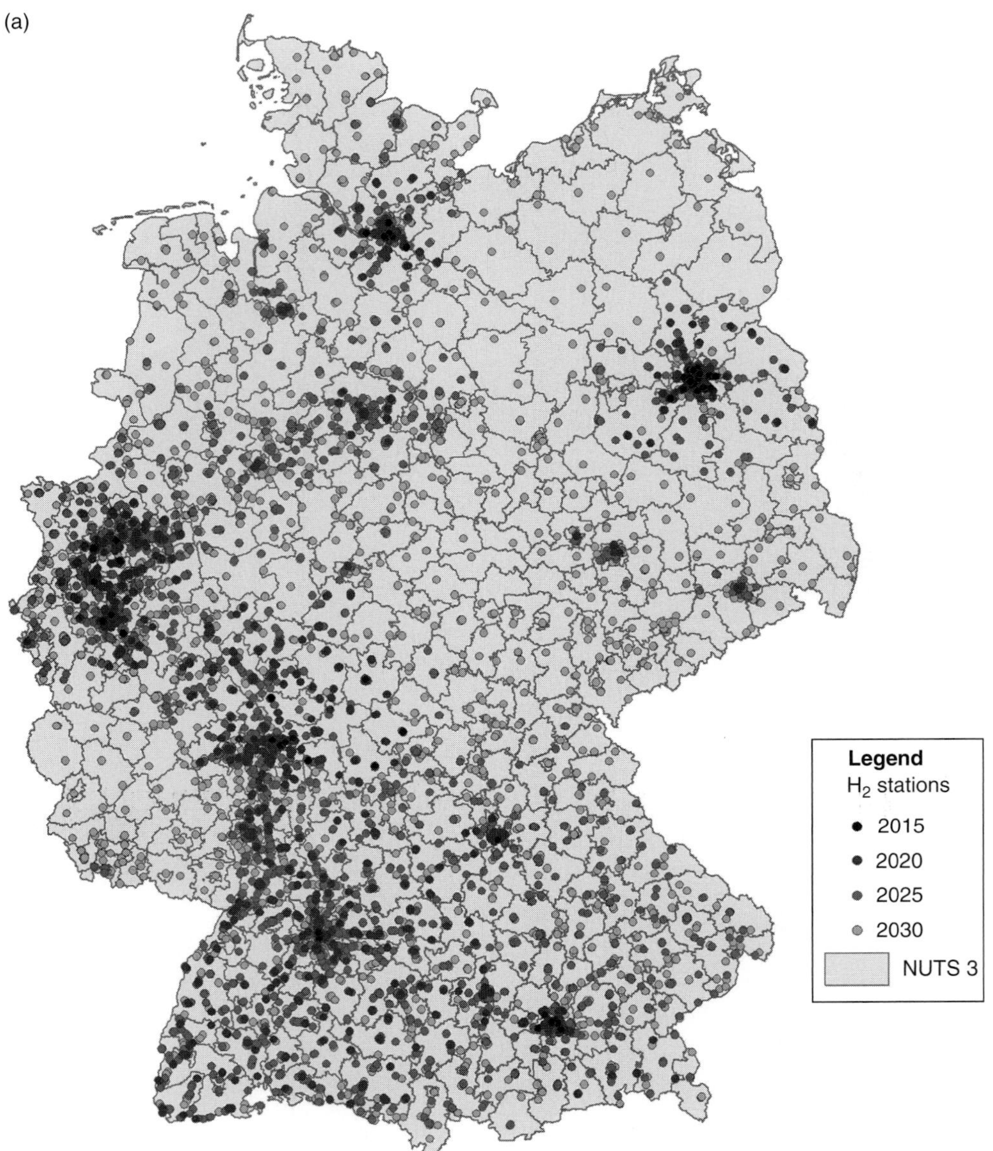

Figure 14.7(a). Fuelling stations in Germany for the 'distributed users' scenario: all hydrogen stations until 2030.

(b)

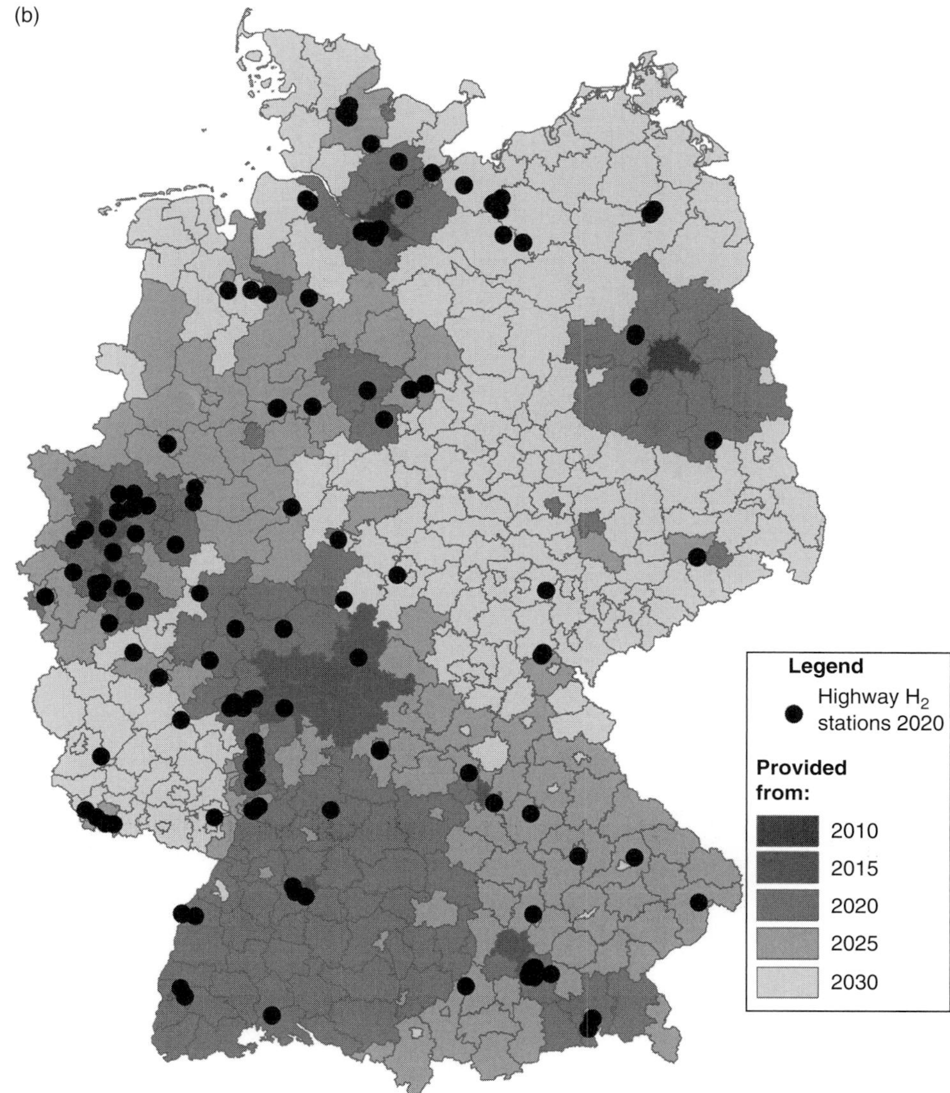

Figure 14.7(b). Fuelling stations in Germany for the 'distributed users' scenario: highway hydrogen station in 2020.

emissions for the power sector. There are no bounds set with regard to the delivery of liquid hydrogen and the supply of renewable hydrogen. Moreover, CCS is not available in the reference scenario. The use of coal gasification is further restricted to hard coal only.

Three other so-called 'infrastructure scenarios' are defined, which are based (1) on a fuelling-station location strategy tailored for maximum user convenience ('FS'),

Table 14.7. *Definition of model scenarios*

	Infrastructure scenarios				Technology scenarios		Energy-price scenarios
Parameter	**Reference scenario (REF)**	**Fuelling stations (FS)**	**Road networks (RN)**	**Concentrated demand (CON)**	**Renewable (REN)**	**(REN-CCS)**	**Low natural gas price (LNGP)**
Geographical distribution of hydrogen demand	**'Distributed users' scenario**	*See Reference scenario*	*See Reference scenario*	**'Concentrated users' scenario**	*See Reference scenario*	*See Reference scenario*	*See Reference scenario*
Fuelling stations	**'FS demand'**	**'FS user'**	*See Reference scenario*	*See Reference scenario*	*See Reference scenario*	*See Reference scenario*	*See Reference scenario*
Road networks	**'Late network'**	*See Reference scenario*	**'Early network'**	*See Reference scenario*	*See Reference scenario*	*See Reference scenario*	*See Reference scenario*
Liquid hydrogen	**No bound**	*See Reference scenario*	*See Reference scenario*	*See Reference scenario*	*See Reference scenario*	*See Reference scenario*	*See Reference scenario*
Renewable energies	**No bound**	*See Reference scenario*	*See Reference scenario*	*See Reference scenario*	**50% renewable H_2 from 2020 on**	**50% renewable H_2 from 2020 on; CCS**	**50% renewable H_2 from 2020 on**
CCS	**No CCS**	*See Reference scenario*	*See Reference scenario*	*See Reference scenario*	*See Reference scenario*	**CCS**	*See Reference scenario*
Energy prices	**High natural gas price**	*See Reference scenario*	*See Reference scenario*	*See Reference scenario*	*See Reference scenario*	*See Reference scenario*	**Low natural gas price**

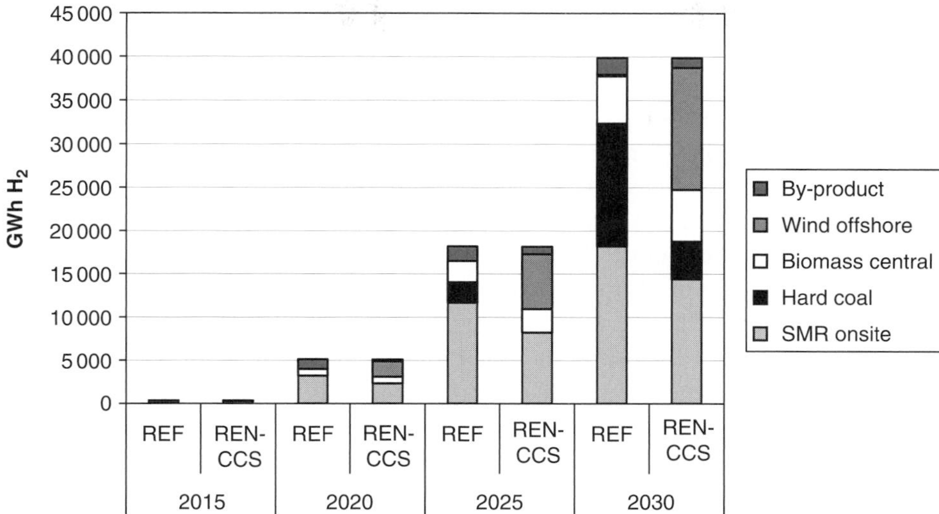

Figure 14.8. Comparison of hydrogen production mix ('REF') and ('REN-CCS') scenarios.

(2) an early road network ('RN') and (3) a 'concentrated users' scenario ('CON'), assuming that the hydrogen infrastructure will be developed first in highly populated areas on the basis of locally bound traffic, whereas rural areas follow later.

Furthermore, two technology-driven scenarios have been developed. In the first scenario, it is assumed that renewable hydrogen production must have a share of at least 50% from 2020 onwards ('REN'). In the second scenario, CCS technologies are additionally available ('REN-CCS').

Owing to uncertainties in the developments of the oil and gas markets, an alternative gas-price scenario is defined with a lower gas-price forecast than the ('REF') scenario, referred to as a 'low natural gas price' scenario ('LNGP').

The following sections focus on the description of these model scenarios and the major findings of the model application regarding the build-up of a hydrogen supply infrastructure. While the scenarios ('REF') and ('REN-CCS') are quantitatively analysed in more detail, the other scenarios are more qualitatively described.

14.5.3 Scenario analysis

Figure 14.8 shows the hydrogen production mix for the Reference scenario ('REF') and the renewable scenario ('REN-CCS') between 2015 and 2030. First of all, it can be seen that during the start-up phase, onsite steam reforming (SMR) at the fuelling stations dominates hydrogen production, as it is more expensive to transport small amounts of hydrogen than to produce it in small-scale units. But the choice between onsite production and central production is also sensitive to the assumed specific investments in onsite technologies and the utilisation of the refuelling stations; if, during the early phase, central SMR production is to play a role, because of lower specific investments, the hydrogen is trucked in liquid form to the fuelling stations. In addition, industrial by-product hydrogen plays a remarkable role, particularly during the first few years; however, hydrogen by-product utilisation is very much dependent on the local conditions, for instance the distance to the refuelling station, as the transport of small quantities over large distances is very costly.

With increasing demand, larger-scale production capacities with subsequent hydrogen transport turn out to be more economic. In 2030, central coal gasification plants gain significant market shares, owing to the assumed increasing gap between gas and coal prices; if lignite was available to the model, the switch to coal in the production mix would happen even earlier and be more dominant, because of the lower price of this domestic resource than of hard coal. Generally, the production mix in the reference case is largely dominated by fossil fuels until 2030. Despite the availability of CCS from 2025 on, the assumed CO_2 price does not justify the CO_2 transport and storage in this period. In addition, the interplay between the location of the coal gasification plant, the hydrogen user centres and the CO_2 storage site has to be taken into consideration.

In the renewable scenario, 50% of the hydrogen must come from renewable sources from 2020 on. Biomass is the cheapest renewable option, but has a limited potential, as the competition between hydrogen, biofuels and other uses has to be considered. Offshore wind via electrolysis could, therefore, play a very important role for hydrogen production after 2020. Onsite SMR also dominates here in the early phase.

Figures 14.9 and 14.10 show the geographical locations of major hydrogen production plants for the 'REF' and 'REN-CCS' scenarios, respectively.

Figure 14.11 shows the development of the cumulative investments for the entire hydrogen supply chain (production, transport and distribution, liquefaction and refuelling stations) for the 'REF' and 'REN-CCS' scenarios. The cumulative investments in these as well as in the other scenarios are clearly dominated by the investment in production technologies. It can be seen that in scenarios with a high renewable share in the hydrogen production mix, total investments increase significantly: for the above scenarios, from 12 to 21 billion euros until 2030. On the basis of the underlying energy price scenario, the resulting specific supply costs in 2030 are about 11 ct/kWh for the Reference scenario and about 12 ct/kWh for the renewable scenario.

The specific costs are relatively high at the beginning of the vehicle uptake, since an initial infrastructure (e.g., in the form of filling stations) will be required, which necessarily exceeds the hydrogen demand. They decline rapidly in the following years and stay more or less constant in the last decade of the analysed period. Assuming a specific hydrogen consumption of 26 kWh_{H_2}/100 km for fuel-cell passenger cars and a specific gasoline or diesel consumption of 5 l/100 km for conventional cars and a

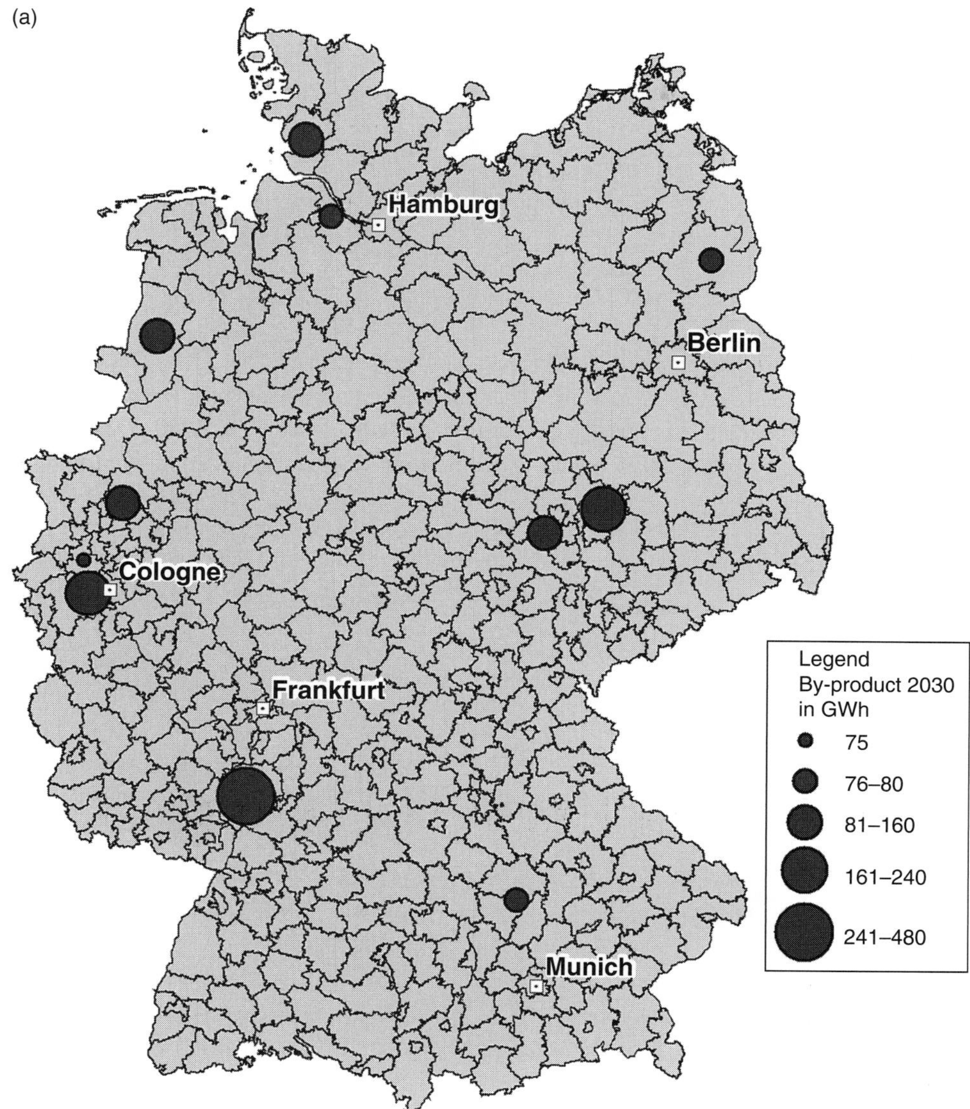

Figure 14.9(a). Hydrogen-production plants in the 'REF' scenario for by-product in 2030.

refinery efficiency of 90%, hydrogen will be competitive at crude oil prices beyond $80–100/barrel (based on before-tax costs of both fuels).

Figure 14.12 shows the shares of the various hydrogen transport and distribution means for the above scenarios. It can be stated that liquid hydrogen is a preferred option for medium amounts of hydrogen dispersed over a large area, which is mainly

(b)

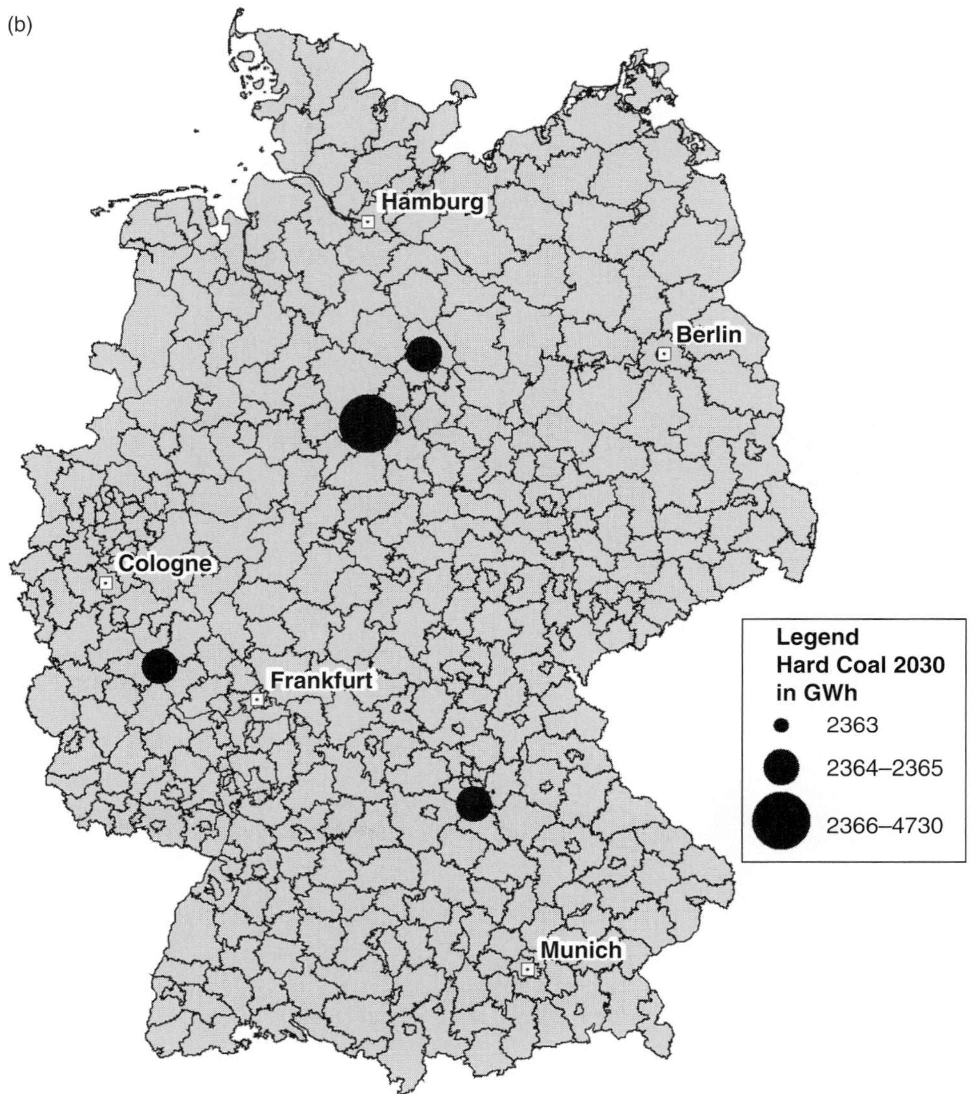

Figure 14.9(b). Hydrogen-production plants in the 'REF' scenario for coal gasification in 2030.

the case during the infrastructure build-up phase until around 2025, in particular in the 'distributed users' scenario. If a more concentrated demand is assumed, the picture turns, so that a pipeline network for hydrogen starts to develop before 2030. Gaseous trailers could play a role for small under-utilised filling stations and very low demand but, in absolute terms, hydrogen transport with gaseous trailers has a negligible influence. The energy price has a great influence not only on the

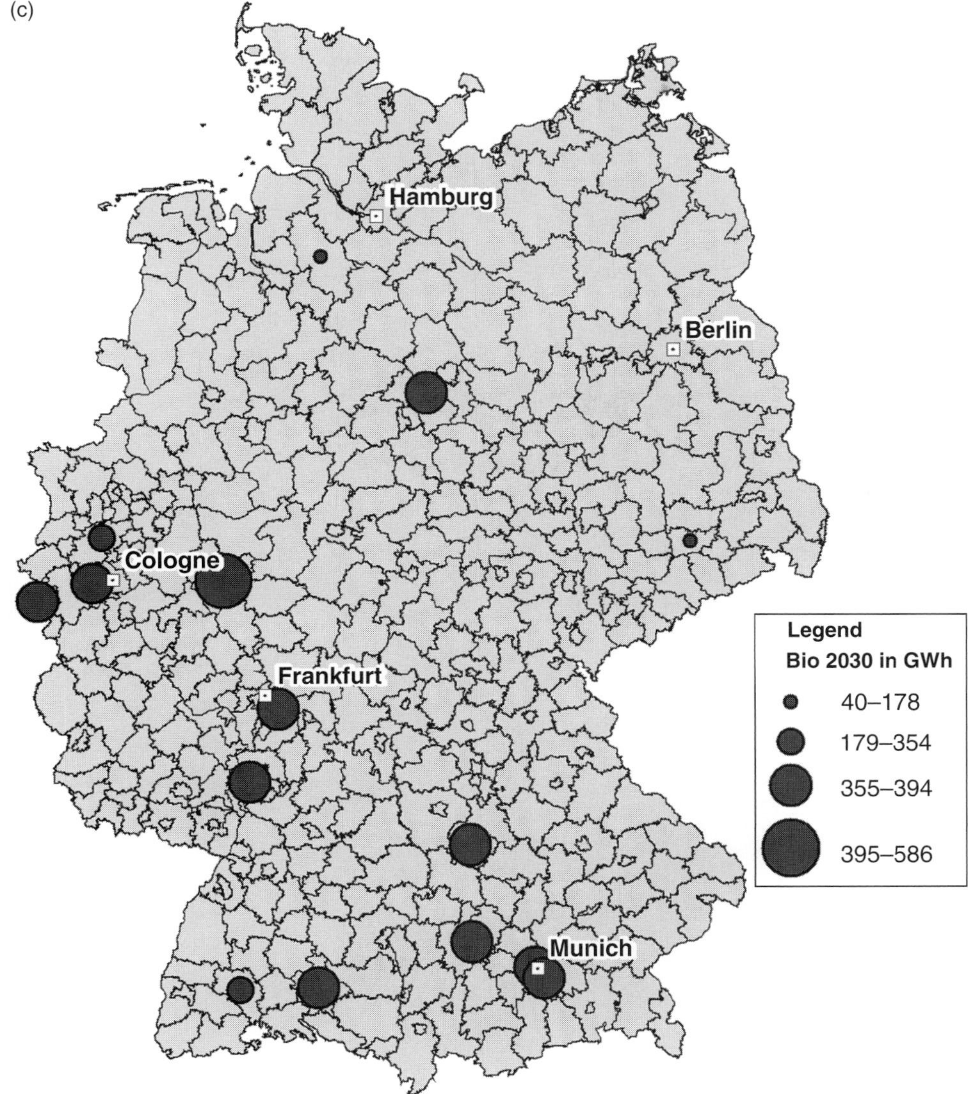

Figure 14.9(c). Hydrogen-production plants in the 'REF' scenario for biomass gasification in 2030.

production of hydrogen but also on transport and distribution. For example, under the 'low natural gas price (LNGP)' scenario, onsite reformers also play a larger role in later periods and, therefore, transportation is less important for distributed hydrogen demand. Also, liquid hydrogen transport is very sensitive to electricity prices, as liquefaction is an energy-intensive process.

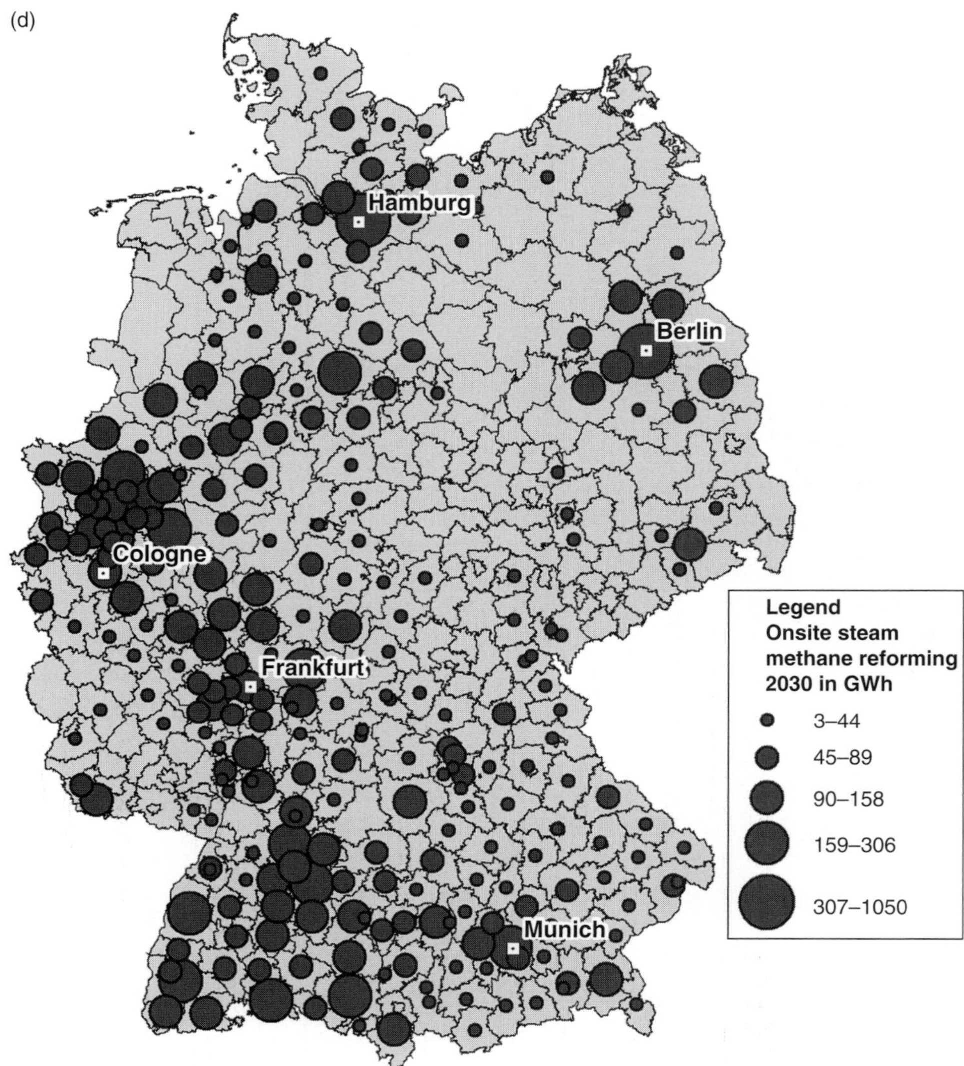

Figure 14.9(d). Hydrogen-production plants in the 'REF' scenario for onsite steam methane reforming in Germany in 2030.

Figure 14.13 exemplifies the location of hydrogen liquefaction plants in Germany in 2030, as well as the transport flows of hydrogen by trailer and pipeline, respectively.

In the following, the major impacts of the other scenarios on infrastructure build-up, production mix, supply costs or choice of transport options will briefly be addressed.

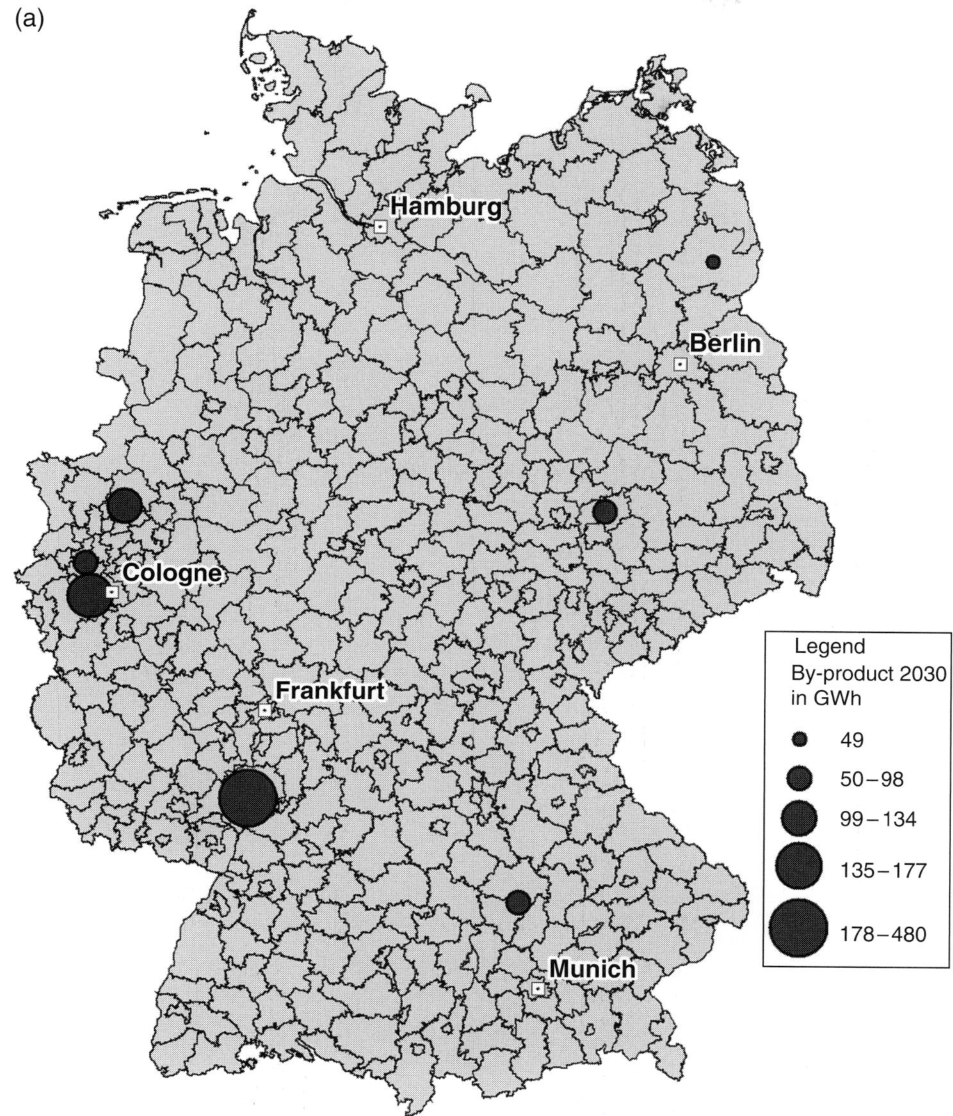

Figure 14.10(a). Hydrogen-production plants in the 'REN-CCS' scenario for by-product in Germany in 2030.

14.5.3.1 Scenario 'fuelling stations (FS)' – ('FS user')

If in the early phase, 15% to 20% of the fuelling stations offer hydrogen, for a given demand, less hydrogen would be supplied by each fuelling station than in the reference case. This would lead to higher investments in fuelling stations. If onsite

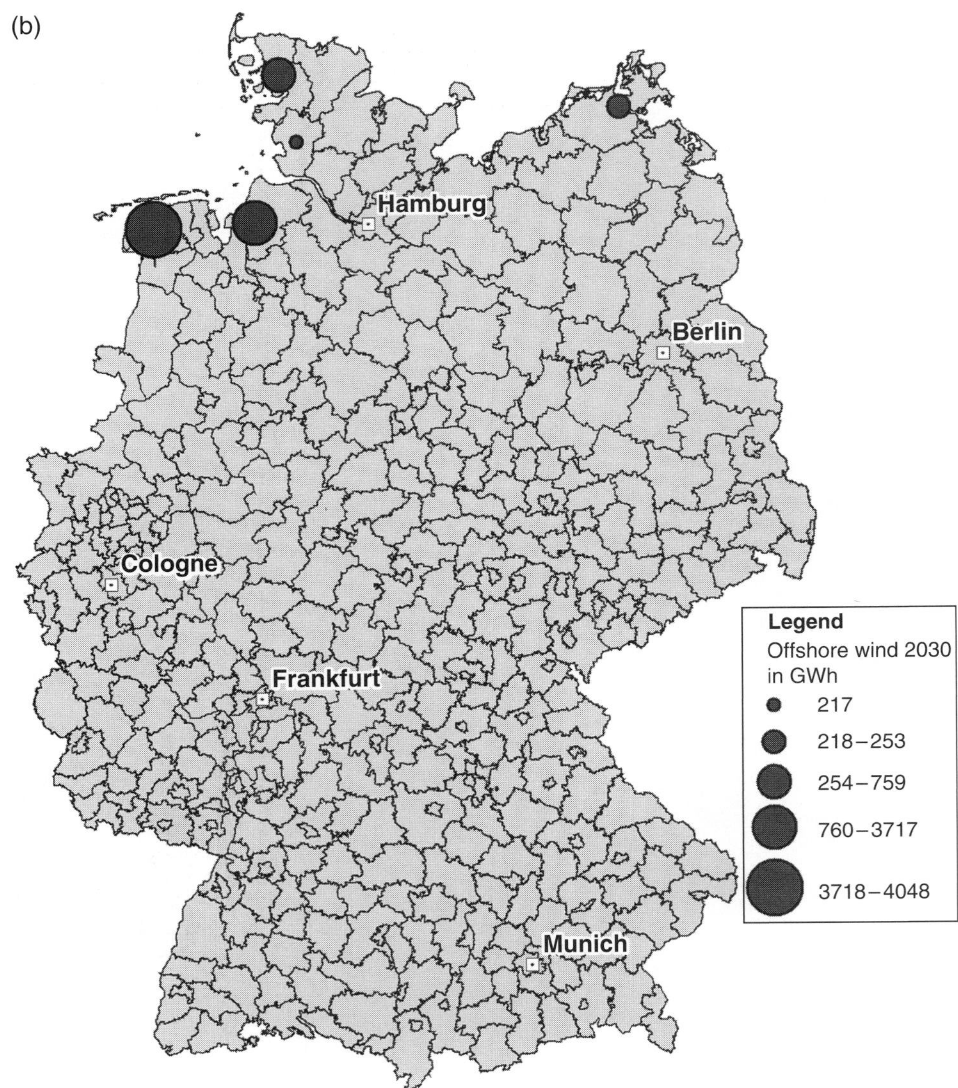

Figure 14.10(b). Hydrogen-production plants in the 'REN-CCS' scenario for offshore wind in Germany in 2030.

production technologies cannot be scaled down to match the hydrogen demand, supply costs would further increase, owing to under-utilisation of the fuelling stations; if hydrogen is produced in central production plants or is available as a by-product and transported in liquefied form, the increase of supply costs would be less drastic. In this scenario, the hydrogen costs at the fuelling station would increase by around 5%.

(c)

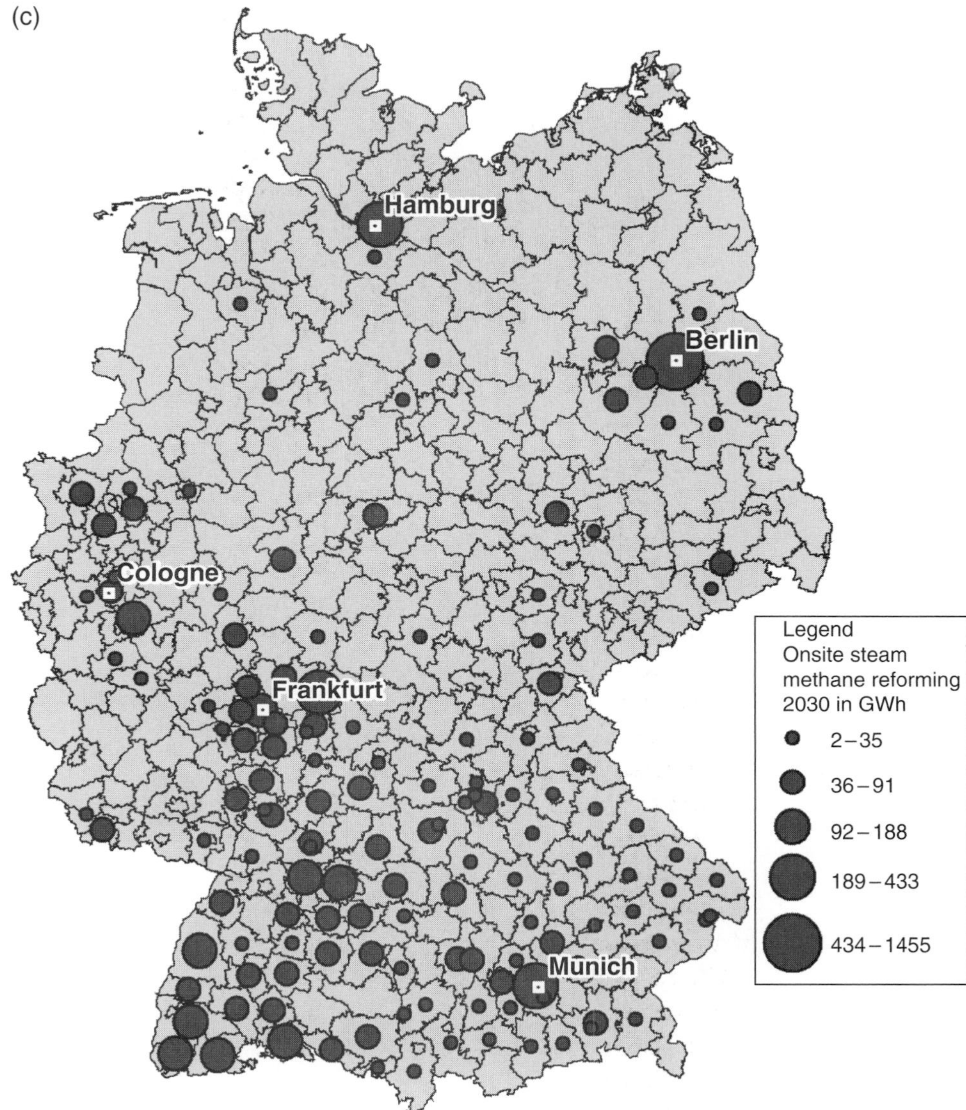

Figure 14.10(c). Hydrogen-production plants in the 'REN-CCS' scenario for onsite steam methane reforming in Germany in 2030.

14.5.3.2 Scenario 'road networks (RN)' – ('early network')

The early road network scenario has a similar effect on the hydrogen infrastructure to the fuelling stations scenario. More fuelling stations will be in place in the early phase. However, in this scenario the fuelling stations will be located on the highways to allow travelling between the user centres. This will surely increase the acceptance of

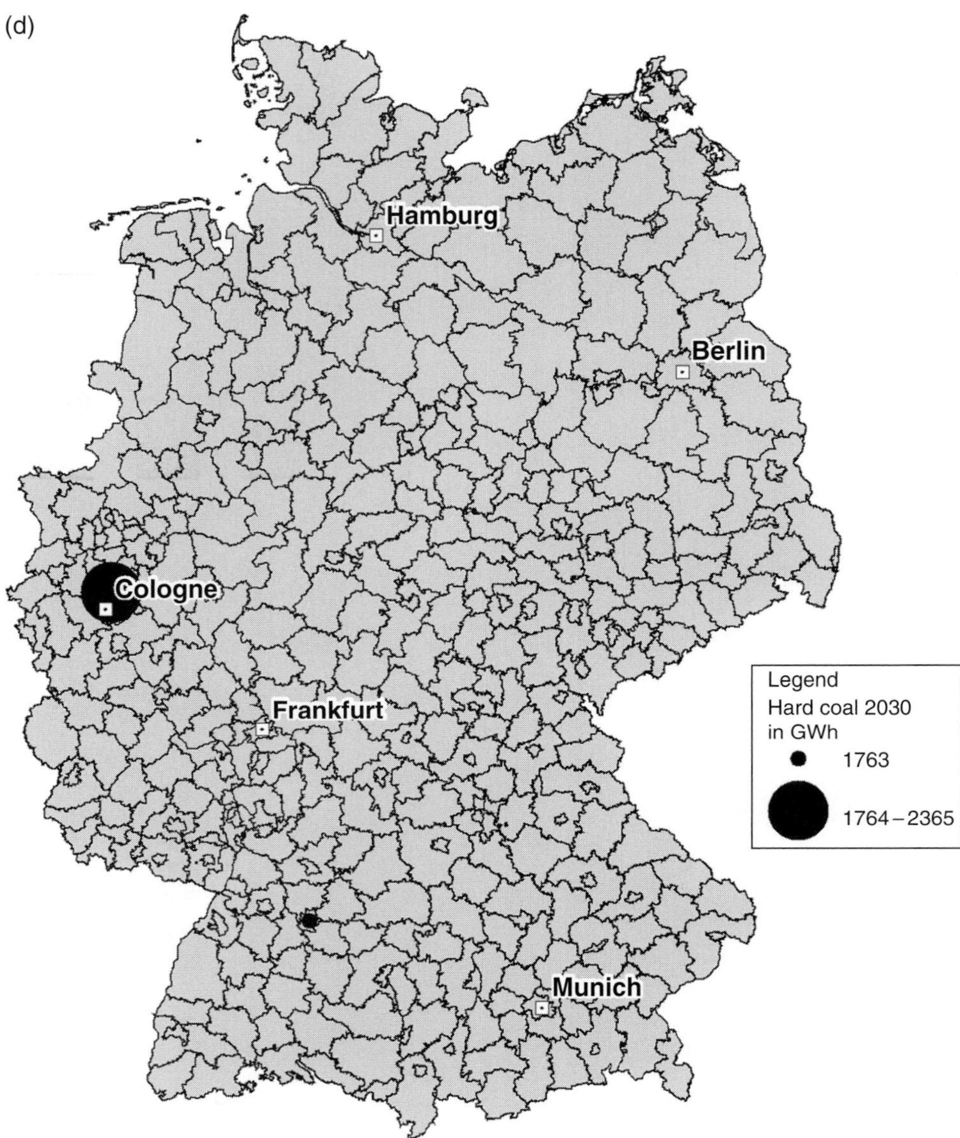

Figure 14.10(d). Hydrogen-production plants in the 'REN-CCS' scenario for coal gasification in Germany in 2030.

the new fuel. But focusing on the challenges for the infrastructure development, this scenario will also lead to higher costs of the fuelling stations. Onsite production on highways might be difficult, if no natural gas infrastructure exists. Also, the specific investments might be very high, if there is only a very low utilisation of these fuelling stations in the early phase. Therefore, liquid hydrogen is dominating the supply of the

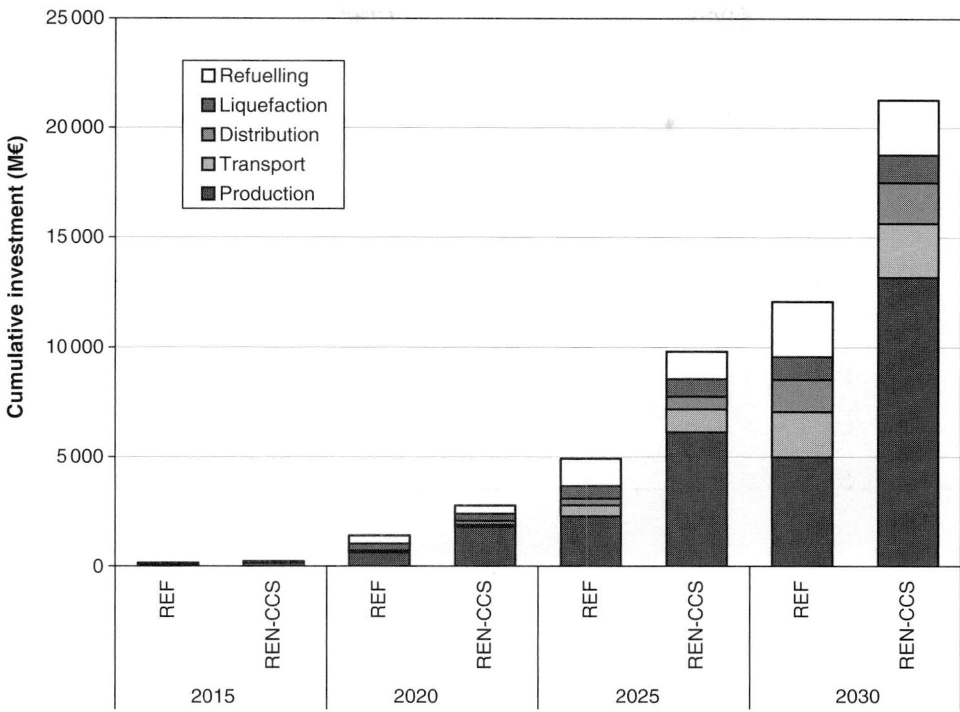

Figure 14.11. Cumulative investment until 2030 for the 'REF' and 'REN-CCS' scenarios.

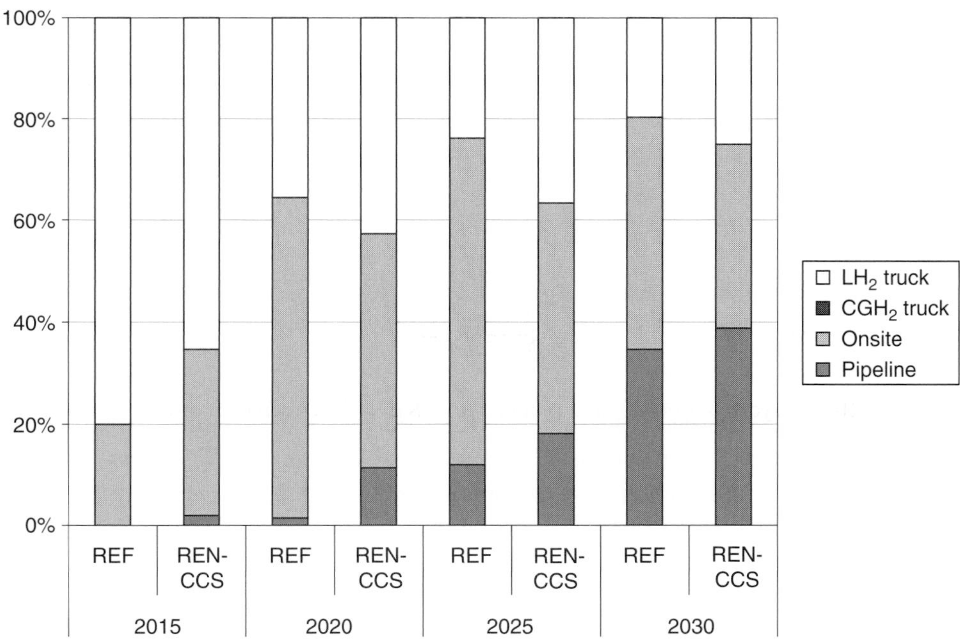

Figure 14.12. Shares of hydrogen distribution for the 'REF' and 'REN-CCS' scenarios.

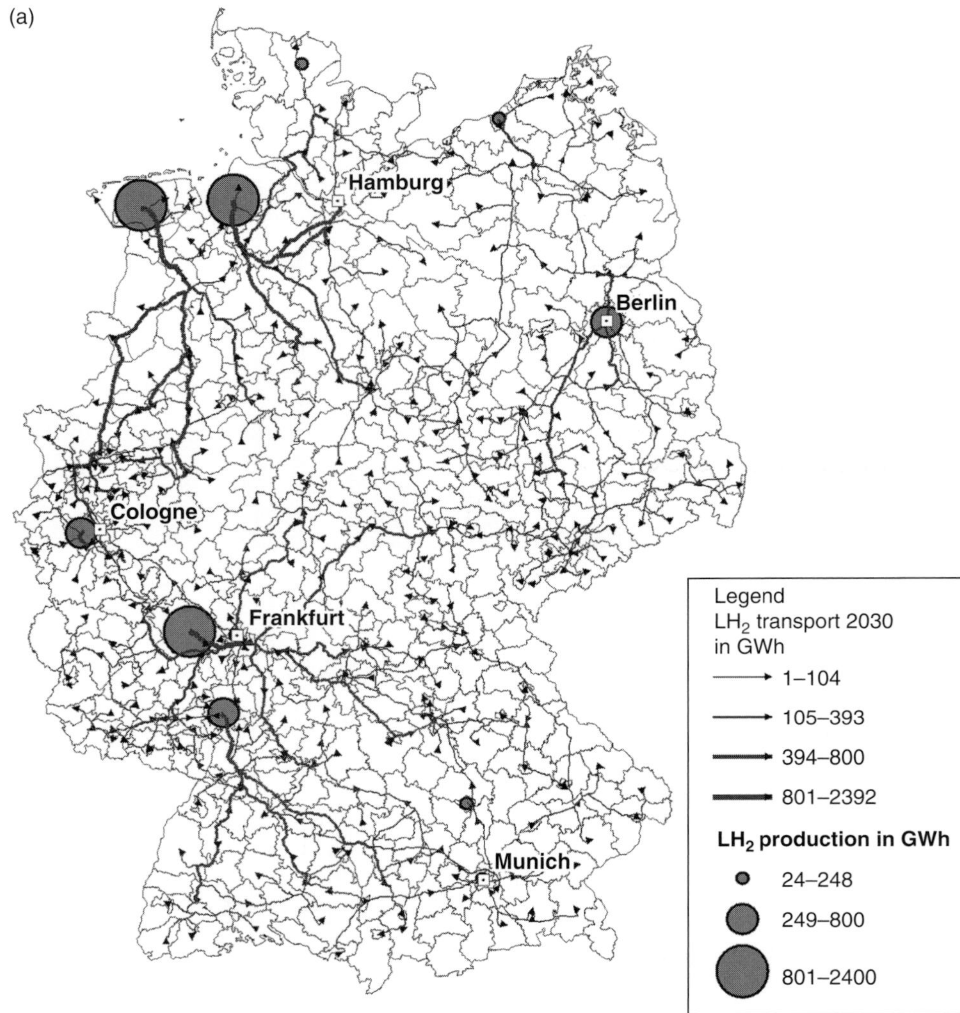

Figure 14.13(a). Liquid hydrogen transport in the 'REN-CCS' scenario in 2030.

highway fuelling stations, if the local fuelling stations are also supplied by liquid hydrogen. The transportation costs of liquid hydrogen are high in general but less sensitive to the distance compared to compressed hydrogen, and also cheaper than onsite production with very low utilisation.

14.5.3.3 Scenario 'concentrated demand (CON)'

The roll-out strategy for hydrogen from 2015 to 2030 ('concentrated users' scenario or 'distributed users' scenario) substantially shapes the development of the hydrogen

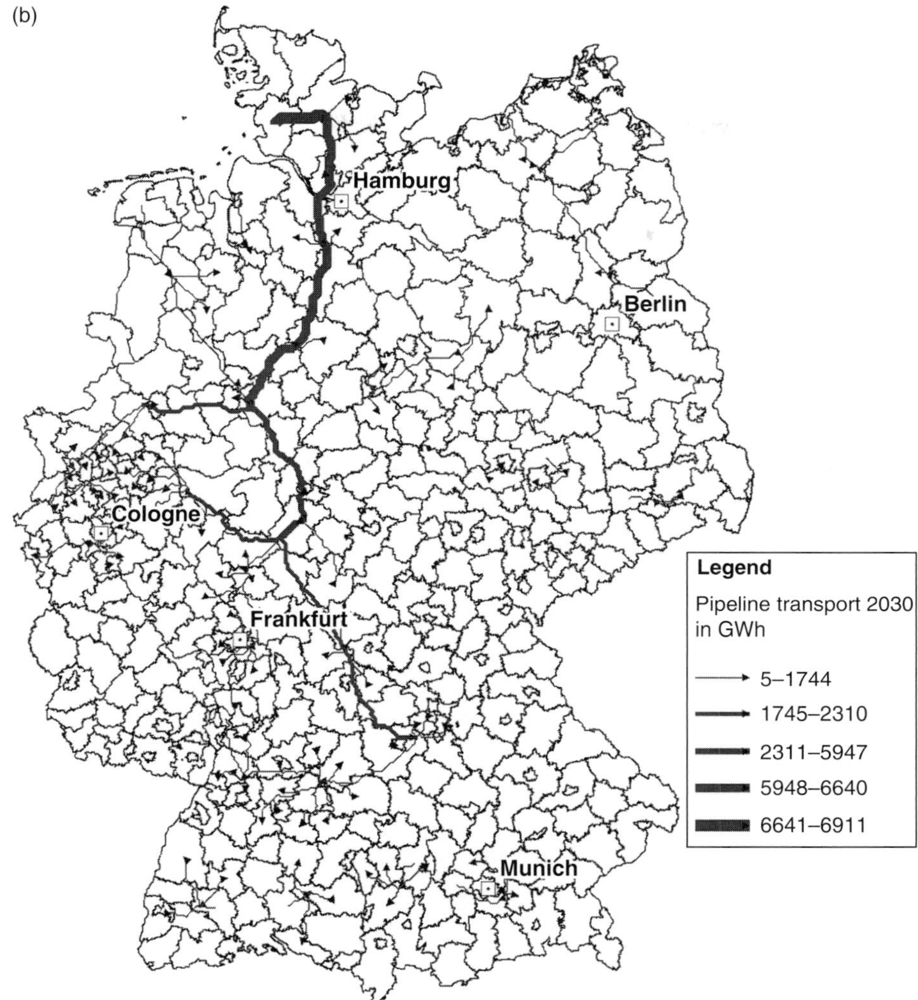

Figure 14.13(b). Pipeline transport in the 'REN-CCS' scenario in 2030.

landscape; the distributed user strategy, with its early widespread use, leads to a more even penetration of hydrogen because more users will have early access to hydrogen. In the concentrated user strategy, areas supplied later will have a backlog in penetration compared with areas supplied earlier. This also affects the fuelling stations: if a certain hydrogen demand is 'spread' over a larger area, a greater number of smaller fuelling stations will be required than if the same demand is concentrated in denser areas. The concentrated scenario with hydrogen demand concentrated in first-user centres tends more to larger-scale production units than the distributed scenario, which results in lower production costs. As a consequence, the concentrated user strategy leads to 10%–20% lower specific hydrogen costs in the early phases (2015–2020), which level out after 2030.

14.5.3.4 Scenario 'CCS'

Under the CCS scenario assumptions, CCS is cost-competitive due to the increasing CO_2 certificate prices. Therefore, the fossil hydrogen production from central production technologies with CCS dominates. In the long run, especially, coal gasification with CCS could gain remarkable hydrogen production shares. However, the total storage potential available beyond 2050 (and its potential competition with the power sector) must also be taken into account and, therefore, CCS might only be a transition technology to other hydrogen-production pathways.

14.5.3.5 Scenario 'low natural gas price (LNGP)'

It is interesting to note that despite the switch from natural gas to coal-based hydrogen production, no major changes in total supply costs can be observed. This fact illustrates that the competitive advantage of natural gas over coal in terms of production costs is based on a very narrow margin and is very sensitive to slight increases in gas prices. If lower natural gas prices are assumed, onsite SMR technologies dominate the hydrogen production for a longer period, if cost reduction through mass production of standardised onsite SMR modules is reached. Otherwise, central SMR plants will play a dominant role under these assumptions.

14.5.4 Hydrogen supply outlook until 2050

In the period after 2030, two dominant technologies could be identified as the main source of hydrogen in Germany: coal gasification with CCS and offshore wind parks, with CCS being incentivised by the assumed increase in CO_2 prices (see Fig. 14.14). If CCS is not available, coal gasification might not be an option after 2030 at all, which leaves only natural gas technologies for fossil hydrogen production, which significantly increase production costs in the long run. In general, the timeframe from 2030 till 2040, with very high rates of growth for hydrogen demand, determines, to a large extent, the production of hydrogen for the next 30 to 40 years, owing to the longevity of the hydrogen production plants once put in place. Therefore, the production options and their consequences for the environment have to be carefully considered. Only in combination with CCS and renewable technologies could fossil fuels play a remarkable role for hydrogen production from a climate perspective in the long term.

Pipelines dominate the transport and distribution of hydrogen in this period and an area-wide pipeline network covers nearly the whole of Germany in 2050. If wind offshore technologies are integrated in the production mix, the pipeline network

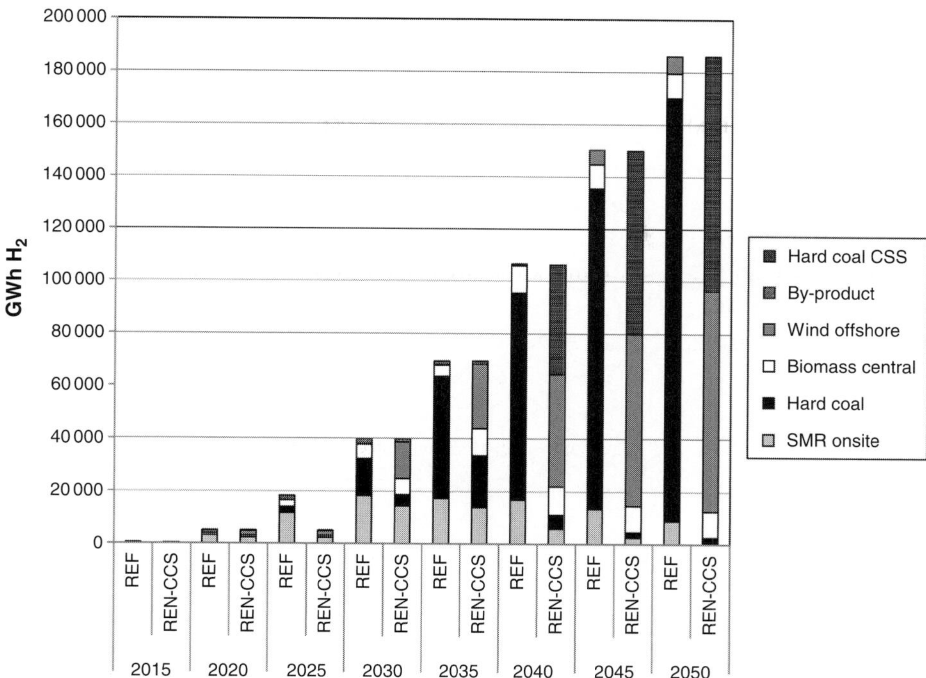

Figure 14.14. Hydrogen production 2015 to 2050 in the reference and renewable scenarios in Germany.

transports the hydrogen from the northern regions of Germany to the demand centres in western and southern Germany, as depicted in Fig. 14.15.

14.5.5 Impact of hydrogen on CO_2 emissions

To get an understanding of the potential CO_2 reduction achievable in the transport sector from the introduction of hydrogen vehicles, CO_2 emissions from hydrogen supply are compared on a well-to-wheel (WTW) basis (i.e., taking into account the entire supply chain) with conventional gasoline or diesel fuel. As a baseline, average WTW emissions for gasoline or diesel-fuelled passenger cars of 160 g CO_2/km are assumed; around 85% of these emissions result from the fuel use in the vehicle.

If the hydrogen is used in fuel-cell vehicles and no CCS is applied, the WTW emissions from hydrogen supply on the basis of natural gas are roughly 100 g CO_2/km, from coal 200 g CO_2/km, from biomass 15 g CO_2/km and from wind power almost negligible (see also Chapter 7). Under this scenario, hydrogen from coal results in significantly higher CO_2 emissions than for conventional cars; in fact an increase of CO_2 emissions of 25% compared with the baseline. If CCS is applied, CO_2 emissions can be reduced to about 30 g CO_2/km for natural gas and 50 g CO_2/km for

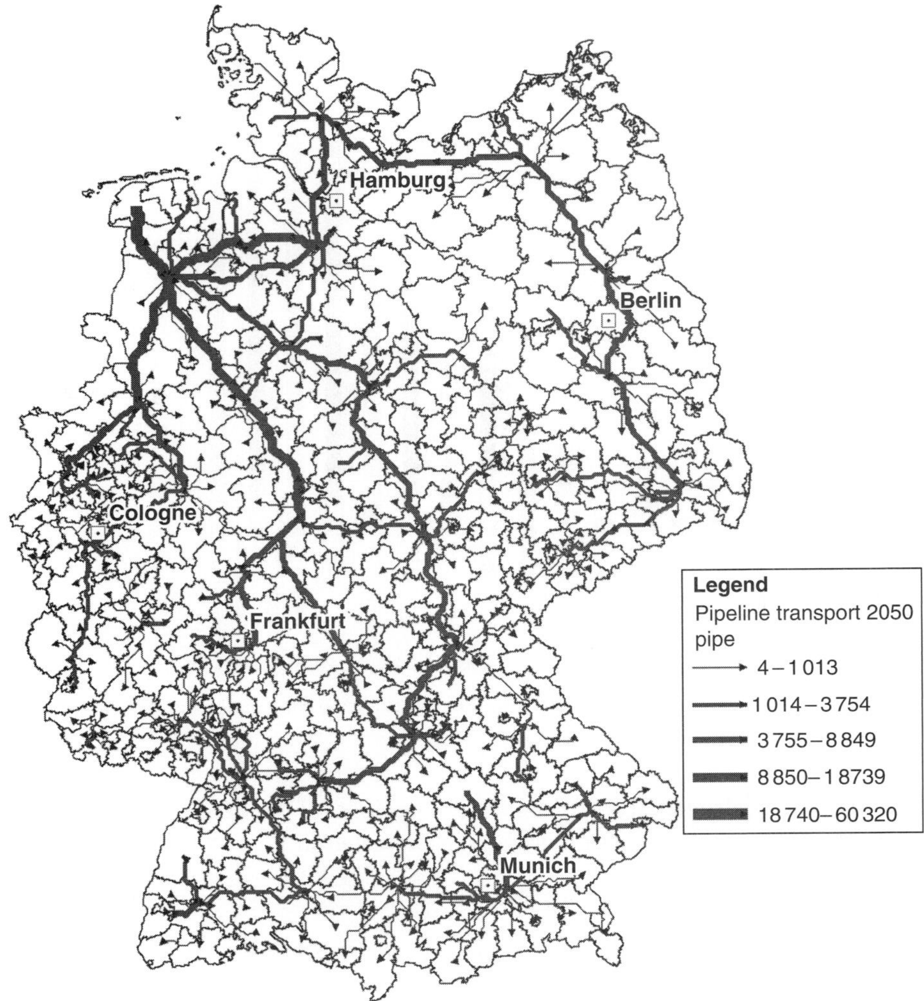

Figure 14.15. Hydrogen pipeline network in the renewable scenario 2050 in Germany.

coal. For hydrogen ICE cars, WTW emissions are between 60% and 80% higher, owing to the lower efficiency of the combustion engine than of the fuel cell (up to 360 g CO_2/km, in the case of coal).

While hydrogen is emission-free at final use, the above analysis clearly shows that where hydrogen is produced from coal without CCS, no overall CO_2 reduction in the energy system is achievable. Given that coal is, in the medium to long term, the most economic feedstock as natural gas prices increase, CCS becomes an inevitable prerequisite for the supply of hydrogen. If CCS was applied for hydrogen production from biomass, a net CO_2 reduction would be possible. The analysis further reveals that hydrogen ICE vehicles should only be used during the transition phase to help

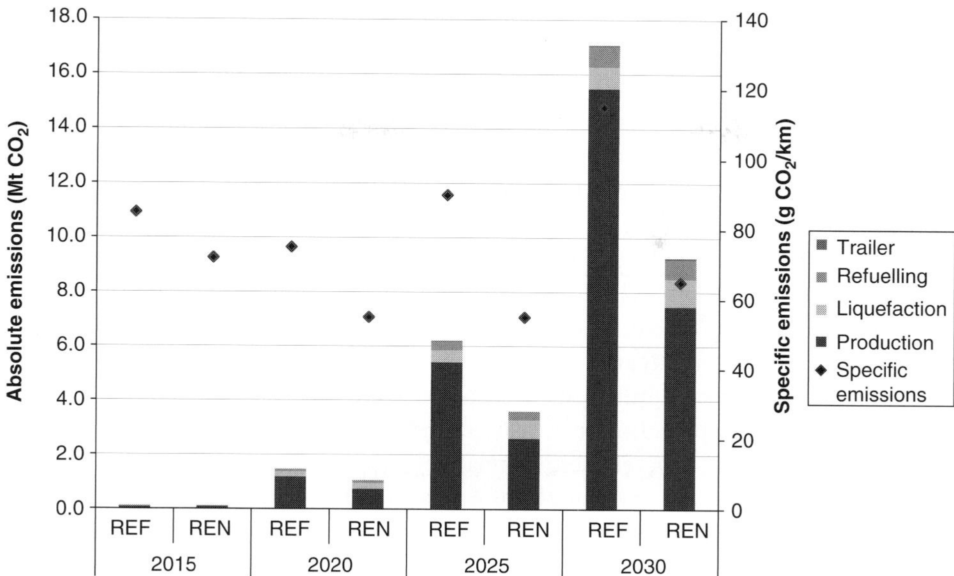

Figure 14.16. Balance of CO_2 emissions.

overcome the initial lack of an area-wide refuelling infrastructure, because they have no advantages in terms of energy efficiency and CO_2 emissions reduction over conventional cars (unless produced from renewable energies). To calculate the absolute CO_2 emissions reduction of the vehicle fleet, the relative reduction (per km) needs to be further multiplied by the hydrogen vehicle penetration rate.

Figure 14.16 compares the absolute and specific CO_2 emissions resulting from hydrogen supply for the 'REF' and 'REN' scenarios (without CCS). In the 'REF' scenario, the absolute CO_2 emissions from hydrogen supply increase to about 17 Mt in 2030; setting a 50% renewable target for hydrogen would almost cut them in half. The WTW CO_2 emissions of the 'REF' scenario are between 80 and 90 g CO_2/km in the initial phase, when hydrogen is produced mainly from natural gas and supplied as an industrial by-product; emissions increase to about 120 g CO_2/km in 2030, owing to the growing share of hydrogen from coal; however, this is still below the baseline of 160 g CO_2/km for conventional cars. In the 'REN' scenario, WTW CO_2 emissions are reduced to 50–70 g/km around 2030.

Under the assumption that CCS will be widely applied from 2025 on, a remarkable reduction of CO_2 emissions could be achieved in the transport sector by hydrogen fuel-cell vehicles over the next few decades, even if hydrogen was produced mainly on the basis of fossil fuels. In the case of a predominantly coal-based production mix without CCS, CO_2 emissions from hydrogen production will exceed those saved through its use in the transport sector. If around 50% of total hydrogen production was to depend on CCS from coal gasification, the total CO_2 volumes to be stored in

Germany and the EU would amount to roughly 10/55 Mt per year by 2030 and 70/350 Mt per year by 2050; accumulating globally to up to 1300 Mt per year by 2050. However, the total available CO_2 storage potential and its potential competition with the power sector must also be considered and could constrain such a scenario.

A common metric used to compare different alternative fuels and power trains to a benchmark fuel and vehicle are CO_2 avoidance costs, which take into account cost of fuel, infrastructure and vehicles. However, depending on the calculation methods, and system boundaries, as well as the assumptions about CO_2 emissions reduction achievable and vehicle prices relative to the benchmark, significant variations in CO_2 avoidance costs are possible. For this reason, the discussion about CO_2 avoidance costs is not addressed in great detail.

Assuming that a premium will have to be paid for fuel-cell vehicles at the beginning, and as a new supply infrastructure needs to be put in place, the specific avoidance costs of hydrogen fuel-cell vehicles are relatively high compared with other alternative fuels and propulsion systems (€600–€1800/t CO_2 avoided), while biofuels for instance are in the range of €170–€350/t CO_2 avoided (JEC, 2007). However, considering increased CO_2 emissions resulting from land use change for first-generation biofuels, this picture might change. Assuming that hydrogen is produced from low-carbon sources and fuel-cell cars are cost-competitive with conventional cars, the avoidance costs of a hydrogen fuel system can be calculated to be in the range of €200 to €500/t CO_2 based on the scenarios presented in this chapter.[13]

Besides the specific avoidance costs, the total potential of CO_2 avoidance of a specific alternative fuel is also important. Here, hydrogen has a vital advantage over biofuels: while the absolute potential of biofuels is limited by the availability of biomass feedstock, hydrogen can be produced from virtually any low-carbon source and the total emission reduction potential is much higher.[14] Equally, owing to their inherent advantages with respect to feedstock base, infrastructure requirements and vehicle efficiency, plug-in hybrid-electric vehicles and battery-electric vehicles have the potential for achieving very low CO_2 avoidance costs in the transport sector.

14.6 Constructing a pan-European hydrogen infrastructure

The following section presents the major outcomes of the HyWays project, whose overarching aim was to develop a validated European hydrogen roadmap and an action plan for introducing hydrogen in transport as well as stationary applications, and to demonstrate how hydrogen can contribute to sustainability. HyWays

[13] These avoidance costs (€/t CO_2) have been calculated according to the following formula:

$$H_2 \text{ supply costs (ct/kWh)} \times H_2 \text{ consumption (kWh/km)}/CO_2 \text{ reduction relative to baseline } (\Delta g\ CO_2/\text{km}).$$

[14] The HyWays project has concluded that hydrogen can reduce the CO_2 emissions from road transport by over 50%; much of the remaining emissions come from goods transport where no hydrogen fuel was assumed (HyWays, 2007). In contrast, biofuels can only supply a fraction of today's transportation energy demand (6% to 15% within the EU), if the competing use of biomass in the stationary sector is taken into account (JEC, 2007) (see also Chapter 7).

(a)

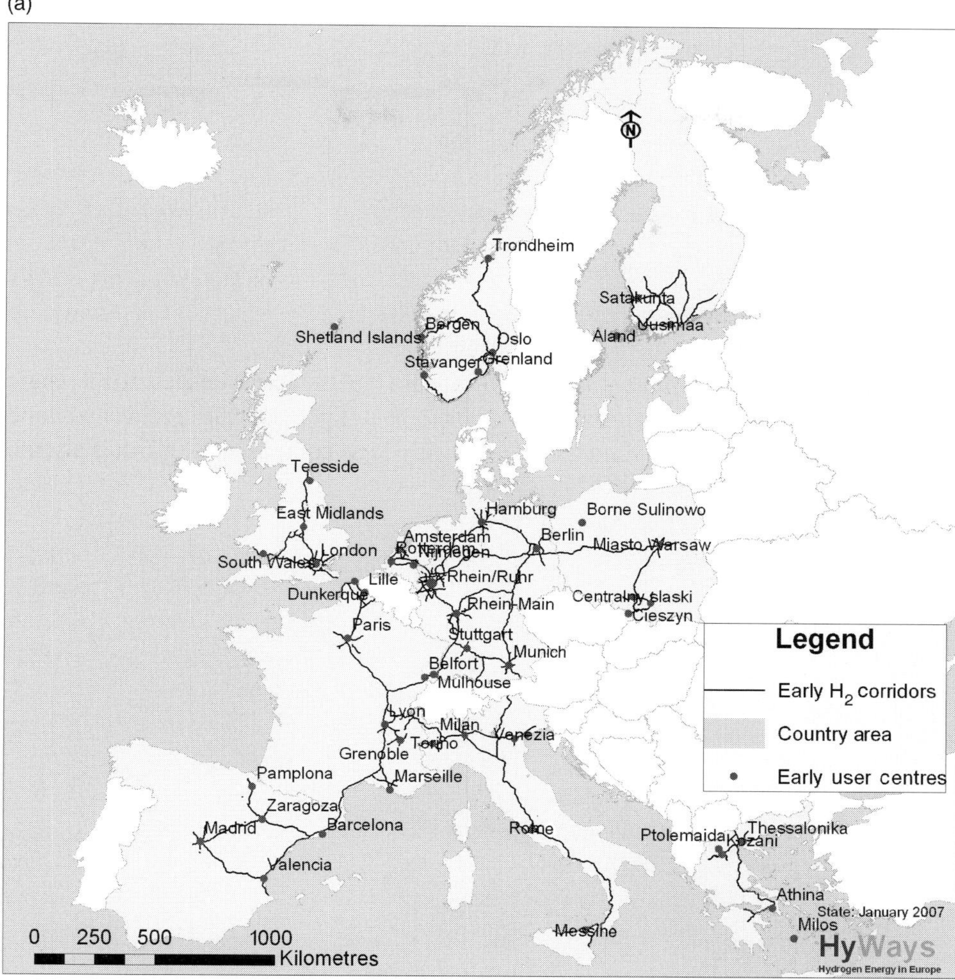

Figure 14.17(a). Early-user centres and early transit road network for hydrogen in the ten HyWays countries (HyWays, 2007).

comprised ten EU member states, who made up 76% of the population and 72% of the land area of the EU27.[15]

Figure 14.17(a) shows the *early-user centres* of all ten HyWays countries and the early hydrogen corridors. Most countries focus on densely populated areas for the early adoption of hydrogen, because of the obvious infrastructure advantages arising from a high density (shorter distribution distances, more users reached per fuelling station). Besides size, indicators like the availability of hydrogen and technology experts, political commitment, existing demonstration projects and, to some extent,

[15] France, Germany, Greece, Italy, the Netherlands, Norway, Finland, Poland, Spain and the United Kingdom. For more information about the HyWays project see www.hyways.de.

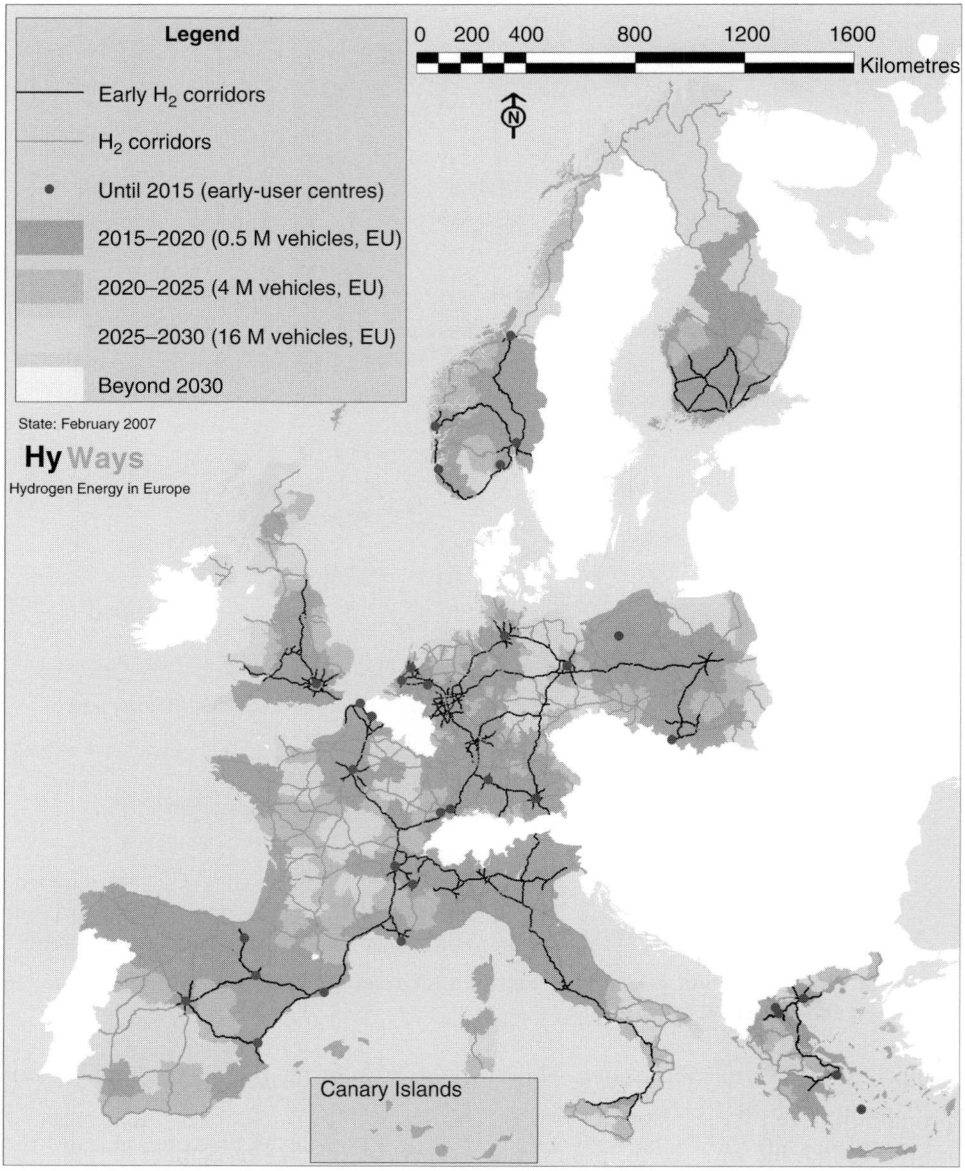

Figure 14.17(b). Time of first hydrogen supply in the ten HyWays countries.

the availability of resources play a major role in selecting the early-user centres. Some countries also include remote areas in the early-user centre portfolio. This is mainly because stranded renewable energy resources can be tapped, and the need for a transit road network is lower due to the remoteness of these regions leading to a stronger focus on the local use of road vehicles.

The resulting early transit road network focuses on connecting early-user centres within the HyWays countries, but also on international links. Furthermore, the motorways around early-user centres with high population densities should be equipped with hydrogen fuelling stations to facilitate daily commuting in these regions.

Figure 14.17(b) shows the subsequent regional deployment of hydrogen use in the later periods (the lighter the shading, the later the region is supplied with hydrogen) for the 'concentrated users' scenario, together with the full network of hydrogen corridors. This roll-out was estimated by the above-described chronological deployment order, which was determined based on purchasing power, catchment population and cars per person in the regions, each with weights decided by the national stakeholders. Accordingly, in the later phases, the existing user centres are extended and simultaneously new user centres are developed until almost the entire area of the countries is covered.

Approximately 500 fuelling stations are required to supply the early transit road network in the ten countries (assuming one fuelling station every 80 km; multilane roads have one on each side). For the supply of all dedicated transit roads (motorways and E-roads), between 1500 and 2000 fuelling stations are required. To supply the early-user centres locally until 2015, approximately 400 additional fuelling stations are sufficient for the ten countries. The number of local fuelling stations increases with the demand increase and the regional expansion of the hydrogen supply and amounts to between 17 000 ('concentrated users' scenario) and 25 000 ('distributed users' scenario) in the timeframe 2025–2030. This corresponds to 20%–30% of all conventional fuelling stations currently existing in these countries.

Figure 14.18 shows the average specific hydrogen costs (including feedstock, production, transport and refuelling), and the cumulated investment in hydrogen infrastructure aggregated for all countries for the HyWays base case scenario.

While refuelling dominates infrastructure investments in the early phases, in the later phases it is superseded by production. The total investment of the ten countries until 2025–2030 (i.e., to reach a hydrogen vehicle penetration rate of approximately 12%) is around €60 billion. However, conventional fuels also require large investments: e.g., the IEA recently assumed that a global investment of as much as US $4300 billion will be required in the oil sector until 2030, to maintain current production levels (IEA, 2006a). Even though a direct comparison of these numbers is not valid, this may be helpful for a placement of the investment needed for hydrogen infrastructure.

The early phase of hydrogen deployment (i.e., approximately 10 000 hydrogen vehicles EU-wide) would show high specific costs of hydrogen of above 50 ct/kWh (intentionally left out in Fig. 14.18). The main reason is the under-utilisation of the production and supply infrastructure due to technology-related capacity thresholds and the overcapacity of refuelling infrastructure required for user convenience. The overall hydrogen costs during this phase are also very sensitive to the required

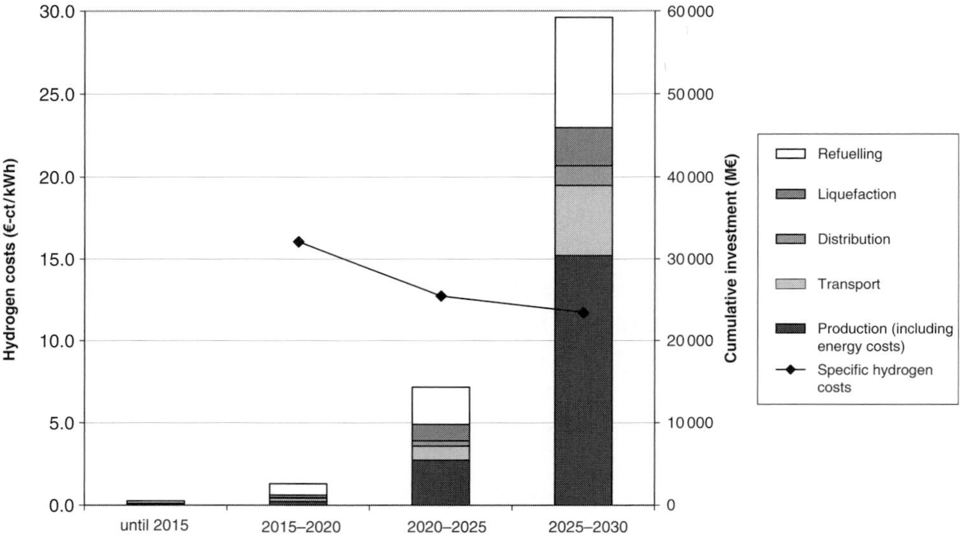

Figure 14.18. Specific hydrogen supply costs and cumulative investments in the 10 HyWays countries (base case scenario with country-specific feedstock bounds and 20% LH_2 demand) (HyWays, 2007).

number of fuelling stations; establishing an early transit corridor network *(long-distance road network)* therefore leads to a cost increase, all the more in the case of a high under-utilisation of fuelling stations. Furthermore, if liquid hydrogen is to be available in all supplied areas (e.g., 20% of the total demand), or if other large-scale production technologies are enforced by the bound setting of the country stakeholders, this will result in a further cost increase due to plant under-utilisation. Cost differences between countries can mainly be explained by the use of different feedstocks and differences in population density.

Still, the total investment at a country level for the early infrastructure until 2015 is limited to €30–120 million. Assuming approximately 1000 vehicles per country, this represents a high specific infrastructure investment per vehicle because the fuelling station utilisation is assumed to be very low. However, this is believed to be required for the initialisation of hydrogen deployment and must be overcome by adequate policy measures. Substantially higher vehicle penetration rates will level out the costs to values between 11 and 16 ct/kWh hydrogen in the medium term until 2030. When comparing these numbers to today's fuel costs, the substantially reduced consumption through improved fuel efficiency must be taken into account.

Table 14.8 summarises the specific features of hydrogen production for each of the ten countries. It must be mentioned that the different price levels are not only derived from the choice of feedstock for hydrogen production but also from regional differences in distribution costs.

Table 14.8. *Characteristics for each of the ten HyWays countries in the timeframe 2025–2030*

Country	Hydrogen demand (GWh)	Relevant feedstocks, roughly in order of declining importance	Hydrogen costs (€ct/kWh), range of scenarios
Finland	1.7	Natural gas (NG), biomass, hard coal, wind, grid electricity, nuclear	10–11
France	25.8	Nuclear, grid electricity, wind power, NG, biomass (electricity dominated)	9–11
Germany	26.1	Hard coal, biomass, wind, by-product, NG, grid electricity	8–11
Greece	4.6	Wind, biomass, lignite, NG (strong focus on domestic energy sources)	9–16[a]
Italy	17.8	Wind, biomass, NG, coal, waste, solar	10–14[a]
Netherlands	6.2	NG, hard coal, biomass, by-product (focus on central production)	10–13
Norway	1.6	Wind, biomass, by-product, grid electricity, NG (no existing NG grid)	11–12
Poland	9.6	Biomass, hard coal, lignite, NG, wind (*in-situ* coal gasification considered)	8–13
Spain	14.9	Wind, biomass, solar, hard coal, NG (high renewable share)	12–16[a]
UK	21.1	NG, coal, wind, nuclear, waste	10–13

Note:
[a]The high maximum prices mainly result from scenarios with a high share of renewables (particularly wind).
Source: (HyWays, 2007).

14.7 Global hydrogen scenarios

The extent to which hydrogen is considered to play a role in the global energy system in the future covers a broad spectrum (see Márban and Valdés-Solís, 2007; Martinot *et al.*, 2007; McDowall and Eames, 2006). In the official world energy reference scenarios of the European Commission (WETO, 2006) (until 2050) and the IEA (2006a) hydrogen is not included as an important energy carrier; hydrogen penetration is only assumed in scenarios with a strict climate policy, high oil and gas prices, and moreover, a breakthrough in the technology of fuel cells and hydrogen storage. According to the IEA (2005), under the most favourable assumptions, hydrogen fuel-cell vehicles will contribute to up to 30% of the vehicle stock by 2050 ('Scenario D'; in the following referred to as the *'low' scenario*). In the most optimistic HyWays scenario, hydrogen vehicles reach a share of about 70% by 2050 (in the following referred to as the *'high' scenario*) (HyWays, 2007). The development of the

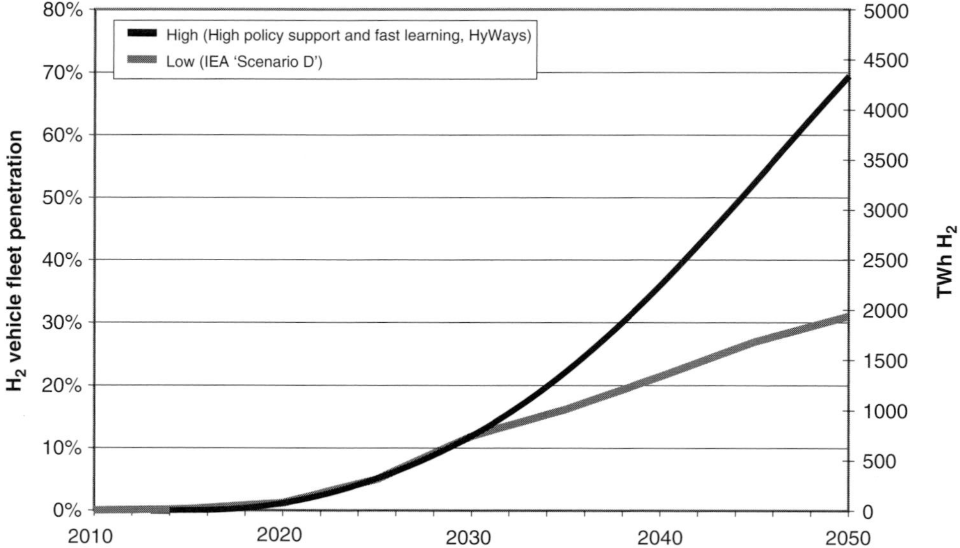

Figure 14.19. Global hydrogen penetration scenarios and related hydrogen demand.

penetration rates and the resulting hydrogen demand of these two scenarios are depicted in Fig. 14.19 (see also Fig. 14.5).

Assuming that the global vehicle stock grows from about 800 million today to 2300 million in 2050 (IEA, 2005), out of which 80% are passenger cars or light-duty vehicles with an annual hydrogen consumption of 3.5 MWh/vehicle, the *'low' scenario* results in a hydrogen demand by 2050 of around 2000 TWh (7.2 EJ) and the *'high' scenario* of 4400 TWh (15.8 EJ).[16] On the basis of a fuel consumption of 6 l/100 km of petrol engine cars and a driving range of 12 000 km per year, the *'low' scenario* would reduce the oil consumption by 7 million barrels per day by 2050, the *'high' scenario* by 16 million barrels per day.

The specific investments for implementing a complete hydrogen supply infrastructure (production plants, transport infrastructure (pipelines) and refuelling stations) vary between €530 million and €670 million/TWh until 2050 across the various scenarios analysed. Assuming an average investment of €600 million/TWh (€167 million/PJ), the necessary (cumulative) investments until 2050 for putting the required supply capacity in place would sum up to around €1200 billion in the *'low' scenario* and €2700 billion in the *'high' scenario*. Although these estimates bear considerable uncertainties and should only be interpreted as indicative orders of magnitude, they have to be reflected in the context of the overall global investments required in the energy sector,

[16] In contrast, the IEA (2005) derives for its most optimistic scenario ('Scenario D'), the *'low' scenario*, a total hydrogen demand for the transport sector of 12.4 EJ in 2050 (this is also similar to (IEA, 2006b)). In this scenario, approximately 80% of the hydrogen demand in 2050 is split into largely equal parts among Europe, North America and China.

Table 14.9. *Cumulative primary energy use for the production of hydrogen until 2050 for various feedstocks*

	Low scenario	High scenario
Natural gas	135 EJ	233 EJ
Coal	179 EJ	309 EJ
Biomass	179 EJ	309 EJ
Wind power (installed) (2050)	1 536 GW	3 439 GW
Nuclear	606 kt U	1 048 kt U

which according to the IEA (2006a) amount to just over US$20 000 billion over the period until 2030 only; of this, it is estimated that the cumulative investments required in the oil and gas sector alone will be in the order of $4300 billion and $3900 billion respectively. (The capital expenditures for all announced oil-sands projects over the period 2006 to 2015 already total US$105 billion, see Chapter 3.) The transition to hydrogen in the transport sector would represent a maximum of 1.7% of the expected GDP growth until 2050. While the investments for introducing hydrogen would add substantially to those already required for the energy system, they should not be considered an insurmountable barrier for implementing a hydrogen fuel system. Further on, to the extent that hydrogen would substitute oil-based transportation fuels, the investments foreseen in the oil sector would be diminished.

To close the loop with the resources assessments in Chapters 3 to 5, in the following, what the cumulative resource depletion or the requirement for the supply of hydrogen until 2050 would be for the above scenarios is analysed. As the global hydrogen production mix in 2050 is hard to predict, for simplicity it is assumed that hydrogen is produced entirely by one particular feedstock; this reflects the maximum impact on resources. Table 14.9 shows the cumulative primary energy use for various primary energy sources that would be required to produce the entire hydrogen demand until 2050 by one feedstock alone.

For the *'high' scenario*, in the case of natural gas, around 3.8% of the natural gas reserves would have been used up by 2050, while for hard coal the cumulative resource depletion amounts to only 1.7% of the current reserves. Assuming that the possible bioenergy potential in 2050 is in the range of 100 EJ to 400 EJ (see Chapter 5), it would either not be sufficient or almost entirely consumed for hydrogen production. As for wind power, in the *'low' scenario*, the currently installed capacity would have to increase 20-fold to meet the hydrogen demand in 2050; if hydrogen vehicles were to reach a share of 70% of the global vehicle stock by 2050, the current wind capacity would have to increase 45-fold. In the case of nuclear hydrogen supply, up to 33% of the RAR130 uranium resources, and 23% of the RAR+IR130 resources would be required. To conclude, while the impact on the depletion of fossil resources is rather marginal, a significant extension of today's

Table 14.10. *Annual primary energy use of passenger cars for various fuels*

Feedstock	Fuel	Energetic efficiency of fuel production process	Fuel consumption (MJ/100 km)	Primary energy use (GJ_{prim}/(year car))[c]
Crude oil	Gasoline	90%[a]	190	**23.3**
	Diesel	90%[a]	177	**21.6**
Oil-sands mining	Gasoline	90%[a]	190	23.3
Natural gas use[b]		–	33.4	3.7
				27.0
Oil sands *in situ*	Gasoline	90%[a]	190	23.3
Natural gas use[b]		–	68.9	7.7
				31.0
Natural gas	CNG	98%	187	**21.0**
	Hydrogen (FC)	73% (71–75)	94	**14.2**
	GTL[d]	63% (58–65)	177	**30.9**
	Methanol	63% (62–67)	200	**34.9**
Hard coal	Hydrogen (FC)	55% (51–63)	94	**18.8**
	CTL[d]	45% (35–50)	177	**43.2**
Biomass	Hydrogen (FC)	55% (50–66)	94	**18.8**
	Biodiesel	43%	177	**45.2**
	BTL[d]	40% (30–50)	177	**48.7**
	Ethanol	44% (40–60)	190	**47.5**
Grid mix	Electricity (BEV)	38%	46	**13.3**

Notes:
[a] Refining.
[b] For upgrading (hydrogen production) and/or steam generation.
[c] Annual driving range: 11 000 km.
[d] FT diesel.

deployment of renewable energies will be required to be able to meet the hydrogen demand by 2050 to a large extent.

Table 14.10 shows, for selected fuel options, the energetic efficiency of the fuel manufacturing process,[17] the average fuel consumption per 100 km and the annual primary energy use per passenger car. Hydrogen shows the lowest annual primary energy use, with the exception of electricity in battery-electric vehicles, assuming it is used in fuel-cell vehicles. Compared with synthetic fuels, this is true on the one hand, because of the 'shorter' process chain, because the complex syngas preparation step is eliminated, and, on the other hand, because the Fischer–Tropsch process itself

[17] The energetic efficiency of the process refers to the energy input required to produce one unit of energy of fuel output. While this approach is rather simplistic, as it does not take into account upstream emissions, for instance from crude oil production or biomass supply, the general conclusions do not change.

requires hydrogen for hydrogenation and hydro-cracking. Because of the more favourable hydrogen-to-carbon ratio, the highest efficiencies for the production of hydrogen as well as Fischer–Tropsch fuels are achieved by the use of natural gas; depending on the chemical composition, either coal and biomass follow next. Even if hydrogen is produced by steam reforming of natural gas, a lower energy use results than when using the natural gas directly in a combustion engine. Biofuels are essentially among the fuels with the highest primary energy use for a given driving range. It must be pointed out, however, that the advantage of hydrogen in terms of primary energy use relative to other alternative fuels largely disappears if the hydrogen is used in internal combustion engines.

14.8 Summary and conclusions

The need for a sustainable energy supply is becoming more pressing in the light of declining fossil energy resources, environmental pollution and climate change. In this context, controversial discussions about hydrogen as an energy carrier are taking place more and more often, with hydrogen infrastructure build-up taking a vital role.

The importance of a hydrogen infrastructure analysis has been outlined. The MOREHyS model shows a novel modelling approach to assess the geographical and temporal set-up of an infrastructure for a hydrogen-based transport system in any given geographical context and to analyse the effects on the corresponding national energy system. The approach introduces hydrogen infrastructure technologies into a linear optimisation model of the energy system and combines this with a geographical information system. By modelling the interaction between the hydrogen supply and the electricity sector, an integrated system analysis can be performed, so that an optimal supply structure for both systems can be derived.

The exact hydrogen supply infrastructure build-up depends strongly on regional or country-specific particularities, such as available feedstocks (like renewable energies), population density, geographical factors and policy support, and must therefore be assessed individually on a country-by-country basis. But there are some cross-national communalities and robust results that can be derived from the hydrogen infrastructure analysis presented in this chapter, which are summarised in the following. It has to be pointed out that these results are based on the assumption that the technical and economic challenges with fuel cells and hydrogen storage will be overcome and that fuel-cell vehicles will be cost-competitive with conventional vehicles.

14.8.1 General conclusions and strategies for the roll-out of a hydrogen infrastructure

- For the introduction of hydrogen, two broad phases can be distinguished: the *infrastructure build-up phase* (2015–2030) and the *hydrogen diffusion phase* (2030–2050). The former can further be subdivided into the *early implementation phase* (2015–2020) and the *transition phase* (2020–2030). Scenario results generally show the greatest differences during the transition phase (mainly depending on the geographical demand distribution and the number and utilisation of filling station), when the initial infrastructure is being developed, but converge in later periods.

- As buses and fleet vehicles operate locally to a large extent and return to a central depot for refuelling and maintenance, they are advantageous during the early implementation phase. Hydrogen use will take off predominantly in densely populated areas or urban environments with favourable political commitment and, during the transition phase, gradually expand towards rural areas. Initially, government support may be necessary, such as funding for pilot and demonstration projects, changes to fuel and vehicle taxation or establishment of low-emission zones in urban centres. A gradual build-up of the infrastructure with an initial concentration in agreed user centres efficiently diminishes the often cited chicken-and-egg problem.
- The introduction of hydrogen in areas with high population density minimises infrastructure costs, but might lead to less optimal hydrogen penetration rates. It also leads earlier to economies of scale in hydrogen production and transportation and its leads to the earlier development of a pipeline network. In addition, a higher utilisation of fuelling stations can be reached in densely populated areas compared with less densely populated areas with the same number of fuelling stations.
- Onsite hydrogen production at the fuelling station (from natural gas) is the preferred option during the early implementation phase (first decade) and also onwards in areas where there is too sparse demand for more centralised schemes (in later periods). With a distribution infrastructure gradually building up and demand for hydrogen rising, onsite production becomes less preferable and there is a trend towards centralised production in later phases; this trend is also supported by the fuel switch from natural gas to coal. During the hydrogen diffusion phase, large-scale central production plants dominate hydrogen supply, mainly on the basis of coal (where available).
- Carbon capture and storage is the key to avoiding an overall increase of CO_2 emissions through fossil hydrogen production, primarily from coal. Except for biomass, renewable hydrogen is only an economic choice under certain circumstances, unless renewable targets are set.
- Hydrogen transport is expensive and should be logistically optimised. Owing to the dominating onsite production, there is little hydrogen transport during the first decade, if so mainly by liquid hydrogen trailers, but also under specific circumstances by compressed gaseous trailers. Liquid hydrogen plays an important role during the transition phase and also to connect outlying areas, at the same time as a pipeline network is being constructed. During the diffusion phase, pipeline transport clearly dominates.
- To allow a sufficient geographical coverage of fuelling stations for user acceptance during the early phase an overcapacity of the supply and refuelling infrastructure (both in terms of number and capacity) is necessary. Generally, between 10% and 30% of existing fuelling stations should be equipped with hydrogen to provide sufficient geographical coverage and achieve user acceptance.

14.8.2 Hydrogen production mix

- The hydrogen production mix is very sensitive to the country-specific context and strongly influenced by the assumed feedstock prices; resource availability and policy support also play a role, in particular for hydrogen from renewable and nuclear energy.

- During the early phase of infrastructure build-up (the first decade), onsite production of hydrogen by small-scale natural gas reforming is clearly the preferred technology (using the existing natural gas pipeline network); to a lesser extent, depending on country-specific conditions, onsite electrolysis and by-product hydrogen are also used. Such a transition, emphasising distributed hydrogen production, better matches the development of hydrogen demand and thus avoids high initial investments in underused large-scale production and distribution facilities.
- The size, number and utilisation of fuelling stations have an impact on the production mix, transport options and infrastructure cost. Larger and well utilised fuelling stations in densely populated areas could be supplied via pipeline. Therefore, central production plays a key role in the production mix here. Small and well utilised fuelling stations are preferably supplied by onsite production, but also mid-sized and large fuelling stations could be supplied by onsite production, if no space limitations are given and the utilisation justifies the investments in onsite production technologies. In these cases, onsite SMR dominates the production mix. However, if the utilisation is lower it might be better to supply the fuelling stations with liquid hydrogen via trucks, because this allows economies of scale in central production to be generated, and saves local investments, which might be under-utilised. This concept seems to be interesting especially for the transition phase, when utilisation might be low in the beginning because of the necessity of a widespread fuelling station network to satisfy customer needs.
- Practical problems may hinder the application of onsite technology. For instance, in densely populated areas, the space requirement can be disadvantageous and onsite SMR might not be an option for fuelling stations with very low initial utilisation, owing to its limited part-load ability and very high specific investments compared with central plants with higher utilisation.
- As hydrogen demand grows and natural gas becomes more expensive, the share of onsite production diminishes and more large-scale central production plants are built, which can also increasingly rely on a growing transport infrastructure (particularly pipeline network) put in place for distribution to the fuelling stations. The fuel switch to coal in later phases, which is only viable with large production plants, further triggers a more centralised production scheme.
- The fossil hydrogen production option dominates during the first two decades while the infrastructure is being developed, and also in later periods if only economic criteria are applied: initially on the basis of natural gas, later with increasing gas prices more and more on the basis of coal (where available). Carbon capture and storage will be critical for these pathways, if hydrogen is to contribute to an overall CO_2 reduction in the transport sector. The production mix between gas and coal is highly sensitive to the ratio of feedstock prices; a switch occurs at a gas:coal price ratio of about 2.5.
- Until around 2030, steam reforming of natural gas plays a role for central production (with CCS), but in the long term this option becomes less attractive owing to the assumed increase of gas prices.
- Hard coal (or lignite) gasification is only economic in large-scale central plants. Owing to low initial demand, this option is restricted to the later phases. Coal gasification is also triggered by a relative increase in natural gas prices. The use of coal heavily depends on the large-scale deployment of CCS, if hydrogen is to be supplied without contributing to a significant rise in CO_2 emissions.

- Hydrogen occurring as a by-product of the chemical industry (mainly chlorine–alkali electrolysis), which is being used thermally today is a (cheap) option (where available), especially for the initial phase because it can be substituted by natural gas. However, investments in purification might be needed. This hydrogen is mainly used where user centres are nearby. It will also contribute to a certain extent during later phases, but with a lower share, owing to its limited potential.
- Renewable hydrogen is mainly an economic option in countries with a large renewable resource base or a lack of fossil resources, for remote and sparsely populated areas (such as islands) or if surplus electricity from intermittent renewable energies must be stored. Otherwise, renewable hydrogen needs to be incentivised or mandated.
- If renewable targets are set, biomass gasification is the cheapest renewable hydrogen supply option; however biomass has restricted potential and competition of end-use, for instance, with other biofuels or stationary heat and power generation. Biomass gasification is applied in small decentral plants during the early phase of an infrastructure roll-out and in central plants in later periods.
- Wind energy (on- and offshore) can, in the early phases, be transmitted through the electric grid and utilised for onsite hydrogen production. In later phases, beyond 2030, it may also be utilised directly for large-scale electrolysis, making use of hydrogen as a storage medium for surplus wind power and furthermore using the electrolyser capacity for wind power regulation. The use of offshore wind energy for hydrogen production generally results in higher hydrogen transport costs, owing to the inherent geographical concentration of the offshore wind potentials, while on the other hand it relieves the electric grid, which could result in savings and higher transmission efficiency. But in the long run, the electricity grid might also be extended, so that onsite electrolysis could be an option beyond 2030 as well. The specific costs of hydrogen production cannot compete with most other options in the period until 2030 under the assumptions made here, while the picture might change with further depletion and price increase of fossil fuels until 2050.
- Nuclear power plants dedicated to hydrogen production result in high infrastructure costs due to large-scale, central production, and hence the need for long-distance transport and distribution. Nuclear power plants are rather an option for later phases with high hydrogen demand. New nuclear technology, such as nuclear thermocycles (for instance the sulphur–iodine cycle) might be an interesting option for the future as well. However, nuclear hydrogen production is likely to face the same public acceptance concerns as nuclear power generation.
- Thermochemical cycles based on solar energy are another long-term option for hydrogen production in countries with favourable climatic conditions.
- Hydrogen production costs are between 8 and 12 ct/kWh during the market introduction phase until around 2030 and between 6 and 14 ct/kWh in the later phase until 2050, mostly depending on the energy feedstock and energy price developments. Feedstock prices and CO_2 costs account for about two-thirds of production costs under the assumed energy price scenarios. If hydrogen were to be supplied by renewable energies, production costs would almost double.

14.8.3 Hydrogen transport options

- The choice of transport option, basically pipeline and liquid hydrogen trailer, depends on transport volumes and distances, as well as on liquefaction energy costs. For large transport volumes and long distances pipelines are exclusively used. Also, with smaller volumes and shorter distances, pipeline distribution is more economic than LH_2 trailers, because of their lower capital intensity at shorter delivery distances. For long distances and dispersed hydrogen demand, such as along motorways or in rural areas, LH_2 trailers are more economic than pipelines. Compressed gaseous hydrogen trailer transport is only economic in situations where very limited demand must be supplied. What makes this option additionally expensive is that extra CGH_2 trailers have to be placed at the fuelling stations as hydrogen storage tanks.

- During the early infrastructure phase until 2020, only minor amounts of hydrogen are transported; mainly in liquid form, and to a lesser extent in gaseous (CGH_2) trailers. Liquid hydrogen trailers mainly play a role during the transition phase. They are mostly used to supply small and medium fuelling stations, both outlying ones (e.g., along motorways) and in centres where onsite production is not a preferable option, because of space restrictions.
- After 2030, with a fast growth in hydrogen demand, pipelines are increasingly the prime choice for hydrogen transport and distribution, all the more as an initial pipeline infrastructure is already in place. In less-populated and remote areas, LH_2 trailers as well as onsite production (depending on feedstock price) remain the most economic option, even in later phases.
- Supplying hydrogen from central plants yields economies of scale, but generally implies higher transport costs, as, e.g., in the case of nuclear production plants. A high share of hydrogen from geographically concentrated renewable energy sources (such as offshore wind) may also result in significantly increased transport costs.
- Small-scale liquefaction is not economically viable. Liquefied hydrogen is only economic with large-scale plants and dispersed hydrogen demand and thus competes with onsite production for outlying fuelling stations with smaller demand. If gaseous hydrogen is needed, LH_2 can be vaporised. Liquefaction is a very capital and energy-intensive process and, with 30%–60% of liquefaction costs coming from electricity, its cost competitiveness is also very sensitive to electricity prices. Depending on the electricity source, LH_2 may also bear a considerable CO_2 footprint. In the very early phase, the capacity of existing liquefiers (where available) may be sufficient.
- Pipeline delivery implies high investments, which are proportional to the delivery distance, but low variable costs. This makes pipelines economically attractive for relatively short distances (<10 km). Pipeline distribution is mainly an option for densely populated areas and larger fuelling stations, as the higher utilisation rate leads to lower transport costs. This indicates that they will become more attractive as hydrogen penetration advances. As pipelines can carry large volumes, they are also the preferred option for long distances and high throughput. A positive side effect may be that their intrinsic storage capacity may facilitate the use of intermittent renewable energy sources.
- The choices of transport options are exposed to many sensitivities (e.g., transport distances, volumes, fuelling station utilisation, cars served, demand for LH_2, energy prices, density of fuelling stations in a region) and the conclusions drawn here should not be regarded as an 'ultimate strategy'. The results show that each of the transport options may play a role under specific conditions. The distance to be covered has the strongest impact on transport costs, which have a much larger impact on the total supply costs of hydrogen than is the case for today's liquid fuels. The supply of hydrogen should thus be close to the user centres, as transport over large distances will increase costs. The primary optimisation goal should,

therefore, be to minimise the average hydrogen transport distances through well planned and distributed siting of the production plants.

- Hydrogen transport costs are in the range of 1 to 4 ct/kWh, with a downward trend in later periods due to the economies of scale achieved by large scale pipeline transport.

14.8.4 Hydrogen supply costs

- Hydrogen supply costs are extremely sensitive to the underlying assumptions about the development of feedstock prices, as these have a decisive impact on production costs. With around 12–14 ct/kWh (€4–€4.6/kg) the specific hydrogen supply costs in the early phase are high, owing to the required overcapacity of the supply and refuelling infrastructure and the higher initial costs for new technologies because of the early phase of technology learning. Around 2030, hydrogen costs range from 10 to 16 ct/kWh ($3.6–$5.3/kg), mainly depending on the feedstock. In the long-term, until 2050, hydrogen supply costs stabilise around this level, however, with an upward trend due to the assumed increase in energy prices and CO_2 certificate prices. Total supply costs remain relatively unchanged when switching from natural gas to coal, in consequence of an increase in gas prices in later periods.
- With a share of 60%–80%, hydrogen production dominates total supply costs. The installation of refuelling stations makes up about 10%. The rest is for transport (including liquefaction).
- At these supply costs, hydrogen becomes competitive in the long run at crude oil prices of beyond $80–$100/barrel (no taxes, no vehicle costs included).[18]
- The specific investments for implementing a complete hydrogen supply infrastructure (production plants, transport infrastructure (pipelines) and refuelling stations) vary between €530 million and €670 million/TWh until 2050 across the various scenarios, with a tendency towards higher numbers in later periods due to higher feedstock prices. Assuming an average of €600 million/TWh (€167 million/PJ), cumulative investments to put the supply capacity in place for reaching a share of 30% (70%) until 2050 of the total car fleet in the EU would amount to roughly €300 billion (€650 billion), globally to €1200 billion (€2700 billion).
- To facilitate the deployment of hydrogen, hydrogen supply along an early road network may be required, but this keeps the total initial investment in infrastructure comparatively small. The risk of high initial investments in a hydrogen infrastructure that may lead to under-used capital can be mitigated by a gradual local build-up.
- It must be pointed out that the energy price scenarios and economic figures for hydrogen technologies, which form the basis for the above estimation of infrastructure build-up costs, date back mostly to the years 2005 and 2006. The drastic increase in energy, as well as equipment, steel and metal, prices experienced in recent years is not reflected in the hydrogen supply costs. For instance, technologies with a high share of feedstock costs in total production costs are very much affected; other great uncertainties exist around the true costs for CCS, which generally tend to be underestimated. Hence, the absolute economic figures presented here have to be taken with caution, as they represent a rather optimistic estimate, assuming that the recent trend of high prices will be sustained for a prolonged period.

[18] This calculation is based on hydrogen fuel-cell cars with a hydrogen consumption of 26 kWh/100 km, gasoline or diesel cars with a consumption of 5 l/100 km and a refinery efficiency of 90%. In addition, $–€ parity is assumed.

14.8.5 Impact of hydrogen supply on CO_2 emissions

- Until 2030, the specific CO_2 emissions over the entire hydrogen supply chain (production, liquefaction, trailer transport, refuelling) are in the range of 60–80 g CO_2/km (compared with 160 g/km for today's gasoline or diesel vehicles). After 2030, with an increasing shift to hydrogen production from coal, *without CCS and renewable targets*, CO_2 emissions generally increase to between 120 and 140 g/km. With 50% renewable hydrogen, the specific emissions could decrease to less than 90 g CO_2/km; adding CCS could further bring them down to as low as 30 g CO_2/km.
- As for hydrogen ICE vehicles, specific emissions are approximately 80% higher, owing to their lower efficiency compared with fuel-cell drives, their application should be limited to the transition phase only.
- Generally, close to 90% of total CO_2 emissions from hydrogen supply result from hydrogen production. In scenarios with a high share of liquid hydrogen transport, hydrogen liquefaction – depending on the electricity mix – could contribute significantly to the total emissions (up to 40 g CO_2/km for an average CO_2 intensity of the electricity mix of 550 g CO_2/kWh).

14.8.6 Integration of hydrogen with the electricity sector (see also Chapter 16)

- In the long run, the co-production of electricity and hydrogen in IGCC plants (with CCS) is a promising option, especially with high gas prices and restrictive CO_2 regimes. Depending on the electricity price (load curve), either electricity or hydrogen can be produced by the IGCC.
- Under strong CO_2 restrictions, renewable energies should be used in the electricity sector to substitute fossil power plants. However, if renewable electricity has to be stored, it could make sense to use this energy for hydrogen production.

References

Almansoori, A. (2006). Design and operation of a future hydrogen supply chain. *Chemical Engineering Research & Design*, **84** (A6), 423–438.

Anderson, D. (1972). Models for determining least-cost investments in electricity supply. *Bell Journal of Economics*, **3** (1), 267–299.

Ball, M. (2006). *Integration einer Wasserstoffwirtschaft in ein nationales Energiesystem am Beispiel Deutschlands*. Dissertation, VDI Fortschritt-Berichte Reihe **16**, No. 177. Düsseldorf: VDI Verlag.

Ball, M., Wietschel, M. and Rentz, O. (2006). Integration of a hydrogen economy into the German energy system: an optimising modelling approach. *International Journal of Hydrogen Energy*, **32** (10–11), 1355–1368.

Caramanis, M. C. (1983). *Electricity Generation Expansion Planning in the Eighties: Requirements and Available Analysis Tools, Energy Models and Studies*. North-Holland Publishing Company, pp. 541–562.

CASCADE MINTS (2005a). *Summary and Objectives*. National Technical University of Athens (NTUA). www.e3mlab.ntua.gr/cascade.html.

CASCADE MINTS (2005b). *Case study comparisons and development of energy models for integrated technology systems*. National Technical University of Athens (NTUA). www.e3mlab.ntua.gr/cascade.html.

Cremer, C. (2005). *Integrating Regional Aspects in Modeling of Electricity Generation – the Example of CO_2 Capture and Storage*. Dissertation No. 16119. ETH Zurich.

DWV (Deutscher Wasserstoff- und Brennstoffzellen-Verband (German Hydrogen and Fuel Cell Association)) (2005). *H_2-Roadmap*. Berlin: DWV. www.dwv-info.de.

Enzensberger, N. (2003). *Entwicklung und Anwendung eines Strom- und Zertifikatmarktmodells für den Europäischen Energiesektor*. Dissertation, VDI Fortschritt-Berichte Reihe 16, No. 159. Düsseldorf: VDI Verlag.

Fischedick, M., Esken, A., Luhmann, H. J., Schüwer, D. and Supersberger, N. (2007). *CO_2 Capture and Geological Storage as a Climate Policy Option. Technologies, Concepts, Perspectives*. Wuppertal Spezial 35e. Wuppertal: Wuppertal Institute for Climate, Environment and Energy.

Forsberg, P. and Karlström, M. (2006). On optimal investment strategies for a hydrogen filling station. *International Journal of Hydrogen Energy*, **32** (5), 647–660.

FORUM (Forum für Energiemodelle und Energiewirtschaftliche Systemanalysen) (1999). *Energiemodelle zum Klimaschutz in Deutschland: Strukturelle und gesamtwirtschaftliche Auswirkungen aus nationaler Perspektive*. Heidelberg: Physica-Verlag.

Gately (1970). *Investment Planning for the Electric Power Industry: An Integer Programming Approach*. Research Report 7035. Department of Economics, University of Western Ontario.

Gielen, D. and Taylor, M. (2007). Modelling industrial energy use: the IEA's energy technology perspectives. *Energy Economics*, **29** (4), 889–912.

Gonzalez-Monroy, L. and Cordoba, A. (2002). Financial costs and environment impact optimisation of the energy supply systems. *International Journal of Energy Research*, **26** (1), 27–44.

Greene, D. L. (1998). Survey evidence on the importance of fuel availability to the alternative fuels and vehicles. *Energy Studies Review*, **8** (3), 215–231.

Greene, D. L. (2005). *HyTrans Model Development*. FY 2005 Progress Report. DOE Hydrogen Program. Knoxville: Oak Ridge National Laboratory, National Transportation Research Center. www.hydrogen.energy.gov/pdfs/progress05/iii_3_greene.pdf.

Hafkamp, W. A. (1984). *Economic-Environmental Modelling in a National-Regional System*. New York: North-Holland.

Hart, D. (E4tech) (2005). *The Economics of a European Hydrogen Automotive Infrastructure*. A study for Linde AG. www.hydrogenday.de/International/Web/Hydrogenday2005.nsf/e4tech%20hydrogen%20study.pdf.

HFP (European Hydrogen and Fuel Cell Technology Platform) (2007). *Implementation Plan, Status 2006*. www.hfpeurope.org.

Höllermann, T. (2004). *Märkte für Wasserstofffahrzeuge in Europa*. Diploma thesis. FhG ISI. Karlsruhe.

Hugo, A., Rutter, P., Pistikopoulos, S., Amorelli, A. and Zoia, G. (2005). Hydrogen infrastructure strategic planning using multi-objective optimisation. *International Journal of Hydrogen Energy*, **30** (15), 1523–1534.

HyWays (2007). *Hydrogen Energy in Europe. Integrated Project under the 6th Framework Programme of the European Commission to Develop the European Hydrogen Energy Roadmap*. Contract No. 502596, 2004–2007. www.hyways.de.

H2A (2008). *US Department of Energy – Hydrogen Program*. www.hydrogen.energy.gov/h2a_analysis.html.

IAEA (International Atomic Energy Agency) (1995). *Computer Tools for Comparative Assessment of Electricity Generation Options and Strategies*. DECADES Project Series Publication. Vienna.

IEA (International Energy Agency) (2005). *Prospects for Hydrogen and Fuel Cells*. IEA Energy Technology Analysis Series. Paris: OECD/IEA.

IEA (International Energy Agency) (2006a). *World Energy Outlook 2006*. Paris: OECD/IEA.

IEA (International Energy Agency) (2006b). *Energy Technology Perspectives 2006. Scenarios and Strategies to 2050*. Paris: OECD/IEA.

JRC (Joint Research Centre, EUCAR, CONCAWE) (2007). *Well-to-Wheels Analysis of Future Automotive Fuels and Powertrains in the European Context; Well-to-Wheels Report*, Version 2c. http://ies.jrc.ec.europa.eu/wtw.html.

Joffe, D. Strachan, N. and Balta-Ozkan, N. (2007). *Representation of Hydrogen in the UK, US and Netherlands MARKAL Energy Systems Models*. UK Sustainable Hydrogen Energy Consortium (UKSHEC), Social Science Working Paper No. 29. London: Policy Studies Institute.

Karlsson, K. and Meibom, P. (2008). Optimal investment paths for future renewable based energy systems – using the optimisation model Balmorel. *International Journal of Hydrogen Energy*, **33** (7), 1777–1787.

Kienzle, S. (2005). *Einbindung einer Wasserstoffinfrastruktur in das Energiesystemmodell Balmorel. Optionen der Bereitstellung von Wasserstoff als Kraftstoff für den deutschen Straßenverkehr*. Diploma Thesis. University of Karlsruhe.

Krzyzanowski, D., Kypreos, S. and Barreto, L. (2008). Supporting hydrogen based transportation: case studies with global MARKAL model. *Computational Management Science*, **5** (3), 207–231.

Leiby, P. N., Greene, D. L., Bowman, D. and Tworek, E. (2006). Systems analysis of hydrogen transition with HyTrans. *Transportation Research Record* (1983), 129–139. Transportation Research Board of the National Academies.

Lev, B. (1983). *Energy Models and Studies*. New York: North-Holland.

Lin, D. Z., Ogden, J., Fan Y. and Sperling, D. (2006). *The Hydrogen Infrastructure Transition (HIT) Model and its Application in Optimising a 50-Year Hydrogen Infrastructure for Urban Beijing*. Research Report UCD-ITS-RR-06–05. Davis: Institute of Transportation Studies, University of California.

Márban, G. and Valdés-Solís, T. (2007). Towards the hydrogen economy? *International Journal of Hydrogen Energy*, **32** (12), 1625–1637.

Martinot, E., Dienst, C., Weiliang, L. and Qimin, C. (2007). Renewable energy futures: targets, scenarios, and pathways. *Annual Review of Environment and Resources*, **32**, 205–239.

McDowall, W. and Eames, M. (2006). Forecasts, scenarios, visions, backcasts and roadmaps to the hydrogen economy: A review of the hydrogen futures literature. *Energy Policy*, **34** (11), 1236–1250.

Melaina, M. W. (2003). Initiating hydrogen infrastructures: preliminary analysis of a sufficient number of initial hydrogen stations in the US. *International Journal of Hydrogen Energy*, **28** (7), 743–755.

Mintz, M., Gilette, J., Elgowainy, A. *et al.* (2006). HDSAM: hydrogen delivery scenario analysis model to analyse hydrogen distribution options. *Transportation Research Record* (1983), 114–120. Transportation Research Board of the National Academies.

Nouweland, A., Borm, P., Brouwers, W., Bruinderink, R. and Tijs, S. (1996). A game theoretic approach to problems in telecommunication. *Management Science*, **42** (2).

Ogden, J. M. (1999). Prospects for building a hydrogen energy infrastructure. *Annual Review of Energy and the Environment*, **24**, 227–279.

Ogden, J. (2006). High hopes for hydrogen. *Scientific American*, **295** (3).

Ogden, J. M., Yang, C. and Johnson, N. (2005). *Technical and Economic Assessment of Transition Strategies Towards Widespread Use of Hydrogen as Energy Carrier*. Report to the United States Department of Energy. Hydrogen, Fuel Cells and Infrastructure Technologies Program, for Phase I of NREL. Technical Report No. UCD-ITS-RR-05–13. University of California Davis, Institute of Transportation Studies.

Parks, K., Milbrandt, A. and Davies, K. (2006). *Energy Systems Analysis: HyDS Modeling Environment*. FY 2006 DOE Hydrogen Program Progress Report, Washington, DC. www.hydrogen.energy.gov/analysis_repository/project.cfm/PID=100.

Prekopa, A. (1995). *Stochastic Programming*. Dordrecht: Kluwer.

Prognos/EWI (2005). *Energiereport IV. Die Entwicklung der Energiemärkte bis zum Jahr 2030. Energiewirtschaftliche Referenzprognose*. Munich: Oldenbourg Industrieverlag.

Roads2HyCom (2007). *European Hydrogen Infrastructure Atlas and Industrial Excess Hydrogen Analysis*, ed. Steinberger-Wilckens, R. and Trümper, S. C. Roads2HyCom. www.roads2hy.com.

Ruth, M. (2007). *MacroSystem Model Overview*. National Renewable Energy Laboratory, US DOE Hydrogen Program. www.hydrogen.energy.gov/macro_system_model.html.

Schindler, J. (2008). *E3database – An Introduction into the Life-Cycle Analysis Tool*. Ottobrunn: Ludwig-Bölkow-Systemtechnik GmbH. www.e3database.com.

Schwoon, M. (2006). *A Tool to Optimise the Initial Distribution of Hydrogen Filling Stations*. Working Paper No. 110. Hamburg: Forschungsstelle für Nachhaltige Umweltentwicklung (FNU).

Shell (2004). *Shell Pkw-Szenarien bis 2030*. Shell Deutschland Oil. www.shell.de.

Smit, R., Weeda, M. and de Groot, A. (2007). Hydrogen infrastructure development in the Netherlands. *International Journal of Hydrogen Energy*, **32** (10–11), 1387–1395.

Smith, A. (2001). *Cleaner Vehicles in Cities. Guidelines for Local Governments*. UTOPIA Project, European Commission, Transport RTD Programme, Contract UR-97-SC-2076.

Song, Y. H. (1999). *Modern Optimisation Techniques in Power Systems*. Dordrecht: Kluwer.

Stephan, C. and Sullivan, J. (2004). An agent-based hydrogen vehicle/infrastructure model. *Evolutionary Computation (CEC) 2004*, **2**, 1774–1779.

Stiller, C. (2008). H2INVEST – Hydrogen infrastructure venture support tool. www.h2invest.com.

Stiller, C., Seydel, P., Bünger, U. and Wietschel, M. (2007). Assessment of the regional hydrogen demand and infrastructure build-up for 10 European countries. www.hyways.de.

Tolley, G. (2005). *Analysis of the Hydrogen Production and Delivery Infrastructure as a Complex Adaptive System*. www.hydrogen.energy.gov/pdfs/review05/

anp_4_tolley.pdf; www.dis.anl.gov/projects/hydrogen_transition_modeling.html.

TREMOVE (2007). *A Policy Assessment Model to Study the Effects of Different Transport and Environment Policies on the Transport Sector for all European Countries*. Version v2.52. www.tremove.org.

Tseng, P., Lee, J. and Friley, P. (2003). *Hydrogen Economy: Opportunities and Challenges*. Washington: Office of Energy Efficiency and Renewable Energy, US Department of Energy. Upton, NY: Brookhaven National Laboratory.

Tzimas, E., Castello, P. and Peteves, S. (2007). The evolution of size and cost of a hydrogen delivery infrastructure in Europe in the medium and long term. *International Journal of Hydrogen Energy*, **32** (10–11), 1369–1380.

Uyterlinde, M. A., Martinus, G. H. and van Thuijl, E. (2004). *Energy Trends for Europe in a Global Perspective. Baseline Projections by Twelve E3-Models in the CASCADE MINTS Project*. ECN Report C–04–094. www.energytransition.info/cascade_mints.

Van Benthem, A. A., Kramer, G. J. and Ramer, R. (2006). An options approach to investment in a hydrogen infrastructure. *Energy Policy*, **34** (17), 2949–2963.

Wang, M. (2008). Overview of GREET model development at Argonne. *GREET user workshop, Sacramento (CA), March 18*. Center for Transportation Research, Argonne National Laboratory. www.transportation.anl.gov/modeling_simulation/GREET/index.html.

Welch, C. (2007). HyDIVE (Hydrogen Dynamic Infrastructure and Vehicle Evolution) model analysis. *Hydrogen Analysis Workshop, August 9–10*. Washington: National Renewable Energy Laboratory (NREL). www1.eere.energy.gov/hydrogenandfuelcells/analysis/pdfs/welch_hydive.pdf.

WETO (2006). *World Energy Technology Outlook 2050*. WETO H_2 Report EUR 22038. Brussels: European Commission, DG Research.

Wietschel, M. (2000). *Produktion und Energie: Planung und Steuerung industrieller Energie- und Stoffströme*. Frankfurt a. M.: Habilitation, Technical University Karlsruhe.

Yang, C., Nicholas, M. A. and Ogden, J. (2006). Comparison of idealized and real-world city station sitting models for hydrogen distribution. UCD-ITS-RP-06–09, Institute of Transportation Studies, University of California, Davis. In *Proceedings of the National Hydrogen Association (NHA) Annual Conference*. Long Beach, California.

Yang, C. and Ogden, J. (2007). Determining the lowest-cost hydrogen delivery mode. *International Journal of Hydrogen Energy*, **32** (2), 268–286.

15

Building a hydrogen infrastructure in the USA

Joan Ogden and Christopher Yang

While the previous chapter addressed the build-up of a hydrogen infrastructure in Europe, this chapter focuses on implementing a hydrogen infrastructure in the USA, where, over the last decade, the vision of hydrogen as a future transportation fuel has gained remarkable momentum.

15.1 Introduction – transportation-energy context in the USA

A large part of primary energy use, greenhouse-gas emissions and air pollution in the United States comes from direct combustion of fuels for transportation and heating. Reducing emissions and energy use from this multitude of dispersed sources (250 million vehicles and perhaps 100 million households and businesses) will mean replacing today's cars and heating systems with higher efficiency, low-emission models, and, ultimately, adopting new fuels that can be produced cleanly and efficiently from diverse sources. This is particularly crucial for transportation, where the number of light-duty passenger vehicles could grow 50% by 2050, and where 97% of fuel comes from petroleum, 60% of which is imported into the United States.

A variety of alternative fuels, including LPG, CNG, ethanol, methanol, as well as electricity, have been implemented on a small scale in the USA, but with limited success – the total number of alternative fuelled vehicles remains less than 1% of the total fleet (Davis and Diegel, 2007). The largest alternative fuel used in the USA is ethanol derived from corn, which is currently blended with gasoline up to 10% by volume in some regions, and accounts for 3% of US transportation energy use.

The context for alternative fuels is rapidly changing in the United States, driven by growing interest in low-carbon, domestically produced fuels and zero-emission vehicles. A host of policy initiatives nationally and in a number of states, such as California, are driving towards lower carbon fuels and zero-emission vehicles (Table 15.1). Among the various alternative fuels, hydrogen stands out as offering zero tailpipe emissions, good performance, fast refuelling time, high efficiency when

The Hydrogen Economy: Opportunities and Challenges, ed. Michael Ball and Martin Wietschel. Published by Cambridge University Press.

Table 15.1. *Policy measures in the United States encouraging hydrogen vehicles*

Vehicle efficiency and emissions policies
- Zero emission vehicle (ZEV) mandate (California and north-eastern states)
  - 250 FCVs by 2008; 2500 by 2011, 25 000 by 2014 in California
  - North-eastern states have followed California's lead
- California AB1493: 30% GHG reduction for new vehicles by 2016
- Future Corporate Average Fuel Economy (CAFE) standards (35 mpg by 2020)

Vehicle-fuel policies
- California Low Carbon Fuel Standard (reduce carbon intensity of fuel by 10% by 2020)
- California SB1505: clean H_2 (hydrogen for transportation must be produced with low GHG emissions)
- California AB32: cap and trade GHG emissions

Hydrogen-specific policies
- H_2 station and vehicle funding incentives (local, state, federal); H_2 programmes in 30 states

used in a fuel cell and the possibility of production from a range of widely available zero carbon resources.

Among the countries that may be the first to implement hydrogen and fuel-cell vehicle technologies, the United States has a unique energy context for a number of reasons. The USA has a lower population density and land-use patterns that result in higher vehicle ownership rates and significantly greater miles driven per vehicle annually. The lower driving range of fuel-cell vehicles could be an impediment to consumer acceptance of these vehicles in the USA, given these driving patterns and current consumer expectations. Another challenge for implementation of alternative fuels, such as hydrogen, is the historically low fuel price in the United States, as compared with many of the industrialised nations of the world. One of the results of the US's low fuel prices is that vehicles typically have lower fuel economy as compared with vehicles in Europe or Japan. The challenge and the opportunity this provides for fuel-cell vehicles is whether they can offer significant increases in fuel economy while maintaining the level of performance and power that US consumers have come to expect. As a result, robust policies that can help enable a transition to sustainable vehicle and fuel technologies are critical in the USA, where fuel prices alone have not been able to motivate consumers and industry adequately.

15.2 Hydrogen-energy policy in the USA

In the past decade, the vision of a hydrogen-fuelled future has gained remarkable momentum with US policymakers and industry. The US Federal government pledged its support for hydrogen research, development and implementation when it announced a five-year, $1.2 billion programme that included the FreedomCar and

Hydrogen, Fuel Cell and Infrastructure Technologies programmes, managed by the Department of Energy (USDOE, 2006).

Along with the Federal government, over 30 US states are developing regional 'roadmaps' or 'hydrogen highways', committing over a billion dollars in public funds since 2002. Private investment may be even larger. Most major car manufacturers are designing, building and demonstrating hydrogen vehicles, investing hundreds of millions of dollars. Daimler, Honda, Toyota and GM have announced plans to commercialise hydrogen fuel-cell vehicles in the timeframe between 2011 and 2020. Car manufacturers and energy companies, like Shell, Chevron and BP, are working with governments to introduce the first fleets of hydrogen vehicles and refuelling mini-networks in California and the north-eastern states. There are currently over 20 hydrogen stations in California, with many more planned by 2010, and 50 stations around the country.

Table 15.1 shows a list of policy measures implemented in the United States (either federally, or more frequently within a state or group of states) that directly or indirectly encourage the development of hydrogen vehicles and fuel. The most direct are policies that set aside funding for research, development and demonstration programmes for hydrogen vehicles, refuelling stations or production and delivery technologies. However, there are also a number of policies that are not hydrogen-specific but that are still favourable to the development of a hydrogen economy. These can be distinguished by their policy focus on either vehicle or fuel characteristics. The first category focuses specifically on vehicles and can lead to emissions reductions or fuel-economy improvements. The other category attempts to focus on the fuel side of the equation by reducing the level of emissions associated with the use of fuel.

The widespread interest in hydrogen rests not only in its long-term social benefits, but also its potential for innovation. Several auto companies have embraced fuel cells as a superior zero-emission technology, and are racing to develop the fuel-cell car. Fuel-cell cars are efficient, clean, quiet and powerful, and open new avenues for vehicle design (Burns *et al.*, 2002). Hydrogen and fuel cells are a logical progression of ongoing technical developments, building on efficiency and increasing electrification of cars, such as hybrid-electric drive trains. (The most efficient hydrogen cars are fuel-cell–battery hybrids.) They could offer new energy services, such as mobile electricity and the ability to provide power to the grid. Some see hydrogen and fuel cells as 'disruptive' technologies that could transform the way we produce, distribute and use energy.

Studies by the National Academies (NRC, 2004) and the International Energy Agency (IEA, 2005), among others, affirm the long-term promise of hydrogen to all but eliminate oil dependence, and emissions of greenhouse gases and air pollutants from the transport sector, beyond what might be achieved by energy efficiency alone. They also highlight the complex challenges that must be overcome before a hydrogen transportation system could become a reality in the United States.

In this chapter, we describe some possible strategies for building up hydrogen infrastructure in the United States over the next several decades. We concentrate on the timeframe from the present to 2030, recognising that hydrogen's full benefits may take several more decades beyond 2030 to be realised, because of the long time constants inherent in changing large well-established energy systems and infrastructure. We discuss the longer term prospects for hydrogen as well.

15.3 Resources for hydrogen production in the USA

Primary energy supply is a key issue for any new transportation fuel. Constraints on domestic resources and production of petroleum have led to increasing levels of imports. An important policy driver for alternative fuels for transportation in the USA is that they will alleviate some of the issues surrounding dependence and economic vulnerability related to petroleum imports, especially from politically unstable parts of the world. This raises some important questions for hydrogen as a future alternative fuel: will the USA have the resources to produce hydrogen for vehicles at low cost and with low environmental impact?

15.3.1 Hydrogen-supply technologies

Like electricity, hydrogen can be produced from a wide variety of primary sources (NRC, 2004; see also Chapter 10). Technologies exist today to make hydrogen via high temperature thermal processing (reforming or gasification) of fossil fuels or biomass to make a synthesis gas or 'syngas', which is further processed to produce hydrogen. Commercially available technologies exist that use electricity to 'split' water into hydrogen and oxygen via electrolysis. Complex processes are being developed that use high-temperature solar or nuclear heat to accomplish water splitting through a series of coupled 'thermochemical' reactions, but these are still in the laboratory stage.

There are several ways of delivering hydrogen to vehicles. Hydrogen can be produced regionally in large plants, stored as a liquid or compressed gas and distributed by truck or gas pipeline. It is also possible to produce hydrogen locally at refuelling stations (or even homes) from natural gas or electricity. Advanced storage technologies are also being developed to improve energy density of hydrogen storage, including use of metal hydrides and alanates and cryocompression that could also improve delivery energy use and costs.

Currently, large amounts of hydrogen are produced from natural gas for oil refining and chemical applications in the USA. These industries could provide a springboard to a hydrogen economy.[1] But ultimately, to obtain the most

[1] The world and the USA already have a significant fossil hydrogen energy economy. Hydrogen is used in oil refining, and contributes to the energy content of petroleum-derived fuels, such as gasoline. Hydrogen-rich synthetic gases (from coal or heavy oil) are used for electric generation. (BP has projects in Scotland and California, making hydrogen in refineries, to run an electric power plant, while capturing CO_2 for enhanced oil recovery.) Hydrogen production currently consumes two per cent of global primary energy, and is growing rapidly. While most hydrogen is produced and used inside refineries or chemical plants, some 5%–10% is delivered to distant users by truck or pipeline. In the United States, this 'merchant hydrogen' delivery system carries enough energy to fuel several million cars. In the near term, excess capacity from refineries and industrial hydrogen plants might fuel up to 100 000 cars in California alone (Ritchey, 2008).

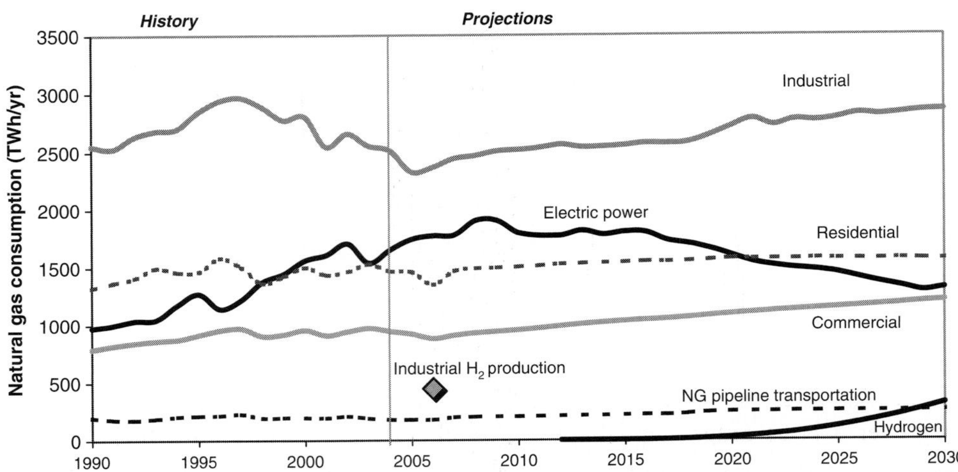

Figure 15.1. Projected natural gas use in the United States by sector (USDOE, 2007a), and projected demand for natural gas to make hydrogen for fuel-cell vehicles until 2030 (based on extension of USDOE's most aggressive scenario for introducing hydrogen fuel-cell vehicles (Gronich, 2006)).

environmental and energy security benefits from a hydrogen economy, a major goal for the USA would be to use diverse low- or zero-carbon domestic resources to make the hydrogen.

15.3.2 Near-term resources for hydrogen production in the USA

In the near term, in the United States, it is likely that hydrogen will continue to be produced via steam reforming of natural gas (approximately 95% of hydrogen currently produced in the USA is from natural gas), either in small distributed reformers at refuelling stations or, potentially, in large central reformers with hydrogen delivery by truck or pipeline. Most of the natural gas the USA uses is domestic with significant imports from Canada (approximately 16% of consumption). However, an increasing amount of natural gas is imported as liquefied natural gas (LNG), approximately 3% of consumption in 2006, estimated to be 20% by 2030 (USDOE, 2007a). The potential increase in imports for a primary energy resource, given the abundance of other energy sources in the country is an important issue for energy policy.

Some analysts have raised concerns that making hydrogen for vehicles will put unacceptable demands on natural gas supplies (Romm, 2004). However, natural gas demand to make fuel for hydrogen vehicles is unlikely to become an issue for many decades.

Figure 15.1 shows projected uses of natural gas for industrial, commercial and residential energy to 2030 (USDOE, 2007a). Current use of natural gas for industrial hydrogen production is shown as a diamond. The most aggressive USDOE scenario for hydrogen vehicle introduction is also shown, with ten million hydrogen vehicles by 2025 (see also Section 15.5). Even under this rapid deployment scenario, the natural gas to make hydrogen for ten million fuel-cell vehicles in 2025 would be less than half the amount of hydrogen used in the refining and chemical industries today, and less than 2% of the total natural gas demand. Recent studies indicate that this could be handled with minor expansions of the existing natural gas system (Vidas, 2007).

Another concern about making hydrogen in the near term from fossil fuels is whether it will increase greenhouse-gas emissions compared with using gasoline. A number of studies have calculated the well-to-wheel greenhouse-gas (GHG) emissions for a variety of vehicle types and fuel supply pathways and the emissions can vary over a large range (see Chapter 7). Hydrogen, when used in a fuel-cell car, generally provides a reduction in greenhouse-gas emissions. Hydrogen, when made from natural gas, the most common method today, and used in an efficient fuel-cell car, would produce greenhouse-gas emissions that are 10%–40% less than those from a gasoline hybrid and 40%–60% less than those from today's conventional gasoline cars (JEC, 2007; NRC, 2004; Wang, 2005). With low-carbon hydrogen supplies, well-to-wheel GHG emissions can approach zero.

15.3.3 Long-term resources for hydrogen production in the USA: realising zero emissions

One of hydrogen's major attractions is tapping vast new resources for transportation fuels that are domestically abundant and converted to H_2 with zero or near-zero emissions. To fully realise the benefits of a hydrogen economy, cost-effective zero-emission fuel supply pathways are needed. Several have been proposed, and each has unique challenges.

The USA has abundant renewable resources, in the form of biomass, wind and solar resources. Producing hydrogen from these sources via renewable electrolysis or biomass gasification is limited mainly by near-term cost rather than technical feasibility or resource adequacy. According to the National Academies study (NRC, 2004), biomass and wind hydrogen might become cost competitive with gasoline in the long term, on a cents-per-kilometre basis. In the very long term, advanced renewable pathways employing direct conversion in photo-electrochemical or photo-biological systems might become practical for production of hydrogen or other fuels.

Hydrogen production from nuclear energy faces several critical challenges. Key among these are cost (for electrolytic hydrogen), technical feasibility (for thermochemical water-splitting systems powered by nuclear heat) and public acceptance. In the USA, no new nuclear power plants have been built for over a decade, though a number of new plants have been proposed and approved. Nuclear hydrogen has the same waste and proliferation issues as nuclear power.

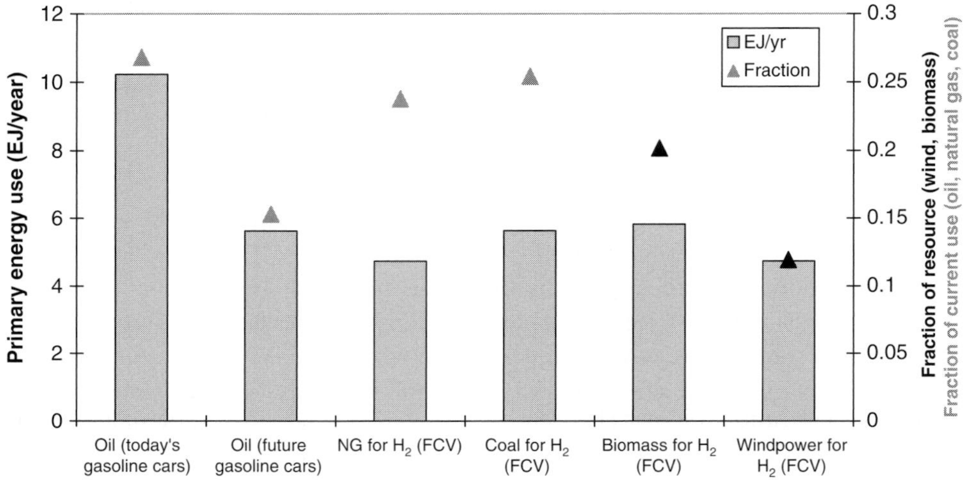

Figure 15.2. Estimated primary energy use and resources needed to produce hydrogen for 100 million fuel-cell vehicles in the United States. The biomass resource is assumed to be 800 million tonnes of biomass per year, and the wind resource is assumed to be 11 000 TWh of electricity per year.

Large-scale production of hydrogen from fossil fuels with carbon capture and sequestration offers near-zero emissions and relatively low cost, assuming suitable carbon disposal sites are nearby (NRC, 2004; Ogden, 2002). Establishing the technical and economic viability and public acceptability of carbon capture and sequestration (CCS) is crucial for long-term use of hydrogen from fossil resources, especially coal. The USA has abundant coal resources – over 200 years of recoverable resources (at current consumption rates). Several industrial-scale demonstration projects are being conducted around the world to validate the concept of CCS. Estimates of the potential resources for carbon storage in underground geological formations in the United States are very large, capable of storing perhaps 100 years of carbon emissions from the energy system (Dooley *et al.*, 2005).

Figure 15.2 shows the amount of primary energy needed to make hydrogen for 100 million fuel-cell cars in the United States (about 50% of the current US fleet or 33% of the projected US fleet in 2050). The amount of primary energy required is given in exajoules (10^{18} joules) per year on the left-hand *y* axis. The fraction of the available annual resource (for biomass and wind) or the current use (for coal or natural gas) are shown on the right-hand *y* axis. For reference, we also plot the energy use for 100 million current gasoline vehicles and 100 million gasoline hybrids. With hydrogen fuel cells the amount of primary energy required is similar to that for gasoline hybrids, and considerably less than for conventional gasoline cars.

Ultimately, we would like to make hydrogen from zero or near-zero carbon sources. There are plentiful near-zero carbon resources for hydrogen production in the United States. For example, a mix of low-carbon resources, including natural gas, coal (with carbon sequestration), biomass and wind power, could supply ample hydrogen for vehicles (Fig. 15.2). With 20% of the biomass resource, plus 15% of the wind resource, plus 25% added use of coal (with sequestration), 300 million hydrogen vehicles (approximately the entire US fleet projected in 2030) could be served with near-zero greenhouse-gas emissions.

The technologies and processes involved in the production and delivery of hydrogen have a significant impact on its cost to the consumer. For example, a recent assessment by the National Academies estimates that for a large-scale future energy system, hydrogen at the pump would cost from \$2 to \$4 per kilogram (\$0.06–\$0.12 per kWh). Given the higher fuel economy of hydrogen fuel-cell vehicles, the fuel cost per kilometre would be similar to or less than today's gasoline vehicles. Several near-zero-carbon hydrogen supplies, notably hydrogen from natural gas or coal with carbon sequestration, biomass hydrogen and wind hydrogen, could become roughly competitive with gasoline, once a large demand is in place.

Renewables and other low-carbon technologies will probably be utilised first in the electricity sector, a development that could help enable zero-carbon hydrogen. Hydrogen might be co-produced with electricity in energy complexes.[2] Hydrogen should be seen in the context of a broader transition to low-carbon sources across the energy system. Public policy will be needed to ensure that low-carbon sources are used for hydrogen. The state of California recently adopted a requirement that hydrogen transportation fuel produced in the state be made using low-carbon sources (emissions must be at least as low as those from steam reforming of natural gas).

15.4 Scenario analysis of US hydrogen infrastructure and vehicle costs

In the next few sections we present a scenario analysis[3] that estimates the cost of producing and supplying hydrogen to fuel-cell light-duty vehicles in the USA over the course of a transition.[4] The scenarios (and the subsequent results of modelling based on these inputs) are not intended to be a prediction of the future, but rather an estimate of likely costs and impacts given the scenario assumptions. We concentrate on the timeframe between the present (2007) and 2030. However, some results are shown to 2050, recognising that the uncertainties grow as we move further into the future.

[2] Researchers at the Environmental Institute of Princeton University suggest that coal or biomass gasification power plants could offer a small 'slipstream' of hydrogen to early hydrogen users (T. Kreutz, private communication, 2006).

[3] One method for understanding the economic and environmental costs and benefits associated with the transition to a hydrogen economy is through the use of scenarios. Scenarios are one way of providing a set of quantitative inputs for use in a computer model or analysis that can be used to calculate the potential impacts associated with these assumptions.

[4] Although buses, delivery vans and other fleet and specialty vehicles could be very important in the early years of a hydrogen transportation system, and in enabling survival and success of companies developing hydrogen and fuel-cell technologies, the focus of this study is on the light-duty vehicle market (personal passenger cars and light trucks) in the USA.

Table 15.2. *Average energy prices for the USA in $2005 (assumed constant for entire scenario). The 2005 and 2030 energy prices from the Annual Energy Outlook 2007 'high-price case' are not significantly different (> 1%).*

	Commercial price[a] $/GJ ($/kWh)	**Industrial price**[b] $/GJ ($/kWh)
Natural gas	10.99 (0.040)	8.11 (0.029)
Coal	N/A	1.43 (0.005)
Electricity	26.37 (0.095)	19.95 (0.072)
Diesel	12.49 (0.045)	N/A
Biomass[c]	N/A	2.51 (0.09)

Notes:
[a] Commercial prices are used for onsite production.
[b] Industrial prices are used for central production.
[c] Biomass prices are assumed to vary as a function of demand; other prices are fixed regardless of demand.

Our scenario analysis is based on two Microsoft Excel®-based models, which we developed for hydrogen infrastructure costs and vehicle scenario analysis.

1. The SSCHISM hydrogen infrastructure model (Yang and Ogden, 2007a; b): designs and costs of an H_2 infrastructure to meet a specified market penetration for FCVs.
2. The STM (Simplified Transition Model): estimates investment to bring hydrogen FCV costs to competitive levels, investment costs for building H_2 infrastructure, GHG-emission reductions and oil savings over time.

Based on input assumptions to be described below, these two models are used to calculate the costs of fuel-cell vehicles and hydrogen infrastructure and estimate the benefits in terms of reductions in CO_2 emissions and oil consumption.

Projections for future numbers of vehicles, vehicle miles travelled, and national energy prices are based on the US Department of Energy's High Price Case projections to 2030 (USDOE, 2007a), while regional feedstock prices are assumed to be constant at 2006 levels.[5] (Assumed prices for gasoline, natural gas and coal are listed in Table 15.2.) The hydrogen system is 'embedded' in this economic future, in that the analysis assumes that the introduction of hydrogen does not change energy prices. In economic calculations, we assume a real discount rate of 15%, and all costs are given in 2005 constant dollars.

15.5 Projections of US hydrogen demand for vehicles to 2030

There have been several recent studies that present scenarios for the future market adoption of hydrogen vehicles and the resulting demand for hydrogen in the United

[5] Detailed state-by-state feedstock and energy prices were not available as projections to 2030, though the average national prices for 2030 seen in the 'high energy price' case are almost identical to their prices in 2006.

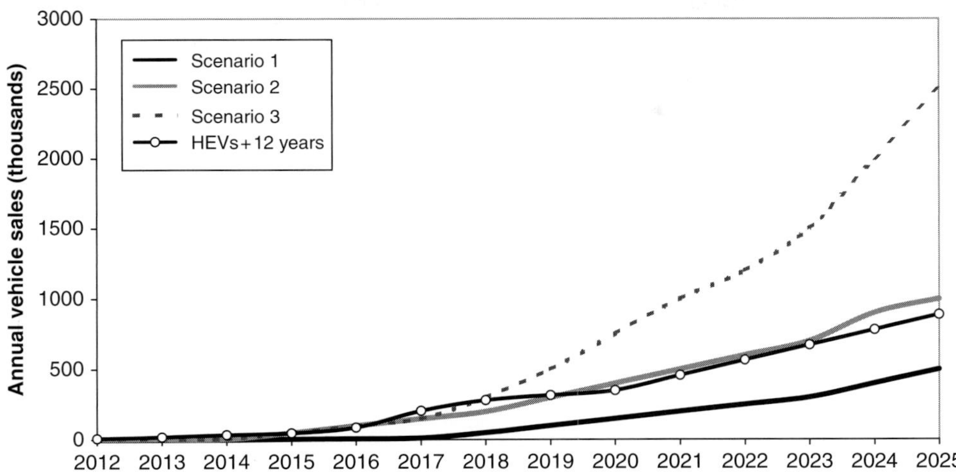

Figure 15.3. Three USDOE scenarios for hydrogen FCV market penetration (Gronich, 2006), and historical market penetration rates for gasoline hybrid vehicles displaced by 12 years.

States (Greene *et al.*, 2007; Gronich, 2006; Joseck and Kapoun, 2007; Kalhammer *et al.*, 2007; McDowall and Eames, 2006; Melaina, 2007; NRC, 2004; Plotkin, 2007).

In 2006 and 2007, the United States Department of Energy convened a series of Hydrogen Transition Analysis Workshops, led by DOE analysts (Gronich, 2006). These were attended by automotive and energy industry participants, as well as other industry, academic and government energy experts. Through a series of meetings, this working group developed three possible scenarios for the introduction of hydrogen vehicles into US markets between 2012 and 2025.

Scenario 1 *Hundreds* to *thousands* of FCVs per year by 2012 and by 2018 *tens of thousands* of FCVs per year. This option is expected to lead to a national market penetration of 2.0 million FCVs by 2025 (1% of total fleet), with a total demand of 26 000 GWh/year (2100 tonnes/day).

Scenario 2 *Thousands* of FCVs by 2012, *tens of thousands* by 2015 and *hundreds of thousands* by 2018. This option is expected to lead to a national market penetration of 5.0 million FCVs by 2025 (2% of total fleet), with a total demand of 65 000 GWh/year (5300 tonnes/day).

Scenario 3 *Thousands* of FCVs by 2012, and *millions* by 2021 such that national market penetration is 10 million by 2025 (4% of total fleet), with a total demand of 90 000 GWh/year (7000 tonnes/day).

Figure 15.3 shows the number of hydrogen vehicle *sales* per year for each scenario, and compares this with the historical market penetration of gasoline hybrid-electric vehicles (HEVs) in the USA, with the HEV curves displaced by 12 years, to reflect hydrogen's later introduction into the market. Scenarios 1 and 2 are similar to the market introduction rate of gasoline hybrid vehicles in the United States, but

	2012	2013	2014	2015	2016	2017	2018	2019	2020	2021	2022	2023	2024	2025
Log Angeles	1	2	2	25	40	50	85	120	160	190	210	250	270	300
New York, Chicago				25	40	50	85	120	150	175	185	225	240	270
San Francisco, Washington/Baltimore					20	30	55	85	120	140	160	190	210	230
Boston, Philadelphia, Dallas						20	50	85	120	145	165	195	210	220
Detroit, Houston							25	50	80	120	140	160	190	210
Atlanta, Minneapolis, Miami								40	75	100	115	130	160	180
Cleveland, Phoenix, Seattle									45	70	90	120	150	170
Denver, Pittsburgh, Portland, St. Louis, Cincinnati, Indianapolis, Kansas City										60	80	110	130	150
Milwaukee, Charlotte, Orlando, Columbus, Salt Lake City											55	80	110	130
Nashville, Buffalo, Raleigh												40	70	90
Nationwide													260	540

Figure 15.4. USDOE plan for the number of light-duty H_2 vehicles sold annually in 27 'lighthouse' cities, given in 1000s of vehicles per year introduced between 2012 and 2025. The overall build-up rate corresponds to DOE Scenario 3. The total number of hydrogen vehicles in 2025 is 10 million, and 2.5 million vehicles are sold that year. Reprinted with permission from Gronich (2006).

displaced by 12–15 years. Scenario 3 is more rapid, and is the working group's view of the fastest rate at which hydrogen vehicles could be introduced.[6] Scenario 3 is similar to the market penetration rate in the National Academies' 2004 study *The Hydrogen Economy*, which was presented as an *optimistic* scenario. We extend the DOE's Scenario 3 to the year 2050 for our analysis.

The DOE scenarios describe the total number of vehicles that are introduced annually, but these vehicles could be distributed in many different ways over different locations. To design and cost hydrogen infrastructure, it is necessary to specify where hydrogen demand would occur. We assume that early hydrogen infrastructure is likely to be built in a phased or regionalised manner where hydrogen vehicles and stations are initially introduced in selected large cities, beginning with those cities like Los Angeles and New York (with interest and motivation to implement hydrogen) and moving to other cities over time. This so-called 'lighthouse' concept reduces infrastructure costs by concentrating development in relatively few key areas. A schedule for phased introduction of hydrogen vehicles in various US cities is shown in Fig. 15.4. The list of 27 cities was chosen based on hydrogen scenario development work by the US DOE (Gronich, 2006; Melendez, 2007).

[6] Most of the emphasis in the United States has been on introduction of hydrogen fuel-cell vehicles, although several initiatives in California (California Hydrogen Highway) have proposed introduction of hydrogen internal combustion engine cars, as a way of getting started earlier with hydrogen. For example, the California Hydrogen Highway Network Blueprint plan suggests introduction of 2000 hydrogen light-duty vehicles in the state by 2010, 650 hydrogen ICEVs, and 1350 FCVs, plus 10 buses and five specialty vehicles, such as fork-lifts or other off-road applications.

15.6 Hydrogen technology and cost assumptions

Our scenarios consider hydrogen and fuel-cell technologies based on foreseeable near- to mid-term extensions of current technology. The hydrogen supply pathways considered are listed in Table 15.3. We do not consider advanced hydrogen production or storage technologies that would require fundamental scientific breakthroughs (for example, hydrogen storage in carbon nanostructures or biological production of hydrogen by algae). Costs and technical information are based on the United States Department of Energy's H2A models, which are a well reviewed source of cost and performance data for current (2005) and mid-term (2015–2030) hydrogen infrastructure technologies (Paster, 2006; USDOE, 2007b).

The assumed capital costs of different hydrogen production systems are summarised in Table 15.4, based on H2A's future (2015) technology assumptions (USDOE, 2007b).

15.7 Fuel-cell-vehicle cost assumptions

Hydrogen infrastructure development is one aspect of a hydrogen transition. Another crucial factor is the introduction of hydrogen vehicles, and bringing the vehicle costs to competitive levels, where they capture market share from gasoline vehicles. As part of our transition scenario analysis, we estimate how the costs of hydrogen FCVs might decrease over time, with R&D learning and mass production. We then compare these costs to those of a reference gasoline vehicle over time, and estimate the investment needed to bring FCVs to cost competitiveness with conventional vehicles.

Table 15.5 lists our cost and performance assumptions for hydrogen fuel-cell vehicles (FCVs) and a gasoline internal combustion engine reference vehicle. FCV

Table 15.3. *Hydrogen supply pathways considered in this analysis*

Resource	H_2 production technology	H_2 delivery method to station (for central plants)
Central production		
Natural gas	Steam methane reforming	Liquid-H_2 truck
Coal	Coal gasification with carbon capture and sequestration	Compressed-gas truck H_2-gas pipeline
Biomass (agricultural, forest and urban wastes)	Biomass gasification	
Onsite production (at refuelling station)		
Natural gas	Steam methane reforming	
Electricity (from various generation resources)	Water electrolysis	

Table 15.4. *Assumed capital costs for hydrogen production systems*

	Plant size tonne/day (MW LHV)	H2A 2015 technologies capital cost (\$ million/MW H_2)
Central NG SMR	50 (70)	0.45
(production plant only)	300 (417)	0.29
	400 (556)	0.27
Central coal with CCS	250 (348)	0.92
(production plant only)	400 (556)	0.84
	1200 (1669)	0.68
Central biomass	30 (42)	0.91
(production plant only)	155 (216)	0.62
	200 (278)	0.58
Onsite SMR (station)	0.1 (0.14)	2.9 (\$0.4 million/station)
	0.5 (0.7)	1.3 (\$0.9 million/station)
	1.5 (2.1)	1.0 (\$2.2 million/station)
Onsite electrolysis (station)	0.1 (0.14)	3.1 (\$0.4 million/station)
	0.5 (0.7)	1.5 (\$1.0 million/station)
	1.5 (2.1)	1.2 (\$2.5 million/station)

Table 15.5. *Assumed cost and performance of hydrogen FCVs and gasoline vehicles*

	H_2 fuel-cell vehicle	Gasoline internal combustion engine reference vehicle
FC drive-train retail price (including fuel cell and H_2 storage)	Costs fall with learning and manufacturing scale to \$100/kW	\$54/kW
H_2 FC vehicle retail price increment compared with gasoline reference vehicle	>\$100 000 (initially) → \$3600 (learned out)	–
FCV market introduction	2012	–
Fuel economy, litres gasoline equivalent per 100 km (miles per gallon gasoline equivalent) (on road = 80% of USEPA fuel economy)	Increasing new car fuel economy 4.6 litre/100 km (51 mpgge) (2015) → 2.8 litre/100 km (84 mpgge) (2050)	*Reference case*: 11.6 litre/100 km (20.2 mpgge) (2005) → 10.0 litre/100 km (23.4 mpgge) (2015) → 8.5 litre/100 km 27.5 mpg (2050)

costs are based on an 80 kW fuel-cell 'engine' with 5 kg (165 kWh) of compressed hydrogen gas stored on-board.

Initially, only a few thousand fuel-cell vehicles are manufactured each year, and the cost is high (>\$100 000 per vehicle). As more vehicles are produced, the cost comes

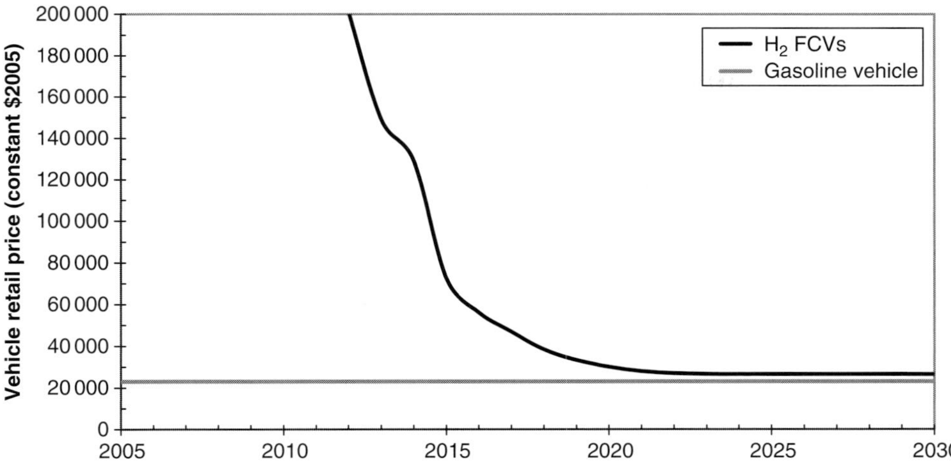

Figure 15.5. Vehicle retail prices versus time according to a learning curve model in Greene *et al.* (2007), assuming hydrogen FCVs are introduced according to the DOE's Scenario 3.

down with learning and scale of manufacturing. We assume that the vehicle price drops according to a learning curve model developed by researchers at Oak Ridge National Laboratory (Greene *et al.*, 2007), based on automobile manufacturers' estimates of fuel-cell vehicle costs in mass production (see Fig. 15.5). Based on a model by Kromer and Heywood (2007), we estimate that the 'learned out' price of a fuel cell vehicle is $3600 more than that of a comparable gasoline vehicle.[7] The on-road fuel economy of the hydrogen FCV is assumed to increase over time from about 4.6 to 2.7 litres per 100 km (51 to 84 miles per gallon gasoline equivalent). The fuel economy of the gasoline car (following the USDOE assumptions) is assumed to increase only modestly from 11.6 to 8.6 litres per 100 km (20.2 to 27.5 mpgge).

15.8 Modelling hydrogen-infrastructure build-up using the SSCHISM model

We use the UC Davis SSCHISM steady-state hydrogen supply pathway model (Yang and Ogden, 2007b) to design hydrogen infrastructure and estimate delivered hydrogen costs. Hydrogen equipment costs and performance are from the H2A model developed by the US Department of Energy (Paster, 2006). The SSCHISM model combines H2A component-level data into complete hydrogen supply pathways from hydrogen production through refuelling, using an idealised spatial model of infrastructure layout in cities (Yang and Ogden, 2007a). Inputs include information about the level of demand (market fraction of hydrogen vehicles), the city

[7] This is consistent with a fuel-cell drive train (the fuel cell, hybrid battery, motor and auxiliaries) manufacturing cost of about $50/kW, and hydrogen-storage cost of $10/kWh, and a retail price mark-up factor of 1.4 (i.e., the retail price is 1.4 times the manufacturing cost).

population and size, the number of stations, local feedstock and energy prices and constraints on viable types of supply. Outputs include the delivered hydrogen cost to the vehicle, hydrogen infrastructure costs, and energy use and CO_2 emissions for each supply pathway in 73 US cities. The SSCHISM model can then determine which pathway is the cheapest method for supplying hydrogen to a particular city at a given market penetration.

Initially, when hydrogen is introduced in each 'lighthouse city', we assume that some minimum number of hydrogen stations is needed to assure adequate coverage and consumer convenience. This constraint is imposed to help deal with the 'chicken-and-egg' problem, of assuring hydrogen fuel availability to early non-fleet vehicle owners. This is assumed to be 5% of existing gasoline stations in cities, based on work by Nicholas *et al.* (2004) and Nicholas and Ogden (2007). For the initial introduction of hydrogen vehicles, it is assumed that we start with 100 kg/day stations at 5% of gasoline stations, for the first several years. In our scenario, these early stations are generally served by onsite natural gas reformer stations, because SSCHISM estimates that it is the cheapest method for such low demand.[8] As more vehicles are introduced and demand grows in a particular city, eventually 100 kg/day stations are not sufficient and additional station capacity is added, increasing station size up to a maximum of 1500 kg/day, while keeping the number of stations constant. About a decade after vehicles are first introduced in a particular city, additional new 1500 kg/day stations are built, and the fraction of gasoline stations offering hydrogen increases over time. To account for under-utilisation of the evolving hydrogen infrastructure as demand grows, we assume a relatively low system-capacity factor of 70%.

15.9 US hydrogen-infrastructure results

The SSCHISM infrastructure model calculates the cost of the potential hydrogen pathway-supply options shown in Table 15.3 for 73 of the largest US urbanised areas and selects the cheapest supply pathway in each city at a specified market penetration. The cheapest pathway choice for any given city depends on the size of the city, level of demand, demand density, and local energy and feedstock prices.

At low market penetration, we find that hydrogen costs are high, because stations are small and under-utilised. This is due to the model requirement that a minimum of 5% of existing gasoline stations supply H_2, even when the number of vehicles is well below 5%. Hydrogen infrastructure at low to moderate demand (up to 5%–10% of the market) is dominated by onsite steam methane reformers, because there is not yet enough demand to justify the large investments required for central production and hydrogen delivery. As hydrogen demand in a particular city grows, it eventually

[8] In the real world, it is possible that these early stations could be supplied using excess hydrogen from industrial or refinery sources, rather than dedicated hydrogen production facilities, but this option is not included in the model.

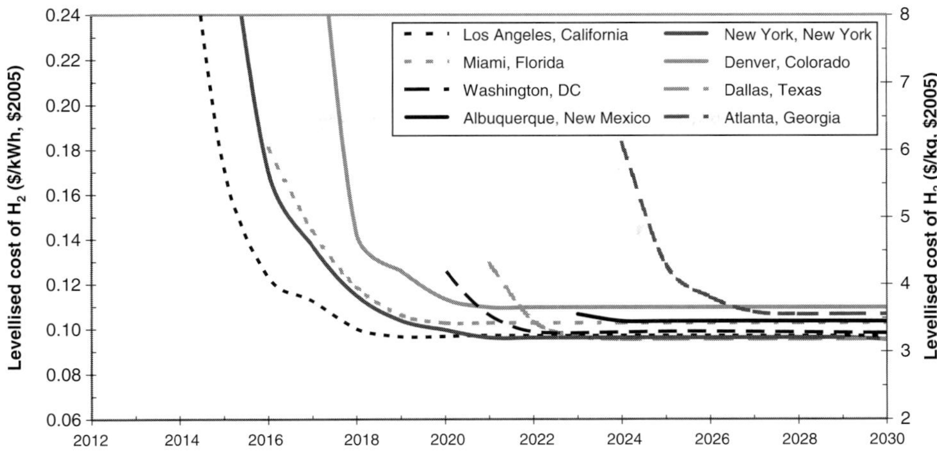

Figure 15.6. Delivered hydrogen cost in US cities for phased introduction of hydrogen cars.

makes sense to build central production plants and delivery systems, because the economies of scale associated with large production plants overcome the additional cost associated with pipeline or truck delivery. This sequence is played out in each of the 73 urban areas in the model. However, the point at which this switch from distributed to central production occurs, and the cheapest central pathway, differs depending on the key factors described above: the size of the city, level of demand, demand density, and local energy and feedstock prices. The switch to central plants tends to occur at a lower market penetration for larger cities, because the actual hydrogen demand is larger for these cities, while onsite SMRs tend to persist in smaller cities for longer.

As new cities are phased in over time, hydrogen is initially costly because of the low demand in the new cities, but costs fall as demand grows. The phased introduction of hydrogen infrastructure and vehicles leads to differences in hydrogen market penetration and also hydrogen cost for different cities. The range and progression of hydrogen costs over time is shown for selected cities in Fig. 15.6, which takes into account not only the staggered introduction time (affecting the market penetration and demand level) but also the local city size and density, and local feedstock and energy prices. These costs are aggregated into an average delivered US hydrogen cost, which is plotted in Fig. 15.7, along with the assumed gasoline price.

Beginning at roughly 10% market share, central plants start to become competitive in many cities. Figure 15.8 shows the distribution of central hydrogen plants by type.[9] A surprising outcome of the analysis is the very low penetration of natural gas-based

[9] All coal and central natural-gas hydrogen plants are assumed to have carbon-capture and sequestration (CCS). Biomass hydrogen plants are assumed to be smaller (30–200 tonnes/day), compared with 50–400 tonnes/day for natural gas central SMRs, and 250–1200 tonnes/day for coal plants. We use a regional biomass supply curve (which specifies the amount of biomass available at a certain $/tonne) (Walsh *et al.*, 1999), to reflect biomass feedstock cost increases as demand grows.

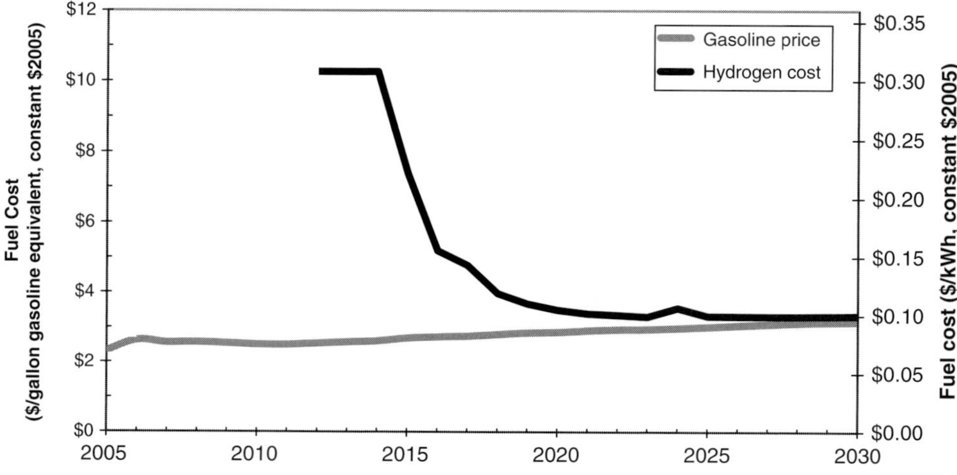

Figure 15.7. Average US delivered hydrogen cost and gasoline price at different years.

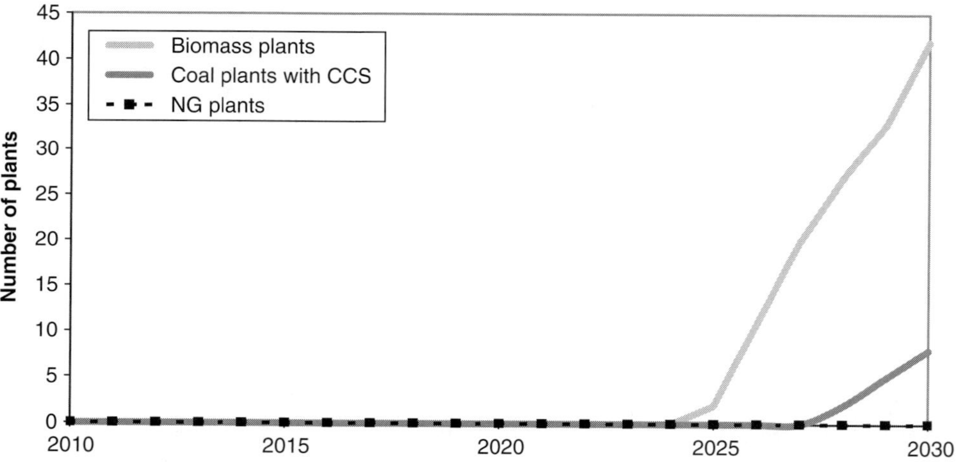

Figure 15.8. Numbers of hydrogen plants by type.

central hydrogen production, even though this is the most common technology for large-scale hydrogen production in the US today. This is a direct result of the relatively high natural gas prices assumed in the study (an average of \$8.6/GJ) as compared with prices for biomass (\$3–5/GJ) and coal (\$1–2.5/GJ). Even though the hydrogen production plant capital cost is higher for coal and biomass than for natural gas (see Table 15.4), the feedstock cost is significantly lower, yielding a lower hydrogen production cost.

Given the low cost and abundance of coal, we find that coal-based hydrogen with CCS is the cheapest central production technology in many parts of the USA.

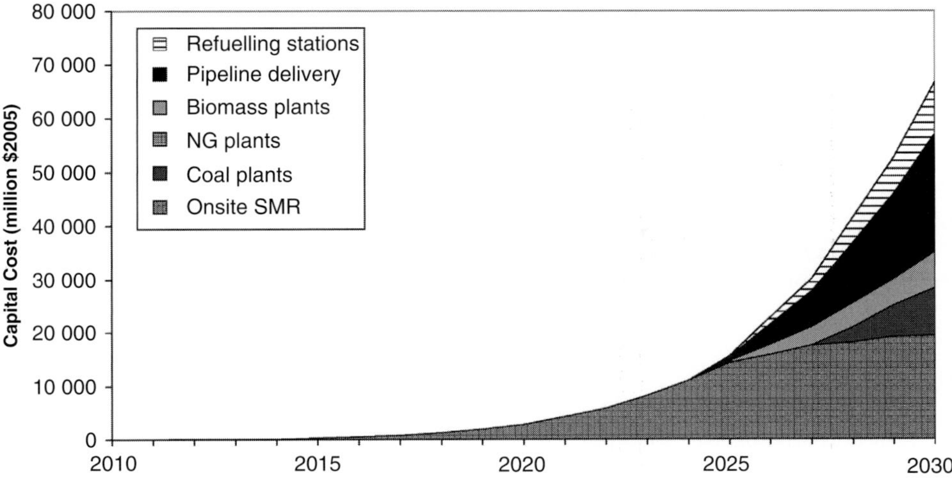

Figure 15.9. Hydrogen infrastructure capital costs to 2030.

However, biomass plants appear earlier than coal and more biomass plants are built, because they are smaller and can become central supplies at smaller market penetration. It is important to note that the delivered cost of hydrogen from coal, biomass and natural gas central plants are typically quite close (within \$0.5/kg). Thus, the choice of a feedstock may be determined by other factors, such as state policies favouring renewables and the availability of carbon-sequestration sites.

Figure 15.9 shows the capital investments required for hydrogen infrastructure up to 2030. Onsite SMRs dominate in the early years. After 2025, central production, from biomass and coal, becomes significant, accompanied by pipeline delivery systems and stations. Later on, central production dominates in large cities, although onsite reformers persist in other areas. By 2030, the majority of capital investment is in central plants and pipeline delivery.

The total infrastructure capital cost is about \$2000–2500 per car served by the system. The total capital costs to build a 'steady-state' hydrogen infrastructure to serve the demands in 2015, 2020 and 2030 are estimated in Table 15.6.

These results suggest that a variety of hydrogen supply pathways will be used in the USA, and the choice will depend on the level of market penetration, and local energy and feedstock prices, as well as the size of the city and the geographical demand density. There is a trend from distributed production towards central production as demand grows beyond about 10%–25% of the fleet. Pipelines begin to appear in a few cities as early as 2025. Because central coal with carbon sequestration and biomass become the major sources of central hydrogen production, the CO_2 emissions from the hydrogen infrastructure system drop over time as these lower carbon supplies are phased in. (The equivalent value would be about 11 kg CO_2/kg H_2 for gasoline (on the same energy basis).)

Table 15.6. *Type of hydrogen supply over time*

	2015	2020	2030
Number of cars served (% total fleet)	70 000	2.0 million (0.7%)	31 million (9%)
Infrastructure capital cost	\$0.3 billion	\$2.7 billion	\$67 billion
Levelled hydrogen cost \$/kWh (\$/kg)	\$0.222 (\$7.41)	\$0.105 (\$3.49)	\$0.099 (\$3.30)
Total number of stations	686 (all onsite SMR)	1970 (all onsite SMR)	18 000 (69% onsite SMR)
H_2 demand GWh/yr (tonne/day)	542 (45)	15 542 (1277)	243 965 (20 052)
Number of central plants	0	0	50 (8 coal w/CCS, 42 biomass)
H_2 production breakdown	100% onsite SMR	100% onsite SMR	69% onsite SMR, 21% biomass, 10% coal with CCS
Pipeline length (km)	0	0	18 000
kg CO_2/kg H_2 produced	10.7	11.2	8.3

More detailed modelling of regional hydrogen systems using geographical information system techniques and optimisation (Johnson *et al.*, 2006; Lin *et al.*, 2008; Parker, 2007) reveal the same trend – the optimal H_2 supply begins with distributed generation at refuelling stations and, as demand grows, central plants with pipelines become the lowest cost supply pathway. Some examples of these detailed regional supply analyses are presented in Section 15.12.

15.10 Benefits of hydrogen vehicles: modelling US fleet reductions in greenhouse-gas emissions and gasoline use

Using the scenarios described above for hydrogen vehicle introduction and infrastructure build-up, we can estimate the savings in national petroleum usage and greenhouse-gas emissions. These estimates are made using the UC Davis Simplified Transition Model or STM. The STM is a vehicle stock model adapted from the Argonne VISION model (Singh *et al.*, 2003; 2005) to keep track of the number of hydrogen and gasoline cars over time. This allows us to estimate the costs, fuel use and GHG emissions over time. The hydrogen vehicle introduction scenario is based on DOE's Scenario 3 (described in Section 15.5) and Fig. 15.10 shows the number of hydrogen FCVs and gasoline vehicles modelled in the STM. Figures 15.11 and 15.12 show the aggregate reductions in oil use and greenhouse gases due to the introduction of H_2 vehicles for the USA to the year 2050.

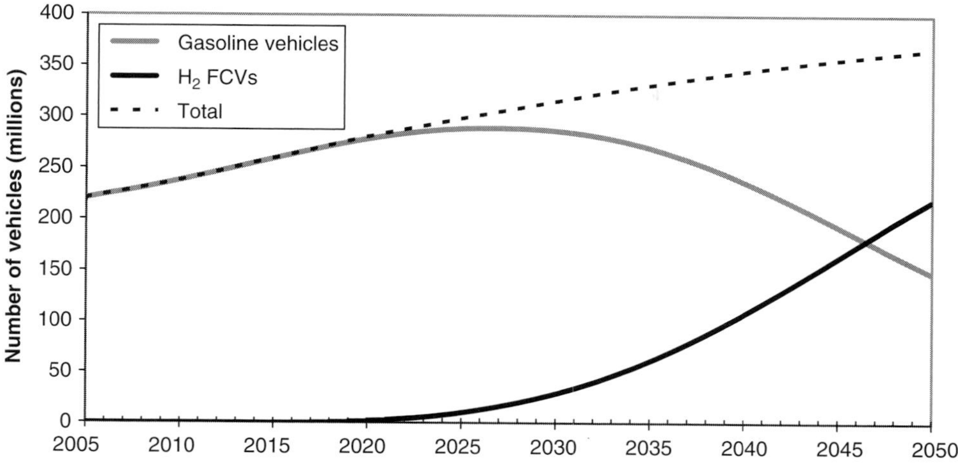

Figure 15.10. Millions of light-duty vehicles in the US fleet.

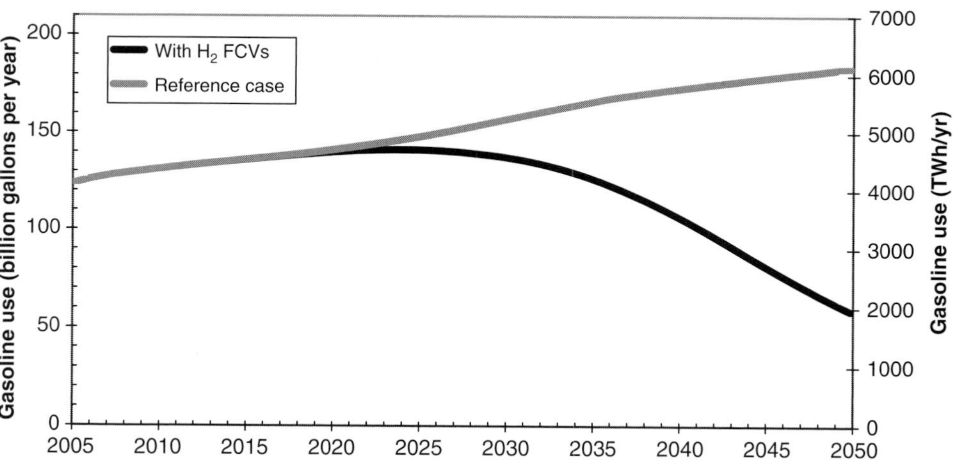

Figure 15.11. US fleet annual gasoline consumption: top, gasoline only; bottom, with H_2 FCVs.

These figures illustrate the long timeframe needed to change the energy system. By 2030, hydrogen is starting to have a relatively small, but increasing effect. The total fraction of hydrogen vehicles in the fleet is only about 9% by 2030 (although the fraction of new vehicle sales is 25%). By 2050, hydrogen's impact on CO_2 emissions is much larger, owing to an increasing percentage of hydrogen vehicles and the declining carbon intensity of the hydrogen fuel. This is illustrated in Fig. 15.12, where we assume that hydrogen vehicles reach about 80% of the new vehicle sales and 60% of the fleet by 2050, improving in efficiency to 80 miles per kg (128 km per kg H_2), and that central coal with CCS and biomass become the dominant hydrogen supply routes.

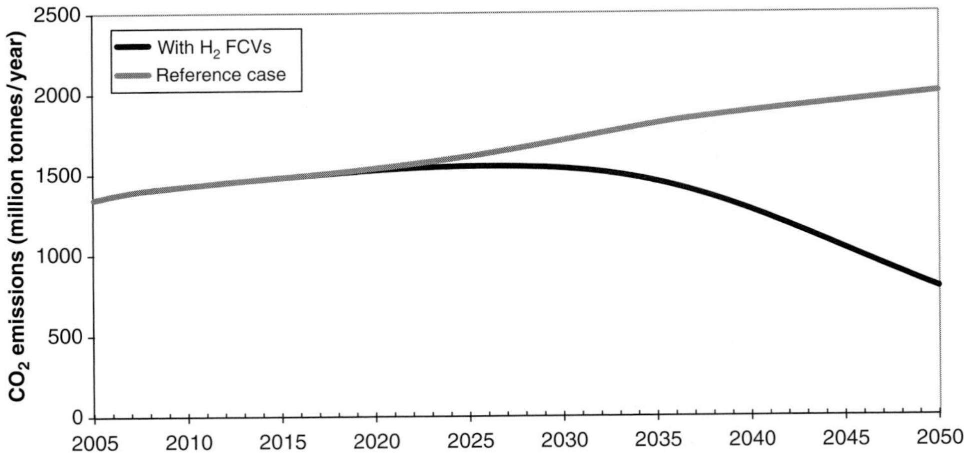

Figure 15.12. US fleet annual GHG emissions: top, gasoline only; bottom, with H_2 FCVs.

15.11 Transition modelling: estimating the investments required to bring hydrogen and fuel-cell vehicles to cost competitiveness

One of the major challenges facing any new vehicle type and alternative transportation fuel is reaching economic competitiveness with gasoline vehicles. Initially, the new vehicles are manufactured in small quantities and the cost of purchasing a vehicle is much higher than a comparable gasoline vehicle, which is a major disincentive to consumers. Getting enough hydrogen fuel-cell vehicles on the road to bring down costs is a key issue. For infrastructure, the analogous problem is putting in enough hydrogen stations for consumer convenience, while bringing down the cost of hydrogen via scale-up. The question is how much money must be invested in the first million or so vehicles and the early infrastructure to reach cost competitiveness.

To study this 'buy-down' process for hydrogen and fuel-cell vehicles, the Simplified Transition Model (STM) was used to aggregate costs over the entire fleet, based on the fuel-cell vehicle and hydrogen infrastructure costs previously described:

- Hydrogen FCV costs come down with learning and scaled-up manufacturing, according to a model developed by Greene *et al.* (2007). Figure 15.5 shows the estimated retail price trajectory for a hydrogen vehicle versus a gasoline vehicle, assuming that FCVs are introduced according to DOE's Scenario 3.
- The delivered hydrogen cost over time is given by the aggregate hydrogen cost curve developed from our infrastructure modelling for this same vehicle introduction rate (Fig. 15.6).

To estimate the overall transition cost, the STM tracks the incremental costs of the hydrogen and fuel-cell vehicles sold in the market relative to the same number of gasoline vehicles. Two components of this 'transition cost' are estimated over time. First is the incremental price of buying hydrogen fuel-cell vehicles each year,

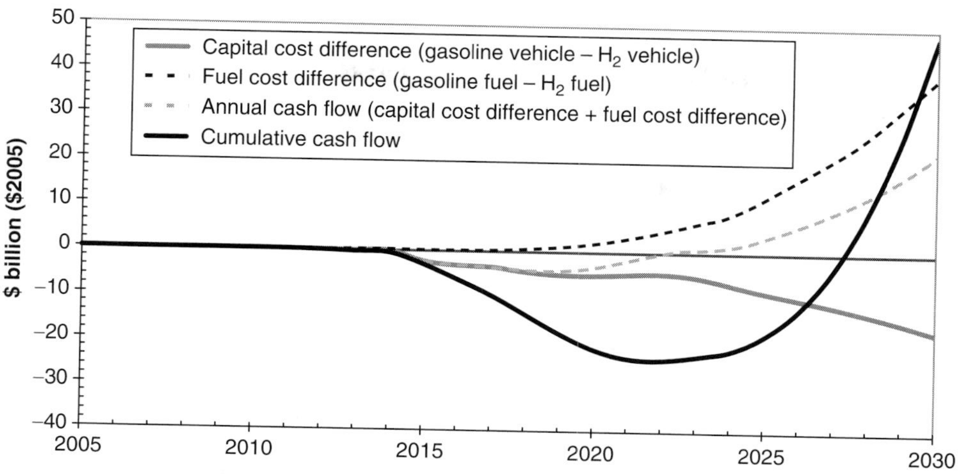

Figure 15.13. Cash flows for introduction of hydrogen fuel-cell vehicles. Positive cash-flow values indicate that the cost of H_2 vehicles and/or fuel is lower than the cost of gasoline vehicles or fuel.

instead of gasoline cars. This is summed over all the hydrogen vehicles sold in a given year and is the aggregated extra cost paid by consumers to buy hydrogen fuel-cell cars instead of gasoline cars. The second cost is the net difference between the annual cost of hydrogen to fuel these cars and the annual cost of gasoline to go the same number of kilometres. The annual cash flow or cost difference between a transition (where hydrogen is introduced) and 'business as usual' (all gasoline cars) is the sum of the vehicle first cost increment and the fuel cost increment.

These annual and cumulative 'cash flows' are shown in Fig. 15.13. The two components of the annual cash flow (the incremental annual vehicle first cost ($/year) and the incremental annual fuel cost ($/year)) are shown, along with the total incremental annual difference or cash flow ($/year) and cumulative cash flow. For the first ten years, the annual cash flow is negative, because the cost of the H_2 FCV is much greater than of the gasoline car. But the annual fuel cost for hydrogen soon becomes less than that for gasoline, because of the higher fuel economy of the H_2 FCV compared with a gasoline car. This saving in fuel costs eventually outweighs the extra cost of buying FCVs, so overall the annual cash flow becomes positive. The year when the cash flow goes positive is termed the 'break-even year'. The break-even year in this analysis is 2023, 11 years after the introduction of H_2 FCVs, when some six million FCVs are on the road (about 3% of the US light-duty fleet). Summing the annual cash flow over time, we find that the cumulative investment needed to reach this 'break-even' year is about $23 billion. (After this time, the cash flow is positive, so the net effect on the economy is positive. Thus $23 billion could be seen as the amount of support that would be needed to bring the H_2 FCVs to economic parity with gasoline vehicles.) The total investment in extra vehicle first costs over this

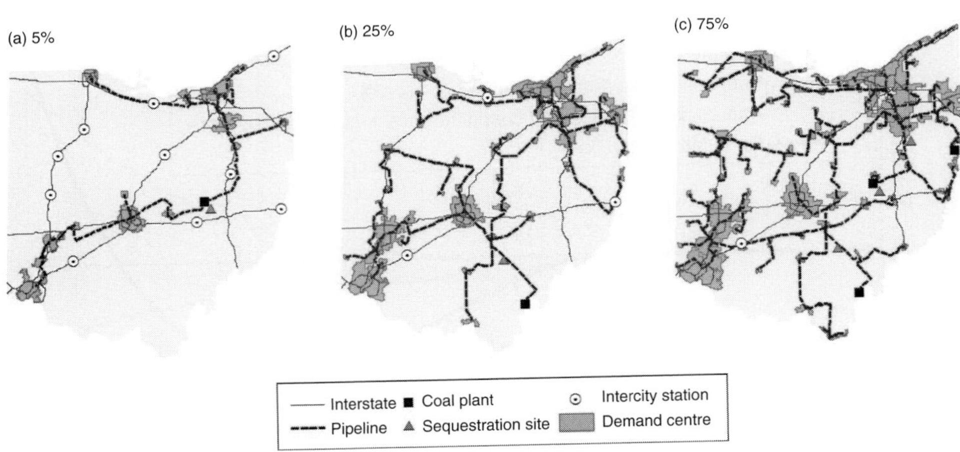

Figure 15.14. Optimal infrastructure configuration at different market penetration levels for a coal-based hydrogen supply system with carbon capture and sequestration in Ohio, USA (Johnson *et al.*, 2006).

11-year period is about \$40 billion, while the total capital investment in hydrogen infrastructure to 2023 is about \$8 billion.[10]

Clearly, there are many uncertainties and assumptions in this analysis, and the numbers should not be taken as precise. From this analysis and others (Greene *et al.*, 2007), it appears that an investment of the order of tens of billions of dollars would be needed to bring hydrogen FCVs to cost competitiveness with gasoline. The largest part of the buy-down cost is attributed to bringing down the cost of vehicles. The cost for early hydrogen infrastructure is many times lower than the cost for early hydrogen vehicles.

15.12 Regional case studies

In this section, we present summary results from several studies conducted at UC Davis on regional hydrogen infrastructure designs in different parts of the USA. These illustrate the diversity of possible solutions for hydrogen supply in the United States.

Johnson *et al.* (2006) studied the design and cost of a coal-based hydrogen infrastructure in Ohio, a coal-rich state, where 90% of electric generation is currently from coal and good geological sequestration sites for CO_2 are available. They used GIS data to estimate demand and choose rights of way and plant locations. A spatially optimised system is shown in Fig. 15.14. The optimal solution tends to favour one or two large coal plants, rather than many smaller ones. The overall capital cost is approximately \$2000 per car served, and the delivered hydrogen cost is \$2.5–\$3.5/kg. This study found

[10] However, investments for hydrogen to reach cost competitiveness with gasoline on a cents-per-kilometre basis are even lower. This happens in about 2018 when hydrogen costs about \$6/kg, because H_2 FCVs are about twice as efficient as gasoline cars. Required infrastructure costs are perhaps \$1–2 billion.

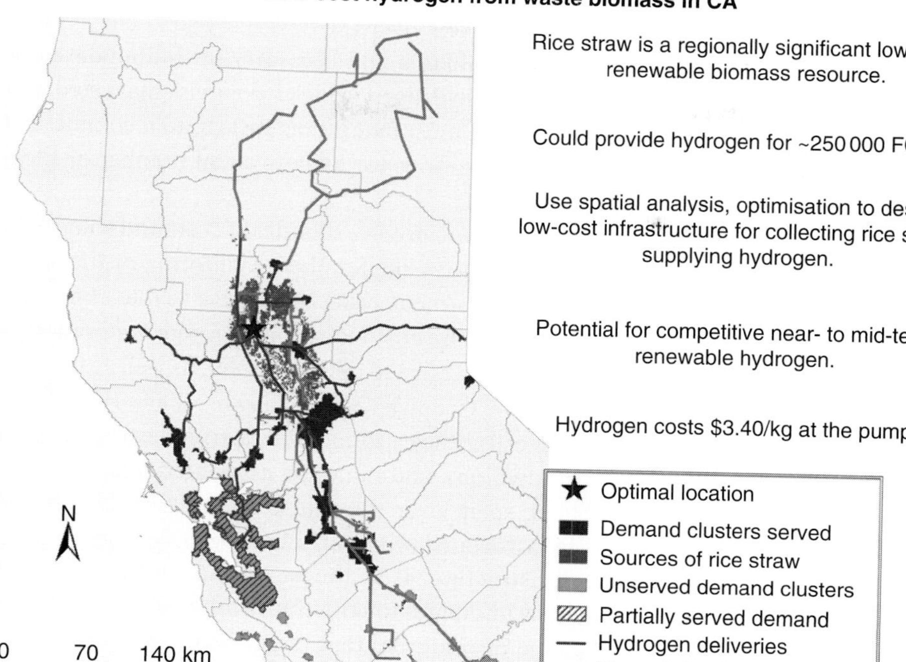

Figure 15.15. Biomass hydrogen infrastructure design, using rice straw as feedstock for hydrogen production (Parker, 2007).

that there is a switch from distributed to central production as the lowest cost infrastructure option at about 25% market penetration.

Parker (2007) examined the possibility of using agricultural waste to make biomass hydrogen. He found that under certain circumstances it would be possible to reduce the costs of biomass hydrogen through optimal location of production plants and design of delivery systems (Fig. 15.15) to minimise costs related to biomass and hydrogen delivery. His best designs yielded delivered hydrogen costs of \$3.5–\$4/kg, competitive with onsite natural gas reforming. The choice of delivery mode (pipeline vs. truck) depended on the market fraction and the type of waste (dense versus more dispersed).

These regional results illustrate the geographically specific nature of hydrogen supply design in the United States. As with the US electricity system, it is likely that hydrogen will be produced from a variety of feedstocks.

15.13 Summary

Building a hydrogen infrastructure in the Unites States will be a decades-long process in concert with growing vehicle markets. The first steps are providing hydrogen to test fleets, and demonstration of refuelling technologies in mini-networks. These are being planned now through projects like the California Hydrogen Highways Network, and several such projects seem likely over the next ten years. The system-level learning from these programmes is valuable and necessary, including development of safety codes and standards. When hydrogen vehicles are mass marketed and sold to consumers in 10–15 years, hydrogen must make a major leap to a commercial fuel available at perhaps 5% of refuelling stations (or an equivalent number of sites) and must be offered at a competitive price.

Getting through this hydrogen transition will involve significant costs and some risk. Concentrating hydrogen projects in key regions like Southern California or the north east corridor will focus efforts, lower investment costs to achieve viable consumer refuelling availability and hasten infrastructure cost reductions through faster market growth and economies of scale. The results presented in this chapter, as well as those of several recent studies (Greene *et al.*, 2007; Gronich, 2006; Lin *et al.*, 2008; IEA, 2005) indicate the costs to 'buy-down' fuel-cell vehicles to market clearing levels (through technological learning and mass production), and build the associated infrastructure might cost tens of billions of dollars, spent over the course of one to two decades. The majority of the cost would be associated with early hydrogen vehicles, with a lesser amount needed for early infrastructure. It is almost certain that government policy will be needed to bring these technologies to cost-competitive levels.

The United States has a wide range of resources that might be used for hydrogen production. In the near term (up to 2025), hydrogen will probably be produced from natural gas, via distributed production at refuelling stations, or, where available, excess industrial or refinery hydrogen. Beyond 2025, central production plants with pipeline delivery will become economically viable in urban areas in the USA. The cheapest low-carbon hydrogen supply pathways appear to be biomass gasification and hydrogen from coal with carbon capture and sequestration. Each could contribute significantly to the long-term hydrogen supply. The cheapest option depends on the market penetration of H_2 FCVs, the local feedstock and energy prices, as well as geographical factors, such as city size and density of demand. Detailed regional studies reveal possibilities for further optimising the hydrogen supply system at the regional level. It appears that hydrogen could be delivered to consumers for about \$3–4/kg, with near-zero emissions of greenhouse gases, on a well-to-wheels basis, which leads to a reduction in fuel cost per mile compared to gasoline vehicles, given the increased efficiency of H_2 FCVs.

How might policy and business strategy affect the future of hydrogen in the US energy system? The context for considering future alternative fuels is dynamic and uncertain. While hydrogen and fuel-cell technologies are progressing, there is continuing technical progress in a variety of other alternative fuel and efficient vehicle technologies, such as hybrid and plug-in electric vehicles and liquid biofuels. At the same time, there is a growing imperative for alternative fuels driven by concerns about oil supply, rising fuel costs and climate change, and the search by politicians for a technical fix to solve these problems quickly.

In the early 2000s, hydrogen and fuel cells were widely seen as the 'end game' in the USA. Over the past few years, it has become apparent that hydrogen will take more time to develop and implement than was previously assumed. Several other alternative fuel options have been proposed in the USA as 'nearer term' or more compatible with the existing energy system, especially liquid biofuels and plug-in hybrid electric vehicles. Many still see hydrogen as a long-term option for the USA, but seek nearer-term strategies.

In the USA, vehicle efficiency is the first step along the road towards a hydrogen economy or any sustainable transportation future. (This is even more true in the USA than in Europe or Japan, where cars are more fuel efficient.) Streamlined, lightweight cars, more efficient engines, and hybrid drive trains are viable near-term technologies that could reduce carbon emissions and oil use over the next few decades. These developments are not in competition with longer-term alternatives like hydrogen, biofuels or battery cars; on the contrary, they are strongly synergistic. Hydrogen and fuel cells are part of a technical progression, building on efficiency, and increasing electrification of cars that encompasses hybrid-electric drive trains, plug-in hybrids and improved batteries.

To realise hydrogen's full benefits in the US context will require making hydrogen from domestic and widely available zero-carbon or decarbonised primary energy supplies. Hydrogen can benefit from ongoing efforts to develop biomass and coal gasification with carbon sequestration for electric power, as well as wind and solar. Hydrogen should be seen as one aspect of a broad move towards lower carbon energy.

Finally, public policy is needed to move towards a goal of zero-emission, low-carbon transportation with diversification away from oil-derived transportation fuels. This calls for a comprehensive strategy, based on developing and encouraging the use of clean, efficient internal combustion engine vehicles in the near term, coupled with a long term strategy of research, development and demonstration of advanced transportation technologies including hydrogen and fuel cells, advanced batteries and biofuels. Even under the most optimistic assumptions, it would be several decades before hydrogen fuel-cell vehicle technologies could make a globally significant impact on reducing emissions and oil use. Beyond this, hydrogen could yield significant benefits, greater than those possible with efficiency alone. This underscores the importance of research, development and demonstration of hydrogen technologies now, so they will be ready when we need them.

References

Burns, L. D., McCormick, J. B. and Borroni-Bird, C. E. (2002). Vehicle of change. *Scientific American*, **287** (4), 64–73.

Davis, S. C. and Diegel, S. W. (2007). *Transportation Energy Data Book*. 26th edn. Oak Ridge National Laboratory.

Dooley, J. J., Dahowski, R. T., Davidson, C. L. *et al.* (2005). *A CO_2 Storage Supply Curve for North America and its Implications for the Deployment of Carbon Dioxide*

Capture and Storage Systems. Joint Global Change Research Institute, University of Maryland.

Greene, D., Leiby, P. and Bowman, D. (2007). *Integrated Analysis of Market Transformation Scenarios with HyTrans*. Oak Ridge National Laboratory Report, ORNL/TM-2007/094.

Gronich, S. (2006). *Hydrogen & FCV Implementation Scenarios, 2010–2025*. Presentation at the USDOE Hydrogen Transition Analysis Workshop, Washington, DC, August 9–10, 2006. www1.eere.energy.gov/hydrogenandfuelcells/analysis/scenario_analysis_mtg.html.

IEA (International Energy Agency) (2005). *Prospects for Hydrogen and Fuel Cells*. IEA Energy Technology Analysis Series. Paris: OECD/IEA.

JEC (Joint Research Centre, EUCAR, CONCAWE) (2007). *Well-to-Wheels Analysis of Future Automotive Fuels and Powertrains in the European Context*. Well-to-Wheels Report, Version 2c. http://ies.jrc.ec.europa.eu/wtw.html.

Johnson, N., Yang, C. and Ogden, J. (2006). *Build-out Scenarios for Implementing a Regional Hydrogen Infrastructure*. Presented at the National Hydrogen Association meeting, Long Beach, CA, March 11–16, 2006.

Joseck, F. and Kapoun, K. (2007). *Fuel Pathways Integration Technical Team*. Presentation at the USDOE Hydrogen Analysis Deep Dive meeting, San Antonio, TX, March 22, 2007.

Kalhammer, F. R., Kopf, B. M., Swan, D. H., Roan, V. P. and Walsh, M. P. (2007). *Status and Prospects for Zero Emissions Vehicle Technology*. Report of the ARB Independent Expert Panel 2007. Prepared for State of California Air Resources Board Sacramento, California. www.arb.ca.gov/msprog/zevprog/zevreview/zev_panel_report.pdf.

Kromer, M. A. and Heywood, J. B. (2007). *Electric Powertrains: Opportunities and Challenges in the US Light-Duty Vehicle Fleet*. MIT Report. LFEE 2007–02 RP.

Lin, Z., Chen, C.-W., Ogden, J. and Fan, Y. (2008). The least-cost hydrogen for Southern California. *International Journal of Hydrogen Energy*, **33** (12), 3009–3014.

McDowall, W. and Eames, M. (2006). Forecasts, scenarios, backcasts and roadmaps to the hydrogen economy: a review of hydrogen futures literature. *Energy Policy*, **34** (11), 1236–1250.

Melaina, M. W. (2007). Turn of the century refueling: a review of innovations in early gasoline refueling methods and analogies for hydrogen. *Energy Policy*, **35** (10), 4919–4934.

Melendez, M. (2007). *Geographically Based Hydrogen Demand & Infrastructure Roll-out Scenario Analysis*. Presented at the 2010–2025 Scenario Analysis for Hydrogen Fuel Cell Vehicles and Infrastructure, January 31, 2007, Washington, DC. www1.eere.energy.gov/hydrogenandfuelcells/analysis/pdfs/scenario_analysis_melendez1_07.pdf.

Nicholas, M. A., Handy, S. L. and Sperling, D. (2004). Using geographic information systems to evaluate siting and networks of hydrogen stations. *Transportation Research Record*, **1880**, 126–134.

Nicholas, M. A. and Ogden, J. (2007) Detailed analysis of urban station siting for California Hydrogen Highway Network. *Transportation Research Record 2006*, **1983**, 129–139.

NRC (National Research Council), Committee on Alternatives and Strategies for Future Hydrogen Production and Use (2004). *The Hydrogen Economy:*

Opportunities, Costs, Barriers, and R&D Needs. Washington, DC: The National Academies Press.

Ogden, J. M. (2002). Modeling infrastructure for a fossil hydrogen energy system with CO_2 sequestration. *Proceedings of the 6th International Conference on Greenhouse Gas Control Technologies (GHGT6)*, October, 2002, Kyoto, Japan.

Parker, N. (2007). *Optimizing the Design of Biomass Hydrogen Supply Chains Using Real-World Spatial Distributions: A Case Study Using California Rice Straw*. Master's thesis. University of California, Davis.

Paster, M. (2006). *Hydrogen Delivery Options and Issues*. Presented at the USDOE Hydrogen Transition Analysis Workshop, Washington, DC, August 9–10, 2006.

Plotkin, S. (2007). *Examining Hydrogen Transitions*. Systems Division, Argonne National Laboratory. Report No. ANL-07/09.

Ritchey, S. (2008). *The Role of Hydrogen in Oil Refining and Implications for Fueling Hydrogen Vehicles*. Master's thesis, University of California, Davis.

Romm, J. (2004). *The Hype About Hydrogen*. Island Press.

Singh, M., Vyas, A. and Steiner, E. (2003). *VISION Model: Description of Model Used to Estimate the Impact of Highway Vehicle Technologies and Fuels on Energy Use and Carbon Emissions to 2050*. Center for Transportation Research. Argonne National Laboratory Report. ANL/ESD/04–1.

Singh, M., Moore, J. and Shadis, W. (2005). *Hydrogen Demand, Production, and Cost by Region to 2050*. Argonne National Laboratory Report. ANL/ESD/05–2.

USDOE (United States Department of Energy) (2006). *Hydrogen Posture Plan. An Integrated Research, Development and Demonstration Plan, December 2006*. www.hydrogen.energy.gov/pdfs/hydrogen_posture_plan_dec06.pdf.

USDOE (United States Department of Energy) (2007a). *Annual Energy Outlook 2007 with Projections to 2030*. Energy Information Administration. www.eia.doe.gov/oiaf/aeo/gas.html.

USDOE (United States Department of Energy) (2007b). *Hydrogen Fuel Cells and Infrastructure Technologies Program Website; H2A Analysis Tools*. www.hydrogen.energy.gov/h2a_analysis.html.

Vidas, H. (2007). *Natural Gas Infrastructure Requirements for Hydrogen Production*. Energy and Environmental Analysis. Presentation at the USDOE Hydrogen Analysis Deep Dive meeting. San Antonio, TX, March 22, 2007.

Walsh, M. E., Perlack, R. L., Turhollow, A. *et al.* (1999). *Biomass Feedstock Availability in the United States: 1999 State Level Analysis*. Knoxville, TV: Oak Ridge National Laboratory.

Wang, M. (2005). *Well-to-Wheels Analysis with the GREET Model*. Argonne National Laboratory. Presentation at the 2005 DOE Hydrogen Program Review, May 26, 2005.

Yang, C. and Ogden, J. (2007a). Determining the lowest-cost hydrogen delivery mode. *International Journal of Hydrogen Energy*, **32**(2), 268–286.

Yang, C. and Ogden, J. (2007b). *US Urban Hydrogen Infrastructure Costs Using the Steady State City Hydrogen Infrastructure System Model (SSCHISM)*. Presented at the 2007 National Hydrogen Association Meeting, San Antonio, TX, March 18–22, 2007. A beta copy of the model is posted on Christopher Yang's website at UC Davis Institute of Transportation Studies: www.its.ucdavis.edu/people.

16

Hydrogen and the electricity sector

Martin Wietschel, Clemens Cremer and Michael Ball

If hydrogen production at a large scale is to be integrated into the energy system, a more holistic view needs to be applied, in particular, with respect to its interactions with the electricity sector. These concern, for instance, the ensuing competition for renewable energies as, in the long term, only hydrogen production using renewable energy sources offers the possibility of reducing dependence on fossil fuels and enhancing security of supply. Other examples are the dispatch of electrolysers or the possible co-production of electricity and hydrogen in IGCC plants (with CCS), which is an important aspect because such a plant design offers the opportunity of producing for two different markets, depending on the market prices for the products. Hydrogen can also be used as a storage medium for electricity from intermittent renewable energies, such as wind energy. The various aspects of the interplay between hydrogen production and electricity generation are addressed in this chapter.

16.1 Hydrogen from intermittent renewable-energy sources

16.1.1 Fluctuating renewable energies and hydrogen

The markets for wind power and also for photovoltaic or solar thermal power are rapidly growing (for details on renewable energies and their market development, see Chapter 5). Despite clear advantages (renewable, CO_2-lean or free), the inherent characteristics of wind- and solar-generated electricity lead to several challenges. These resources are intermittent, differ in their seasonal availability and secure capacity is low, which makes it more difficult to predict power output than for conventional power plants. One additional barrier for these resources is that they depend on local conditions, like wind and place, and, therefore, the transport of electricity over long distances to demand centres could be necessary. One problem is that the capacity of our electricity grids is often restricted, because today's grids are not designed for the transport of electricity over long distances. In regulated electricity markets, the power plants were constructed near the demand centres, since the

The Hydrogen Economy: Opportunities and Challenges, ed. Michael Ball and Martin Wietschel. Published by Cambridge University Press.

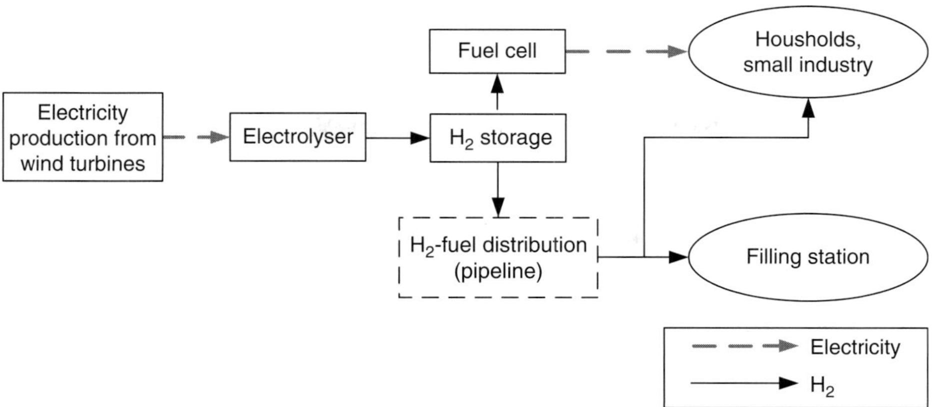

Figure 16.1. Example for the design of a self-sufficient energy system based on hydrogen.

transport of electricity is more expensive than the transport of primary energy carriers, like gas, coal or uranium.

Different options for solving these kinds of problem, like extension of power grids or storage in compressed air systems, exist. Also hydrogen could be a solution here, since it offers the opportunity for storing the energy, transporting the energy, and using the energy universally. Large amounts of hydrogen can be stored underground. Additionally, hydrogen pipelines offer the possibility of storing relevant amounts of hydrogen, as today the natural-gas grid fulfils two requirements: the transport and the storage issue. Furthermore, hydrogen can be stored in tanks of vehicles. Here the principal idea is that if electricity has to be stored it should be stored in the application field (the transport sector), where storage is inherently necessary.

Among others, a flexible solution for isolated or weak grids would be to invest in energy-storage systems using hydrogen as the universal energy vector (for electricity, heating and transport). This option will be discussed briefly in Section 16.1.2.

Closely linked to this discussion is the hydrogen corridor question, where hydrogen produced from renewable energies is transported over long distances and country borders. Owing to a lot of specific questions, this issue is dealt with in a stand-alone chapter (see Chapter 17).

16.1.2 Self-sufficient energy systems using hydrogen

In remote areas, such as low-populated islands without any access to the electricity grid, fluctuating energy resources, like wind and sun, need an energy storage or a complementary electricity generation unit to guarantee power also in periods of no or low wind and sunshine. Normally, diesel generators or, in some cases also batteries, take this job. Production of hydrogen via electrolysis, storage of hydrogen, and re-electrification of hydrogen through fuel cell or gas turbines could be an alternative. Figure 16.1 shows a possible layout for such a self-sufficient energy system using hydrogen.

From 1990 onwards, a number of demonstration projects have been carried out, which both demonstrate the technical feasibility of wind-hydrogen systems and identify areas where further research is needed.[1]

The demonstration projects are designed as stand-alone solutions on islands or other remote areas for supplying households and transport or for testing purposes (e.g., component integration or intermittent electrolyser optimisation). Owing to the growing significance of wind energy in many countries in the last decade, most of the demonstration projects focus on wind-hydrogen solutions. Wind turbines with smaller capacities (10 to 600 kW) and electrolysers with a capacity between 2 and 5 kW are used. Norsk's Utsira project in Norway is showcasing hydrogen's ability to store intermittent wind energy, albeit with just ten homes connected to the generating turbines (www.hydrogen.no/hydrogenaktiviteter/prosjekter/utsira-vind-hydrogenprosjektet?set_language = en). Hydrogenics' Prince Edward Island Hydrogen Village in Canada is a more ambitious effort, and uses the company's electrolysers to create an energy source that can power vehicles as well as store the often unpredictable surges and drops in wind-generated electricity (www.hydrogenics.com/in_newsdetail.asp?RELEASEID = 161016). A hybrid power plant is also scheduled to start operation in Germany in 2010, involving a 6 MW wind park to produce electricity, and to store excess energy in the form of hydrogen produced by a 500 kW electrolyser (www.enertrag.com).

The market of self-sufficient energy systems can be seen as one of the first niche markets of hydrogen. In this context, the market potential is high. In Greece alone, more than 150 populated islands exist and in remote areas of developing countries without a well developed energy infrastructure, like the Philippines, this could be a promising option. However, as for other applications, the costs, especially for the fuel cell, have to be reduced significantly to be cost-competitive with today's conventional solution. One specific economic problem with very-small-scale solutions is the influence of missing economies of scale, as well as low operating hours of the equipment (particularly for electrolyser and transport equipment). In such cases, high investments have a significant influence. Additionally, more conceptual work is required for optimised system configuration.

A more detailed discussion of self-sufficient energy-systems using hydrogen can be found in Schönharting and Nettesheim (2006) and Altmann *et al.* (2000).

16.1.3 Large stranded renewables and hydrogen

Much of the world's richest wind and other renewable resources are stranded. It is necessary to build many large, new transmission systems to bring this energy to distant markets. For example the wind resources in the Russian Far East and the

[1] For an overview on actual demonstration projects, see Altmann and Gamallo (2000) and Geer (2005), and for additional demonstration projects, see PURE (2006) and Schatz (2006).

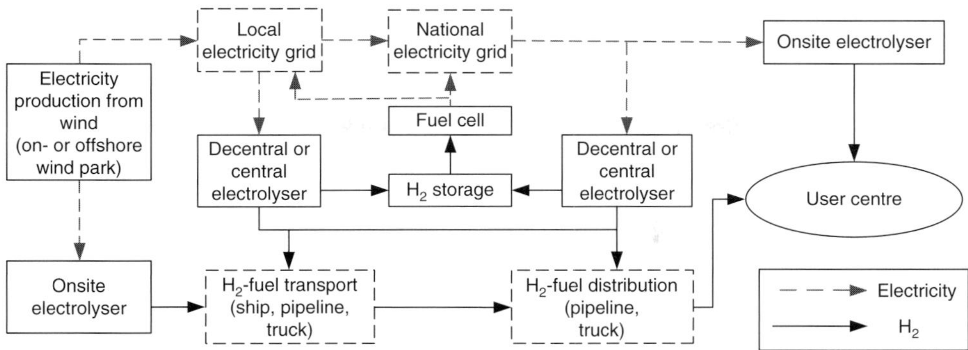

Figure 16.2. Example for the design of an energy system based on hydrogen.

Great Plains of North America are stranded, without means for gathering and transmitting the energy to distant markets. The wind resources of the twelve Great Plains states in the USA, if fully harvested annually, would equal the entire energy consumption of the USA in 2002 (around 10 000 TWh) (see Leighty *et al.* (2006) and the discussion in Chapter 17). Earth's richest biomass and direct-insolation resources, like those in North Africa, are also stranded.

Hydrogen, as a storage and transport medium for long distances between large renewable production locations and user centres, where today's electricity grids are not able to transport the electricity because of capacity restrictions, could be an interesting option (see Fig. 16.2). However, relevant barriers also exist. The question of creating large hydrogen corridors, with a focus on renewables, but also discussing it for other feedstock is handled in a separate chapter of this book (Chapter 17).

16.1.4 Case study, Germany: comparison of different storage options for surplus wind electricity

16.1.4.1 Storage of electricity from wind energy

Introduction Nowadays, the electricity produced from wind energy is directly fed into the grid in most countries. As long as thermal electricity production is substituted, mainly fuel savings are realised in the conventional electricity system, which leads to emissions and financial savings. If the production of electricity via renewables and the demand are harmonised in time and location, storage of electricity is, in most cases, not attractive from an economical viewpoint. Another situation occurs, when high wind penetration rates are reached and so the wind generation exceeds actual load in a specific area. One option is then the transmission of electricity to other regions. If grid extension measures are necessary, the storage of electricity from wind energy can be an attractive financial option to avoid this. If penetration rates increase further, situations can appear where all demand is already covered by renewable generation. When using storage options, a curtailment of renewable generation is not required and further potential can be exploited. For both circumstances, the production of hydrogen can be a valuable storage option. However, hydrogen as a storage option stands in competition with other storage options, such as compressed air.

In a case study for the north-western region of Germany, different storage options for surplus wind generation will be analysed. The north-western region has a huge wind generation potential, especially in the offshore region of the North Sea (see Fig. 16.3). On the other hand, only a few load centres are located in that area, making it necessary to build new grid lines for electricity transmission or storage options, if the full potential is to be exploited.

Simulation of wind power generation To analyse the fluctuating wind generation from on- and offshore locations in the region, a simulation model for wind generation is used (developed by Sensfuß *et al.* (2003; 2004)). The simulation is based on wind-speed data from the German weather service (DWD) that are converted into power output. Offshore locations are simulated with wind-speed data from the islands in the North Sea. For the conversion into power output, the power characteristics of typical wind turbines are used. Further input parameters are the air temperature and density. Each wind-speed measurement point serves as a representative location that has a specific power capacity installed, a specific kind of turbine and a certain roughness of the surrounding. The power output P mainly depends on the third power of the wind speed v. P in one time interval i for one location (Loc) and one turbine type (Typ) is then calculated by:

$$P_{i,\mathrm{Loc},\mathrm{Typ}} = 0.5 \cdot \rho_{S_0} \cdot A_{\mathrm{Rot},\mathrm{Loc},\mathrm{Typ}} \cdot c_P(v_{i,k,\mathrm{Loc},\mathrm{Typ}}) \cdot v^3_{i,k,\mathrm{Loc},\mathrm{Typ}}$$

with

ρ_{S_0} = Standard air density,
$A_{\mathrm{Rot},\mathrm{Loc},\mathrm{Typ}}$ = Rotor area,
c_P = Power value depending on wind speed,
$v_{i,k,\mathrm{Loc},\mathrm{Typ}}$ = Hub wind speed depending on air density.

The wind speed is given for a height of 10 m above ground and is converted into wind speed at hub height using a logarithmic height profile of the wind speed. The main parameter of the height profile is the roughness of the surrounding area. Furthermore, the wind speed is altered by air density, because the power output of wind turbines depends on the air density.

Input data Wind-speed measurements for 176 locations are used for the simulation covering a time period of one year. The data were collected in 1998, which was an average wind year. The surplus electricity of wind energy is calculated in four

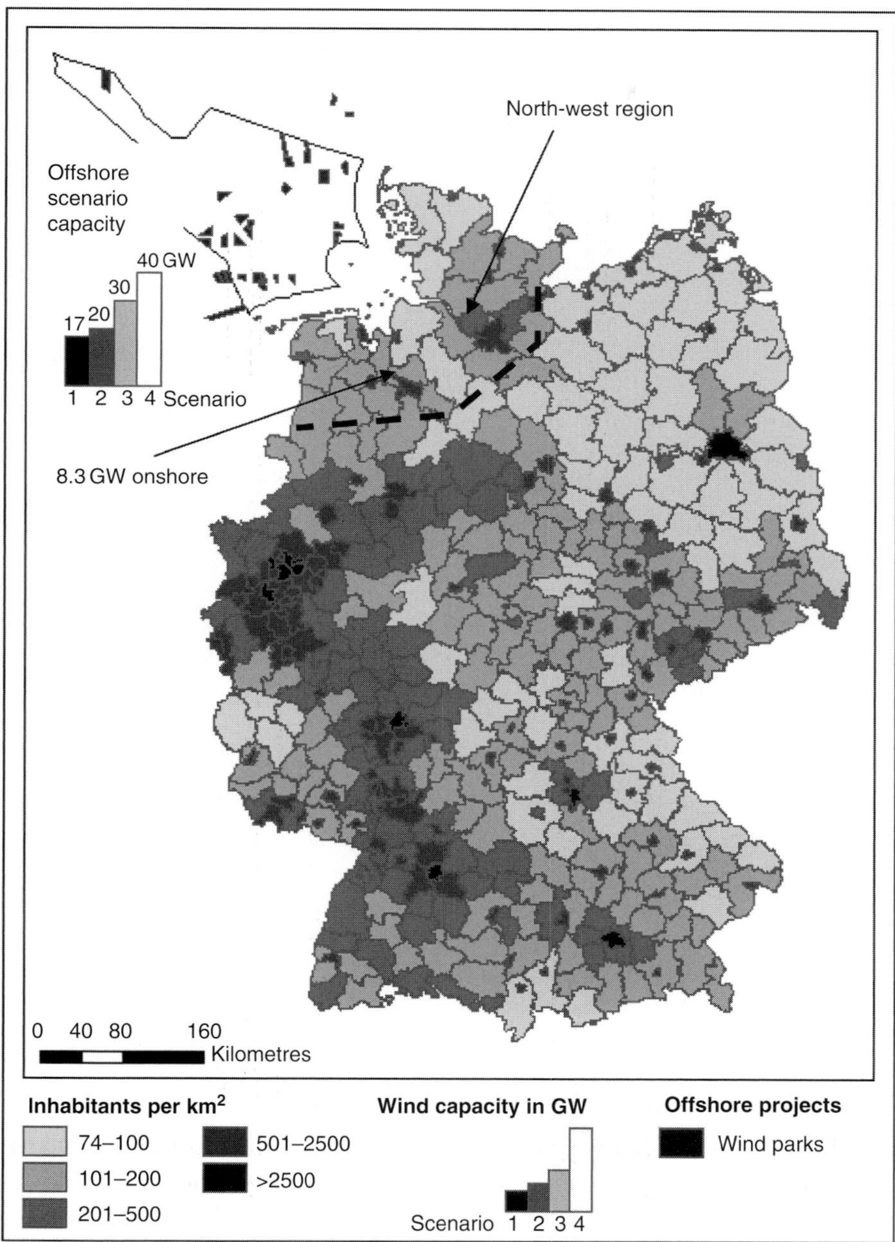

Figure 16.3. Case study for integrating offshore wind energy in the north-western region in Germany – location of wind capacities and demand centres (own illustration based on DENA (2005; 2007)).

scenarios assuming increasing offshore wind capacity. In the simulation, only part of Germany, the north-western region, is considered (see Fig. 16.3), analogical to the classification in DENA (2005).

The onshore capacity stays constant in all scenarios at 8.3 GW. The offshore capacity in the North Sea increases from 17 GW in the first scenario to 40 GW in the fourth scenario, and is 20 and 30 GW in the second and third scenarios, respectively (first scenario based on DENA (2005); Ragwitz *et al.* (2005)). In 2007, Germany's installed wind capacity was 20 GW, generating 5% of Germany's electricity demand. (The total power generation capacity of coal, gas, nuclear and hydropower amounts to about 110 GW.) In Scenario 1, the north-western region would only provide 13% of today's electricity demand, which increases to 26% in Scenario 4. The potential for offshore wind generation is huge. Planned projects have a capacity of more than 40 GW. It is expected that 20 GW can be realised until 2020 and a further 16.5 GW in the following years (DENA, 2005). The four scenarios imply a strong development of offshore capacity. For this reason, a high share of surplus wind generation is expected.

The turbine technology is represented by five different turbine types in the simulation covering turbines between 500 kW and 4.5 MW. The highest shares have 1.5 and 2.3 MW turbines on onshore locations. For the offshore locations, the characteristic of a 4.5 MW turbine (ENERCON E112) is used.

Electricity demand in the north-west region For the determination of the dynamic electricity demand in the north-western region of Germany, the load curves for Germany are used as a basis (UCTE, 2000; 2005). The load curves on the electricity transmission grid are published by the Union for the Co-ordination of Transmission of Electricity (UCTE) for every third Wednesday in a month. Data for the weekends are published for 2000. Using this approach 90% of electricity demand is considered. Only electricity generation that is not fed into the grid is not covered by these statistics. The generators are in general industrial enterprises that use the electricity for their own purposes. In the later analyses, this demand is not considered in the calculation of surplus wind energy.

The load curve of Germany is then scaled by the peak demand in the north-western region to get the local load curve. In DENA (2005), the peak demand is indicated at 7.3 GW for the region, which is much under the potential of wind capacities.

Calculation of surplus wind generation Hourly time series of wind generation and electricity demand are used for the calculation of the surplus wind generation. Furthermore, some assumptions are made following the arguments in DENA (2005). At least 700 MW of nuclear power plants and 160 MW of coal power plants are necessary to secure grid stability. The transmission capacity of the grid is

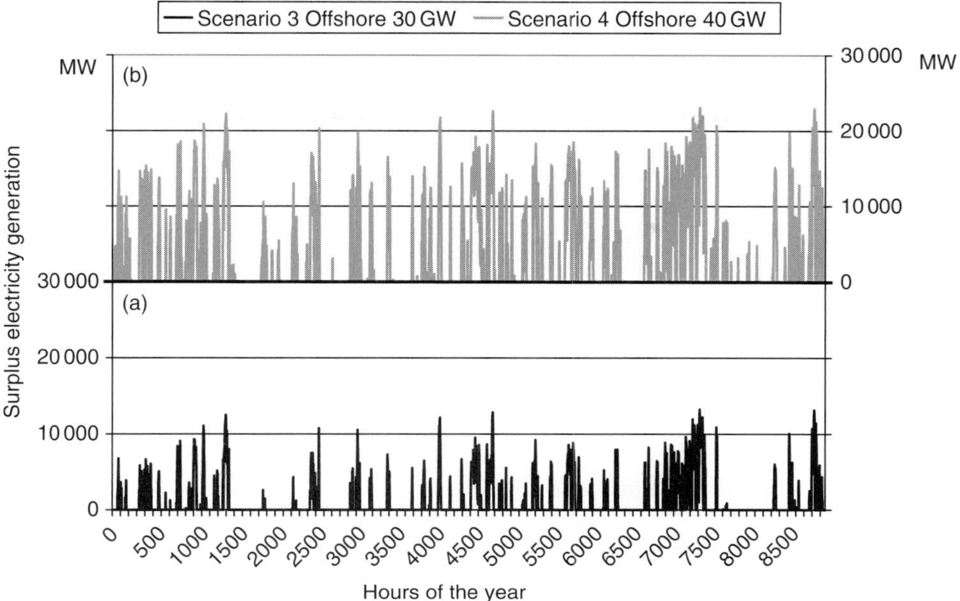

Figure 16.4. Time series of surplus electricity generation in (a) Scenario 3 and (b) Scenario 4 (simulation results).

extended to export electricity out of the region. It is limited to 19 GW, which constitutes an extended grid infrastructure compared with the situation today.

The surplus electricity is then calculated from the electricity generation that does not supply local demand or that cannot be exported to other regions. The time series of surplus electricity generation is shown in Fig. 16.4 for Scenarios 3 and 4. The fluctuating character of the wind generation can also be observed in the surplus electricity generation. In the case of high installed offshore capacity, high surplus peaks also occur, but only a few times in the year. The peaks can reach up to 23 GW in the fourth scenario. On the other hand, there are also longer time periods with no surplus electricity generation.

If the time series is sorted in order of the surplus electricity generation, a power duration curve can be obtained (see Fig. 16.5). It can be seen that in the fourth scenario, almost 3000 h per year surplus electricity is generated. In the other scenarios, the duration of surplus electricity decreases to 1800 h and to 180 h, respectively. In the first scenario, no surplus electricity is generated.

A summary of the simulation results is shown in Table 16.1. In the first scenario, no surplus electricity is generated, because the grid is extended to cope with this amount of wind energy. In the fourth scenario, the surplus electricity increases to 17.3% of the wind generation. If the surplus electricity should be used in a storage system, it is necessary to dimension the storage option. If the storage option should

Table 16.1. *Simulation results for wind and surplus electricity generation in four scenarios (own simulations)*

		Installed wind capacity scenarios			
		1	2	3	4
Onshore wind capacity	GW	8.3	8.3	8.3	8.3
Offshore wind capacity	GW	17	20	30	40
Wind electricity generation	GWh	77 431	88 394	124 935	161 476
Surplus electricity	GWh	0	244	8 825	27 965
Share of wind generation	%	0	0.3	7.1	17.3
Peak surplus generation	GW	0	3.5	13.3	23.1
Surplus electricity used in storage option (min. operation 2000 h/year)	GWh	0	0	0	14 348

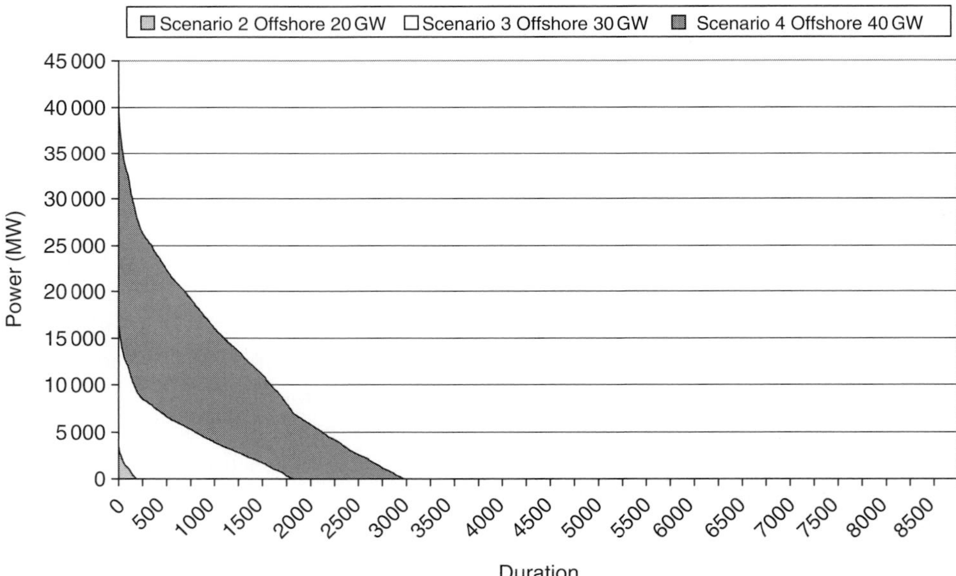

Figure 16.5. Power duration curve of surplus electricity generation.

be used, for economic reasons, for at least 2000 h per year, in the fourth scenario, only, a storage option with a capacity of 5.8 GW can be installed. The 2000 hours seem to be necessary because the investments in storage options (compressed air or electrolysers for hydrogen production) are capital intensive, so that a minimum operation time is required. In this case, it can be operated at full capacity 2000 h per year and uses 51% of the surplus electricity.

16.1.4.2 Analysed storage options

There are different possibilities for storing and using large amounts of surplus wind electricity. In the German case study, compressed-air energy storage (CAES) and storage in hydrogen are compared with respect to carbon emission reduction and costs. For geographical reasons, pump storage is no option and other storage options do not fit the requirements of storing large amounts of fluctuating energy. The comparison is made for different application fields: the re-electrification and network supply (stationary application) and the use of hydrogen in passenger cars (mobile application).

Compressed-air energy storage versus hydrogen in the stationary sector In the compressed-air energy storage (CAES) case, surplus wind electricity is used to compress air, which is then stored in subterrestrial caverns. At peak load, the compressed air is used in a conventional gas turbine to produce electricity. The gas-turbine process is still dependent on fuels (e.g., natural gas); compressed air is merely used to improve the efficiency of the process. In the study, the analysis is based on two existing CAES power plants: Huntorf (Germany, 1978) and McIntosh (USA, 1991) and one fictive CAES plant that reflects state-of-the-art technology (GT-26).

In the case of the hydrogen path, it is assumed that surplus wind electricity is used to produce hydrogen via electrolysis (efficiency 70%). Hydrogen is then stored in pressure tanks at 50 bar and is re-electrified at peak load in gas turbines (GT, efficiency 40%) or gas–steam turbines (GST, efficiency 60%), with a hydrogen-to-natural-gas ratio of 8:2.

The carbon emission reduction of the two considered storage options is calculated with reference to a conventional gas turbine or gas and steam turbine. The result is shown in Fig. 16.6. The emission reduction refers to 1 kWh surplus wind electricity. The black bars reflect the reference emissions of the conventional gas turbine (GT) and gas–steam turbines (GST). The other bars show the figures for the CAES and the hydrogen paths. The emissions that occur during the storage paths are marked in grey; the emission reduction is visualised in grey and white stripes.

In this comparison, the biggest emission reduction is achieved by the CAES paths. They are of the order of 317 to 373 g/kWh wind electricity, while the emission reduction of the hydrogen pathways is between 159 and 169 g/kWh wind electricity.

For the economic comparison of the two storage options, the specific re-electrification costs of the stored wind energy are calculated. These costs are made up of the investment, operation and maintenance costs, input electricity costs (wind electricity) and the fuel costs (natural gas). As hydrogen technologies are not in a commercial state, the calculation is also performed with target costs for electrolysers. Carbon emissions are also monetarily included, assuming a certificate price of € 20/t. Table 16.2 summarises the major economic assumptions.

Comparing the real costs of hydrogen and CAES storage paths on the basis of re-electrification with gas turbines (Fig. 16.7) on a basis of 2000 operating hours, the

Table 16.2. *Assumptions for economic calculation*

	GT-26	Huntorf	Electrolyser	Unit
Investment (real)	500	600	1100	€/kW
Investment (target)			400	€/kW
O&M (real)	2	4	3	% of investment
O&M (target)			1.5	% of investment
Surplus wind-electricity (off-peak)	4	4	4	ct/kWh
Natural gas price	2.35	2.35	2.35	ct/kWh
CO_2 certificate price	20	20	20	€/t

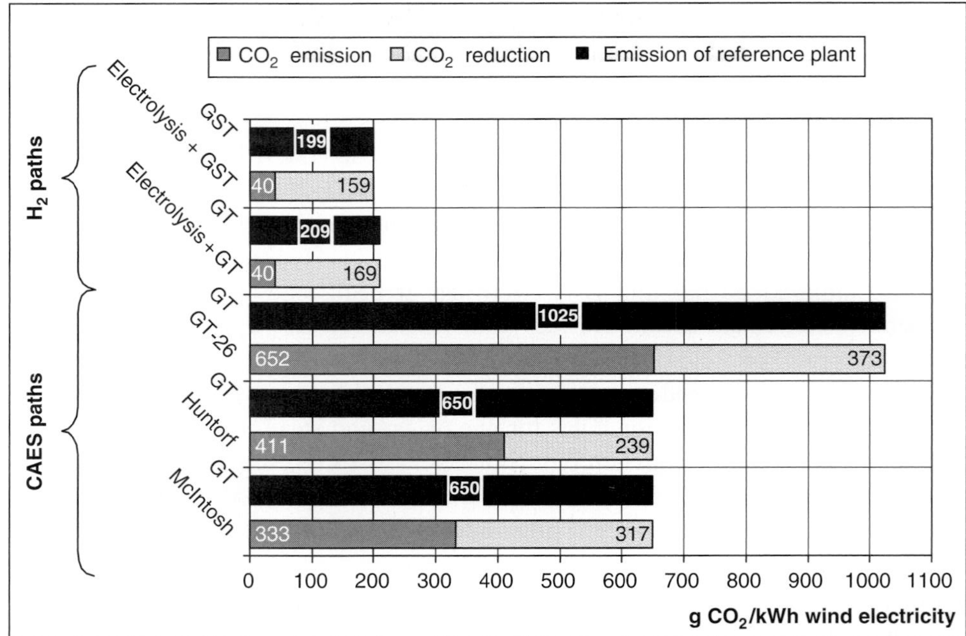

Figure 16.6. Carbon emission reduction for 1 kWh stored and re-electrified wind electricity for CAES and hydrogen storage options (Wietschel *et al.*, 2006).

CAES paths perform significantly better with output electricity costs between 21 and 28 ct/kWh (hydrogen path: 40 ct/kWh). Only if better conversion technologies are applied for the hydrogen paths (GST instead of GT), or the calculations are based on target costs, do the economics become significantly better. The CAES paths would also perform better with GST plants and may also exploit further cost reduction potential.

Automotive paths It has been shown that hydrogen as a storage option for surplus wind electricity has no advantages; neither with respect to carbon emission reduction

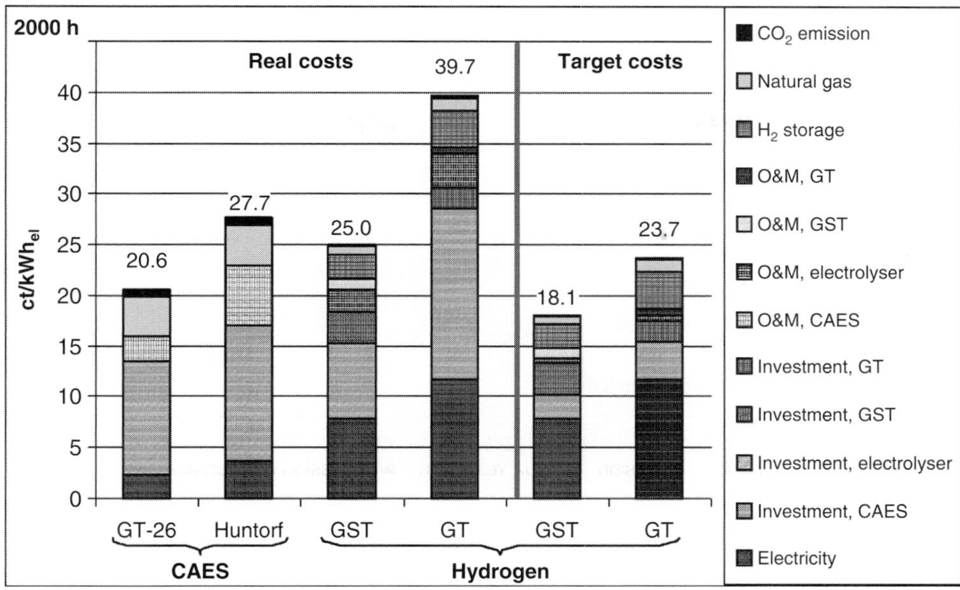

Figure 16.7. Specific costs per kWh output-electricity of CAES and hydrogen storage paths.

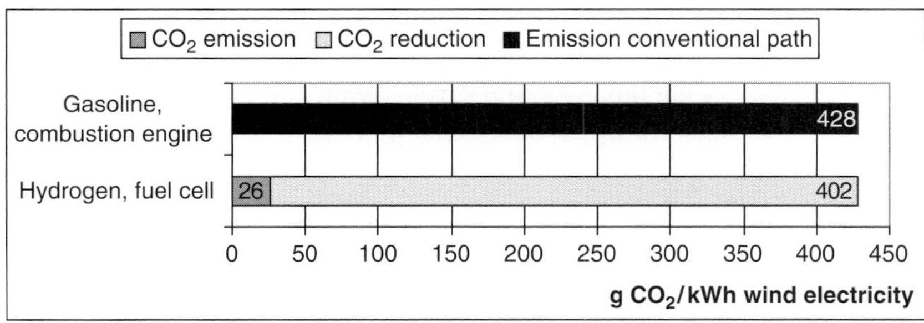

Figure 16.8. Carbon emission reduction by using hydrogen from surplus wind electricity in the automotive sector.

nor to economics when the purpose is re-electrification and grid integration. In this case, compressed-air energy storage is the favourable option. However, if hydrogen is used in the automotive sector, the situation is different. In this case, wind electricity is used to produce hydrogen via electrolysis (efficiency, 70%) at the filling station. It is compressed to 880 bar (using grid electricity), refuelled into the vehicle and converted in a fuel cell (vehicle efficiency, 44%). The reference system for this option is gasoline in a combustion engine (vehicle efficiency, 23%). The result is shown in Fig. 16.8: the carbon emission reduction is 402 g/kWh for one kWh wind electricity.

Compared with the previous results – the carbon emission reduction by storing and re-electrifying surplus wind electricity, it can be seen that the highest carbon emission

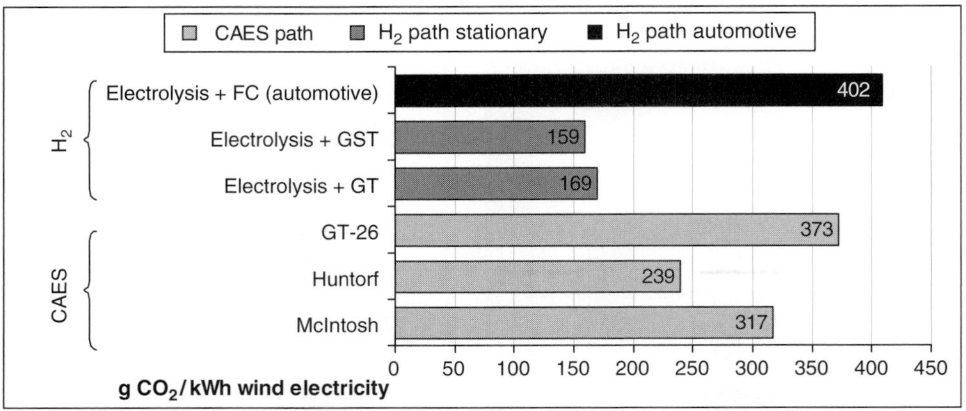

Figure 16.9. Carbon emission reduction per kWh surplus wind electricity used in the stationary and automotive sectors.

reduction per kWh surplus wind electricity is achieved, if the electricity is used to produce hydrogen for automotive applications (compare this with Fig. 16.9).[2]

Hydrogen from surplus wind electricity as vehicle fuel is fairly expensive. According to the calculations and assumptions of Wietschel *et al.* (2006) the price is 9.5 ct/kWh hydrogen. This is calculated with the assumptions shown in Table 16.2, compression costs of 1.5 ct/kWh and electricity costs from wind power of 4 ct/kWh. However, the 4 ct/kWh, which has a major influence on the total cost (see Fig. 16.10) is debatable, because we are talking about surplus wind and the question is what other alternatives for use exist. An opportunity cost approach may be useful, e.g., taking into account the gains of other uses, like electricity production via the compressed air option, which will lead to a much lower price for the electricity.

For the economic evaluation, this price is compared with the costs of cheaper hydrogen production options, like natural gas reforming. With an operating time of 6750 h and a natural gas price of 2.35 ct/kWh, hydrogen costs are at 5.3 ct/kWh. This is much lower than the surplus wind pathway, if an electricity price of 4 ct/kWh is assumed. Further calculations have been performed, to show at what natural gas price natural gas reforming would reach hydrogen costs from surplus wind electricity: the hydrogen price of surplus wind electricity is only reached at a natural gas price of 5.5 ct/kWh. If a carbon tax of €20/t is introduced, the necessary natural gas price is 5 ct/kWh (compare Fig. 16.10).

For comparison, the average gasoline price in 2006 was 14.5 ct/kWh with and 4.1 ct/kWh without taxes (in Germany (MWV, 2006)). The price of hydrogen from surplus wind electricity is thus more than double the price of gasoline (assuming that

[2] The calculations of the carbon emission reduction are based on the deviation from the reference-system gasoline in a combustion engine. If reference systems with higher efficiencies are applied, e.g., hybrid-electric vehicles, the total carbon emission reduction of the hydrogen path is lower. In this case, a modern CAES power plant might have benefits.

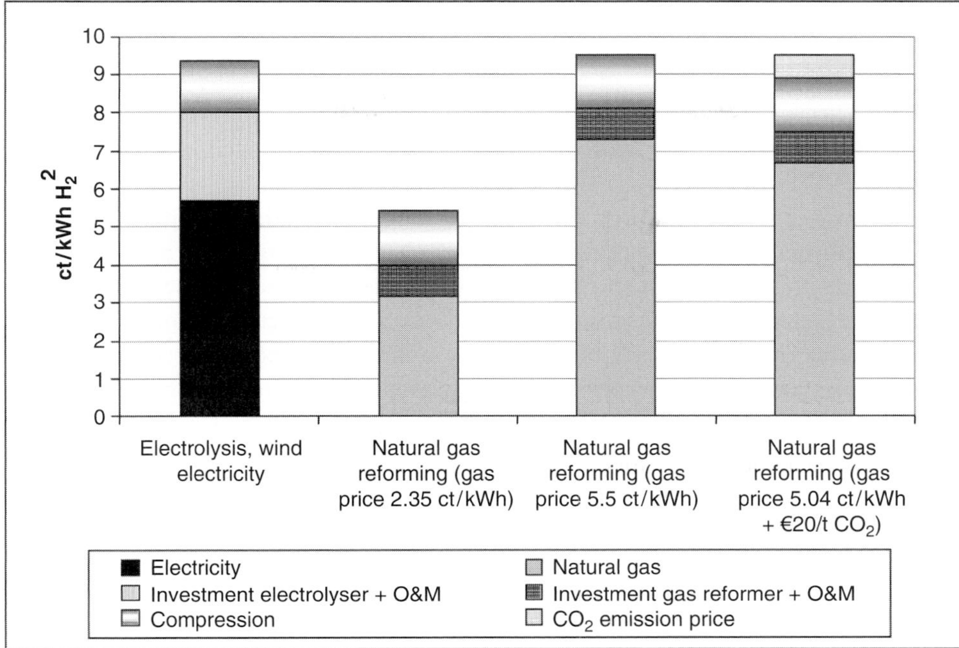

Figure 16.10. Cost comparison: hydrogen from surplus wind electricity and from natural gas reforming at different natural gas prices and with and without carbon taxes.

cost targets of hydrogen technologies are realised). On the other hand, the efficiency of fuel cells is about twice as high, compared with internal combustion engines. For a complete economic evaluation, the costs of the propulsion system are also relevant. For fuel cells, however, these costs are hard to define (see Chapter 13). They are currently far more expensive than conventional drive concepts, but they do have a large cost reduction potential in technological improvements and mass production. Some experts believe that the costs of fuel-cell vehicles may drop as low as or even below those of conventional vehicles.[3]

16.1.4.3 Summary of case-study results

In the scope of this case study, alternative storage options for surplus wind electricity production in the region of north-western Germany are analysed. One assumption is that today's limited grid capacity could be extended only by a certain amount. The comparison of the storage option is based on economic and ecological criteria.

For alternative scenarios of capacity extension (from 25 GW to 48 GW installed wind capacity), the amount of surplus electricity that has to be stored is calculated on

[3] Compare Smekens *et al.* (2002); Tsuchiya and Kobayashi (2002); Blesl and Ohl (2003); the IEA (2005) and, for a more pessimistic view, Gielen and Simbolotti (2005).

the basis of a simulation model. The results show that from 38 GW installed capacity onwards, surplus wind electricity could be identified. In the scenario with 38 GW, 8.8 TWh/year (7.1% of all produced electricity) and in the scenario with 48 GW, 30 TWh/year (17.3%) is surplus electricity that has to be stored, when a use is desired. Compressed air or hydrogen are storage options for such large amounts of electricity.

In the case of re-electrification of the stored energy for stationary applications, the storage option of using compressed air has a clear advantage over hydrogen. The CO_2 savings are higher and the costs are lower. In the next case, the use of the surplus electricity for mobile application via hydrogen is analysed. It could be shown that here the CO_2 savings are higher than for the case of compressed air with re-electrification for stationary applications. However, taking into account other hydrogen production options, like steam methane reformers, the cost of surplus electricity is a barrier, if the surplus wind electricity production is calculated on a full-cost basis. But it could be argued that an alternative use of this wind surplus electricity now exists and, therefore, the costs are near zero (only operation costs and costs for lifetime shortening are relevant). In such a case, this option is cost-competitive compared with other hydrogen production options.

16.2 Co-production of hydrogen and electricity

16.2.1 Plant concept for co-production of hydrogen and electricity

Today, different processes (steam reforming, autothermal reforming, partial oxidation, gasification) are available and commercially mature for hydrogen production from natural gas or coal. These processes would have to be combined with technologies for CO_2 capture and storage (CCS), to keep the emissions profile low. A power plant that combines electricity and hydrogen production can be more efficient than retrofitted CO_2 separation systems for conventional power plants.

Conceptually, these plants could be designed to deliver only hydrogen, only electricity through combined cycle plants, or a mix of both. The main underlying principle is to convert any carbon-containing fuel into a syngas that usually contains a mixture of carbon monoxide (CO) and hydrogen (H_2). When derived from a solid fuel such as coal, cleaning of the syngas especially of dust is required prior to further conversion. With a water-gas shift reaction, the CO together with water is converted into CO_2 and more hydrogen. From this gas stream, the CO_2 can be separated in a gas-separation process.[4]

The use of hydrogen for electric energy production from fossil fuels in large centralised plants will contribute positively to achieve important reductions of CO_2

[4] Several studies have looked in detail at the concept of co-generation of electricity and hydrogen, e.g., Chiesa *et al.* (2005); Kreutz *et al.* (2005); IE/IPTS, (2005); Yamashita and Barreto (2003), as well as the EU R&D project Dynamis (www.dynamis-hypogen.com).

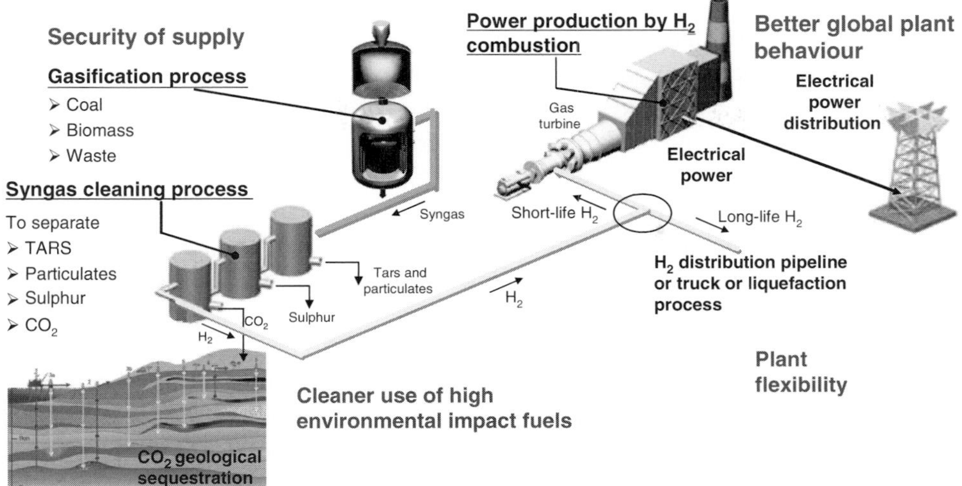

Figure 16.11. Combined production of hydrogen and electricity (HyWays, 2007).

emissions if combined with CO_2 capture and sequestration processes. Such plants could also help to increase the diversification of resources, since a variety of fossil feedstocks, including resources, such as coal and waste that otherwise cause major impacts on the environment, as well as biomass can be used as fuel. The general principle of co-producing hydrogen and electricity from coal is illustrated in Fig. 16.11.

Within such a plant, depending on the pressure of the syngas, the separation can be performed by chemical absorption (usually with amine solvents) under lower pressure conditions or by physical absorption (e.g., with methanol) under higher pressure conditions (see also Chapter 6). Likewise, pressure-swing absorption can be employed. With the special properties of hydrogen, membrane separation processes could also be a very promising solution for the separation task.

The separation process yields an energy-rich gas stream of hydrogen that can be used without generating further greenhouse gases. Furthermore, a stream of highly concentrated CO_2 is produced that can be taken to storage after being cleaned of major impurities.

The described process is very much comparable to the already employed technology of integrated gasification combined-cycle power (IGCC) plants. The IGCC plants first make use of a gasification process to convert the feedstock into a syngas. The syngas is then cleaned of undesired components, like dust and sulphur. Finally, the syngas is used as a fuel in a combined-cycle process for electricity generation. The advantages of an IGCC lie in the possibility of making use of the highly efficient combined-cycle process while using fuels like coal or petroleum coke that normally cannot be used with a gas turbine. Further, the production and subsequent cleaning of a syngas allows the emission of air pollutants to be reduced very effectively and very efficiently. When comparing the already existing IGCC plants with a future co-production of hydrogen and electricity, it should be noted though, that the IGCC lacks the steps of the water-gas shift and of CO_2 removal. Instead, the cleaned syngas, containing CO and H_2, is taken directly to a gas turbine for electricity production. The additional treatment of the syngas for the CO_2 separation can be viewed as further development of the process to allow a reduction of greenhouse-gas emissions.

With an intermediate stream of hydrogen being part of the concept of an IGCC with CO_2 capture, the idea of designing plants producing both hydrogen and electricity is evident. Technically, this would mean diverting part of the hydrogen stream coming from the CO_2-separation plant and marketing it directly as a product. The remainder of the hydrogen would then be used as a fuel for electricity generation in a combined-cycle gas turbine plant.

Hydrogen originating from a gasification process requires additional treatment to remove impurities. The type and effort of the treatment depends on the purity requirements of the possible target market. Large users of hydrogen today are the refinery industry, ammonia producers and methanol producers. There are many other industrial processes where hydrogen is used, such as in the electronics, food and glass industries.

The concept of co-production of hydrogen and electricity is also interesting in that it allows investment in a plant that can operate in a mature electricity market. At the same time, it can participate in an emerging hydrogen market at reduced economic risks. Worldwide, several projects have been announced that aim at the realisation of demonstration plants producing electricity with CO_2 capture and storage by use of IGCC technology. These demonstration plants could produce hydrogen for external use at comparatively low efforts, and some projects explicitly aim at the co-production of hydrogen and electricity. The currently most visible projects are:

- The Futuregen initiative in the United States (see www.futuregenalliance.org; at the time of writing, it was not yet clear whether the project would go ahead),
- The HYPOGEN programme of the European Union,
- The GreenGen initiatve in China (see www.greengen.com.cn/en),
- The ZeroGen project in Australia (see www.zerogen.com.au),
- The IGCC demonstrator in Japan.

The HYPOGEN programme of the European Union could stand as an example of the current projects aiming at the co-production of hydrogen and electricity from fossil fuels with CO_2 capture and storage. The programme in its original scope envisages the large-scale demonstration of this co-production in a project organised as a public–private partnership. According to the programme layout from 2001, the demonstrator plant should start operation by 2013. Within the European Union research project 'Dynamis', a pre-engineering of possible plants and an investigation of sites for storage locations was undertaken (see www.dynamics-hypogen.com). The principal plant concepts investigated in this project were coal-based IGCC plants with hydrogen production and CO_2 capture. On the other hand, a plant concept based on the use of natural gas was analysed with a post-combustion capture of CO_2 associated with a combined-cycle plant for electricity production. The hydrogen generation in the gas-based concept is envisaged by a parallel gas reforming unit.

16.2.2 Results from MOREHyS model runs

As described in Chapter 14, one essential aspect of the MOREHyS model is that the modelling approach integrates optimisation of the hydrogen infrastructure build-up with system planning and power-station management and takes into account interactions between the supply of hydrogen and the existing energy (electricity) system. This section briefly illustrates, on the basis of the MOREHyS model, how the concept of co-producing hydrogen and electricity can be integrated into the existing energy system (for the example of Germany; for more information, see Ball (2006)). In the following, the influence of high natural gas prices and CO_2 restrictions for the power sector on the hydrogen production mix, as well as the interactions with the electricity generation, are discussed, in particular with respect to the deployment of IGCC plants (with CCS). (In accordance with Chapter 14, it is assumed that CCS is available from 2020 onwards.) Accordingly, the MOREHyS reference scenario is extended by two scenarios: one *with* CO_2 constraints for power generation and one *without* CO_2 constraints for power generation.

For both scenarios, a clear shift in hydrogen production to coal-fired IGCC plants could be identified, as compared with the Reference scenario. In the scenario *with* CO_2 capping, IGCC plants (with CCS) are triggered by the CO_2 constraint on electricity generation as well as the high gas prices, as, besides hydrogen, electricity can be provided (nearly) CO_2-free (see Fig. 16.12b). Without such a constraint, more conventional coal gasification plants are installed to produce the relevant amount of hydrogen. But also in the scenario *without* CO_2 restrictions, owing to the high gas price, IGCC plants are economically attractive because of the co-production of electricity in connection with hydrogen.

Moreover, for the scenario *without* CO_2 capping, electrolysis enters the model from about 2025 on, as a result of the lower electricity prices (owing to the high gas prices and the lack of CO_2 constraints, a clear switch to more economic coal-fired power generation is observed) and the relatively higher hydrogen production costs for steam methane reformers (owing to the high natural gas price).[5]

Figure 16.12a illustrates, for the scenario *without* CO_2 capping, and as a result of model optimisation, for the year 2030, the load curve of electricity generation for a typical working day and a typical weekend day for Germany, and the resulting

[5] An important aspect of electrolysers is also that they can, in principle, follow the load curve of fluctuating energy production capacities without any time delay (owing to the electrochemical process).

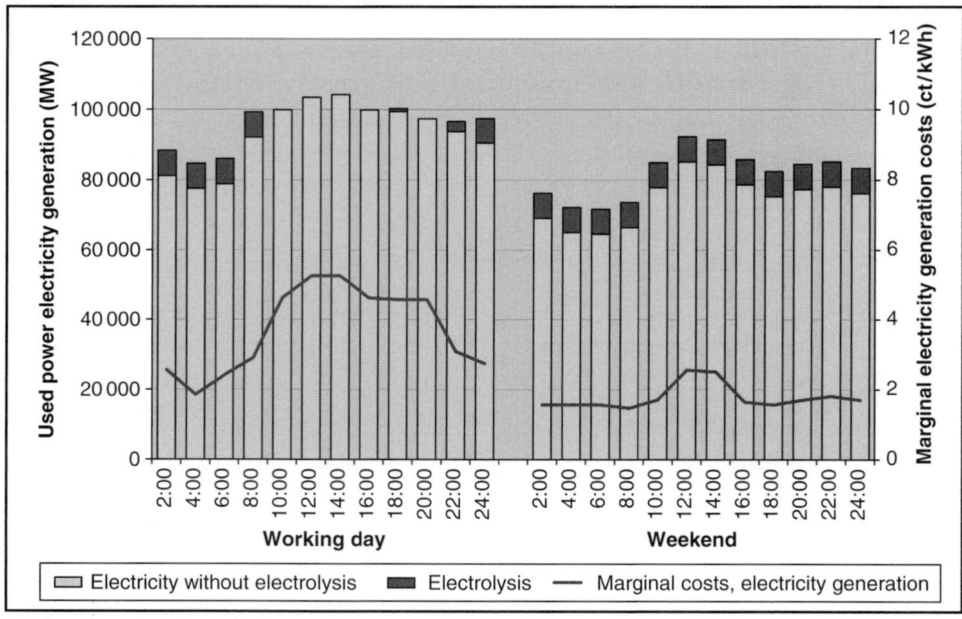

(a) Load curve, electrolysis

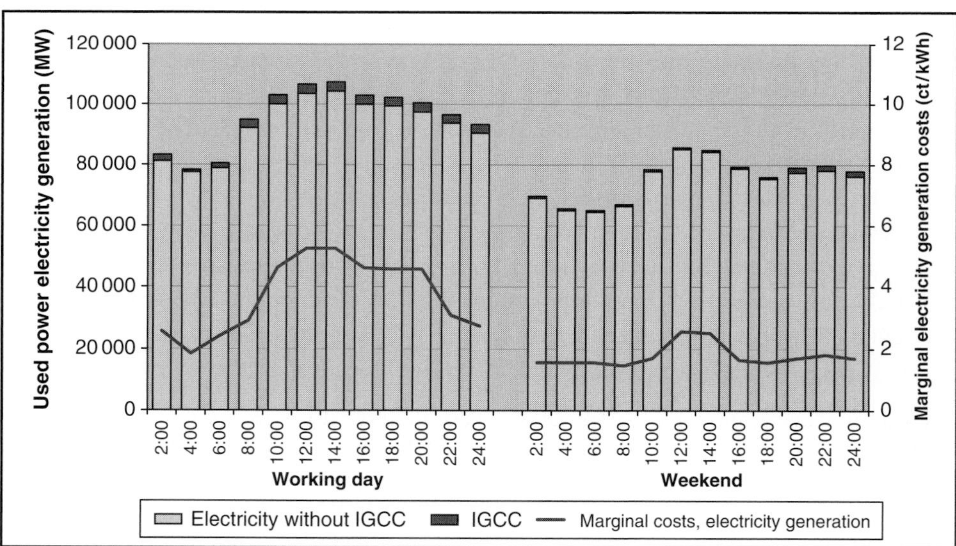

(b) Load curve, IGCC

Figure 16.12. Load curves of electrolysers and IGCC plants.

marginal electricity generation costs, as well as the load curve for electrolysers. It can be seen that the electrolysers produce hydrogen mainly in weak and middle load hours (i.e., during night-times or weekends), when the marginal electricity generation costs (electricity prices) are comparatively low. At the same time, this result shows the effect that electrolysers could potentially be used for load levelling in power grids. It can further be concluded that because of limited electrolyser capacities – owing to the requirement of a minimum number of full load hours for their economic operation (>2000 hours) – not all available excess power generation capacities are utilised. Under the assumption of an existing hydrogen market, it could principally turn out to be an economic option for utilities to use cheap electricity during weak load hours for hydrogen production: the hydrogen could then be either produced by the utilities themselves or the electricity could be sold at higher margins to external hydrogen producers. However, while this might be attractive from an economic point of view, from an energy-efficiency perspective it does not make sense to use fossil-based electricity for the production of hydrogen, which is then converted back into electricity in fuel cells. If produced from off-peak electricity onsite, hydrogen storage at the power plant would also be required.

In Fig. 16.12b, the electricity production of IGCC plants (with CCS) is shown for the scenario with high gas price and a CO_2 emissions cap. Especially during the peak load time, this type of plant is rather used to produce electricity, whereas during the weak and medium load periods more hydrogen is produced. This result shows one important advantage of IGCC plants: they can deliver to two markets, the electricity market and the transport market, depending on the price signals. This could be one major driver for IGCC in the near future, as outlined in the next section.

16.2.3 Market implications of co-production of hydrogen and electricity

A large energy conversion plant for the co-production of hydrogen and electricity will operate in two distinct markets. From an operator's perspective it is definitely advantageous not to be fully dependent on a single market, as adverse conditions in the one market could be at least partially compensated for in the other market. To make full use of this option, the plant could be designed with a flexible yield of hydrogen and electricity (see, e.g., Starr *et al.* (2007) who have analysed in detail the implications of a product flexibility for such plants). With a flexible share of the two products, the operator of a co-production plant can deliver to the market offering the higher margin of profits. For example, the plant could supply electricity only at peak load on the electricity market and hydrogen when the electricity market is at base load or during night-time.

The flexibility of a co-production plant, however, is associated with costs that have to be balanced against the advantages to operate in the two markets. First, additional costs arise from the fact that the combined-cycle power plant or the hydrogen purification plant will have to share the hydrogen from the gasification plant and hence will have a reduced time of usage. The reduced time of usage will produce a higher share of capital costs. Second, when designed for full market flexibility, neither the power plant nor the hydrogen purification plant will operate at the design point of optimal efficiency (see also Chapter 10). Consequently, the energy costs of such a plant will be higher than for a plant with a fixed output ratio.

The additional costs for the plant flexibility have not only to be compared to the potential market income but the effects on the operation mode of the syngas plant should also be accounted for. It is possible that the market flexibility will allow a higher usage time of the entire plant with profitable operation. This given, designing a plant with flexible outputs would contribute positively to the time of use of the gasification plant and of the CO_2-capture plant.

The plant concept for co-production of hydrogen and electricity is applicable to a very broad range of fossil fuels and also biomass without paying tributes to climate change. At the same time energy supply security is improved, as a result of the diversification of (fossil) feedstock options.

The main risk lies in the potential failure of permanent underground storage of CO_2. This requires that special attention be paid to demonstrate the economic and technical feasibility of such processes and the availability of sites to sequester all CO_2 produced (see also Chapter 6). The CCS technologies currently under development could extend the time available to develop a full and durable solution for a sustainable power and fuel provision based on renewable sources.

The use of hydrogen derived from fossil fuels in electricity production will broaden the sectors where such carriers can be used in a sustainable way. There it will provide the opportunity to utilise the advantages offered by hydrogen as demonstrated in the transport sector, enabling the power sector to diversify its feedstocks with very low CO_2 emissions.

16.3 Where best to use renewables – electricity sector or transport sector?

If hydrogen usage is to take off and hydrogen is to contribute significantly to a reduction of transport-related CO_2 emissions and a diversification of energy sources, it must be produced from renewable energies in the long term. In this respect, however, hydrogen is increasingly in competition with stationary electricity generation for feedstock availability. In particular, with respect to the most effective CO_2 reduction the question arises; which is the best end-use sector for (the still-limited) renewable energies: the electricity sector or the transport sector? This point is addressed by means of some simple efficiency calculations.

Figure 16.13 shows the amount of CO_2 that can be replaced by 1 kWh of renewable electricity, when used to replace both fossil electricity and fossil fuels in the transport

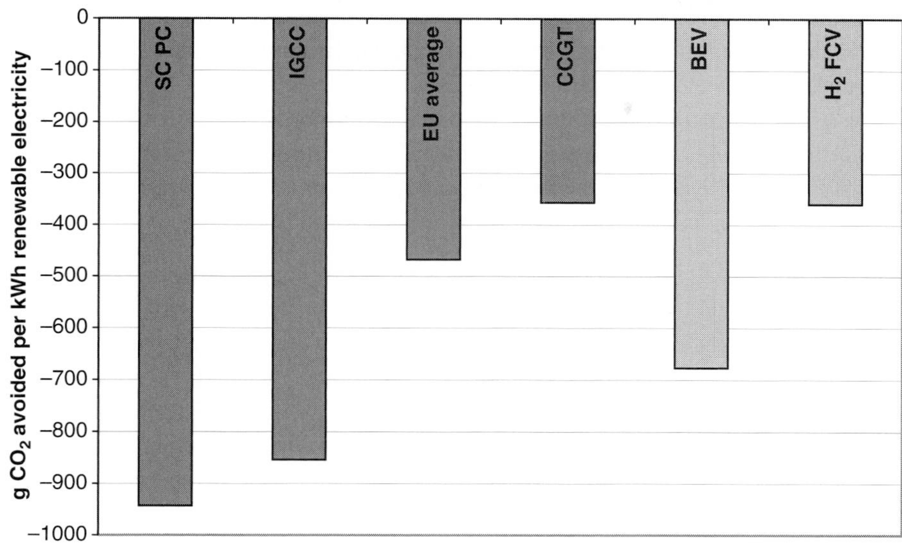

Figure 16.13. Comparative CO_2 reduction of renewables from power generation and transport fuels.

sector; for the transport sector a further distinction is made between use in battery-electric vehicles (BEV) and hydrogen fuel-cell vehicles (FCV).[6]

These results show that renewables should be used first to replace coal-fired power plants in a fossil-based power system (as in Germany, for instance), if CO_2-emission targets exist. Moreover, if produced by renewable energies, hydrogen adds to security of supply and reduced CO_2 emissions only in so far as the non-fossil fuel source is additional to what would otherwise be used in electricity production. This picture may change if electricity from fluctuating renewable sources, like wind or the Sun has to be stored (as shown before). In this case, the use of hydrogen as a storage option could have some advantages, because the potential of cheap electricity options like pump storage plants is limited in Germany (and in Europe). Further analysis should focus on this specific topic.

When looking at the transport sector alone, it becomes clear that the use of renewable electricity in battery-electric vehicles is by far the most efficient application and yields a much higher CO_2 reduction than hydrogen fuel-cell vehicles. This is because of the high discharge rate of batteries, which is almost double the efficiency of a fuel cell. However, battery vehicles still face significant technical hurdles (for details, see Chapter 7). Meanwhile, plug-in hybrids (if commercialised) with CO_2 emissions as low as 40 g/km could significantly help reduce CO_2 emissions from the

[6] For the transport sector, the comparison is made on the basis of the following assumptions: 130 g CO_2/km per vehicle for conventional fuel, a consumption of 0.6 MJ/km for BEV and 1.0 MJ/km for hydrogen FCV, and an electrolyser efficiency of 80% (see also Hammerschlag and Mazza, 2005).

transport sector, while renewable energies could be employed in the power sector. These analyses underline that, in the long term, electricity is a major competitor for hydrogen.

16.4 Summary

Alongside the option of producing hydrogen from electricity using electrolysers, there are other direct links between hydrogen and electricity production in the power sector. A rapid build-up of wind power, and also photovoltaic and solar thermal electricity generating capacities, is expected. Despite some clear advantages (renewable, CO_2 free), the intermittency of wind- and solar-generated electricity poses a challenge with regard to load levelling when capacity grows. Hydrogen could be one solution to meeting this challenge, as it offers the opportunity to store and transport the energy. Especially, for mobile applications, where energy has to be stored under any circumstances, this could be a more promising option instead of storing and re-electrification for stationary applications.

Also in remote areas, such as low-populated islands, without any access to the electricity grid and for using large renewable resources, which are far away from any user centres (so-called stranded resources), hydrogen could be an attractive solution (see also the discussion in Chapter 17).

Using hydrogen to produce electrical energy from fossil fuels in large centralised plants will contribute positively to achieving important reductions of CO_2 emissions, if this is combined with CO_2 capture and sequestration processes. Such plants will also help to increase the diversification of resources, since a variety of fossil feedstocks can be used, including resources such as coal and waste that otherwise cause major impacts on the environment, as well as biomass.

References

Altmann, M. and Gamallo, F. (2000). *Autarke Wind-Wasserstoff-Systeme*. www.hyweb.de/Wissen/autarke.htm.

Altmann, M., Niebauer, P., Pschorr-Schoberer, E. and Zittel, W. (2000). WHySE wind-hydrogen supply of electricity markets – technology – economics. *Wind Power for the 21st Century Conference*, September 25–27, 2000, Kassel (Germany). www.hyweb.de/Wissen/pdf/windpw00.pdf.

Ball, M. (2006). *Integration einer Wasserstoffwirtschaft in ein nationales Energiesystem am Beispiel Deutschlands*. Dissertation, VDI Fortschritt-Berichte 16, No. 177. Düsseldorf: VDI Verlag.

Blesl, M. and Ohl, M. (2003). *Fuel Cells: Bottom-up Interpretation of the Experience Curve*. Presentation during the Workshop of EU-EXTOOL and IEA EXCEPT 2003.

Chiesa, P., Consonni, S., Kreutz, T. and Williams, R. (2005). Co-production of hydrogen, electricity and CO_2 from coal with commercially ready technology.

Part A: performance and emissions. *International Journal of Hydrogen Energy*, **30** (7), 747–767.

DENA (2005). *Energiewirtschaftliche Planung für die Netzintegration von Windenergie in Deutschland an Land und Offshore bis zum Jahr 2020*. Cologne: Energiewirtschaftliches Institut. Commissioned by Deutsche Energie Agentur (DENA), Berlin.

DENA (2007). *Interactive Map of the North Sea. Planned Offshore Projects in the North Sea*. Deutsche Energie Agentur (DENA). www.offshore-wind.de/page/index.php?id = 2620.

Geer, T., Manwell, J. F. and McGown, J. G. (2005). A feasibility study of a wind/hydrogen system for Martha's Vineyard, Massachusetts. *American Wind Energy Association; Windpower 2005 Conference*, May 2005. www.ceere.org/rerl/publications/published/2005/AWEA05_Wind-Hydrogen.pdf.

Gielen, D. and Simbolotti, G. (2005). *H_2 Policy Analysis using the ETP model*. Paris: IPHE Task Force on Socio-Economics.

Hammerschlag, R. and Mazza, P. (2005). Questioning hydrogen. *Energy Policy*, **33** (2005), 2039–2043.

HyWays (2007). Hydrogen Energy in Europe. www.hyways.de.

IEA (International Energy Agency) (2005). *Prospects for Hydrogen and Fuel Cells*. IEA Energy Technology Analysis Series, Paris: OECD/IEA.

IE/IPTS (Institute for Energy and Institute for Prospective Technological Studies) (2005). *Hypogen Pre-Feasibility Study*. Report EUR 21512 EN. European Commission, Directorate-General, Joint Research Centre.

Kreutz, T., Williams, R., Consonni, S. and Chiesa, P. (2005). Co-production of hydrogen, electricity and CO_2 from coal with commercially ready technology. Part B: economic analysis. *International Journal of Hydrogen Energy*, **30** (7), 769–784.

Leighty, W. C., Hirata, M., O'aAshl, K. and Benoit, J. (2006). *Large Stranded Renewables: The International Renewable Hydrogen Transmissions Demonstration Facility (IRHTDF)*. www.hydrogennow.org/Facts/Pipeline/Leighty/WEC-Sydney-Sept04-30Apr-Final-Rev12May.pdf.

Mineralölwirtschaftsverband e.V. (German Association of Mineral Oil Economy; MWV) (2006). www.mwv.de.

PURE Energy Centre (2006). *PURE Project*. www.pure.shetland.co.uk/html.

Ragwitz, M. *et al.* (2005). *Analysis of the EU Renewable Energy Sources' Evolution up to 2020 (FORRES 2020)*. Report for the European Commission, Directorate General for Enterprise and Industry. Karlsruhe: Fraunhofer IRB Verlag.

Schatz project (2006). *The Schatz Solar Hydrogen Project*. www.schatzlab.org/projects/real_world/schatz_solar.html.

Schönharting, W. and Nettesheim, S. (2006). *RES2H2 a European Hydrogen Project: Hydrogen Generation from Wind Energy*. www.res2h2.com/goals0.htm.

Sensfuß, F., Ragwitz, M. and Wietschel, M. (2003). Fluktuationen der Windenergie und deren Vorhersagbarkeit bei einem verstärkten Ausbau des Offshore Anteils in Deutschland bis 2020. Fraunhofer Institute for Systems and Innovation Research, In *Proceedings Internationale Energiewirtschaftstagung (IEWT) 2003*. Vienna.

Sensfuß, F., Klobasa, M., Ragwitz, M. and Wietschel, M. (2004). Energiemodelle zum europäischen Klimaschutz – der Beitrag der deutschen Energiewirtschaft. In *Forum für Energiemodelle und Energiewirtschaftliche Systemanalysen in Deutschland*. Münster: LIT-Verlag, pp. 745–753.

Smekens, K. E. L., De Feber, M. A. P. C. and Seebregts, A. J. (2002). Learning in clusters: methodological issues and lock-out effects. *International Energy Workshop*. EMF/IIASA, 18–20 June 2002, Stanford University.

Starr, F., Tzimas, E. and Peteves, S. (2007). Critical factors in the design, operation and economics of coal gasification plants: the case of the flexible co-production of hydrogen and electricity. *International Journal of Hydrogen Energy*, **32** (10–11), 1477–1485.

Tsuchiya, H. and Kobayashi, O. (2002). Fuel cell cost study by learning curve. *International Energy Workshop*. EMF/IIASA, 18–20 June 2002. Stanford University. www.iiasa.ac.at/Research/ECS/IEW2002/docs/Paper_Tsuchiya.pdf.

UCTE (2000). *Hourly Load Values of a Specific Country Every Weekend of the Year 2000*. The Union for the Co-ordination of Transmission of Electricity (UCTE). www.ucte.org.

UCTE (2005). *Hourly Load Values of a Specific Country, Every 3rd Wednesday of a Specific Year*. The Union for the Co-ordination of Transmission of Electricity (UCTE). www.ucte.org.

Wietschel, M., Hasenauer, U., Vicens, N. J., Klobasa, M. and Seydel, P. (2006). Ein Vergleich unterschiedlicher Speichermedien für überschüssigen Windstrom. *Zeitschrift für Energiewirtschaft*, **2**, 103–114.

Yamashita, K. and Barreto, L. (2003). *Integrated Energy Systems for the 21st Century: Coal Gasification for Co-producing Hydrogen, Electricity and Liquid Fuels*. Interim Report IR-03-039. Laxenburg, Austria: International Institute for Applied Systems Analysis (IIASA).

Further reading

Crotogino, F. and Huebner, S. (2008). *Energy Storage in Salt Caverns – Developments and Concrete Projects for Adiabatic Compressed Air and for Hydrogen Storage*. www.kbbnet.de.

VDE (Association for Electrical, Electronic and Information Technologies) (2009). *Energiespeicher in Stromversorgungs system mit hohen Anteil erneuerbarer Energieträger*. ETG Task Force Energiespeicher. Frankfurt: VDE.

Yang, C. (2008). Hydrogen and electricity: parallels, interactions, and convergence. *International Journal of Hydrogen Energy*, **33** (8), 1977–1994.

17

Hydrogen corridors

Martin Wietschel and Ulrike Hasenauer

The discussions in Chapters 3 and 5 have shown that conventional fossil as well as renewable resources are limited and that strong competitions between the future uses of the primary energy carriers exist. In Chapter 16 the necessity of long-term transport of hydrogen for such reasons as relevant geographical distances between location of resources and demand centres was discussed in detail in the geographical scope of a country. Now, this chapter deals with questions of import or export of hydrogen. While import strategies – or energy corridors – are common and well known for natural gas, oil and electricity, the idea of hydrogen corridors is foreign to many people.

Such hydrogen corridors offer, among other things, in the long term the possibility of coping with the energy resource limitations for hydrogen production and improving energy supply security. Domestic energy resources are limited within the most industrial countries and, therefore, one open research question is whether it is a sustainable option to produce hydrogen outside and import it.

17.1 What is a hydrogen corridor?

In the following, we understand as a hydrogen corridor the import-based supply chain of hydrogen, including production in the country of origin and transport to the border of the country of destination. For a better understanding of what is a hydrogen corridor and which elements can be involved, an example for a concept of a hydrogen corridor between Iceland and the EU25 is shown in Fig. 17.1. Here, the principle is as follows. Geothermal or hydroenergy in Iceland is used to produce electricity, which is transmitted via electric current lines to a hydrogen production plant, close to the harbour. There, hydrogen is produced via electrolysis and liquefied. Liquid hydrogen is than transported with tank ships to the EU25, where it is distributed with trucks. A possible utilisation of the hydrogen can be as fuel for buses and passenger cars. It is clear that this is only one possible solution. As hydrogen can be produced from any primary energy source and with very different geographical

The Hydrogen Economy: Opportunities and Challenges, ed. Michael Ball and Martin Wietschel. Published by Cambridge University Press.

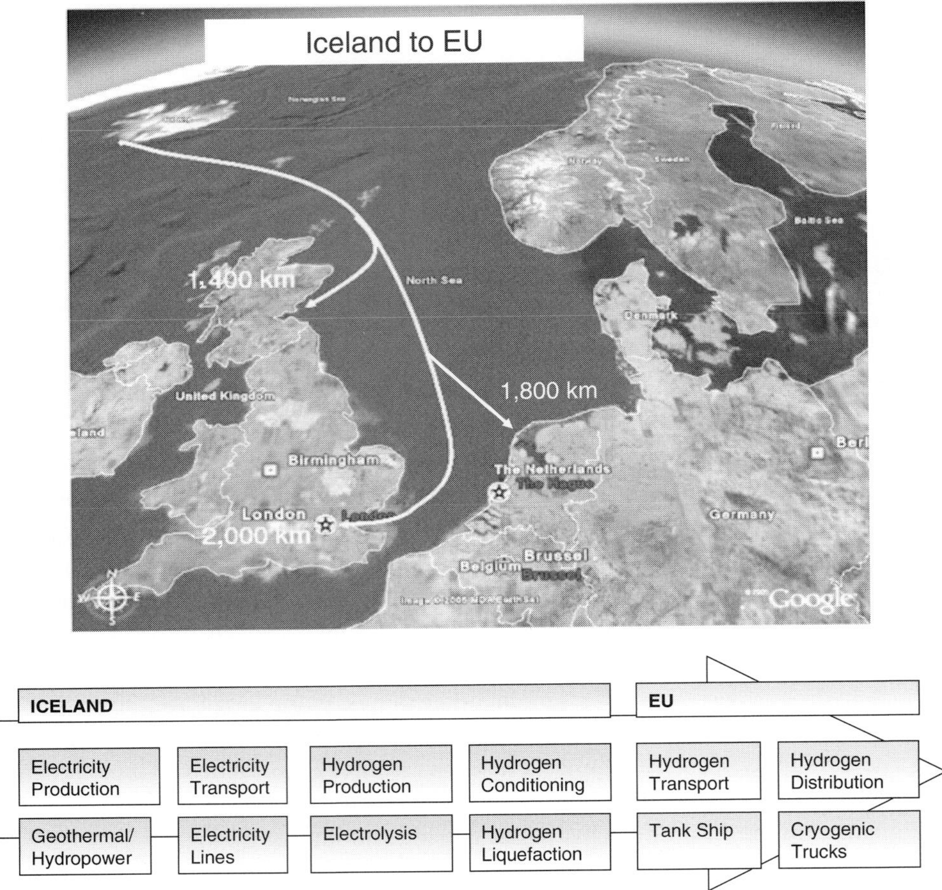

Figure 17.1. Elements of a hydrogen corridor between Iceland and the EU, based on hydro- or geothermal power.

and political pre-conditions and thus transport options, each hydrogen corridor is very different and needs case-specific in-depth analysis.

17.2 Why hydrogen corridors?

Assuming that hydrogen will be an important energy carrier in all parts of daily life, one of the most important questions is where the hydrogen will come from. Global efforts are to design an energy system that is more sustainable than that of our time. In this respect, sustainability means providing energy that pollutes our environment as little as possible at reasonable costs and with a high security of supply. Hydrogen produced by renewable feedstock is, on a long-term perspective, a very promising option and plays a prominent role in most stakeholders' visions about hydrogen as an energy carrier (see Chapter 8). However, renewable hydrogen produced near to

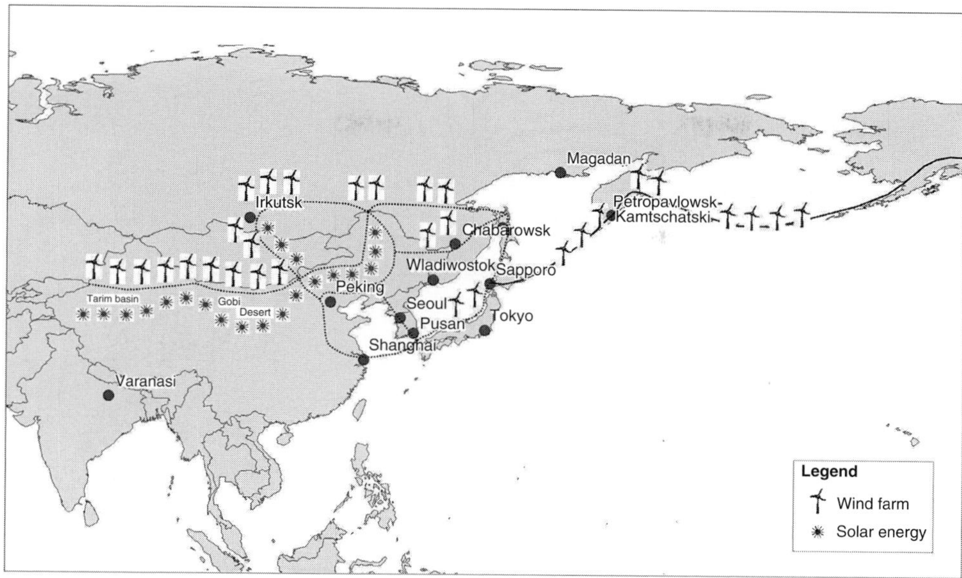

Figure 17.2. Natural gas pipeline in east Russia, which is fed by renewably produced hydrogen along its route to large demand centres; see Leighty *et al.* (2004) for more project information.

consumer centres is often very cost-intensive today. Moreover, the availability is usually restricted and competition with other uses is high (see Chapter 16). A solution to this dilemma is seen in the establishment of hydrogen corridors. Earth's richest renewable energy sources are often stranded, in the sense that they are not close to the demand centres (see also Chapters 5 and 16). Here, renewable energy sources are often comparatively cheap, owing to extreme climate conditions (radiation, water flow, wind conditions) or geological particularities (geothermal abnormalities). Gathering this renewable energy and using hydrogen as transport medium to supply distant markets could be a step towards a more sustainable energy supply in future. As an example, in the project: 'Large Stranded Renewables: International Renewable Hydrogen Transmission Demonstration Facility (IRHTDF)' the use of such large stranded renewable energy and the transmission via pipelines is studied (Leighty *et al.*, 2004). The consortium proposes to construct new natural gas pipelines in such a way that they are also capable of transmitting natural gas–hydrogen blends or even pure hydrogen. The focus is on a natural gas pipeline project in east Russia. The idea is to add hydrogen from diverse renewable sources all along the route to the natural gas pipeline and subsequently replace natural gas with hydrogen completely, once the gas fields are exploited (compare Fig. 17.2).

17.3 Overview of hydrogen-corridor studies

Although the concept of hydrogen corridors is widely unknown, the idea is not new. During the last decades, the feasibility and prospects of exporting large amounts of

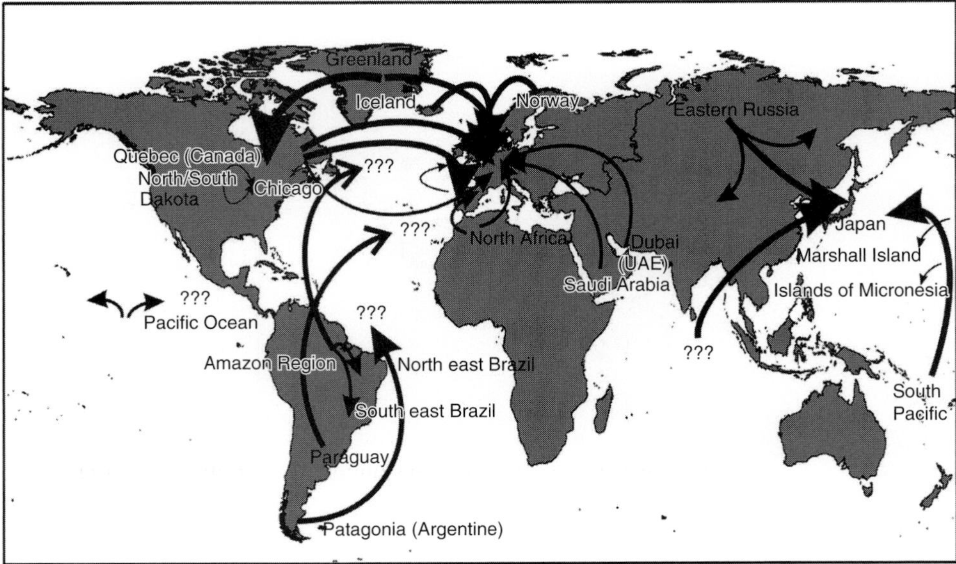

Figure 17.3. Geographical overview of investigated hydrogen corridors.

energy via hydrogen from regions where energy is abundantly available was the subject of various studies. A comprehensive overview of worldwide hydrogen-corridor studies was performed by the Ludwig-Bölkow-Systemtechnik (LBST, Germany) in the frame of the project 'Encouraged: Energy Corridor Optimisation for European Markets of Gas, Electricity and Hydrogen' (detailed information on this project is given in Section 17.4). The focus was on project goals, main results and lessons learned. The considered supply countries and destinations are shown in Fig. 17.3. Of the 26 projects analysed, the main destinations are Europe, the USA, Canada and Japan. Potential supply countries are Greenland, Patagonia, Paraguay, Saudi Arabia, Norway, Iceland and North Africa.

The hydrogen-corridor studies have considered various energy sources for the production of hydrogen. These energy sources are listed in Table 17.1.

The studies have also considered different transport options for their respective hydrogen corridors. Some of the studies also compare alternative methods of energy transport (e.g., electricity or chemical compounds). Table 17.2 contains a list of the different transport options.

As Table 17.1 and Table 17.2 show, most projects focus on renewable energy sources for hydrogen production and subsequent transport. This is because for fossil energy sources, like natural gas or coal, with high (volumetric) energy content, feedstock transport is preferable to hydrogen transport, owing to the low volumetric energy content of hydrogen and because the infrastructure is already in place. With respect to long-distance hydrogen transport, no clear favourable solution can be identified. The choice depends very much on the specific local circumstances. In all

Table 17.1. *Overview of hydrogen sources considered in various studies*

Hydrogen source	Studies (examples)	Literature
Hydropower	A Study for the Generation, Inter-Continental Transport, and Use of Hydrogen as a Source of Clean Energy, on the Basis of Large-scale and Cheap Hydro-Electricity (Hydrogen Pilot Project – Canada)	(DECHEMA, 1987)
	Euro–Quebec Hydro-Hydrogen Pilot Project (EQHHPP)	(Drolet, 1994)
	Norwegian Hydro Energy in Germany (NHEG)	(Andreassen *et al.*, 1993)
	Feasibility Study for Export of Hydrogen from Iceland to the European Continent (EURO-HYPORT)	(Nordic H_2 Energy Foresight, 2001)
Glacier power	Power from Glaciers: The Hydropower Potential of Greenland's Glacial Waters	(Partl, 1977)
On- and offshore wind energy	Large Scale Wind Hydrogen Production in Argentine Patagonia	(Raballo and LLera, 2004)
	Wasserstofferzeugung in Offshore-Windparks – 'Killer-Kriterien', grobe Auslegung, Kostenabschätzung – Hydrogen Production in Offshore-Windparks – 'Killer Criteria', Basic Design, Cost Estimation	(Altmann *et al.*, 2001)
Solar thermal power plants	Comparison between High-Voltage Direct-Current Transmission and Hydrogen Transport	(Kaske and Plenard, 1985)
Photovoltaic energy	HYSOLAR – A German–Saudi Arabian Partnership	(Abaoud and Steeb, 1998; Siegel and Phan, 1990)
Geothermal power	Feasibility Study for Export of Hydrogen from Iceland to the European Continent (EURO-HYPORT)	(Nordic H_2 Energy Foresight, 2001)
Ocean power, thermal energy conversion	The Hawaiian International Hydrogen Energy Pilot Project: A Concept for the Commercial Implementation of Non-Fossil Hydrogen Energy in the Pacific Ocean Area (HIHEPP)	(Krock and Zapka, 1991)
Natural gas	A Perspective of Renewable Energy and New Technology in Northeast Asia	(O'Hashi and Hiraishi, 2001)

Table 17.2. *Overview of transport options considered in various corridor studies*

Transport option	Studies (examples)	Literature
Comparison of transport of methylcyclohexan (MCH), ammonia (NH_3) and LH_2 by ship and LH_2 transport by aircraft	A Study for the Generation, Inter-Continental Transport, and Use of Hydrogen as a Source of Clean Energy, on the Basis of Large-scale and Cheap Hydro-Electricity (Hydrogen Pilot Project – Canada)	(DECHEMA, 1987)
Comparison of transport of methylcyclohexan (MCH) and LH_2 by ship	Euro–Quebec Hydro-Hydrogen Pilot Project (EQHHPP)	(Drolet *et al.*, 1994)
Comparison between HVDC and hydrogen pipeline transport	Comparison between High-Voltage Direct-Current Transmission and Hydrogen Transport	(Kaske and Plenard, 1985)
	Power from Glaciers: The Hydropower Potential of Greenland's Glacial Waters	(Partl, 1977)
	Wasserstofferzeugung in offshore Windparks – 'Killer-Kriterien', grobe Auslegung, Kostenabschätzung – Hydrogen Production in Offshore Wind Parks – 'Killer Criteria', Basic Design, Cost Estimation	(Altmann *et al.*, 2001)
	Transmitting Wind Power from the Dakotas to Chicago: A Preliminary Analysis of a Hydrogen Transmission Scenario	(Gibbs and Biewald, 2000)
LH_2 transport by ship	Norwegian Hydro Energy in Germany (NHEG)	(Andreassen *et al.*, 1993)
	Wind-produced hydrogen exported from Patagonia	(Siteur, 1993)
	HYSOLAR – A German–Saudi Arabian Partnership	(Abaoud and Steeb, 1998; Siegel and Phan, 1990)
	Comparative Study Between the HYSOLAR Project and a Hypothetical International Project in Brazil for Hydrogen Production and Exportation (BHP) from Photovoltaic Energy and Secondary Hydroelectricity Combined Supply	(Soltermann and Da Silva, 1998)

Table 17.2 (*cont.*)

Transport option	Studies (examples)	Literature
	International Clean Energy Network Using Hydrogen Conversion (WE-NET)	(Hijikata, 2002)
	Power from Glaciers: The Hydropower Potential of Greenland's Glacial Waters	(Partl, 1977)
	BMW Feasibility Study on Hydrogen Production in Dubai	(Hoffmann, 2001)
Containerised LH_2 transport	Intercontinental Liquid Hydrogen Delivery System	n/a
Pipeline transport	A Perspective of Renewable Energy and New Technology in Northeast Asia	(O'Hashi and Hiraishi, 2001)
Usage of the existing or an expanded NG pipeline network	Transport, Speicherung und Verteilung von Erdgas heute – von Wasserstoff morgen – Transport, Storage and Distribution of Natural Gas Today – of Hydrogen Tomorrow	(Fasold, 1987)

studies, large hydrogen volumes are assumed. This is because large infrastructure investments are justifiable only in connection with a high capacity utilisation, which in turn requires a significantly high demand.

All currently existing hydrogen-corridor studies are pre-feasibility studies, which means that they are far from realisation. It is too early to establish hydrogen corridors. There is no demand for large amounts of hydrogen at present and sources of cheap hydrogen are already available for start-up scenarios. Furthermore, the industry is not yet prepared for large investments in a hydrogen infrastructure because it is still uncertain whether hydrogen will become accepted as an important energy carrier at all.

Among the most important hydrogen-corridor studies of the last years is the Euro–Quebec Hydro-Hydrogen Pilot Project (EQHHPP). In 1988, the European Commission and the Government of Quebec came to an agreement to investigate jointly the perspectives of renewably produced hydrogen as clean fuel. Together with European and Canadian industrial companies and research organisations, the various steps of the project were carried out. The project lasted from 1989 to 1998 and involved around 80 companies from seven countries. The goal was to investigate the feasibility and perspectives of producing hydrogen from Quebecois surplus hydropower (around 140 MW), transporting it to Europe and using it in various end-use technologies.

By 1991, the general feasibility could be proven with a $\pm15\%$ cost accuracy. The development, realisation, testing and demonstration of key hydrogen application and infrastructure technologies, such as buses, aircraft, transport containers and co-generation units has generated plentiful experiences and technological improvements. In a scaled-down model of the full-size barge tank investigated in EQHHPP it could be proven that large-scale vacuum super-insulated storage and transport tanks can be built and behave very much as forecast in relevant mathematical and thermodynamic models. However, the approach to delivering rather large quantities of LH_2 into a not yet prepared and developed market did not turn out to be practical. Therefore the approach was modified to develop advanced lightweight containers for initial start-up markets (Drolet *et al.*, 1994).

17.4 Energy-corridor optimisation for a European market of hydrogen

17.4.1 Introduction

One recent study of the role of hydrogen corridors for Europe is 'Encouraged: Energy Corridor Optimisation for European Markets of Gas, Electricity and Hydrogen', funded by the European Commission within the Sixth Framework Programme (FP6).[1]

The project aimed to identify economically optimal and viable hydrogen corridors between the EU25 (as of 31st December, 2005) and neighbouring countries and to assess the feasibility and necessity of such corridors. The analysis is based on consistent hydrogen scenarios and looks at the barriers to and the benefits from establishing a pan-European 'energy network' for hydrogen. Collecting state-of-the art insights and involving stakeholders for consensus building are additional goals of the study.

This section is structured as follows: first, an overview of the scenario framework is given for the possible development of hydrogen demand in the next decades in the EU25. These figures have to be determined first to see when and to what extent hydrogen corridors may play a role.

Next, 12 hydrogen production options outside the EU are selected and their potential for importing hydrogen is assessed. Subsequently an analysis of the economics of these hydrogen corridors is performed, taking into account the hydrogen production and long-distance transport costs. The results are also compared with the costs of EU domestic hydrogen production. The domestic production option is used as a benchmark for the evaluation of the corridor options.

Then the main results with respect to adequate corridor possibilities, timing and cost-competitiveness are drawn and the potential of hydrogen that could be imported to the EU via such corridors is estimated.

17.4.2 Scenario framework

When analysing the prospects of hydrogen corridors between neighbouring countries and the EU, it is important to estimate the future development of hydrogen demand

[1] The analysis of hydrogen corridors was performed by the Fraunhofer Institute for System and Innovation Research (ISI) in Germany, with input from DFIU (Germany), LBST (Germany), INE (Iceland) and ENEA (Italy). For further information on the Encouraged project, contract No. 006588, see www.bsrec.bg/newbsrec/encouraged.htm and the EC (2007).

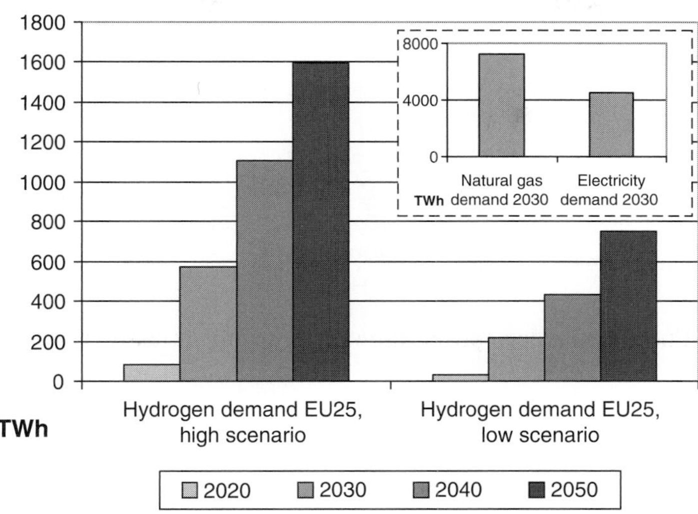

Figure 17.4. Different long-term hydrogen, natural gas and electricity demand projections for the EU25.

in the EU, taking into account the future development of the energy system, too. The hydrogen corridor analysis is, therefore, embedded within the framework of two penetration scenarios: one with low and one with high hydrogen penetration with a time horizon up to 2050. The assumptions are based on the HyWays project phase I, an integrated project of the EU aiming at developing a European roadmap for hydrogen (www.hyways.de). In HyWays, the hydrogen demand in phase I is assessed for six European countries: Germany, France, Italy, the United Kingdom, the Netherlands and Norway (see also Chapter 14). In the framework of the project 'Encouraged', the hydrogen demand of the six countries is extrapolated to the hydrogen demand of the EU25, proportional to the population ratio. It is assumed that hydrogen is mainly used in the transport sector. Figure 17.4 shows the hydrogen demand assumed for the EU25 according to the high and low penetration scenarios up to 2050. For comparison, the forecasts of the European Commission (Mantzos *et al.*, 2003) for natural gas and electricity demand for EU25 in 2030 are also included. From this comparison, it becomes clear that the assumed hydrogen demand is fairly large from 2030 onwards (around 30% market penetration for passenger cars, buses, and light-duty vehicles).

Of course, assumptions about future hydrogen demand are subject to uncertainties because it is not yet clear whether hydrogen will become part of the energy system at all.

17.4.3 Selection of hydrogen feedstock and production centres outside the EU

Based on the experiences made in other hydrogen corridor studies and looking at the hydrogen vision of stakeholders and policy makers, eight hydrogen production

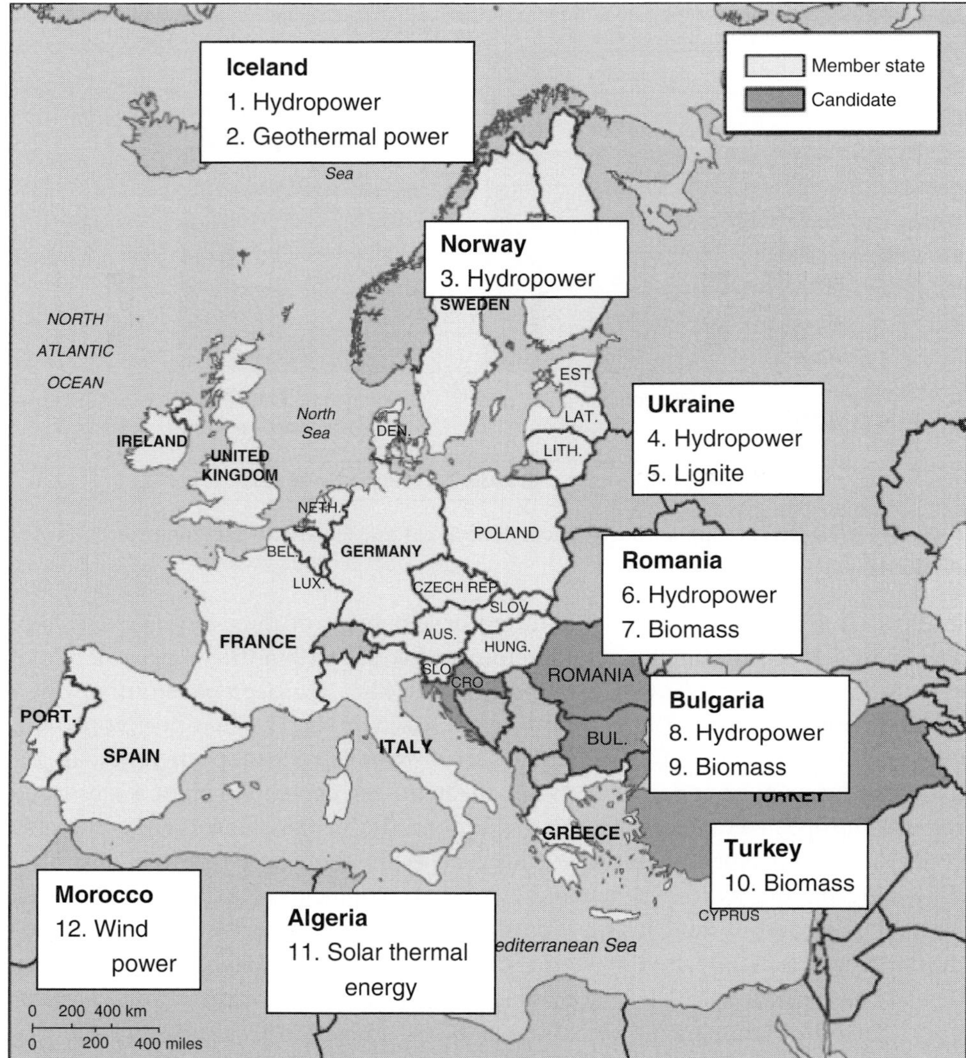

Figure 17.5. Selected hydrogen production centres outside the EU25.

centres outside the EU25 and six feedstocks are selected for further in-depth analysis. The feedstock selection focuses on renewable energy sources, namely solar thermal energy, wind power, geothermal power, hydropower and biomass. As an exception, one hydrogen corridor based on lignite is also assessed. This is because significant sources of cheap lignite exist and it makes no sense to transport the lignite itself, owing to its low heating value. Converted to hydrogen, lignite might contribute to an increase in the security of energy supply in the EU. Among the selected hydrogen production centres outside the EU25 are: Morocco, Algeria, Iceland, Norway, Romania, Bulgaria, Turkey and the Ukraine. The corridor options are shown in Fig. 17.5.

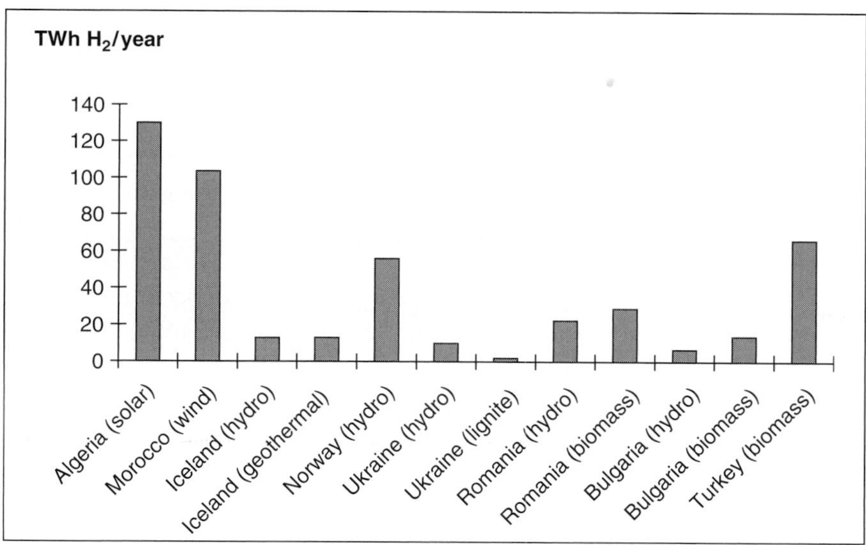

Figure 17.6. Maximum theoretical annual volume of hydrogen by feedstock and country in 2040.

Long-distance transport has a large impact on the total costs. Compared with other fuels, the volumetric energy content of hydrogen is comparatively low, which makes long-distance transportation less efficient. Therefore, if natural gas or hard coal are the considered feedstock for hydrogen production, it would be better to import the feedstock and produce hydrogen at (or near to) the place of use.

For the selected 12 hydrogen production options outside the EU, a detailed analysis is performed of the additional feedstock or electricity potential that – if exploited – might be used for hydrogen production in the supplying country and the amount of hydrogen that could be produced annually is calculated (see Fig. 17.6) – depending on the conversion efficiency of the selected production technology and feedstock. The additional potential is defined as the realisable potential (i.e., the theoretically feasible potential in a certain year under the assumption that all existing barriers are overcome and all drivers are active) minus the achieved potential (i.e., today's gross inland production of the considered energy source) (see Chapter 5).

The chosen reference year is 2040. The biomass potential is taken from the Green-X project on the additional renewable potential in the EU (see www.green-x.at/Green-X%20viewer.htm). The time horizon of the Green-X forecasts is 2020. It is assumed that the potential remains constant after 2020. North Africa has the largest additional potential (wind and solar), followed by Turkey (biomass) and Norway (hydro).

17.4.4 Economic evaluation of hydrogen corridors

17.4.4.1 Database, methodology and calculation results

The economics of importing hydrogen from neighbouring countries to the EU depend on the hydrogen feedstock and production costs on the one hand, and on the long-distance transport costs on the other. Another aspect is the price difference in comparison with European hydrogen-production costs using domestic energy carriers. The technical and economic data of the hydrogen production and transportation technologies are taken from a Fraunhofer ISI internal database, for which extended literature surveys and manufacturer interviews were performed. The technology data reflect a future perspective with a time horizon up to 2040. This means that technical development processes, learning effects and potential cost reductions from mass production and economies of scale are included.

17.4.4.2 Production

Table 17.3 shows the hydrogen-production costs in the selected neighbouring countries projected for 2040. Two figures are given in the column for feedstock and electricity costs. While the first figure reflects the specific feedstock and electricity costs per kWh hydrogen produced and thus includes the efficiency of the hydrogen production process, the figure in parentheses refers to the feedstock or electricity costs in ct/kWh_{el}. During hydrogen production via gasification, electricity is co-produced and credits are given at the level of $4\,ct/kWh_{el}$.

By far the cheapest hydrogen production option is lignite gasification in the Ukraine at only 1.7 ct/kWh hydrogen. This is because lignite is a very cheap feedstock (0.4 ct/kWh), and the production process is also comparatively inexpensive (plus excess electricity can be sold). However, of the options considered, lignite gasification is the only one that releases significant amounts of CO_2 emissions: around 640 g CO_2 are emitted to produce 1 kWh hydrogen. With respect to the renewable production options, hydropower in Iceland, Norway, Romania and Bulgaria, as well as geothermal power in Iceland, perform very well, followed by biomass reforming. Hydrogen production in North Africa from wind power or solar radiation is more expensive.

17.4.4.3 Transport

Figure 17.7 compares different hydrogen transport options. It is assumed that technical barriers are overcome (especially for ship transport) and learning effects are used. Exemplarily, an annual hydrogen transport volume of 10 TWh is assumed over a distance of 500 kilometres, to compare different hydrogen transport modes in terms of costs per kWh transported hydrogen (the numbers, however, exclude liquefaction, a pre-condition for hydrogen transportation in ships or trucks). Liquid hydrogen transport in trucks is generally not an economically reasonable option

Table 17.3. *Hydrogen production costs from different feedstocks in EU25 neighbouring countries in 2040 (prices in €2005)*

Location	Hydrogen production	Feedstock cost (electricity cost) (ct/kWh_{H_2}) (ct/kWh_{el})	Plant-related costs[a] (ct/kWh_{H_2})	Total production costs (ct/kWh_{H_2})
Algeria	Solar thermal water splitting	–	8.2	8.2
Morocco	Wind electrolysis	8 (5.6)	1.8	9.8
Iceland	Geothermal electrolysis	3.0 (2.1)	0.7	3.7
Iceland	Hydro electrolysis	3.4 (2.4)	0.7	4.1
Norway	Hydro electrolysis	3.9 (2.7)	0.7	4.5
Ukraine	Hydro electrolysis	2.9 (2.1)	0.7	3.6
Ukraine	Lignite gasification	0.3 (0.4)	1.4	1.7
Romania	Hydro electrolysis	2.9 (2.1)	0.7	3.6
Romania	Biomass staged reforming	3.4 (1.6)	2.5	5.9
Bulgaria	Hydro electrolysis	2.9 (2)	0.7	3.5
Bulgaria	Biomass gasification (staged reforming)	3.6 (1.7)	2.5	6.1
Turkey	Biomass gasification (staged reforming)	3.6 (1.7)	2.5	6.1

Note:
[a] Plant-related costs include the annualised investment of the plant and operation and maintenance costs (not included are feedstock and electricity costs).

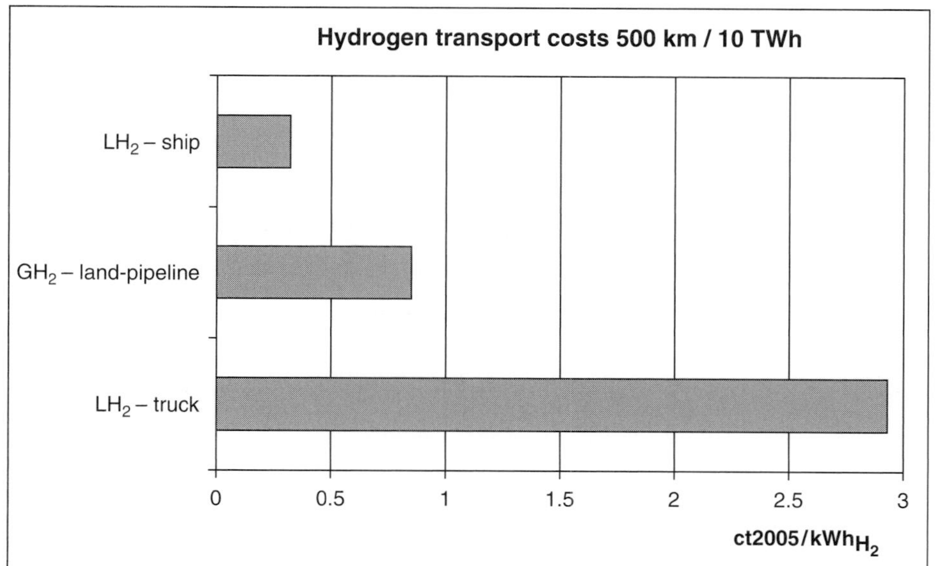

Figure 17.7. Comparison of different hydrogen transport options. Time perspective till 2040.

for long-distance transportation of large hydrogen volumes. It is more economic to transport such volumes in pipelines (over land) or ships (over sea) (see also Chapter 12). Compared with hydrogen transport, overland HVDC (high-voltage direct-current) electricity transmission is cheaper, but hydrogen transportation could be the more favourable option for distances over water.

17.4.4.4 Cost analysis of corridors

Figure 17.8 illustrates the costs of supplying hydrogen via corridors and within the EU25 from domestic feedstocks with a time perspective of 2040. The EU data are used as a benchmark for hydrogen corridors. The corridor options include production in the country of origin and long-distance hydrogen transportation to the border of the neighbouring EU25 countries. Solar hydrogen production in Algeria is considered with solar thermal water splitting and not with electrolysis. Thus, no electricity is needed and no feedstock costs are incurred in this case. In the case of transportation by ship, liquefaction costs are also included, as they are part of the supply chain.

The EU hydrogen production options include production costs only. Because the EU is very large and, thus, there is a high cost variability of feedstocks and electricity sources, low and high cost ranges are assumed. The distribution in the EU and the conditioning at the place of use (compression or liquefaction) are not included because these are not decision-relevant when comparing hydrogen corridors with the option of domestic hydrogen production.

In Fig. 17.9, the costs of selected hydrogen corridors are compared with conventional gasoline fuel. The costs are shown without taxes in €2005, to have a fair basis for comparison; it can be assumed that the taxes and earnings of different fuels are very similar and, therefore, not decision-relevant.

For the comparison, the costs of hydrogen distribution and compression at the filling station are added, assuming that hydrogen is delivered in gaseous form. Furthermore, the negative effects of carbon emissions are included in monetary terms for fossil fuel-based paths at a cost assumption of € 20/tCO_2. The hydrogen supply costs are within a range of about double the cost of gasoline.

17.4.5 Results

Based on the analysis of the potentials and the economics of hydrogen corridors from neighbouring countries and a cost comparison with domestic hydrogen production in the EU25, which is used as a benchmark, the following conclusions can be drawn:

- Long-distance transport costs play a relevant role when evaluating corridors: compared with feedstock and production costs, the transport costs of hydrogen from neighbouring countries may lead to cost increases of 17 to 65%.
- The cheapest hydrogen-corridor options are hydrogen based on hydropower from Romania, the Ukraine and Bulgaria, followed by hydrogen from hydropower and geothermal power in

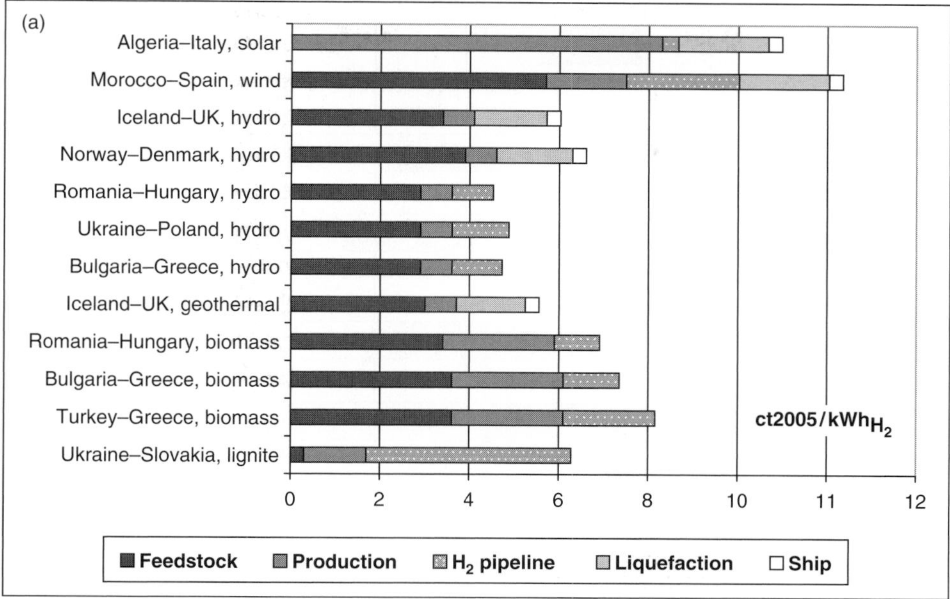

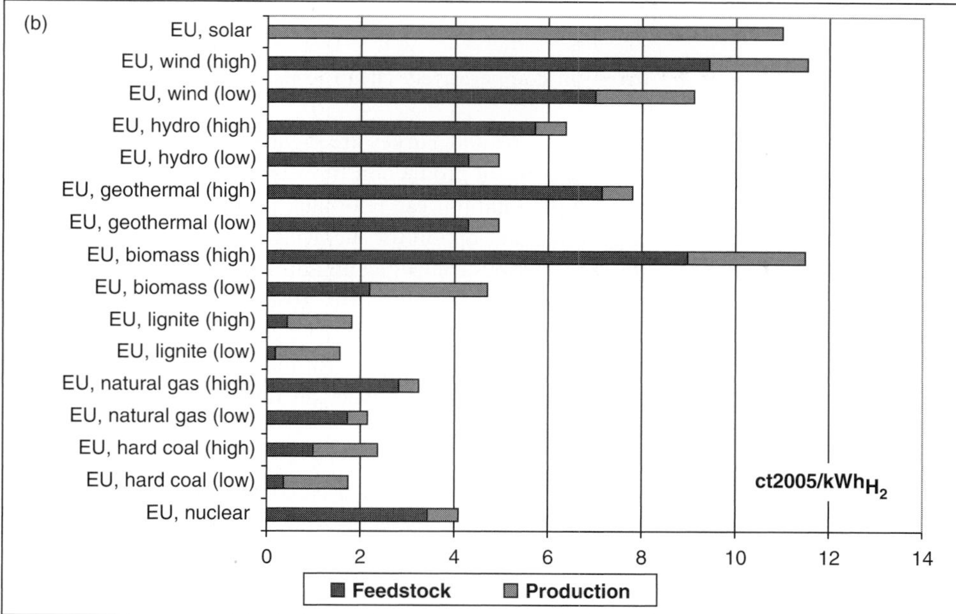

Figure 17.8. Comparing the costs of hydrogen production and long-distance transportation from selected neighbouring countries to the EU25 with EU hydrogen production from domestic sources with a time perspective of 2040.

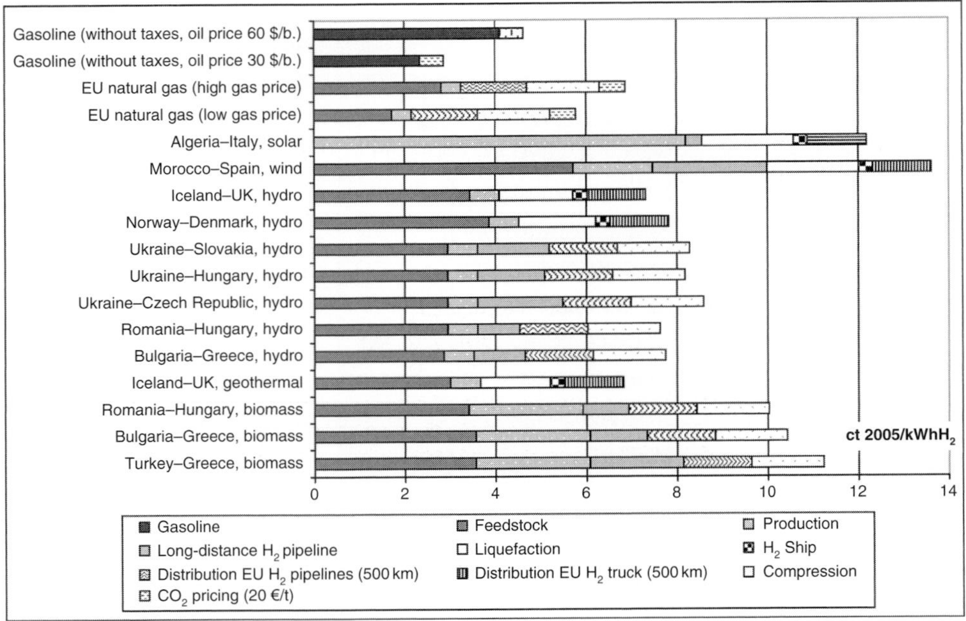

Figure 17.9. Comparison of hydrogen with conventional gasoline. Time perspective: 2040.

Iceland. These options are based on comparatively cheap feedstocks and moderate long-distance transport costs.

- Biomass corridors from Romania, Bulgaria and Turkey are slightly more expensive but still well within the range of European hydrogen production from biomass. Biomass corridors are up to 1.5 times more expensive than natural gas reforming in the EU25 itself (next to hydrogen by-products, the cheapest relevant option for supplying hydrogen with acceptable carbon emissions). However, there are various competing utilisation possibilities for biomass.
- Hydrogen from wind power and solar radiation in North Africa are the most expensive solutions, owing to the high costs of production and long-distance transport. These options are three to five times more expensive than natural gas reforming in the EU. Introducing a CO_2 price of €20/t CO_2 does not have a major impact on this result. However, these corridors have a very large renewable potential!
- Compared with the production in the EU25, the costs of some of the renewable corridors are in an acceptable range. Hydrogen imports perform well, even compared with conventional fuels like gasoline, if the conversion in the vehicle is also taken into account. The high efficiency of fuel cells, which is twice that of conventional internal combustion engines, can outweigh the higher fuel costs. However, it should be kept in mind that today's fuel cells are still much more expensive than conventional power trains.
- The prospects of hydrogen corridors based on lignite are questionable because the EU25 has large sources of lignite itself and the total costs of these corridors are disproportionately high because of the added costs of long-distance hydrogen transport. Second, a corridor based on lignite has no positive environmental side effect – unlike the corridors based on renewable feedstock or electricity sources – that might justify higher hydrogen costs.

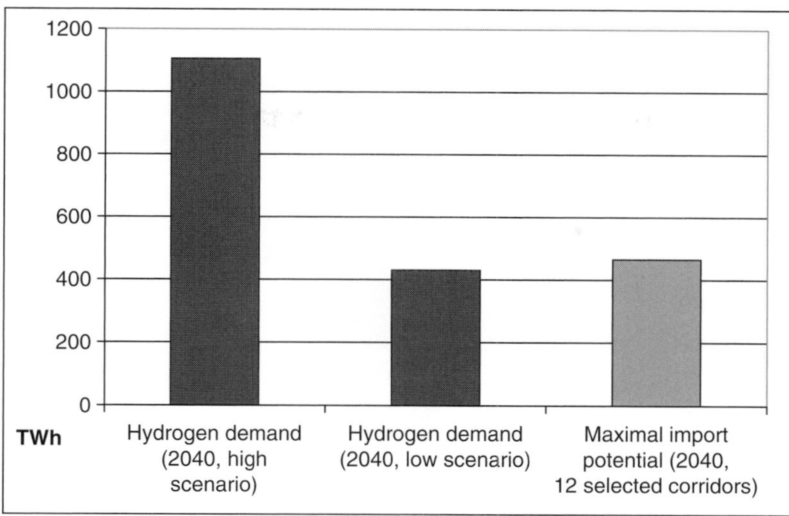

Figure 17.10. Hydrogen demand according to HyWays scenarios and maximal import potential in 2040 for 12 selected hydrogen corridors.

- Hydrogen is not an attractive intermediate transport medium for electricity. If electricity is the desired end product, transmitting electricity is usually more beneficial than the production, transport and re-electrification of hydrogen, e.g., in fuel cells, owing to electricity's better overall efficiency and lower costs.
- If a fossil feedstock, like gas or coal, is the energy source for hydrogen production then the fossil energy carrier should be transported, since, owing to its higher volumetric energy content this is easier than transporting the hydrogen itself (because of the low volumetric energy density of hydrogen and because the infrastructure for gas and coal transport is already in place).
- Nuclear power is not a promising option in the context of hydrogen corridors. First, it is cheaper to transport uranium or enriched uranium, or even electricity, instead of hydrogen. Second, there may be acceptance problems related to nuclear power in some countries (on both sides: the 'production' country as well as the 'consumer' country).
- There is a large potential for importing renewable hydrogen from neighbouring countries. As shown in Fig. 17.10, the potential of the 12 hydrogen production centres outside the EU25 of around 465 TWh/year meets the total hydrogen demand of the low hydrogen penetration scenario in 2040 and nearly half the demand of the high hydrogen penetration scenario. Expressed in different terms, it could power half the EU25 vehicle fleet (if driven by fuel cells).

The analysis of the potential and the economic performance of hydrogen supply routes from neighbouring countries to Europe performed in this study are mainly based on techno-economic criteria. In practice, there are many factors that might adversely influence the emergence of hydrogen corridors, among others, competition with alternatives in a free market economy. In contrast to these factors, a 'barrier' is

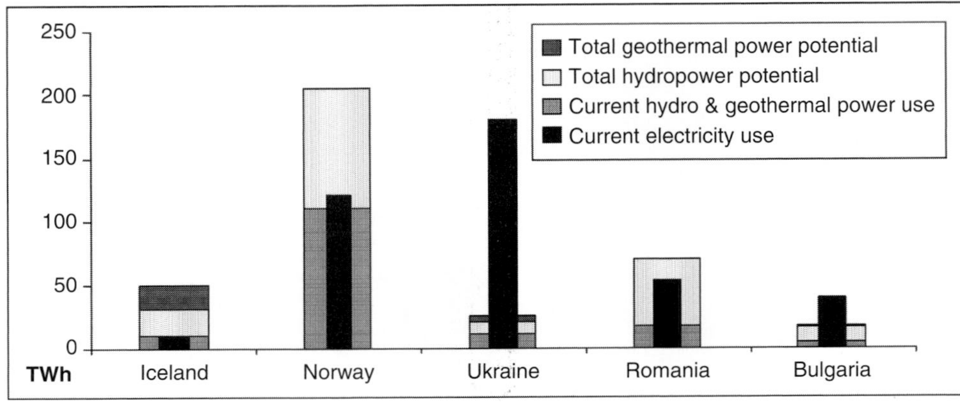

Figure 17.11. Comparison of renewable electricity potentials and current electricity consumption in selected countries.

defined as a technical or market imperfection or unfair support of alternatives that prevents the implementation of a hydrogen corridor that has been proven beneficial.

17.4.5.1 Competing utilisation possibilities of additional feedstock or electricity potential

The nature of renewable energy is such that it should be used as close to its origin as possible, since energy efficiency may drop significantly if the energy is transmitted over long distances. This is especially the case if non-fluctuating energy sources are considered, which do not pose the same difficulties concerning grid integration as wind or solar power. Corridors based on renewable electricity for hydrogen production and export are, therefore, not eligible if more electricity is used nationally than the total realisable renewable potential available. In these cases, grid integration is the better solution. With regard to the analysed corridors, this is true for the Ukraine and Bulgaria (see Fig. 17.11). Owing to the intermittent nature of wind and solar power, grid integration might cause difficulties with respect to grid stability, and in this case hydrogen production could serve as storage for the stochastic energy sources.

In Norway and Romania, hydrogen production and export is in direct competition with electricity transmission via high-voltage direct-current lines (HVDC). This solution is particularly attractive because hydropower is a non-fluctuating renewable energy source and does not destabilise the grid, as, for example, wind or solar power do.

In Turkey, Romania and Bulgaria, biomass could be used for the production and export of biofuels or biogas, for local heating or decentralised electricity generation.

Hydrogen from hydro or geothermal power from Iceland offers the lowest barriers with respect to competing utilisation possibilities; electricity export via electricity grid is less attractive, owing to Iceland's isolated position.

Table 17.4. *Eligibility criteria for renewable-based hydrogen corridors*

- Countries that have large potentials of cheap **non-fluctuating renewable electricity** sources where the domestic use of renewable electricity is close to 100% and potential still remains.
- Countries that have large potentials of cheap **fluctuating renewable electricity** sources (solar or wind) that cannot be integrated into the national electricity grid (owing to problems of grid stability).
- Countries that have a large potential of cheap **biomass** that exceeds what can be used locally (biomass is usually not transported over large distances, owing to its low energy density).
- Countries that can supply hydrogen at a comparable cost to conventional fuels (maximally twice the price of gasoline without taxes, owing to the double efficiency of fuel cells compared with internal combustion engines).

17.4.5.2 Competing utilisation of hydrogen within the country of origin

When contemplating the prospects of importing hydrogen from neighbouring countries to Europe, the following has to be considered: importing hydrogen is more economic, the larger the quantities of hydrogen transported via one specific corridor route. The premise for transporting large quantities is a large potential of the hydrogen production source in the corridor area, on the one hand, and a large demand for hydrogen in Europe, on the other. But if hydrogen demand increases in Europe, it is likely to also increase on a worldwide basis and within the corridor areas, too. In this case, it is probably more reasonable to use hydrogen locally rather than exporting it.

The checklist in Table 17.4 specifies the criteria that make a country eligible for exporting hydrogen.

17.4.5.3 Hydrogen corridors are not suitable for the start-up phase of hydrogen use in the energy sector (no small-scale solutions)

Introduction schemes for hydrogen typically start on a small (i.e., local or regional) scale: compared with the late 1980s, the focus has shifted from 'global link' concepts to more localised production and supply concepts. Today's stakeholder view is that hydrogen infrastructure build-up will start with small-scale hydrogen use, which will then increase smoothly. Furthermore, the start-up will occur at many places in Europe. In contrast, creating hydrogen corridors requires large-scale solutions (large infrastructure installations, e.g., hydrogen liquefaction plants, hydrogen ships and CGH_2 transport pipelines), to exploit economy-of-scale effects and thus high investments. Investors have to be found to carry the risk of such investments under uncertain market conditions. An incremental development of hydrogen corridors is not possible, because pipelines, tank ship concepts or liquefaction plants require a certain minimum size to work efficiently and economically.

Sources of cheaper hydrogen for the start-up phase are already available (e.g., as a chemical by-product, mostly fossil-based) and are very often used only thermally and, thus, suboptimally. Some of these could be made available at a reasonable cost for new applications (e.g., captive vehicle fleets), as long as the hydrogen can be substituted easily by, e.g., natural gas.

17.4.5.4 Hydrogen corridors are not a major option for portable applications and stationary applications

Most of the relevant stakeholders believe that hydrogen use will take off first in early markets like the '4Cs' (computers, camcorders, cellular phones, cordless tools) and will then penetrate the stationary sector before becoming widespread in the transport sector. However, hydrogen as a large-volume transport vector for electricity is justifiable only if the hydrogen enters the transport sector as a fuel. Further, as mentioned before, the re-electrification of 'electrolytic' hydrogen is not economic and is less energy-efficient than the HVDC long-distance transmission of electricity.

17.4.5.5 Competing energy supply structures

Existing, competing conventional energy supply structures (electricity and natural gas grids) may rule out the establishment of hydrogen corridors.

17.4.5.6 Conflict of interests concerning utilisation of the natural gas grid

Using existing natural gas pipelines has the potential for significantly lowering transport costs, but the utilisation of the natural gas pipeline network for hydrogen transport requires the consensus of all parties involved and a concerted effort to start its implementation.

17.5 Summary

This chapter deals with questions of import or export of hydrogen. While import strategies are common and well known for natural gas, oil and electricity, the idea of hydrogen corridors is foreign to many people. Hydrogen produced by renewable feedstock is, on a long-term perspective, a very promising option. Earth's richest renewable energy sources are often stranded, in the sense that they are not close to the demand centres, and, therefore, hydrogen corridors could play an important role in the future.

Analyses on cost and potentials using domestic production of hydrogen as a benchmark conclude that hydrogen import supply routes are particularly attractive if based on renewable energy sources and can significantly contribute to securing energy supply and reducing greenhouse-gas emissions. Owing to the relevant influence of transport costs on the economic assessment of hydrogen corridors, it is important to consider large-scale solutions to exploit economies of scale and to lower the specific costs by large throughputs. Hydrogen corridors are no suitable option at the start-up phase of hydrogen use, when only small quantities are required.

Hydrogen can be produced from environmentally benign sources (both inside and outside the EU), and hydrogen can also be a useful energy vector to bring environmentally benign energy into the EU. The potential of renewable hydrogen corridors is remarkable; however, they are more an option for the very long term. Besides the relatively high costs of the long distance transport, competing domestic utilisation possibilities for renewable resources (such as for power and heat generation) and the potentially increasing demand for hydrogen within the export countries of hydrogen themselves may limit this option.

Hydrogen corridors based on hard coal or natural gas are not suitable. It is better to transport these feedstocks directly instead of hydrogen, because of their much higher energy content and consequently lower transport cost and because the infrastructure for it is already in place.

References

Abaoud, H. and Steeb, H. (1998). The German-Saudi HYSOLAR program. *International Journal of Hydrogen Energy*, **23** (6), 445–449.

Altmann, M., Gaus, S., Landinger, H., Stiller, C. and Wurster, R. (2001). *Wasserstofferzeugung in offshore Windparks. 'Killer-Kriterien', grobe Auslegung und Kostenabschätzung*. Ottobrunn: L-B-Systemtechnik GmbH. www.hyweb.de/Wissen/pdf/GEO_Studie_Wasserstoff_oeffentlich.pdf.

Andreassen, K., Henriksen, N., Øyvann, A., Bünger, U. and Ullmann, O. (1993). Norwegian hydro energy in Germany (NHEG). *International Journal of Hydrogen Energy*, **18** (4), 325–336.

DECHEMA (Deutsche Gesellschaft für Chemisches Apparatewesen, Chemische Technik und Biotechnologie e.V.) (1987). *A Study for the Generation, Inter-Continental Transport and Use of Hydrogen as a Source of Clean Energy, on the Basis of Large-Scale and Cheap Hydro-electricity (Hydrogen Pilot Project – Canada)*. Translation of the German Final Report. Frankfurt am Main: DECHEMA.

Drolet, B., Gretz, J., Kluyskens, D., Sandmann, F. and Wurster, R. (1994). The Euro-Québec Hydro-Hydrogen Pilot Project [EQHHPP]: Demonstration Phase. Hydrogen Energy Progress X. *Proceedings of the 10th WHEC, Cocoa Beach*. Florida: Florida Solar Energy Center–University of Central Florida, vol. 1, pp. 23–46.

EC (2007). *Energy Corridors – European Union and Neighbouring Countries*. Final Report of Encouraged Project. Luxembourg: Office for Official Publications of the European Communities.

Fasold, H. (1987). Transport, Speicherung und Verteilung von Erdgas heute – von Wasserstoff morgen. *Tagung der VDI-GET, 11–12 March 1987*, München. Hamburg: Urban, vol. 104, (4), pp. 165–171.

Gibbs, B. and Biewald, B. (2000). *Transmitting Windpower from the Dakotas to Chicago: A Preliminary Analysis of a Hydrogen Transmission Scenario*. Cambridge, Massachusetts: Synapse Energy Economics. www.synapse-energy.com/Downloads/SynapseReport.2000-09.ELPC-Leighty-Foundation.Hydrogen-Transmission-Scenario-Analysis.00-21.pdf.

Hijikata, T. (2002). Research and development of international clean energy network using hydrogen energy (WE-NET). *International Journal of Hydrogen Energy*, **27** (2), 115–129.

Hoffmann, P. (2001). *International Directory of Hydrogen Energy & Fuel Cell Technology*. 2nd edn. Rhinecliff, NY: The Hydrogen Letter Press.

Kaske, G. and Plenard, F. J. (1985). High-purity hydrogen distribution network for industrial use in Western Europe. *International Journal of Hydrogen Energy*, **10** (7–8), 479–482.

Krock, H. and Zapka, M. (1991). The Hawaiian International Hydrogen Energy Pilot Project (HIHEPP): a concept for the commercial implementation of non-fossil hydrogen energy in the Pacific Ocean area. *OCEES International, 1–3 October 1991, Hawaii*. New York: IEEE.

Leighty, W., Hirata, M., O'Hashi, K. and Benoit, J. (2004). Large stranded renewables: the International Renewable Hydrogen Transmission Demonstration Facility (IRHTDF). *World Energy Congress*, 4th September, 2004. Sydney.

Mantzos, L., Capros, P. and Kouvaritakis, N. (2003). *European Energy and Transport Trends to 2030*. Athens: National Technical University of Athens (NTUA). http://europa.eu.int/comm/dgs/energy_transport/figures/trends_2030/1_pref_en.pdf.

Nordic H_2 Energy Foresight. (2001). *Euro – HYPORT*. www.h2foresight.info/NordicR + DActivities/NordicActivities_demonstrations_Iceland_ Hyport.htm.

O'Hashi, K. and Hiraishi, K. (2001). A perspective of renewable energy and new technology in northeast Asia. *World Energy Council, 18th Congress, Buenos Aires, October 2001*. Tokyo: Asian Pipeline Research Center of Shibaura Institute of Technology, Northeast Asian Gas and Pipeline Forum.

Partl, R. (1977). *Power From Glaciers: The Hydropower Potential of Greenland's Glacial Waters*. Final report. Laxenburg, Austria: IIASA.

Raballo, S. and LLera, J. (2004). Large scale wind hydrogen production in Argentine Patagonia. *International Conference for Renewable Energies, 1–4 June 2004, Bonn, Germany*. Buenos Aires: C.A.P.S.A. – Capex S.A. Group.

Siegel, A. and Phan, Q. (1990). *HYSOLAR-Programme zur Simulation und Auslegung von PV-Elektrolyseanlagen*. Report of DLR (Deutsches Zentrum für Luft- und Raumfahrt) and KACST (King Abdulaziz City for the Science and Technology). Stuttgart: DLR.

Siteur, F. (1993). Wind energy in South America. The role of the Argentinean Wind Energy Association as a consensus builder. *Windtech International*. www.windtech-international.com/content/view/166/68.

Soltermann, O. E. and Da Silva, E. P. (1998). Comparative study between the hysolar project and a hypothetical international project in Brazil for hydrogen production and exportation (BHP) from photovoltaic energy and secondary hydroelectricity combined supply. *International Journal of Hydrogen Energy*, **23** (9), 735–739.

18

Macroeconomic impacts of hydrogen

Martin Wietschel, Sabine Jokisch, Stefan Boeters, Wolfgang Schade and Philipp Seydel

Often only technical aspects are considered when looking at hydrogen as an energy carrier. However, the introduction of hydrogen could have relevant implications for GDP, welfare and job developments in a nation or region. The competitiveness of a nation could be one major driver for hydrogen use as an energy carrier. These issues are discussed in the following. Among other things, possible economic effects are shown on the basis of a quantitative model analysis and assessed for relevant EU member states.

18.1 Introduction

A lot of research in the hydrogen and fuel-cell field has a strong technology focus (e.g., technology research, application of technologies, technology roadmap and infrastructure build-up). The analysis of economic impacts tends to concentrate on the necessary investments for hydrogen infrastructure build-up (see Chapters 14 and 15). Other very important impacts of hydrogen as an energy fuel, such as those on employment, gross domestic product (GDP), international competitiveness or welfare, are often mentioned as important benefits of a hydrogen use as an energy vector but they are usually not well analysed.

For such kinds of analysis, four different methodological approaches are normally used for analysis in the energy sector: input–output, general equilibrium, system dynamics and econometric models.

Such kinds of model, emphasising an aggregate description of the overall economy, are general economic models with a rather rudimentary treatment of the energy system. Following the top-down approach, they describe the energy system (similar to the other sectors) in a highly aggregated way. Three different approaches will be presented in this chapter[1] with a focus on the following two topics:

[1] This work was part of the EU project HyWays (The Development and Detailed Evaluation of a Harmonised 'European Hydrogen Energy Roadmap', see www.hyways.de). HyWays aimed at developing a validated and well accepted roadmap for the introduction of hydrogen in the energy system. Furthermore, the MATISSE and TRIAS projects developed and assessed scenarios of alternative energy supplies for transport including hydrogen (see www.matisse-project.net/projectcomm and www.isi.fhg.de/TRIAS).

The Hydrogen Economy: Opportunities and Challenges, ed. Michael Ball and Martin Wietschel. Published by Cambridge University Press.

- Competitiveness of the EU in the field of hydrogen compared to different world regions,
- Employment and GDP impacts of hydrogen as an energy fuel.

To analyse the second topic, it is helpful first to analyse the international competitiveness of countries because import–export effects trigger many of the positive or negative impacts of hydrogen not only on employment figures, but also on other economic values.

Section 18.2 describes the competitiveness of economies in the field of hydrogen using a so called lead market approach. Section 18.3 deals with hydrogen penetration scenarios, which are the underlying scenarios for the different model applications in Sections 18.4–18.6.

The following three sections present different model applications to analyse the impacts of hydrogen to the economies using the scenarios described in Section 18.3. In Section 18.4 employment effects for ten European countries will be exemplarily analysed with an input–output model. In Section 18.5, GDP effects for different European countries will be analysed with a general equilibrium model. Section 18.6 presents a system dynamic model, which deals with GDP and employment effects. Section 18.7 summarises the different model approaches, presents and discusses the results, and draws overall economic conclusions.

18.2 International competitiveness of economies in the field of hydrogen

18.2.1 Theory of lead market

Trade developments have a significant influence on the economic impacts of using hydrogen as an energy vector in a national economy. Therefore, the so-called lead-market concept is used to estimate the international competitiveness of a country and to forecast the international trade development (import–export shares) of hydrogen technologies. Lead markets are defined as the regional markets to adopt an innovative design first in a technology field that has specific characteristics that enhance the probability that the same design will be adopted broadly in other countries as well (Beise, 2005; Meyer-Krahmer, 2003). Different theoretical approaches can be used to explain such lead markets.

In traditional (neoclassical) trade theory, the theorem of factor proportionality plays an important role in explaining trade flows (Ohlin, 1933; Samuelson, 1948). This theorem postulates that there will be gains from trade through increased specialisation in sectors that make intensive use of those factors with which countries are relatively well endowed (e.g., capital availability or labour cost). However, the well known Leontief paradox (against the theorem of factor proportionality, the USA actually import labour- and capital-intensive products, (Leontief, 1956)) and other empirical research[2] opened up a debate on the value of this theorem and new

[2] Kaldor (1978) and Robson (1987) showed that market shares of exports and increasing relative unit costs went hand in hand for a number of countries.

hypotheses were developed on factors influencing international competitiveness. One idea was that it is technological rather than price competition in capitalist economies that matters most. Therefore, technology factors, like productivity and research and development, were taken into account to explain market success (Fagerberg, 1996; Sell, 2001; Wakelin, 1997). The technology-gap theory (e.g., Posner, 1961) and the product-cycle theory (e.g., Vernon, 1966) postulated that especially well developed countries introduce new products and can win innovation yields for a limited time. Explanations for this are differences in the technological capabilities of countries (technology gap) and higher know-how requirements during the market introduction phase of a technology.

Next to the superior technological capability of firms or industry branches, other aspects are also important: market introduction strategies for new technologies, market introduction advantages due to access to regular customers, the internal capability of a firm to identify and use the market chances of new developments and reputation (Beise, 2004, 2005; Meyer-Krahmer, 2004).

Whereas these approaches are supply-oriented, others also take the demand side into account (Blümle, 1994; Fagerberg, 1995; Linder, 1961; Sell, 2001). They argue that the origins of innovations are based on the development of a new demand that cannot be satisfied by existing products. Innovations require lead users (von Hippel, 1988; Morrison *et al.*, 2004), who expect novel and high benefits from the new products, are trendsetters for commercial developments and radical innovations and have connections with producers. Other market factors are also important (Dosi *et al.*, 1990; Porter, 1990). High competition between firms leads to increased inventions (this has been validated empirically) and the development of alternative technology designs, which increase the probability of selecting the most appropriate technology design (market structure advantages). Market size, market dynamics and market growth all have a significant impact on cost reduction and international competitive advantages. The transfer (export) to other regions is advantageous if the risks of technology adoption are minimised via demonstration effects and similar consumer preferences.

Apart from supply-side and demand-side factors, the regulative framework (e.g., standardisation, approval requirements) may also be relevant in explaining the success of innovations (Blind and Jungmittag, 2005; Beise and Rennings, 2005; Porter, 1990).

18.2.2 Indicators for the identification of lead markets for hydrogen and fuel cells

To identify lead markets for hydrogen and fuel cells, a holistic approach is chosen which takes supply- and demand-side factors into account as well as regulative impacts. Quantification of lead market factors is often not possible (e.g., how are technology capabilities or transfer advantages measurable?).

Most of the empirical research in technology-dominated sectors focuses on the correlation of international trade and technological capability. Research and development budgets and patents are mainly used as indicators to explain the success of a country in international trade figures (see Fagerberg (1995) for a comprehensive overview of relevant studies). In general, the empirical results of these studies support a positive correlation between technology (measured by R&D intensity and patents) and export performance for a large number of industrial branches and countries. For the research work on hydrogen and fuel cells, it is important to notice the strong evidence linking export success and technology in the car, chemical and machinery industries (Fagerberg, 1996), because these sectors are relevant for hydrogen and fuel cells.

Some empirical research also includes a variable assumed to reflect the size of the country or its domestic market. The results are not uniform for countries and branches, but two studies identify a significant correlation between scale and trade performance for the car industry, among others (Fagerberg, 1995; Soete, 1981). With regard to the importance of price competition, the evidence is less clear cut. As might be expected, price competition seems to be important in many low-tech industries (e.g., textiles and clothes). The investment variable measured per worker fails to have a significant impact in all but a few cases.

For the following study, which was carried out in the HyWays project (HyWays, 2006a), more than 40 indicators are used to analyse the lead market potential of the different countries in the hydrogen and fuel-cell fields. Next to the well accepted indicators of patents (see Fig. 18.1) and R&D budget (see Fig. 18.2), others are also selected. These should be treated carefully because of the lack of statistical evidence for the influence of these indicators on trade volume. However, the indicator choice was discussed with numerous experts working on macroeconomic and innovation research questions. Among others, the following indicators were chosen: demonstration projects, foreign trade in hydrogen-relevant industrial branches, domestic-energy resources and electricity prices, venture capital investments and current production of hydrogen for industry applications.

Next to absolute figures, relative figures (e.g., patent activity in relation to inhabitants) are also chosen to compensate for the fact that larger economies have a clear advantage if only absolute figures are employed. To come to a final evaluation of the lead-market potential, the indicators have to be weighted and summarised to one value. Here, a scoring approach is used with weighting factors fixed in collaboration with experts. This scoring approach is a pragmatic and 'simple' approach, which allows the evaluation of alternatives on the basis of different quantitative and non-quantitative criteria. However, some disadvantages and methodical difficulties can be identified. A detailed discussion of this method and also a presentation of more elaborated multicriteria methods can be found in Geldermann and Rentz (2005) and Keeney and Raiffa (1976). The weighting of criteria in the scoring approach depends on the evaluation of experts and, therefore, the experts' selection (and their internal knowledge) has a strong influence on the results.

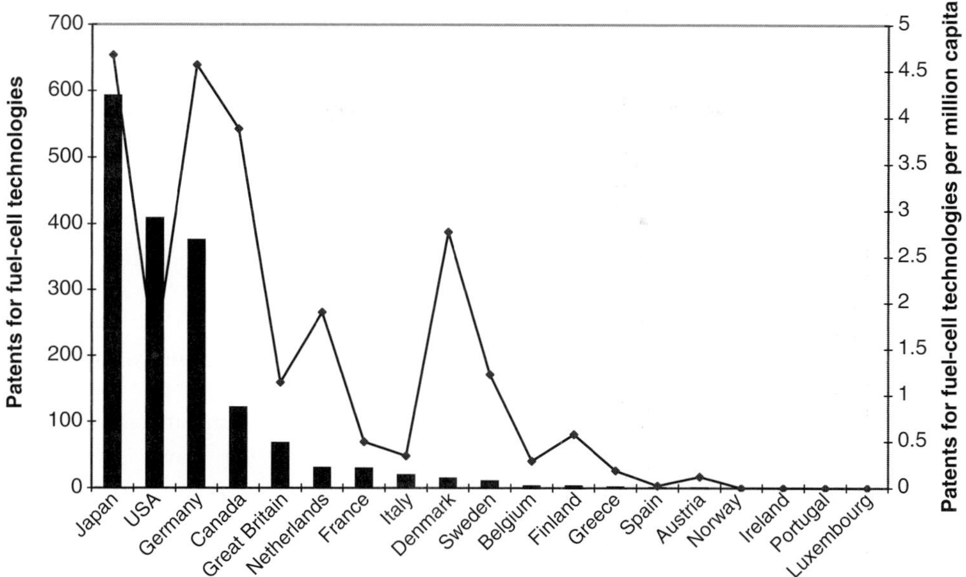

Figure 18.1. Total number of fuel-cell patents and fuel-cell patents per million capita (1990–2002) (world patents (WO) and European patents (EP)), source: own patent analysis; more background information can be found in Wietschel (2005).

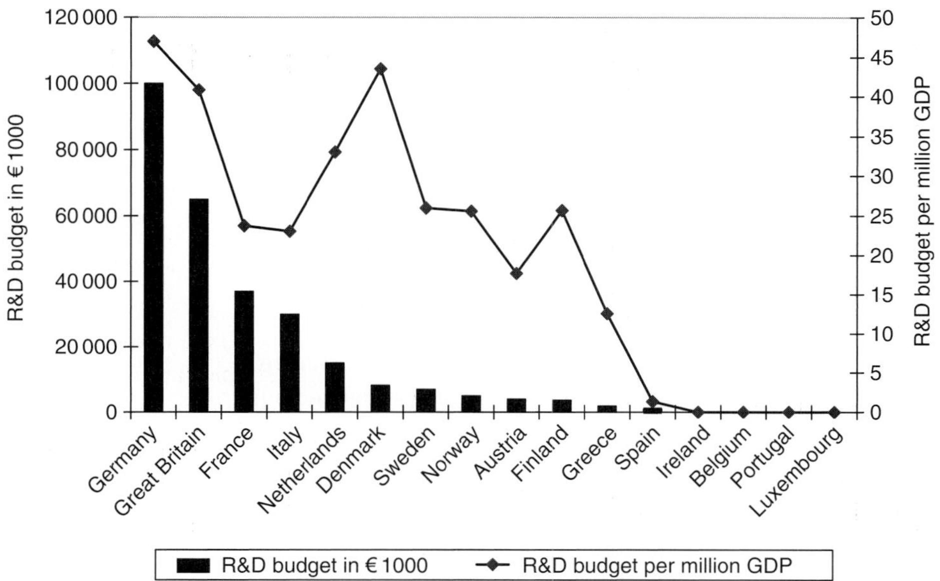

Figure 18.2. Public R&D budget in fuel cells and hydrogen (Amorelli *et al.*, 2004).

It should be mentioned that the quantitative figures for the indicators are based on the available historical data and that very recent developments may not be reflected in this analysis. In addition to the quantitative indicator approach, results of other studies were also taken into account (Amorelli *et al.*, 2004; FHIRST, 2004; HySociety, 2004; OECD, 2005).

18.2.3 Conclusions

The main outcomes of the analysis are:

- European industry is well placed with regards to conventional generation of hydrogen from fossil fuels and via water electrolysis.
- Europe is behind the USA and Japan for industrial development and manufacturing of fuel cells.
- In the fuel-cell research Europe, and the USA, have a broader approach. At present it cannot be estimated if this is an advantage or a disadvantage, especially against Japan, where the focus is on PEMFC.
- Although Europe is at the forefront in many fields, a lack of co-ordination and of well defined targets could make it less suited to commercial and industrial exploitation. But at present there are some efforts to fund and develop a scientifically sound and broadly accepted European roadmap.
- In Europe, the ecological targets and the existing environmental problems and targets, as well as the population density offer good conditions for the implementation of a hydrogen and fuel-cell economy.
- The overall competitiveness of the industries in Europe, Japan and the USA can be regarded as equivalent. But the better access to venture capital in the USA, and also in Canada, can give these countries an advantage, especially in developing innovative small and medium-sized companies.
- Concerning law regulations, there are efforts in Europe, but the USA, Japan and Canada have the leading position.
- Overall, Europe can be regarded as well positioned, but Europe runs the risk of falling further behind the USA and Japan, which are in the leading position right now and likely to enlarge their lead through specific improvements.

18.3 Hydrogen-penetration scenarios and investments in hydrogen technologies

Owing to still existing uncertainties on a technical level, a market development forecast for mobile and stationary hydrogen and fuel-cell applications cannot be made with any reasonable probability. Crucial subsystems, such as fuel-cell stacks or hydrogen-storage systems, and also key components, like catalysts and membrane-electrode assemblies, have already achieved significant progress, but still require further breakthroughs on their way towards mass commercialisation, so that a more

Table 18.1. *Scenarios for the potential development of hydrogen vehicles; share in vehicle stock (for more details of the scenarios, see HyWays, 2006c)*

Total share of fleet	2010	2020	2030	2040	2050
High penetration	–[a]	3.3%	23.7%	54.4%	74.5%
Medium penetration	–[a]	1.2%	11.9%	35.9%	69.4%
Low penetration	–[a]	0.1%	2.8%	12.9%	36.0%

Note:
The scenarios 'high', 'medium' and 'low' penetration correspond to the scenarios 'very high policy support; fast learning', 'high policy support; fast learning' and 'moderate policy support; moderate learning', respectively, as used in Chapter 14.

[a] Demonstration vehicles and fleets only.

Table 18.2. *Scenarios for the potential development of stationary hydrogen applications in the residential sector (for more information, see HyWays, 2006c)*

Total share of households	2010	2020	2030	2040	2050
High penetration	–	1%	4%	8%	10%
Medium penetration	–	1%	4%	8%	10%
Low penetration	–	0.1%	0.5%	2%	5%

Table 18.3. *Scenarios for the possible development of stationary hydrogen applications in the commercial and services sector (for more information, see HyWays, 2006c)*

Total share of commercial demand	2010	2020	2030	2040	2050
High penetration	–	0.3%	1.3%	2.7%	3.3%
Medium penetration	–	0.3%	1.3%	2.7%	3.3%
Low penetration	–	>0%	0.2%	0.7%	1.7%

evolutionary based forecast is not within the scope of this work. Considering the key findings of the European Hydrogen and Fuel Cell Technology Platform (2005) Deployment Strategy and the Strategic Research Agenda (European Hydrogen and Fuel Cell Technology Platform, 2005), in HyWays, a set of different penetration rates reflecting either a very optimistic or a more conservative development that has been adapted to the specific needs of both mobile and stationary applications was developed, see Tables 18.1 to 18.3. (For more background information on the scenarios, see HyWays, 2006b.)

Having exogenously fixed the demand for hydrogen, the infrastructure build-up necessary to satisfy this demand then has to be determined (in terms of numbers of

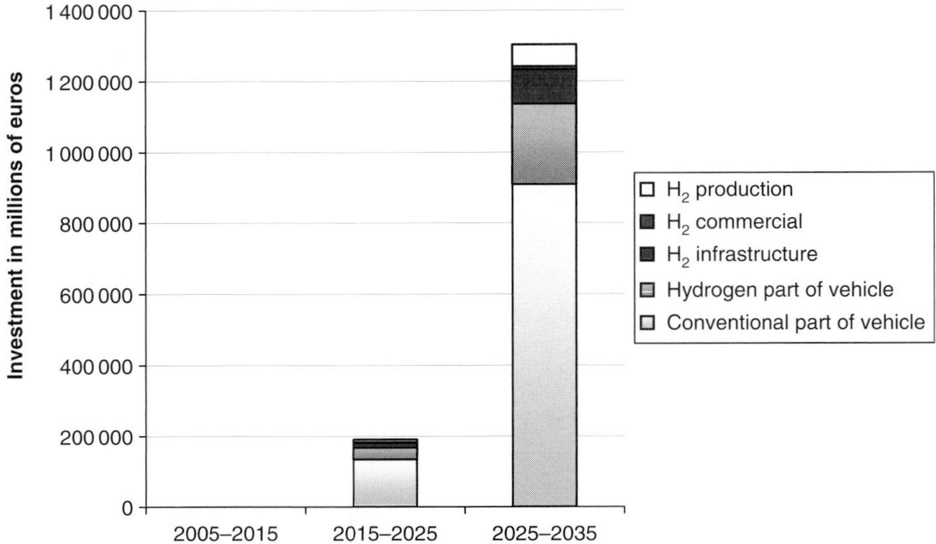

Figure 18.3. Structure of the investments of the six HyWays countries (France, Germany, Greece, Italy, The Netherlands and Norway) in a hydrogen economy (cumulative for a ten-year period, hydrogen high-penetration scenario).

vehicles, pipelines and fuelling stations, as well as of hydrogen production plants across the country). To do this, an energy system model (MARKAL) is used (see Seebregts *et al.* (2001) for a description of MARKAL). This is a typical bottom-up engineering-based linear activity model with a large number of energy technologies to capture the substitution of energy carriers at the primary and final energy levels, process substitution, process improvements (gross efficiency improvements, emission reductions), technology learning, spill-over effects and energy savings. It is mostly used to compute the cheapest way of meeting a given demand for final energy or energy services subject to various system constraints, such as exogenous emission reduction targets, or prescribed energy technology paths (such as an administered phase-out of nuclear power or phase-in of green energy). The use of mathematical programming in the context of energy planning is a well known field of research (Gately, 1970), but is still of great interest (Gonzalez-Monroy and Cordoba, 2002; Song, 1999; Wietschel, 2000), especially due to the new situation in liberalised energy markets. The extension or adaptation of such models to hydrogen use in the energy system is a recent research development in this field (CASCADE MINTS, 2006; HyWays, 2006b).

For the following analysis, the investment figures and the selected technologies from MARKAL are an important input. The structure of the investments necessary for a transition to a hydrogen economy is clearly dominated by the expenditures for hydrogen vehicles (see Fig. 18.3). If a hydrogen vehicle is imported, not just the hydrogen drive system will be imported, but very likely the whole vehicle. Therefore, the structure of the domestic vehicle industry sector turned out to be one of the key factors for an employment analysis, and further for GDP and welfare development.

In the start-up phase, decentralised production options play an important role. Later on, central production options from fossil fuels with carbon capture and storage (CCS) are expected to be the leading production source in Europe, with renewable hydrogen then being phased-in. In this later stage, hydrogen pipelines also seem to be the most important transport mode. For more details on the methodology, database and results see HyWays (2006b).

18.4 Input–output model: ISIS

18.4.1 General modelling approach

The ISIS (Integrated Sustainability Assessment System) model belongs to the class of input–output models. Studies conducted with input–output models are mesoeconomic analyses in which macroeconomic impacts, such as changes in income-flow effects, only play a subordinate role. The spotlight of ISIS here is on impacts triggered by structural shifts. These types of analysis constitute a classical field of application for an input–output model (Duchin, 1994; Leontief, 1986), which aims to map fully the flow of goods between the economic sectors for an economic domain, such as Germany, on a mesoeconomic aggregation level (Meyer-Krahmer, 1999; Petit, 1995).

Based on the sectoral interrelations, which are illustrated in the input–output model and which form the status quo, the impulses triggered by the closed-loop strategies can be modelled. These either affect upstream production (e.g., increased secondary raw materials and a parallel decrease in the use of primary raw materials) or are linked with changes in the final demand for goods (e.g., a decrease of cars with conventional combustion engines and an increase of cars with fuel cells, as drive systems lead not only to changes in the automotive sector, but also to different supplier structures for the automotive sector). With the help of the information about the interrelations contained in the input–output model, the effects triggered upstream by these impulses can be determined right up to the material suppliers.

The main elements of the used input–output model ISIS are described in the following. At the core of ISIS is a statistical input–output model (IO model) used to examine the structural impacts of the various strategies. Other modules for employment effects, qualification structure and job conditions, regional effects and environmental effects were developed or added to analyse other dimensions of sustainability. The results of the scenario calculations from the IO model, i.e., production changes as a result of the different strategies, serve as inputs for the other modules.

The IO model used for ISIS is based on the most recent input–output tables of the European Statistical Office (Eurostat, 2004) for the year 2000. It can be assigned to

Table 18.4. *Layout of an input–output table*

		Manufacturing sector	Final demand sector				Production value
		Sectors 1–59	Private consumption	Public consumption	Fixed investments	Exports	
Manufacturing sector	Sectors 1–59	Interrelation matrix: supplies of goods and services among the sectors (intermediate demand) (millions of euros)					
Imports							
Gross value added							
Production value							

the group of static, open Leontief models (Leontief, 1956). It divides the single economies into 59 manufacturing sectors and six sectors of final demand and illustrates both the supplies of goods and services between the manufacturing sectors (intermediate demand) and supplies from these to the final demand sectors (see Table 18.4 for the structure of input–output tables). This IO model is used to calculate sectoral differences in production between the reference and the transition scenarios. These results form the basis for further analyses in the additional modules. In these additional analyses, therefore, not only the direct but also the indirect effects from all input relations are taken into account.

Assuming that a linear approximation can be made for the correlation between the sectoral employment level and the sectoral production level, the quantitative impacts on employment are calculated using job coefficients. Of course, the constant input–output coefficients are a strong assumption and could be criticised (Zhang and Folmer, 1998). However, structural effects could be analysed in a *ceteris-paribus* analysis with the chosen approach.

Which employment effects can be listed in the individual sectors are analysed. In an *impulse-based approach*, the total employment effects are assigned to the economic impulses that trigger them. These impulses consist of the demand shifts triggered by a hydrogen scenario, e.g., the rise in producing hydrogen vehicles on the one hand (positive impulse) and the drop in production of conventional vehicles on the other hand (negative impulse). The employment effects assigned to one economic impulse include the workers who are directly or indirectly involved in the production of goods linked with the impulse. They thus specifically incorporate upstream production. The ratio relating the change in the number of employed to the impulse triggering the change is characterised as the specific total employment effect.

The projection analyses are based on the years 2020 and 2030. To avoid overestimating the employment effects it is necessary to take into account the growth in productivity that will have taken place by this time. To do so, sector-specific productivity indices are determined, which indicate the ratio of specific employment (e.g., employment or gross output value) in the projected year to specific employment in the base year. These indices are based on data in *The Impact of Technological and Structural Change on Employment* (Christidis *et al.*, 2002).

18.4.2 Modelling a hydrogen economy in ISIS

An important step is to integrate the new technologies like fuel-cell vehicles and electrolysers into the input–output model, because in a standardised input–ouput table they are missing. Therefore, an in-depth technological analysis is necessary to integrate new technologies into the input–output disaggregation (including the investments and also the operation and maintenance costs of technologies).

Different stakeholders have been interviewed about future technology developments. Over 20 industry partners have been interviewed and given input to assemble a technology picture of the future hydrogen technologies (e.g., for hydrogen pipelines, stacks or electrolysers).

The different technologies and tasks are separated into single parts. Some parts that are similar to existing technologies could easily be integrated into the sectoral form of an input–output analysis. Other new parts, such as fuel cells, are split up in more detail. On the basis of single components, a sectoral analysis is then carried out by an expert group composed of engineers, industrial engineers and economists to define the possible economic sectors which ‘produce’ this kind of economic commodity.

Of course, the analysis is afflicted with uncertainties about the future development of single technologies. Some breakthroughs in single technology developments could not be foreseen within this approach. This may lead to some wrong allocation of technology components to industrial sectors.

As an example of a sectoral technology split, see the central electrolysis technology in Fig. 18.4.

The structure of the investments necessary for a transition to a hydrogen economy is clearly dominated by the expenditure on hydrogen vehicles (see Fig. 18.3). If a hydrogen vehicle is imported, it is very likely that not just the hydrogen drive system,

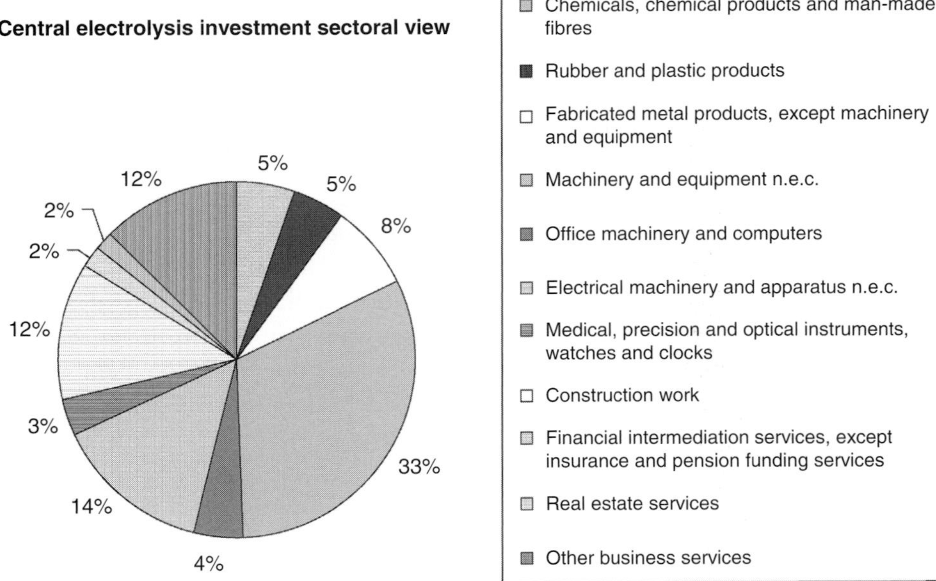

Figure 18.4. Example of technology analysis, e.g., 5% in rubber and plastic products means that 5% of the investments in a central electrolyser are produced in the rubber and plastic products sector in economic terms.

but the whole vehicle will be imported. Therefore, the structure of the domestic-vehicle industry turned out to be one of the key factors for the employment analysis, and also for GDP and welfare development.

Three import–export scenarios have been analysed. Each scenario describes a possible future for the competitiveness of hydrogen technologies produced within the EU.

The so-called 'structural identity scenario' is based on the assumption that the international competitiveness of domestic hydrogen technologies is mainly influenced by today's competitiveness of industrial sectors producing goods that are very similar to hydrogen technologies. For example, if a country makes and exports conventional cars, this country is likely to do so in the future as well. Or if a country has a leading position at present in producing gas turbines, this country is likely to produce hydrogen turbines in the future.

These assumptions are open to criticism because today's domestic industry based on conventional technologies does not automatically equate to a leading position for hydrogen technologies in the future. For example, if a country has the current manufacturing capacity to construct conventional internal combustion engines, this does not necessarily mean that it will also have a relevant industry for stack production in the future, owing to the technological differences between the technologies

Table 18.5. *Import–export scenario assumptions (e.g., 1.4 for cars means that for every car sold in the domestic market, the domestic producer exported an additional 1.4 cars) (FC = fuel cell, ICE = internal combustion engine)*

H_2 car production (FC, ICE)	Structural identity	Pessimistic	Optimistic
Finland	0.4	0.0	0.4
France	1.4	0.0	2.0
Germany	2.0	0.0	3.0
Greece	0.1	0.0	0.1
Italy	0.9	0.0	1.3
Netherlands	0.8	0.0	1.2
Norway	0.2	0.0	0.2
Poland	1.1	0.0	1.7
Spain	1.9	0.0	2.9
United Kingdom	1.0	0.0	1.6

(e.g., different materials, production processes and education). Two alternative scenarios were developed to take this into account.

The ‘pessimistic scenario’ shows what could happen if other regions of the world assume the leading position and Europe has to import hydrogen vehicles. In this scenario, it was assumed that all hydrogen-vehicle technology will be imported. In contrast, the ‘optimistic scenario’ assumes that major efforts will be undertaken, which result in increased EU exports in hydrogen vehicles and technologies. Indeed, it is supposed that hydrogen vehicles, as well as stacks and hydrogen production technology, could be exported even more than conventional technology (see Table 18.5).

An overview of the assumed import–export ratios is shown in Fig. 18.5.

18.4.3 Results

Figure 18.6 shows the employment development for the ten countries analysed. Small gains can be achieved if the import–export shares for hydrogen technologies are similar to conventional technologies. This result is mainly influenced by a lower level of automation and standardisation for hydrogen technologies in the start-up phase. However, the same level of competitiveness as for conventional technologies must be reached on world markets first and this will be a challenging task, looking at the results of the lead market analysis. Based on the situation at present, a small drop in employment figures can be expected.

The largest direct effects on employment due to the transition to a hydrogen economy are seen for the automotive industry, and to a lesser extent for the plant and equipment sector. This is because losing (or winning) market shares to (or from) a foreign competitor may lead to lost (or gained) sales of the complete vehicle, not

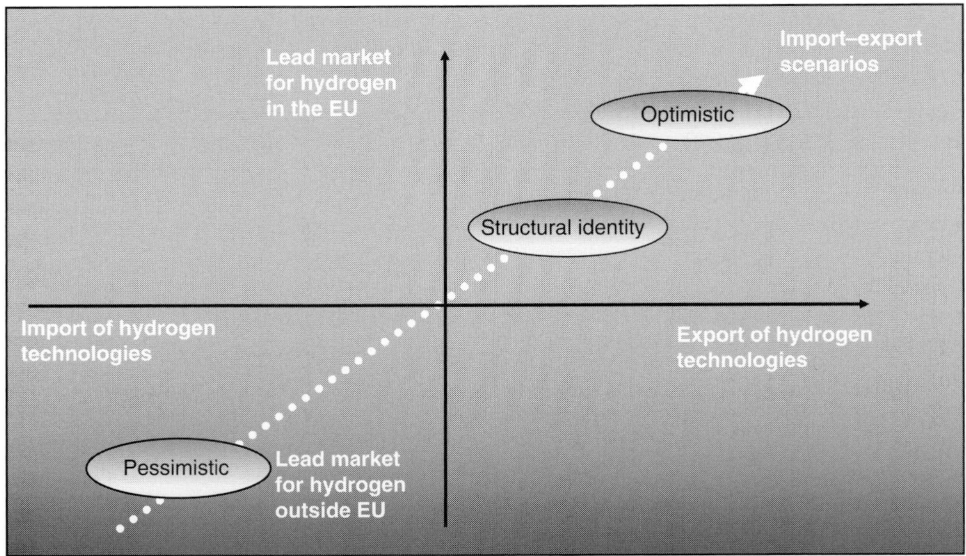

Figure 18.5. Import–export scenarios.

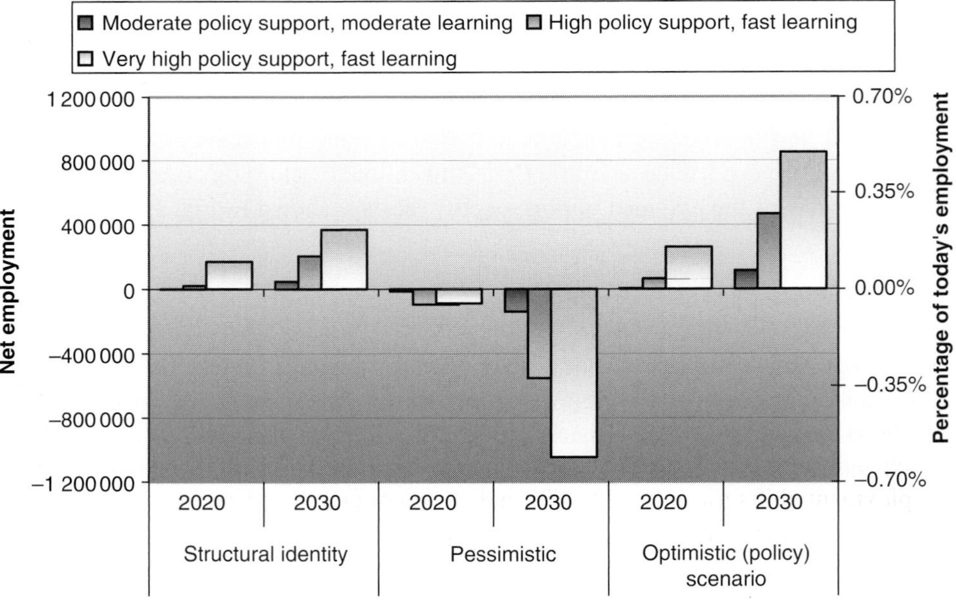

Figure 18.6. Net employment effects for the hydrogen high penetration, and medium penetration scenarios with optimistic learning rates for hydrogen passenger cars, as well as hydrogen low penetration scenarios with moderate learning rates for hydrogen passenger cars for the years 2020 and 2030. The overall net employment effects for the ten HyWays countries in three import and export scenarios are shown.

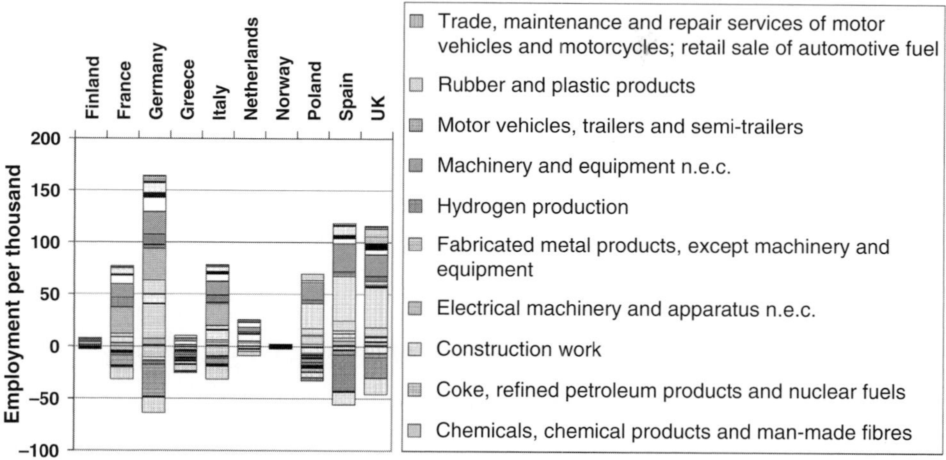

Figure 18.7. Employment effects for the hydrogen high penetration in the structural identity scenario for the different economic sectors of the ten HyWays MS countries.

just the hydrogen drive system. Whether the impact is negative or positive depends strongly on the effort Europe puts into consolidating or improving its current position in the car market. This holds even stronger for the current car manufacturing countries, which therefore face the dilemma: should they invest in a new technology, possibly losing many billions of R&D and infrastructure build-up investments, or not, at the possible expense of even higher losses in GDP and jobs. In particular, France, Germany, Spain, the UK and Italy could be seen as candidates for the dilemma situation. The differences between the winnings and losses of these countries could be traced back to the differences in the structure of the automotive sector. Germany, for example, has very high export shares in the automotive sector, compared with other countries. Therefore, drastic changes in the share of hydrogen vehicles could also have a stronger effect on the production for export and not only on the production for domestic use.

Replacing conventional vehicles by fuel-cell vehicles (FCVs) induces sectoral employment shifts away from traditional car manufacturing (see Fig. 18.7).

The most important shift is from the historical automotive sector to the new hydrogen-vehicle technology sectors. This conglomerate consists mainly of fabricated metals, the electrical and the machinery plastic sector and the chemical sector.

A small shift away from the refinery sector towards the new and more labour-intensive hydrogen production sector could also be identified. Employment gains could also be made in the construction and the machinery and equipment sectors because of hydrogen infrastructure investments.

This shift requires considerable education and training of the workforce and, in combination with possible mass production by 2020, necessitates early political action because of the required gradual build-up of manufacturing capacity and skilled labour. Further research work is necessary to analyse today's weak and strong aspects of the education system in European countries and to develop an action plan for adaptation to the new requirements. This analysis should take into account the demographics of the European nations, which is an important factor. Will Europe be able to find the necessary skilled labour such as engineers and technicians or will they 'import' skilled labour from countries like China and India?

18.5 Computable general equilibrium model: PACE-T(H_2)

18.5.1 General modelling approach

The PACE-T(H_2) model belongs to the class of dynamic computable general equilibrium (CGE) models,[3] which are based on microeconomic theory. Dynamic CGE models explain the origination and spending of income for all major economic agents (households, firms, government, abroad) simultaneously and numerically solve the path for all endogenous variables over several periods. These models thus provide a consistent framework for determining economy-wide repercussion effects of different policy scenarios and their distributional consequences.

The current version of the PACE-T(H_2) model of the world economy contains 12 regions. These are the ten countries featured in the integrated EU project HyWays[4] (Finland, France, Germany, Greece, Italy, the Netherlands, Norway, Poland, Spain, the United Kingdom), the rest of the EU (plus Switzerland and Iceland), and the rest of the world. We distinguish between nine production sectors (transport, energy intensive production, five energy sectors, the rest of production, and an aggregate of investments). The model is set up as a perfect foresight model with optimal savings and investment decisions of the representative agent, who maximises the present value of her lifetime utility given her intertemporal budget constraint. The period length is chosen to be ten years, with the model horizon extending from 2000 to 2050. International trade is modelled in the Armington fashion (Armington, 1969), where goods produced in different countries are treated as imperfect substitutes and import shares depend on their relative prices. An exception is the international crude oil market, which is homogenised so that every country either imports or exports crude oil.

[3] The general equilibrium structure of these models is based on Arrow and Debreu (1954). An introduction to applied general equilibrium analysis can be found in Shoven and Whalley (1984). CGE models are used in almost all economic fields to quantify the impact of various policy reforms. For an overview of these studies, see, e.g., Bhattacharyya (1996), Conrad (2002), Gottfried *et al.* (1990), or Gunning and Keyzer (1995).

[4] The PACE-T(H_2) model was first applied for the analysis of introducing hydrogen in the transport sector within HyWays. For more information on HyWays, see www.hyways.de.

The production of commodities in each region is captured by aggregate production functions, which characterise technology through substitution possibilities between various inputs (capital, labour, energy, non-energy, intermediate inputs). Domestic demand is composed of the final demand of private and public consumers and the intermediate demand from other sectors. The final demand of the representative agent is given as a composite of transport services and an aggregate of other consumption commodities, which combines consumption of an energy aggregate with a non-energy consumption bundle. All essential functions in the model are of the nested constant elasticity of substitution (CES) type.

A special feature of PACE-T(H_2) is the implementation of passenger cars as a durable consumption good, i.e., there is a separate stock of automobile capital. Households do not consume cars, as such, but transport services, which are produced with various inputs. The value of these services is composed of capital services of the automobile stock present in the respective economy, fuel and expenditure for repair and maintenance. Car lifetime is assumed to be 12 years for all cars. The model distinguishes between three different size classes for cars. These are small, medium and large cars, which are powered by either a conventional technology or hydrogen.

The following equilibrium conditions constitute the essential parts of the model: (i) zero profit conditions for all production sectors (under the assumption of perfect competition), (ii) market clearance on all markets (perfectly adjusting prices) and (iii) exhaustion of the representative consumer's budget through consumption purchases and savings.

18.5.2 Modelling a hydrogen economy in PACE-T(H_2)

Hydrogen enters PACE-T(H_2) as a fuel for passenger cars. Consumers in the model choose between transportation services produced by either a conventional or a hydrogen technology. Both car technologies are assumed to be perfect substitutes. The baseline simulation in PACE-T(H_2) assumes the hydrogen technology to be inactive. In the scenario analysis, the share of hydrogen cars in overall new passenger cars (penetration rates) is set exogenously. Since consumers always choose the cheapest alternative, the more expensive technology would simply be withdrawn from the market. To prevent such a development, both technologies have to be equally expensive. This is reached by either taxing or subsidising the hydrogen technology. The respective amounts are lump-sum transferred to the consumer.

The assumption that conventional and hydrogen cars are perfect substitutes must be considered as a modelling short cut, given the fact that there are further factors affecting car demand, like noise or driving properties. However, owing to the lack of detailed empirical data on hydrogen car demand there is no possibility of calibrating a more flexible demand function.

The use of hydrogen as fuel input in passenger transport demands a hydrogen production and distribution technology. The model therefore features a separate sector for hydrogen production, which is characterised by fixed input shares. These inputs differ between the considered countries because of the different mix of hydrogen production technologies with different fuel or electricity inputs. The aggregate hydrogen production technology in PACE-T(H_2) is characterised by the average costs of hydrogen production and distribution calculated as the sum of costs for investment, fuel inputs and operation and maintenance (O&M) for the various production and distribution technologies. The so-called bottom-up data on car technology and hydrogen production and distribution technology are taken from the energy-system model MARKAL. These data are mainly related to the capacity and activity of various technologies and related parameters, such as activity levels, delivery cost, fuel costs, fuel inputs, installed and new capacity, investment cost, and costs for operation and maintenance. Given the output quantities and the cost structure of the different hydrogen technologies, the aggregate production data of the IO sectors are split up to accommodate a consistent bottom-up representation. The input structure of car and hydrogen production technologies is adopted from the ISIS model.

Since the consumer's car purchasing decision is based on after-tax consumer car prices, PACE-T(H_2) requires further input of tax data for the model regions. These data are taken from DIW (2002) and MWV (2005).

The PACE-T(H_2) model analyses the world economy based on the GTAPinGAMS structure (Rutherford, 1998). It incorporates top-down benchmark data in the form of social accounting matrices to which the model is calibrated. The main data source for the calibration of national and international commodity flows is the GTAP5 database (GTAP, 2002). The trade shares of hydrogen and conventional cars are taken from the sector 'Motor vehicles and parts' of the GTAP database.

Finally, all scenarios analysed in PACE-T(H_2) assume the same international import–export structures of hydrogen and conventional cars (structural identity assumption). In other words, if a country is producing and exporting conventional cars at present, the assumption is made that this country will also produce and export hydrogen cars in the future.

18.5.3 Results

Now we turn to the simulation results of the high-penetration scenario in the HyWays project for selected countries. Starting in 2020, hydrogen penetration rates in the considered scenario gradually increase over time to reach on average almost 75% in the vehicle stock in 2050 (see Table 18.1). The scenario furthermore assumes a fairly steep learning curve for the hydrogen technology, i.e., a high rate of cost decrease. While during the first years of market entrance, lifetime costs of hydrogen cars exceed those of their conventional counterparts, hydrogen cars start to become competitive after 2020 so that their lifetime costs are below those of conventional cars.

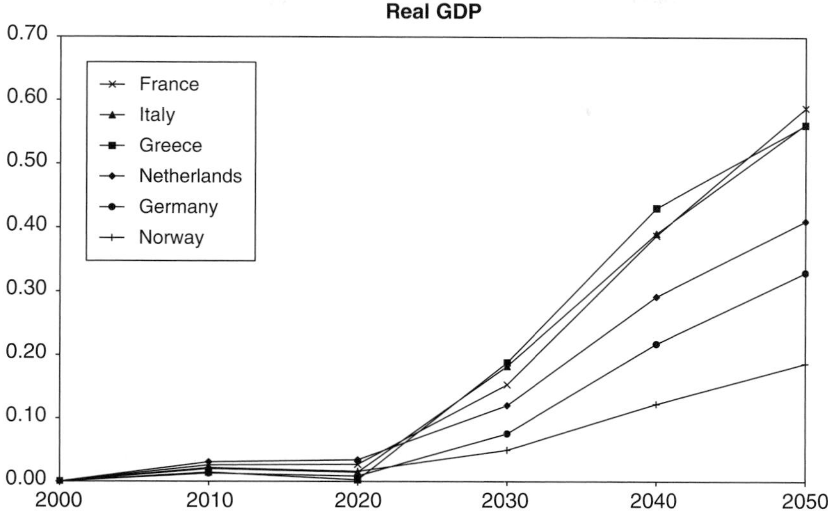

Figure 18.8. Real GDP (changes in % relative to baseline); source: PACE-T(H_2).

Figure 18.8 shows the development of real GDP. In the period before 2020, GDP remains almost unchanged, since there are only a few hydrogen cars in the market. Afterwards, however, GDP starts to rise. Thus, GDP in 2050 exceeds the respective baseline value by approximately 0.6% in France, Italy and Greece; by 0.4% in the Netherlands; by 0.3% in Germany; and by 0.2% in Norway. These differences across the countries mainly stem from the difference between lifetime costs of conventional and hydrogen cars. Lifetime costs of cars depend on their purchase costs, O&M costs and, in the case of conventional technology, on the oil price or, in the case of the hydrogen technology, on hydrogen production costs, which might heavily differ between the countries owing to differing production levels and technologies.

The development of real consumption in the six selected countries, as shown in Fig. 18.9, is slightly different. Obviously, all countries experience positive long-run effects of the introduction of hydrogen cars. This finding shows the medium- and long-run cost savings from hydrogen cars. These release additional resources, which lead to an increase in the representative consumer's real income and thus additional consumption demand. In contrast to the development of real GDP, consumption demand already increases in 2010, owing to consumption smoothing over time of the representative household. Note that the changes in real consumption differ between the considered countries. These differences are mainly driven by hydrogen production costs and the assumed hydrogen penetration rates in the different car classes, which determine the cost savings potential for the introduction of the hydrogen technology. Since in the underlying data hydrogen production costs are lowest in

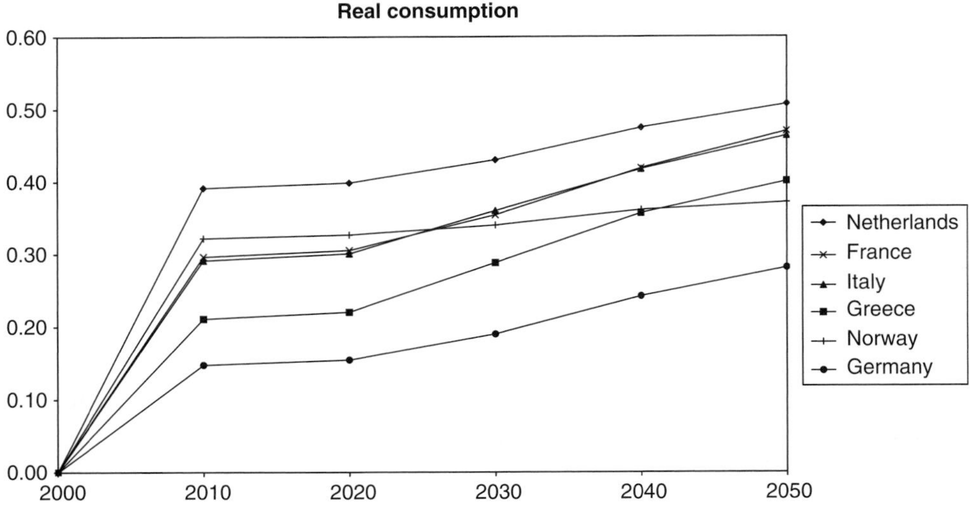

Figure 18.9. Real consumption (changes in % relative to baseline); source: PACE-T(H_2).

the Netherlands, the cost savings potential due to the introduction of hydrogen cars is highest. Consequently, consumption in the Netherlands also increases the most. The opposite holds true for Germany. Here hydrogen production costs are highest, which leads to much smaller cost savings and thus the lowest rise in consumption across the six considered countries.

The observed increase in real consumption translates directly into welfare gains, which are measured as weighted averages of real consumption over time. Not surprisingly, welfare increases are largest in the Netherlands, France and Italy and more modest in Norway, Greece and Germany.

Assuming high hydrogen penetration rates and a large decrease in hydrogen production costs seems to be a very optimistic scenario. As Table 18.1 indicates, the HyWays project analysed further scenarios with lower penetration rates combined with smaller rates of cost decrease. The macroeconomic analysis with PACE-T (H_2) revealed comparable findings for these alternative scenarios, i.e. (small) improvements in real consumption, welfare and GDP from the introduction of the hydrogen vehicles (for more details, see the reports posted on www.hyways.de). The magnitude of these effects in the alternative scenarios is dampened since lower penetration rates and shallower learning curves mean smaller future cost reductions for hydrogen cars and thus smaller cost savings.

All in all, if with increasing maturation the hydrogen technology gains a cost advantage over the conventional technology, as assumed here, the associated cost savings potential leads to positive effects on the consumer's budget and thus on consumption demand. As soon as the hydrogen technology becomes competitive, GDP is positively affected.

18.6 System-dynamics model: ASTRA

18.6.1 General modelling approach

Assessment of transport strategies (ASTRA) is a system-dynamics model generating time profiles of variables and indicators needed for policy assessment. System-dynamics models are built of systems of non-linear differential equations. However, because of the size of complex models of social systems, analytical solutions will usually not be found, so results have to be computed by numerical integration, replacing the differential equations with difference equations and leading to a simulation approach. Construction of system-dynamics models assumes that the behaviour of systems is primarily determined by their feedback mechanisms. 'The central concept that system dynamicists use to understand system structure is the idea of two-way causation or feedback.' (Meadows, 1980). Two classes of feedback loop can be distinguished: (1) negative feedback loops that are target seeking and stabilise system development, and (2) positive feedback loops that drive systems towards accelerated growth or decline, such that systems consisting only of positive feedbacks would either explode or implode.

Originally ASTRA was developed on the base of existing models that have been converted into a dynamic formulation feasible for implementation in system dynamics and allowing for closure of the feedbacks between the models. Among these models have been the macroeconomic model, ESCOT (Schade *et al.*, 2002) and the classical four-stage transport model, SCENES (ME&P, 2000). The ASTRA model then runs scenarios for the period 1990 until 2030 using the first 12 years for calibration of the model. Data for calibration stem from various sources, with the bulk of data coming from the EUROSTAT (2005) and the OECD online databases (OECD, 2005). A detailed description of ASTRA is provided by Schade (2005).

The ASTRA model consists of eight modules and the version described in this section covers the 27 European Union countries (EU27) plus Norway and Switzerland. The major interlinkages and feedback loops between the eight modules are shown in Fig. 18.10.

Each module has a specific purpose:

- The population module (POP) calculates the population development for the EU29 countries with one-year age cohorts, i.e., the ageing problem is considered.
- The macroeconomics module (MAC) provides the national economic framework. The MAC combines different theoretical concepts, as it incorporates neoclassical elements, like production functions; Keynesian elements, like the dependency of investments on consumption extended by influences from exports or government debt; and elements of endogenous growth theory, like the implementation of endogenous technical progress, as one important driver for long-term economic development.
- The regional-economics module (REM) mainly calculates the generation and distribution of freight transport volume and passenger trips. Transport generation is performed individually for each of the 76 zones of the ASTRA model. Distribution splits trips of each zone into three distance categories of trips within the zone and two distance categories crossing the zonal borders and generating origin–destination (OD) trip matrices.

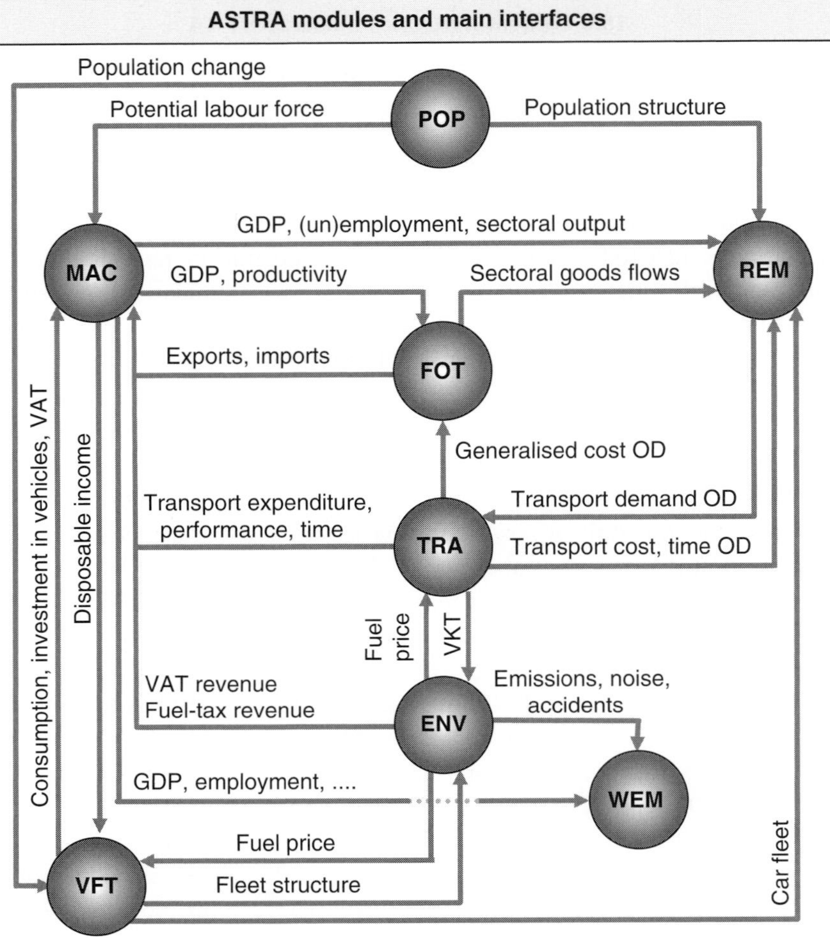

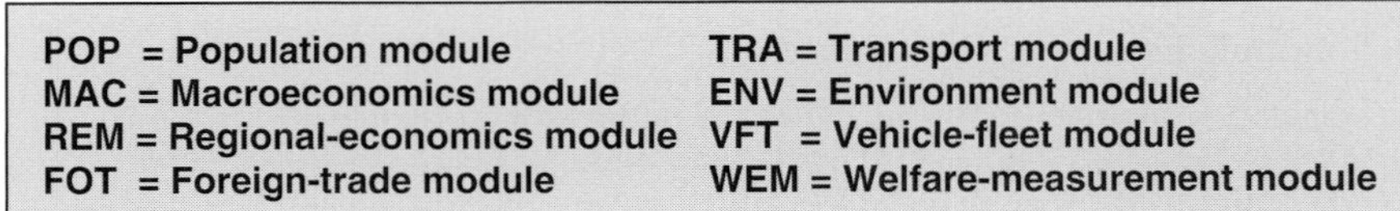

Figure 18.10. Overview of modules and feedback loops of the ASTRA model.

- The foreign-trade module (FOT) is divided into two parts: trade between the included EU29 countries (intra-EU model) and trade between the EU29 countries and the rest of the world (RoW), which is divided into nine regions (EU-RoW model). Trade flows generate freight transport flows.
- The transport module (TRA) performs the mode split for passenger and freight transport and models travel times using a simplified network-capacity approach. The modal split

determines the structure of transport expenditures and taxes, feeding back into the macro-economics module and the distribution models.

- The environment module (ENV) estimates transport emissions, fuel consumption and transport accidents based on vehicle-km-travelled (VKT) and emission factors, which are determined by the vehicle fleet structure.
- The vehicle-fleet module (VFT) describes the vehicle fleet composition for all road modes. Vehicle fleets are differentiated into age classes, based on one-year age cohorts, and emission standard categories. Additionally, vehicle fleets are differentiated into gasoline and diesel-powered cars of different cubic capacity and into hybrid vehicles, hydrogen internal combustion engine vehicles and hydrogen fuel-cell vehicles. The vehicle fleet develops according to income changes, development of population and fuel prices. The purchase of hydrogen vehicles is taken exogenously as input of the absolute values from the high-penetration scenario of the EU HyWays project.

18.6.2 Modelling a hydrogen economy in ASTRA

Modelling a hydrogen economy in ASTRA concentrates on the adaptation of the transport system and its related energy supply. It requires input from other studies concerning (1) the penetration of hydrogen cars and the build-up of hydrogen supply infrastructure, and (2) the build-up of renewables capacity to produce hydrogen from non-fossil energy sources.

The basic framework of the analysis is provided by the business-as-usual (BAU) scenario of the ASTRA model, which is extended in particular by inputs on hydrogen technologies from the HyWays project (HyWays, 2006b) and a study on growth and employment impacts of renewables where ASTRA is connected with the GreenX model (Schade, 2006). For the ASTRA BAU scenario, the model endogenously determines economic variables (e.g., GDP, employment, investment, trade flows), transport variables (e.g., passenger and freight transport performance per mode divided into trip purposes and distance classes, vehicle fleets) and environmental variables (e.g., consumption of the different types of fuels, emissions, accidents). Trends of the major variables from the different fields are shown in Fig. 18.11.

Further variables that determine the scenario are given exogenously. This includes the prices for crude oil (see also Fig. 18.11) and natural gas, which are taken from the WETO-H_2 reference case (WETO, 2006).

Market entering of hydrogen cars is taken from the HyWays project, which involved an intense stakeholder process with car manufacturers and fuel suppliers to develop a scenario for market penetration of hydrogen cars (HyWays, 2006b). For the presented analysis, the HyWays high-penetration scenario was taken. For simplification in the ASTRA model, hydrogen-ICE cars and hydrogen-ICE hybrids were aggregated into one category (H_2-ICE), as were hydrogen-FCs and hydrogen-FC-hybrids (hydrogen-FC). The applied market development of these two categories is shown in Fig. 18.12. In terms of implementation, ASTRA estimates the total purchase of new cars endogenously and then subtracts the exogenously provided

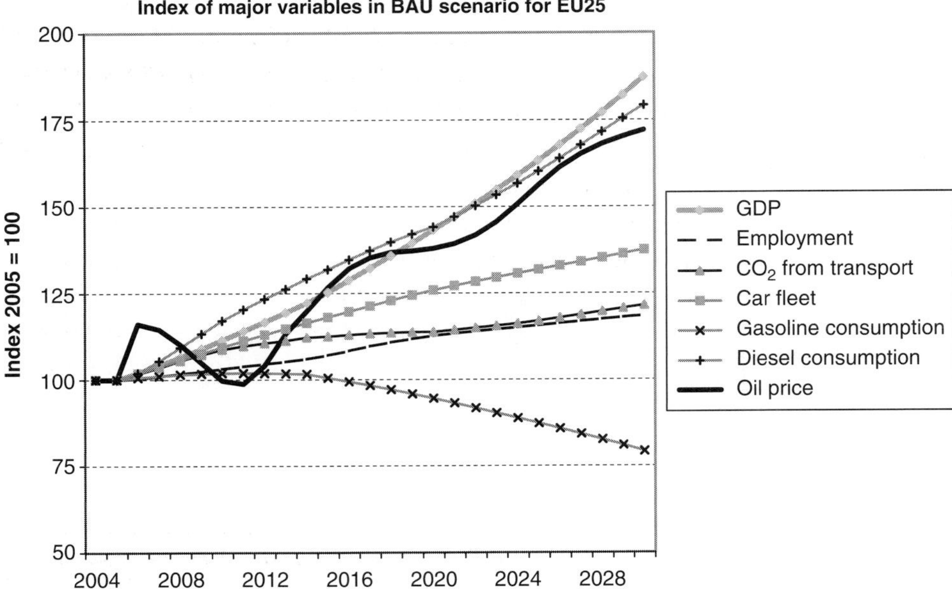

Figure 18.11. Overview on the trends in the BAU scenario (source: ASTRA results, oil price from WETO-H_2 study).

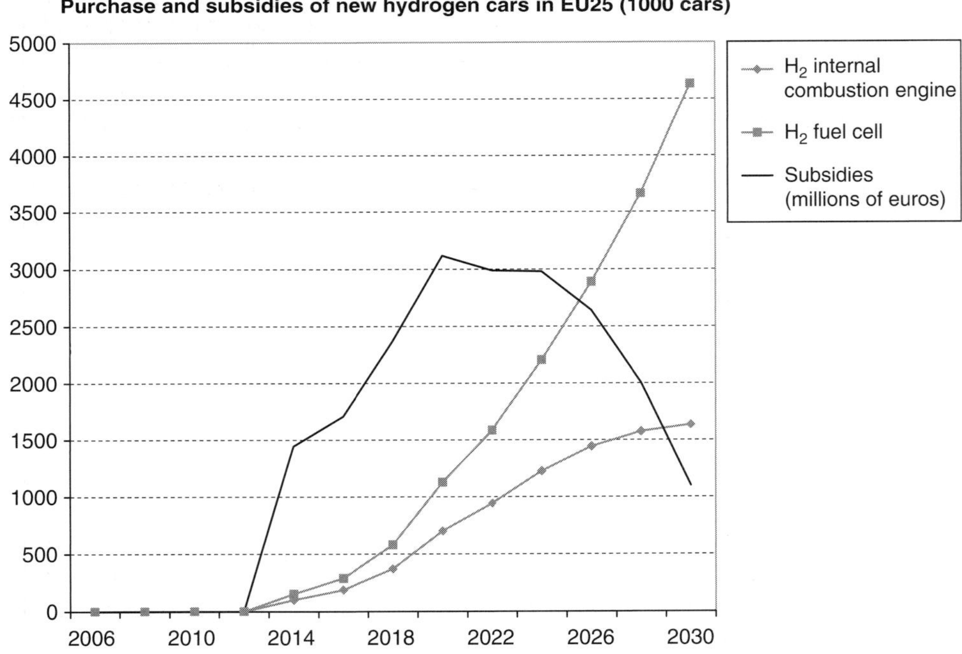

Figure 18.12. Subsidies and diffusion of hydrogen cars into car fleet of EU25 (source: ASTRA results based on HyWays high penetration scenario).

numbers of the hydrogen cars to get the distribution between hydrogen-powered cars and conventional cars. In 2030, this leads to shares of hydrogen cars of about 30% of all new purchased cars. In terms of production location of vehicles, the structural identity scenario is taken, implying that hydrogen cars are manufactured with the same spatial distribution as conventional cars.

It is expected in HyWays that at the time of introducing the first hydrogen cars in 2013, subsidies by the government have to be provided, owing to the high cost of the fuel cells. These subsidies diminish over time, such that the peak of absolute subsidisation amounting to €3 billion is reached in 2020, though the number of hydrogen cars sold increases continuously (see Fig. 18.12).

The higher prices of cars, which is balanced by subsidies, has two impacts in ASTRA: first, car manufacturers increase their revenues and output, compared with BAU, and second, a few other sectors that manufacture significant shares of the fuel cell also benefit. HyWays estimates that about one third of a car's price is related to the drive train. For hydrogen-fuel-cell cars, out of this one third about 30% is assumed to be provided by the chemical sector and 40% by the electronics sector in ASTRA. The remaining 30% is still manufactured by the vehicle sector. Hence, the according shares of demand for H_2 fuel-cell vehicles are shifted from the vehicles sector, which before produced 100% of the drive train, to the chemicals and electronics sectors, respectively. This affects the sectoral final demand and the input–output table calculations in ASTRA.

Analyses of the cost of producing hydrogen conclude that some production pathways, even today, are competitive compared with fossil fuels for transport (Hilkert, 2003). Under this hypothesis, it is feasible to build up the infrastructure for hydrogen production and fuelling from revenues generated by hydrogen sold. Consequently, the infrastructure investments required to build up the fuelling infrastructure for H_2 cars are calculated endogenously from the hydrogen fuel demand of the hydrogen cars in service using efficiency values from HyWays (25.9 kWh H_2/100km for H_2-FCs and 46.4 kWh H_2/100km for H_2-ICEs) in 2010 and an efficiency improvement curve that reduces this H_2 consumption between 2010 and 2050 by 30%.

The calculated demand for hydrogen can be satisfied by ten different production pathways in ASTRA: five renewable pathways (biomass, wind, solar-thermal, geothermal and hydro) and five other pathways (natural gas, coal, electrolysis with electricity from average grid mix, nuclear, by-product hydrogen). For a number of countries, a specific mix of pathways was developed in HyWays, based on potentials for renewables and policy approaches (e.g., high share of nuclear in France, high share of CCS in Poland). These mixes are transferred to the remaining EU27 member states according to similarities to countries analysed in HyWays. Based on the hydrogen demand and the strategies of the individual countries for considering renewable pathways, the required investments into additional capacity for renewables to produce hydrogen are derived considering the technology split onto different economic sectors and learning curves depending on already installed capacity (see Fig. 18.13).

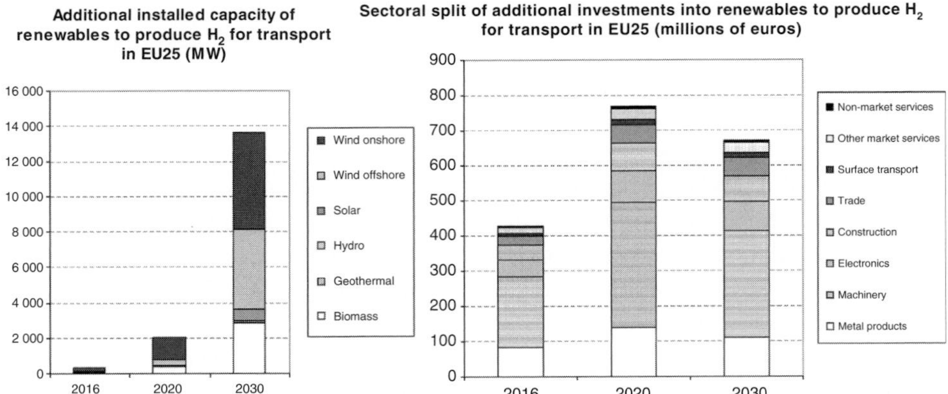

Figure 18.13. Additional installed capacity and investments into renewables to produce hydrogen for transport in the EU25. (Source: ASTRA results based on HyWays country approaches to applying renewable technologies.)

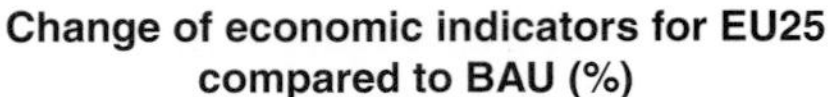

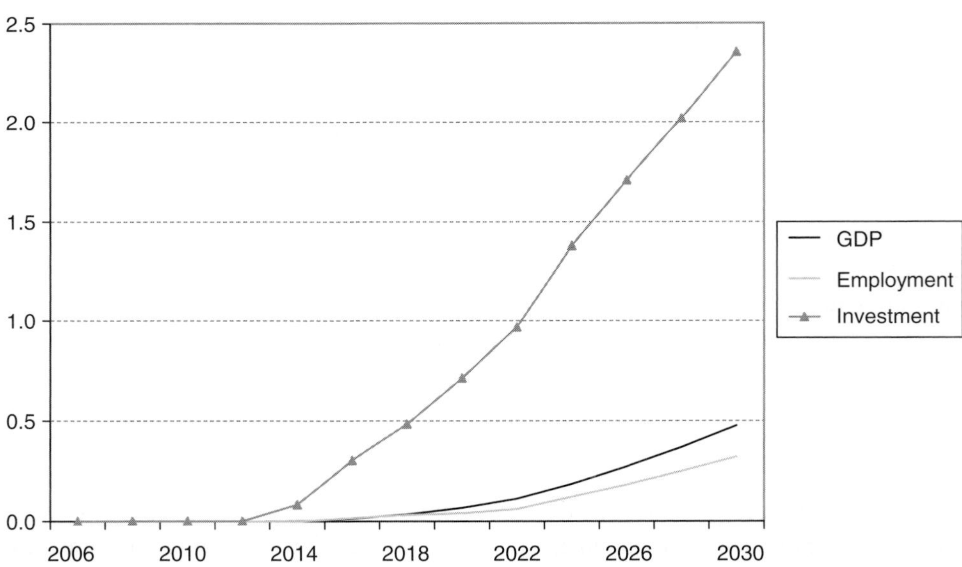

Figure 18.14. Impact on economic indicators through introduction of hydrogen cars for EU25.

18.6.3 Results

Based on the framework of economic development, energy prices, hydrogen car penetration and the structure of renewable hydrogen production described in the previous sections, the hydrogen car's high-penetration scenario of HyWays is simulated with the ASTRA model and the results are compared with the BAU scenario. Figure 18.14

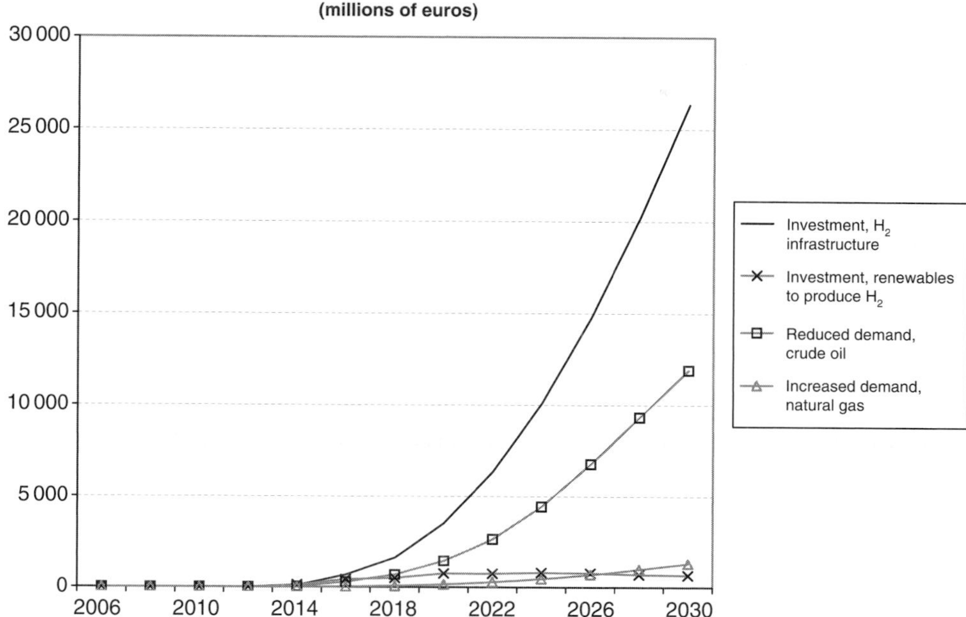

Figure 18.15. Changed investments and resource demand for EU25.

presents the changes in the major economic variables for the EU25. Overall, the economic development proves to be positive with a growth of close to 0.5% of GDP in 2030, a growth of 0.3% of employment and a stronger increase of investment by 2.4%. This increase of investment has several reasons: first, the above explained additional investment in hydrogen production and fuelling infrastructure, as well as for the additional renewable capacities required to produce 'renewable' hydrogen (see also Fig. 18.15), which are both funded by revenues of selling hydrogen as a fuel, and, second, the changed structure of final demand, reducing non-transport related consumption by 0.5% and increasing it for the purchase of vehicles, which trigger stronger investments in other sectors than the sectors losing consumption shares. Third, the wider economic effects following these additional investments, i.e., effects such as increased employment and income, leading to higher GDP, leading to increased demand and, hence, more investment in the second round.

As expected, major environmental indicators are affected positively by the introduction of hydrogen cars. Demand for gasoline drops by more than 13% until 2030, compared with BAU, and demand for diesel by about 2%. The difference is significant, as in this scenario only passenger cars are equipped with fuel cells and H_2-ICE engines, but neither buses nor light-duty vehicles are expected to be equipped with fuel cells. This means only a small share of diesel fuel consumers is affected, i.e., diesel cars, while buses, light- and heavy-duty vehicles (LDV, HDV) continue to run on diesel.

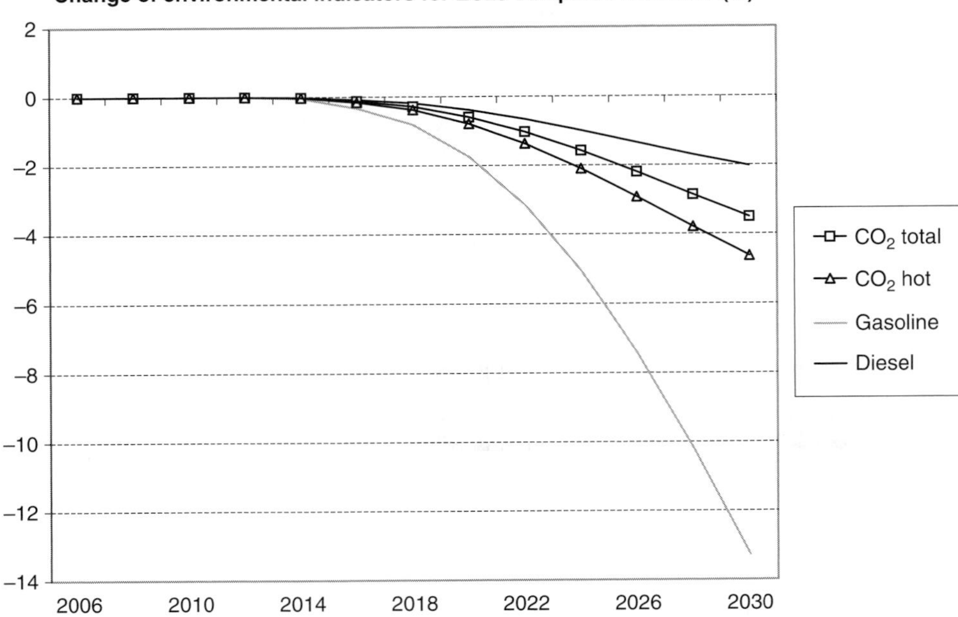

Figure 18.16. Impact on environmental indicators through introduction of H_2 cars for EU25.

Also, as GDP grows a bit stronger, freight transport will be increasing, thus raising demand for diesel from freight transport compared to BAU.

Total CO_2 emissions from transport are reduced by about 3.5% in 2030. However, emissions from the driving activity decrease by 4.6% (CO_2 hot in Fig. 18.16), which is significantly stronger than the reduction for total transport CO_2. The reason is that ASTRA calculates the life-cycle emissions for the total transport CO_2 emissions and these include upstream emissions, i.e., those emissions that are generated during the production of fuel. Since, to some extent, hydrogen is produced by non-renewable energies, e.g., natural gas or by-product hydrogen, some upstream emissions occur, such that the change of CO_2 emissions while driving and of total CO_2 emissions differ.

A further positive economic impact, besides increased investment, is the change of imports of fossil fuels. For crude oil, this amounts to a value of € 12 billion of savings in the year 2030, with a minor compensation of increased imports of natural gas reaching more than € one billion in 2030 (see Fig. 18.15).

The 25 economic sectors of ASTRA are affected differently by the hydrogen-penetration scenario. Looking at the changes of sectoral final demand compared with BAU (see Fig. 18.17), both winners and losers can be identified. As can be expected, winning sectors are those contributing to vehicle and fuel-cell production, as well as those satisfying the additional demand for investment goods to produce vehicles and renewable installations. Hence, the largest change in 2030 can be observed for electronics and machinery, both sectors contributing to FC-drive-train production

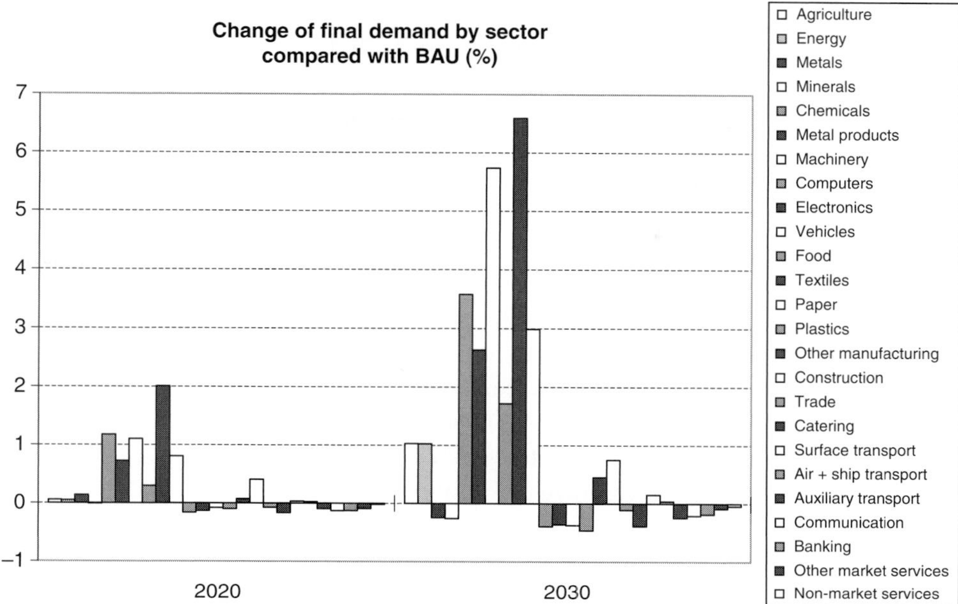

Figure 18.17. Sectoral changes of final demand for EU25 due to hydrogen car penetration and renewable investments.

and renewable technologies. Also on the significantly positive side are the vehicles and chemicals sectors, both contributing more to the new vehicle technologies, and the metal-products sector, which is more relevant for investments into renewables. Losing sectors would be those providing consumption goods, where demand is reduced owing to the shift towards transport consumption. These sectors include food, textiles, paper, plastics and catering.

18.7 Summary and conclusions

Research in the hydrogen and fuel-cell field has a strong technology focus. The analysis of economic impacts tends to concentrate on the necessary investments for hydrogen infrastructure build-up. However, the presented work shows that hydrogen as an energy carrier also has impacts on employment, gross domestic product (GDP) and international competitiveness. To analyse these impacts, the so-called lead market analysis has been conducted with the conclusion that Europe can be regarded as well positioned. But Europe also runs the risk of falling further behind the USA and Japan, which are in the lead right now. Different hydrogen penetration scenarios from the HyWays project have been analysed, with the result that the automotive industry seems to be very important as a sector for the transition phase from a macroeconomic perspective. Three different macroeconomic models have been applied to analyse economic impacts: the ISIS input–output model focuses on sectoral changes and resulting employment effects; PACE-T(H_2) is a computable general equilibrium model that has special emphasis on GDP and welfare effects, including price effects, and the system dynamic model, ASTRA, additionally gives special insights in investment effects.

The major conclusions from all economic modelling activities could be drawn as follows: introduction of hydrogen leads, under the assumption of a similar competitiveness as today for non-hydrogen technologies, to significant positive effects for long term GDP, employment and investment.

Using a more Keynesian approach for GDP and employment analysis, for a scenario with a relevant market penetration of hydrogen vehicles, the following could be stated (till 2030 23.7% of vehicle stock): overall, the economic development proves to be positive with a growth of close to +0.5% of GDP in 2030, a growth of +0.3% of employment, and a stronger increase of investment by +2.4%, over the baseline. This increase of investment has several reasons. First the above-explained additional investment into hydrogen production and fuelling infrastructure, as well as for the additional renewable capacities required to produce 'renewable' hydrogen, is funded by revenues of selling hydrogen as a fuel, and, second the wider economic effects following these additional investments, i.e., such effects as increased employment and income leading to higher GDP, leading to increased demand and hence more investment in the second round. Further, the development of new technologies requires a higher education level, needs more workers due to a lower automation level, and is a barrier for off-shoring of production.

The conclusions above are all based on the assumption that no shifts in export–import relations occur (e.g., same market share of fuel-cell vehicles for European car manufactures as today for conventional vehicles). However, the mid-term employment effects in automotive sectors could be drastic if the assumption of similar competitiveness is rejected for hydrogen technologies. It could also be assumed that in such a case relevant negative consequences for GDP would occur. However, with great efforts in the field of hydrogen, exports might be possible with additional positive effects for employment and growth.

Therefore, for the large automotive countries, the following dilemma can be identified. On the one hand, job losses could be drastic if these countries were to lose market shares due to late market entry. On the other hand, there are uncertainties regarding the market success of hydrogen cars and the potential risk of losing several billion euros, owing to investments in premature hydrogen infrastructure and hydrogen car development. Compared with large automotive countries, the economic risks of a hydrogen economy are much smaller for countries without a large automotive industry and promise significant increases in employment here if the right strategy is pursued. Similar conclusions can be drawn for the plant and equipment branches.

In contrast to the Keynesian approach, CGE models assume perfectly adjusting prices, given scarce resources and cover all economy-wide repercussion effects of different policy scenarios. Applying a CGE approach for analysing the economic effects of introducing hydrogen cars, the findings are the following: macroeconomic variables are positively affected from the introduction of hydrogen cars with real consumption growing up to 0.5% and real GDP up to 0.6% in 2050 compared with the business-as-usual scenario, depending on the considered country. This finding is a result of the assumed cost advantage of hydrogen technology over conventional technology. Positive GDP effects are observed as soon as the hydrogen technology becomes competitive. Decreasing costs of the hydrogen technology mean lower use of resources in the production of hydrogen cars. The released additional resources lead to an increase in the consumers' real budget and thus their consumption demand. However, the magnitude of macroeconomic changes differs across countries, mainly driven by country-specific assumptions on cost developments for hydrogen technology and the different mix of hydrogen-production technologies.

Contrary to the model with fixed prices and infinitely elastic supply factor, the CGE model proves considerably less sensitive to assumptions about international competitiveness. If the international production structure for hydrogen cars deviates from that of conventional cars, prices and production structures are adjusted. The consequences for aggregate variables, like real consumption and GDP, are small.

References

Amorelli, T., Hernández, H., Lonza, L. *et al.* (2004). *Assessing the International Position of EU's Research and Technological Development & Demonstration (RTD&D) on Hydrogen and Fuel Cells*. ESTO Publications, Joint Research Centre. http://esto.jrc.c.europa.eu/publication/pub.cfm?id = 1298.

Armington, P. S. (1969). A theory of demand for products distinguished by place of production. *IMF Staff Papers*, **16**, 159–178.

Arrow, K. J. and Debreu, G. (1954). Existence of an equilibrium for a competitive economy. *Econometrica*, **22**, 265–290.

Beise, M. (2004). Lead markets: country-specific drivers of the global diffusion of innovations. *Research Policy*, **33**, 997–1018.

Beise, M. (2005). Lead markets, innovation differentials and growth. *International Economics and Economic Policy*, **1** (4), 305–328.

Beise, M. and Rennings, K. (2005). Lead markets and regulation: a framework for analyzing the international diffusion of environmental innovations. *Ecological Economics*, **52** (1), 5–17.

Bhattacharyya, S. C. (1996). Applied general equilibrium models for energy studies: a survey. *Energy Economics*, **18** (3), 145–164.

Blind, K. and Jungmittag, A. (2005). Trade and the impact of innovations and standards: the case of Germany and the UK. *Applied Economics*, **37** (12), 1385–1398.

Blümle, G. (1994). The importance of environmental policy for international competitiveness. In *Interactions between Economy and Ecology*, ed. Matsugi, T. and Oberhauser, A. Berlin: Duncker & Humblot, pp. 35–57.

CASCADE MINTS (2006). *Summary and Objectives and Case Study Comparisons and Development of Energy Models for Integrated Technology Systems.* NTUA (National Technical University of Athens). www.e3mlab.ntua.gr/cascade.html.

Christidis, P., Hernandez, H. and Lievonen, J. (eds.) (2002). *Impact of Technological and Structural Change on Employment: Prospective Analysis 2020*. Sevilla: IPTS (Institute for Prospective Technological Studies) (EUR 20258 EN).

Conrad, K. (2002). Computable general equilibrium models in environmental and resource economics. In *The International Yearbook of Environmental and Resource Economics 2002/2003*, ed. Tietenberg, T. and Folmer, H. Cheltenham: Edward Elgar, pp. 66–114.

DIW (Deutsches Institut für Wirtschaftsforschung) (2002). Europäischer Vergleich der besonderen Steuer- und Abgabensysteme für den Erwerb, das Inverkehrbringen und die Nutzung von Kraftfahrzeugen. *Endbericht*. Berlin: DIW.

Dosi, G., Pavitt, K. and Soete, L. (1990). *The Economics of Technical Change and International Trade*. New York: Springer.

Duchin, F. (1994). Input–output analysis and industrial ecology, In *The Greening of Industrial Ecosystems*, ed. Allenby, B. R. and Richards, D. Washington, DC: National Academy Press, pp. 61–68.

European Hydrogen and Fuel Cell Technology Platform (2005). *Deployment Strategy Foundation Report and Strategic Research Agenda Foundation Report.* https://www.hfpeurope.org/hfp/keydocs.

EUROSTAT (2004). *Input Output Tables – Data*. http://epp.eurostat.ec.europa.eu.

EUROSTAT (2005). *Themes (various) – Data*. http://epp.eurostat.ec.europa.eu.

Fagerberg, J. (1995). *Is there a large-country advantage in high-tech? Working Paper No. 526*. Oslo: Norwegian Institute of International Affairs.

Fagerberg, J. (1996). Technology and competitiveness. *Oxford Review of Economic Policy*, **12** (3), 39–51.

FHIRST (2004): Fuel cells and hydrogen improved R&D strategy for Europe. www.ec.europa.eu/research/energy/pdf/efchp_hydrogen18.pdf.

Gately, D. (1970). *Investment Planning for the Electric Power Industry: An Integer Programming Approach*. Research Report 7035. University of Western Ontario, Department of Economics.

Geldermann, J. and Rentz, O. (2005). Multi-criteria analysis for the assessment of environmentally relevant installations. *Journal of Industrial Ecology*, **9** (3), 127–142.

Gonzalez-Monroy, L. and Cordoba, A. (2002). Financial costs and environment impact optimisation of the energy supply systems. *International Journal of Energy Research*, **26** (1), 27–44.

Gottfried, P., Stöß, E. and Wiegard, W. (1990). Applied general equilibrium tax models: prospects, examples, limits. In *Simulation Models in Tax and Transfer Policy*, ed. Petersen, H.-G. and Brunner, J. K. Frankfurt, pp. 205–344.

GTAP (Global Trade, Assistance, and Production) (2002). *The GTAP5 Data Base, Center for Global Trade Analysis*. West Lafayette: Purdue University.

Gunning, J. W. and Keyzer, M. A. (1995). Applied general equilibrium models for policy analysis. In *Handbook of Development Economics*, vol. 3A, pp. 2025–2107.

Hilkert, M. (2003). *Pathways for a transition to a sustainable hydrogen transportation fuel infrastructure in California*. Diploma Thesis at the University of Karlsruhe with support of the University of California, Davis.

HySociety (2004). *Promoting a hydrogen-based society*. www.ec.europa.eu/research/energy/pdf/efchp_hydrogen21.pdf.

HyWays (2006a). *HyWays – Hydrogen Energy in Europe*. Integrated Project under the 6th Framework Programme of the European Commission, 2004–2007. www.hyways.de.

HyWays (2006b). *HyWays A European Roadmap: Assumptions, Visions and Robust Conclusions from Project Phase I*. Germany: L-B-Systemtechnik GmbH, Ottobrunn. www.hyways.de.

HyWays (2006c). *HyWays D3.13 Report on the HyWays Scenario Assumptions*. Ottobrunn, Germany: L-B-Systemtechnik GmbH. www.hyways.de.

Kaldor, N. (1978). The effect of devaluation on trade in manufacture. In *Further Essays on Applied Economics*. London: Duckworth.

Keeney, R. L. and Raiffa, R. (1976). *Decisions with Multiple Objectives; Preferences and Value Tradeoffs*. New York: John Wiley & Sons.

Leontief, W. (1956). Factor proportions and the structure of American trade: further theoretical and empirical analysis. *Review of Economics and Statistics*, **XXVIII**.

Leontief, W. (1986). *Input Output Economics*. New York: Oxford University Press.

Linder, S. B. (1961). *An Essay on Trade and Transformation*. Uppsala: Almqvist and Wiksell.

ME&P (2000). *SCENES European Transport Forecasting Model and Appended Module: Technical Description*. Deliverable D4 of SCENES (Modelling and methodology for analysing the interrelationship between external developments and European transport). Project funded by the European Commission 4th RTD framework.

Meadows, D. H. (1980). The unavoidable a priori. In *Elements of the System Dynamics Method*, ed. Randers, J. Cambridge: Productivity Press.

Meyer-Krahmer, F. (1999). Innovation als Beitrag zur Lösung von Beschäftigungsproblemen? *Mitt Arbeitsmarkt-Berufsforsch*, **32** (4), 402–541.

Meyer-Krahmer, F. (2003). Lead-Märkte und Innovationsstandorte. In *Kunststück Innovation: Praxisbeispiele aus der Fraunhofer-Gesellschaft*, ed. Warnecke, H.-J. *et al*. Berlin: Springer, pp. 23–28.

Meyer-Krahmer, F. (2004). Vorreiter-Märkte und Innovation. In *Made in Germany '21*, ed. Steinmeier, F. W. and Machnig, M. Hamburg: Hoffmann und Campe – Verlag, pp. 96–110.

Morrison, P. D., Roberts, J. H. and Midgley, D. F. (2004). The nature of lead users and measurement of leading edge status. *Research Policy*, **33**, 351–362.

MWV (Mineralölwirtschaftsverband) (2005). *Mineralöl-Zahlen 2004*. Hamburg: MWV.

OECD (2005). *Statistics Subjects (various)*. http://www.oecd.org/statsportal.

Ohlin, B. (1933). *Interregional and International Trade*. Cambridge: Harvard University Press.

Petit P. (1995). Employment and technological change. In *Handbook of the economics of innovation and technological change*, ed. Stoneman, P. Oxford: Blackwell.

Porter, M. E. (1990). *The Competitive Advantage of Nations*. New York: Free Press.

Posner, M. V. (1961). International trade and technical change. *Oxford Economics Papers*, **30**, 323–341.

Robson, P. (1987). *The Economics of International Integration*, 3rd edn. London: Allen & Unwin.

Rutherford, T. F. (1998). *GTAPinGAMS: The Dataset and Static Model*. Working Paper. University of Colorado.

Samuelson, P. (1948). International trade and the equalisation of factor prices. *Economic Journal*, **58**, 163–184.

Schade, B., Rothengatter, W. and Schade, W. (2002). *Strategien, Maßnahmen und ökonomische Bewertung einer dauerhaft umweltgerechten Verkehrsentwicklung*. Final Report. On behalf of the German Federal Environmental Agency. Berlin: Erich-Schmidt-Verlag.

Schade, W. (2005). *Strategic Sustainability Analysis: Concept and Application for the Assessment of European Transport Policy*. Baden-Baden: Nomos Verlag.

Schade, W. (2006). Input to Study on RES-ETH 2020: ASTRA Results. Working Paper (Draft). Karlsruhe: Fraunhofer ISI.

Seebregts, A. J., Smekens, K. E. L. and Goldstein, G. A. (2001). Energy/environmental modelling with the MARKAL family models. *OR2001 Conference, Energy and Environment Session, 3–5 September*, Duisburg, Germany.

Sell, H. (2001). *Einführung in die internationalen Wirtschaftsbeziehungen*, 2nd edn. Munich: Oldenbourg Wissenschaftsverlag.

Shoven, J. B. and Whalley, J. (1984). Applied general-equilibrium models of taxation and international trade: an introduction and survey. *Journal of Economic Literature*, **22**, 1007–1051.

Soete, L. G. (1981). A general test of technological gap trade theory. *Weltwirtschaftliches Archiv*, **117** (4), 638–660.

Song, Y. H. (ed.) (1999). *Modern Optimisation Techniques in Power Systems*. Dordrecht: Kluwer.

Vernon, R. (1966). International investment and international trade in the product cycle. *Quarterly Review of Economics*, **88**, 190–207.

von Hippel, E. (1988). *Sources of Innovation*. New York: Oxford University Press.

Wakelin, K. (1997). *Trade and Innovation*. Cheltenham: Edward Elgar Publishing.

WETO (2006). *World Energy Technology Outlook 2050 (WETO-H_2)* (EUR 22038). Brussels: European Commission.

Wietschel, M. (2000). *Produktion und Energie: Planung und Steuerung industrieller Energie- und Stoffströme*. Frankfurt a. M.: Lang.

Wietschel, M. (2005). Patents in fuel cells and hydrogen production. In *Fuel Cell Today*. www.fuelcelltoday.com.

Zhang, Z. and Folmer, H. (1998). Economic modelling approaches to cost estimates for the control of carbon dioxide emissions. *Energy Economics*, **20** (1), 101–120.

19

Sustainable transport visions: the role of hydrogen and fuel-cell vehicle technologies

Martin Wietschel and Claus Doll

This chapter examines the role of hydrogen, and fuel-cell vehicle technologies in particular, in contributing to a future sustainable transport system and also shows the limitation of such an approach. Particular areas that need to be addressed in this respect include emissions, safety, land use, noise and social inclusion. Vehicle technologies will play a key role in addressing several of these.

19.1 Introduction

Transport systems perform vital societal functions but in their present state cannot be considered ‘sustainable’. Particular concerns in this respect include climate change, local air emissions, noise, congestion and accidents. One of the most controversially discussed long-term solutions to today’s problems of the transport sector is the introduction of hydrogen as an energy fuel and fuel-cell vehicles. In this chapter, we integrate the two debates – one on the transition to a ‘hydrogen economy’ and one on sustainable transport. We try to answer the question on what role hydrogen and fuel-cell vehicle technologies could play in a sustainable transport vision.

This chapter is structured as follows: Section 19.2 introduces the economic relevance of transportation by looking back at the beginning of industrialisation. Section 19.3 then summarises current evidence on the social effects of today’s motor transport with reference to different world regions. Departing from this analysis, the subsequent section briefly derives a set of exit strategies towards a more sustainable transport future, considering a wide range of policy instruments and transport externalities. Being the core element of this chapter, Section 19.5 eventually turns attention to hydrogen-based propulsion systems and their potential contribution to a responsible transport policy. Here, positive as well as negative externalities of producing, providing and using hydrogen fuels are taken into account. The data and information come from an analysis of recent research studies carried out by the authors in collaboration with European partners. In the final section, a summary and conclusions are given.

The Hydrogen Economy: Opportunities and Challenges, ed. Michael Ball and Martin Wietschel. Published by Cambridge University Press.

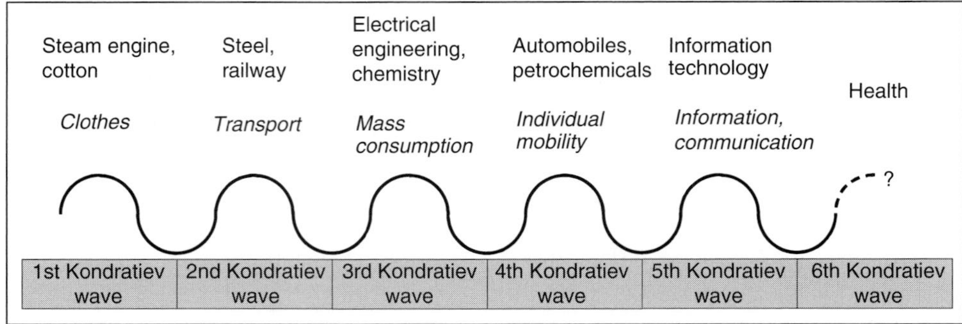

Figure 19.1. 'Schumpeter–Freeman–Perez' paradigm of five waves since the industrial revolution, and the sixth one to come.

19.2 Economic development of the transport sector

19.2.1 Long-term economic cycles and the influence of transport

The development of human society and its economic performance have always been closely linked to breakthrough innovations in products and production technologies. The starting point of the industrial revolution, for example, was set by the invention of the steam engine and its application to the production and sale of large quantities of cotton. The development of sufficiently safe and powerful means to move people and goods have played an essential role for the economic dynamics of these times, as illustrated by Fig. 19.1. Roughly 50 years after the market entry of the steam engine, the railways came, first in Britain, then Europe and then the rest of the world, allowing fast and comfortable mass transportation over long distances. Furthermore, the invention of the steam boat in 1807 and the development of steel hulls for ocean-going vessels now allowed high quantities of bulk to be carried over long distances (Nakamura *et al.*, 2004). The fast expansion of the railway networks all over the world is closely linked to advances in steel production, which have made it possible to produce high quantities of tracks and rolling stock. Eventually, the build-up of large production capacities fostered urbanisation, as the mobility of working class people was still limited to walking distances. The next breakthrough innovations boosting the economic development of Western countries were electrical engineering and the rise of the chemical industry, which had its high phase around 1900.

During that time, in 1885, Karl Benz invented the world's first petrol-powered automobile in Germany. Owing to the initially bad performance of gasoline motors, alternative propulsion concepts emerged: in 1904, the first patent of a petrol–electric hybrid car was applied for and the first vehicle to exceed the 100 kph speed limit was an electric-powered machine. But gasoline soon became the first choice because of its low price and high energy-storage capabilities. Although very expensive and thus reserved for privileged people, the automobile was considered a great step towards more environmentally friendly and safe transport.

Environmental concerns did not play a significant role in the post-war phase of economic prosperity, but they have done so since the 1980s. More recently, politicians have been seeking for ways to decouple economic development from transport growth, to make existing transport systems more environmentally friendly and to look for new options to move people and goods. In all cases, modern computer systems and information technologies play a major role in making transportation more sustainable.

The Russian economist, Kondratiev, has analysed the waves in world economic growth from the invention of the steam engine and cotton production to the rise of the chemical and automotive industries in post-war time and the contemporary emergence of the information society. Figure 19.1 illustrates the five cycles that have been identified since the eighteenth century.

Transport is crucial for our economic competitiveness and commercial, economic and cultural exchanges. This sector of the economy accounts for some 1000 billion euros, or over 10% of the EU's gross domestic product, and employs ten million people. Transport also helps to bring Europe's citizens closer together, and the Common Transport Policy is one of the cornerstones of European development and integration (EC, 2001). However, one of the consequences of this economic structure is that we are much more aware of the long-term consequences of our behaviour. Besides those effects that can be experienced directly by people, such as congestion, accident consequences or noise exposure, our contribution to climate change or the health effects of long-term noise disturbance can be scientifically proven. The magnitude of the external effects of transport and possible strategies to mitigate them will be the subject of the following sections.

19.3 Social effects of transportation

19.3.1 Overview: desired and adverse effects of transport

Transportation in general is not a value on its own, but serves primary economic, social or private interests. Transport provides access to markets and production sites and thus helps to foster productivity and economic development of production, retail and service industries. Providing the possibility for people to move quickly and safely, modern transport systems furthermore help to use human capital more effectively and to create and develop cultural surroundings. Additionally, efficient transport networks and systems give all people access to high-level education facilities and to the health system. By that, transport helps to increase the quality of life.

But transportation also constitutes an industry in itself, providing jobs and actively contributing to economic growth. In the USA, transport industries contributed roughly 4% to employment and 6% to production in 2002. In Europe, the figures are somewhat higher; a contribution of 6.5% to employment and 10% to production is observed. Roughly 40% of these figures are attributable to vehicle production, while the remaining 60% are generated by transport-related services. The pressure towards more fuel-efficient or even alternatively powered vehicles and towards more efficient logistics services will further increase the importance of the sector (Schade *et al.*, 2006).

Besides the certainly positive impacts of high-standard transport facilities on economic development and cultural life, moving towards a long-term sustainable society requires a look at its negative effects. The external social effects of transport have been studied by numerous institutions since the early 1990s in Western countries. An external cost, also known as an externality, arises when the social or economic activities of one group of persons have an impact on another group and when that impact is not fully accounted, or compensated for, by the first group. Since the fourth RTD framework programme, the Commission of the European Communities (EC) has taken a leading role in this branch of research by funding several huge studies on the topic. Important milestones are the ExternE project family (Bickel *et al.*, 2005), the UNITE project (Nash *et al.*, 2003) and the GRACE project (Nash *et al.*, 2007). The EC research is driven by its policy objectives to establish a competitive, socially balanced and sustainable common market. Having identified safety, environment and climate protection as the major challenges of a sustainable European transport policy, the EC promotes the pricing of each mode of transport according to its true marginal social costs (EC, 1992; 1998; 2001). Therefore, the emphasis of EC-funded research is on estimating marginal costs by traffic condition. Marginal costs show in which ways different categories of costs vary with an additional driven kilometre. Full accounting of external cost is only regarded as a monitoring instrument to judge the progress in marginal cost-pricing reforms. Separate accounts for urban and rural areas are not carried out in EC funded studies.

Further monitoring studies on the external costs of transport in Europe have been conducted at a country level and by international organisations, such as the International Union of Railways (Maibach *et al.*, 2004; 2007a). Contrasting the total network-related perspective of European cost estimates, the tradition in North America is somewhat different. The TTI Urban Mobility Report (Schrank and Lomax, 2005) and a comparable report by Transport Canada (2006) exclude highway traffic and compute urban road congestion and environmental costs for selected cities. The studies are hardly comparable, as they differ in scope and assumptions, but the main message is clear: most environmental problems caused by transport are urban problems. While in Western Europe, conditions are reported to have substantially improved over the past decades, thanks to technical standards of vehicles, the ongoing trend towards bigger sports utility vehicles (SUVs) and minivans counterbalances this positive trend (OECD, 2002).

Unfortunately, positive trends could not be seen for the megacities in developing countries. The very low income of people only allows for the purchase and use of old vehicles with low technical standards (Nakamura *et al.*, 2004). This is of major concern, in so far as half of the world population (48% in 2003) live in urban areas.

Table 19.1. *External costs of transport in Western Europe 2000*

	Total transport	Road transport	Share of road transport
Total costs (€ million)			
Air pollution	174617	164282	94%
Accidents	156439	152588	98%
Climate change[a]	97857	56192	57%
Noise	45644	40410	89%
Others	77861	74752	96%
Total	**552418**	**488224**	**88%**
Average			
– €/1000 pkm	55.4	68.2	
– €/1000 tkm	72.5	79.4	

Note:
[a] Calculated with €70/t CO_2; pkm = passenger kilometre, tkm = ton kilometre.
Source: (Maibach *et al.*, 2004).

This corresponds to three billion people, which is an increase of 33% against 1990. According to World Bank research, in 2020 the share of urban residents is expected to be 55% (4.1 billion people), where the biggest share of this increase will take place in developing countries (Léautier, 2006). Although the share of European people living in cities (75%) is well above the world average, transportation problems here are much less pronounced, as high income levels allow for the enforcement of rigid environmental and safety standards.

The effects of transport on economy, people and on the environment are manifold. They include the consequences of transport accidents and fatalities, nuisance and health effects caused by steady noise exposure, air emissions and the exhaust and resuspension of particles, climate impacts by the emission of greenhouse gases, soil and water contamination, and the deterioration of natural habitats. Moreover, the financial burden of infrastructure provision and the additional travel and production costs caused by congestion should be mentioned; but these items are mainly borne by transport users themselves and thus are only partly imposed on society as a whole. Not all of these effects are equally relevant for all means of transport. While accidents constitute the major problem of car travel, the railways definitely face a noise problem and air transport contributes most to the emission of climate gases.

Table 19.1 presents some evidence on the level and the structure of the external costs of various transport modes of the 15 EU member states before the enlargement of the Community in 2004 plus Norway and Switzerland. The figures show the monetary values of the effects that transport imposes on society and that are not directly covered by taxes, charges or other regulations.

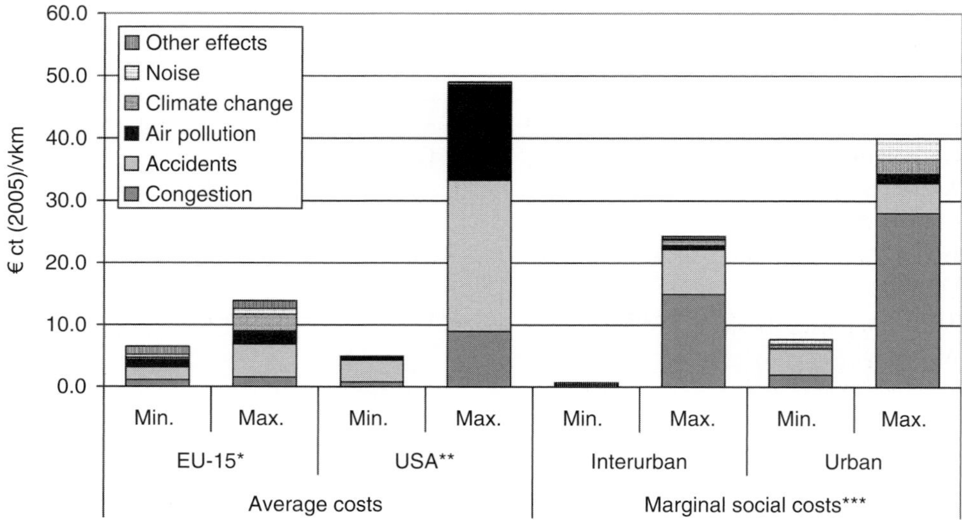

Figure 19.2. Average and marginal external costs of transport, EU15 and USA, by cost category (DOT, 1997; Maibach *et al.*, 2004; 2007a; 2007b; Nash *et al.*, 2003; VTPI, 2007); data have been processed to 2005 euros by average price indices and exchange rates.

Excluding the uncovered costs of infrastructure provision and congestion effects, which are largely carried internally by the transport sector, the social effects of transport are estimated at €552 billion for Western Europe. This constitutes 6.2% of GDP. Using a shadow value of €70 per ton of CO_2, as proposed by the German Federal Environmental Agency (UBA, 2007), climate change costs stand at third place, behind the costs of air pollution and accidents. Using a value of €170 as proposed for the upper limit by Maibach *et al.* (2004), climate change costs would double and thus clearly lead the list. The dominant transport mode is road, causing 88% of external costs, followed by aviation with 8.6% and rail with 1.2% of total costs. Recent estimates for Germany (Maibach *et al.*, 2007a) indicate that the external costs of road transport are roughly four times the costs for providing and maintaining the roadway infrastructure. Thus, pricing for external effects would increase road user charges by a significant proportion.

In the United States and Canada, several estimates of urban and interurban transport externalities have been carried out during the last decades (VTPI, 2007). A summary of national evidence for the most important cost categories was provided by the Federal Highway Administration within the 1997 Highway Cost Allocation Study (DOT, 1997) for total and average costs. Figure 19.2 compares European and US results by vehicle kilometre and sets the average cost estimates in relation to marginal social cost figures. While average costs are usually estimated on a top-down basis for a larger area, marginal social costs result from the analysis of traffic situations and can thus reflect local conditions much better than average costs.

Although the scope of networks and vehicle classes diverges between EU15 and US estimates, some conclusions can be drawn out of Fig. 19.2. The large amount of EU-funded research has led to a broad consensus on values and methods, while the US results, dating back to work of the 1990s, impressively demonstrate the uncertainty of the early estimates in the field across all cost categories. It is further remarkable that the US approaches seem to ignore the climate-change problem, even in their high estimates. The marginal cost values finally demonstrate that all cost categories besides accidents are much more an urban than an interurban problem.

19.3.2 Climate change effects

In one form or another, motorised transport requires fuels to be burned. The fuels' energy might either be directly converted into physical power by thermal engines or might be transformed into intermediate fuels or electric power. In any case, the use of carbon-based fuels entails the emission of carbon dioxide and other air pollutants into the atmosphere. Under constant fuel consumption rates and fixed modal shares, the CO_2 emissions of passenger and goods transport are more or less directly linked to transport volumes and these are, despite all political intentions of decoupling transport activities from economic growth, closely related to the development of economic activities. The effects of greenhouse-gas emissions on climate, weather extremes, nature, biodiversity, access to drinkable water, food supply and other living conditions are broadly discussed in the latest IPCC reports (IPCC, 2007).

Since the beginning of the industrial revolution in the middle of the nineteenth century, man-made CO_2 emissions from burning fossil fuels have risen from zero to around six billion metric tons of carbon dioxide. One third of these can be attributed to the use of liquid fuels, which is to a large extent used by motor vehicles, diesel trains and aircraft. Figure 19.3 shows the worldwide trend from 1850 to 2002, highlighting the emissions from liquid fuels for the two large economies of North America and Western Europe. The graph illustrates the break in the strong rise of the post-war era initiated by the oil price shock in the early 1980s and the breakdown of the former communist economies after 1989.

For transportation analysis, the tons of CO_2 emitted are commonly computed from sales statistics of the various fuels and the data are categorised to vehicle classes or transport sectors by specific mileage and fuel consumption rates. Figure 19.4 presents the total greenhouse-gas emissions for the EU15, the EU25 and the USA for the years 1993 and 2003 distinguished by road transport modes and industry sectors. (EU transport emissions are categorised by energy consumption, as reported by Mantzos and Capros (2006).) The total rise of GHG emissions in the EU25 of 4%

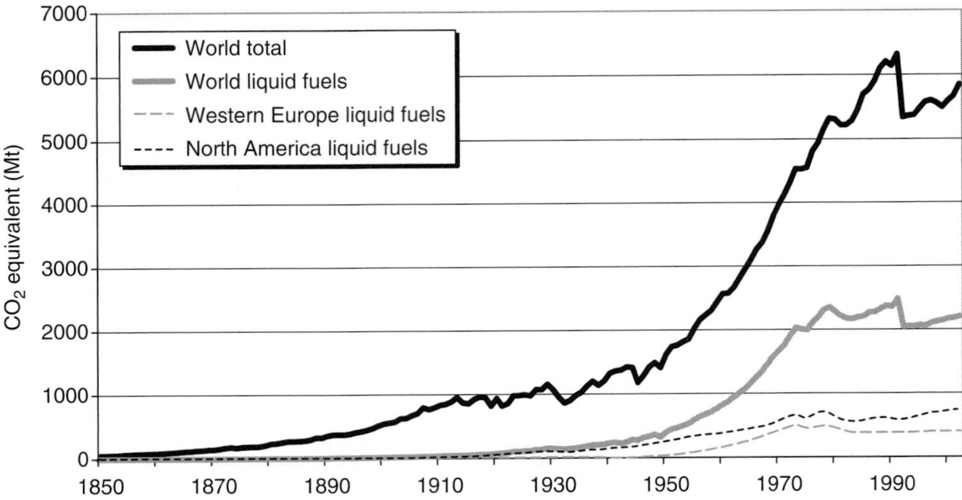

Figure 19.3. Development of CO_2 emissions from fossil fuels worldwide 1850 to 2002 (Marland *et al.*, 2007).

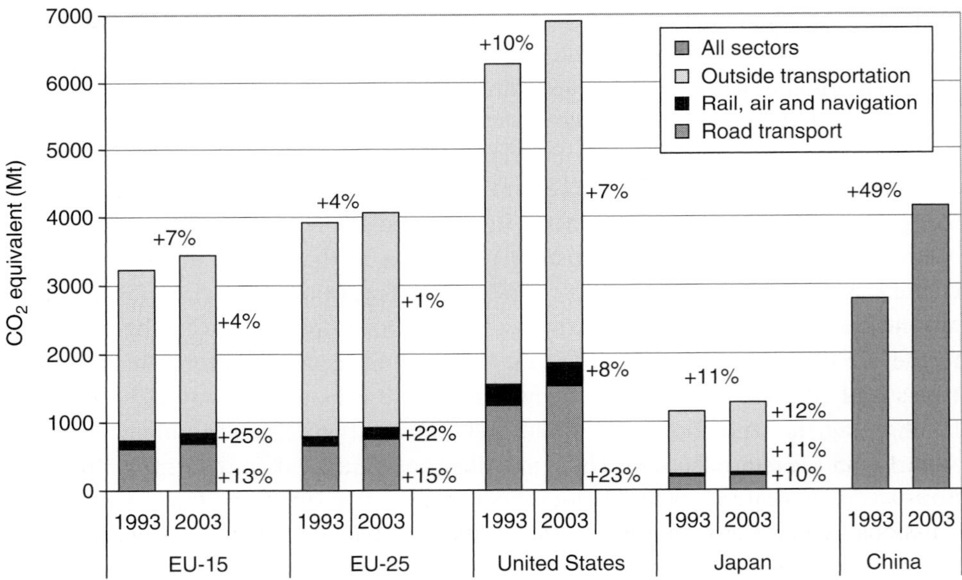

Figure 19.4. Greenhouse-gas emissions by broad sector in selected world regions 1993 and 2003 (own compilation based on EPA (2006); Mantzos and Capros (2006); MLIT (2006); UN (2007)).

between 1993 and 2003 is contrasted with a 7% increase in Western Europe. This increase is largely a result of the growth in transport activities, as the figures for the remaining economic sectors are only 1% (EU25) and 4% (EU15). Furthermore, the figures indicate that the good performance of the EU according to the Kyoto targets is, to a large extent, a result of the breakdown of the eastern European economy in the past decade. Transport emissions are largely from road transport, including bus transit, individual transport and road haulage. Nevertheless, other modes show a much more expressed increase in GHG emissions, which can be attributed to two facts: first, the EC has made enormous efforts to stabilise or even increase rail shares in goods transport, and second, air and maritime transport show constantly high growth rates, far exceeding that of land transport. However, all transport modes violate international reduction targets, which is even more expressed and much more focused on road in the USA than in the EU. In contrast, Japan shows constantly high growth rates across all sectors. The Peoples' Republic of China is not obliged to reduce emissions according to the Kyoto Protocol. Although the country shows, by far, the highest growth rate among the selected world regions, it still ranks far below the industrial states concerning CO_2 rates per capita.

According to the US Bureau of Transportation Statistics (BTS, 2007), in 2003 transport accounted for 27% of CO_2-equivalent of greenhouse-gas emissions, of which more than 80% stemmed from the combustion of fossil fuels. During this decade, road transport showed, by far, the biggest share (82%) and the strongest growth (+23%) of greenhouse-gas emissions. Among these, heavy trucks were the main drivers, with a growth of 51%. Astonishingly, the aviation sector only raised its GHG emissions by 1.9% between 1993 and 2003; this was, however, a direct consequence of the terrorist attacks of 9/11. Between 1993 and 2000 the air transport sector's emissions rose by 15% (BTS, 2007).

Approaches to value greenhouse-gas emissions are difficult because of their long-term impacts on the global environment. Current estimates range between €5 per ton based on the price of emission certificates up to €140 per ton. The values based on emission permits take advantage of cheap compensation strategies by clean development mechanism (CDM) measures, but as soon as the developing countries catch up with industrialisation, these 'low-hanging fruits' will become rare and avoidance costs will increase. Moreover, avoidance costs depend on the agreed targets; these are usually political compromises and in no way reflect the urgency of action.

Climate change is a global phenomenon and thus it does not matter where CO_2 and other GHG are emitted. But it is the driving cycle in urban transport, with frequent stops and accelerations, which is most fuel consuming and produces most exhaust emissions. Accordingly, marginal social costs of climate change for passenger cars is about twice as high in city centres than on rural roads. However, the uncertainty range is high, ranging from €6 to €23/100 vkm (vehicle kilometre) as different cities provide different network qualities (Essen *et al.*, 2007).

In 1995, the European Commission concluded an agreement to stepwise reduce the average CO_2 emission of new passenger cars to 120 g/km and in 1998 a corresponding agreement was concluded with the Association of European Car Manufacturers (ACEA), envisaging a reduction to 140 g CO_2/km until 1998 on a voluntary basis. Although progress has been made, the average emissions did only decrease from 186 g/km in 1995 to 164 g/km in 2004. In its communication of 7th February, 2007, the EC has changed its strategy: until 2012, car manufacturers will be obliged to reduce average CO_2 emissions of their vehicle fleets to 130 g/km by technical measures. A further reduction to 120 g/km shall be achieved by complementary measures, including alternative fuels, energy-efficient tyres and air conditioners, better road and safety management and changes in driving behaviour (compare http://ec.europa.eu/reducing-co2-emissions-from-cars-index_en.htm). Further, the EU and its member states have agreed to promote alternative energy sources and have partly included elements of CO_2-based vehicle taxation systems. But the scene is scattered and clear market signals are still missing in Europe.

Driven by smog problems and rising energy prices, the State of California has supported the development of alternative transportation fuels (fuels other than gasoline or diesel) since the creation of the Energy Commission in 1975. Earlier programmes included demonstration programmes with vehicles using 'neat' ethanol and methanol; infrastructure development for methanol–gasoline blends (M85); support for flexible fuel, natural gas and electric vehicles. In 2006, the Californian Air Resources Board (ARB) established the Global Warming Solutions Act, which presents a set of measures to reduce CO_2 emissions, including an emissions cap until 2020, mandatory reporting rules, research programmes and the establishment of an advisory board.

19.3.3 Air pollution

The emissions of motor vehicles into the air are manifold. Among the most harmful substances are carbon monoxide (CO), hydrocarbons (HC) and formaldehyde (HCHO), nitrogen oxides (NO_x), particulate matter (PM) and organic gases other than methane (NMOG). All these substances are toxic to different degrees or foster the development of cancer, which is particularly true for ultra-small particles with a maximum size of 0.1 μm ($PM_{0.1}$). While particulate matter has attracted increasing attention over the past decade, other substances, such as heavy metals or sulphur oxides (SO_x), responsible for acid rain, have become less important in transportation, since fuels have been improved and catalytic converters have become obligatory in western countries. According to the OECD (2002) direct emissions and the resuspension of particulate matter is considered the biggest problem by Western European cities. In respect of health, particulate emissions and ozone are the main problems, contributing to around 370 000 premature deaths in Europe each year (EEA, 2006a). Although not included in these external-cost studies, increasing dependence on car use has been linked to rises in obesity in Western societies.

Concentrations of particulate matter and SO_2, accordingly, are the highest in cities with extremely high population densities, low technical standards and an excessive use of two wheelers, as is the case for Asian agglomerations. The worst conditions are observed in Delhi, Beijing, Tehran and Mumbai. But Western European cities also

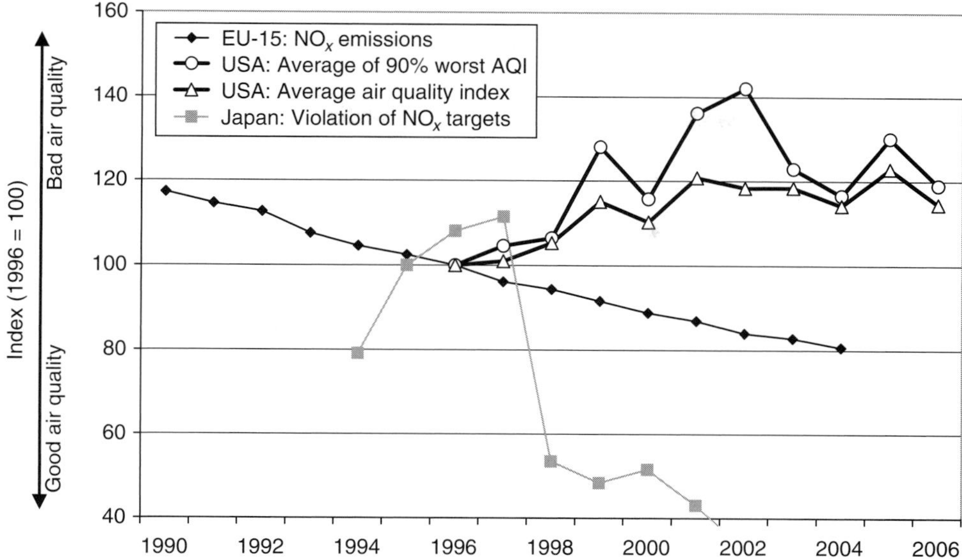

Figure 19.5. Emissions and air quality indicators for the EU15, the USA and Japan (AirNow, 2007; EEA, 2006b; c).

rank highly in this comparison: for example, Milan takes the worst position concerning NO_x concentration. Looking at transport-related air emissions alone, the interrelationship between the cities' population density and the concentration of pollutants gets even more obvious. Here, Bangkok, Hu-Chi-Min City, Tehran and Cairo lead the ranking – followed by Bologna, Manila and Mexico City.

Owing to their toxicity, air pollutants are most harmful in densely populated areas, among which the biggest ones are the megacities in developing or transition countries. In particular, in these countries the motorisation rate grows fastest. From 1991 to 1996 high income countries have faced a 8% rise, while the least developed countries, starting from a very low level, had a 36% increase in motorisation. Linking these figures with the low technical standards of vehicles in developing countries clearly indicates the negative side effects of the third world's struggle to catch up with western economic development. Nakamura *et al.* (2004) reveal this by stating that particulate matter is five times, SO_2 three times and NO_2 slightly above developed country levels for mid- and low-income countries. According to Jun (2007) two-thirds of China's urban residents breathe seriously polluted air resulting from one and a half decades of strong economic growth without taking care of its adverse effects.

Returning the attention to the developed world, Fig. 19.5 shows selected indicators of the emission of NO_x in the EU15 countries, the US air quality index and the violation of NO_x concentration standards in Japan for all emission sources in relation to 1996 values. There is no air quality indicator covering all EU member states, but

Table 19.2. *Transport and total air emissions by pollutant in the EU 2005*

	Transport sector			All sectors		
	1990	2004	1990–2004	1990	2004	1990–2004
NO_x (Mt)	6346	4071	−36%	13504	9289	−31%
CO (Mt)	31425	12542	−60%	51129	25809	−50%
NMVOC (Mt)	5642	1888	−67%	14185	7847	−45%
SO_x (Mt)	559	67	−88%	16491	4951	−70%

Source: EEA (2006c).

according to Larssen *et al.* (2004), concentration levels for most pollutants are broadly in line with the development of emissions into the atmosphere. The US air quality index (AQI) measures the concentration relative to target levels of various pollutants weighted by their toxicity and the Japanese indicator is derived from the number of measurement points not exceeding concentration limits over the whole year. Although the indicators are not really comparable between the regions, they show very positive developments in Europe and in Japan concerning NO_x emissions. The US picture looks less favourable, but air quality does seem to have stabilised since the beginning of the century. The developed countries are far ahead of the still-worsening conditions in the developing world, concerning air quality.

Data from the EEA (2006c) reveal that air emissions from transport follow a slightly less expressed, but well comparable trend to entire emissions between 1990 and 2004. Table 19.2 shows the share for the transport sector of certain pollutants in the EU: 40% to 50% of NO_x and carbon monoxide (CO), 25% of non-methane volatile organic compounds (NMVOC) and just 1% of sulphur oxides (SO_x). In developing countries, and here in particular in the big megacities, the share for transport of SO_x, heavy metal and particulate emissions (not listed here) will – because of the old vehicle fleets – be much higher.

Vehicle emissions into the atmosphere other than CO_2 can be reduced more or less effectively by technical solutions. Alternative or synthetic fuels, filter technologies, engine-internal optimisation measures, such as high pressure injection systems, exhaust gas recirculation or other solutions constitute possible options. Thus, non-CO_2 emissions will be – although they still constitute a top priority – of descending importance when discussing long-term sustainable transport systems in western countries. Technical progress is reflected by the dynamic development of emission standards, as introduced by the European Commission and by the US government. Table 19.3 provides an overview of the progressing emission limits and their structure in the EU and the USA. There are similar standards for goods vehicles. Airlines are increasingly faced with airport-specific emission regulations and emission-specific landing fees. Only those means of transport that are usually considered as environmentally friendly – namely rail and shipping – are not subject to environmental

Table 19.3. *Exhaust emission standards for passenger cars in the EU and the USA*

Exhaust emission standards	**CO** (g/km)		**HC / HCHO**[a] (g/km)		**NO_x** (g/km)		**PM** (g/km)	**NMOG** (g/km)
Fuel type	Gasoline	Diesel	Gasoline	Diesel	Gasoline	Diesel	Diesel	Gasoline
	European Union							
Euro 1 (7/1992)	3.16	3.16	1.13[b]	1.13[b]	[b]	[b]	0.18	
Euro 2 (7/1997)	2.2	1	0.5[b]	0.7[b]	[b]	[b]	0.08	
Euro 3 (4/2000)	2.3	0.64	0.2	0	0.15	0.5	0.05	
Euro 4 (4/2005)	1	0.5	0.1	0	0.08	0.25	0.025	
Euro 5 (2008)	1		0.05	0	0.08	0.08	0.025	
Euro 6 (2010)	1		0.05	0	0.04	0.04	0.025	
	United States							
LEV	2.11		0.009		0.03			0.046
ULEV	1.05		0.005		0.03			0.024
SULEV	0.62		0.002		0.01			0.006

Notes:

[a] EU: HC, USA: HCHO.

[b] Joint limit $HC + NO_x$.

Abbreviations: HCHO = Formaldehyde; NMOG = non-methane organic gas; LEV = low-emission vehicle; ULEV = ultra-low-emission vehicle; SULEV = super-ultra-low-emission vehicle.

Source: Stan (2005).

standards. But, given the rather old locomotive and vessel fleets worldwide, there is growing awareness of the problem as the fast development in road and aviation makes these modes catch up in environmental terms.

In the Asian–Pacific area, only Japan has developed its own vehicle emission standard. Other countries have adopted the US system (Taiwan) or the European standards to different degrees (most other countries). But it should not be forgotten that the share of new cars applying to these standards in developing countries is comparatively low.

Further ways to deal with the problem of air pollution include setting binding rules for fleet emission rates or, as in the case of California, oblige manufacturers to offer a particular share of zero-emission vehicles. Furthermore, access control for old vehicles is applied in European cities primarily to combat particulate matter concentrations. These regulatory actions are increasingly supported by fiscal policies, e.g., rebates or exemptions on vehicle taxes or road-user charges, as in the cases of the German and Swiss motorway tolls and the London congestion charge.

Technical solutions may well tackle the problem. But in terms of global equity, forcing high-end technologies as first choice to combat the adverse environmental impacts of transport will be in favour of the industries in the developed countries. The analysis of world patent applications and global trade flows on behalf of the German Environment Ministry (Edler *et al.*, 2007) demonstrates that developing and transition economies only play a very minor role in the developement and trade of new vehicles and propulsion systems. In the field of sustainable propulsion systems, non-OECD countries participate in world trade with 13% of imports and exports, but contributed only 2% of patent applications in 2005.

19.3.4 Transport noise and vibrations

The emission of noise is different from that of air pollutants or climate gases, as noise effects are restricted to the time of emission; there are no cumulative concentrations developing in the air. Further, the perception of sound as 'disturbing' depends on the level of other sound sources and on what people are currently doing. Consequently, the valuation of noise needs to take into account the type of area and the time of day. Besides disturbance, constantly high noise levels above 70 dB(A) can even lead to physical symptoms, such as increasing blood pressure and an increased risk of cardiac infarctions. People dying prematurely then cause suffering and grief to relatives and friends and economic losses to the private and public sector. In particular, in low-income countries it is not only the disturbance of noise emissions by cars, buses, tramways, trains or aircrafts that harm people, but also vibrations caused by construction works and heavy vehicles, which may cause serious damages to people's homes and historic buildings.

Noise is a purely local and very time-sensitive phenomenon, and is most relevant in urban areas and at night-time. Noise at night-time is perceived at roughly twice the levels during daytime. Depending on population densities and settlement typologies, rural areas may be ten times less sensitive than urban centres. Further, noise is not perceived by the human ear in a linear way; this means that two sound sources of the same volume are not perceived to be twice as loud as one of the sources alone. Accordingly, adding an extra vehicle to a road with dense traffic is much less harmful than driving it along a low frequented road. This fact is underlined by the vastly diverging figures on marginal external noise costs depicted in Fig. 19.2.

In Western Europe in the year 2000, 18 million people were exposed to average noise levels above 70 dB(A) caused by traffic. Of these, 15.9 million were disturbed by road, 0.7 million by railway and 1.4 million by aircraft noise. At lower noise levels, the number of people disturbed is much higher. The economic valuation of noise may be made by willingness-to-pay surveys, including hedonic pricing methods, assessing the development of land or housing prices with varying levels of permanent noise exposure. Although hedonic pricing is widely applied, filtering out the pure effects of noise on the prices of real estate appears difficult, and thus current estimates tend to over-value noise related impacts. A comparison of several international studies in Maibach *et al.* (2004) has revealed that the share of income that people are willing to pay for reducing noise exposure by one decibel is remarkably similar across Europe.

Although much has been done to reduce vehicle noise in the past decades, the strong growth of traffic has counterbalanced this benefit. Motor vehicles emit two types of noise: engine noise and rolling sound. Engine noise can be reduced by encapsulated motors or quieter propulsion systems, such as fuel-cell or electric motors. But, in particular when speeds are high, rolling sound dominates the picture. Thus, alternative motor systems may ease, but will not solve the noise problem of transport.

19.3.5 Transport accidents

Traffic accidents cause suffering and grief to the casualties, and to their relatives and friends, entail production losses to the victim's employer, impose costs to the insurance system, to the health sector and to traffic police and, finally, cause economic losses due to damaged public and private property. Out of this long list of consequences of traffic crashes, the value posed by society on human health and lives accounts by far for the highest share of the social costs of traffic accidents. By applying the contingent valuation method (CVM) or by analysing the benefits of public safety programmes, contemporary European studies arrive at a value of €1.5 million per death case and at roughly 10% of this value for severe injuries, irrespective of whether an accident was self-inflicted or not. Fatal accidents are an urban as well as an interurban problem. While in city traffic the probability of crashes is higher than on rural roads, accident severities increase with travel speeds and are thus less pronounced on urban roads.

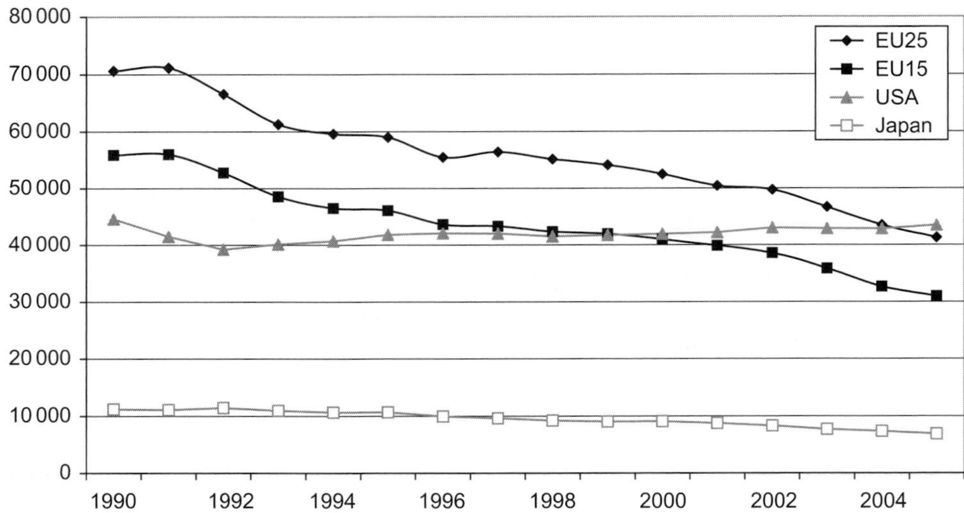

Figure 19.6. Development of fatal accidents since 1990 in Europe, the USA and Japan (BTS, 2007; EC, 2006a; NPA, 2007).

Total fatalities and accident rates per inhabitant or per passenger kilometre are among the most frequently cited and well documented indicators of transport system performance across the world. By drawing comparisons between the EU, the USA and Japan, Fig. 19.6 shows significant differences between single countries. While accident rates have declined in Western Europe (EU15) and Japan by 40% since 1990, strong traffic growth rates in the USA and the new EU member states have entailed stable fatality rates within the past 15 years.

Besides safety-belt regulations, accident rates are influenced by speed-limit standards, drivers' education levels, technical vehicle checks and regulations on driving with lights by day. Propulsion technologies are not relevant in this case and thus the switch from traditional combustion engines to alternative fuel-powered motors will not influence accident rates. Sure, the question can be raised that quieter electric vehicles could increase the risk of accidents because vehicles can not be heard as quickly. But up to now there is no evidence for this assumption and thus the thesis is rejected in the subsequent elaborations.

19.3.6 Congestion

Economists frequently argue that congestion is a transport-sector internal phenomenon imposed by transport users on each other. Impacts on the economy, e.g., commuters arriving late at their workplace or goods deliveries being delayed are well known by the travellers or shippers and thus congestion- or delay-related time costs may be computed, but must not be added up with system-external costs. Moreover,

Table 19.4. *State and projection of traffic congestion in the EU and the USA*

Mode	Europe		USA	
Interurban roads	Mainly Randstad and Ruhr areas and urban access	**C** ↘	Highway intersections and around agglomerations	**B** ↘
Urban roads	Severe congestion in some cities, no general problem	**C** ↘	Steadily increasing but not perceived a major problem	**D** ↘
Rail	Only at port hinterland lines; technical standards	**B** →	Considerable lag in grade-separated facilities in major lines	**D** ↘
Aviation	Problems in major hubs (London, Paris); airspace	**C** ↘	Constant investments, still recovering from 9/11	**B** ↗
Waterborne transport	Only port hinterland transport (Rotterdam)	**B** →	Port capacity and congestion hinterland routes	**D** ↘

Source: (Schade *et al.*, 2006).

shippers' surveys indicate that the unreliability of travel times constitutes a much bigger problem than the average delay, which can be anticipated in the production process. Besides time losses, congestion cost analyses report additionally 10% of costs for excessive fuel use and air pollutant and climate emissions. This leads to the environmental dimension of congestion: owing to less favourable driving cycles for combustion engines, fuel use and air emission factors are roughly twice the emission factors in rural transport. In particular, in the megacities in Latin America and Asia, congestion contributes to a large extent to air-quality problems.

Maibach *et al.* (2004) estimate total congestion costs of €63 billion for the EU15, Switzerland and Norway. This accounts for 0.7% of GDP and is 13% of system-external costs of road transport. According to the marginal cost figures given in Table 19.1 congestion is an agglomeration problem. Schrank and Lomax (2005) estimate delay costs of motorised road traffic in 85 US cities of different size of €48 million and Transport Canada (2006) reports €2.7 million for seven cities. Apart from parts of the USA and the UK, consistent time series of traffic congestion are not available on either side of the Atlantic. To arrive at a practical comparison, Schade *et al.* (2006) have transformed the available evidence and statements of operators and policy makers into a set of qualitative indicators. The intermodal comparison in Table 19.4 reveals that the situation in the USA appears more critical for most means of transport than in the EU, which can be explained by different traditions in establishing long-term transportation plans. By the analysis of studies and statistics and by numerous interviews with national bodies, the study has classified the current state of transport networks from A (no delays) to E (widespread and durable congestion in most days) and the expected future development from ↗↗ (strong improvement) via ↗ (improvement), → (stable situation) and ↘ (worsening) to ↘↘ (rapid deterioration).

It is interesting to note that, despite these severe problems, congestion is not always given the first priority in urban policy. For example, Schade *et al.* (2006) stated that, by comparison with the EU and the EU situation, US citizens value health, security and education problems much higher than pure congestion, although the latter has dramatically increased in severity, duration and spread across all classes of city. It can be assumed that this also holds for many large-sized world cities, which makes the additional environmental burden the real challenge in congestion mitigation.

19.3.7 Other transport externalities

The above effects are entirely related to the operation of vehicles and are therefore closely linked to the volume of transport demand. But there are a number of effects that the production and existence of infrastructures and vehicles impose on society. These include environmental effects caused by the construction or the disposal of infrastructures and vehicles, of energy generation (including the risk of nuclear power plants and the storage of nuclear waste), the deterioration of nature, landscape and natural habitats, the sealing of soil, the separation of urban districts and the visual intrusion caused by the existence of transport infrastructures. The economic assessment of these effects is quite challenging. Repair or replacement cost approaches provide a solution, but the range of possible values remains large. However, the total magnitude of these additional cost items is minor and thus does not justify demanding assessment approaches.

A particular point of discussion in transportation planning is the question of land use. In densely populated agglomerations, different activity patterns of people and firms compete for scarce space: production, retail, culture, recreation, housing – and different modes of transportation. Ways of solving these conflicts are to locate particular destinations, such as shopping areas or production sites outside the city centre. However, these measures create new trips or make the efficient combination of multipurpose journeys more demanding. Other ways are to increase the efficiency of transport systems by traffic demand management (TDM) or mobility management. Modern cities in western countries successfully introduced pedestrian zones and multipurpose city centres. These measures shorten daily trips, promote walking and cycling and should increase the living quality of road space. Latest concepts in urban land-use planning are ‘meeting zones’, where pedestrians and car drivers use the same road without specific regulations or traffic signs.

19.4 Exits towards a sustainable future

The above list of effects reveals that transport and its environment do not suffer from a single problem, but from several heterogeneous effects caused by different modes and affecting different population groups or environmental sectors with individual perspectives for the future. As the problems are heterogeneous, so should the strategies to calm or remove them be. A single instrument healing all of transport's ills does not exist, as the time horizon, the order of magnitude or even the direction of effects of policy measures on different effects may vary considerably.

A good example is kilometre-dependent road pricing, which is strongly promoted as the universal policy tool by the European Commission (EC, 1998; 2001). This instrument is doubtless capable of increasing the capacity utilisation of existing infrastructures and the fluidity of traffic and thus even lowering the emissions of air pollutants and greenhouse gases per vehicle kilometre. As the Swiss and the German examples show, a technology-dependent differentiation of tariffs will substantially increase the renewal of the vehicle fleet and thus support long-term environmental targets. But their introduction will also cause traffic to move to the secondary networks and cause problems with noise disturbance and traffic safety there. Noise and safety, therefore, must be addressed by regulatory instruments, such as speed or weight limits, driving bans and the enforced control of safety regulations.

The example illustrates that the reduction of negative impacts of transport requires decisions on two things: the preferences of goals and the instruments to be applied to reach them. In this work, we define the overall goal of transport policy as to ensure long-term sustainable development with respect to the environment, the public sector and transport users. Environmental sustainability includes the reduction of greenhouse gases, air pollutant and noise emissions below particular target levels, which might be the result of a political compromise or based on scientific studies. While the Kyoto targets to address climate change clearly belong to the first category, noise targets are commonly based on evidence by market or epidemiological studies.

Policy instruments (see Table 19.5) to meet the various sustainability targets may be classified into infrastructure investment, traffic demand management (TDM), regulation, pricing and support research and development (R&D) in new technologies. Each of these measures consists of several variants, which will, however, not be considered in detail in this volume. The following matrix undertakes an attempt to propose the appropriateness of the various classes of policy measures to meet specific subobjectives of a sustainable transport development.

Investment in new infrastructure, in particular in the road sector, still constitutes the first choice for calming congestion problems and for improving the accessibility of remote regions applied by national and trans-national planning authorities. Examples are the TEN-T investment programme, followed by the European Union (EC, 2007), the SAFETEA-LU programme of the US Department for Transport (SAFETEA-LU, 2005) and the German Anti Congestion Programme of the German Federal Government (ABMG, 2002). Besides the primary benefits from capacity extension, the construction of new infrastructure is frequently justified by employment effects during the construction phase. As experiences in many western countries in the past decades have shown though, the problem of new traffic generated by better capacity supply seriously violates the sustainability goals of a balanced

Table 19.5. *Policy goals and instruments for sustainable transport policy*

Sustainability goals	Instruments				
	Investment	TDM	Pricing	Regulation	R&D
Financial viability	–– New investment and maintenance obligations	++ More efficient use of existing infrastructures	++ Revenues, better use of existing capacity	0	0
Congestion mitigation	+ Effective; problem of induced traffic	++Access or speed control, mobility management	++ Time-variant charges	0	+ Telematics, information and TDM systems
Safety	+ Safer roads; problem of induced traffic	+ Smoother traffic flows	+ Risk-dependent insurance premiums	++ Weight and speed limits, social standards	+ Driver assistance systems, tyres and pavements
Noise reduction	+ Noise walls, bypass routes	– Possibly traffic deviation to secondary routes	– Possibly traffic deviation to secondary routes	++ Local or temporal driving bans	+ Quiet engines, tyres and pavements
Air quality	– Better traffic flow but induced demand	+ Better driving conditions, less fuel consumption	++ Tolls, tax by emissions, tax reduction for bio-fuels	++ Exhaust emission standards; fuel composition	++ Fuels and propulsion systems, filter technology
Climate protection	– Better traffic flow but induced demand	+ Better driving conditions, less fuel consumption	+ Fuel tax, circulation tax by fuel consumption	0 Emission trading, max. average fleet emissions by manufacturer	++ Fuels, propulsion systems, fuel-efficient engines
Nature conservation	–– Disruption of habitats, ground sealing	0	0	+ Compensation measures, access control	0

Source: Own compilation.

transport policy. More severe than the social and environmental effects of the induced traffic, however, may be that new concepts of mobility are hindered by such ambitious transport investment plans. Finally, financing goals are often not met as in the case of huge investment projects very frequently a bias towards the underestimation of costs and the over estimation of benefits appears, which can be supposed to be politically desired (Flyvbjerg *et al.*, 2003).

Instruments of traffic demand management address the capacity problem from a different direction. The goal is to match traffic demand more effectively to existing infrastructure capacity. There are basically two approaches: managing traffic flows and influencing user behaviour. Flow management may be achieved by a dynamic regulation of route access, speed limits or charges or providing information on alternative routes. All these variants are applied in practice, but they are restricted to a rather narrow window of capacity utilisation. When sufficient road space is available, traffic demand management is not needed and within an oversaturated network only drastic measures will solve the problem. Positive examples for the latter are the London, Singapore or Norwegian congestion tolls, special lanes for highly occupied vehicles (HOV lanes or bus lanes) or access control, as in the case of Zurich. Mobility management is of a more strategic nature, in that it addresses people's travel choices or transport patterns of firms and freight forwarders. Examples in passenger transport are to make working times more flexible or to spread school start times in order to mitigate traffic peaks, to promote public transport by supporting free job tickets, parking space management in central business districts or the redesign of urban fabrics to shorten the distances people have to travel in daily life. Prominent examples in freight transport are the promotion of combined transport facilities or of goods distribution centres near urban areas. The examples show that there are manifold examples of successful mobility management, but it also has to be stressed that most consumers are hardly willing to change their style of living just for the sake of overall society. The management of traffic demand and mobility thus needs to be supported by more powerful policy measures.

One of these measures is pricing and taxation policy. Price signals can provide incentives for people to behave in a particular way. Commonly used are tax reductions on biofuels or clean vehicles, the differentiation of infrastructure charges by exhaust or noise emission standards, as in the case of Germany, or the reduction of tolls on HOV lanes, according to US practice. Following the policy of the European Commission, differentiated user pricing is the first best alternative to ensure sustainable development within the transport sector. As recommended by neoclassical welfare theory, all users of all modes of transport should, according to neoclassical theory, be priced according to the true marginal social costs that they cause to society (EC, 1998). While this concept constitutes a wonderful closed theoretical model, in practical application it constitutes of a number of pitfalls. For example, the welfare optimum is reached only when (1) all economic sectors are included in the pricing scheme, (2) charges instantly vary by time and location according to the current development of costs, (3) all users and the price-setting institution at any time have perfect information on costs and options, (4) all users behave fully selfishly and (5) the state is a perfect manager always seeking to maximise social welfare (6) without causing any type of transaction cost (Doll, 2005). As none of these criteria is fulfilled in practice, the EC has partly withdrawn the promotion of marginal cost pricing goals. The more pragmatic policy now focuses on the strengths within each mode of transport and looks more to financing goals. With this objective, a number of European countries (Switzerland, Austria and Germany) have recently introduced tolls for heavy-goods vehicles on national networks and others will follow (the Netherlands, the Czech Republic and possibly the UK). An important issue to ensure the acceptability of the tolls is a common agreement on the use of the revenues from transport pricing. As practical examples as well as theoretical analysis reveal they must be used to improve transport conditions.

Pricing cannot convey all of the partly contrary sustainability objectives. In many cases, specific regulations are more effective. Moreover, the presence of transaction costs associated with a sophisticated pricing system (20% of the German HGV motorway tolls are from the operating and enforcement system) makes regulation in some cases cheaper. Examples are driving bans in front of hospitals during night-time, speed and access limits, compulsory emission standards of new vehicles or maximum pollutant concentrations or noise levels.

The policy-oriented instruments discussed so far have the disadvantage that they require a constant periodic review and re-adjustment. People tend to adapt to new regulatory environments and to seek for possibilities to maintain their well known behavioural patterns. When policies introduce hard measures, public resistance may make the undertaking very expensive or even impossible. Furthermore, the temporal success of a regulatory environment may turn as either the political environment or the public awareness for particular problems changes. In contrast, technical innovations, as soon as they have penetrated the market, will not be removed again, so long as they allow people to maintain their lifestyle with less harmful effects on nature and society. New generations of cleaner engines and fuels, and driver-assistance systems provide a platform on which future developments can build. Of course, there are technical solutions that require major initial public support, e.g., to install expensive infrastructures, or which demand for considerable research and development activities. It is difficult to state under which conditions or when these technologies will be accepted by the market. Illustrative examples are fuel-cell vehicles, Maglev trains or the 3-litre car. Nevertheless, the fast development of the economies in Central and Eastern Europe, Asia and South America call for effective technical solutions to the sustainability problem of motorised transport.

To summarise: there are many problems and a variety of instruments available to address them. Investment in new infrastructures may create safer roads or rail tracks, but is counterproductive owing to the creation of new demand. The management of traffic flows and demand can help to use existing infrastructure more efficiently and thus to save financial and natural resources. Further, because of the demographic development in Europe with its shrinking society, excessive traffic demand and congestion will be less of a problem in the future in Europe. But looking to America and Asia shows that congestion is still growing and will remain on the political agenda through the coming decades, raising questions of demand management, pricing, access control, investment and land-use planning. Safety, air pollution and even noise can be addressed by vehicle and infrastructure technologies and regulations and are thus out of the focus of future challenges in transportation. The remaining long-run environmental problem of the transport sector is thus the emission of greenhouse gases. Ideally this issue needs to be addressed from different sides. The development of more energy-efficient engines and fuels needs to be supported by limits on maximum fuel consumption rates, CO_2-emission taxes, emission trading systems, pricing schemes and – most important – the provision of attractive transit systems. Good examples for successful integrated policies are the TransMilenio concept in the Latin American cities of Bogotá and Coritiba and the London congestion charge, the various rail and tram concepts in Germany and Japan and the active promotion of zero-emission vehicles by the State of California.

19.5 Contribution of hydrogen-based propulsion systems

19.5.1 Introduction

The discussions in the preceeding sections have shown the bandwidth of social problems of today's transport system. In the following, it will be discussed which of these impacts are affected by the introduction of hydrogen vehicles. Here, 'impacts' may be positive or negative.

Hydrogen vehicles have no effect on the development of congestion, land use due to transport infrastructure and accidents. Whereas the first two points are behind any discussion, some people argue that the introduction of hydrogen vehicles could increase the level of accidents because these propulsion systems produce less noise than internal combustion engines. Up to now there is no empirical evidence to confirm or reject this point. Further, this argumentation seems to be critical because – as the discussion in Section 19.3 has clearly shown – noise is a serious social problem of transportation and much better technical solutions may exist to reduce the level of accidents instead of producing a permanent level of noise. An example is the development of video-based assistant systems, which warn the driver of pedestrians and bikers.

Positive impacts of hydrogen vehicles can be expected in impacts of climate changes, local emissions and noise. These impacts will be discussed in detail in the following sections. However, some new problems could be identified by the introduction of hydrogen as a fuel in the transport sector, which will be treated later on in this chapter.

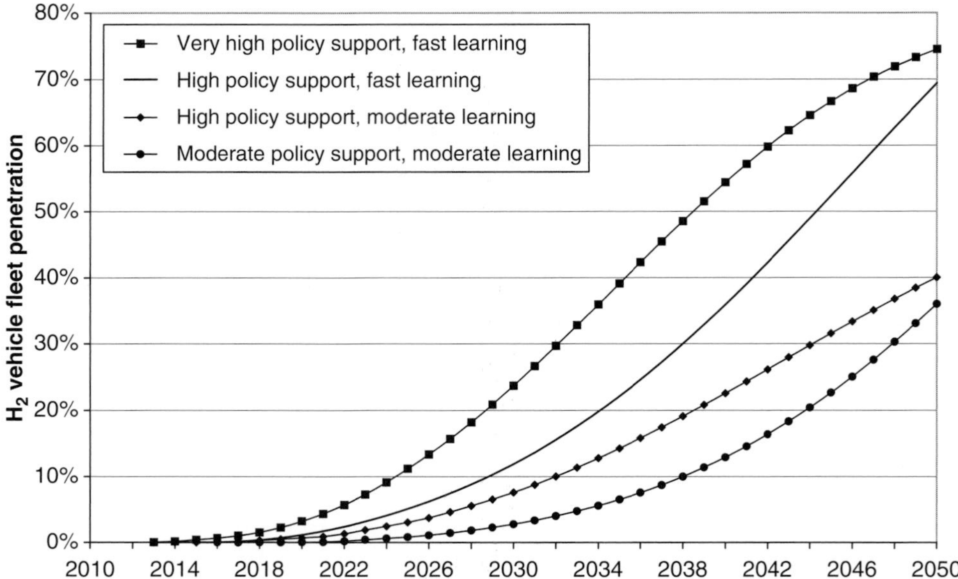

Figure 19.7. Development of the penetration rate of hydrogen vehicles for passenger transport (scenarios of HyWays, 2007, see Chapter 14).

19.5.2 Reduction of external effects by hydrogen use

Whereas the use of hydrogen as a fuel in the transport sector does not lead to relevant greenhouse-gas emissions, the upstream process (production and transport of hydrogen) could be relevant. The detailed energy pathway analysis in Chapter 7 has shown that both effects – more or less greenhouse-gas emissions – could be found, depending, among other things on the feedstock and hydrogen production technology. However, the majority of hydrogen pathways lead to clearly lower greenhouse-gas emissions compared with today's gasoline or diesel vehicles. Also, compared with many other alternative fuels and propulsion systems (e.g., biomass or natural gas as a fuel or conventional hybrid vehicles) most of the hydrogen pathways lead to lower greenhouse-gas emissions (see Chapter 7 for details and references).

The relevant reduction potential due to the introduction of hydrogen vehicles has been analysed in many successful international research projects. In the HyWays project, the impact on CO_2 emissions was analysed for different scenarios for selected European countries (HyWays, 2007). (HyWays was an integrated project where industry partners, research institutes and European member state representatives worked on the development of a European hydrogen roadmap and action plan. As well as models, extensions with stakeholder involvement in more than 50 workshops were carried out.) The impact of hydrogen on CO_2 emission is determined by both the penetration rate of hydrogen end-use applications – see Fig. 19.7 (within road transport, passenger cars, light-duty vehicles and city buses gradually shift to

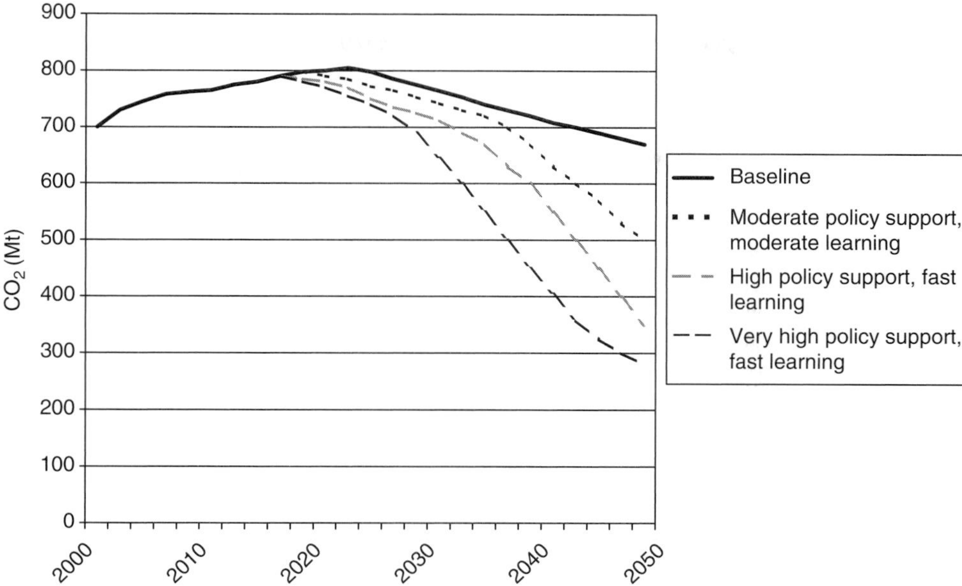

Figure 19.8. Development of total CO_2 emissions for road transport (see Chapter 14 for scenario definitions).

hydrogen; for more background information, see Chapter 14) – and the way the hydrogen is produced. In HyWays, a mix of renewable, fossil energy carriers, mainly with carbon capture and sequestration, and nuclear was assumed as major sources for hydrogen production.

The emissions shown in Fig. 19.8 include emissions during the production process for hydrogen as well as for gasoline and diesel. In the baseline scenario, with the assumption that no hydrogen as a fuel will enter the market, the demand for transport increases substantially, explaining the increase in CO_2 emissions until 2020. In this scenario, total CO_2 emissions in 2050 are 10% below the emission level in 1990 due to higher shares of biofuels in the light of current EU policy goals. As a result of the introduction of hydrogen, total CO_2 emissions from road transport decrease impressively by about 350 Mton in 2050 in the 'fast-learning scenario', reducing emissions by 55%–60% compared with the baseline scenario. In the scenario with moderate policy support and moderate learning, total CO_2 emission in 2050 will decrease slightly by over 30%.

Figure 19.9 shows the development of the marginal abatement costs (cost (€) to reduce the marginal unit (ton) of CO_2). In the baseline scenario, the marginal abatement costs increase to over 100 €/ton of CO_2, in order to meet the assumed 30% CO_2 emission reduction goal in 2050 in every country compared with the base year 1990. As a result of the introduction of hydrogen into the energy system, the marginal abatement costs decrease by 15%–30%, see Fig. 19.9. This means that, in

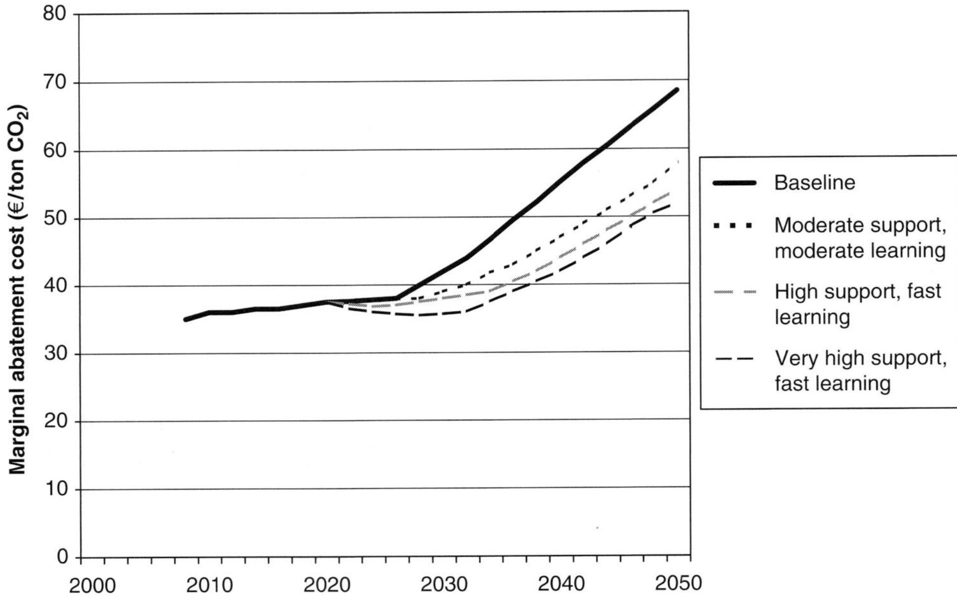

Figure 19.9. Development of the marginal abatement costs for CO_2 reduction for the whole energy system.

time, hydrogen does become a cost-effective emission-reduction option, lowering the costs of meeting future CO_2-emission-reduction targets. Comparable results are found if an 80% CO_2 reduction target is implemented for 2050.

In the WETO-H_2 study (EC, 2006b), supported by the European Commission, hydrogen in the transport sector only plays a role if a very ambitious carbon constraint is assumed, and also a series of technology breakthroughs have to be reached that significantly increase the cost-effectiveness of hydrogen technologies, in particular in end use. (WETO-H_2 has been prepared by a consortuim of six European research organisations. The aim of WETO-H_2 is to analyse the development of the European energy system, embedded in the world energy system, for different scenarios. One is the 'hydrogen' case.) In such a case, hydrogen will be used in 2050 worldwide in 30% of passenger cars and about 80% of these will be powered by fuel cells; 15% will be hydrogen hybrid vehicles and 5% hydrogen internal combustion engines. The share of renewable energy in hydrogen production will be 50% and that of nuclear 40%. World emissions of CO_2 in 2050 will fall by 18 Gt (about 40%) compared with the reference projection. The reference projection describes a continuation of existing economic and technological trends, including short-term constraints on the development of oil and gas production and moderate climate policies, for which it is assumed that Europe keeps the lead. About three-quarters of the reduction achieved in the hydrogen case are achieved in the generation of electricity. This result shows that the deployment of hydrogen in the world energy system is compatible with ambitious climate policies consistent with a trajectory of long-term stabilisation of greenhouse-gas emissions at 550 ppm (see also Chapter 2).

Also the International Energy Agency (IEA) has carried out a comprehensive research study of the worldwide consequences of an introduction of hydrogen (IEA, 2006). (The study is a response to the Group of Eight (G8) leaders at their Gleneagles Summit in July 2005, and to the International Energy Agency Ministers who had met two months earlier. Both groups called for the IEA to develop and advise on alternative scenarios and strategies aimed at a clean, clever and competitive energy future.) If the most favourable, cost-optimal scenario is compared with a similar scenario where hydrogen and fuel cells are not part of the technology portfolio, but energy efficiency measures (hybrids) and other alternative fuels like ethanol enter the market, the net benefit of hydrogen and fuel cells will show a 5% reduction of CO_2 emissions (1.4 Gt CO_2) by 2050. The reduction is mainly based on hydrogen use in the transport sector. As in the afore-mentioned HyWays project, the marginal CO_2 abatement costs are lower in the hydrogen scenario than in the baseline scenario.

All in all, the studies analysed show that the use of hydrogen in the transport sector can significantly reduce the CO_2 emission of the transport sector compared with other scenarios without hydrogen, even when taking into account tailpipe and upstream emissions as well as alternative technology developments. This is an important result because CO_2-emission reduction in the transport sector is a very challenging task. However, the influence on worldwide CO_2 emission is limited. In most studies, hydrogen is mainly introduced for transport applications and here only a share of vehicles are hydrogen vehicles, among other reasons, because for heavy trucks most of the studies see no benefits of hydrogen as a fuel. In many forecasts, this sector shows a particularly strong growth. In other sectors, which are relevant for greenhouse-gas emissions, hydrogen plays no, or at least a very limited role. Therefore, hydrogen could be only one building block among several measures for reduction of greenhouse gases. This topic is also handled in more detail in Chapter 20.

The large-scale deployment of hydrogen in the transport sector (cars, light-duty vehicles and city buses) has a significant impact on the reduction of atmospheric pollutant emissions. Emission reduction of pollutants is one of the main drivers for the introduction of hydrogen. These benefits are often mentioned. However, the number of quantitative analyses is limited. In the following, the results of the HyWays project (HyWays, 2007) will be presented.

In HyWays, the Cobert model, which is designed to calculate emissions of the transport sector, was used. A detailed description of the methodology used to quantify the impact on local air emissions is given in HyWays (2007) and Mattucci (2007). In the baseline scenario, without hydrogen, the local pollutant emissions are generally decreasing owing to more severe legislations on exhaust emissions of vehicles. Two new Euro legislations (V and VI) have been added for cars and light-duty vehicles to

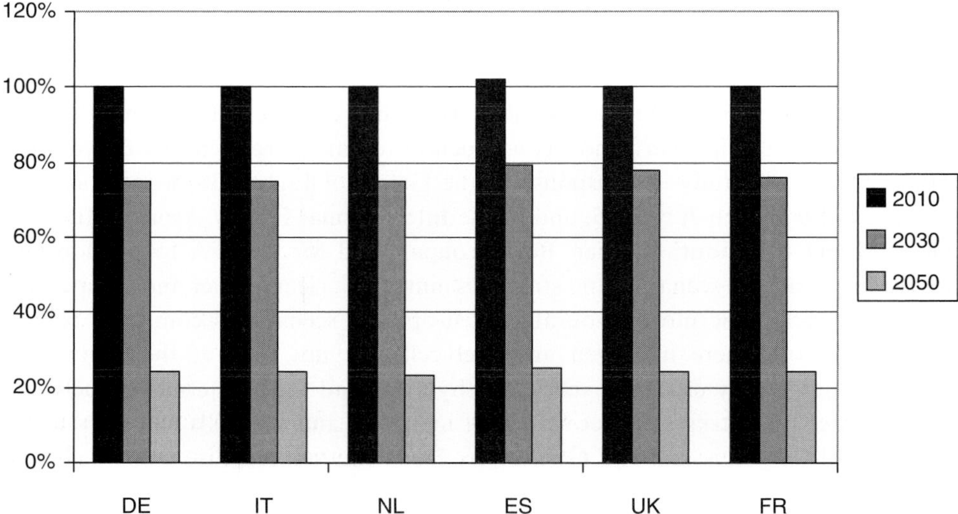

Figure 19.10. Impact on reduction of fine dust emissions as a result of the introduction of hydrogen in road transport: comparison of the high-support and fast-learning scenario with the baseline scenario; scenario data are taken from Mattucci (2007).

ensure that more stringent requirements on vehicle emissions can be imposed by the European Commission. The legislations reduce the acceptable levels on pollutant emissions (Euro V) and impose limitations on fuel consumption for both cars and light-duty vehicles (i.e., as result of voluntary agreements between car manufacturers and the EC, the Kyoto Protocol and post-Kyoto initiatives to counteract climate changes) and start in 2010 and 2015, respectively. The impact of the introduction of hydrogen in road transport has been assessed for three domains; urban, extra-urban and highway. Therefore, the impact in heavily polluted areas is specifically taken into account, since the highest concentration and, therefore, the highest impact on health occurs in densely populated areas, such as urban ones. Projections of emission levels are made for all the pollutants (CO, NO_x, PM, volatile organic compounds (VOC), etc.).

An indication of the environmental effects of hydrogen deployment for each of six European countries (France, Germany, Italy, the Netherlands, Spain, and the UK) is given in Fig. 19.10, where the fine dust emissions are shown for the hydrogen high-penetration scenario. Similar trends are found for other pollutants. The data are normalised in respect to the baseline scenario and show a trend very similar for the analysed countries with a reduction of more than 70% in 2050. The results are an average per country. At a local level, higher reductions can also be achieved if non-technical measures, such as limitation of city centre access for non-zero emission vehicles, are taken.

Fuel-cell vehicles contribute to reduced noise pollution, since the drive system is nearly noiseless. This is especially important in urban areas, where the share of noise coming from the drive system is dominant, compared to other noise sources from vehicles like aerodynamic and rolling sound, depending on tyres and road surface. Noise constitutes a particular problem during sleep periods because noise exacerbates ischaemic heart disease and mental illnesses (see the Health Commission (2003) and Section 19.2).

The advantage of noise reduction for fuel-cell vehicles is often mentioned, among others by test users of such vehicles, but scientific quantitative analysis under real drive conditions are very limited. In TRL (2007) it is assumed that a hydrogen fuel-cell vehicle would produce 50% less noise than a comparable vehicle with a conventional internal combustion engine. Given the dominance of rolling aerodynamic sound at high speeds the noise reduction along motorways, for instance, will be limited.

A rough calculation suggests that the sum of external costs for fuel-cell powered vehicles due to the emission of CO_2, local air pollutants and noise over the vehicle's lifespan are between 1000 and 1500 euros lower than for conventional combustion engines. For conventional vehicles, the average external cost for climate change, noise and air pollution have been calculated (see Fig. 19.2 for data – averages are calculated from Min. and Max. values for EU15 for external costs – and references). In the next step the emission reduction for greenhouse gases, noise, and air pollution for fuel-cell vehicles in comparison to conventional vehicles are calculated for the year 2030 and 2050. For greenhouse gases and air pollution, the data come from the HyWays project, and for noise from TRL (2007). Then the external cost difference for a fuel-cell vehicle in comparison to a conventional vehicle is calculated. For the lifetime of vehicles, 12 years, and for vehicle km per year, 12 000, are assumed (average figures for Germany).

19.5.3 Increase of external effects by hydrogen use

There has been little discussion or analysis, however, of possible negative impacts of hydrogen and fuel-cell diffusion within transport (EC, 2003). One problem mentioned is the use of platinum group metals (platinum, palladium and rhodium, PGM) as catalysts in the fuel-cell stack. The resources are limited, South Africa and Russia dominate the production of platinum, which is a critical issue for supply security, and the mining of platinum is energy intensive and linked with a lot of serious environmental impacts. Next to fuel-cell vehicles today, PGM can also be found in catalytic converters for gasoline and diesel vehicles. The environmental evaluation depends, among other things, on the reduction of the PGM content on fuel-cell stacks and the recycling rate, as well as – for benchmark reasons – on the development of PGM in conventional vehicles.

Based on data of Saurat (2006), where an increase of PGM is predicted for conventional vehicles, owing to higher environmental legislations and a decrease of PGM for fuel cells (the reduction of PGM for fuel cells has made a rapid progress in the past), in Schade *et al.* (2007) the overall development of platinum has been forecast for EU25. Here, the scenario for very high policy support and fast learning

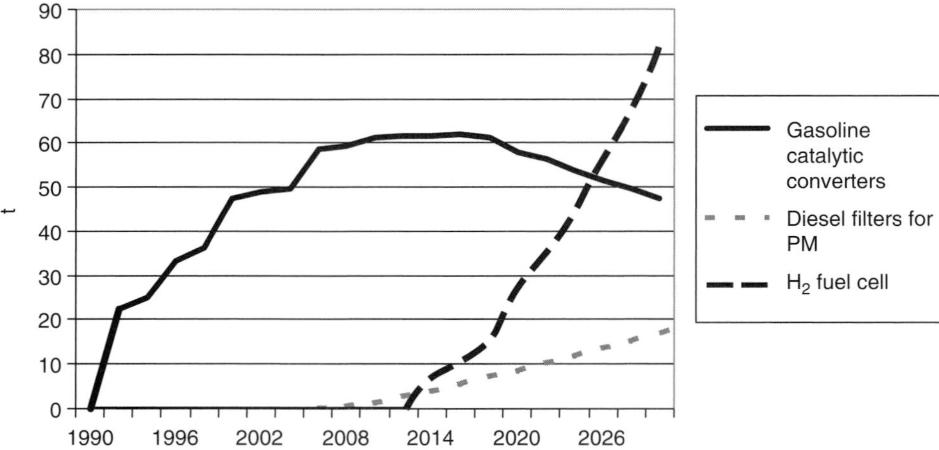

Figure 19.11. Development of platinum demand for new cars in EU25.

(see Fig. 19.7) has been used as a penetration scenario for vehicles. The results show that even though the platinum content in fuel cells is subject to a significant learning process and is hence strongly reduced, it can be expected that with the accelerated diffusion of fuel-cell vehicles into the fleet after 2020, the platinum demand increases to levels higher than current demand, reaching close to 150 tons in the year 2030 (see Fig. 19.11). This would cause concerns about environmental impacts. One option for mitigating these problems is to develop strategies for recycling and reusing the platinum when cars are scrapped.

Also other materials should be included in the assessment. Fuel cells and hydrogen-storage tanks require different materials from internal combustion engines and gasoline- or diesel-fuel tanks. As an example, fuel-storage tanks in hydrogen vehicles are likely to require more steel or alloys than modern plastic gasoline or diesel tanks. If mass-produced hydrogen-powered vehicles were heavier, then they would need stronger brakes, and would wear tyres more rapidly. The net effect on material assets greatly depends on whether mass-produced hydrogen vehicles are more or less durable than gasoline and diesel vehicles, which is not currently known (TRL, 2007).

One other identified negative impact relates to possible public anxieties about the safety of hydrogen technologies, which are viewed as a barrier to be overcome through improved public communication (L-B-Systemtechnik, 2006). For technical aspects of security, see Chapter 11, and for acceptance discussion, see Chapter 8 where the safety issues are analysed in detail and the conclusion is that hydrogen systems of a high regulation standard are not expected to have any negative impacts on accident frequency or severity (see also TRL, 2007). The focus in the hydrogen economy literature on overcoming barriers to a hydrogen transition suggests support for a 'technology push' model of technological development. This linear, deterministic perspective typically focuses on the technological potential of innovation while ignoring its social relevance (Shackley *et al.*, 2004). Furthermore, it reduces the complexity and ignores possible feedbacks and 'rebound effects' associated with innovation (HySociety, 2004).

However, the discussion on social impacts cannot only be limited on the propulsion system but must also include the production, transport, delivery and storage of hydrogen. (For more background information on the discussion of energy production and environmental impacts, see EEA, 2006b; EPA, 2007; IAEA, 2007.) Important new impacts of hydrogen based on hydrogen production include:

1. Severe reactor-accident release, waste-repository release, land disturbance, and others, if hydrogen is produced via uranium feedstock;
2. Global climate change, air-quality degradation (coal, oil), lake acidification and forest damage (coal, oil), land disturbance and others, if hydrogen is produced by fossil fuels;
3. Storage of carbon dioxide by carbon capture and sequestration solutions for fossil fuels: this is often not seen as a long-term solution and exploratory studies indicate that carbon sequestration may not be acceptable to the public (Whitmarsh *et al.*, 2006);
4. Increased demands on land use if hydrogen uses biomass or other renewables as feedstock, which is particularly critical if biomass is imported from countries where the biomass production goes hand in hand with destruction of valuable eco-systems, like jungles.

Looking at the last three problems listed above, it becomes clear that the introduction of hydrogen in transport applications is fundamentally linked to wider energy systems. Discussion of sustainable hydrogen in transport is, therefore, not limited to the context of sustainable transport but also includes discussion of sustainable energy systems. This increases the complexity of the issue, and expands the range of interests and perspectives to be considered in decision making about hydrogen transport technologies. Together, the possible negative societal impacts of hydrogen use in transport and the implications for energy systems suggest a need for more reflexive and inclusive debates about the range of impacts and beneficiaries of hydrogen and fuel-cell technologies. It should be kept in mind that the exploration and transport of oil also lead to relevant negative impacts on the environment.

19.6 Summary and conclusions

Transport systems perform vital societal functions, but in their present state cannot be considered 'sustainable'. Particular concerns in this respect include local air pollution (particulate matter, ozone), accidents, climate change, congestion, land use and noise. Increasing attention is being focused on hydrogen transport technologies as a means to achieving more sustainable transport. Recent estimates for Germany indicate that the external costs of road transport are roughly four times the costs of providing and maintaining the roadway infrastructure and four times that of an untaxed gasoline price.

A lot of research studies have been carried out to analyse the possible CO_2 effects of hydrogen as a fuel in the transport sector. All in all, the analysed studies show that the use of hydrogen in the transport sector can significantly reduce CO_2 emission of the transport sector compared with other scenarios without hydrogen, even if taking into account tailpipe and upstream emissions as well as alternative technology developments (e.g., biomass as a transport fuel, efficiency improvements of the propulsion system) in the baseline scenario without hydrogen. This is an important result because CO_2 emission reduction in the transport sector is a very challenging task. The reduction goes up to 60% compared with a reference development for selected European countries. However, the results are very sensitive to assumptions like feedstock for hydrogen and hydrogen-vehicle penetration rates and show a relevant bandwidth.

Local air emissions, responsible for particulate matter, ozone and acid rain, as well as noise, could also be significantly reduced by the introduction of hydrogen fuel-cell vehicles. Emissions of NO_x, SO_2, and particulates can be reduced by 70% to 80% compared with a reference case. Especially in highly densely populated areas this is one major benefit of hydrogen that is often underestimated. The number of mega-cities worldwide is increasing, which demonstrates the increasing importance of this topic. A rough calculation shows that the CO_2, local emissions and noise benefits of a fuel-cell vehicle lead up to reduction of average external cost by 1000 to 1500 euros per vehicle compared with a conventional vehicle.

However, other transport problems, such as congestion and accidents, are not mitigated – and some may even be exacerbated – by hydrogen use. New problems in the upstream process of hydrogen production, like nuclear waste or land use due to increasing biomass demand, could be possible. These impacts have to be benchmarked against the impacts of upstream processes of conventional fuels. Another important topic is the question of material assessment, e.g., the use of platinum group metals in the fuel-cell stack or increase of material use for tanks, because mining is often linked with considerable environmental impacts. The total net effects of these topics are currently unknown, not least because the prediction of technological developments of fuel-cell vehicles is very uncertain. This is also true for conventional vehicles for benchmarks. Further research is necessary.

Thus, we highlight the need for integrated energy and transport policies and argue for more reflexive and inclusive societal debate about the impacts and benefits of hydrogen transport technologies. Hydrogen could be one relevant element to solve today's unsustainable problem in the transport system. However, next to technical solutions and new fuels, other measures, like infrastructure investment, traffic-demand management, regulation and pricing, like city taxes, are necessary to avoid major external effects of today's transport system.

References

AirNow (2007). *Airnow* EPA, NOAA National Weather Service, NASA Earth Science, National Park Service Air Resources, National Association of Clean Air Agencies, Environment Canada. www.airnow.gov.

ABMG (2002). *Autobahnmautgesetz für schwere Nutzfahrzeuge*. BGB 1 I, pp. 1234, Bundesregierung, Federal Government of Germany.

Bickel, P., Friedrich, R. *et al.* (2005). *Externalities of Energy: Extension of Accounting Frameworks and Policy Applications (ExternE-Pol)*. Final Technical Report. Project funded by the European Commission's EEDS programme, project co-ordinator: IER, University of Stuttgart.

BTS (2007). Bureau of Transportation Statistics. www.bts.gov.

Doll, C. (2005). *Allokation gemeinsamer Kosten der Straßeninfrastruktur. Anwendbarkeit der Konzepte der kooperativen Spieltheorie in der Wegekostenrechnung*. Baden-Baden: Nomos-Verlag.

DOT (1997). *Federal Highway Cost Allocation Study 1997*. US Department of Transportation, in co-operation with the Federal Highway Administration, Federal Transit Administration and the Federal Railroad Administration. Washington, DC: Department of Transportation.

EC (1992). *The Future Development of the Common Transport Policy – A Global Approach to the Construction of a Community Framework for Sustainable Mobility*. COM (1992) 494. Brussels: European Commission.

EC (1998). *Fair Payment for Infrastructure Use: A Phased Approach to a Common Transport Infrastructure Charging Framework in the EU*. COM (1998) 466. Brussels: European Commission.

EC (2001). *European Transport Policy for 2010: Time to Decide*. Luxembourg: Office for official publications of the European Communities, European Commission.

EC (2003). *Hydrogen Energy and Fuel Cells – A Vision of Our Future*. Final Report of the High Level Group. EUR 20719 EN. Brussels: European Commission – Directorate-General for Research & Directorate General for Energy and Transport.

EC (2006a). *Energy and Transport in Figures 2006*. Brussels: European Commission, Directorate General for Energy and Transport, in co-operation with Eurostat.

EC (2006b). *World Energy Technology Outlook – 2050 (WETO-H_2)*. Brussels: Directorate-General for Research, Directorate for Energy, European Commission.

EC (2007). *Trans-European Networks – Towards an Integrated Approach*. Communication from the Commission. Brussels: Commission of the European Communities.

Edler, D., Blazejczak, J., Walz, R. *et al.* (2007). *Wirtschaftsfaktor Umweltschutz. Vertiefende Analyse zu Umweltschutz und Innovation*. Study commissioned by the Federal Ministry for Environment, Nature Conservation and Nuclear Safety (BMU) and the Federal Environment Agency (UBA). Berlin: DIW, Fraunhofer-ISI, Roland Berger.

EEA (2006a). *Transport and Environment: Facing a Dilemma. TERM 2005: Indicators Tracking Transport and Environment in the European Union*. Copenhagen: European Environment Agency. http://europa.eu.int.

EEA (2006b). *Energy and Environment in the European Union, Tracking Progress Towards Integration*. No. 8/2006. Copenhagen: European Environment Agency.

EEA (2006c). *Annual European Community LRTAP Convention Emission Inventory 1990–2004*. Submission to EMEP through the Executive Secretary of the UNECE. EEA Technical Report 8/2006. Copenhagen: European Environment Agency.

EPA (2006). *Inventory of US Greenhouse Gas Emissions and Sinks: 1990–2004*. Washington, DC: Environmental Protection Agency.

EPA (2007). *How Does Electricity Use Affect the Environment?* Washington, DC: Environmental Protection Agency. www.epa.gov/cleanrgy/energy-and-you/affect/index.htm.

Essen, H. v., Boon, B., Maibach, M. and Schreyer, C. (2007). *Methodologies for External Cost Estimation and Internalisation Scenarios*. Discussion paper for the Workshop in Internalisation on 15th March, 2007. Delft: CE Delft, Infras (Zurich).

Flyvbjerg, B., Rothengatter, W. and Bruzelius, N. (2003). *Megaprojects and Risk – An Anatomy of Ambition*. Cambridge University Press.

Health Commission (2003). *Noise and Health: Making the Link*. The London Health Commission. www.londonshealth.gov.uk/pdf/noise_links.pdf.

HySociety (2004). *D7 Report of Social Impacts of Hydrogen. Results of Work Package 2 (Task 2.4) of the HySociety Project*. www.hysociety.net.

HyWays (2007). *Roadmap*. EU Research project (Integrated Project). www.eu.fhg.de/h2database/index.html.

IAEA (2007). *Sustainable Development – Nuclear Power*. Vienna: Division of Public Information, International Atomic Energy Agency. www.iaea.org/Publications/Booklets/Development/index.html.

IEA (2006). *Prospects for Hydrogen and Fuel Cells*. IEA Energy Technology Analysis Series. Paris: IEA.

IPCC (2007). *Climate Change 2007. Impacts, Adaptation and Vulnerability*. Working Group II: Contribution to the Intergovernmental Panel on Climate Change 4th Assessment Report. Summary for Policymakers. Geneva.

Jun, M. (2007). *How Participation can help China's Ailing Environment*. China Dialogue website. Accessed from: www.chinadialogue.net/article/show/single/en/733-How-participation-can-help-China-s-ailing-environment.

Larssen, S., Adams, M. L., Barrett, K. J. *et al.* (2004). *Air pollution in Europe 1990–2000*. Report prepared for the European Environment Agency (EEA). Copenhagen.

L-B-Systemtechnik (2006). *A European Roadmap: Assumptions, Visions and Robust Conclusions from Project Phase I*. Ottobrunn, Germany: L-B-Systemtechnik.

Léautier, F. A. (ed.) (2006). *Cities in a Globalizing World. Governance, Performance & Sustainability*. WBI Learning Research Series. Washington, DC: The World Bank Institute, The World Bank.

Maibach, M., Schreyer, C., Schneider, C. *et al.* (2004). *External Costs of Transport. Update Study*. Final Report to the International Union of Railways (UIC), Paris. Zurich: Infras.

Maibach, M., Schreyer, C., Doll, C. and Bickel, P. (2007a). *Externe Kosten des Verkehrs*. Study commissioned by the Allianz Pro Schiene e.V. Frankfurt.

Maibach, M., Schreyer, C., van Essen, H. P., Boon, B. H. and Doll, C. (2007b). *Internalisation Measures and Policies for all External Costs of Transport (IMPACT): Handbook on Estimation of External Costs in the Transport Sector (2nd Draft)*. On-going study for the European Commission, DG-TREN. CE-Delft.

Mantoz, L. and Capros, M. (2006). *European Energy and Transport, Trend to 2030 – Update 2005*. Report to the European Commission. Brussels.

Marland, G., Boden, T. A. and Andres, R. J. (2007). Global, Regional, and National Fossil Fuel CO_2 Emissions http://cdiac.ornl.gov/trends/emis/em_cont.html.

Mattucci, A. (2007). *Environmental Analysis for Hydrogen Deployment*. HyWays Background Report. ENEA. www.hyways.de.

MLIT (2006). *White Paper on Transport Infrastructure and Land Use in Japan 2005*. Tokyo: Ministry of Land, Infrastructure and Transport. www.mlit.go.jp/english/white-paper/mlit-index.html.

Nakamura, H., Hayashi, Y. and May, A. D. (eds.) (2004). *Urban Transport and the Environment – An International Perspective*. World Conference on Transportation Research Society and Institute for Transport Policy Studies. Amsterdam: Elsevier.

Nash, C. *et al.* (2003). *Unification of Accounts and Marginal Costs for Transport Efficiency (UNITE). Final Report for Publication*. Project funded under the European Commission's 5th RTD programme, project co-ordinator: ITS, University of Leeds.

Nash, C. *et al.* (2007). *Generalisation of Research on Accounts and Cost Estimation (GRACE)*. Ongoing project funded under the European Commission's 6th RTD programme, project co-ordinator: ITS, University of Leeds.

NPA (2007). *Statistics 2005 Road Accidents Japan*. Abridged edition. Tokyo: Traffic Bureau, National Policy Agency.

OECD (2002). *Implementing Sustainable Urban Travel Policies*. Final Report. Paris: OECD.

SAFETEA-LU (2005). *Safe, Accountable, Flexible, Efficient Transportation Equity Act: A Legacy for Users*. Public Law 109–59. Washington, DC: Department of Transportation.

Saurat, M. (2006). *Material Flow Analysis and Environmental Impact Assessment Related to Current and Future Use of Platinum Group Metals in Europe*. Master's Thesis. Chalmers University and Wuppertal Institute.

Schade, W. *et al.* (2006). *COMPETE – Analysis of the Contribution of Transport Policies to the Competitiveness of the EU Economy and Comparison with the United States*. Final Report to the European Commission. Fraunhofer ISI, Karlsruhe.

Schade, W., Wietschel, M. and Weaver, P. (2007). *Reframing Sustainable Transport: Exploring Hydrogen Strategies Using Integrated Sustainability Assessment (ISA)*. Paper presented to L2L – Sustainable Neighbourhood – from Lisbon to Leipzig through Research 4th BMBF-Forum for Sustainability, German EU Council Presidency 8–10th May, 2007, Leipzig.

Schrank, D. and Lomax, T. (2005). *The 2005 Urban Mobility Report*. Texas Transportation Institute, Texas A&M University System.

Shackley, S., McLachlan, C. and Gough, C. (2004). *The Public Perceptions of Carbon Capture and Storage*. Working Paper 44. Manchester: Tyndall Centre for Climate Change Research, UMIST.

Stan, C. (2005). *Alternative Antriebe für Automobile: Hybridsysteme, Brennstoffzellen, Alternative Energieträger*. Berlin: Springer-Verlag.

Transport Canada (2006). *The Costs of Urban Congestion in Canada*. Transport Canada Environmental Affairs.

TRL (2007). *Technical Assistance and Economic Analysis in the Field of Legislation Pertinent to the Issue of Automotive Safety: Evaluation of the Impact (Extended Impact Assessment) of the Introduction of Hydrogen as a Fuel to Power Motor-vehicles Considering the Safety and Environmental Aspects*. Final Report for Enterprise Directorate-General, European Commission. TRL Limited.

UBA (2007). Externe Kosten kennen – Umwelt besser schätzen: Die *Methodenkonvention zur Schätzung externer Kosten am Beispiel Energie und*

Verkehr. Umweltbundesamt. www.umweltbundesamt.de/uba-info-presse/hintergrund/externekosten.pdf.

UN (2007). *Millennium Development Goals Indicators. The Official United Nations Site for the MDG Indicators*. United Nations. http://mdgs.un.org/unsd/mdg/Default.aspx.

VTPI (2007). *Transportation Costs and Benefits*. Resources for measuring transportation costs and benefits. TDM Encyclopedia. Victoria, BC: Victoria Transport Policy Institute. www.vtpi.org/tdm/tdm66.htm.

Whitmarsh, L., Wietschel, M., Jäger, J. *et al*. (2006). *MATISSE WP7.1 – Hydrogen Transport Stakeholder Workshop: Findings from Break-Out Group Discussions and Questionnaires*, MATISSE Internal Report.

20

Energy-efficient solutions needed – paving the way for hydrogen

Eberhard Jochem

Primary and final energy demand per capita or per gross domestic product (GDP) is quite high, which reflects the large losses at each level of energy conversion and use. This section stresses the fact that energy use will have to become much more efficient before hydrogen as a final energy carrier becomes attractive, given its relatively high generation cost. The option of energy and material efficiency is often forgotten, owing to a traditionally supply-oriented energy policy and the fact that efficient solutions of material end-energy use have so far remained without powerful lobbying institutions. The world of energy and material efficiency – which represents the most profitable option for many decades in this century – has to be tackled before hydrogen stands a chance of becoming a major final energy carrier and finds its place within a sustainable energy system in industrialised countries.

20.1 Present energy losses – wasteful traditions and obstacles to the use of hydrogen

In 2003, almost 450 000 PJ of global primary energy demand delivered around 295 000 PJ of final energy to customers, resulting in an estimated 141 000 PJ of useful energy after conversion in end-use devices. Thus, around 300 000 PJ – or two-thirds – of primary energy are presently lost during energy conversion, e.g., in power plants, refineries, kilns, boilers, combustion engines and electrical motors, mostly as low- and medium-temperature heat. These losses also include the small share of losses from the transmission, transformation and distribution of grid-based energies (see Fig. 20.1).

Conversion efficiencies in primary energy conversion are somewhat better in countries with high shares of hydropower (like Norway or Switzerland), but the extensive losses of thermal power plants generally determine the high energy losses in the conversion sector. The efficiency of the conversion from final to useful energy is determined, to a large extent, by the enormous conversion losses in road vehicles

The Hydrogen Economy: Opportunities and Challenges, ed. Michael Ball and Martin Wietschel. Published by Cambridge University Press.

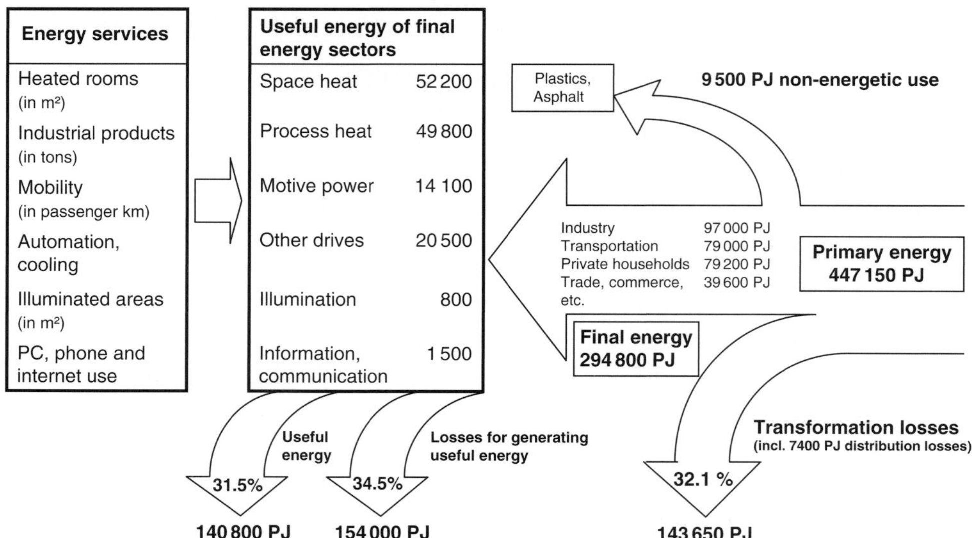

Figure 20.1. The energy system from services to useful, final and primary energy, world 2003 (OECD, 2005).

(around 80%), high-temperature industrial processes (50% losses) and illumination (90% losses, since incandescent light bulbs operate more as a heating system than a lighting source). Finally, the losses at the useful energy level are substantially determined by heat and cooling losses from buildings, whether these are for residential, commercial or factory use (see Fig. 20.1).

Since today it is costly to produce, transport and store hydrogen as a very clean secondary energy source (as compared with other liquid fuels), the amount of energy to be delivered for conversion to any useful energy or final energy will be a cost-determining factor in future energy systems. If the demand for useful energy or final energy in a particular application can be reduced by a factor of five or even ten (e.g., by passive houses, membrane or biotechnology processes instead of thermal separation or synthesis processes in industry), the amount and the cost share of the energy to be supplied can also be drastically reduced relative to that needed today. Given this dependence, it is obvious that more energy-efficient solutions are the prerequisites to promoting the use of hydrogen in several applications and sectors.

This vision cannot wait. Mankind is facing several major energy-related challenges this century: the threat and consequences of climate change, the reconcentration of crude-oil production in the Middle East, and the energy price risks of peaking oil production.

In the light of these challenges, the Board of the Swiss Federal Institutes of Technology (1998) has developed the vision of a ‘2000 Watt per capita society by the middle of the 21st century' (see also Marechal *et al.*, 2005). A yearly 2000 Watt per capita demand for primary energy corresponds to 65 GJ/capita per year, which is equivalent to one-third of today's per-capita primary energy use in Europe of some 170 GJ per capita and year. The greater inefficiencies and per capita incomes in North America mean that this figure is even higher here, around 300 GJ/capita and year, or 10 000 W/capita, while the global average is presently 2000 W per capita (OECD, 2005). Assuming a doubling of GDP (gross domestic production) per capita in Europe within the next 60 to 70 years, the 2000 Watt per capita society implies an improvement in primary energy use by a factor of four to five, admitting some influence of structural change on less energy-intensive industries and consumption patterns. This vision poses a tremendous challenge to research and development, the political system and technology producers to improve energy and material efficiency. It is obvious that completely new technologies and supporting organisational and entrepreneurial measures are needed to meet this goal. This vision could also be labelled the twin technology to hydrogen applications, in order to make the various options affordable and sustainable from the technical, economic and policy points of view.

Highly energy-efficient solutions would bring about a 'double dividend': a lower energy demand and a clean energy supply at eventually the same (or even lower) cost levels as at present. The argument that clean and sustainable energy is too expensive would be gradually eroded, the more the energy losses at all levels could be reduced.

Carbon-emitting fossil fuels such as coal, oil and natural gas presently comprise some 80% of the global primary energy demand, and produce CO_2 emissions of approximately 26 billion t per year (2006). This trend is increasing annually by 1.5%, mostly because of the fast growth of fossil-fuel use in emerging economies. Today's CO_2 emissions are already four times what the atmosphere is able to absorb in this century, assuming that the global mean surface temperature does not increase by more than 2 °C within this period. The adaptation costs and damage costs of extreme climate events have already started increasing to noticeable levels that are not included in the cost of fossil energy use (EEA, 2004; Stern, 2006).

When considering how to improve energy efficiency in the future, the focus has often been (and still is) on energy-converting technologies (e.g., thermal power plants, gas turbines or combustion engines and boilers). There are, however, three additional areas that can play a role in reducing future energy demand, which presently receive little attention:

- Energy losses at the level of useful energy (presently about 34% of the primary energy demand in Europe or almost 32% at the global level) could be substantially reduced or even avoided by using such technologies as low-energy or passive buildings, physico-chemical technologies or biotechnology processes instead of thermal or high pressure processes in industry; they could also be reduced by substantially lighter vehicles, the recuperation or storage of braking energy and more intensified heat recovery in all final energy sectors.

- Second, the demand for energy-intensive materials could be reduced by recycling or substitution of those materials, by improving their design based on new knowledge such as bionics, or by improved material properties that are still emerging from material science and nanotechnology (Stahel, 1997). The partial substitution of plastics and other chemical products based on petrochemical by-products based on biomass is also an option to reduce the dependence on fossil fuels, even though this only represents a small share of 2.6% of total fossil fuel use (Patel and Narayan, 2005; Patel 2005).
- Finally, when measured in their actual annual operating hours, many appliances, machinery, industrial plants and cars are scarcely used – only 200 to 400 hours per year; it would make sense therefore to intensify their use by pooling (e.g., car sharing, leasing of machines, as is already common with harvesting machinery, but is still in its infancy in many other fields, such as construction machines and energy-intensive industrial production processes (Fleig, 2000; Jochem *et al.*, 2002)).

Empirical and theoretical considerations suggest that the overall energy efficiency of today's industrial economies could be improved by some 80% to 90% within this century (e.g., Enquête Commission, 1991). This complies with the vision of the Board of the Swiss Federal Institutes of Technology (1998) and was confirmed by a major analysis of its technical feasibility (Jochem *et al.*, 2002; 2004). Within this context, the authors consider technological advances that lead to highly efficient energy use to be promising investments. Research and development (R&D) that furthers these technologies is a crucial prerequisite. Countries and firms that invest in these technologies and the related R&D are likely to support a sustainable growth of their economies. On top of this, if applied in the transportation sector, they will make a significant contribution to the pressing problems of climate change and the reconcentration of world oil production, help to manage the secure supply of energy and counter the risks of high energy prices, in view of peaking world conventional oil production within the next 10 to 20 years.

To conclude: there are extremely high losses associated with the current energy system in every country in the world. However, there are numerous technological options available to reduce these losses by about 2% per year over the next decades. The sooner major innovations and efficiency improvements can be realised, the earlier the opportunities for hydrogen applications are likely to materialise. The earlier industrialised countries can achieve low-energy solutions, the more likely the double dividend is of reducing greenhouse-gas emissions by reducing energy demand and the increased use of hydrogen as a clean secondary energy source as well as through a greater use of renewables instead of fossil fuels as primary energies.

20.2 How to speed up major energy-efficient innovations in material and energy efficiency

The vision of the Board of Swiss Federal Institutes of Technology (1998) may sound over-optimistic, as it implies doubling the efficiency improvements actually achieved in OECD countries (about 1% per year) over the last 35 years (IEA, 2006) to an overall annual improvement in efficient energy use of 2.0%–2.3%/year. To identify promising activities, innovation policies and research areas for major improvements in efficient energy use and conversion, short-term and long-term options have to be distinguished and the re-investment cycles of the capital stock in the various fields of energy use and conversion have to be considered as one of the major restrictions. The following analytical issues then result:

- *Organisational options and short-term investments* in existing efficiency solutions will contribute to speeding up the efficient use of energy and materials and 'buying down' the cost of new energy-efficient technologies in this and the next decade (e.g., highly efficient electrical motors, components of low-energy and passive houses, condensing boilers, heat pumps, etc.). Almost everyone is aware of the cost-decreasing effects of learning and economies of scale when it comes to renewable energies or fuel cells, but so far analyses have largely ignored the fact that many technologies of efficient energy use follow the same pattern of specific cost reductions (for example, the experience curve of highly efficient window systems or building wall insulation have a coefficient comparable to that of wind power (Jakob and Madlener, 2004)).
- In the longer term, *new technologies* may contribute significantly to a 2000 Watt per capita society in their second-generation phase. During market diffusion, this strategic option *focuses on technology substitution* rather than on technology improvements, which are often minor and incremental. Passive houses, for instance, reduce the final energy demand by a factor of eight to ten relative to the present average final energy demand of the housing stock; light and efficient cars may cut energy demand by more than 50% by recovering brake energy using power electronics and improved efficiencies from the reduced rolling of tyres and air resistance of the vehicle. Membrane technologies can reduce the energy demand of thermal separation processes by 60% to 90% in relevant industrial branches.
- The low rate of 1% annual energy-efficiency improvement over the last 35 years raises more questions. What are the major obstacles and market imperfections that have to be addressed by policy measures in order to speed up the realisation of energy and material efficiency potentials? Since obstacles and market imperfections have been the object of study for a long time (IPCC, 2001), are there any additional concepts of motivation or opportunities which have not yet been applied but which are likely to speed up activities and markets for realising energy efficiency options?

20.3 The focus on major improvements in energy-efficient solutions

Although short-term energy-efficiency improvements may be welcome to reduce energy costs and CO_2 emissions in the next two decades, these still have to be checked against the criteria of sustainable development. On the one hand, minor efficiency improvements today in long-lived capital stocks, such as buildings, railways, roads, or central power plants with re-investment cycles of 40 to 60 years may lead to a lock-in situation (e.g., a building not insulated during refurbishment now will generate high energy costs over the next 50 years); on the other hand, costly efficiency

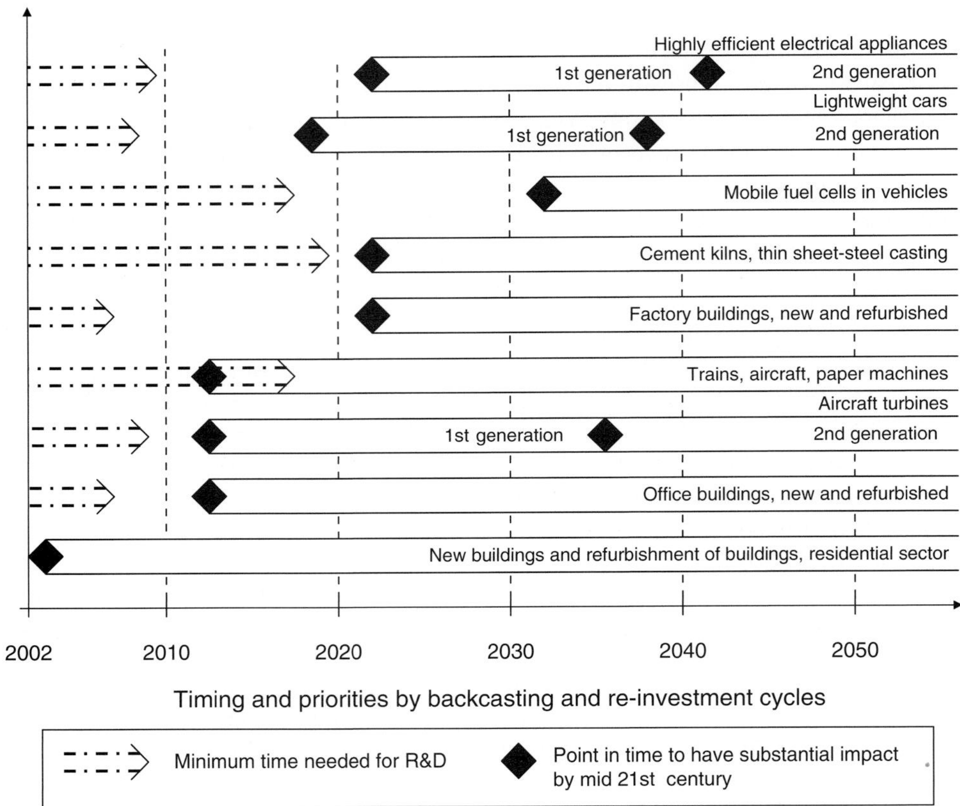

Figure 20.2. Timing and priority setting of efficiency policies and R&D using backcasting and re-investment cycles to avoid lock-in pathways.

improvements – possibly subsidised by tax payers or energy consumers – in product areas with re-investment cycles of three to five years may result in a waste of resources if they are expected to be much less expensive and more profitable in four to six years time.

The technique of backcasting can be used as an aid to prioritise short-term options and long-term innovations based on their re-investment cycles and to time their introduction correctly. This helps to avoid lock-in technology pathways by examining re-investment cycles and the necessary R&D periods that must be considered (see Fig. 20.2). Fuel cells are one example which may ultimately generate too many CO_2 emissions after 2040 if the hydrogen used to power them is still produced from fossil fuels. The concept of backcasting has to be combined with the usual bottom-up models as a new way to identify the role of energy efficiency policy in the context of a sustainable energy system (IEA, 2006). As the challenges posed by dwindling oil resources and climate change are pressing, the strategic concern here is less 'minimal cost at a given level of emissions', but rather a 'minimal time span reducing specific energy demand by a factor of four to five at acceptable cost'.

There are numerous, profitable short-term investments and organisational measures for more efficient energy use in every final energy sector and the transformation sector (IEA, 2006; Romm, 1999; UNDP, 2000). There are even many very simple organisational measures that can avoid losses of useful energy or energy for conversion, which are unknown or not considered in many cases, such as avoiding the loss of compressed air or heat in many applications and industrial sites or the high stand-by losses in many electrical appliances. Realising these potentials in the short term supports the awareness of and search for additional efficiency improvements and new efficient solutions as part of a new cultural tradition targeting a sustainable energy system in the long term. However, the final objective should be to focus on new highly efficient technologies that enable renewable energies and clean secondary energies like hydrogen to penetrate markets at an acceptable social cost. How can this be realised? What are the obstacles to be overcome? How can actors be motivated?

If the efficiency potentials are as high as a factor of four or five in the long term, from a theoretical point of view, then the process of R&D – both its decision making and its performance – has to be improved. This means creating a process that identifies the present technical and cost bottlenecks of a new efficient technology with greater accuracy and which makes the right selection of promising new and efficient technologies (including their acceptance by the target groups involved in the innovation process).

In general, to be selected for further R&D efforts, a technological field has to meet the following selection criteria (Jochem *et al.*, 2006):

- A minimum energy demand of at least 0.2% to 0.3% of a country's primary energy demand at present, or a similar percentage to be realised by the new technology in 2050. This share seems small, but is actually significant given the hundreds of technologies with substantial efficiency potentials.
- An envisaged energy-efficiency potential of at least 20% to 25% in the field of energy-conversion technologies and more than 50% at the level of useful energy and material efficiency. This efficiency step is needed to pay off the R&D efforts, in order to be able to meet the global challenges mentioned in the first section of this chapter.

Beside these selection criteria, the analysis used evaluation criteria to cover quantifiable and non-quantifiable aspects:

- the position in the technology cycle, major technical bottlenecks and the position relative to competing or traditional technologies;
- cost reduction potentials of the new technologies due to learning and economies of scale effects;
- a currently favourable (or achievable in the future) export position of German or European technology producers able to manufacture the new technology;
- perceived favourable acceptance of the new technologies or market obstacles that can be overcome during the next decades; and finally
- the timing of re-investment cycles and the length of R&D time necessary for possible market introduction.

These selection and evaluation criteria were applied systematically to four technological fields, three of which contribute to new energy-efficient solutions. Passive houses, for example, with their major components of insulation solutions, window systems, ventilation and control techniques are close to market diffusion within the next ten years. Fuel cells for mobile uses in vehicles, however, are still a long way from market introduction, for instance, because of unresolved problems regarding the deactivation of the membrane electrode assembly (MEA) and the need for cost reductions by about one order of magnitude. Other types of fuel cells for stationary uses may be closer to market introduction, owing to less severe technical bottlenecks and better economic competitiveness.

20.4 Obstacles and market imperfections – but also motivations and opportunities

Consulting engineers usually return from onsite visits to companies having located substantial energy-efficiency potentials that are easy to realise and usually have high rates of internal return (Romm, 1999). Even the energy managers of large companies are often not informed about all the new innovations of efficient-energy use. The limited realisation of profitable efficiency potentials has already been the subject of discussions about obstacles and market imperfections for a decade (IPCC, 2001), and the heterogeneity of these obstacles and potentials has been tackled by different sets of policy measures and instruments.

Surveys and interviews show that the attention paid to energy-efficiency investments in companies, public administrations and private households is often very low and heavily influenced by the priorities of those responsible for decision making (Ramesohl, 2000; Schmid, 2004; Stern, 1992). In other cases, project-based economic evaluations do not consider the relatively high transaction costs of the investor and also the substantial risks involved in the case of long-term investments; both aspects may be decisive for small efficiency investments (Ostertag, 2003).

Traditional investment priorities steer the motivation and behaviour of staff and determine the careers of young engineers and their efforts; energy engineers often have difficulties in 'making a convincing case' to the management (Schmid, 2004). The co-benefits of energy-efficient new technologies are rarely identified and not included in profitability calculations by the energy or process engineers, because of a lack of a systemic view of the whole production site and possible changes related to efficiency investments.

Besides the economic reasons behind this priority setting of companies, public administrations and private households, there are also psychosocial, motivational and behavioural aspects, which have rarely been analysed except by some sociologists and psychologists in the 1990s (e.g., Jochem *et al.*, 2000; Stern, 1992). Social relations, such as competitive behaviour, mutual estimation and acceptance, not only play a role between, but also within companies. Efforts to improve energy efficiency are influenced by the intrinsic motivation of a company's actors and decision makers, the interaction between those responsible for energy and the management, the internal stimuli of key actors, and their prestige and persuasive power (InterSEE, 1998; Schmid, 2004).

20.4.1 Widening the view from energy-related research to innovation systems

Obstacles and market imperfections still have the flavour of a mechanical concept about them. The implication is that they simply have to be removed and proper frame conditions set. However, accelerated innovation and more effective R&D will only become a reality if the existing innovation system is ready to consider new methods, support first movers and entrepreneurial solutions and invest in R&D and new products and technologies. The existence of a proven technology's energy-saving potential alone does not further the 2000 Watt per capita society. Energy is actually only saved once a technology's (behavioural) potential has materialised as a result of research and development, as well as thorough innovation policies, and the technology is being broadly marketed and used.

Therefore, the research and innovation system of a country has to be analysed and must be convinced by the opportunities and the new vision of a 2000 Watt per capita society. Any recommended efficiency-policy portfolio and R&D efforts have to be evaluated within the context of the relevant research and innovation boundary conditions of the actors and institutions involved (see Fig. 20.3). The research and innovation systems of a country encompass the 'biotopes' of all the institutions that are:

- Engaged in scientific research and the accumulation and diffusion of knowledge (e.g., in Switzerland, the federal research institutions, such as the two Institutes of Technology (ETH Zürich and EPF Lausanne), the national laboratories (Paul Scherrer Institute, EAWAG and the material science laboratory EMPA), universities and schools);
- Engaged in education and professional training as well as the dissemination of new knowledge to a broader audience (i.e., educational institutions, media); and
- Developing and producing new technologies, processes and products; and commercialising and distributing these (e.g., intermediates, infrastructure, technology producers and trade).

An innovation system also comprises relevant policy institutions, in Switzerland, for example, the Office of Energy (BFE) or the National Science Foundation, which are responsible for setting the economic, financial (e.g., venture capital), and legal boundary conditions and regulations (standards, norms) as well as the public and private investments in the appropriate infrastructure. Each innovation system of a country (and even of a sector or a technological field within a country) is unique and has developed its profiles and strengths only over the course of decades. Each is based on stable exchange relationships and interactions among the institutions of science and technology, industry, commerce, and the political system (Edquist, 1997).

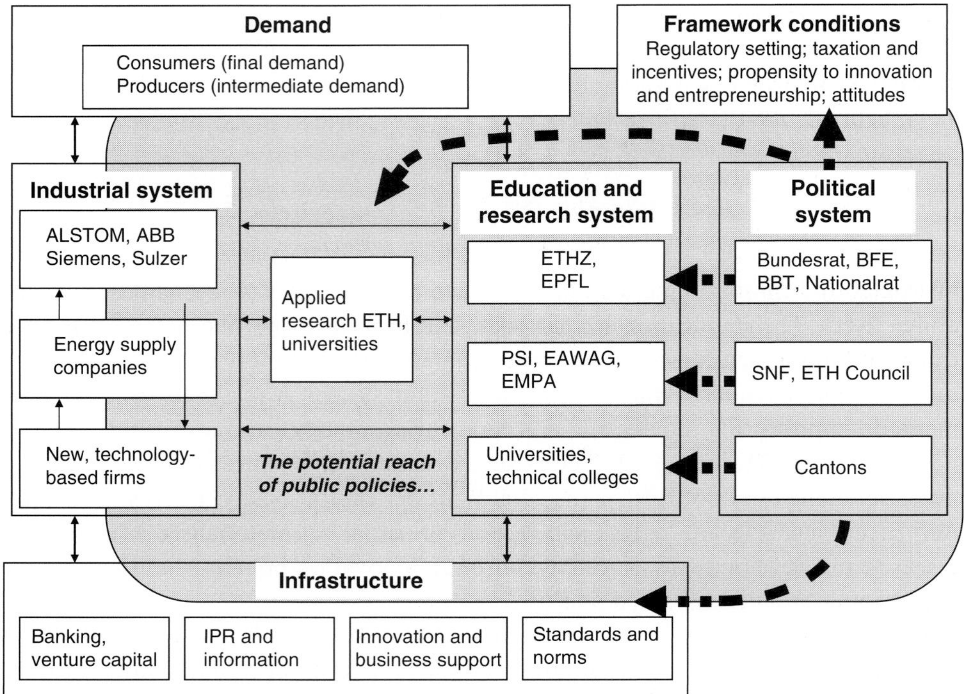

Figure 20.3. Scheme of Swiss energy and energy-efficiency research and the related innovation system (source: Fraunhofer ISI).

Since energy and material efficiency is dispersed over all the sectors of an economy and private households, the efficiency innovation system is characterised by:

- a high degree of compartmentalisation (e.g., buildings, road transportation, industrial branches, energy companies) and a corresponding sectorisation of the political administration with low interdepartmental exchange and co-operation;
- non-interlinked arenas (corporatist negotiation deadlocks involving the sovereignty of regions in federal states (e.g., 26 cantons in Switzerland, in cases such as building codes), or of member states of the European Union and related failed attempts at restructuring responsibilities in governments or at the EU level);
- dominance of a 'linear model' of energy supply in political approaches (and among related technologists, energy economics researchers and consultants), which focuses on energy supply options (such as costly renewables or fusion energy, for which the technical and economic feasibility will remain an open question for many decades to come) and tends to neglect efficiency opportunities at the useful energy and energy-service levels.

These characteristics of the efficiency innovation system are general in nature and indeed almost independent of the specific country considered, but they are also highly dependent on the ubiquity and heterogeneity of energy and material efficiency. The weaknesses of under-co-ordinated innovation policy making, which seem to prevail in the energy and material efficiency field, should be analysed in more detail. Topics here include poorly articulated demand and weak networks that hinder fast knowledge transfer, legislation and market boundary conditions which favour incumbent technologies (with high external costs), flows in the capital markets (focusing on large-scale technologies and players); and poorly organised actors.

The organisational structure of the 'energy-efficiency community' and the related research and innovation system obviously requires vast improvement when compared with energy supply communities, whether thermal power plants, renewables, hydrogen or fusion energy. The community supporting hydrogen might be much more effective if they followed a joint policy strategy with the efficiency community, which still has a very low lobbying capacity.

20.5 Policy aspects supporting the efficiency path to a sustainable energy system

Although the technical feasibility of a 2000 Watt per capita society in the second half of this century has been clarified and its economic feasibility is likely to be achievable at present energy price levels, its practical realisation remains an open question. The external cost of fossil-fuel use and many obstacles at sectoral and technology levels still hinder a fast implementation, as do a limited perception of the motivation and opportunities of possible first movers and supporting actors and a limited understanding of the innovation system.

Pre-conditions for success include research on innovation-focused and co-ordinating roles for government, addressing the large portfolio of technologies and innovations, reinforcing user–producer relations, supporting the formation of new networks, stimulating learning and economy-of-scale effects, as well as the articulation of demand and prime movers. Research in these fields will have to involve evolutionary economics, the sociology of organisation and science, political science and management science.

In 2006, the German government, for instance, made a decision to improve energy productivity by 50% between 1990 and 2020; this is equivalent to an annual 2.8% increase in primary energy productivity. With the selection of this extremely ambitious target, the government has set off along a technological path where hydrogen use could play a role. However, it will require wise decisions to trigger a policy process capable of accelerating the efficiency progress towards the targeted path.

20.5.1 The European Commission's Action Plan for Energy Efficiency

For the first time, I have the impression that energy-efficiency potentials are being clearly recognised and adequately addressed by the *Action Plan for Energy Efficiency* of the European Commission (Commission of the European Communities, 2006). This is a policy programme that supports energy-policy objectives (low energy costs, energy security, environmentally benign energy use), as well as innovation, employment and environmental policies. The Action Plan has the potential to speed up the innovative capacity in efficient energy use in Europe by leading the policy process, assuring a harmonised setting of boundary conditions in Europe and inducing greater competition among administrations of the member states, as well as technology producers and service sectors.

The *Action Plan for Energy Efficiency* clearly indicates that the initiative in energy efficiency issues is being increasingly focused at the EU level. This may be a wise strategic policy process, as many products are traded, many companies have plants in several EU member states and many consumers cross national borders quite often or even regularly. The need for harmonisation of the efficiency markets and technologies within the EU is increasingly being promoted by the Commission. The recently published Action Plan is 'intended to mobilise the general public and policy makers at all levels of government, together with market actors, and to transform the internal energy market in a way that provides EU citizens with the globally most energy-efficient infrastructure, buildings, appliances, processes, transport means and energy systems'.

The Action Plan does not set an efficiency target (like the German government), but cites energy efficiency potentials between 25% and 30% for individual sectors, such as buildings, road and air transportation and industry, and suggests a 2.5% annual improvement of the primary energy intensity between 2005 and 2020. The plan identifies ten priority actions, comprising appliance and equipment labelling and minimum energy-performance standards, building-performance requirements, more efficient power generation and distribution, high fuel efficiency of cars, and supporting small and medium-sized companies with several measures.

The priority actions also include specific actions for energy efficiency in the new member states, a coherent use of taxation (which certainly needs more attention), specific activities in large cities and fostering energy efficiency worldwide through collaboration with international organisations. The Action Plan does not explicitly include research and development issues or a time horizon longer than 15 years. This indicates that even the strategic concept of European energy policy lacks the longer-term perspective of the energy-efficiency path, despite the fact that much improved energy efficiency is actually a prerequisite for future sustainable forms of energy supply such as renewables and hydrogen use. The Chinese government also adopted a strong position late in 2006, when it decided to target a 4% improvement in annual energy productivity for the next five years.

20.6 Summary

The organisational structure of the 'energy efficiency community' and the related research and innovation system obviously have to be much improved when compared to energy supply communities – whether thermal power plants, renewables, hydrogen or fusion energy. The community promoting hydrogen might have greater success if they followed a joint policy strategy with the efficiency community, which still has a very low lobbying capacity. As highly efficient energy use can be considered the twin of hydrogen use in the long term, an energy efficiency policy should be particularly supportive of major reductions of energy losses realised by new technologies, substituting the traditional conversion technologies from final to useful energy. Examples of such technologies include remote, low-energy buildings and highly efficient, light vehicles, particularly in road and air transport applications. Energy efficiency is the key for the solution of our actual energy problems.

References

Board of the Swiss Federal Institutes of Technology (1998). *2000 Watt Society – Swiss Model. Sustainability Strategy within the Swiss Institutes of Technology* (in German). Wirtschaftsplattform. ETH Zurich.

Commission of the European Communities (2006). *Action Plan for Energy Efficiency: Realising the Potential.* COM(2006)545 final, 19th October, 2006. Brussels.

Edquist, C. (1997). *Systems of Innovation: Technologies, Institutions and Organisations.* London: Pinter Publishers.

EEA (European Environmental Agency) (2004). *Impacts of Europe's Changing Climate.* Report No. 2/2004, Copenhagen. ISBN 9789291676927.

Enquête Commission (1991). *Protecting the Earth. A Status Report with Recommendations for a New Energy Policy.* Bonn: Bonner University Press.

Fleig, J. (2000). *Zukunftsfähige Kreislaufwirtschaft. Mit Nutzenverkauf, Langlebigkeit und Aufarbeitung ökonomisch und ökologisch wirtschaften.* Stuttgart: Schäffer-Poeschel.

IEA (International Energy Agency) (2006). *Energy Technology Perspectives 2006 – Scenarios and Strategies – In Support of the G8 Plan of Action.* Paris: OECD.

InterSEE (1998). *Interdisciplinary Analysis of Successful Implementation of Energy Efficiency in Industry, Commerce and Service.* Wuppertal Institut für Klima Umwelt Energie, AKF-Institute for Local Government Studies, Energieverwertungsagentur, Fraunhofer Institut für System- und Innovationsforschung, Institut für Psychologie der Universität Kiel, Amstein&Walthert, Bush Energie. Wuppertal, Kopenhagen, Wien, Karlsruhe, Kiel.

IPCC (Intergovernmental Panel on Climate Change) (2001). *Climate Change 2001 – Mitigation: Contribution of Working Group III to TAR of the IPCC.* Cambridge: Cambridge University Press.

Jakob, M. and Madlener, R. (2004). Riding down the experience curve for energy-efficient building envelopes: the Swiss case for 1970–2020. *International Journal of Energy Technology and Policy* (Special Issue on Experience Curves), **2** (1–2), 153–178.

Jochem, E., Sathaye, J. and Bouille, D. (2000). *Society, Behaviour, and Climate Change Mitigation.* Advances in Global Change Research. vol. 8. Dordrecht: Kluwer Academic Publishers.

Jochem, E., Favrat, D., Hungerbühler, K. *et al.* (2002). *Steps Towards a 2000 Watt Society. Developing a White Paper on Research & Development of Energy-Efficient Technologies*. ETH Zurich: CEPE.

Jochem, E., Andersson, G., Favrat, D. *et al.* (2004). *Steps Towards a Sustainable Development – A White Book for R&D of Energy-Efficient Technologies*. Zurich: CEPE/ETH and Novatlantis.

Jochem, E., Bradke, H., Cremer, C. *et al.* (2006). Developing an assessment framework to improve the efficiency of R&D. Karlsruhe: Fh-ISI.

Marechal, F., Favrat, D. and Jochem, E. (2005). Energy in the perspective of the sustainable development: the 2000 Watt society challenge. *Resources, Conservation and Recycling*, **44** (3), 245–262.

OECD (2005). *Energy Balances of OECD Countries*. Paris: OECD/IEA.

Ostertag, K. (2003). *No-Regret Potentials in Energy Conservation – An Analysis of Their Relevance, Size and Determinants*. Heidelberg: Physica.

Patel, M. (2005). Environmental life cycle comparisons of biodegradable plastics. Chapter 13. In *Handbook of Biodegradable Polymers*, ed. Bastioli, C. Shawbury, UK: Rapra Technology Ltd. pp. 431–484.

Patel, M. and Narayan, R. (2005). How sustainable are biopolymers and biobased products? The hope and the reality. In *Natural Fibres, Biopolymers and Their Biocomposites*, ed. Mohanty, A. K., Misra, M. and Drzal, L. T. Boca Raton (USA): CRC Press, pp. 833–853.

Ramesohl, S. (2000). Social interactions and conditions for change in energy-related decision making in CMCs – an empirical socioeconomic analysis. In *Society, Behaviour and Climate Change Mitigation*, ed. Jochem, E., Sathaye, J. and Bouille, D. Advances in Global Change Research, vol. 8. Dordrecht: Kluwer Academic Publishers, pp. 207–228.

Romm, J. (1999). *Cool Companies*. London: Earthscan.

Schmid, Ch. (2004). *Energieeffizienz in Unternehmen: Eine handlungstheoretische und wissensbasierte Analyse von Einflussfaktoren und Instrumenten*. Dissertation, Vbf Zurich.

Stahel, W. R. (1997). The service economy: wealth without resource consumption? *Philosophical Transactions of the Royal Society A*, **355**, 1386–1388.

Stern, N. (2006). *The Economics of Climate Change. The Stern Review*. New York: Cambridge University Press.

Stern, P. C. (1992). What psychology knows about energy conservation. *American Psychologist*, **47**, 1224–1232.

UNDP, WEC and UNDESA (2000). *World Energy Assessment*. New York: UNDP.

21

The future of hydrogen – opportunities and challenges

Martin Wietschel and Michael Ball

21.1 Context – the energy challenge of the future

Today's energy and transport systems, which are based mainly on fossil energy carriers, can in no way be evaluated as sustainable. Given the continued growth in the world's population as well as the progressive industrialisation of developing nations, particularly in Asia, but also in South America, the global demand for energy is expected to continue to escalate in the coming decades – by more than 50% until 2030, according to the International Energy Agency (IEA) – with fossil fuels continuing to dominate global energy use. At the same time, there is a growing global consensus that greenhouse-gas (GHG) emissions, which keep rising, need to be managed, to prevent dangerous anthropogenic interference with the climate system. Hence, security of supply and climate change represent two major concerns about the future of the energy sector, which give rise to the challenge of finding the best way to rein in emissions while also providing the energy required to sustain economies. Concerns over energy supply security and climate change, as well as local air pollution and the increasing prices of energy services, are having a growing impact on policy making throughout the world.

The transport sector today accounts for some 18% of primary energy use and some 17% of global CO_2 emissions, with the vast majority of emissions coming from road transport. Transport is also responsible for 20% of the projected increase in both global energy demand and greenhouse-gas emissions until 2030. At present, oil is still the largest primary fuel, with a share of more than one third in the global primary-energy mix and more than 95% of transport energy demand. Any oil-supply disruptions would, therefore, hit the transport sector hardest since, worldwide, it is almost entirely dependent on oil. Moreover, there is a high geographical concentration of oil as well as a growing import dependency on a few, often politically unstable countries (at least from the western world's point of view). Mounting anxiety about the economic and geopolitical implications of

The Hydrogen Economy: Opportunities and Challenges, ed. Michael Ball and Martin Wietschel. Published by Cambridge University Press.

possible shortages in the supply of oil as a pillar of our globalised world based on transportation and the need to reduce greenhouse-gas emissions in the transport sector are triggering the search for alternative fuels.

21.2 The challenge for road transport

21.2.1 Shortage of cheap oil – a driver for a paradigm shift in the transport sector?

Oil and gas still make the world work; fossil fuels still account for about 80% of today's global primary energy supply. The worldwide demand for oil has reached new heights, led by China and other rapidly industrialising countries. The shrinking margin between oil production capacity and demand is largely responsible for the rapid rise in oil prices in recent years. Given the extent to which the industrialised world has come to depend on oil as a pillar of its economy, possible shortages in the supply of oil as a consequence of declining production are likely to result in abrupt and disruptive changes.

There will always be considerable uncertainty of how much oil still exists under the Earth's surface and how much can be recovered. There is a long history of failed forecasts regarding the peaking of oil production, and experience shows that reserves are usually underestimated. However, there are compelling reasons why current projections can be taken to be more reliable than previous ones. For instance, global production has been exceeding new discoveries since the 1980s and the size of new discoveries is also decreasing.

The most recent world energy scenarios project that global oil demand will grow by more than one-third until 2030. According to this projection, cumulative oil production will have to almost double to meet the rising demand. The analysis in this book reveals a mismatch between these growth scenarios and the remaining potential of conventional oil: if we continue with business as usual, we are likely to face shortcomings in the supply of oil in the coming decades. The analysis further suggests that conventional oil production will peak some time between 2010 and 2020. Moreover, the dependency on oil imports in all the major importing regions is projected to increase in the future, especially the reliance on OPEC countries, since these hold around 75% of the known remaining reserves.

The analysis of resource potential vs. growth in oil demand clearly indicates that it is time to develop alternatives to oil as the main fuel for the transport sector. This leads to the question of how the growing transport energy demand can be met in the long term when conventional (easy) oil runs out and crude prices remain high. Next to energy-saving strategies, what are the choices? The principal options being discussed include unconventional oil from oil sands or oil shale, synthetic liquid fuels (Fischer–Tropsch fuels) on the basis of gas or coal, biofuels, electricity as a 'fuel' for plug-in hybrid electric vehicles (PHEVs) or battery-electric vehicles (BEVs), and hydrogen for use in fuel-cell vehicles (FCVs).

21.2.2 Climate change and other environmental impacts

Transport systems perform vital societal functions, but in their present state cannot be considered 'sustainable'. Particular concerns in this respect include local air pollution (particulate matter, ozone), climate change, congestion, land use, accidents and noise. Local air pollution, especially from road transport, is quickly becoming a major issue for urban air quality, particularly in the world's growing megacities. At a global level, greenhouse-gas emissions from the transport sector and from fuel production represent another major problem and are increasingly subject to regulation around the world. Since the 1970s, GHG emissions from the transport sector have grown by more than 120% worldwide, and most scenarios predict that this trend will continue in the future.

The increasing global demand for fuel is one of the main reasons for the rise in greenhouse-gas emissions. Emissions of CO_2, the main greenhouse gas from human activities, are the subject of a worldwide debate about energy sustainability and the stability of the global climate. Evidence that human activities are causing the planet to warm up is now unequivocal according to the Intergovernmental Panel on Climate Change (IPCC). To meet stringent climate change targets, such as stabilising CO_2-equivalent concentrations below 550 ppm, or limiting the global temperature rise to 2 °C above pre-industrial levels requires drastic CO_2 reductions of 60% to 80% in 2050 compared with 1990 emissions, which is a daunting challenge. This will require a portfolio of technologies and mitigation activities across all sectors, such as improving energy efficiency, carbon capture and storage (CCS) and the use of renewable energies or nuclear power. Deep cuts in emissions will also be required in the transport sector. Implementing CO_2-emissions-reduction measures in the transport sector is often accompanied by the co-benefits of reducing traffic congestion and/or improving air quality.

21.2.3 The options

Resolving the problems of road transport requires new solutions for transport energy use. The principal options are demand-side measures, more efficient vehicles and cleaner fuels. The former aim at reducing transport demand and using vehicles more efficiently, and primarily include transport demand management (TDM), like city taxes, road and parking pricing, a modal shift from private car use to public transport, park & ride, car sharing or promoting cycling and walking, but also improved freight logistics, shifting freight transport from roads to rail, teleworking as well as improved driving habits. In the near and medium term, smaller cars, more lightweight and aerodynamic construction, improved conventional internal combustion engines (ICEs), hybridisation or dieselisation can all further improve the fuel economy of vehicles and help reduce fuel consumption and emissions. To give a theoretical example of what an improved fuel economy of vehicles could achieve:

a dieselisation of the entire US light-duty vehicle fleet, or likewise replacing the current US gasoline vehicle fleet with more efficient European-like gasoline vehicles, would result in fuel savings of as much as two to three million barrels of oil per day. Nevertheless, longer-term strategies must focus on developing alternative fuels and propulsion systems.

An important differentiator between alternative fuel options is the carbon footprint of the fuel supply chain, when considering their full production life cycle, i.e. the well-to-wheel (WTW) CO_2-eq. intensity. Typically, for today's conventional liquid fuels, 15–20% of WTW CO_2-eq. emissions result from the fuel supply and 80–85% from the fuel use (combustion) in the vehicle. Reducing or even eliminating CO_2 emissions from the vehicle part (as is the case for hydrogen and electricity), therefore generally has the single biggest impact on overall reduction of GHG emissions in the transport sector. Incentives also need to be given to motivate car manufacturers to produce more low-fuel consumption and CO_2-efficient vehicles and to encourage people to buy cars with reduced fuel consumption. However, mobility is one of the major drivers of economic growth and societal development, so reducing energy demand and CO_2 emissions from transport, especially from personal transport, is a particular challenge.

21.3 Alternative fuels and propulsion systems

21.3.1 Getting on with liquid fuels

The present level of oil prices, growing concerns about whether world oil supplies will be able to meet increasing demand, especially from the developing economies of Asia, as well as the mounting number of countries experiencing declines in conventional oil production are prompting significant investments in oil sands and a renewed interest in oil shale, as well as in synthetic fuels from gas (gas-to-liquids, GTL) and coal (coal-to-liquids, CTL). There is also a significant push for biofuels taking place around the world. Interest in these alternatives is also motivated by energy-security concerns, which tend to stimulate a greater reliance on indigenous energy resources, which often result in increased greenhouse-gas emissions. Nevertheless, all these fuels have in common that they are simple to handle, have a high volumetric energy density, are easy to store on board a vehicle and can use the existing distribution and refuelling infrastructure.

Despite the considerable growth of the Canadian oil sands industry in recent years, there are still several difficulties that could impede the future development of this industry; for instance, the heavy reliance on natural gas and water, which are necessary for both the extraction of bitumen from oil sands and its upgrading to synthetic oil, as well as the associated high emissions of CO_2. For nearly a century, the oil shale in the western United States has been considered a possible substitute source for conventional crude oil. If a technology can be developed to recover oil from oil shale, economically the quantities would be in the range of today's conventional oil reserves. But the economics of shale-oil production have persistently remained behind conventional oil. The prospects of oil-shale development are uncertain and many issues related to technology performance, and environmental and socioeconomic impacts remain unresolved. The potential resource base of both oil sands and oil shale is vast, but their extraction generally comes at a much higher energy penalty and CO_2 intensity than conventional oil production – CO_2 emissions from Canadian oil sands production and upgrading, for instance, may be up to three times higher than from conventional crude oil extraction, unless CCS is applied – and may also result in detrimental environmental impacts, such as loss of biodiversity.

While synthetic fuels (Fischer–Tropsch fuels) can be designed for optimal combustion in the engine and thus significantly reduce local emissions (e.g., from low sulphur content and low particle emissions), the production of synthetic fuels from fossil energy sources is much more CO_2 intensive than conventional refining; in the case of CTL, more than ten times as intensive (without carbon management). Moreover, solely from a thermal process efficiency point of view, the syngas route favours the production of hydrogen rather than Fischer–Tropsch fuels (neglecting infrastructure build-up and vehicle availability), as this does not require Fischer–Tropsch synthesis, which itself has substantial hydrogen requirements, especially for the fuel synthesis on the basis of feedstocks with low hydrogen: carbon ratios, such as coal and biomass. At present, the methanol route also has a low energy efficiency, because of the energy-consuming intermediate step of synthesis gas. If a catalyst is discovered that is capable of converting methane directly and easily to methanol, then this would also be a very attractive fuel for use in direct-methanol fuel cells (DMFC). However, as things stand at present, hydrogen can be used directly as a secondary energy carrier in fuel cells without emitting any CO_2, and hence it does not make sense to convert hydrogen to methanol (another secondary energy carrier), which *does* emit CO_2 when used in a fuel cell.

Growth prospects for any unconventional oil source will depend, to a large extent, on the prices for conventional hydrocarbons and on environmental constraints. The potential of synthetic fuels such as those derived from coal or natural gas is also in question due to the highly capital-intensive nature of the Fischer–Tropsch conversion process and these fuels will have difficulty in achieving scale without major innovation and significant advances in technology. If the cost of producing unconventional oil becomes competitive with the cost of oil from conventional sources – either from technological improvements or higher oil prices – and the environmental problems can be overcome, then unconventional fuels will find a place in the fossil-fuel market in the future. If oil prices remain at relatively high levels, unconventional-fuel production (including GTL and CTL) could reach between 7 and 15 Mb/day in 2030, i.e., between 6% and 13% of the total projected oil production at this time, of which around a third would come from oil sands. In the near and medium term, unconventional oil may delay the mid-depletion point of oil production for a short time, but the global decline of production cannot be prevented long term if demand continues to surge.

Today, owing to policy support schemes, 'first-generation' biofuels (biodiesel and bioethanol) are gaining relevant market shares in some world regions, such as Europe or the USA, as a means to reduce transport-related GHG emissions and enhance supply security. However, there are various concerns associated with the supply of biofuels (particularly 'first-generation' biofuels), which challenge their overall sustainability and may constrain large-scale production: net reduction of GHG emissions, competition for water resources, use of pesticides and fertilisers, land use, impacts on biodiversity (such as loss of rainforest) as well as competition with food (crop) production for arable land availability, which may drive up food and fodder prices. As a consequence, biofuel mandates, especially in the EU, are being scrutinised and reviewed from a sustainability perspective. Moreover, biomass use for transportation fuels is increasingly in competition with stationary heat and power generation for feedstock availability. 'Second-generation' biofuels, such as lignocellulosic ethanol, could potentially extend the feedstock base and avoid interference with the food chain. However, more R&D is needed before these 'second-generation' biofuels become commercially available.

Tailpipe CO_2 emissions from biofuels are not much different from those for gasoline and diesel, but as the CO_2 released has previously been fixed by photosynthesis in the plants, biofuel combustion is generally considered to be CO_2 neutral. But the overall balance of GHG emissions over the entire supply chain of biofuels depends on several factors, such as crop type and yield, the amount and type of energy embedded in the fertiliser production and related emissions, the emissions impact of land-use and land-use change as well as the (fossil) energy used to harvest and transport the feedstock to the biorefinery and the energy intensity of the conversion process. In the case of biodiesel from palm oil from Indonesia for example, the GHG emissions are several times higher than is the case for conventional diesel production. Besides this, from the overall perspective of CO_2 reduction in the energy sector, the massive extension of biofuel production must be critically considered, as biomass can be used up to three times more efficiently in heating and combined heat and power than in producing the currently used biodiesel and bioethanol. While this also holds true for hydrogen, biomass yields more kilometres when used via hydrogen in fuel-cell cars than liquid biofuels in ICE cars; moreover, as hydrogen is produced via gasification, it can be considered a second-generation biofuel.

Biofuels are appealing as, once produced, they require only limited changes in infrastructure and the performance and costs of a vehicle powered by biofuel are not substantially different from those of a fossil-fuel powered vehicle. But biofuels alone cannot solve the dual problem of meeting a growing transport energy demand and reducing emissions. Biofuels (including biohydrogen) have only a limited ability to replace fossil fuels and should not be regarded as a 'silver-bullet' solution to reducing transport emissions. Biomass availability, competition for end uses as well as socioeconomic and environmental implications all place limits on biofuel use. The emergence of the 'food-versus-fuel' debate adds further constraints, At a global level, it is estimated that biomass-derived fuels could substitute a maximum of 20%–30% of today's total vehicle fuel consumption.

Another alternative (though non-liquid) fuel already being used in many countries is (compressed) natural gas (CNG). Of all the alternative fuels (apart from hydrogen and electricity), natural gas achieves the greatest reduction, of 20%–25%, in vehicle emissions of CO_2. The primary requirement for CNG is the implementation of new refuelling stations, as a natural gas distribution infrastructure is already largely in place in many countries. As an interim solution, in response to the lack of a widespread availability of fuelling stations, bi-fuel vehicle concepts are being introduced. Nevertheless, as the benefits of CNG, as well as of LPG (liquefied petroleum gas), are unlikely to offset the costs associated with further development of the refuelling infrastructure, vehicle conversions and safety issues, they will only play a limited role in countries with favourable market conditions. In addition, in the long term, natural gas will face the same resource–economic constraints as crude oil.

21.3.2 The prospects of electric mobility

Triggered, among other things, by the development of hybrid vehicles, there is renewed interest from both car manufacturers and electric utilities in electric vehicles as a means to reduce emissions and a lot of research is being done on the development of new battery types. It is possible to rank these vehicles by increasing battery involvement in the propulsion system and thus extended electric driving range as follows: hybrid-electric vehicles (HEV), plug-in hybrid-electric vehicles (PHEV), both of which incorporate an ICE (or maybe later on a fuel cell), and pure battery-electric vehicles (BEV) without an ICE.

Plug-in hybrid-electric vehicles and battery-electric vehicles are major competitors of hydrogen fuel-cell vehicles. Like hydrogen vehicles they shift emissions 'upstream' to the 'fuel' production, and when driven in the electric mode, they are zero-emission vehicles too. Kilometres driven electrically are also lower in well-to-wheel CO_2 emissions than those driven by conventional fuels (even if the electricity is taken from new coal-fired power stations). The switch to electricity also reduces the oil dependency of the transport sector by opening it up to the much wider portfolio of primary energy sources of the power sector. If renewable or carbon-neutral electricity is used for recharging, this is by far the cheapest option to reduce GHG emissions from road transport taking only fuel costs into account. Hydrogen FCVs and electric vehicles can thus help reduce CO_2 emissions and local air pollution and alleviate energy security concerns, while at the same time offering a potential storage option for surplus electricity from intermittent renewable energies.

The main attraction of this option is that the 'fuel supply' infrastructure, i.e., the electricity grid, already exists, although an extensive network of recharging stations would still need to be implemented, with all potential recharging options – slow charging (at home or in public), fast charging and battery swapping stations – having their own specific challenges, not least with respect to practicability and customer acceptance. In addition, the impact of large-scale electrification of the transport sector on the electricity system must be analysed and understood. This includes determining the resulting electricity demand and load profile, and impacts on generation capacity as well as on the transmission and distribution capacity of the grid; for instance, if network extensions are necessary, this could be a major economic barrier.

Battery-electric vehicles are apparently the most energy-efficient solution and superior to fuel-cell vehicles, as the high discharge rate of a battery is almost double the efficiency of a fuel cell, but major technical and economic breakthroughs for vehicle batteries will have to be realised first. The most promising battery technology today is the lithium-ion battery. The size and weight, as well as the low energy density, of (existing) batteries are, at present, a constraint on the range of purely battery-powered cars, limiting their suitability to largely short-distance urban operation. Currently, long recharging times of 3–8 hours, the high cost of batteries – all the more if they should allow for extended driving ranges – and scarcity of some metals, among others, lithium, are further restrictions. But battery-electric vehicles are still a promising option for niche markets such as small city cars, which have to fulfil strict environmental requirements in highly polluted urban areas and two-wheelers (scooters), which for instance are becoming increasingly popular in Asian megacities.

If battery performance were to improve markedly and, at the same time, costs could be reduced, BEVs could represent a complete solution to decarbonising the transport sector, thus making the discussion about hydrogen largely obsolete. But given the above constraints, it seems likely that the major impact of batteries on the transport sector will be through PHEVs, which have lower requirements on battery performance as they are only partly dependent on battery power. If developed with an electric range of 50–80 km, plug-in hybrids could 'fuel' 60%–80% of their energy demand from the power grid (as, on average, less than 20% of trips exceed 60 km), thereby drastically reducing the liquid-fuel demand of the vehicle. Even PHEVs with only a 30 km range would be a good match for urban driving patterns and could displace between one and two thirds of liquid fuel. However, the extent to which PHEVs can contribute to a decarbonisation of the transport sector depends on the utility factor, i.e. the fraction of driving that is performed by electricity, and whether emissions can be reduced from the generation sector: for example, PHEVs offer no CO_2 savings if charged using the fossil-fuel dominated electricity mix of most countries; if charged using electricity from coal, they would be even more CO_2-intensive than the average vehicle fleet, all the more when compared against evolving efficiency standards for new cars (as proposed in the EU, for instance). In this respect, the CO_2 performance of PHEVs relative to standard ICE vehicles will critically depend on the increase in conventional vehicle efficiency and related lower tailpipe CO_2 emissions as a result of very stringent emission targets. Ultimately, the CO_2 benefits of PHEVs would also depend on the marginal power plant during the charging period.

It has to be understood that, due to the battery chemistry, there will always be inherent trade-offs among power density, energy density, longevity, safety and the cost of batteries, and hence some of the development targets will have to be compromised. For instance, due to the low energy density and the high costs of batteries, increasing the electric range of vehicles above 150 km will be very costly. For this reason, the primary application of full electric vehicles is seen in 'niche markets' only, such as public city transport (buses), short-distance individual transport (second vehicles and city cars) in cities or urban agglomerations, as well as two-wheelers (still having to rely on alternative solutions beyond the range limits of these vehicles). However, the size of the 'niche' market can be 'significant', as a 60 km all-electric range covers up to two thirds of annual mileage (which would, for instance, be very attractive for daily commuters) and because the concept of electric vehicles helps to address some of the challenges accompanying the global urbanisation trend. PHEVs overcome some of the shortcomings of full electric vehicles and can act as bridging technology for passenger transport, but they are only a partial solution when it comes to decarbonising transport, because other solutions are still needed for long-distance driving as well as road freight transport, and there remains the long term necessity to develop alternatives to replace today's conventional liquid fuels, for the reasons outlined previously.

As for hydrogen vehicles, market acceptance will also be a key enabler for electric vehicles: the inherent limitations of PHEVs/BEVs need to be accepted by customers, as they have to adjust their mobility behaviour to the technical conditions of the vehicle. Customer acceptance particularly correlates with driving range and costs; the inconvenience of possibly long recharging times and restrictive charging patterns also need to be accepted. The fact that BEVs will only cover a fraction of the driving range – although sufficient for typical daily driving needs – poses a particular challenge to their attractiveness, as consumers may have difficulty accepting a vehicle that is limited in range and would probably have to afford two cars to overcome this range issue.

Today, there is a general consensus that in the coming two decades electric vehicles, i.e. PHEVs and BEVs, will gain a material share of the vehicle fleet in many countries. However, the upsides as well as limitations of electric mobility need to be addressed realistically. In this respect it helps to consider that previous battery-electric car 'hypes' – in the 1970s and at the beginning of the 1990s – failed, as batteries fell short of achieving their development targets. A number of uncertainties remain that could substantially affect the take-up of electric vehicles, such as customer acceptance (related to the limited driving range as well as availability, accessibility and user convenience of recharging stations), doubts about battery technology development, electricity infrastructure bottlenecks with significant car penetration, future regulation and taxation regimes (regarding vehicles and electricity price) and increasingly efficient ICE engines.

21.3.3 Making a choice

Owing to its chemical and physical properties, oil is an excellent energy carrier for the transport sector, despite the associated environmental effects like the greenhouse-gas emissions and local emissions released when burning it in combustion engines. But the total energy efficiency of this carrier is low (overall it is around 20%, and in this sense, today's vehicles are actually producing more heat than propulsion energy). A range of alternatives to today's liquid fuels exist which all exhibit constraints and drawbacks of some kind at present. No other fuel will ever be as easy (and cheap) to produce and handle as gasoline and diesel from conventional crude oil. While these fuels can be produced from crude oil with high efficiency, the production of any alternative fuel will generally incur higher conversion losses, across the supply chain and – in the case of gaseous fuels – be more difficult to handle and require a new distribution and refuelling infrastructure. Hydrogen and electricity even require new propulsion systems (fuel cells, batteries).

Despite the ongoing efforts to develop new solutions for transport energy use, liquid petroleum-based fuels will retain their dominant role in the transport sector for the coming decades. Particularly road freight transport (which may account to up to 50% of total road transport fuel demand and related emissions in some countries) as well as shipping and aviation will continue to rely on liquid fuels for some time, as their inherently advantageous properties make them the preferred choice for these sectors. The lack of 'readily' available alternatives is likely to put pressure on biofuels to be used as 'low-carbon' fuel for freight transport, predominantly for trucks, and hence underpin the role of hydrogen and electricity for passenger transport. In whichever case, however, for being an environmentally benign alternative, in order for biofuels deployment to take off at a large scale requires that the criteria of sustainability be met first; to this end technologies must be developed that allow biofuels to avoid both the competition with food and land-use conversion.

The alternative fuels and drive systems available only seem to be viable on the mass market, if the oil price stays above \$60 to \$70 /bbl for a sustained period. Oil prices peaked above \$140 /bbl in summer 2008 and many experts believe that stable oil prices over \$100 /bbl could be reached in the next one or two decades. The higher the market prices of fossil fuels, the more competitive low-carbon alternatives will become: The principal choice here is between biofuels, electricity and hydrogen, provided that they are produced either from low/zero-carbon feedstock or that the CO_2 generated during their production is captured and stored. But higher priced conventional oil resources, on the other hand, can also be replaced by high-carbon alternatives such as oil sands, oil shale or synthetic fuels from coal and gas.

Among the various low-carbon options, hydrogen seems especially promising at the current level of knowledge, as it can contribute to the three most important targets with respect to transportation energy use, which are being increasingly favoured by policy makers around the world: GHG-emissions reduction, energy security and reduction of local air pollution. The major competitor to hydrogen in this respect is electricity, as it potentially offers the same benefits with respect to these energy policy objectives. Of all the alternatives, full electric vehicles have the potential for the lowest fuel costs and GHG emissions, owing to their higher efficiency throughout the fuel supply chain and the vehicle's lower fuel consumption. From today's perspective, it seems to be a technology race between the battery and the fuel cell. While the challenges for batteries are technical and economic in nature, for fuel cells, they are rather economic. If a 'miraculous' battery is invented, it is obvious that this will outweigh any other alternative fuel. But current battery technology suggests that pure battery-electric vehicles will only be an option for short-distance transport (in urban areas), due to their limited driving range; for extended range, electric vehicles will need to rely on the combination with plug-in hybrid technology, and therefore be a partial solution only for CO_2 reduction in the transport sector. Nevertheless, as a bridging technology, plug-in hybrids still have considerable potential to reduce the demand for liquid fuels and provide cleaner mobility.

What would kill the prospects for hydrogen is the 'ideal battery' offering 'unlimited range' (as hydrogen is less efficient than electricity) and/or 'unlimited' supply of 'low-carbon fuels' (i.e. in principle second generation biofuels), because hydrogen is more cumbersome to distribute and use than liquid fuels. However, it is wise to assume that neither of these will come true. While the fraction of driving performed by electricity will undoubtedly grow, there is unlikely to be a 'silver bullet' in the coming decades and the transport sector will witness a much more diversified portfolio of fuels in the future. In the short to medium term, hydrogen will be additional to what biofuels and electricity can offer for energy security and CO_2 emissions reduction. In the long run, however, hydrogen holds promise to overcome some of the limitations of biofuels and electricity, allowing for further decarbonisation of transport. Particularly in this respect, when compared with electricity, hydrogen is more promising, as fuel-cell vehicles cover the entire driving spectrum, allow fast refuelling and have the same potential for reduction of CO_2 and local pollution as electric vehicles. In addition, longer term, batteries could act as potential range extenders for fuel-cell vehicles, thus benefiting from the development of PHEVs. Recognising the limitations of electric vehicles, hydrogen fuel-cell vehicles are also generally seen as the long-term solution by major car manufacturers.

21.4 Why hydrogen?

Neither the use of hydrogen as an energy vector nor the vision of a hydrogen economy is new. Until the 1960s, hydrogen was used in many countries in the form of town gas for street lighting as well as for home energy supply (cooking, heating, lighting), and the idea of a hydrogen-based energy system was already formulated in the aftermath of the oil crises in the 1970s. Moreover, hydrogen is an important chemical feedstock, for instance for the hydrogenation of crude oil or the synthesis of ammonia. The breakthroughs in fuel-cell technology in the late 1990s are the main reason behind the revival of interest in hydrogen. While hydrogen can be utilised in different applications (mobile, stationary and portable), the transport sector is going to play a crucial role for the possible introduction of hydrogen, as outlined in the previous section. This is also where fuel cells can make the most of their high conversion efficiencies, compared with the internal combustion engine.

Hydrogen offers a range of benefits as a clean energy carrier (if produced by 'clean' sources), which are receiving ever greater attention as policy priorities. Creating a large market for hydrogen as an energy vector offers effective solutions to both the control of emissions and the security of energy supply. Hydrogen is emission-free at the point of final use and thus avoids the transport-induced emissions of both CO_2 and air pollutants. Being a secondary energy carrier that can be produced from any (locally available) primary energy source (unlike other alternative fuels, except electricity), hydrogen can contribute to a diversification of automotive fuel sources and supplies and offers the long-term possibility of being solely produced from renewable energies. Hydrogen could also be used as a storage medium for electricity from intermittent renewable energies, such as wind power. Assuming that CCS is eventually realised on a large scale, clean power generation from fossil fuels would be possible via the production of hydrogen. Moreover, there is also the possibility of co-producing electricity and hydrogen in IGCC plants.

However, it is hard to justify the use of hydrogen solely from a climate-policy perspective. It must be stressed that hydrogen is not an energy source in itself but a secondary energy carrier, in the same way as electricity. Like electricity, as far as the security of supply or greenhouse-gas emissions are concerned, any advantage from using hydrogen as a fuel depends on how the hydrogen is produced. If produced from coal, it augments the security of supply, but causes much higher CO_2 emissions (unless the CO_2 is captured and stored, a critical prerequisite for this pathway). If produced using non-fossil fuels (nuclear or renewable), it adds to the security of supply and reduces CO_2 emissions, but only in so far as the non-fossil fuel source is additional to what would otherwise be used in electricity generation. This means that any assessment of the virtues of switching to hydrogen as a transportation fuel involves a number of assumptions about long-term future energy-policy developments.

Local air emissions, responsible for particulate matter, ozone and acid rain, as well as noise, could be significantly reduced by the introduction of hydrogen fuel-cell vehicles. Emissions of NO_x, SO_2 and particulates can be reduced by 70% to 80% compared to a case without hydrogen. Especially in densely populated areas this is one major benefit of hydrogen, which is often underestimated. As there are a growing number of megacities worldwide, the importance of improving urban air quality is also increasing. A calculation shows that the CO_2, local emissions and noise benefits of a fuel-cell vehicle can reduce the average external cost of a vehicle by US$ 1000 to US$ 1500 compared with a conventional vehicle.

21.5 The role of fuel cells

In fuel cells, electricity and water are usually produced from hydrogen and oxygen in an electrochemical reaction that also releases heat. In contrast to conventional electricity generation, which takes place in a three-stage conversion process (chemical energy – thermal energy – mechanical energy – electricity), in a fuel cell, chemical energy is directly converted into electrical energy. A fascinating point is the potential theoretical efficiency of fuel cells.

A lot of different fuel-cell types exist, which do not require hydrogen as fuel. Therefore, fuel cells could enter the market independently of hydrogen production or infrastructure build-up. This is especially valid for portable applications and stationary applications. Stationary (high-temperature) fuel cells – and hence distributed heat and power generation – are not necessarily a market for hydrogen because they can use, e.g., natural gas from the gas mains directly; conversion to hydrogen would only reduce their overall efficiency (although it would allow central removal and storage of the CO_2). The situation is different for mobile applications, where the dominant fuel-cell type is the proton-exchange-membrane (PEM) fuel cell, which only functions with pure hydrogen.

On the other hand, it makes no sense to introduce hydrogen in the transport sector without fuel cells in the long run because of the high electricity to heat ratio and the high overall conversion efficiency of fuel cells powered by hydrogen: today, the efficiency of the fuel-cell system for passenger cars is around 40% (in the future maybe 50%) compared with 25%–30% for the gasoline or diesel powered internal combustion engine under real driving conditions. Fuel-cell systems have a higher efficiency at partial load than full load, which also suggests their suitability for application in motor vehicles, which are usually operated at partial load, e.g., during urban driving. In addition, the fuel cell's exhaust produces zero emissions when fuelled by hydrogen. Road-transport noise in urban areas would also be significantly reduced. Furthermore, fuel-cell vehicles could possibly even act as distributed electricity generators when parked at homes and offices and connected to a supplemental fuel supply. From this perspective, the use of hydrogen in internal combustion engines can only be an interim solution.

Today, the power train costs of fuel-cell vehicles are still far from being competitive. They have the largest influence on the economic efficiency of hydrogen use in the transport sector and the greatest challenge is to drastically reduce fuel-cell costs from currently more than \$2000/kW to less than \$100/kW for passenger cars. On the other hand, fuel-cell drive systems offer totally new design opportunities for vehicles: because they have fewer mechanical and hydraulic subsystems compared with combustion engines, they provide greater design flexibility, potentially fewer vehicle platforms and hence more efficient manufacturing approaches, which may lead to additional cost reductions. Nevertheless, this cost-reduction potential has to be realised first and is in a continuous interplay with the requirements for efficiency and lifetime. This is the major source of uncertainty for the market success of fuel-cell vehicles. Additional technical challenges like hydrogen storage and safety issues have to be solved as well.

To achieve a relevant market success, it is essential to meet the fuel-cell targets set for costs, lifetime and reliability. These technology developments obviously always take longer than planned by industry. However, preparation for the structural changes in industry is just as important as the technical optimisation of fuel cells. Qualified service technicians and skilled workers must be available to ensure that the introduction of fuel-cell technology is managed as smoothly as possible. The success of hydrogen in the transport sector will crucially depend on the development and commercialisation of competitive fuel-cell vehicles. The current time line for mass roll-out of fuel-cell vehicles as envisaged by major car manufacturers is between 2015 and 2020.

21.6 Hydrogen storage

Hydrogen storage is regarded as one of the most critical issues, which must be solved before a technically and economically viable hydrogen fuel system can be established. In fact, without effective storage systems, a hydrogen economy will be difficult to achieve.

Considerable progress has been achieved over the past few years concerning hydrogen-propelled vehicles. Most development efforts have concentrated on the propulsion system and its vehicle integration. At present, there is a general agreement in the automotive industry that the on-board storage of hydrogen is one of the critical bottleneck technologies for future car fleets. Still, no approach exists as yet that is able to comply with the technical requirements for a range greater than 500 km while meeting all the performance parameters regardless of costs. The physical limits for the storage density of compressed and liquid hydrogen have more or less been reached, while there is still potential in the development of solid materials for hydrogen storage, such as systems involving metal hydrides.

21.7 Supply of hydrogen

21.7.1 Hydrogen production

Hydrogen occurs naturally in the form of chemical compounds, most frequently in water and hydrocarbons. Hydrogen can be produced from fossil fuels, nuclear and renewable energy sources by a number of processes, such as natural gas reforming, gasification of coal and biomass, water electrolysis, water splitting by high-temperature heat, photo-electrolysis and biological processes. The global hydrogen industrial gas business is significant and total production amounts to around 700 billion Nm^3 (enough to fuel more than 600 million fuel-cell cars) and is based almost exclusively on fossil fuels: roughly half on natural gas and close to one-third on crude oil fractions in refineries. Most of this hydrogen is produced onsite for captive uses. The major use of hydrogen is as a reactant in the chemical and petroleum industries: ammonia production has a share of around 50%, followed by crude oil processing with slightly less than 40%.

Natural gas reforming, coal gasification and water electrolysis are proven technologies for hydrogen production today and are applied on an industrial scale all over the world. Steam reforming of natural gas is the most used process in the chemical and petrochemical industries; it is currently the cheapest production method and has the lowest CO_2 emissions of all fossil production routes. Electrolysis is more expensive and only applied if high-purity hydrogen is required. With an assumed increase in natural gas prices, coal gasification becomes the most economical option from around 2030 onwards. Biomass gasification for hydrogen production, still at an early stage today, is expected to become the cheapest renewable hydrogen supply option in the coming decades, although biomass has restricted potential and competes with other biofuels, as well as for heat and power generation. Biomass gasification is applied in small decentralised plants during the early phase of infrastructure roll-out and in centralised plants in later periods. Steam reformers and electrolysers can also be scaled down and implemented onsite at fuelling stations (although still more expensive), while coal gasification or nuclear energy are for large-scale, central production only and therefore restricted to later phases with high hydrogen demand.

In the medium to long term, hydrogen may be produced by natural gas reforming or coal gasification in centralised plants with CCS. Carbon capture and storage is essential to avoid an overall increase in CO_2 emissions through fossil hydrogen production, primarily from coal. Carbon abatement for hydrogen production is relatively cheap, since the (additional) costs of CO_2 capture in connection with hydrogen production from natural gas or coal are mainly the costs for CO_2 drying and compression, since CO_2 and hydrogen are already separated as part of the hydrogen-production process (even if the CO_2 is not captured). Taking the costs for CO_2 transport and storage into account, total hydrogen production costs increase by about 3%–5% in the case of natural gas reforming and 10%–15% in the case of coal gasification.

Hydrogen also occurs as a by-product of the chemical industry (for instance, chlorine–alkali electrolysis) and is already being used thermally. This represents another (cheap) option (where available), because it can be substituted by natural gas, although investments in purification might be necessary. This option is relevant for supplying hydrogen during the initial start-up phase in areas where user centres are nearby.

Nuclear power plants dedicated to hydrogen production are an option for later phases with high hydrogen demand. Thermochemical cycles based on nuclear energy or solar energy are a long-term option for hydrogen production with new nuclear technology (for instance the sulphur–iodine cycle) or in countries with favourable climatic conditions. However, nuclear hydrogen production is likely to face the same public acceptance concerns as nuclear power generation. The production of hydrogen from photo-electrolysis (photolysis) and from biological processes is still at the level of basic research.

Generally, the hydrogen production mix is very sensitive to the country-specific context and strongly influenced by the feedstock prices; resource availability and policy support also play a role, in particular for hydrogen from renewable and nuclear energy. The fossil hydrogen production option dominates the first two decades, while an infrastructure is being developed, and also during later periods if only economic criteria are applied: initially on the basis of natural gas, subsequently, with increasing gas prices, more and more on the basis of coal. However, as fossil production gradually shifts from distributed natural gas-based production to more centralised (coal)-based production, this offers the opportunity for capturing and sequestering the CO_2 generated. In the short to medium term, renewable hydrogen will mainly be an economic option in countries with a large renewable resource base and/or a lack of fossil resources, for remote and sparsely populated areas (such as islands) or for storing surplus electricity from intermittent renewable energies. Otherwise renewable hydrogen needs to be incentivised or mandated.

It is evident that hydrogen needs to be produced in the long term from processes that avoid or minimise CO_2 emissions. Renewable hydrogen (made via electrolysis from wind or solar generated electricity or via biomass gasification) is surely the ultimate vision (particularly from the viewpoint of mitigating climate change), but not the pre-condition for introducing hydrogen as an energy vector. Until this goal is reached, hydrogen from fossil fuels will prevail, but the capture and storage of the produced CO_2 then becomes an indispensable condition if hydrogen is to contribute to an overall CO_2 reduction in the transport sector. The expected dominance of fossil hydrogen during the introduction phase (the period until around 2030) is reflected in the various hydrogen roadmaps, as is the more prevalent role of renewable energies for hydrogen production in later periods. With the exception of biomass, the specific costs of hydrogen production from renewable energies are not considered to be competitive with most other options during this period. In particular, for hydrogen from renewable electricity to be economically viable, the cost of electrolysers must come down sharply.

Hydrogen production costs depend, to a very large extent, on the assumed feedstock prices. The typical range until 2030 is between 8 and 12 ct/kWh ($2.6–$4/kg). In the long term, until 2050, with an expected increase in feedstock prices (fossil fuels) and CO_2 prices, hydrogen production costs will increase as well.

21.7.2 Hydrogen distribution

Different options are available for hydrogen transport and distribution: delivery of compressed gaseous and liquid hydrogen by trucks and of gaseous hydrogen by pipelines. Pipelines have been used to transport hydrogen for more than 50 years, and today there are about 16 000 km of hydrogen pipelines around the world that supply hydrogen to refineries and chemical plants; dense networks exist, for example, between Belgium, France and the Netherlands, in the Ruhr area in Germany and along the Gulf coast in the United States.

The technical and economic competitiveness of each transport option depends on transport volumes and delivery distances. Pipelines are the preferred option for large quantities and long distances. Liquid hydrogen trailers are for smaller volumes and long distances, and compressed gaseous hydrogen trailers are suitable for small quantities over short distances. Pipelines are characterised by a very low operating cost, mainly for compressor power, but high capital costs. Liquid hydrogen has a high operating cost due to the electricity needed for liquefaction (which accounts for 30%–60% of the total liquefaction costs and may also represent a significant CO_2 footprint), but lower capital costs, depending on the quantity of hydrogen and the delivery distance. Distance is also the deciding factor between liquid and gaseous trailers. Hydrogen transport costs are typically in the range of 1–4 ct/kWh ($0.3–$1.3/kg).

Because of the specific physical and chemical properties of hydrogen, pipelines must be made of non-porous, high-quality materials, such as stainless steel; therefore, the investments in a hydrogen pipeline for a given diameter are up to twice those for natural gas pipelines. The costs could be considerably reduced if the natural gas infrastructure could be adapted to hydrogen. As hydrogen can diffuse quickly through most materials and seals and can cause severe degradation of steels, mainly due to hydrogen embrittlement, the use of existing natural gas pipelines could be problematic and has to be investigated on a case-by-case basis. Coating or lining the pipelines internally, or adding minor amounts of oxygen, could solve the problems in using existing long-distance transmission pipelines made from steel. Next to embrittlement, hydrogen diffusion, however, would prohibit the transport of hydrogen in low-pressure, natural gas distribution pipelines, which are often made of plastic materials. In addition, valves, manifolds and, in particular, compressors would need to be modified, as they are optimised to work under a certain range of conditions, such as gas composition. Another possibility could be to blend hydrogen with natural gas up to a certain extent and either separate the two at the delivery point, or use the mixture, e.g., in stationary combustion applications. However, to what extent this is feasible and reasonable is still a matter of debate, given that hydrogen is an expensive and valuable commodity, and because admixture to and extraction from natural gas is no solution for fuel-cell vehicle owners. To conclude, the introduction of hydrogen would largely require a new dedicated pipeline transportation and distribution infrastructure.

21.8 Hydrogen infrastructure build-up

How the hydrogen supply infrastructure would develop and what this would look like depends heavily on country-specific conditions, such as the available feedstock (like renewable energies), population density, geographic factors and policy support, and must, therefore, be assessed on a country-by-country basis. Nevertheless, based on the hydrogen infrastructure analysis presented in this book, it is possible to derive some robust strategies and cross-national communalities for the introduction of hydrogen in the transport sector. It is important to bear in mind that the technical and economic challenges concerning fuel cells and hydrogen storage are assumed to have been resolved and that fuel-cell vehicles are cost-competitive with conventional vehicles.

Assuming mass roll-out of fuel-cell vehicles starting around 2015, for the introduction of hydrogen, two broad phases can be distinguished: the infrastructure build-up phase (2015–2030) and the hydrogen diffusion phase (2030–2050). The former can further be subdivided into an early implementation phase (2015–2020) and the transition phase (2020–2030). Scenario results generally show the greatest differences during the transition phase, when the initial infrastructure is being developed, but tend to converge in later periods. Generally, there is a transition from small-scale distributed production at the beginning to more larger-scale centralised production in later phases, as demand picks up.

Hydrogen use will take off predominantly in densely populated areas and urban environments with favourable support policies and, during the transition phase, will then gradually expand outwards into rural areas. As buses and fleet vehicles, such as delivery vans, operate locally to a large extent, run on short, regular routes and return to a central depot for refuelling and maintenance, they are ideal candidates for hydrogen during the early implementation phase, as they do not need an extensive network of refuelling stations. Hydrogen ICE vehicles with bi-fuel conversions are advantageous during the early phase as well, as they avoid the necessity of having an area-wide coverage of refuelling stations in place right from the beginning. Strong policy measures, such as zero-emission mandates or tax incentives are essential to encourage the early adoption of hydrogen vehicles.

The introduction of a hydrogen fuel system is best accomplished initially through the distributed production of hydrogen, mainly onsite at the fuelling stations. This is the most economic approach, as it avoids constructing an extensive and costly transport and distribution infrastructure, which accompanies centralised production. This could be deferred until the demand for hydrogen is large enough. A distributed production system during the transition phase can be installed rapidly as the demand for hydrogen expands, thus allowing hydrogen production to grow at a pace that is reasonably matched to hydrogen demand. This approach gives the market time to develop and diminishes the risks for investors, as it avoids fixing large amounts of capital in under-used large-scale production and distribution facilities, while it is still unclear how hydrogen demand will develop. The preferred technology is small-scale natural gas reforming (using the existing natural gas pipeline network), followed, to a lesser extent, by gasification of biomass, onsite electrolysis (from grid electricity or wind or solar energy) and by-product hydrogen.

Onsite hydrogen production at the fuelling station is not only the preferred option during the early implementation phase (first decade), but also in areas where demand is too low for more centralised schemes (in later periods). Onsite production becomes less preferable once a distribution infrastructure has gradually been built up and demand for hydrogen has risen and there is, therefore, a trend towards centralised production in later phases. In remote areas with a high energy demand, however, it is possible that local energy resources may be exploited by manufacturing hydrogen to meet local transport energy demands.

Owing to the dominant onsite production, there is little need for hydrogen to be transported during the first decade. If this takes place, it is mainly by liquid hydrogen trailers, but also by compressed gaseous trailers under specific circumstances. Liquid hydrogen plays an important role during the transition phase (until 2030) and in connecting outlying areas, such as along motorways or in rural areas. At the same time a pipeline network is being constructed, and pipelines clearly dominate hydrogen transport in the diffusion phase (after 2030).

Transport options are exposed to many sensitivities (e.g., distances, volumes, fuelling station utilisation, demand for liquid hydrogen, energy prices, density of fuelling stations in a region) and there is no ‘ultimate best strategy’, as each of the options can play a role under specific conditions. The distance to be covered has the strongest impact on transport costs, which influence the total supply costs of hydrogen to a much larger extent than is the case for today’s liquid fuels. As transport is so expensive, hydrogen should be produced close to the user centres. The primary optimisation goal is, therefore, to minimise the average hydrogen transport distances through well planned siting of the production plants.

Projected hydrogen supply costs are highly sensitive to underlying assumptions about the development of feedstock prices, as these have a decisive impact on production costs; uncertainty, therefore, increases significantly with longer-term projections. Being representative for both the European Union and North America, at around 12–14 ct/kWh ($4–$4.6/kg), the specific hydrogen supply costs in the early phase are high, owing to the required overcapacity of the supply and refuelling infrastructure and the higher initial costs for new technologies because of the early phase of technology learning. Around 2030, hydrogen costs range from 10 to 16 ct/kWh ($3.6–$5.3/kg) in the above-mentioned regions, mainly depending on the feedstock. In the long term, until 2050, hydrogen supply costs stabilise around this level, but with an upward trend, due to the assumed increase in energy prices and CO_2 certificate prices. Also, while fossil hydrogen costs will rise in accordance with the expected increase in fossil-fuel prices, at the same time, costs for renewable hydrogen are expected to go down, ultimately reaching a break-even point.

With a share of 60%–80%, hydrogen production dominates total supply costs. The installation of refuelling stations makes up about 10%; the remainder is for transport and liquefaction. At these supply costs, hydrogen becomes competitive in the long run at crude oil prices above $80–$100/barrel (no taxes, no vehicle costs included). The specific investments for implementing a complete hydrogen supply infrastructure (production plants, transport infrastructure (pipelines) and refuelling stations) vary between 150 and 190 M$/PJ until 2050 across the various scenarios, with a tendency towards higher numbers in later periods, owing to higher feedstock prices.

21.9 Hydrogen and the electricity sector

If hydrogen production on a large scale is to be integrated into the energy system, a more holistic view needs to be applied to its interactions with the electricity sector. Alongside the option of producing hydrogen from electricity via electrolysers, there are other direct links between hydrogen and the power sector. These concern, for instance, the ensuing competition for renewable energies as, in the long term, only hydrogen production from renewable energy sources can reduce the dependency on fossil fuels and enhance the security of supply. Hydrogen can also be used as a storage medium for electricity from intermittent renewables, such as wind energy. Finally, the co-production of 'clean' electricity and 'clean' hydrogen in IGCC plants (with CCS) is potentially a promising option for the future.

The uses of renewable energies will face increasing competition with respect to feedstock availability for electricity generation and fuel production and, in the case of biomass, also for food production. Today, since renewable energy supplies are still limited in most countries, the question is where they are best employed in order to achieve the largest CO_2 reduction for the energy system as a whole: in the power sector or the transport sector. New renewable energy sources will reduce CO_2 emissions to a greater extent if they are used to generate power that displaces grid-mix electricity, than if applied to producing vehicle fuels like hydrogen (or other renewable fuels) and thus substituting conventional fuels. This reduction will be even greater if there is a high share of fossil fuels in the power mix. Moreover, if produced by renewable energies, hydrogen supplements the security of supply and reduces CO_2 emissions only to the extent that the renewable energy source is additional to what would otherwise be used for power generation. This picture may change if electricity from fluctuating renewable sources, such as wind or solar, has to be stored. Hence, from the overall perspective of CO_2 reduction, renewable energies should only be deployed in the transport sector in large amounts after they have achieved significant penetration in the power sector.

As for the competition between hydrogen and electricity from renewable energies in the transport sector, it is clear that the use of renewable electricity in battery-electric vehicles is by far the most efficient application and yields a much higher CO_2 reduction than hydrogen fuel-cell vehicles, owing to the high discharge rate of batteries. However, full-battery vehicles still face significant technical and economic hurdles and while plug-in hybrids can significantly reduce the demand for liquid fuels, any large reduction in CO_2 emissions would require the batteries to be charged using low- or zero-carbon electricity, particularly in the light of proposed future vehicle standards.

A rapid build-up is expected in wind power and also in photovoltaic and solar-thermal electricity-generating capacities. Despite some clear advantages – renewable and CO_2-free – the intermittency of wind- and solar-generated electricity poses a challenge with regard to load levelling when capacity grows. Hydrogen could be one solution to this problem, as it offers the possibility of storing and transporting the energy. In particular, for mobile applications, where energy has to be stored anyway, this could be a more promising option than storage and re-electrification for stationary applications. Hydrogen production could also be an attractive option for remote areas, such as low-populated islands without access to the main grid and for exploiting large renewable resources, which are far away from user centres (so-called stranded resources).

Using hydrogen to produce electrical energy from fossil fuels in large centralised plants will contribute to achieving significant reductions of CO_2 emissions, if combined with CO_2 capture and storage. Such plants will also help to increase the diversification of resources, since a variety of fossil feedstocks can be used, including resources, such as coal and waste, that otherwise cause major impacts on the environment, as well as biomass. In addition, it is possible to co-produce hydrogen and electricity in these plants, which can contribute positively to load levelling. For instance, IGCC plants can be used to produce more electricity during peak periods, and to produce more hydrogen during off-peak periods. This demonstrates the important advantage of IGCC plants: they can deliver to two markets, the electricity market as well as the transport market, depending on the price signals.

As already mentioned, the high temperatures of the exhaust gases produced from the high temperature (stationary) molten-carbonate fuel cell (MCFC) and solid-oxide fuel cell (SOFC) are ideal for co-generation and combined-cycle power plants, reaching an overall system efficiency of up to 90%, or for providing heat for space and hot-water heating. But since they can be fuelled directly by hydrocarbons because of their high operating temperatures, natural gas or biogas would be the fuel of choice here, rather than hydrogen.

21.10 International competitiveness and economic impacts

The transport sector is of high economic relevance for some world regions, for instance in Europe and the USA, where it contributed between 4% and 6.5% to employment, and between 6% and 10% to production in 2002. Roughly 40% of these figures are attributable to vehicle production. Therefore, the international competitiveness of the transport sector is also of high political relevance to some regions. Looking at hydrogen, the structure of the necessary investments in hydrogen as an energy vector is clearly dominated by the expenditures on hydrogen vehicles. If a hydrogen vehicle is imported, it is very likely that this involves the whole vehicle, not just the hydrogen-drive system. Therefore, the structure of the domestic vehicle industry is one of the key factors for the development in employment and gross domestic product (GDP).

A macroeconomic analysis for Europe reveals that the introduction of hydrogen results in positive effects for long-term employment – assuming a similar degree of competitiveness as today's for the non-hydrogen technologies that will be replaced with hydrogen technologies. The development of new technologies requires a higher qualification level, needs more workers due to a lower automation level, and is a barrier to off shoring production. The overall impact on economic growth (GDP) from introducing hydrogen into the energy system is small. Besides the fact that net changes in expenditure patterns are small, the fact that hydrogen is introduced in only part of the energy system also helps to explain the relatively small impacts on GDP. Overall, however, the economic development proves to be positive, partly, because of an increase in investments. This increase in investments is for several reasons: first, additional investments in hydrogen production and fuelling infrastructure as well as in the additional renewable capacities required to produce 'renewable' hydrogen (both funded by revenues from selling hydrogen as a fuel), and second, the wider economic effects following these additional investments, i.e., increased employment and income resulting in higher GDP, which leads to increased demand and hence more investment in the second round.

It should be pointed out that the economic analysis is based on the assumption that, after an initial period of support, hydrogen technologies show cost advantages over conventional technologies. In addition, the above conclusions are all based on the assumption that there are no shifts in exports or imports (e.g., same market share of fuel-cell vehicles for European car manufactures as today for conventional vehicles). However, mid-term employment effects in automotive sectors could be drastic if the assumption of similar competitiveness is rejected for hydrogen technologies. Therefore, the following dilemma can be identified for the major car-making countries. On the one hand, job losses could be drastic if these countries lose market shares as a result of late market entry. On the other hand, there are uncertainties regarding the market success of hydrogen cars and the potential risk of losing several billion dollars from investments in premature hydrogen infrastructure and hydrogen-car development. Compared with these countries, the economic risks of a hydrogen economy are much smaller for countries without a large domestic automotive industry and promise significant increases in employment here if the right strategy is pursued. Similar conclusions, but with lower total effects, can be drawn for the plant and equipment branches.

Replacing conventional vehicles with fuel-cell vehicles induces a sectoral employment shift away from traditional automobile manufacturing to the fabricated metal, electrical, machinery and rubber and plastic sectors among others. Because of the required gradual build-up of manufacturing capacity and a skilled labour force, preparing for expected mass production makes early political action essential.

21.11 Global hydrogen scenarios

The extent to which hydrogen is considered to play a role in the global energy system in the future ranges widely across the various world energy scenarios. Hydrogen is not included as an important energy carrier in the official reference scenarios. Hydrogen penetration is only assumed in scenarios with a strict climate policy, high oil and gas prices and, moreover, a technology breakthrough in fuel cells and hydrogen storage. Under the most favourable assumptions, hydrogen vehicles are projected to reach shares of 30%–70% of the global (light-duty) vehicle stock by 2050, resulting in a hydrogen demand of around 7 EJ or 16 EJ, respectively, with the vast majority split to largely equal parts among Europe, North America and China. The resulting reduction of oil consumption would be in the range of 7 to 16 million barrels per day.

For an average investment of M$170/PJ needed to implement a complete hydrogen supply infrastructure (i.e., production plants, transport pipelines, refuelling stations, etc.), the (cumulative) investments required to put the necessary supply capacity in place by 2050 would add up to around US$1200 billion to US$2700 billion, respectively, for the above demand scenarios; this is equivalent to up to $75 billion per year (only around 6% of worldwide armament expenditure in 2006). Although these estimates harbour considerable uncertainties and should only be interpreted as indicative orders of magnitude, they should be considered in the context of the overall global investments required in the energy sector. According to the IEA, these are already estimated to amount to around US$20 000 billion in the period up to 2030. The cumulative investments required in the oil and gas sectors alone total $4300 billion and $3900 billion, respectively, in this period. The transition to hydrogen in the transport sector would represent a maximum of 1.7% of the projected GDP growth until 2050. While the investments for introducing hydrogen would add substantially to those already required for the energy system, they should not be considered an insurmountable barrier to implementing a hydrogen-fuel system. Moreover, to the extent that hydrogen would substitute oil-based transport fuels, the investments foreseen in the oil sector would be diminished.

The global hydrogen mix in 2050 is hard to predict. In the scenario where the cumulative hydrogen demand until 2050 is met by one primary energy source alone – which reflects the maximum impact on resources – it can be concluded that the impact on the depletion of fossil resources would be marginal (up to 4% of the current natural gas reserves and up to 2% of hard coal reserves). In contrast, a significant expansion of renewable energies dedicated to hydrogen production would be required: up to a 40-fold increase of the current global installed wind capacity and up to six times the current global biomass use. It is not possible to conclude any preference for hydrogen (from fossil fuels) over oil sands and oil shale from a resource point of view, because the primary energy expended for their production – although significantly higher than for the recovery of conventional oil – yields more 'mobility' than when it is used for hydrogen production.

To get an idea of the potential CO_2 reduction achievable in the transport sector from the introduction of hydrogen vehicles, CO_2 emissions from hydrogen supply have to be compared on a well-to-wheel (WTW) basis (i.e., taking into account the entire supply chain) with conventional gasoline and diesel fuels. Generally, close to 90% of total CO_2 emissions of the hydrogen supply chain result from hydrogen production. Assuming average WTW emissions for conventional passenger cars of 160 g CO_2/km (although recognising that more stringent thresholds are being introduced), hydrogen fuel-cell vehicles achieve – *without CCS* – a CO_2 emissions reduction of around 35% for hydrogen from natural gas, 90% for biomass and almost 100% for wind energy. Under this scenario, hydrogen from coal results in a 25% increase in CO_2 emissions (which would even exceed WTW emissions from fuels derived from oil sands). If CCS is applied, CO_2 emissions for fossil hydrogen can be reduced by up to 80%. For hydrogen ICE cars, WTW emissions are, in any case, 60%–80% higher, owing to the lower efficiency of the combustion engine than of the fuel cell; their application should, therefore, be largely constrained to the transition phase.

While hydrogen is emission-free at final use, it is evident that if hydrogen is produced from coal without CCS, no overall CO_2 reduction in the energy system is achievable. Given that coal will become the most economic feedstock in the medium to long term as natural gas prices increase, CCS becomes an inevitable prerequisite for the supply of hydrogen.

Both the production of hydrogen from coal and the production of oil from unconventional resources (oil sands, oil shale, CTL, GTL) result in high CO_2 emissions and substantially increase the carbon footprint of fuel supply, unless the CO_2 is captured and stored. While the capture of CO_2 at a central point source is equally possible for unconventionals and centralised hydrogen production, in the case of hydrogen, a CO_2-free fuel results, unlike in the case of liquid hydrocarbon fuels. This is all the more important, as around 80% of the WTW CO_2 emissions result from the fuel use in the vehicles. If CCS were applied to hydrogen production from biomass, a net CO_2 removal from the atmosphere would even be achievable.

21.12 Perspectives

It has taken more than a century for the existing transportation system to evolve, which today still relies almost entirely on one energy source, crude oil. Recently, the growing anxiety about energy security against the background of increasing import dependency has been shaping the discussion about the future supply of energy and, in particular, the supply of fuels. To what extent the decline in (conventional) oil production will become a problem largely depends on to what extent it is possible to find substitute energy carriers in due time. In this respect, industry and policy makers are increasingly being challenged to develop alternatives to oil. It will only be possible to manage shortages in the supply of oil and realise the necessary transition to alternative fuels by setting the direction at an early stage, as any change in the energy system will take a long time.

While the enthusiasts can legitimately call hydrogen the 'ultimate fuel', it is also legitimate to see hydrogen as a compromise: it is more cumbersome to distribute and use than liquid fuels; it is less efficient than electricity, even when used in a fuel cell. Consequently, hydrogen would not have a significant role as an energy carrier, if the scope for biofuels were 'unlimited', or if the elusive 'better battery' were invented. However, it is prudent to consider the case that neither of these will come to pass.

Hydrogen offers the possibility of responding to all the major energy policy objectives in the transport sector at the same time, i.e., GHG-emissions reduction, energy security and reduction of local air pollution and noise. Hydrogen could provide the link between renewable energy and the transport sector, transforming biomass, wind and solar energy into transport fuel and reducing oil dependence, as well as CO_2 emissions. Moreover, hydrogen could play an important role as a means of storing surplus electricity from intermittent renewable energies. Nevertheless, for hydrogen from fossil fuels, especially coal, CCS is a critical prerequisite if an overall CO_2 reduction over the entire supply chain is to be achieved. Moreover, if CCS is deployed on a large scale, fossil fuels could be decarbonised via the production of hydrogen, which could then be used as a clean fuel for power generation.

Energy systems and technologies evolve slowly – the combustion engine took more than a century to be developed and improved. Hydrogen and fuel cells will be no different, and it will take several decades for the build-up of a hydrogen infrastructure and for hydrogen to make a significant contribution to the fuel mix. However, threats such as dwindling energy resources or climate change may lead to a faster market penetration of hydrogen vehicles than anticipated in general.

Hydrogen should not be evaluated in isolation, but benchmarked against its main competitors, as assessing its potential without taking competing options into account would result in misleading conclusions. The introduction of hydrogen should also be analysed in the context of the development of the energy system as a whole. In the transport sector, in a long-term perspective, alongside hydrogen, only electricity seems to have the potential to fulfil all the above transport energy requirements, too.

The widespread introduction of hydrogen as a vehicle fuel faces three major technical challenges: developing cost-competitive and efficient fuel cells for vehicles, designing safe tanks to store hydrogen on-board with an acceptable driving range and developing an infrastructure for hydrogen production, distribution and refuelling. Both the supply side (the technologies and resources that produce hydrogen) and the demand side (the hydrogen conversion technologies) must simultaneously undergo a fundamental transformation, as one will not work without the other.

However, shifting transport to hydrogen is not only a technical issue. It would also induce structural economic changes through the build-up of manufacturing capability and the development of a large-scale industry, producing and distributing hydrogen. In addition, there would be trade-flow changes, owing to reduced trade in fossil fuels and increased trade in feedstocks for hydrogen production, and changes in employment opportunities.

Technology breakthroughs that substantially reduce the costs of the whole supply chain are essential for the successful take-off of a hydrogen market. But it has been shown that the introduction of hydrogen in the transport sector seems feasible from an economic viewpoint; for instance, the cumulative capital needed to develop a hydrogen infrastructure should not be considered a deterrent when considered relative to the estimated investments required over the next decades in the energy sector, in general, and in the oil sector, in particular, to keep up production levels. Hydrogen production and infrastructure costs are not an economic barrier at today's prices of conventional energy carriers. The critical element is the cost development of the fuel-cell propulsion system, whose forecasts are a major source of uncertainty here.

The introduction of hydrogen requires a joined-up approach between all relevant stakeholders. Hydrogen and fuel cells (but also biofuels and electricity) are unlikely to emerge in future energy markets without decisive and favourable policy support and incentives. Measures need to be put in place and upheld long enough to create public awareness and stimulate consumer acceptance of hydrogen and to guarantee investment security for entrepreneurs, since significant industry investments are required for vehicle manufacturing and infrastructure build-up well in advance of market forces. Moreover, regulations, codes and standards (RCS) are required for the production, distribution, storage and use of hydrogen (especially with regard to vehicle and on-board storage system safety). International co-operation will also be crucial to establish trans-boundary hydrogen infrastructures, because vehicles are driven, imported and exported across country borders. Last, but not least, the public will need to be trained and educated in the use of hydrogen technologies, for instance, the refuelling of hydrogen cars.

Hydrogen will probably mainly replace oil-based fuels in the transport sector while other energy carriers like electricity will continue to play a role. Using the term 'hydrogen economy', therefore, may be misleading. Via renewable energies and CCS, hydrogen has the potential to solve some of our energy problems, but improving energy efficiency also plays a vital role in tackling climate change as well as contributing to the security of energy supply.

The discussion about the sense and nonsense of hydrogen as an energy carrier has been controversial in the past and this is likely to continue in the foreseeable future, as will be the case for any of the alternative fuels. A lot of this controversial dispute can be explained by the fact that the parameters and assumptions for the evaluation are often not laid down clearly. The evaluation of hydrogen worldwide is positive if:

- the oil price remains above \$80 to \$90/barrel in the medium and long term and other conventional energy carrier prices are also high,
- renewable energies and CCS are deployed on a large scale,
- the transport sector has to reduce greenhouse-gas emissions significantly, and
- there is no major technological breakthrough in vehicle batteries.

This book has considered in detail the potential for and costs of technologies and measures to introduce hydrogen, recognising that this is subject to significant uncertainties. These include the difficulties of estimating the costs of technologies several decades into the future, as well as how fossil fuel prices will evolve in the future. It is also difficult to predict what the public acceptance of hydrogen will be.

The analysis presented here does not attempt to give a definitive answer. Instead, this book aims to shed some light on the challenges and opportunities that lie ahead for countries seeking to develop hydrogen energy policies. However, the authors would like to stress that hydrogen should not be seen as the all-encompassing solution to the world's energy problems and, in particular, not as the one and only response to the challenges faced by the transport sector. It is also highly unlikely that any single technology or fuel has the potential to be this 'silver bullet', able to meet the energy challenge and all the other criteria for improving energy security and mitigating the effects of climate change and other harmful environmental impacts, because all the options are subject to constraints of some kind.

While the transport sector will witness a much more diversified and regionally fragmented portfolio of fuels in the coming decades, there is a general consensus that the share of electric mobility, in its broadest sense, electric-drive vehicles powered by a battery or fuel cell (either directly or through a hybrid drive train), will increase markedly. It is clear that in the short term, hydrogen will be additional to what biofuels and electric mobility can offer for energy security and CO_2 mitigation. In the long run, however, hydrogen has the potential to overcome the limitations of biofuels and electricity and allow for further decarbonisation of transport.

Today, there are a growing number of public–private partnerships aimed at accelerating the commercialisation of hydrogen and fuel-cell technologies, as well as hydrogen demonstration projects around the globe. The critical question is whether these partnerships will be able to pave the way for the commercial introduction of hydrogen vehicles. Will hydrogen remain the fuel of the future? The coming decade will provide the answer.

Further reading

Barreto, L., Makihira, A. and Riahi, K. (2002). The hydrogen economy in the 21st century: a sustainable development scenario. *International Journal of Hydrogen Energy*, **28** (3), 267–284.

Bose, T. and Malbrunot, P. (2007). *Hydrogen. Facing the Energy Challenges of the 21st Century*. John Libbey Eurotext.

Busby, R. L. (2005). *Hydrogen and Fuel Cells: A Comprehensive Guide*. Penn Well Corporation.

Drennen, T. E. and Rosthal, J. E. (2007). *Pathways to a Hydrogen Future*. Elsevier.

Deutscher Wasserstoff- und Brennstoffzellen-Verband (DWV; German Hydrogen and Fuel Cell Association) (2006). *Woher Kommt die Energie für die Wasserstofferzeugung? Status und Alternativen*. Authored by Ludwig-Bölkow-Systemtechnik (LBST) (Schindler, J., Wurster, R., Zerta, M., Blandow, V. and Zittel, W.). Berlin: DWV.

Ewing, R. A. (2007). *HYDROGEN – Hot Stuff Cool Science: Discover the Future of Energy*. 2nd edn. PixyJack Press.

Hammerschlag, R. and Mazzap, P. (2005) Questioning hydrogen. *Energy Policy*, **33** (16) 2039–2043.

Holland, G. and Provenzano, J. (2007). *The Hydrogen Age*. Gibbs Smith.

Lovins, A. B. (2003). *Twenty Hydrogen Myths*. Rocky Mountain Institute. www.rmi.org.

Peavey, M. A. (1998). *Fuel from Water: Energy Independence with Hydrogen*. 11th edn. Merit Products Inc.

Rand, D. A. J. and Dell, R. M. (2007). *Hydrogen Energy. Challenges and Prospects*. Royal Society of Chemistry (RSC) Publishing.

Romm, J. J. (2005). *The Hype About Hydrogen: Fact and Fiction in the Race to Save the Climate*. Island Press.

Romm, J. J. (2006). *Der Wasserstoff-Boom. Wunsch und Wirklichkeit beim Wettlauf um den Klimaschutz*. Weinheim: WILY-VCH.

Schulte, I., Hart, D. and van der Vorst, R. (2004). Issues affecting the acceptance of hydrogen fuel. *International Journal of Hydrogen Energy*, **29** (7), 677–685.

Schultz, M. G., Diel, T., Brasseur, G. P. and Zittel, W. (2003). Air pollution and climate-forcing impacts of a global hydrogen economy. *Science*, **302** (24th October, 2003).

Science (2004). Toward a hydrogen economy. *Science*, **305** (13th August, 2004).

Scientific American Magazine (2007). *Oil and the Future of Energy: Climate Repair, Hydrogen, Nuclear Fuel, Renewable and Green Sources, Energy Efficiency.* The Lyons Press.

Scott, D. S. (2007). *Smelling Land: The Hydrogen Defense Against Climate Catastrophe.* Canadian Hydrogen Association.

Wald, M. L. (2004). Questions about a hydrogen economy. *Scientific American*, May (2004), 66–73.

Index